# Handbook of Electric Machines

## Other McGraw-Hill Reference Books of Interest

### Handbooks

*Baumeister* • Marks' Standard Handbook for Mechanical Engineers
*Bovay* • Handbook of Mechanical and Electrical Systems for Buildings
*Brady and Clauser* • Materials Handbook
*Brater and King* • Handbook of Hydraulics
*Croft, Carr, and Watt* • American Electricians' Handbook
*Fink and Beaty* • Standard Handbook for Electrical Engineers
*Harris* • Shock and Vibration Handbook
*Hicks* • Standard Handbook of Engineering Calculations
*Hicks and Mueller* • Standard Handbook of Professional Consulting Engineering
*Juran* • Quality Control Handbook
*Kurtz* • Handbook of Engineering Economics
*Maynard* • Industrial Engineering Handbook
*Pachner* • Handbook of Numerical Analysis Applications
*Parmley* • Mechanical Components Handbook
*Parmley* • Standard Handbook of Fastening and Joining
*Perry and Green* • Perry's Chemical Engineers' Handbook
*Raznjevic* • Handbook of Thermodynamic Tables and Charts
*Rohsenow, Hartnett, and Ganic* • Handbook of Heat Transfer Applications
*Rohsenow, Hartnett, and Ganic* • Handbook of Heat Transfer Fundamentals
*Rothbart* • Mechanical Design and Systems Handbook
*Seidman and Mahrous* • Handbook of Electric Power Calculations
*Smeaton* • Motor Application and Control Handbook
*Smeaton* • Switchgear and Control Handbook
*Tuma* • Engineering Mathematics Handbook
*Tuma* • Handbook of Physical Calculations
*Tuma* • Technology Mathematics Handbook

### Encyclopedias

Concise Encyclopedia of Science and Technology
Encyclopedia of Electronics and Computers
Encyclopedia of Energy
Encyclopedia of Engineering

### Dictionaries

Dictionary of Electrical and Electronic Engineering
Dictionary of Electronics and Computer Science
Dictionary of Mechanical and Design Engineering
Dictionary of Scientific and Technical Terms

# HANDBOOK OF ELECTRIC MACHINES

**Syed A. Nasar**
**Editor in Chief**
*Professor of Electrical Engineering,*
*University of Kentucky*

**McGRAW-HILL BOOK COMPANY**

**New York St. Louis San Francisco Auckland Bogotá**
**Hamburg Johannesburg London Madrid Mexico**
**Milan Montreal New Delhi Panama**
**Paris Sâo Paulo Singapore**
**Sydney Tokyo Toronto**

**Library of Congress Cataloging-in-Publication Data**
Handbook of electric machines.

Includes index.
1. Electric machinery—Handbooks, manuals, etc.
I. Nasar, S. A.
TK2181.H26 1987 621.31'042 86-10348
ISBN 0-07-045888-X

Copyright ©1987 by McGraw-Hill, Inc. All rights reserved.
Printed in the United States of America. Except as permitted
under the United States Copyright Act of 1976, no part of this
publication may be reproduced or distributed in any form or by
any means, or stored in a data base or retrieval system, without
the prior written permission of the publisher.

1234567890 DOC/DOC 893210987

ISBN 0-07-045888-X

*The editors of this book were Harold B. Crawford and Edward N. Huggins,
the designer was Mark E. Safran, and the production supervisor
was Thomas G. Kowalczyk. It was set in Times Roman by Denver Data.*

*Printed and bound by R. R. Donnelly & Sons Company.*

# CONTENTS

# CONTRIBUTORS

**K. J. Binns, D.Sc.** *Professor, Department of Electrical Engineering and Electronics, The University of Liverpool, England* (CHAP. 9).

**I. Boldea, Ph.D** *Associate Professor, Department of Electrical Machines, Polytechnical Institute "TR. Vuia," Timisoara, Romania* (CHAP. 7).

**T. W. Dakin, Ph.D.** *Consulting Scientist, Boca Raton, Fla.* (CHAP. 13).

**S. B. Dewan, Ph.D.** *Professor, Department of Electrical Engineering, University of Toronto, Canada* (CHAP. 8).

**A. J. Ellison, D.Sc.** *Professor Emeritus, Department of Electrical and Electronic Engineering, The City University, London, England* (CHAP. 14).

**C. Flick, M.S.** *Advisory Engineer, R&D Center, Westinghouse Electric Corporation, Pittsburgh, Pa.* (CHAP. 10).

**H. E. Jordan, Ph.D.** *Manager Corporate R&D, Reliance Electric Company, Cleveland, Ohio* (CHAP. 4).

**B. C. Kuo, Ph.D.** *Professor, Department of Electrical Engineering, University of Illinois at Urbana-Champaign, Urbana* (CHAP. 11).

**S. A. Nasar, Ph.D.** *Professor, Department of Electrical Engineering, University of Kentucky, Lexington* (CHAPS. 1, 2, AND 7).

**M. G. Say, D.Sc.** *Professor Emeritus, Heriot-Watt University, Edinburgh, Scotland* (CHAPS. 1 AND 5).

**G. R. Slemon, D.Sc.** *Dean, Faculty of Applied Science and Engineering, University of Toronto, Canada* (CHAP. 8).

**A. J. Spisak, M.S.** *President, Electro Mechanical Engineering Associates, Pittsburgh, Pa.* (CHAP. 12).

**A. Straughen, M.A.Sc.** *Professor Emeritus, Department of Electrical Engineering, University of Toronto, Canada* (CHAP. 8).

**L. E. Unnewehr, M.S.** *Director of Advanced Electronics Group, Allied Automotive, Troy, Mich.* (CHAPS. 2 AND 3).

**C. G. Veinott, D.Eng.** *Consulting Engineer, Sarasota, Fla.* (CHAP. 6).

**S. J. Yang, Ph.D.** *Reader, Department of Electrical and Electronic Engineering, Heriot-Watt University, Edinburgh, Scotland* (CHAP. 14).

# PREFACE

The purpose of this book is to provide, under one cover, authoritative information on electric machine theory, performance, tests, and design for practicing engineers, consultants, researchers, and educators. The presentation is such that the book will aid in solving on-the-job problems as well as in reviewing various types of electric machines—their characteristics, analysis, and design. The various machine types presented have information on: physical description, equivalent circuits, performance calculations, design equations, winding details, and standards for ratings (NEMA) and testing (IEEE, ANSI). Thus, the book contains a wealth of useful information on nearly all commonly used electric machines.

The book begins with the basic concepts and physical laws relating to electric machines and magnetic circuits. Subsequently, large synchronous, induction, and dc commutator machines are discussed in considerable detail. Topics covered in later chapters include: small electric motors; linear motors; electronic control of motors; permanent magnet and superconducting machines; electric motors for control applications; heating, cooling, and ventilating of electric machines; electric machine insulation; and noise and vibration encountered in the rotating machines.

To keep the presentation to the point, derivations of analysis and design equations are not included. The contributors are recognized authorities in their fields and thus bring up-to-date information to their respective chapters. Lists of references are included for further study.

*Syed A. Nasar*

# Handbook of Electric Machines

CHAPTER 1

# BASIC CONCEPTS AND TECHNOLOGY

M. G. Say
S. A. Nasar

## 1.1 THE ELECTRIC MACHINE

The electric machine is a converter of energy—mechanical into electric energy as a generator, and electric into mechanical energy as a motor. Generators produce electric energy for power-supply networks, ships, aircraft, and road transport vehicles. Motors drive industrial, processing, and domestic plants; mining equipment; trains; office machines; and many other devices. Electric machines range in power capability from a few milliwatts to a thousand megawatts or more.

A machine links an electric energy system (e.g., a power network or a battery) to a mechanical energy system (e.g., a prime mover or a mechanical load) through the terminal connections and the shaft. The machine's behavior is influenced by the properties of the attached systems. In particular, these systems affect its transient response to a change in terminal conditions; typical responses include faulty starting, stalling, and load fluctuation.

## 1.2 BASIC PRINCIPLES

Mechanical energy $w_m$ involves a force $f_m$ acting over a distance $x$ or a torque $M_m$ acting over an angle $\theta$. Electric energy $w_e$ involves the displacement of a charge $q$ (a current $i$ for a time $t$) through a potential difference $v$. The energies $w$ and the corresponding powers $p = dw/dt$ are therefore

Mechanical: $$w_m = f_m x \qquad p_m = f_m \frac{dx}{dt} = f_m u \tag{1.1}$$

or $$w_m = M_m \theta \qquad p_m = M_m \frac{d\theta}{dt} = M_e \omega_r \tag{1.2}$$

Electric: $$w_e = vq \qquad p_e = v \frac{dq}{dt} = vi \tag{1.3}$$

where $u = dx/dt$ is the translational speed and $\omega_r = d\theta/dt$ is the rotational speed. In the electric machine, the physical mechanism of interchange between electric and mechanical energy is the magnetic field, a characteristic property of electric current.

## Magnetic Field

Experimentation with magnets establishes that there is a "field of force" between them in a pattern suggesting "lines" of magnetic flux along which mechanical forces act (Fig. 1.1*a*). The lines can be taken as endowed with an "elastic thread" property, a tendency to shorten in length and to thicken laterally. The flux plot of Fig. 1.1*a* gives an appreciation of the force between the magnets.

***Magnetic Circuit.*** Magnetic flux is solenoidal; i.e., it occupies a magnetic circuit linked by the associated current. Around any path linking it, a current $i$ develops a magnetomotive force (mmf) $\mathcal{F} = i$. When the electric circuit is coiled into $N$ turns, a path linking all $N$ turns has the mmf $\mathcal{F} = Ni$ distributed around the magnetic circuit to give a path element of length $l$ the magnetic field strength $H$. Summation of $H \cdot dl$ around a closed path is $\mathcal{F}$. At any point, $H$ gives rise to a magnetic flux of density $B = \mu H$, where $\mu$ is the magnetic property of the material medium called its absolute permeability. Summation of $B$ over the area of the flux path gives the total flux $\phi$.

***Maxwell Stresses.*** Mechanical force is transmitted as the result of two stresses. At a point where the flux density is $B$ and the field strength is $H = B/\mu$, there is (1) a tensile stress $\frac{1}{2}BH$ along a flux line and (2) a compressive stress $\frac{1}{2}BH$ in all directions normal to a flux line. This Maxwell concept underlies the elastic thread analogy.

***Magnetic Circuit Law.*** The total flux $\phi$ is related to the total mmf $\mathcal{F}$ by $\phi = \mathcal{F}/R_m = \mathcal{F}\Lambda$, where $R_m$ is the reluctance and $\Lambda$ the permeance of the magnetic circuit. For a path element of length $l$ and uniform cross section $a$ over which the flux density is $B$, the mmf required is $Hl$. Then

$$R_m = \mathcal{F}/\phi = Hl/\mu Ha = l/\mu a \qquad \Lambda = 1/R_m = \mu a/l \tag{1.4}$$

For fields in free space (or air), $\mu = \mu_0 = 4\pi/10^7 \simeq 1/800{,}000$, and $\mu_0$ is the *magnetic space constant*. For fields in ferromagnetic material, the absolute permeability is $\mu = \mu_r\mu_0$, where $\mu_r$ (the permeability relative to free space) is a numeric ranging from about 10 to 400,000; thus, a flux of density $B$ can be set up in ferromagnetic

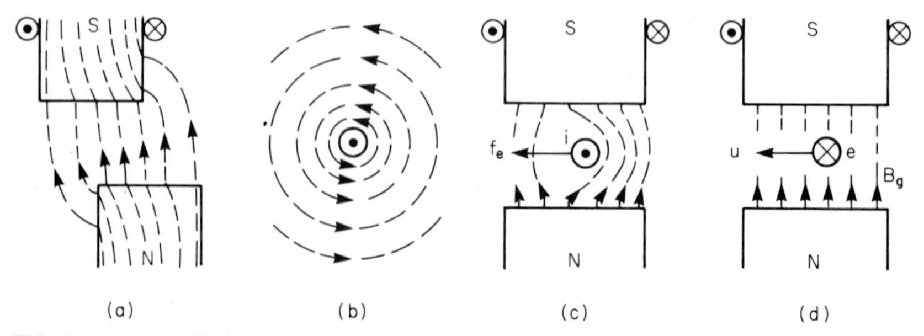


**FIG. 1.1** Magnetic field effects: (*a*) alignment of flux lines; (*b*) field due to a straight conductor; (*c*) force on a conductor; (*d*) voltage induced in a conductor.

materials with a smaller mmf than in air. Consequently, the magnetic circuit of a practical electric machine comprises a ferromagnetic structure with an essential airgap in which useful force effects can be exploited.

## Magnetic Field Effects

These are summarized in Fig. 1.1.

***Alignment.*** The flux plot (Fig. 1.1*a*) of the field across an airgap between two highly permeable ferromagnetic poles shows that the Maxwell tensions tend to draw the poles together. The force is so directed as to reduce the reluctance of the magnetic path in air: it is an *alignment*, or *reluctance*, force.

***Interaction.*** In accordance with Ampère's law, an isolated conductor in air (Fig. 1.1*b*), carrying current $i$ and of active length $l$, has an associated magnetic field in concentric circular paths. In a path of radius $r$, the field strength is $H = i/2\pi r$ and the flux density is $B = \mu_0 H$. Now let the conductor be set in the airgap between two magnetic poles that provide a gap flux density $B_g$: the resultant flux (the superposition of $B_g$ and $B$) has the configuration of Fig. 1.1*c*. Maxwell stresses result in a force $f_e$ on the current in a direction at right angles to both $i$ and $B_g$; this force is given by

$$f_e = B_g l i \tag{1.5}$$

This is an *interaction* force, $f_e$ reversing if either $i$ or $B_g$ is reversed.

***Induction.*** Let a conductor be moved at a speed $u$ across a magnetic field of density $B_g$, as in Fig. 1.1*d*. In accordance with Faraday's law, the conductor is the seat of a motional electromotive force (emf)

$$e_r = B_g l u \tag{1.6}$$

in the direction indicated. Reversal of $u$ reverses $e_r$.

***Energy Conversion.*** If the conductor carrying $i$ moves in the direction of $f_e$ at speed $u$, mechanical work is done at the rate $p_m = f_e u$. But the motion generates $e_r$ opposing the current, which must be maintained by an equal and opposing source voltage; there is thus an electric power input to the conductor of $p_e = e_r i$. Combining the effects of Figs. 1.1*c* and *d*,

$$p_m = f_e u = (B_g l i)u = (B_g l u)i = e_r i = p_e \tag{1.7}$$

An electric input power $p_e$ develops a mechanical output $p_m$, the machine working in the *motor* mode. If $i$ (and thus also $f_e$) is reversed, $e_r i$ becomes an electric output into an external circuit, and with the same direction of motion the machine operates in the *generator* mode, converting a $p_m$ input to a $p_e$ output. The machine acts in either mode simply by reversal of current direction.

***Energy Transfer.*** A principle invoked in some machines enables the current $i$ to be furnished inductively, with no external connection. This follows from a full statement of Faraday's law of electromagnetic induction: An emf $e$ occurs in an electric circuit whenever its flux linkage changes. The linkage is $\psi = N\phi$, and $e = d\psi/dt$, with components $e_r$ of motion and $e_p$ of pulsation:

$$e = \frac{d\psi}{dt} = \frac{\partial\psi}{\partial\theta}\frac{\partial\theta}{\partial t} + \frac{\partial\psi}{\partial t} = \frac{\partial\psi}{\partial\theta}\omega_r + \frac{\partial\psi}{\partial t} = e_r + e_p \tag{1.8}$$

for a rotary machine. If a winding carries a current $i$, then $e_r i$ is the rate of electromechanical energy conversion and $e_p i$ is the rate of electric energy transfer.

***Linked Electromechanical System.*** In Fig. 1.2 the electrical system is represented by a voltage source $v_g$ and a source impedance $Z$. A mechanical system attached to the shaft of the machine and rotating at angular speed $\omega_r$ has a load or a prime mover of torque $M_m$ together with a frictional torque $M_f$, an elastic torque $M_s$ (arising from spring members or shaft distortion), and an inertia $J$ that demands or supplies a torque when the machine changes speed. The electric machine has an input voltage $v$ and an input (or output) current $i$. In the conversion-transfer region, $e = e_r + e_p$ is generated by a magnetic field. Resistor $R$ and inductor $L$ account, respectively, for $I^2R$ loss and nonuseful ("leakage") flux linkage.

***Prototype Machines.*** The necessary condition for conversion is that the working flux density $B_g$, the conductor current $i$, and the angular (or linear) motion shall be mutually at right angles. The arrangements in Fig. 1.3 conform to this requirement. The flux and current directions are shown for homopolar (unipolar) machines; for heteropolar (alternate N- and S-pole) machines, the current direction is reversed in regions of opposite polarity in order to preserve unidirectional interaction force.

***Heteropolar Cylindrical Machine.*** Shown for a two-pole magnetic circuit in Fig. 1.4*a*, this is by far the most common form. The active region is the airgap between a fixed member (stator) and a rotatable member (rotor). Both are ferromagnetic and, with the airgap, complete the magnetic circuit. Conductors are attached axially to the stator bore surface in two bands, the conductors on one band being joined to those on the other by end connectors to form turns. The resulting stator current-sheet pattern develops a distributed mmf, with its peak $\mathcal{F}_1$ magnetizing on the axis of the winding. The rotor is provided with a similar current-sheet pattern of peak mmf $\mathcal{F}_2$ on the winding axis.

Let $\mathcal{F}_2$ be displaced from $\mathcal{F}_1$ by an angle $\lambda$: then a torque is developed that tends to realign the axes. The system seeks a minimum reluctance position, with $\mathcal{F}_1$ and $\mathcal{F}_2$ aligned. Alternatively, either member can be considered to produce the gap flux density $B_g$ in which the currents $i$ of the other lie, developing an interaction torque. If the mmf distributions around the airgap are sinusoidal, the torque is

$$M_e = k\mathcal{F}_1\mathcal{F}_2 \sin \lambda \tag{1.9}$$

where $\lambda$ is the *torque angle* and $k$ depends on the airgap dimensions. For $\lambda = 0$, the torque vanishes; displacement increases the torque to a maximum for $\lambda = \pi/2$ rad; and for further displacement, the torque falls to zero for $\lambda = \pi$ rad (at which the system is in unstable equilibrium).

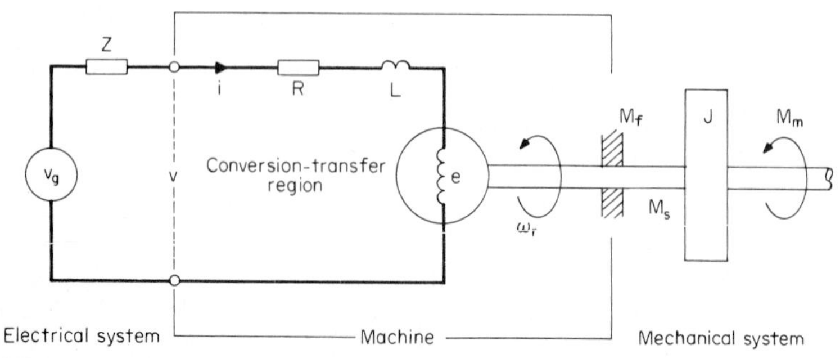


**FIG. 1.2** Linked electromechanical system.

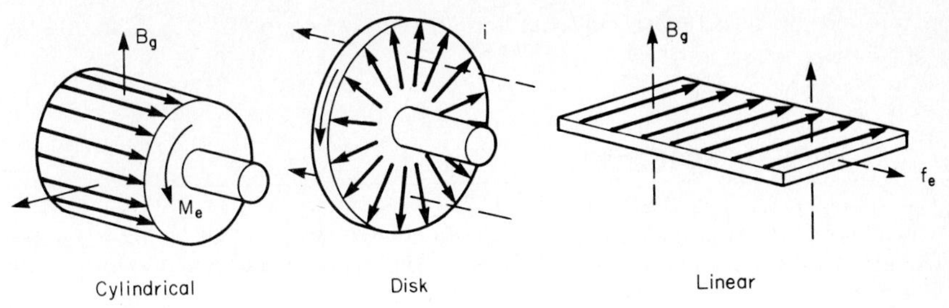


**FIG. 1.3** Machine morphology.

The machine in Fig. 1.4*b* exploits the maximum torque condition by making $\lambda = \pi/2$ rad permanently. This requires a special form of commutator winding on the rotor. A salient-pole stator construction is normally employed.

Let an ideal machine operating at optimum torque angle have a stator current-sheet density $A$ and a gap flux density $B$, both being *mean* values: then the interaction force per unit of gap surface area is $BA$. With a stator bore diameter $D$ and active length $l$, the total interaction force is $f_e = BA\pi Dl$ and the torque $f_e \times \frac{1}{2}D$ is

$$M_e = \tfrac{1}{2}BA\pi D^2 l \tag{1.10}$$

At angular speed $\omega_r$ the converted power is $P_e = M_e\omega_r$. Then, for a rotational speed $n = \omega_2/2\pi$,

$$P_e = \pi^2 BAD^2 ln \tag{1.11}$$

Let $A = 79$ kA/m, $B = 0.85$ T, $D = 1.20$ m, $l = 3.5$ m, and $n = 60$ r/s: then $M_e =$ 530 kN·m and $P_e = 200$ MW, typical of a high-speed turbogenerator of moderate rating. For a small motor with $A = 27$ kA/m, $B = 0.48$ T, $D = 0.15$ m, $l = 0.23$

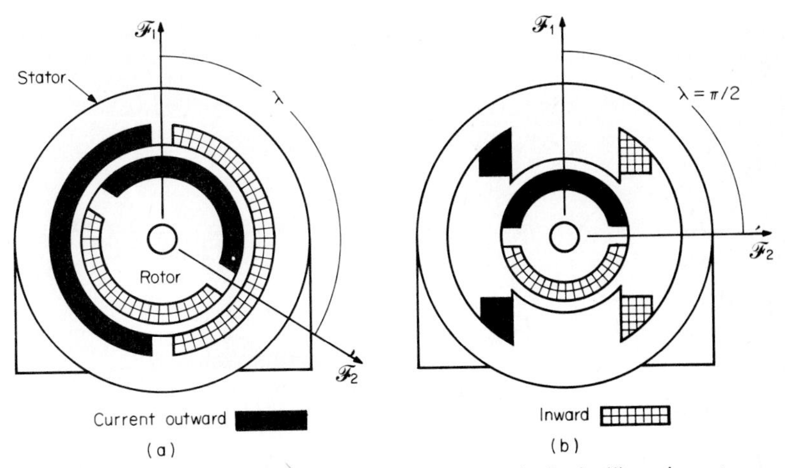


**FIG. 1.4** Prototype two-pole machines: (*a*) a two-pole magnetic circuit; (*b*) maximum torque condition.

m, and $n = 30$ r/s (1800 r/min), the torque and power are 105 N·m and 20 kW, respectively.

## 1.3 TYPES OF MACHINE

A machine requires a unidirectional torque to be maintained when the rotor rotates. The electrical supply system has a dominant influence on the physical arrangement. The three most common machines are all heteropolar and have at least one member cylindrical; they are distinguished by the nature of the airgap flux:

1. *Three-phase ac:* Traveling-wave gap flux, one member dc-excited (*synchronous* machine) or fed by energy transfer from the other (*induction*, or *asynchronous*, machine).
2. *DC or single-phase ac:* Fixed-axis flux, the rotor having a commutator winding (*commutator* machine).

### Traveling-Wave Gap Flux

The stator current-sheet pattern in Fig. 1.4*a* is set up by three similar windings (*A*, *B*, and *C*) displaced (in a two-pole machine) by an angle $2\pi/3$ rad, as in Fig. 1.5*a*. The conductors in phase bands *A* and *A'* are connected to form turns, and similarly for *B* and *C*. With the phase windings excited by balanced three-phase currents of frequency $f_1$, the sequential cyclic reversal of currents in the displaced phase windings shifts the current-sheet pattern—as shown in Fig. 1.5*b* for peak current (i) in phase A and for peak current (ii) in phase B (one-third of a period later). Thus, the stator mmf $\mathcal{F}_1$ produces a *traveling wave* of mmf and gap flux, and hence a "rotating field" moving at *synchronous* speed [$n_s = f_1$ (r/s), or $\omega_1 = 2\pi f_1$ (rad/s), for a two-pole machine]. (In general, the synchronous speed of a machine with $p$ pole pairs is $n = f_1/p$, or $\omega_1 = 2\pi f_1/p$.)

Suppose the rotor, rotating at angular speed $\omega_r$, to have a three-phase winding carrying currents of frequency $f_2$: then it has a current-sheet pattern of mmf $\mathcal{F}_2$ rotating at angular speed $\omega_2 = 2\pi f_2$ with respect to the *rotor* body, and therefore

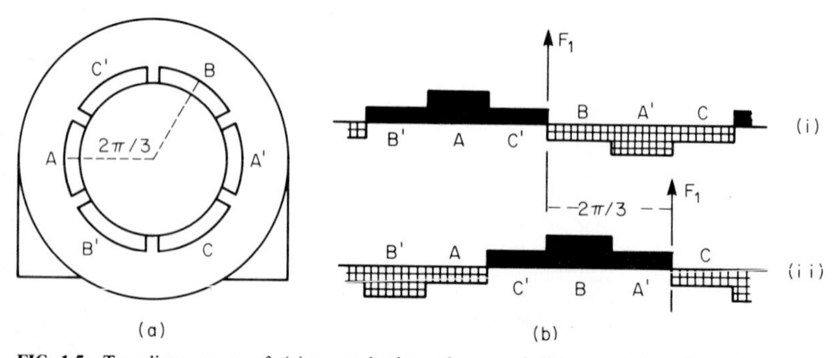


**FIG. 1.5** Traveling-wave mmf: (*a*) two-pole three-phase mmf; (*b*) current sheet for peak current in phase A and in phase B, one-third of a period later. The height of the shaded blocks is proportional to the instantaneous phase current.

at $\omega_r \pm \omega_2$ with respect to the *stator*. For unidirectional torque, $\mathcal{F}_1$ and $\mathcal{F}_2$ must rotate in synchronism, and $\omega_r \pm \omega_2 = \omega_1$ is the essential running condition.

***Synchronous Machine.*** Because the rotor is dc-excited, $\mathcal{F}_2$ is "fixed" to the rotor body, i.e., $\omega_2 = 0$ and $\omega_r = \omega_1$. Thus, the rotor must rotate in synchronism with the traveling-wave mmf. The torque angle accommodates to the torque demand up to a maximum for the torque angle $\lambda = \pi/2$ rad. The machine can operate in both generator and motor modes by a reversal of the torque angle.

***Induction Machine.*** The rotor winding, isolated and closed on itself, derives its current inductively from the stator. If the rotor spins at synchronous speed, its conductors move with the stator field and no current is induced. But if $\omega_r$ is less than $\omega_1$ by a fractional "slip" $s = (\omega_1 - \omega_r)/\omega_1$, then the rotor conductors lie in a field changing at slip frequency $s\omega_1$, and currents of this frequency are induced to provide an mmf $\mathcal{F}_2$ traveling around the rotor at this frequency. This gives $(\omega_r + \omega_2) = \omega_1$, the required condition. Thus, torque is developed for any slip $s$ other than zero (synchronous speed). When the machine is driven above synchronous speed, the torque is reversed and the machine generates.

### Fixed-Axis Gap Flux

In the usual structural form, Fig. 1.4*b*, the gap flux is produced by the fixed-axis stator mmf $\mathcal{F}_1$. The rotor mmf $\mathcal{F}_2$ has the optimum torque angle $\lambda = \pi/2$ rad. As the rotor spins, the current of each individual conductor is reversed as it passes from the outward- into the inward-directed region of the current sheet, a *commutation* process involving sliding contacts or sequence switching. In consequence, the machine can develop torque at standstill and, ideally, at any speed.

***DC Commutator Machine.*** Except for miniature machines with $\mathcal{F}_1$ produced by permanent magnets, both stator and rotor windings are dc-excited, with individual currents ("shunt" or "separately excited") or with the same current ("series"). The torque is smooth and continuous, with both motor and generator modes obtainable.

***Single-Phase Commutator Machine.*** As simultaneous reversal of $\mathcal{F}_1$ and $\mathcal{F}_2$ leaves the torque direction unchanged, the dc motor can, without essential change, be operated on one-phase ac with the stator and rotor windings in series. The torque has a double-frequency pulsation about a unidirectional mean.

***Other Forms.*** There are many variants, especially in small and miniature machines. Single-phase induction motors require special techniques ("split-phase," "shaded-pole"). Some operate on alignment ("reluctance," "hysteresis," "stepper"). The linear machine is usually based on the three-phase rotary motor. A few large homopolar dc machines have been built. These variants lack the functional simplicity of those described above, but the basic principles are still valid.

## 1.4 THREE-PHASE SYSTEMS

Polyphase circuits—three-phase circuits in particular—are used in the majority of transmission, distribution, and energy conversion systems where the power, or voltampere (VA), levels are above 10 to 20 kilowatts (kW) or kilovoltamperes (kVA). The basic reasons for using polyphase systems are related to *power density*, which is defined as the ratio of either power to weight or power to volume of the device. Thus, an electric machine of a given weight is capable of delivering more power in polyphase than in single-phase designs. In some electrical applications the voltampere rating is often substituted for the power rating. For example, the specific weight of an ac motor is generally less for a three-phase motor than for a single-phase motor at output power ratings above 1 horsepower (hp). Above approximately 5 hp, a three- or two-phase motor will always have a lower specific weight than a single-phase motor. Because three-phase systems are by far the most common, the following discussion will be restricted to them.

Suppose we have a system of three ac voltages of a certain frequency such that their amplitudes are equal but these voltages are displaced from one another by 120° in time. We may mathematically express this system of voltages as

$$v_{a'a} = V_m \sin \omega t \tag{1.12}$$

$$v_{b'b} = V_m \sin (\omega t - 120°) \tag{1.13}$$

$$v_{c'c} = V_m \sin (\omega t - 240°) \tag{1.14}$$

These voltages are graphically depicted in Fig. 1.6. In terms of their root-mean-square (rms) values, these voltages may be written in phasor notation as

$$V_{a'a} = V\underline{/0°} \tag{1.15}$$

$$V_{b'b} = V\underline{/-120°} \tag{1.16}$$

$$V_{c'c} = V\underline{/-240°} \tag{1.17}$$

Figure 1.7*a* shows the phasor representation of Eqs. (1.15) to (1.17). Notice that Fig. 1.7*b* also shows (hypothetically) three voltage sources corresponding to Eqs. (1.15) to (1.17). Consequently, a three-phase (voltage) source having three equal voltages 120° out of phase with one another may be defined; this system is called a *three-phase balanced system*—in contrast to an unbalanced system, in which the magnitudes may be unequal and/or the phase displacements may not be 120°. For a balanced three-phase system, the phasor sum of the three voltages is zero.

Let us abbreviate $v_{a'a}$, $v_{b'b}$, and $v_{c'c}$ as $v_a$, $v_b$, and $v_c$, respectively. Now, referring to Figs. 1.6 and 1.7*a*, we observe that the voltages attain their maximum values in the order $v_a$, $v_b$, and $v_c$. This order is known as the phase sequence *abc*. A reverse

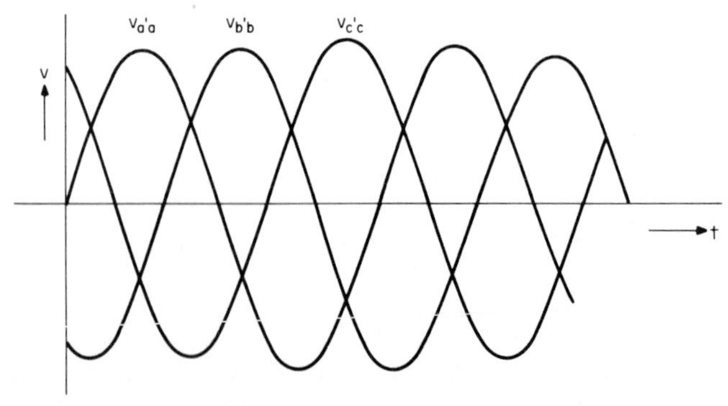


**FIG. 1.6** A system of three voltages of equal magnitude, but displaced from each other by 120° in phase.

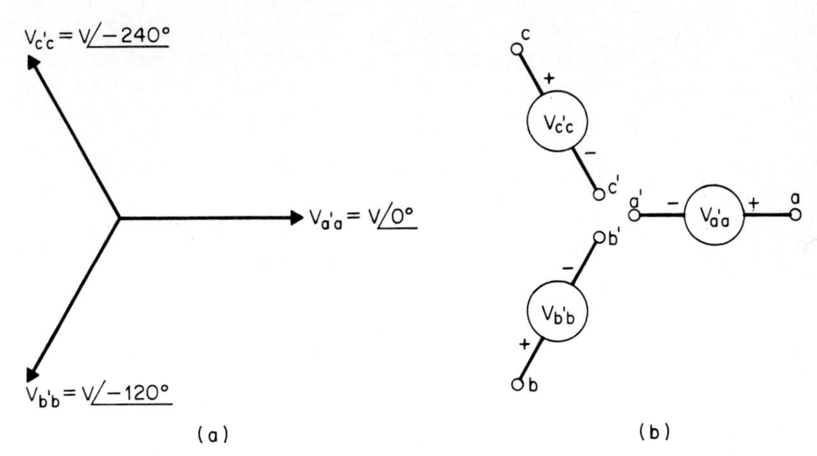


**FIG. 1.7** (*a*) Balanced three-phase phasor representation; (*b*) three-phase voltage source.

phase sequence will be *acb*, in which case the voltages $v_c$ and $v_b$ lag $v_a$ by 120° and 240°, respectively.

## Three-Phase Connections

There are two common, practical ways to interconnect the three voltage sources shown in Fig. 1.7*b*. These two forms of interconnection are illustrated in Figs. 1.8*a* and *b*, respectively labeled the "wye" and "delta" connections. The result, for either type of connection, is a three-terminal ac source of power supplying a balanced set of three voltages to a load. In the wye connection, the terminals $a'$, $b'$, and $c'$ are joined together to form the neutral point *o*. If a lead is brought out from the neutral point *o*, the system becomes a *four-wire three-phase* system. The terminals *a* and $b'$, *b* and $c'$, and *c* and $a'$ are joined individually to form the delta connection.

Notice from Fig. 1.8*a* that two types of voltages can be identified: voltages $V_{a'a}$, $V_{b'b}$, and $V_{c'c}$ across the three individual phases, known as the *phase voltages*; and voltages $V_{ab}$, $V_{bc}$, and $V_{ca}$ across the lines *a*, *b*, and *c* (or *A*, *B*, and *C*), known as the *line voltages*. The line voltages are related to the phase voltages such that

$$V_l = \sqrt{3}V_p \tag{1.18}$$

The relationships of all the phase voltages and line voltages are illustrated in the phasor diagram of Fig. 1.9.

For the wye connection, it is clear from Fig. 1.8*a* that the line currents $I_l$ and phase currents $I_p$ are the same. Thus, we may write

$$I_l = I_p \tag{1.19}$$

For the delta connection, Fig. 1.8*b* verifies that the line voltages $V_l$ are the same as the phase voltages $V_p$. Hence,

$$V_l = V_p \tag{1.20}$$

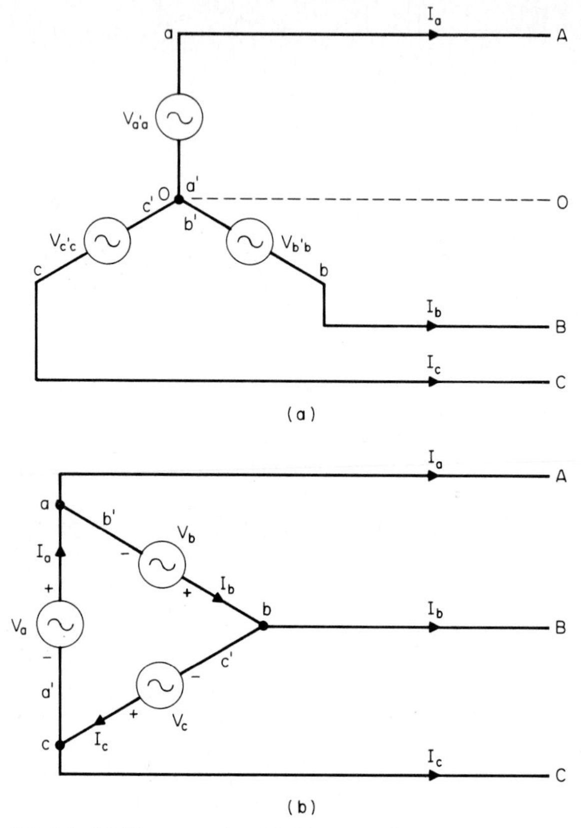


**FIG. 1.8** (*a*) Wye connection; (*b*) delta connection.

Figure 1.10 shows the phase currents and line currents for the delta-connected system of Fig. 1.8*b*. The phase currents and line currents are related to each other by

$$I_l = \sqrt{3} I_p \tag{1.21}$$

## Power in Three-Phase Systems

In contrast to the fact that the instantaneous power in a single-phase ac circuit is pulsating in nature, it is interesting to note that the instantaneous power in a balanced three-phase circuit is a constant.

If the three-phase circuit is not purely resistive and has a power-factor angle $\theta_p$, then the average power delivered by each phase is given by $V_p I_p \cos\theta_p$. Thus, the total power delivered to the load by this balanced system is given by

$$P_T = 3(V_p I_p \cos\theta_p) \tag{1.22}$$

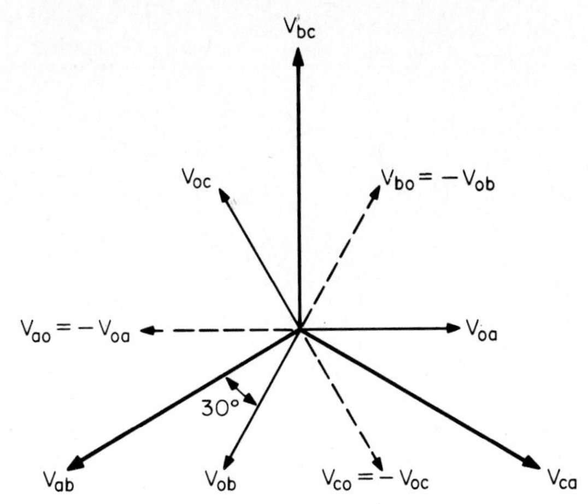


**FIG. 1.9** Voltage phasors for Y connection.

For wye or delta connection, the total power in terms of line voltages and currents can be expressed as

$$P_T = \cos \sqrt{3} V_l I_l \cos \theta_p \tag{1.23}$$

It should be noted that the angle $\theta_p$ is still the angle between the phase voltage and phase current.

A graphical representation of the instantaneous power in a three-phase system is given in Fig. 1.11. It is seen that the total instantaneous power is constant and is equal to 3 times the average power. This feature is of great value in the operation of three-phase motors, for the constant instantaneous power implies an absence of torque pulsations and consequent vibrations.

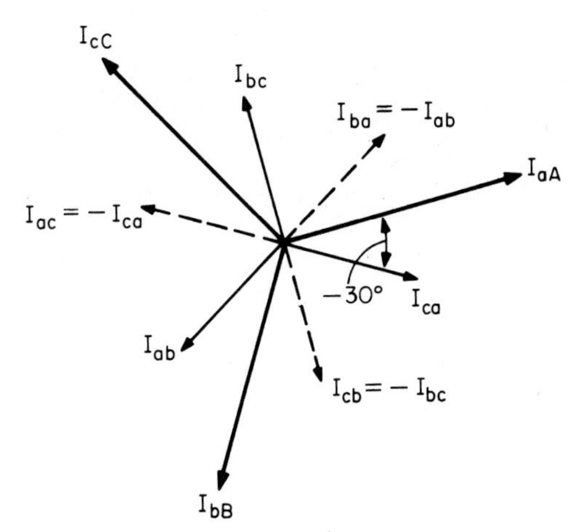


**FIG. 1.10** Current phasors for Δ connection.

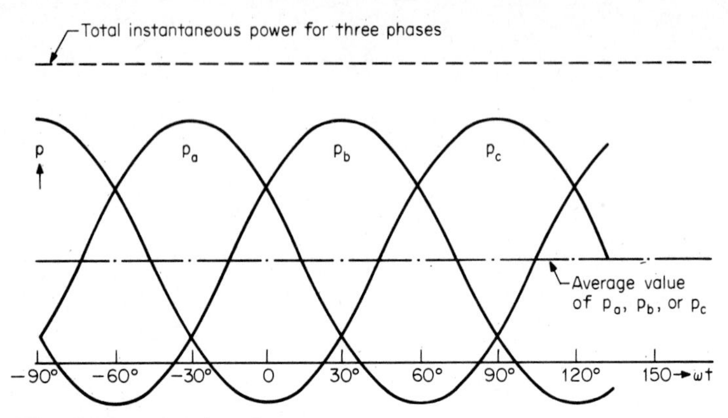


**FIG. 1.11** Power in a three-phase system.

## Wye-Delta Equivalence

In working with *balanced* three-phase systems, one usually makes computations on a *per-phase* basis first and then obtains the total results for the entire circuit on the basis of the symmetry and other factors that must apply. Thus, if a set of three identical wye-connected impedances is connected to a wye-connected three-phase source (as shown in Fig. 1.12), then, because of the symmetry, point $o'$ is at the same potential as point $o$; therefore, the two may be connected to each other. Thus, each generator supplies only its own phase, and the computations on a per-phase basis are therefore legitimate.

Sometimes it is necessary, or desirable, mathematically to convert a set of identical impedances connected in wye to an equivalent set connected in delta, or vice versa (Fig. 1.13). In order for this transformation to be valid, the impedances seen between any two of the three terminals must be the same. For this condition, the wye impedances in terms of the delta impedances, or vice versa, are such that

$$Z_w = \frac{Z_d}{3} \tag{1.24}$$

and

$$Z_d = 3Z_w \tag{1.25}$$

Hence, we can switch back and forth between balanced wye and delta as desired.

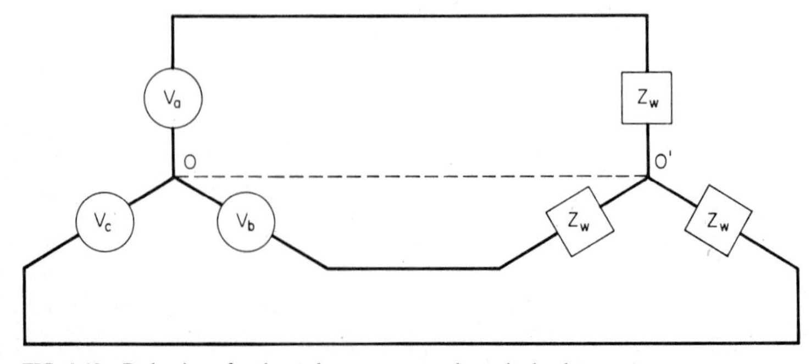


**FIG. 1.12** Reduction of a three-phase system to three single-phase systems.

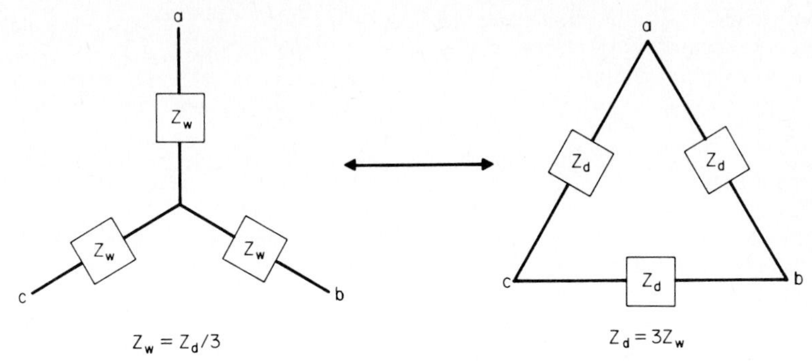


**FIG. 1.13** Equivalence of Y and Δ impedances.

### Harmonics in Three-Phase Systems

The term *harmonics* in power circuitry refers to components of current or voltage the frequency of which is greater than the *fundamental*, or *source*, frequency. Voltage and current components the frequency of which is *less* than the fundamental, or applied, frequency are known as *subharmonics*. Voltage and current harmonics, as generally used in power circuit theory and in this handbook, are found by means of a Fourier series analysis of the voltage or current waveforms or, experimentally, by waveform analyzers or similar instrumentation. In power circuits, the source harmonic current is due to the nonlinear characteristic of the magnetic circuit of an electromechanical device, such as a transformer or alternator. Both current and voltage harmonics also result from the application of nonsinusoidal voltage or current sources. In general, sources involving power semiconductors are nonsinusoidal.

The role of harmonic voltages and currents in polyphase power systems is a fascinating subject and involves far more theory, analysis, and experimentation than could possibly be covered in this handbook. There are, however, several simple concepts worth noting. In a three-phase system, the phase shift of the third harmonic of the second phase is $3 \times 120°$, or $360°$; likewise, in the third phase the phase shift is $3 \times 240°$, or $720°$ or $360°$. This tells us that in a three-phase system the third harmonics are all shifted in such a manner as to be in phase with each other. In a wye connection, this may add up to considerable current flowing in the neutral wire; in a delta, this amounts to a "circulating" current around the delta. The delta connection is often used for just this purpose—to "trap" the third harmonic currents—that is, to contain third harmonic currents as circulating currents around the delta windings. The "price" for trapping third harmonic currents in a delta winding is, of course, increased ohmic losses—and hence increased temperature rise—in the windings. The benefits are improved voltage and current waveforms on the three-phase lines. Trade-offs such as this are usually made in power systems on the basis of cost, efficiency, and complexity versus the penalties for nonsinusoidal waveforms.

## *1.5 NAMEPLATE RATINGS*

Each commercial rotating machine has a nameplate attached to it in some manner. Besides the legal and warranty considerations related to the numerical values listed on the nameplate and the usefulness of these values in applying a machine to a specific load, the nameplate describes the following: the type of machine; its power capability; its speed, voltage, and current characteristics; and some of its environmental limitations. The nameplate gives the set of parameters most often used analytically or experimentally to determine machine performance and is widely used in college laboratory experiments and in machine applications. Therefore, a brief description of the meaning of machine nameplate parameters follows.

**1.** *Power:* This is expressed in terms of the unit horsepower in motors and of watts or kilowatts in generators. It refers to the *continuous output* power that the machine can deliver; in machines with variable loading, continuous power is equivalent to average load-cycle power. The continuous power rating is primarily a function of the thermal capacity of the machine, which in turn depends on the frame configuration, sometimes called the machine "package." There are two basic types of packages in commercial machines, with a number of variations within each type determined by the type of environment and external heat-transfer equipment in which the machine will be operated: *open* (drip-proof, splashproof, externally ventilated, etc.) and *totally enclosed* (nonventilated, fan-cooled, dustproof, water-cooled, encapsulated, etc.).

**2.** *Speed:* This is expressed in the unit revolutions per minute (rpm, or r/min). The speed listed on the nameplate depends on the general type of machine:

Synchronous—synchronous speed

Induction—speed at rated power (synchronous minus slip speed at rated power)

DC—usually base speed, the maximum speed at which rated torque can be supplied

Universal—usually no-load or light-load speed

DC control—usually no-load speed

Many commercial motors are designed to operate from two different voltage-source levels or at two different synchronous speeds, and in such cases there will be two sets of speed ratings.

**3.** *Voltage:* This is the nominal *voltage* rating of the source to which the machine windings are to be connected; the units are volts. In polyphase machines the rated voltage is always expressed as a *line-to-line* voltage. The rated voltage also gives the *insulation* level at which the machine has been constructed and tested. Insulation levels are standardized in commercial machines.

**4.** *Current:* This is the *steady-state current* in the armature or power circuit at rated power output and rated speed, expressed in rms in ac machines and average amperes in dc commutator machines. In polyphase machines this current is always a *line current*. In dc commutator machines the *field-circuit* current rating is the current's "full field," i.e., the field required for maximum torque at base speed.

**5.** *Voltamperes:* In ac motors the input voltampere rating is derived from the current and voltage ratings. In single-phase motors this is

$$\text{Rated voltamperes} = (\text{rated volts})(\text{rated amperes})$$

In three-phase motors it is

$$\text{Rated voltamperes} = \sqrt{3}\,(\text{rated volts})(\text{rated amperes})$$

In most ac generators the output voltampere rating, from which the output power factor can be derived, is stamped on the nameplate in place of the current rating. Both the current and the voltampere ratings of machines are determined primarily by the thermal characteristics of the windings.

**6.** *Temperature rise:* This is the maximum safe temperature rise (in degrees Celsius) in the "hot spot" of the machine, which is usually the armature or power windings.

**7.** *Service factor:* This number indicates how much over the nameplate power rating the machine can be continuously operated without overheating. It has a value of 1.15 for many commercial motors. In addition, there is a series of *short-time overload ratings* for most commercial motors, which can be obtained from the manufacturer.

**8.** *Frequency:* This is the frequency of the supply voltage, in hertz.

**9.** *Efficiency index:* A recent addition to the nameplates of some manufacturers, this index gives an indication of minimum and nominal efficiency.

**10.** *Torque:* Torque is generally listed on nameplates only for control and torque motors. The units vary considerably, although ounce-inches are most common. The rated steady-state torque in power motors can be derived from nameplate power and speed. Two important torque parameters in ac machines—pull-out and starting torques at rated voltage—can be obtained through the manufacturer or by means of relatively simple laboratory tests.

**11.** *Inertia:* This parameter, the inertia of the rotating member, is a nameplate item only for control motors.

**12.** *Manufacturer:* The manufacturer's name also appears on the nameplate.

**13.** *Frame size:* The frame size is the manufacturer's indication of the physical size of the motor and its mounting dimensions. For motors of ratings greater than 1 hp, the National Electrical Manufacturers Association (NEMA) has developed a standard system of frame numbering and the standardized dimensions of these frame sizes.

A typical nameplate of a dc motor is shown in Fig. 1.14. Some other information given on some nameplates includes the serial number, the style, whether the motor is open or totally enclosed, and a code letter to indicate locked-rotor kVA.

## 1.6 EFFICIENCY AND DUTY CYCLE

This section discusses efficiency and losses in a broad and general sense. (Later chapters discuss them specifically in terms of various machine types.)

### Efficiency

For the electric machines discussed in this handbook, *efficiency* has the following meaning:

$$\eta = \frac{\text{output power or energy}}{\text{input power or energy}} \tag{1.26}$$

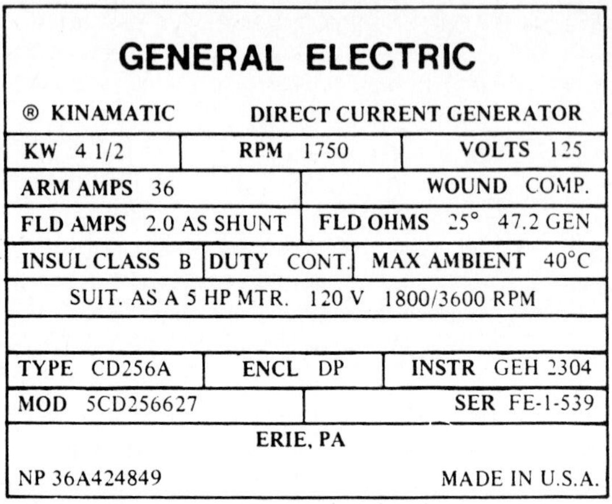

| GENERAL ELECTRIC | | |
|---|---|---|
| ® KINAMATIC | DIRECT CURRENT GENERATOR | |
| KW 4 1/2 | RPM 1750 | VOLTS 125 |
| ARM AMPS 36 | | WOUND COMP. |
| FLD AMPS 2.0 AS SHUNT | FLD OHMS 25° 47.2 GEN | |
| INSUL CLASS B | DUTY CONT. | MAX AMBIENT 40°C |
| SUIT. AS A 5 HP MTR. 120 V 1800/3600 RPM | | |
| | | |
| TYPE CD256A | ENCL DP | INSTR GEH 2304 |
| MOD 5CD256627 | | SER FE-1-539 |
| ERIE, PA | | |
| NP 36A424849 | | MADE IN U.S.A. |

**FIG. 1.14** Typical nameplate of a dc motor.

This can also be expressed in terms of mechanical and electrical losses in either energy or power terms as

$$\eta = \frac{\text{output}}{\text{output} + \text{losses}} = \frac{\text{input} - \text{losses}}{\text{input}} \tag{1.27}$$

In the International System of Units (SI), the units of power are watts (W); the SI units of energy are joules (J) and wattseconds (Ws) or watthours (Wh), and 1 J = 1 Ws.

In electric machines, either the numerator or denominator of Eqs. (1.26) and (1.27) is a mechanical power or energy. The mechanical power of a rotating machine can be expressed as

$$P_m = T_{\text{av}} \Omega_{\text{av}} \tag{1.28}$$

where $T_{\text{av}}$ = shaft torque, N·m
$\Omega_{\text{av}}$ = shaft speed, rad/s

On the electrical side of a machine, power is expressed as

$$P_e = VI \cos \theta \qquad \text{sinusoidal} \tag{1.29}$$

or

$$P_e = V_{\text{av}} I_{\text{av}} \qquad \text{dc or pulse} \tag{1.30}$$

where $V$ = terminal voltage, V
$I$ = terminal current, A
$\theta$ = power-factor angle

In these equations the rms parameters have been designated by uppercase, unsubscripted symbols; time-average parameters are designated by uppercase symbols and the subscript "av." The power calculated by these equations is *average power*. It is also common to have instantaneous quantities on the right side of these equations, in which case lowercase symbols would be used and the power on the left would be referred to as instantaneous power. The use of both average and instantaneous power is common in the analysis of electromechanical systems. Energy $W$ is the time integral of power, i.e.,

$$W = \int_0^t p\,dt \tag{1.31}$$

### Duty Cycle

In many applications, electric machines are operated at power levels that change with time instead of being at constant power. When the variation in power level can be described as a periodic function, it is known as a *duty cycle.*

Duty-cycle variations are used primarily to describe the variations of loading of a motor, but the concept is equally valid for generators, transformers, and other power devices under such conditions. A duty cycle may be expressed in terms of input or output power, shaft torque, armature current, or other machine parameters that describe the load variation. Examples of applications in which the load characteristics can be so described include automatic washing machines, refrigerator compressors, many types of machine tool processes, and electric vehicles traveling through a prescribed driving cycle.

The principal purposes for the use of duty-cycle analysis are (1) to determine the size or the *rating* of a machine to satisfy a particular load requirement and (2) to determine *energy efficiency*. The latter purpose is important for machines operating at several power levels since the *power efficiency* is generally different at each power level in the duty cycle.

## 1.7 RATINGS AND NEMA CLASSIFICATIONS

Although the ratings discussed here are closely related to the nameplate ratings discussed in Sec. 1.5, they are restricted to those which specify the operating limitations on a machine. As in all physical systems, there are physical limitations on electromechanical devices that set bounds on the operation of these devices. One obvious limitation on an electric machine is some maximum speed, operation beyond which will result in damaged bearings, disintegration of the rotating member, or some other such catastrophe. Another type of bound is *saturation*, the limit on the performance of the magnetic members of the device; exceeding the saturation seldom results in any physical damage, but it limits device performance in many ways. Two other significant bounds on an electric machine's operation that result in physical damage or hazardous operation when exceeded are thermal bounds and commutation bounds.

Obviously, a machine should never be operated anywhere near bounds that may cause damage to the machine or injury to operating personnel. Therefore, another set of bounds, known as *machine ratings*, has been developed to guide the machine owner or operator in the proper and safe operation of the machine. Machine ratings are always considerably below the parameter values that might result in hazardous operation or machine damage. Originally, the ratings were somewhat subjective and based largely on past experience and long periods of testing under load; more recently, they have sometimes been determined by analytical models of machines and machine subsections. There is a definite legal connotation to a machine rating, and by the user. Like most rules, however, machine ratings are designed to be broken or exceeded under certain conditions, and most manufacturers will supply the user with "overload" ratings. This implies that there is a considerable margin between a machine rating and the safety bounds just discussed. Most rotating machines designed for power applications are rugged, long-lived devices and can stand some abuse from the user or the environment; however, this margin of safety built into most rotating machines should be known before using it and should be used only when absolutely essential. In other words, for normal applications, a machine should be operated only at or below its ratings.

It is not the purpose of this handbook to give a detailed explanation of the scope and meaning of machine ratings. This complex and everchanging subject is described in a number of publications available in most libraries and from machine manufacturers and users. The *National Electrical Code* describes the use and application of all types of machines, including safety requirements.

The sizes of commercial machines are designated by a parameter known as *frame size*. Frame-size standards are determined by NEMA and are described in its literature. NEMA has also established a designation, known as *motor classes*, for the general efficiency and starting-torque characteristics of ac motors. Motor classes A to D exhibit decreasing running efficiency and increasing starting torque with successive alphabetical letter designations. The NEMA classification of the most common types of three-phase induction motors is as follows:

*NEMA design B:* Normal torques; normal slip; normal locked amperes

*NEMA design A:* High torques; low slip; high locked amperes

*NEMA design C:* High torques; normal slip; normal locked amperes

*NEMA design D:* High locked-rotor torque; high slip

Figure 1.15 shows typical NEMA design B, C, and D motor torque-speed characteristics.

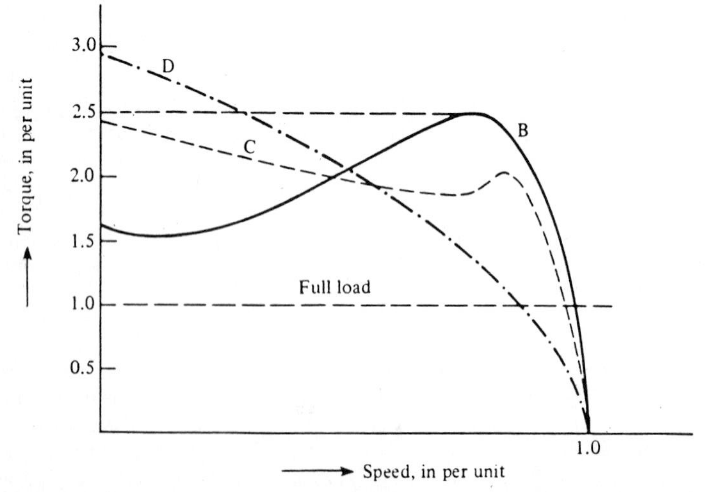


**FIG. 1.15** Torque-speed characteristics of NEMA design B, C, and D motors.

The wire used in electric machines is known as *magnet wire.* Wire sizes are most commonly described in terms of numerical designations known as the American Wire Gauge (AWG). Electrical insulation for magnet wire is designated by another alphabetical listing of A to H. The insulation class refers primarily to the safe operating temperature of the insulation and, to a lesser degree, to the structural characteristics and physical strength of the insulation; the temperature range of this designation is from 90°C (class A) and 110°C (class B) to 180°C (class H).

# CHAPTER 2
# MAGNETIC CIRCUITS

**S. A. Nasar**
**L. E. Unnewehr**

## *2.1 INTRODUCTION*

The essential element in all electric machines and electromechanical devices of the electromagnetic type is an electromagnetic system. The function of this system is to establish and control electromagnetic fields in order to accomplish the desired process of energy conversion, energy transfer, or energy processing. In the simplest sense, an electromagnetic system consists of electric circuits located in a region of space and having a very specific geometry designed to establish the required electromagnetic field relationships.

It is possible to use conventional electric circuit theory to describe and analyze many of the functions and performance characteristics of an electromagnetic system in terms of these electric circuits. However, to understand the basic energy conversion process and to be able to determine the electric circuit parameters of an electric machine, it is necessary to understand the electromagnetic field of the machine and to become familiar with the terms and analytical expressions used to describe this field. Also, much of the process of designing a machine is centered around the design of the magnetic system. This chapter thus reviews some basic concepts of magnetic field theory and shows how they are of value in the design and analysis of electric machines.

The term *field* is a concept used to describe a distribution of forces throughout a region of space. The electric field describes the force on a unit of charge of electricity, the electron. The magnetic field describes the force on a magnetic dipole. Electric machines of the *electromagnetic type* produce forces or torques resulting from the presence of the magnetic force field. There is a class of electric machines known as *electrostatic machines* whose forces result from the presence of electric fields; these are much less common than electromagnetic machines.

Fields are three-dimensional spatial phenomena. A truly three-dimensional analysis of a field becomes very complex and time-consuming and will tie up tremendous blocks of computer storage when used in computer analysis methods. Fortunately, three-dimensional analysis is seldom necessary because of the property of fields known as *symmetry*. Symmetry considerations allow us to resolve the three-dimensional problem into one of two dimensions or even into one dimension within a limited region of space, thus simplifying the analysis and conceptual difficulties. Much of the task of analyzing machines and electromechanical devices rests on identifying the symmetry of its fields.

Tests for symmetry simplification revolve around answers to two questions: (1) What dimensional coordinate components of the field do *not* exist? (2) With which dimensional coordinate components does the field *not* vary? Probably the most "symmetrical" of electromagnetic devices is a transformer with a toroidal (doughnut-shaped) core and distributed windings (i.e., windings wound uniformly around the circumference of the toroid). Envision taking a cross-sectional "slice" perpendicular to the core. No matter where this slice is taken around the circumference, one would expect the magnetic field relationships across the cross section of this slice to be the same, since there is no change in the geometry or in the winding as one moves around the circumference of the toroid; therefore, the magnetic field can be examined on the basis of this two-dimensional cross section of the slice.

## 2.2 REVIEW OF ELECTROMAGNETIC FIELD THEORY

The analysis of electric machines appropriately begins with the adaptation of Maxwell's equations to the specific spatial symmetry and materials' coefficients associated with this class of system. Vector notation will be used to begin the analysis, since this notation is extremely valuable in determining various directional parameters of induced voltage, force, torque, etc.; subsequently, after these directional considerations have been established, scalar notation will be used. In all equations, the International System of Units (SI), sometimes termed the *rationalized mks* (or meter-kilogram-second) system, is assumed. In magnetic systems two other systems of units have been widely used by practitioners: the centimeter-gram-second (cgs) system and the English system. To ensure understanding, boldface symbols indicate vectors. Unless otherwise indicated, cartesian coordinates are assumed. Maxwell's equations, which govern the electromagnetic phenomena at any point in space, are given as

$$\nabla \times \mathbf{E} = -\frac{\partial \mathbf{B}}{\partial t} \tag{2.1a}$$

$$\nabla \times \mathbf{H} = \mathbf{J} + \frac{\partial \mathbf{D}}{\partial t} \tag{2.1b}$$

$$\nabla \cdot \mathbf{B} = 0 \tag{2.1c}$$

$$\nabla \cdot \mathbf{D} = \rho \tag{2.1d}$$

For almost all electromagnetic systems, the charge distribution $\rho$ and the electric field flux density $D$ can be assumed to be negligibly small. By means of Stokes' theorem, it is possible to transform the first and second of Maxwell's equations into their integral form, applicable over a region in space:

$$\oint \mathbf{E} \cdot dl = -\int_s \frac{\partial \mathbf{B}}{\partial t} \cdot d\mathbf{s} \tag{2.2}$$

$$\oint \mathbf{H} \cdot dl = \int \mathbf{J} \cdot d\mathbf{s} \tag{2.3}$$

These two equations are known, respectively, as Faraday's law and Ampère's law, named for the persons who first experimentally verified these relationships.

It is most important to observe the directional parameters described by the vector notation in these equations because these are the basis of the left-hand and right-hand rules used in machine analysis.

A third significant field relationship is the force equation

$$d\mathbf{F} = I d\mathbf{l} \times \mathbf{B} \tag{2.4}$$

where $I$ is the current flowing in a differential conductor of length $d\mathbf{l}$. One simple application of Eq. (2.4) is to integrate this differential force over a volume in which the current flows in a conductor and the flux density $\mathbf{B}$ is uniform:

$$\mathbf{F} = \oint I d\mathbf{l} \times \mathbf{B} = I\mathbf{l} \times \mathbf{B} = \mathbf{B}I\mathbf{l} \sin \theta \mathbf{a_F} \tag{2.5}$$

where $\theta$ is the angle between the direction of the conductor and the magnetic field. In many machine configurations this angle is 90°, giving

$$\mathbf{F} = \mathbf{B}\mathbf{l}I \tag{2.6}$$

which is the well-known $\mathbf{B}\mathbf{l}I$ rule used in machine analysis.

## 2.3 MAGNETIC MATERIALS

In free space, $\mathbf{B}$ and $\mathbf{H}$ are related by the constant $\mu_0$, known as the *permeability of free space*:

$$\mathbf{B} = \mu_0 \mathbf{H} \tag{2.7}$$

and

$$\mu_0 = 4\phi \times 10^{-7} \mathrm{H/m} \tag{2.8}$$

The value of $\mu_0$ is given in SI units; the SI unit of $\mathbf{B}$ is tesla and of $\mathbf{H}$ is ampere per meter. It is still common for material characteristics to be given in cgs units and sometimes in English units.

Within a material, Eq. (2.7) must be modified as follows in order to describe a different magnetic phenomenon than that occurring in free space:

$$\mathbf{B} = \mu \mathbf{H} \tag{2.9}$$

$$\mu = \mu_R \mu_0 \tag{2.10}$$

where $\mu$ is termed *permeability* and $\mu_R$ *relative permeability*, a nondimensional constant. Permeability in a material medium as defined by Eq. (2.10) must be further qualified as applicable only in regions of materials that are *homogeneous* (uniform in quality) and *isotropic* (having the same properties in any direction). In materials not having these characteristics, $\mu$ becomes a vector (instead of a tensor). Finally, note that for some common materials, Eq. (2.9) is *nonlinear* and $\mu$ varies with the magnitude of $\mathbf{B}$. This results in several subdefinitions of permeability related to the nonlinear *B-H* characteristic of the material (discussed below under "The *B-H* Curve").

A material is classified according to the nature of its relative permeability $\mu_R$, which is actually related to the internal atomic structure of the material and will not be discussed further at this point. Most "nonmagnetic" materials are classified as either *paramagnetic*, for which $\mu_R$ is slightly greater than 1.0, or *diamagnetic*, in which $\mu_R$ is slightly less than 1.0. However, for all practical purposes, $\mu_R$ can be considered as equal to 1.0 for all these materials.

There is one interesting case of diamagnetism that is becoming of interest in certain types of electromagnetic devices. This is "perfect diamagnetism" (Meissner effect), which occurs in certain types of materials known as *superconductors* at temperatures near absolute zero. In such materials $B \rightarrow 0$ and $\mu_R$ is essentially zero; that is, no magnetic field can be established in the superconducting material. This phenomenon has several potential applications: for instance, in several types of rotating machines and in switching devices.

Two classes of magnetic materials of use in electric machines are ferromagnetic and ferrimagnetic materials. (1) Ferromagnetic materials are subgrouped into hard and soft materials, this classification roughly corresponding to the physical hardness of the materials. Soft ferromagnetic materials include the elements iron, nickel, cobalt, and one rare earth element; most soft steels; and many alloys of the four elements. Hard ferromagnetic materials include the permanent magnet materials (such as the alnicos), several alloys of cobalt with the rare earth elements, chromium steels, certain copper-nickel alloys, and many other metal alloys. (2) Ferrimagnetic materials are the ferrites and are composed of iron oxides that have the formula $MeOFe_2O_3$; Me represents a metallic ion. Ferrites are subgrouped into hard and soft ferrites, the former being the permanent magnetic ferrites, usually barium or strontium ferrite. Soft ferrites include the nickel-zinc and manganese-zinc ferrites and are used in microwave devices, delay lines, transformers, and other generally high-frequency applications. (3) A third class of magnetic materials of growing importance is made from powdered iron particles or other magnetic materials suspended in a nonferrous matrix, such as epoxy or plastic. Sometimes termed *superparamagnetic*, powdered-iron parts are formed by compression or injection molding techniques and are widely used in electronics transformers and as cores for inductors. Permalloy (molybdenum-nickel-iron powder) is one of the earliest and best known of the powdered materials.

Several magnetic properties of magnetic materials are important in the study of electromagnetic systems: permeability at various levels of flux density, saturation flux density, $H$ at various levels of flux density, temperature variation of permeability, hysteresis characteristic, electrical conductivity, Curie temperature, and loss coefficients. Because of the nonlinear characteristics of most magnetic materials, graphical techniques are generally valuable in describing their magnetic characteristics. The two graphical characteristics of most importance are known as the *B-H* curve, or magnetization characteristic, and the hysteresis loop.

## The *B-H* Curve

Figure 2.1 shows a typical *B-H* characteristic. This characteristic can be obtained in two ways: the virgin *B-H* curve, obtained from a totally demagnetized sample, or the normal *B-H* curve, obtained at the tips of hysteresis loops of increasing magnitude. (There are slight differences between the two methods that are not important for our purposes.)

In Fig. 2.1, notice the nonlinear regions designated as I and III. In region III *magnetic saturation* occurs, and the flux density within the material cannot increase beyond the *saturation density* $B_s$. The small increase that occurs beyond

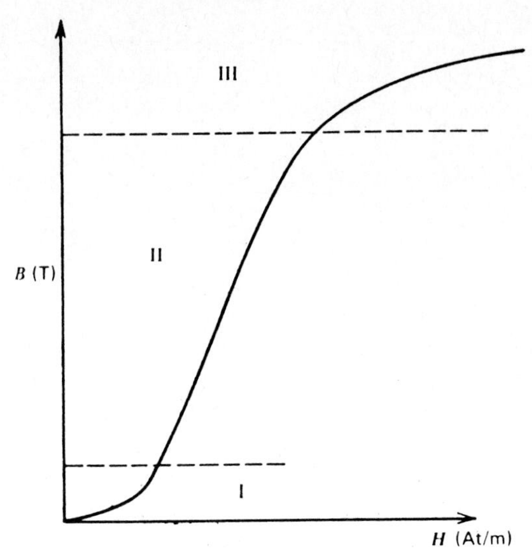


**FIG. 2.1** A typical *B-H* curve.

this condition is due to the increase in the space occupied by the material according to the relationship $B = \mu_0 H$. It is often convenient to subtract out this component of "free space" flux density and observe only the flux density variation within the material. Such a curve is known as the *intrinsic magnetization* curve and is of use in the design of permanent magnet devices.

The regions shown in Fig. 2.1 are also of value in describing the nonlinear permeability characteristic. From Eq. (2.9) it is seen that permeability is the slope of the *B-H* curve. In the following discussion relative permeability is assumed. The slope of the *B-H* curve is actually properly called *relative differential permeability*, or

$$\mu_d = \frac{1}{\mu_0}\frac{dB}{dH} \tag{2.11}$$

*Relative initial permeability* is defined as

$$\mu_i = \lim_{H \to 0} \frac{1}{\mu_0}\frac{B}{H} \tag{2.12}$$

and is seen to be the permeability in region I. This is important in many electronics applications where signal strength is low. It can also mislead one in measuring the inductance of a magnetic core device with an inductance bridge because the low signal strength in most bridges will often magnetize the sample only in region I, where the permeability is relatively low. In region II the *B-H* curve for many materials is relatively straight, and if a magnetic device is operated only in this region, linear theory can be used. In all regions the most general permeability term is known as *relative amplitude permeability* and is defined as merely the ratio of *B-H* at any point on the curve, or

$$\mu_a = \frac{1}{\mu_0}\frac{B}{H} \tag{2.13}$$

## The Hysteresis Loop

The second graphical characteristic of interest is the hysteresis loop. A typical sample is shown in Fig. 2.2; this is a *symmetrical hysteresis loop*, obtained only after a number of reversals of the magnetizing force between plus and minus $H_s$. This characteristic illustrates several parameters of most magnetic materials, the most obvious being the property of hysteresis itself. The area within the loop denotes a nonreversible energy and results in an energy loss known as the *hysteresis loss*. This area varies with temperature and with the frequency of reversal of $H$ in a given material.

The second quadrant of the hysteresis loop is very valuable in the analysis of devices containing permanent magnets. An example of this portion of hysteresis loop for alnico V is shown in Fig. 2.3. The intersection of the loop with the horizontal ($H$) axis is known as the *coercive force* $H_c$ and is a measure of the magnet's ability to withstand demagnetization from external magnetic signals. Often shown on this curve is a second curve, known as the *energy product*, which is the product of $B$ and

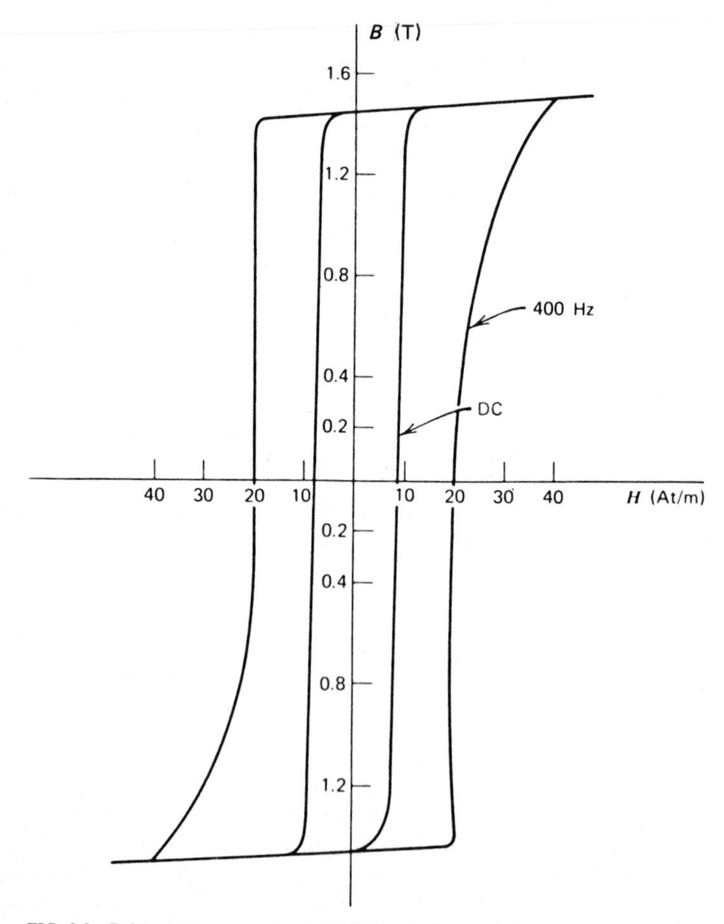


**FIG. 2.2** Deltamax tape-wound core 0.002-in strip hysteresis loop.

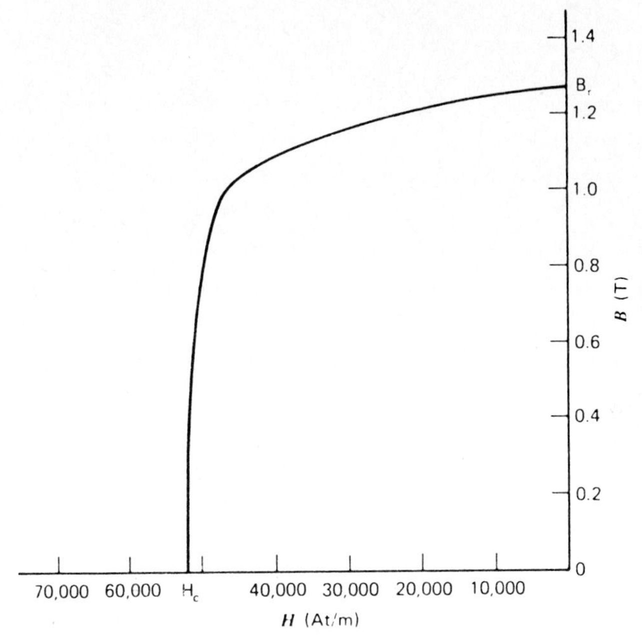


**FIG. 2.3** Demagnetization curve of alnico V.

$H$ plotted as a function of $H$ and is a measure of the energy stored in the permanent magnet. The value of $B$ at the vertical axis is known as the *residual flux density*.

Table 2.1 gives the parameter values for several common magnetic materials; the *Curie temperature* (or Curie point) $T_c$ is the critical temperature above which a ferromagnetic material becomes paramagnetic. Several $B$-$H$ curves are given in Fig. 2.4.

Some magnetic materials, such as Permalloy, supermendur, and other nickel alloys, have a *maximum* relative permeability of over 100,000, giving a ratio to the permeability of a nonmagnetic material, such as air or free space, of $10^5$. High permeability of this magnitude can be realized only in a few materials and only over a very limited range of operation. The permeability ratio between good and poor magnetic materials over a typical working range of operation is more like $10^4$ at best.

## 2.4 MAGNETIC LOSSES

A characteristic of magnetic materials that is very significant in the energy efficiency of an electromagnetic device is the energy loss within the magnetic material itself. In electric machines and transformers, this loss is generally termed the *core loss*, or sometimes the *magnetizing loss* or *excitation loss*.

Traditionally, core loss has been divided into two components: *hysteresis loss* and *eddy-current loss*. The hysteresis loss component (mentioned in the previous section) is generally held to be equal to the area of the low-frequency hysteresis loop times the frequency of the magnetizing force in sinusoidal systems.

**TABLE 2.1** Characteristics of Soft Magnetic Materials

| Trade name | Principal alloys | Saturation flux density, T | $H$ at $B_{sat}$, A/m | Amplitude permeability max. $\mu_m$ | Coercive force $H_c$, A/m | Electrical resistivity, $\mu\Omega$-cm | Curie temperature, °C |
|---|---|---|---|---|---|---|---|
| 48NI | 48% Ni | 1.25 | 80 | 200,000 | | 65 | |
| Monimax | 48% Ni | 1.35 | 6,360 | 100,000 | 4.0 | 65 | 398 |
| High Perm 49 | 49% Ni | 1.1 | 80 | | | 48 | |
| Satmumetal | Ni, Cu | 1.5 | 32 | 240,000 | | 45 | 398 |
| Permalloy (sheet) | Ni, Mo | 0.8 | 400 | 100,000 | 1.6 | 55 | 454 |
| Moly Permalloy (powder) | Ni, Mo | 0.7 | 15,900 | 125 | | | |
| Deltamax | 50% Ni | 1.4 | 25 | 200,000 | 8 | 45 | 499 |
| M-19 | Si | 2.0 | 40,000 | 10,000 | 28 | 47 | |
| Silectron | Si | 1.95 | 8,000 | 20,000 | 40 | 50 | 732 |
| Oriented T | Si | 1.6 | 175 | 30,000 | | 47 | |
| Oriented M-5 | Si | 2.0 | 11,900 | | 26 | 48 | 746 |
| Ingot iron | None | 2.15 | 55,000 | | 80 | 10.7 | |
| Supermendur | 49% Co, V | 2.4 | 15,900 | 80,000 | 8 | 26 | |
| Vanadium Permendur | 49% Co, V | 2.3 | 12,700 | 4,900 | 92 | 40 | 932 |
| Hyperco 27 | 27% Co | 2.36 | 70,000 | 2,800 | 198 | 19 | 925 |
| Flake iron | Carbonal power | $\simeq$0.8 | 5,200 | 5–130 | | $10^5$–$10^{15}$ | |
| Ferrotron (powder) | Mo, Ni | (Linear) | (Linear) | 5–25 | | $10^{16}$ | |
| Ferrite | Mg, Zn | 0.39 | 1,115 | 3,400 | 13 | $10^7$ | 135 |
| Ferrite | Mn, Zn | 0.453 | 1,590 | 10,000 | 6.3 | $3\times10^7$ | 190 |
| Ferrite | Ni, Zn | 0.22 | 2,000 | 160 | 318 | $10^9$ | 500 |
| Ferrite | Ni, Al | 0.28 | 6,360 | 400 | 143 | | 500 |
| Ferrite | Mg, Mn | 0.37 | 2,000 | 4,000 | 30 | $1.8\times10^8$ | 210 |

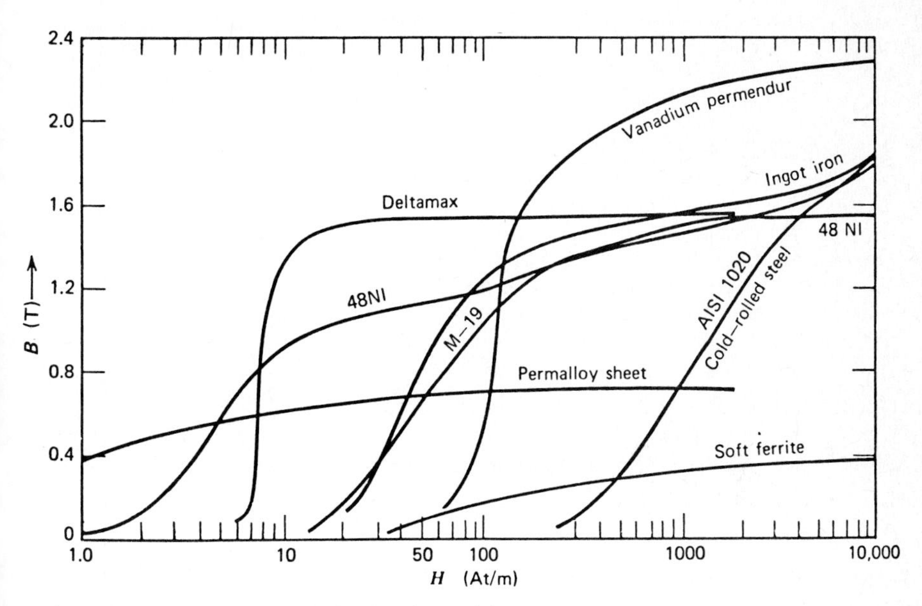


FIG. 2.4 *B-H* curves of selected soft magnetic materials.

## Eddy-Current Loss

Eddy-current losses are caused by induced electric currents (called eddies because they tend to flow in closed paths within the magnetic material itself). The eddy-current loss in a sinusoidally excited material, neglecting saturation, can be expressed by the relationship

$$P_e = k_e f^2 B_m^2 \qquad \text{W/kg} \tag{2.14}$$

where $B_m$ is the maximum value of the flux density, $f$ is the frequency, and $k_e$ is the proportionality constant, depending on the type of material and the lamination thickness.

To reduce the eddy-current loss, the magnetic material is *laminated*, that is, divided into thin sheets with a very thin layer of electrical insulation between the sheets. The sheets must be oriented in a direction parallel to the flow of magnetic flux. The eddy-current loss is roughly proportional to the square of the lamination thickness and inversely proportional to the electrical resistivity of the material. The lamination thickness varies from about 0.5 to 5 mm in electromagnetic devices used in power applications and from about 0.01 to 0.5 mm in devices used in electronics applications. Laminating a magnetic part usually increases its volume. This increase may be appreciable, depending on the method used to bond the laminations together. The ratio of the volume actually occupied by magnetic material to the total volume of a magnetic part is the *stacking factor*. This factor is important in accurately calculating flux densities in magnetic parts. Table 2.2 gives typical stacking factors for the thinner lamination sizes.

**TABLE 2.2** Stacking Factor for Laminated Cores

| Lamination thickness, mm | Stacking factor |
|---|---|
| 0.0127 | 0.50 |
| 0.0254 | 0.75 |
| 0.0508 | 0.85 |
| 0.1–0.25 | 0.90 |
| 0.27–0.36 | 0.95 |

The stacking factor approaches 1.0 as the lamination thickness increases. In powdered-iron and ferrite magnetic parts, there is an "equivalent stacking factor" that is approximately equal to the ratio of the volume of the magnetic particles to the overall volume.

## Core-Loss Values

Manufacturers of magnetic materials have obtained core-loss data under the condition of sinusoidal excitation for most of their products. Figures 2.5 and 2.6*a* show measured core-loss values for two common magnetic materials: M-15, a 3 percent silicon steel widely used in transformers and small motors, and 48NI, a nickel alloy used in many electronics applications. Fig. 2.6*b* shows measured core loss in a ferrite material.

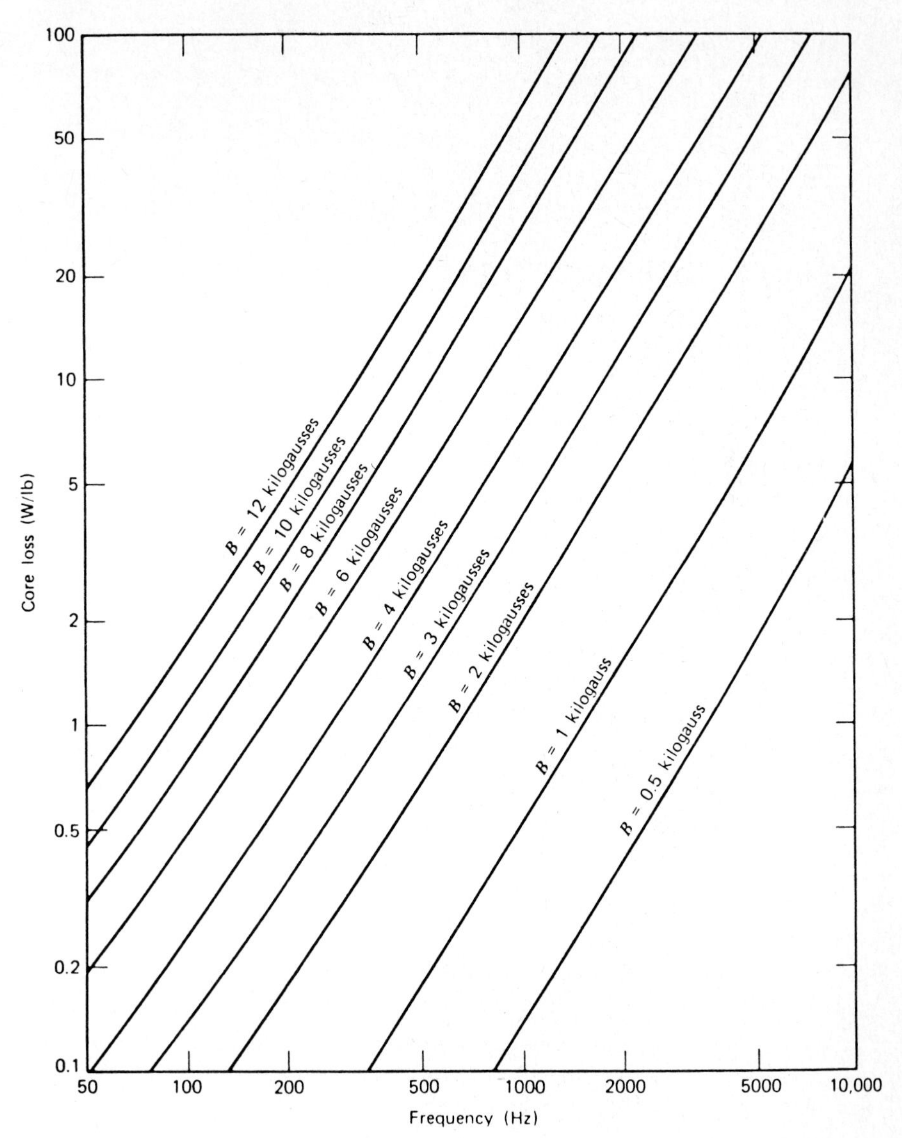


**FIG. 2.5** Core loss for nonoriented silicon steel 0.019-in-thick lamination. (*Courtesy of Armco Steel Corporation.*)

## Apparent Core Loss

This term describes the *total* excitation requirements of an electromagnetic system, including core loss. It is defined as the product of the root-mean-square (rms) exciting current with the rms value of induced voltage in the excitation winding. The SI unit of apparent core loss is the voltampere.

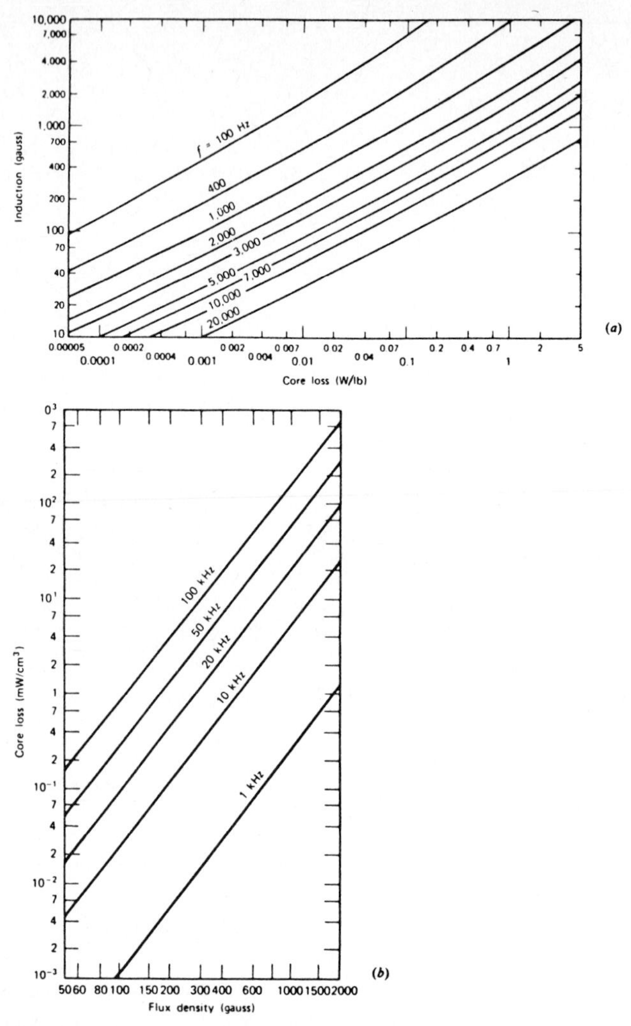


**FIG. 2.6** (***a***) Core loss for typical 48 percent nickel alloy 4 mils thick. (*Courtesy of Armco Steel Corporation.*) (***b***) Core loss for Mn-Zn ferrites.

## 2.5 MAGNETIC CIRCUIT ANALYSIS

It is important to emphasize that a magnetic field is a *distributed parameter* phenomenon; that is, it is distributed over a region of space. As such, rigorous analysis requires the use of the distance variables as contained in the divergence and curl symbols of Eq. (2.1). However, under certain conditions it is possible to apply *lumped parameter* analysis to certain classes of magnetic field problems just as it is applied in electric circuit analysis. The accuracy and precision of such analysis in the magnetic circuit problem is much less, however, than in electric circuit problems.

This section briefly describes lumped circuit analysis as applied to magnetic systems, often called *magnetic circuit analysis*. Magnetic circuit analysis follows the approach of simple dc electric circuit analysis and applies to systems excited by dc signals or, by means of an incremental approach, to low-frequency ac excitation. Its usefulness is in sizing the magnetic components of an electromagnetic device during design stages, calculating inductances, and determining airgap flux densities for power and torque calculations. Let us begin with a few definitions.

## Definitions

**1.** *Magnetic potential:* For regions in which no electric current densities exist, which is true for the magnetic circuits discussed in this handbook, the magnetic field intensity **H** can be defined in terms of *scalar* magnetic potential $\mathcal{F}$ as

$$\mathbf{H} = \nabla \mathcal{F} \qquad \mathcal{F} = \int \mathbf{H} \cdot d\mathbf{l} \tag{2.15}$$

It is seen that $\mathcal{F}$ has the dimension of amperes, although ampere-turn (At) is frequently used as a unit for $\mathcal{F}$. For a potential rise or source of magnetic energy, the term *magnetomotive force (mmf)* is frequently used. As a potential drop, the term *reluctance drop* is often used. There are two types of sources of mmf in magnetic circuits: electric current and permanent magnets. The current source usually consists of a coil of a number of turns $N$ carrying a current known as the *exciting current*; note that $N$ is nondimensional.

**2.** *Magnetic flux:* Streamlines, or flow lines, in a magnetic field are known as lines of magnetic flux, denoted by the symbol $\phi$ and having the SI unit weber. Flux is related to **B** by the surface integral

$$\phi = \int_{\mathbf{S}} \mathbf{B} \cdot d\mathbf{S} \tag{2.16}$$

**3.** *Reluctance:* Reluctance is a component of magnetic impedance, somewhat analogous to resistance in electric circuits except that reluctance is not an energy loss component. It is defined by a relationship analogous to Ohm's law:

$$\phi = \frac{\mathcal{F}}{\mathcal{R}} \tag{2.17}$$

The SI unit of magnetic reluctance is henry$^{-1}$. If we assume that the flux density has only one directional component $B$ and is uniform over a cross section of area $A_m$ taken perpendicular to the direction of $B$, Eq. (2.16) becomes $\phi = BA_m$. We also assume that $H$ is nonvarying along the length $l_m$ in the direction of $B$, and Eq. (2.17) becomes, with some rearranging,

$$\mathcal{R} = \frac{\mathcal{F}}{\phi} = \frac{Hl_m}{BA_m} = \frac{l_m}{\mu A_m} \tag{2.18}$$

which is similar to the expression for electrical resistance in a region with similary uniform electrical properties.

**4.** *Permeance:* The permeance $\mathcal{P}$ is the reciprocal of reluctance and has the SI unit henry. Permeance and reluctance are both used to describe the geometric characteristics of a magnetic field, mainly for the purpose of calculating inductances.

**5.** *Leakage flux:* Between any two points at different magnetic potentials in space, a magnetic field exists, as shown by Eq. (2.15). In any practical magnetic circuit there are many points (or, more generally, planes) at magnetic potentials different from each other. The magnetic field between these points can be represented by flow lines, or lines of magnetic flux. Where these flux lines pass through regions of space (generally air spaces, electrical insulation, or structural members of the system) instead of along the main path of the circuit, they are termed *leakage flux lines*. In coupled circuits with two or more windings, the definition of leakage flux is specific: The flux links one coil but not the other.

Leakage is a characteristic of all magnetic circuits and can never be completely eliminated. At dc or very low ac excitation frequencies, magnetic shielding consisting of thin sheets of high-permeability material can reduce leakage flux. This is done not be eliminating leakage but by establishing new levels of magnetic potential in the leakage paths, the better to direct the flux lines along the desired path. At higher frequencies of excitation, electrical shielding (such as aluminum foil) can reduce leakage flux by dissipating its energy as induced currents in the shield.

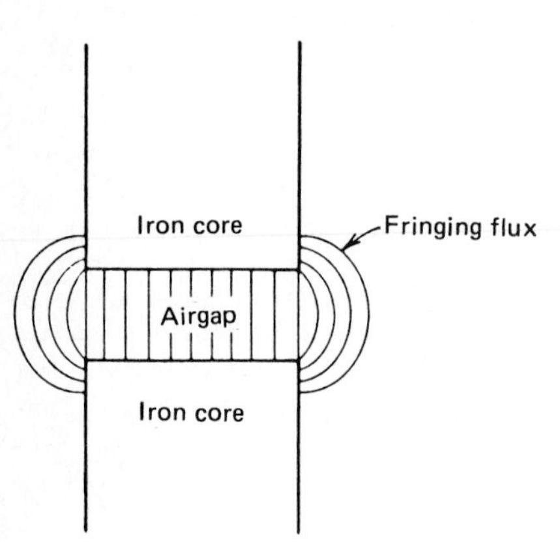


**FIG. 2.7** Flux fringing at an airgap.

**6.** *Fringing:* Fringing is somewhat similar to leakage; it describes the spreading of flux lines in an airgap of a magnetic circuit. Fig. 2.7 illustrates fringing at a gap. Fringing results from lines of flux that appear along the sides and edges of the magnetic members at each side of the gap, which are at different magnetic potentials. Fringing has the effect of increasing the effective area of the airgap, which must be considered with the length of the airgap.

## Ampère's Law Applied to a Magnetic Circuit

According to Eq. (2.3), the integral around any closed path of the magnetic field intensity **H** equals the electric current contained within that path. In using this integral expression, note that *positive current* is defined as flowing in the direction of the advance of a right-handed screw turned in the direction in which the closed path is traversed.

Let us apply Ampère's law to the simple magnetic circuit whose cross section is shown in Fig. 2.8. Note that for the data and the directions shown, the current's direction is into the plane of the paper for the conductors enclosed by the integration path for coils 1 and 3 and out of the plane of the paper for coil 2. From the left side of Eq. (2.3) we obtain

$$\oint \mathbf{H} \cdot dl = \int_0^{l_m} \mathbf{H}_m \cdot dl + \int_0^{l_g} \mathbf{H}_g \cdot dl \tag{2.19}$$

If the magnetic material is linear, homogeneous, and isotropic and if leakage flux is neglected, Eq. (2.19) becomes

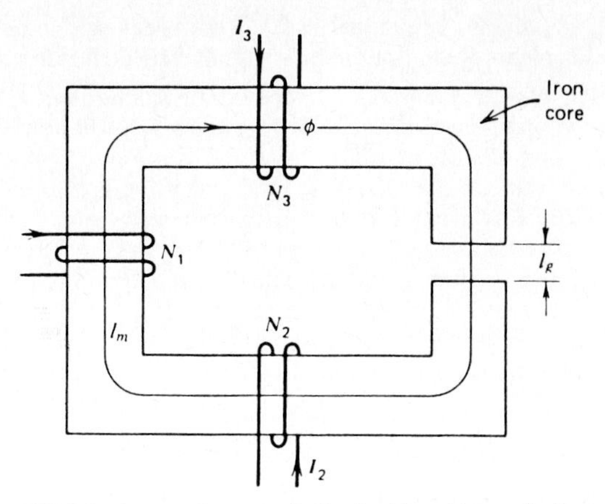


FIG. 2.8 A composite magnetic circuit with multiple excitation (mmf's).

$$\oint \mathbf{H} \cdot d\mathbf{l} = H_m l_m + H_g l_g = \phi(\mathcal{R}_m + \mathcal{R}_g) = \mathcal{F}_m + \mathcal{F}_g \tag{2.20}$$

where $\mathcal{R}_m$ and $\mathcal{R}_g$ are the reluctances of the magnetic member and gap, respectively, and $\mathcal{F}_m$ and $\mathcal{F}_g$ represent the magnetic potential or reluctance drop across these two members of the magnetic circuit. The right side of Eq. (2.3) gives

$$\oint \mathbf{H} \cdot d\mathbf{l} = N_1 I_1 + N_3 I_3 - N_2 I_2 \tag{2.21}$$

Combining Eqs. (2.20) and (2.21) yields

$$N_1 I_1 + N_3 I_3 - N_2 I_2 - \mathcal{F}_m - \mathcal{F}_g = 0 \tag{2.22}$$

We may generalize Ampère's law on the basis of this simple example to state that "The sum of the magnetic potentials around any closed path is equal to zero," which is analogous to Kirchhoff's voltage law in electric circuits. From the preceding equations, note that:

1. A magnetic core with an airgap is analogous to a simple series dc circuit, as shown in Fig. 2.9.

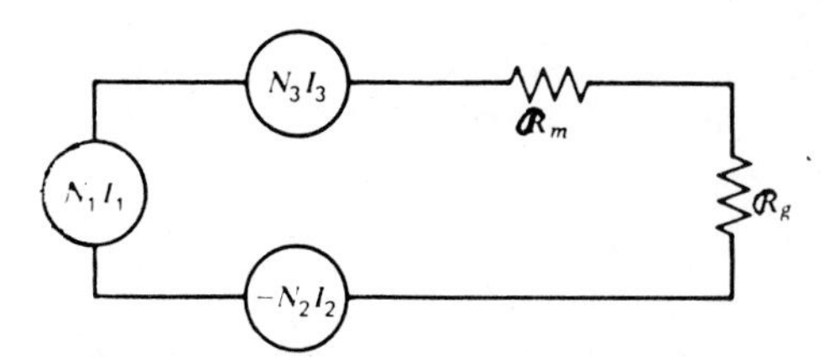


FIG. 2.9 Approximate equivalent circuit for Fig. 2.8.

2. Because of the symmetry about the plane of the paper of the system of Fig. 2.8, a two-dimensional representation of the magnetic field is acceptable.
3. The calculation of reluctances is a more cumbersome approach for determining reluctance drops than is the use of magnetic field intensities; therefore, for numerical solutions, the use of $(H_m l_m + H_g l_g)$ is always preferred. The calculation of magnetic reluctances for magnetic sections requires the determination of amplitude permeability from *B-H* curves (such as shown in Fig. 2.4), the same process required to determine $H_m$; however, if the reluctance technique is continued, a great deal of additional mathematical manipulation is required that often introduces trivial errors. The reluctance technique is, however, most valuable in the qualitative analysis of magnetic circuits, such as in developing equivalent circuits, in comparing different geometric sections, or in calculating inductance.
4. The reader may wonder about solving the inverse problem; that is, given the exciting ampere-turns, determine the flux (or flux density) in the airgap. A little thought will show that there is no direct analytical solution to this problem because of the nonlinear *B-H* characteristic of the magnetic material. The flux density must be known before the field intensity or reluctance of the magnetic member can be determined. This type of problem is amenable to iterative computer techniques, although considerable computer storage space may be required, since the *B-H* characteristics of the magnetic material must be stored for repeated access during the iterative process.

## Limitations of the Magnetic Circuit Approach

The number of problems in practical magnetic circuits that can be solved by the approach outlined above is limited, despite the similarity of this approach to simple dc electric circuit theory. The main purpose of introducing magnetic circuits is to state some very fundamental principles and definitions necessary to understanding electromagnetic systems, not to offer a problem-solving technique. The limitations of magnetic circuit theory rest primarily on the nature of magnetic materials as contrasted with conductors, insulators, and dielectric materials. Most of these limitations have already been introduced as assumptions in the discussion of magnetic circuits. Let us assess the significance of these assumptions.

1. *Homogeneous magnetic material:* Most materials used in practical electromagnetic systems can be considered homogeneous over finite regions of space, allowing the use of the integral forms of Maxwell's equations in calculating reluctances and permeances.
2. *Isotropic magnetic materials:* Many sheet steels and ferrites are oriented by means of the metallurgical process during their production. Oriented materials have a "favored" direction in their grain structure, giving superior magnetic properties when magnetized along this direction.
3. *Nonlinearity:* This is an inherent property of all ferromagnetic and ferrimagnetic materials. However, there are many ways of treating this class of nonlinearity analytically:
   - *a.* As can be seen by observing the *B-H* curves shown in this chapter, a considerable portion of the curve for most materials can be approximated as a straight line, and many electromagnetic devices operate in this region.
   - *b.* Numerous analytical and numerical techniques have been developed for describing the *B-H* and other nonlinear magnetic characteristics.

*c.* The nonlinear *B-H* characteristic of magnetic materials manifests itself in the relationship between flux and exciting current in electromagnetic systems; the relationship between flux and induced voltage is a *linear* relationship, as given by Faraday's law, Eq. (2.2). It is possible to treat these nonlinear excitation characteristics separately in many systems, such as is done in the equivalent-circuit approach to transformers and induction motors.

*d.* An inductor whose magnetic circuit is composed of a magnetic material is a nonlinear electric circuit element, such as a coil wound on a magnetic toroid. With an airgap in the magnetic toroid, however, the effect of the nonlinear magnetic material on the inductance is lessened. Rotating machines and many other electromechanical devices have airgaps in their magnetic circuits, permitting the basic theory of these devices to be described by means of linear equations.

4. *Saturation:* All engineering materials and devices exhibit a type of saturation when output fails to increase with input (for instance, in the saturation of an electronic amplifier). Saturation is very useful in many electromagnetic devices, such as magnetic amplifiers, saturable reactors, and other types of magnetic switching devices.

5. *Leakage and fringing flux:* This is a property of all magnetic circuits. It is best treated as a part of the generalized solution of magnetic field distribution in space, often called a boundary-value problem. In many rotating-machine magnetic circuits, boundaries between regions of space containing different types of magnetic materials (usually, a boundary between a ferromagnetic material and air) are often planes or cylindrical surfaces that, in a two-dimensional cross section, become straight lines or circles. Leakage inductances can frequently be determined in such regions by calculating the reluctance or permeance of the region using fairly simple integral formulations. The spatial or geometric coefficients so obtained are known as permeance coefficients. An example of this approach is given under "Inductance" at the end of this section.

## The Ideal Magnetic Circuit

The ideal magnetic circuit can be defined by using the assumptions discussed above in this section (see "Ampère's Law Applied to a Magnetic Circuit"). It is composed of magnetic materials which are homogeneous, isotropic, and linear and which have *infinite permeability*. Airgap fringing is usually neglected.

Figure 2.10 shows the boundary between two regions of different magnetic permeability. Let region 1 be free space with $\mu_1 = \mu_0$ and let region 2 be the magnetic material with $\mu_2 \to \infty$. The boundary conditions, in terms of field components nor-

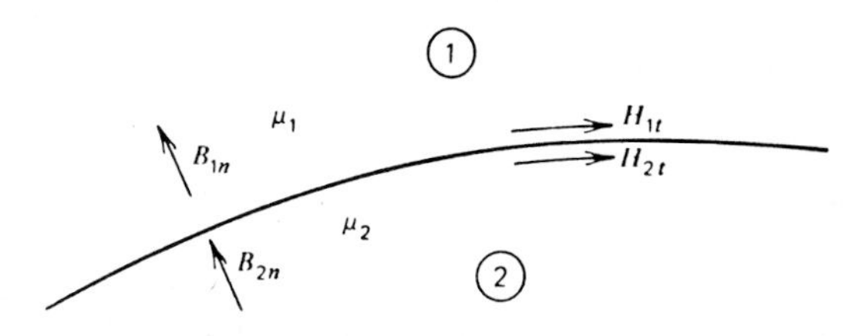


**FIG. 2.10** Magnetic materials interface.

mal to the boundary ($H_n$ and $B_n$) and tangential to the boundary ($H_t$ and $B_t$), can be shown to be (assuming zero current density on the boundary surface)

$$B_{1n} = B_{2n} \qquad H_{1t} = H_{2t} \tag{2.23}$$

where the subscripts 1 and 2 refer to the respective regions. Within the magnetic material (region 2),

$$H_{2t} = \frac{B_{2t}}{\mu_2} \rightarrow 0(\mu_2 \rightarrow \infty) \tag{2.24}$$

showing that the tangential component of field intensity in the magnetic material is zero as the permeability approaches infinity. Therefore, from Eq. (2.23) the tangential component of the magnetic field in region 1 at the boundary is also zero. Also, from Eq. (2.15) it can be seen that $\mathcal{F}$, the magnetic potential, is zero along a path parallel to the tangential field within the magnetic material. There are two important conclusions from this analysis:

1. Flow lines, or lines of magnetic flux, are perpendicular to the surface of a perfect magnetic conductor.
2. There is no potential difference, or reluctance drop, between points or planes at different locations within a perfect magnetic conductor.

These characteristics of an ideal magnetic circuit are frequently used in magnetic circuit analysis.

## Faraday's Law and Induced Voltage

To relate a time variation of the magnetic flux to an electric field variation around a closed path, recall Faraday's law, Eq. (2.2):

$$\oint \mathbf{E} \cdot d\mathbf{l} = -\int_{\mathbf{S}} \frac{\partial \mathbf{B}}{\partial t} \cdot d\mathbf{s} \tag{2.2}$$

To determine the direction of the electric field vector **E** in this expression, use the fingers of the right hand to indicate the positive direction about the closed path. The thumb indicates the direction of $d\mathbf{s}$; a flux density **B** in the direction of $d\mathbf{s}$, *increasing with time*, results in a direction for **E** *opposite* to the positive direction about the closed path. The left side of Eq. (2.2) is termed *electromotive force (emf)*, or induced voltage $e$; the right side has been defined as the magnetic flux in Eq. (2.16). Using these definitions, the scalar form of Faraday's law is

$$e = -\frac{d\phi}{dt} \tag{2.25}$$

The flux linkage $\lambda$ is defined as the product of a number of turns $N$ and the flux $\phi$ linking $N$, or

$$\lambda = N\phi \tag{2.26}$$

If the line integral in Eq. (2.2) is made about $N$ closed paths representing $N$ series turns, Eq. (2.25) becomes

$$e = -\frac{d\lambda}{dt} \tag{2.27}$$

Faraday's law is applicable to cases in which the time variation of voltage results from a time-varying flux linking a stationary coil (as in a transformer), or from a coil or conductor physically moving through a static flux, or from any combination of the two situations. For the case of a conductor moving in a magnetic field, the induced voltage is often termed *motional emf* and is defined as

$$\text{Motional emf} = \oint (\mathbf{U} \times \mathbf{B}) \cdot d\mathbf{l} \tag{2.28}$$

where $\mathbf{U}$ is the velocity vector of the conductor. A special application of this relationship is useful in rotating-machine analysis: Assume a conductor of length $\boldsymbol{l}$ moving perpendicularly to the direction of a uniform magnetic field $\mathbf{B}$ at a constant velocity $\mathbf{U}$; Eq. (2.28) becomes (with $B$, $l$, and $U$ mutually perpendicular)

$$\text{Motional emf} = BlU \tag{2.29}$$

The direction associated with motional emf can be worked through the integral and vector processes on the right side of Eq. (2.28) in a process similar to that described after Eq. (2.2), but a simple rule, known as the *right-hand rule*, is much easier for cases in which Eq. (2.29) applies. Extend the thumb and the first and second fingers of the right hand so that they are mutually perpendicular to each other. If the thumb represents the direction of $U$ and the first finger the direction of $B$, the second finger represents the direction of the emf along $l$.

## Energy Relations in a Magnetic Field

The potential energy in a magnetic field is defined throughout space by the volume integral

$$W = \frac{1}{2}\int_{\text{vol}} \mathbf{B} \cdot \mathbf{H}\, dv = \frac{1}{2}\int_{\text{vol}} \mu H^2\, dv = \frac{1}{2}\int_{\text{vol}} \frac{B^2}{\mu}\, dv \tag{2.30}$$

This equation is valid only in regions of constant permeability; therefore, its usefulness is limited to *static* linear magnetic circuits.

## Inductance

*Inductance* is defined as flux linkage per ampere:

$$L = \frac{\lambda}{i} = \frac{N\phi}{i} \tag{2.31}$$

Consider the magnetic toroid around which are wound $n$ distinct coils electrically isolated from each other, as shown in Fig. 2.11. The coils are linked magnetically by the flux $\phi$, some portion of which links each of the coils. A number of inductances can be defined for this system: $L_{km}$ = (flux linking the $k$th coil due to the current in the $m$th coil)/(current in the $m$th coil). Mathematically, this can be stated as

$$L_{km} = \frac{N_k(K\phi_m)}{i_m} \tag{2.32}$$

where $K$ is the portion of the flux due to coil $m$ that links coil $k$ and is known as the *coupling coefficient*. By definition, its maximum value is 1.0. A value of $K$ less

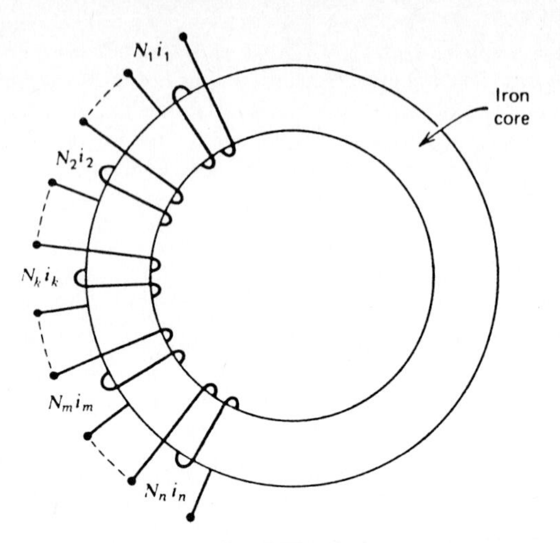


**FIG. 2.11** A toroid with $n$ windings.

and of coil $m$. When the two subscripts in Eq. (2.32) are identical, the inductance is termed *self-inductance*; when different, the inductance is termed *mutual inductance* between coils $k$ and $m$. Mutual inductances are symmetrical, i.e.,

$$L_{km} = L_{mk} \tag{2.33}$$

Inductance can be related to the magnetic parameters derived earlier in this section. In Eq. (2.32), using Eq. (2.17), $\phi_m$ can be replaced by the magnetic potential $\mathcal{F}_m$ of coil $m$ divided by the reluctance $\mathcal{R}$ of the magnetic circuit; the magnetic potential of coil $m$, however, is $N_m I_m$. Making these substitutions in Eq. (2.32) gives

$$L_{km} = \frac{K N_k N_m}{\mathcal{R}} = K N_k N_m \mathcal{P} \tag{2.34}$$

where $\mathcal{P}$ is the permeance, the reciprocal of the reluctance. In a simple magnetic circuit (such as the toroid of Fig. 2.11), the reluctance from Eq. (2.18) can be substituted into Eq. (2.34), giving

$$L_{km} = \frac{K N_k N_m A_t \mu_t}{l_t} \tag{2.35}$$

where $\mu_t$, $A_t$, and $l_t$ are the permeability, cross-sectional area, and mean length of the toroid, respectively.

Stored energy can be expressed in terms of inductance:

$$W = {}^1\!/_2 L i^2 \tag{2.36}$$

By substituting for $L$ from Eq. (2.31) and for $Ni$ (magnetic potentials) from Eq. (2.17), Eq. (2.36) can be expressed as

$$W = {}^1\!/_2 R \phi^2 \tag{2.37}$$

## 2.6 MAGNETIC CIRCUITS CONTAINING PERMANENT MAGNETS

The second type of excitation source commonly used for supplying energy to magnetic circuits in rotating machines and other types of electromechanical devices is the permanent magnet. In a circuit excited by a permanent magnet, the operating conditions of the permanent magnet are largely determined by the external magnetic circuit. Also, the operating point and subsequent performance of the permanent magnet are a function of how the magnet is physically installed in the circuit and whether it is magnetized before or after installation. In many applications, the magnet must go through a stabilizing routine before use. These considerations are, of course, meaningless for electrical excitation sources.

For permanent magnet excitation, the object is to determine the size (length and cross section) of the permanent magnet. The first step in this process is to choose a specific type of permanent magnet, since each type of magnet has a unique characteristic that will partially determine the size of the magnet required. In a practical design this choice will be based on cost factors, availability, mechanical design (hardness and strength requirements), available space in the magnetic circuit, and the magnetic and electrical performance specifications of the circuit. Except for the new neodymium-iron-boron magnet, most permanent magnets are nonmachinable and must usually be used in the circuit as obtained from the manufacturer. Table 2.3 summarizes some of the pertinent characteristics of common permanent magnets.

**TABLE 2.3** Characteristics of Permanent Magnets

| Type | Residual flux density $B_r$, G | Coercive force $H_c$, Oe | Maximum energy product, G-Oe × $10^6$ | Average recoil permeability |
|---|---|---|---|---|
| 1% Carbon steel | 9,000 | 50 | 0.18 | |
| 3 1/2% Chrome steel | 9,500 | 65 | 0.29 | 35 |
| 36% Cobalt steel | 9,300 | 230 | 0.94 | 10 |
| Alnico I | 7,000 | 440 | 1.4 | 6.8 |
| Alnico IV | 5,500 | 730 | 1.3 | 4.1 |
| Alnico V | 12,500 | 640 | 5.25 | 3.8 |
| Alnico VI | 10,500 | 790 | 3.8 | 4.9 |
| Alnico VIII | 7,800 | 1,650 | 5.0 | — |
| Cunife | 5,600 | 570 | 1.75 | 1.4 |
| Cunico | 3,400 | 710 | 0.85 | 3.0 |
| Vicalloy 2 | 9,050 | 415 | 2.3 | — |
| Platinum-cobalt | 6,200 | 4,100 | 8.2 | 1.1 |
| Barium ferrite-isotropic | 2,200 | 1,825 | 1.0 | 1.15 |
| Oriented type A | 3,850 | 2,000 | 3.5 | 1.05 |
| Oriented type B | 3,300 | 3,000 | 2.6 | 1.06 |
| Strontium ferrite | | | | |
| Oriented type A | 4,000 | 2,220 | 3.7 | 1.05 |
| Oriented type B | 3,550 | 3,150 | 3.0 | 1.05 |
| Rare earth–cobalt | 8,600 | 8,000 | 18.0 | 1.05 |
| NdFeB | 11,200 | 8,500 | 30.0 | 1.05 |

Permanent magnet excitation is chosen for a specified airgap flux density with the aid of the second quadrant *B-H* curve, often called the *demagnetization curve*, for a specific type of magnet. This curve has been introduced in Sec. 2.3 as Fig. 2.3. The *B-H* characteristics of a number of alnico permanent magnets are shown in Fig. 2.12; Fig. 2.13 shows the characteristics of several ferrite magnets. Not included in either of these sets of curves is the characteristic of the recently introduced neodymium-iron-boron (NdFeB) permanent magnet. Figure 2.14 compares the *B-H* characteristic of the NdFeB magnet with that of some other commonly used magnets. In addition to being superior to most permanent magnets, NdFeB magnets are less expensive compared to samarium-cobalt (SmCo) magnets and have strength, hardness, and machining characteristics comparable to those of iron and steel. Also shown on these figures are curves of *energy product*, the product of $B$ in gauss (G) and $H$ in oersteds (Oe), and of *permeance ratio*, the ratio of $B/H$. The energy product is a measure of the magnetic circuit as a function of its flux density and field intensity; in general, a permanent magnet is used most efficiently when operated at conditions of $B$ and $H$ that result in the maximum energy product. Permeance coefficients are useful in the design of the external magnetic circuit; this parameter is actually relative

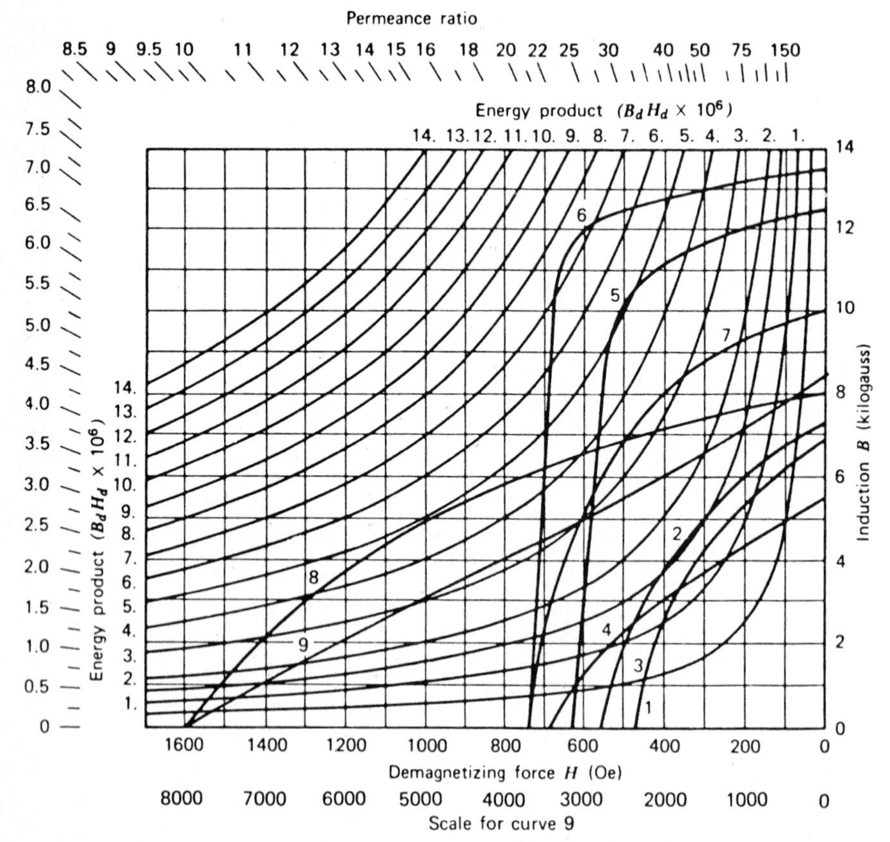


**FIG. 2.12** Demagnetization and energy product curves for alnicos 1 to 8. *Key*: 1, alnico I; 2, alnico II; 3, alnico III; 4, alnico IV; 5, alnico V; 6, alnico V-7; 7, alnico VI; 8, alnico VIII; 9, rare earth–cobalt.

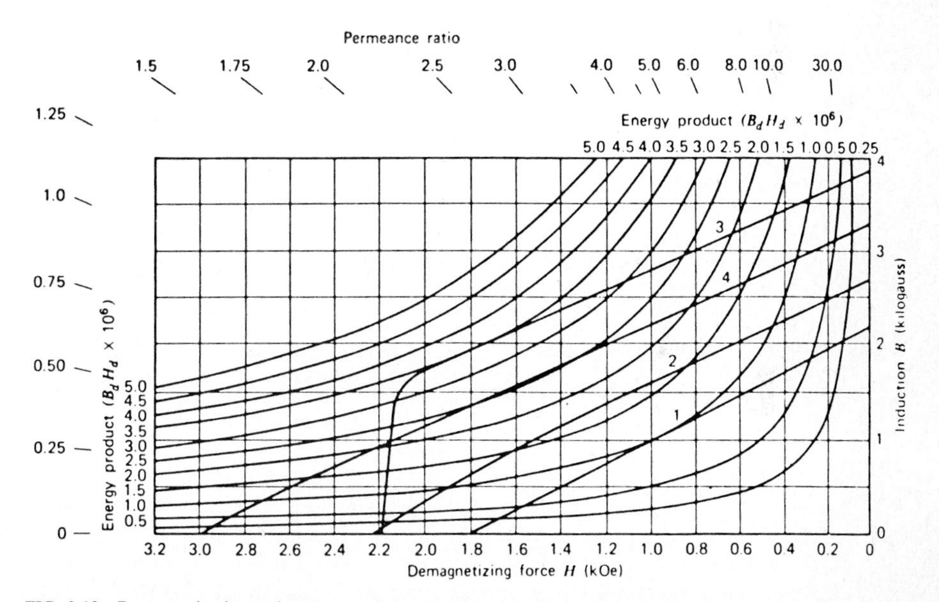


**FIG. 2.13** Demagnetization and energy products curves for Indox ceramic magnets. *Key*: 1, Indox I; 2, Indox II; 3, Indox V; and 4, Indox VI-A.

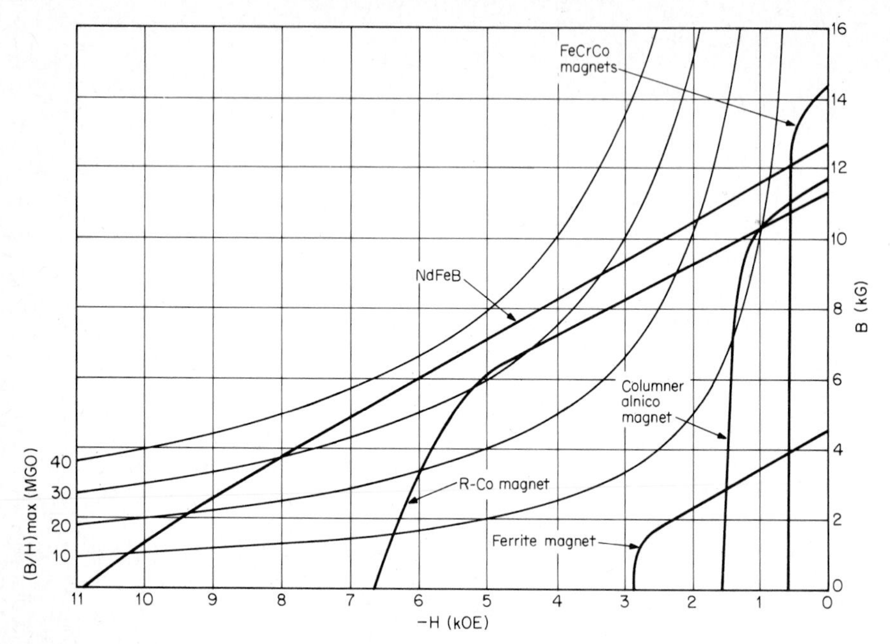


**FIG. 2.14** Demagnetization curves of certain permanent magnets.

permeability, as defined in Sec. 2.3, since $\mu_0$ is 1.0 in the cgs system of units. The symbols $B_d$ for flux density and $H_d$ for field intensity are used to designate the coordinates of the demagnetization curve.

Once the permanent magnet type has been chosen, the design of the magnet's size follows the general approach taken in Sec. 2.5. From Ampère's law, Eq. (2.3),

$$H_d l_m = H_g l_g + V_{mi} \tag{2.38}$$

where $H_d$ = magnetic field intensity of magnet, Oe
$l_m$ = length of magnet, cm
$H_g$ = field intensity in gap = flux density in gap (in cgs units)
$l_g$ = length of gap, cm
$V_{mi}$ = reluctance drop in other ferromagnetic portions of the circuit, Gb

The cross-sectional area of the magnet is calculated from the flux required in the airgap as follows:

$$B_d A_m = B_g A_g K_1 \tag{2.39}$$

where $B_d$ = flux density in magnet, G
$A_m$ = cross-sectional area of magnet, $\text{cm}^2$
$B_g$ = flux density in gap, G
$A_g$ = cross-sectional area of gap, $\text{cm}^2$
$K_1$ = leakage factor

The leakage factor $K_1$ is the ratio of the flux leaving the magnet to the flux in the airgap. The difference between these two fluxes is the leakage flux in the regions of space between the magnet and the airgap. The leakage factor can be determined by a number of formulas developed for the common configurations used in permanent magnet magnetic circuits; these formulas can be found in Refs. 1, 2, and 3. From Ref. 2, a leakage factor for the configuration of Fig. 2.15 is given by

$$K_1 = 1 + \frac{l_g}{A_g}\left(1.7C_G\frac{G}{G+l_g} + 1.4W\sqrt{\frac{C_w}{H}} + 0.67C_H\right) \tag{2.40}$$

where $C_H$ = perimeter of cross section of circuit of length $H$
$C_w$ = perimeter of cross section of circuit of length $W$
$C_G$ = perimeter of cross section of circuit of length $G$

and where $l_g$ is shown in Fig. 2.15.

It may be shown that the leakage factor for the configuration of Fig. 2.15 is high. Thus, this is not a very efficient magnetic circuit. Stated another way, the permanent magnet is in the wrong location within the magnetic circuit. A much more efficient use of the permanent magnet in this configuration is to locate it adjacent to the airgap, as shown in Fig. 2.16. From Ref. 2, the leakage factor for Fig. 2.16 is

$$K_1 = 1 + \frac{l_g}{A_g}0.67C_G\left(1.7\frac{0.67G}{0.67G+l_g} + \frac{l_g}{2G}\right) \tag{2.41}$$

It is interesting to observe the volume of permanent magnet material required to establish a given flux in an airgap. Solving for $A_m$ in Eq. (2.39) and for $l_m$ in Eq. (2.38) (neglecting $V_{mi}$) and noting that in the cgs system $H_g = B_g$, we obtain

$$\text{Volume} = A_m l_m = \frac{B_g A_g l_g K_1}{B_d H_d} \tag{2.42}$$

It is seen that magnet volume is a function of the square of the airgap flux density. The importance of the leakage factor in minimizing the required magnet size is also apparent from this equation. The denominator of Eq. (2.42) is the energy product that is a function of the permanent magnet material and the operating point on the demagnetization curve of the magnet.

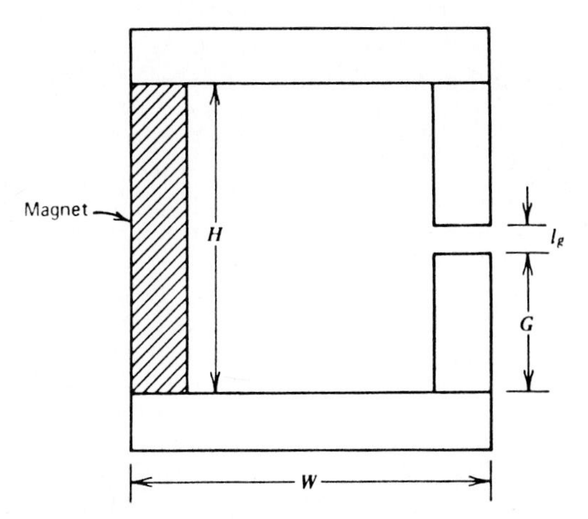


**FIG. 2.15** A magnetic circuit with a permanent magnet.

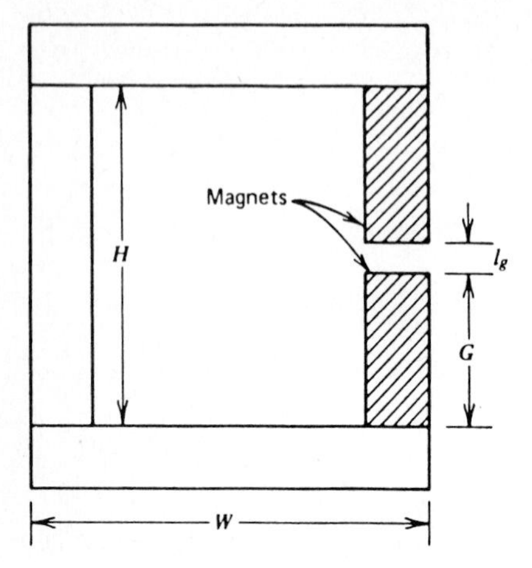


**FIG. 2.16** An efficient use of permanent magnets in a magnetic circuit.

The permeance-ratio parameter shown in Figs. 2.12 and 2.13 is the ratio of the equivalent permeance of the external circuit, $A_g K_1/l_g$, to the permeance of the space occupied by the permanent magnet, $A_m/l_m$, in the cgs system of units. This can be seen by solving for $B_d$ from Eq. (2.39) and for $H_d$ from Eq. (2.38) (neglecting $V_{mi}$) and taking the ratio:

$$\frac{B_d}{H_d} = \frac{A_g l_m K_1}{A_m l_g} = \frac{P_{ge}}{P_M} = \tan \alpha \tag{2.43}$$

Equation (2.43) is deceptively simple in appearance because the task of obtaining analytical expressions for $K_1$ is very difficult, as has been seen. Also, the reluctance drop in the soft-iron portions of the magnetic circuit, $V_{mi}$, must be included somehow in Eq. (2.43). This is even a more difficult task, since the reluctance drop is a function of both the permanent magnet's operating point $B_d H_d$ and the effects of leakage flux in the iron. The reluctance drop is usually introduced by means of a factor similar to the leakage factor and is based on measurements in practical circuit configurations. The various expressions that make up Eq. (2.43) are of value in observing general relationships among the magnetic parameters as the permeance of the external circuit is varied. This leads us to the second type of permanent magnet circuit.

Circuits with a varying airgap will be briefly described with the aid of Fig. 2.17. Keeping Eq. (2.43) in mind, let us observe the variation of $B$ and $H$ of a permanent magnet as the external circuit's permeance is varied. Figure 2.17 shows a typical second-quadrant $B$-$H$ characteristic of a permanent magnet. Theoretically, it is possible to have infinite permeance in the external magnetic circuit that would correspond to $\alpha = 90°$ in Eq. (2.43), and the magnet operating point would be at $B_d = B_r$ and $H_d = 0$ in Fig. 2.17. This situation is approximated by a permanent magnet having an external circuit consisting of no airgap and a high-permeability soft-iron member, often called a "keeper." In practice, however, there is always a small equivalent airgap and a small reluctance drop in the keeper, the operating point is to the left of $B_r$, and $\alpha$ is less than 90°.

For circuits with a finite airgap, the operating point will be at some point $A$ on the $B$-$H$ curve, and the permanent magnet will develop the magnetic field

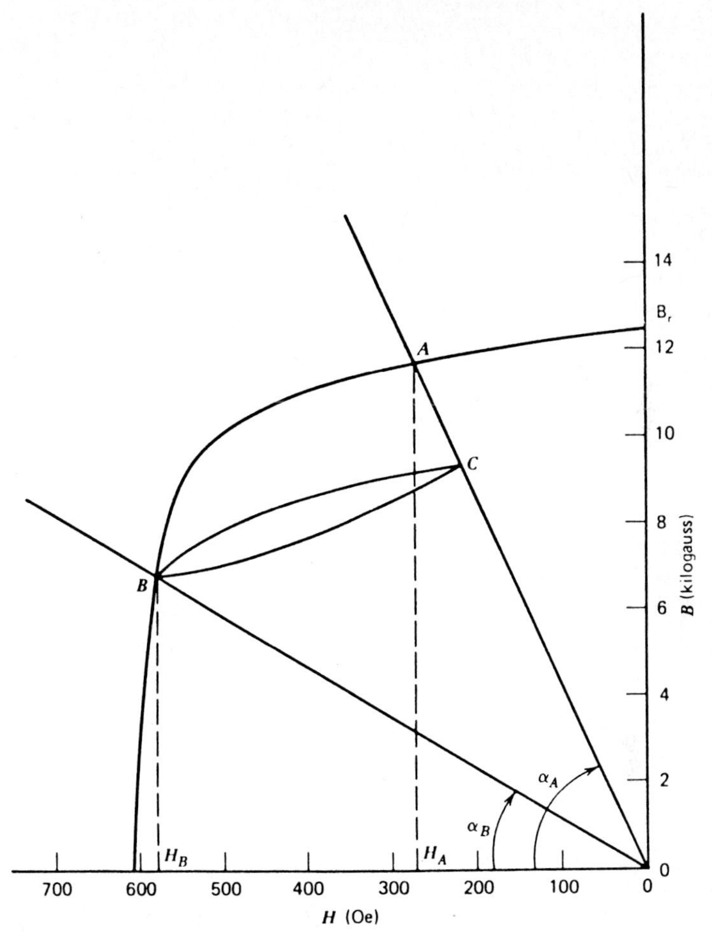


**FIG. 2.17** Second-quadrant *B-H* characteristic of a permanent magnet.

intensity $H_A$ to overcome the reluctance drop of the airgap and other portions of the external magnetic circuit. If the airgap is increased, $P_{ge}$ decreases, and the magnet must develop a larger magnetic field intensity $H_d$. From Eq. (2.43), it is seen that $\alpha$ decreases and that the operating point in Fig. 2.17 will move farther to the left to some point, say *B*, at $\alpha_B$. If the airgap is subsequently returned to its original value, the operating point will not return to *A* but, instead, to *C*. If the airgap is successively varied between the two values, the operating point will trace a "minor hysteresis loop" between *B* and *C*, as shown in Fig. 2.17. The slope of this loop is known as *recoil permeability*; since it is a slope on the *B-H* plane, it is also sometimes called incremental permeability, as defined in Eq. (2.11). Recoil permeability is an important parameter of a permanent magnet for applications with varying airgaps; values of this parameter are given in Table 2.3 for the permanent magnets shown.

## REFERENCES

1. H. C. Roters, *Electromagnetic Devices*, Wiley, New York, 1941.
2. *Design and Application of Permanent Magnets*, Indiana General Corporation Manual, no. 7, Valparaiso, Ind., 1968.
3. *Permanent Magnet Design*, Bulletin M303, Thomas and Skinner, Indianapolis, 1967.

# CHAPTER 3
# SYNCHRONOUS MACHINES

**L. E. Unnewehr**

## 3.1 GENERAL CONSIDERATIONS

A machine is classified as "synchronous" if under normal operation it operates at a fixed speed determined by the frequency of the applied external signal. This normal operating speed is known as *synchronous speed*; in revolutions per minute, it is given by

$$n_s = 120\frac{f}{P} \qquad \text{r/min} \tag{3.1}$$

where $f$ = frequency of applied signal, Hz
$P$ = number of poles of the synchronous machine

One of the principal merits of synchronous machines is this *invariant* relationship between machine speed and the frequency of the external source.

Synchronous machines are used in an extremely wide range of output power applications, a range probably greater than for any other class of rotating machine. On the low end are the clock and timing motors and control alternators in the milliwatt range; at the high end are the giant alternators used in electric power generation in the 50- to 750-MW range. Synchronous machines also have a greater diversity of physical configurations than does any other class of rotating machine. The principal configuration and the one to which the term *synchronous machine* usually applies is the subject of this chapter.

In generator applications, synchronous machines are frequently called alternators. The terms *alternator* and *synchronous generator* will be used interchangeably in this chapter. In terms of physical size and power ratings, alternators represent the largest class of rotating electric devices in existence. More than 90 percent of the electric energy used in the world is generated by alternators. However, alternators are also made in much smaller sizes and power ratings. The smallest is probably the device known as an ac tachometer, a tiny alternator used as a speed sensor.

Synchronous motors are by far the most common type of motor, given their widespread use in clock and recording applications. The types of synchronous motors that are counterparts to the alternators mentioned in the previous paragraph are built in power ratings of from a few hundred watts to approximately 100 MW and are used in applications where constant speed is important, including pumps, compressors, and drives for textile mills.

## 3.2 PHYSICAL DESCRIPTION

The type of synchronous machine to be discussed in this chapter contains two electrical windings and is excited from both windings.

One winding, termed the *armature winding*, is that in which electric energy is developed. It is a power winding and must be constructed of conductors capable of carrying current densities derived from the power rating of the machine. The excitation produced by this winding is normally called *armature reaction*.

The second winding of a synchronous machine is called the *field winding*. Its function is to set up the magnetic excitation for the machine. In a very important class of synchronous machines of relatively small power ratings, this winding is replaced by permanent magnets; the resulting configuration is known as a permanent magnet synchronous machine.

Physically, a synchronous machine consists of a stationary member, called the *stator*, and a rotating member, called the *rotor*. In general, the armature and field windings can be located on either the rotor or the stator, but the most common configuration in large synchronous machines is to have the armature winding on the stator. The rotating member, or rotor, thus houses the field winding. A subclassification of synchronous machines is based on the physical construction of the rotating member and field winding: A *salient-pole* structure is one in which individual electrical windings are wound around an even number of poles or projections of magnetic materials (called saliencies) mounted around the shaft of the rotor; the second type of rotor structure is called a *nonsalient smooth rotor*, or *cylindrical rotor*. Figures 3.1 and 3.2 illustrate the rotor-stator configuration of primitive salient-pole and cylindrical-rotor synchronous machines, respectively.

In the most common type of synchronous machine, the magnetic circuit is constructed with an even number of poles, whether in the salient-pole or cylindrical-rotor construction. Therefore, $P$, as used in Eq. (3.1), is an even number. In some methods of analyzing synchronous and induction machines, the pole-pair parameter

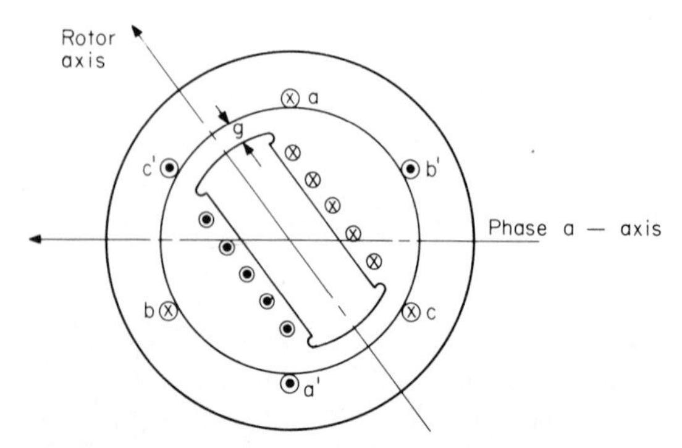


**FIG. 3.1** Cross section of a primitive two-pole salient-pole synchronous machine.

is used, which is always half the magnitude of $P$. In the following analysis, however, $P$ will always represent the total number of poles.

### Pole Pitch: Electrical Degrees

In radial-airgap machines, one *pole pitch* $\lambda$ is defined as

$$\lambda = \frac{360}{P} \tag{3.2}$$

Pole pitch is an arc expressed in mechanical degrees or radians, and $P$ such arcs make up the circumference of a machine. By definition, a pole pitch may also be expressed in terms of electrical degrees as

$$\lambda = 180 \text{ electrical degrees} \tag{3.3}$$

or $\pi$ electrical radians. Thus

$$\text{Electrical degrees} = \frac{P}{2} \quad \text{mechanical degrees} \tag{3.4}$$

about any circumference of a synchronous machine.

### Airgap and Magnetic Circuit of a Synchronous Machine

In Figs. 3.1 and 3.2, the separation between the rotor and the stator is represented by the symbol $g$ and is called the airgap. In cylindrical machines, the airgap is the *radial distance* between the two magnetic surfaces of the rotor and the stator; hence, cylindrical machines are frequently called radial-airgap machines. In this chapter the symbol $g$ represents a distance along the radius of a machine; in some cases the distance along the machine diameter is used, which is twice the value of the radial airgap and is termed the *diametrical airgap*.

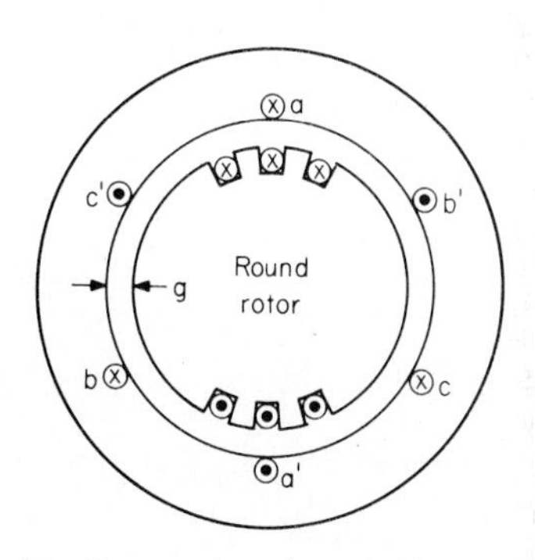


**FIG. 3.2** Cross section of a primitive two-pole cylindrical-rotor synchronous machine.

Synchronous machines in relatively small power ratings are often constructed in disk, or "pancake," geometries rather than in cylindrical geometries. This form of geometry is usually dictated by structural or geometric considerations and is more common in machines with permanent magnet excitation than in wound-field machines. Figure 3.3 illustrates a primitive disk-shaped synchronous machine. The disk (or pancake) machines are *axial-airgap machines*. In general, the magnetic relationships that define machine performance are similar in radial- and axial-airgap machines; therefore, once the magnetic circuit parameters have been defined in terms of the respective geometries, there is little difference in the analysis of radial- or axial-airgap machines. Nevertheless, note that the analysis presented in this chapter assumes radial-airgap, or cylindrical, geometry.

The magnetic circuit of a conventional synchronous machine is similar to that

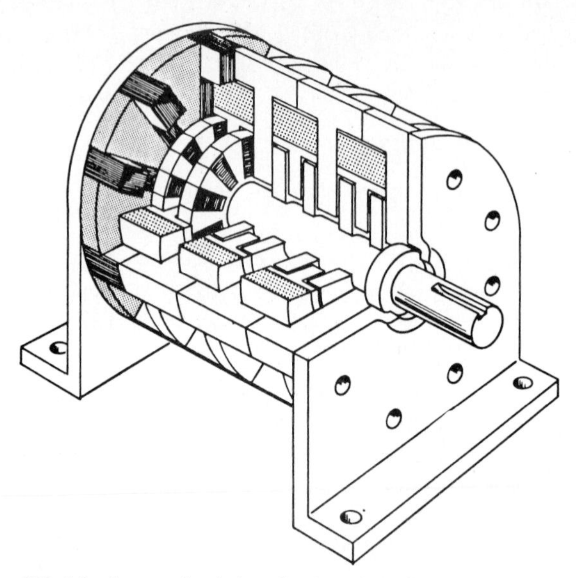

**FIG. 3.3** Cross-sectional view of a three-phase disk motor.

of conventional induction and dc machines. It consists of three basic elements: the airgap, the stator, and the rotor.

***Airgap.*** The airgap exists between the rotor and the stator. The basic dimension affecting the magnetic and electrical properties of a synchronous machine is the radial distance $l_g$. In most practical machines, however, the surfaces separated by $l_g$ are not homogeneous but are constructed of alternate sections of magnetic and nonmagnetic materials knows as *teeth* and *slots*, respectively. Such a pattern may exist on the airgap surface of either the stator or the rotor. Magnetically, a nonhomogeneous surface on the rotor or the stator results in an effective increase in airgap length. This increased length is known as the *effective airgap* $l_{ge}$ and is defined as

$$l_{ge} = K_{cs} K_{cr} l_g \tag{3.5}$$

where $K_{cs}$ and $K_{cr}$ are known as Carter coefficients.[1-3] The Carter coefficients are a function of the slot and/or tooth geometry and are defined below in Eq. (3.6). The physical airgap length $l_g$ is constant around the circumference of a cylindrical-rotor machine. In salient-pole machines the physical airgap is nonuniform. The nonuniformity is caused by increasing the airgap at the outer portions of the salient pole in order to make the pole flux distribution more sinusoidal, a process known as *pole shaping*.

***Stator.*** Stators are constructed of magnetic laminations stacked or assembled in the axial direction and separated by thin films of electricity insulating materials, either oxides formed in the heat-treating process or films applied in liquid form, such as varnishes. The laminations are held together to form a rigid body by means of through bolts, by welds applied on the outer circumference of the laminations, or (in the case of some small-machine stators) by pressure bonding, with the insulating material as the bonding agent. The stator stack is further strengthened by the stator housing at its outer periphery. The stator stack is secured to the housing by means of radial bolts or, in the case of smaller machines, by pressure-fit techniques. A stator lamination consists of three basic magnetic sections: the slots, the teeth, and the stator yoke, or *back iron*. A typical stator lamination is shown in Fig. 3.4. There are many geometries of stator slots; three common geometries for large synchronous machines are shown in Figure 3.5. The Carter coefficient for these geometries is

$$K_c = \frac{(b_1 + b_{tt})(b_1 + 5l_g)}{(b_1 + b_{tt})(b_1 + 5l_g) - b_1(b_1 + 0.5l_g)} \tag{3.6}$$

In Eq. (3.5), $K_{cs}$ is obtained by using the stator slot dimensions in Eq. (3.6) and $K_{cr}$ is obtained by using the rotor slot dimensions in Eq. (3.6).

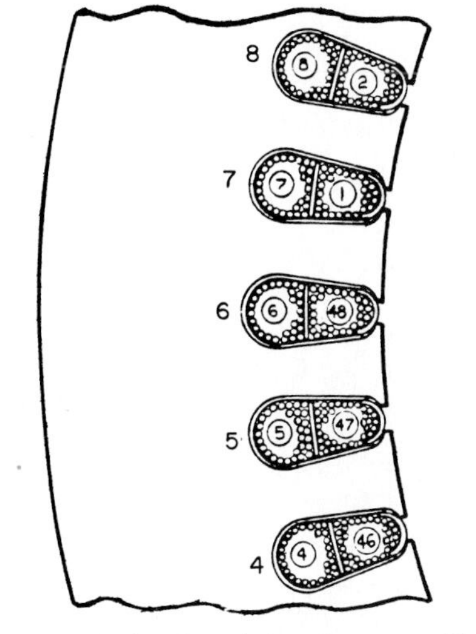

**FIG. 3.4** Portion of a typical synchronous machine stator lamination.

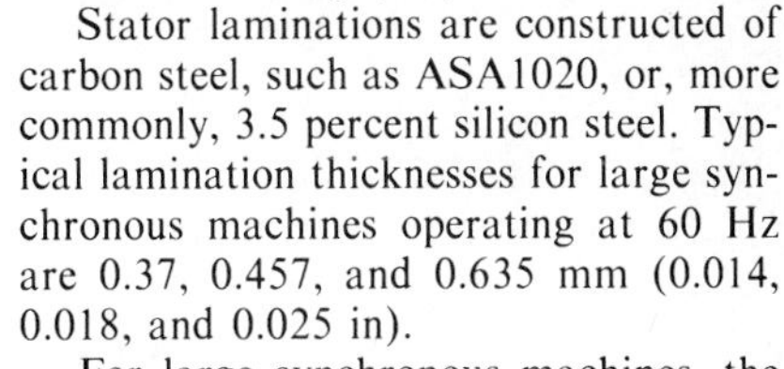

Stator laminations are constructed of carbon steel, such as ASA1020, or, more commonly, 3.5 percent silicon steel. Typical lamination thicknesses for large synchronous machines operating at 60 Hz are 0.37, 0.457, and 0.635 mm (0.014, 0.018, and 0.025 in).

For large synchronous machines, the stacking factor is generally between 0.92 and 0.98.

***Rotor.*** Synchronous machines are constructed with two classes of rotors, salient-pole and cylindrical rotors, as noted above. The former is commonly used in synchronous motors and in slow-speed alternators driven by hydroturbines or waterwheels; the latter construction is common in higher-speed alternators driven by steam turbines, diesel engines, or gas turbines.

There are two basic electrical windings on synchronous machine rotors: (1) the field, or excitation, winding and (2) the damper, or amortisseur, winding. The function of the former has been discussed above. The damper winding is identical in structure and in function to the squirrel-cage winding on brushless induction machines. Its function in synchronous machines is to damp rotor mechanical oscillations by supplying positive or negative induction motor torque; in some cases, it is also used as a means for starting synchronous machines as induction motors.

The damper winding is constructed in a manner similar to that used in squirrel-cage–winding construction on large induction machines (see Chap. 4); i.e., copper or aluminum bars are pushed into the slots rather than being installed by a casting process (as is done on smaller machines). The bars are shorted together electrically at each end by one of two techniques: a shorting ring around the outer periphery constructed of the same material as the bars, or a lamination matching the steel laminations in shape but constructed of the same material as the bars. The bars must be brazed to the end ring or end lamination in order to achieve good electrical and structural connections at each bar connection.

The field winding's construction and shape depend upon whether the rotor is

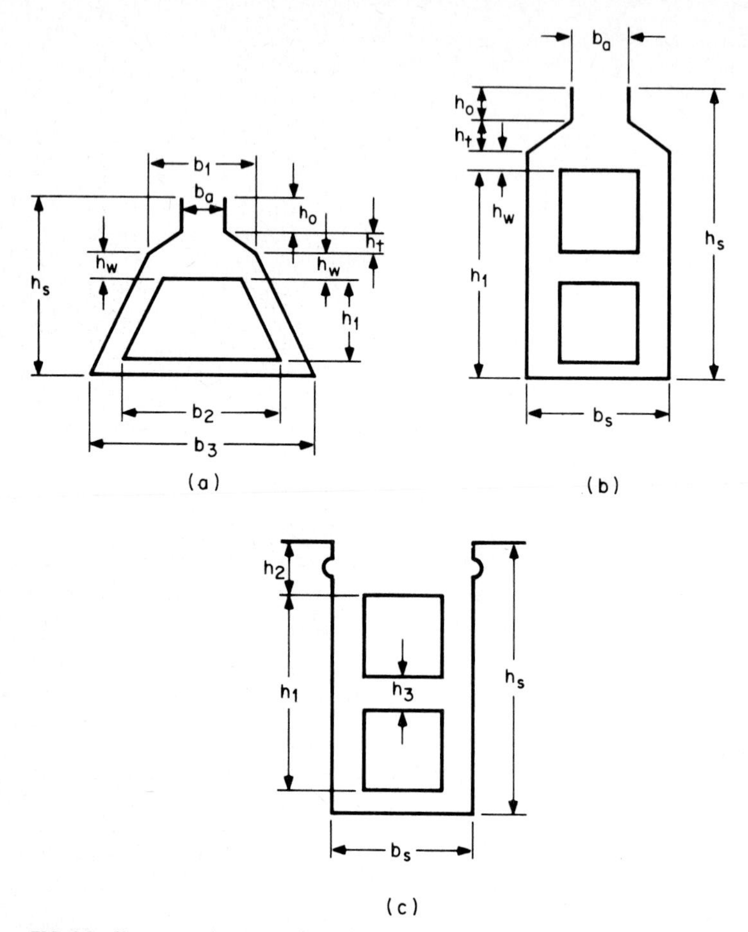


**FIG. 3.5** Slot geometries: ***(a)*** semiclosed slot with one-layer winding; ***(b)*** semiclosed slot with two-layer winding; ***(c)*** open slot.

of the salient-pole or cylindrical type. The field winding is located under the pole face and wound around a magnetic section of constant cross-sectional area. In some cases, given that the magnetic pole-face sections are removable, the field winding can be preformed and slipped over the inner pole section before the pole-face section is mounted. This results in a considerable construction-cost savings, although it adds another airgap at the joint between the pole-face and inner-pole sections, thus requiring additional magnetic potential in the field winding. In cylindrical-rotor machines, the field winding is laid in slots and wound in a distributed manner.

Rotors of either type are generally constructed in stacks of magnetic laminations. Whereas the field flux is a unidirectional (or dc) quantity, there are several components of bidirectional (or alternating) flux induced in portions of the rotor (namely, the pole-face regions) by stator currents and by the effects of variable reluctance in the airgap regions. Therefore, laminated pole sections are generally desirable in order to reduce magnetic losses caused by these effects. The cylindrical-rotor magnetic circuit consists of the same components as the stator. The salient-pole magnetic circuit may be divided into three sections: pole face, inner pole, and rotor yoke, or "inner iron." Materials used in rotor laminations are normally either carbon steel or silicon steel, as in stator laminations.

Slip rings are mounted at one end of the rotor, to which the field winding is electrically connected. Copper-graphite or liquid-metal brushes form the electrical connection to the external power supply for the field winding.

***Structural Members.*** The principal structural members of a synchronous machine, in addition to the electrical and magnetic members discussed above, are the shaft, the bearings and bearing housing, the stator housing, and the mounting structure. The latter forms the mechanical connection to the floor or pedestal on which the machine is mounted and must be capable of transmitting the rated reaction torque of the machine. Engine-driven alternators are often flange-mounted to the engine in a cantilevered manner.

## 3.3 SYNCHRONOUS MACHINE WINDINGS

The functions of synchronous machine windings are (1) to provide a magnetic field of the proper orientation and (2) to provide paths for current conduction. There are three basic types of windings used in synchronous machines: distributed, solenoid, and damper windings.

**1.** *Distributed windings:* This style of winding is laid in slots distributed around the rotor or stator airgap surface and is constructed of insulated wire or bundles of wire.

In large machines the conductor is composed of square or rectangular copper or aluminum bars. The bars are formed into the shape of the elemental unit of a distributed winding, known as a coil, and are insulated with a cloth of nylon, Mylar, or other material; they are then coated with an insulated varnish and thermally cured or baked to form a rigid unit. The coil is placed as a unit into the proper slots—which requires "open slots" of the geometry illustrated in Fig. 3.5*c*. These coils are known as preformed coils.

In medium- and smaller-size machines, conductors are formed from one or more insulated wires known as magnet wire. These bundles of wire are forced into the slots through the slot openings at the airgap surface, coated with an insulating varnish, and thermally cured or baked for structural rigidity and high insulation resistance. Such coils are called random-wound, or mesh, coils.

Distributed windings, whether of the preformed or random-wound type, are placed in the slots according to specific rules to achieve the proper magnetic field relationships. Distributed windings are used for armature windings and for cylindrical-machine field windings.

**2.** *Solenoid windings:* The windings of salient-pole fields are of the solenoid style, very similar to windings used in dc machine fields, many electromagnets, power relays, etc. Field windings are generally multilayer windings, and insulating strips are often used to separate the layers. The solenoid must also be isolated from the magnetic pole structure itself by a hard material with good qualities of electrical insulation, mechanical strength, and puncture or tear resistance to prevent tearing at the sharp corners of the pole structure.

**3.** *Damper windings:* The damper windings of synchronous machines are located at the outer surface of the rotor in a manner similar to squirrel-cage bars on induction machines. In general, these windings are formed and installed in a manner identical to those used in induction motor construction.

## Polyphase Distributed Windings

Synchronous machine armature windings are generally connected to external polyphase systems. Even in single-phase synchronous machines, a pseudo-polyphase external system is created by means of energy-storage circuit elements. The phase relationships of the external source are imposed upon the armature windings, and these relationships are a significant factor in setting up the required magnetic field orientation of the machine. Two- and three-phase armature windings are common for synchronous machines, but the following analysis and notation will be based upon the latter. Several definitions used in distributed winding analysis are:

*Coil:* The basic element of a winding.

*Turn, T:* A coil may consist of one or more turns of conductor connected in series.

*Active conductor, Z:* That portion of a turn in the slot which is acted upon by the excitation flux. Note that there are twice as many active conductors as total turns.

*Parallel paths, A:* The number of electric circuits in parallel in a phase winding.

*Coil pitch, $\beta$:* The arc, measured in electrical degrees, between the two sides of a coil.

*Coil throw:* When the slots are numbered consecutively around the circumference of a machine, the coil throw is the difference between the numbers of the slots embraced by the coil.

*Number of slots, S:* The slot pitch is the electrical arc between slot centers, or

$$\alpha = \frac{180}{S} \qquad \text{electrical degrees} \tag{3.7}$$

*Phase belt, $S_{pm}$:* The number of coils per phase per pole in a double-layer winding, or

$$S_{pm} = S/P/M \tag{3.8}$$

***Double-Layer Lap Winding.*** This is the most common layout for armature windings in synchronous machines. *Double layer* refers to two layers of coil sides in each slot, i.e., two coil sides per slot or one coil per slot. *Lap* refers to the manner of connecting one coil in a phase belt to the other coils in that group: the end of one coil is "lapped back" and connected to the beginning of another coil in the slot adjacent to the beginning of the first coil. This is seen in the layout of lap coils in Fig. 3.6. In a manner similar to that in dc armature windings, a coil is placed in the slots with one coil side in the upper part of one slot and the other coil side in the lower part of another slot.

One pole is made up of a phase belt $S_{pm}$ from each phase. Thus, in a three-phase machine, the portion of the winding that constitutes one pole consists of three phase belts; in a two-phase machine, there are two phase belts per pole. The *P* phase belts (one from each pole) that make up a phase winding may be connected to each other in series or in parallel or in appropriate series-parallel combinations; the choice of series, parallel, or series-parallel connections of the phase belts is related to many

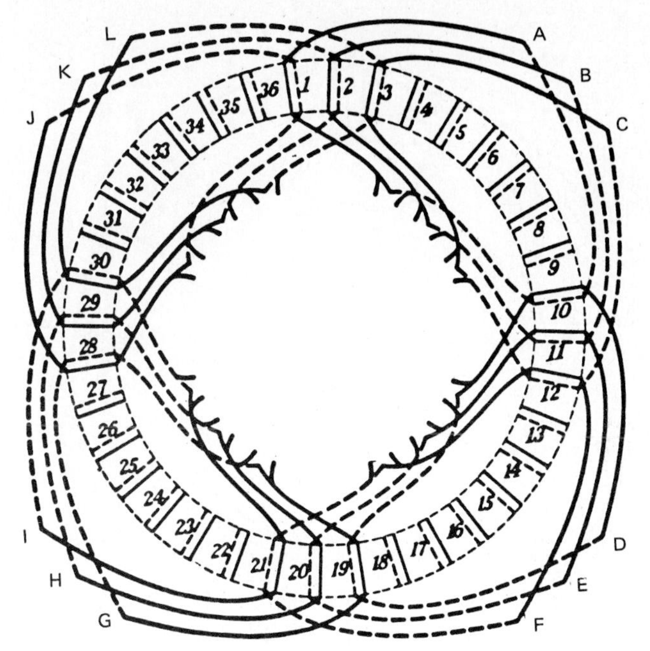


**FIG. 3.6** Lap coils arranged in 12 slots of a 36-slot core for one of the phases of a three-phase winding.

choices available in designing a machine as well as to the machine voltage and current ratings. It is seen that phase-belt coil sides occupy one-third of a pole pitch; from the definition of pole pitch in electrical degrees [Eq. (3.3)], the coil sides of a phase belt therefore occupy an arc of 60 electrical degrees. This type of winding is often called a 60° winding. Since the time-phase relationships of the windings must be the same as their spatial relationships, the currents in the three phase belts making up a pole must have a phase relationship of 60 electrical degrees rather than the characteristic 120° relationship. This requires that the middle phase belt making up a machine pole (usually designated as phase B) be reversed in its electrical connection with respect to the other two. This is shown in the complete schematic layout of a double-layer three-phase lap winding in Fig. 3.7. It is seen from Fig. 3.7 that the number of phase belts in each phase winding equals the number of poles $P$; thus, in a six-pole winding, there are six phase belts in each phase and a total of 18 phase belts. It is also seen that half of these phase belts are so connected as to result in poles of one polarity, say north, and that the other half are so connected as to result in the opposite, or south, polarity. Such a winding as has been described—a winding for which $S_{pm}$ is an integer—is known as an integral slot winding.

It is entirely possible to connect a winding so that the poles of only one polarity are formed, say north poles. Such a winding is known as a 120° winding. In such a winding, equivalent poles of the opposite polarity are formed by the return flux paths that occur in between the created polarities. Such a winding results in a less regular airgap flux pattern and is not as popular as the 60° winding.

***Pitch Factor.*** The coil pitch (as described later in this chapter) may be equal to or less than the pole pitch [Eqs. (3.2) and (3.3)]. The coil pitch influences

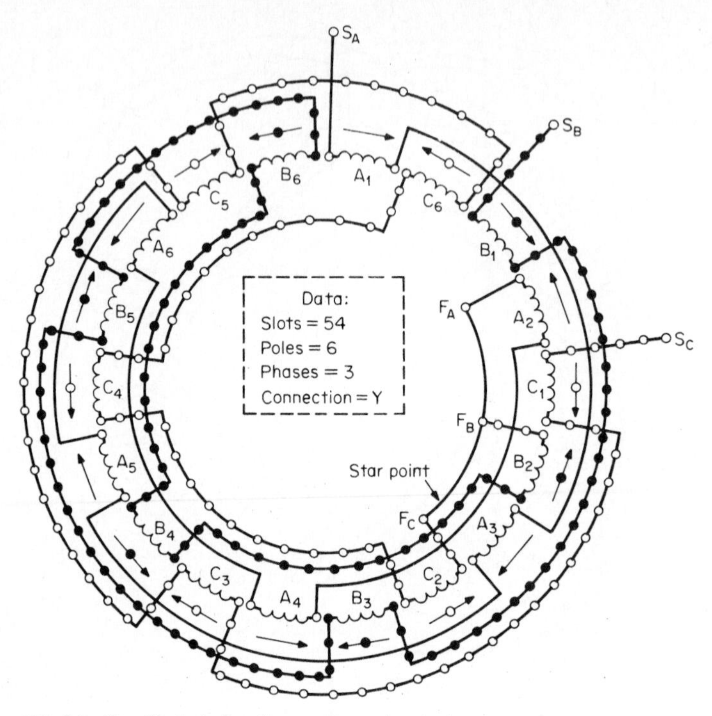


**FIG. 3.7** Simplified winding diagram illustrating the bottom-top, top-bottom connection method for a lap winding.

the magnitude of the voltage induced in a coil. Refer to Fig. 3.8; it illustrates a linear layout of a pole pair, the stator slots in the region of the pole pair, and one armature coil laid in a pair of armature slots. A salient-pole structure is shown, but the conclusions are identical for a cylindrical-rotor field structure with distributed windings.

The pitch factor is determined by the spatial location of the coil sides with respect to the magnetic axes of the poles $P_n$ and $P_s$, as shown in Fig. 3.8. As the field structure rotates, a voltage is induced in each coil side. If the coil pitch is equal to the pole pitch, each coil side will be in an identical location with respect to a north and south pole, respectively, as shown by the coil drawn in solid lines in Fig. 3.8; therefore, the two coil-side induced voltages will be in time phase with each other and will be additive, giving a coil voltage of twice the voltage induced in one coil side. If the coil pitch is less than the pole pitch by one slot, as shown by the dotted coil side in Fig. 3.8, the two coil-side voltages will still be of the same magnitude but will be out of phase with each other, and the total coil voltage will be less than twice the coil-side voltage. Such a coil is called a *short-pitched*, or chorded, coil and is commonly used in most classes of rotating machines. The *pitch factor* $K_p$ is defined as the ratio of the short-pitched voltage to the full-pitched voltage. It may be shown that

$$K_p = \cos \frac{n\alpha}{2} \tag{3.9}$$

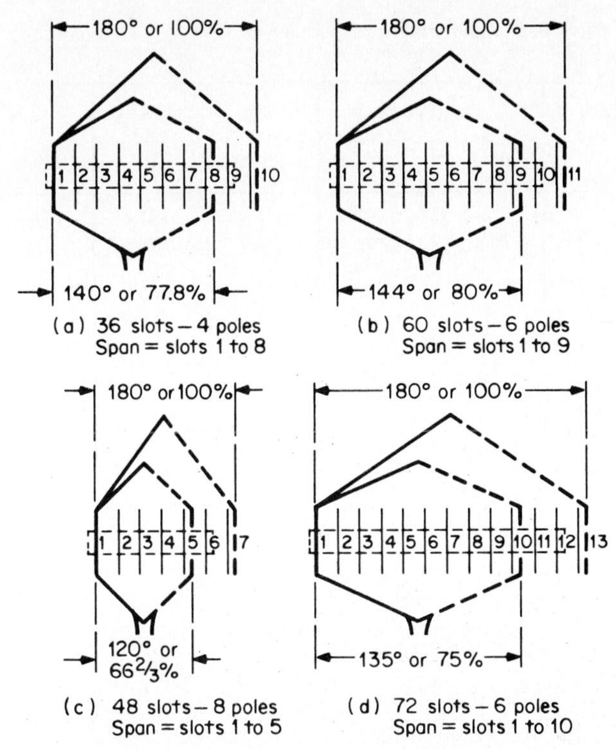


**FIG. 3.8** Fractional-pitch lap coils and their corresponding degrees and percentage values. The 100 percent pitch coils are indicated for comparison.

where $n$ = number of slots lost by short pitching. The coil pitch as a decimal or fraction is expressed as

$$\lambda_c = \frac{\text{coil throw (slots)}}{\text{pole pitch (slots)}} \tag{3.10}$$

or, in terms of slots,

$$\lambda_c = \frac{S/P - n}{S/P} \tag{3.11}$$

For example, in Fig. 3.8 there are eight slots per pole and $n$ is one. The pitch, from Eq. (3.11) is 7/8; the slot pitch, from Eq. (3.7), is 22.5 electrical degrees; and the pitch factor $K_p$, from Eq. (3.9), is cos (11.25), or 0.981. The actual coil voltage is therefore $2 \times 0.981E_c$. Table 3.1 gives the pitch factors for many slot combinations, expressed in terms of fractional pitch.

Short-pitched coils are used in synchronous machines for two major reasons: (1) reduction of harmonic induced voltage and (2) economy, since a short-pitched coil uses less conductor and insulation materials and is generally easier to place in slots than is a full-pitched coil.

***Distribution Factor.*** The phase belt, the basic unit of distributed lap windings, consists of one or more coils, with the coil sides generally located in adjacent slots.

**TABLE 3.1** Pitch Factors for Fundamental Components in a Three-Phase Armature Winding

| Pitch | Pitch factor $k_p$ | Pitch | Pitch factor $k_p$ | Pitch | Pitch factor $k_p$ | Pitch | Pitch factor $k_p$ |
|---|---|---|---|---|---|---|---|
| 1 | 1 | 13/15 | 0.978 | 11/15 | 0.914 | 7/12 | 0.793 |
| 23/24 | 0.998 | 18/21 | 0.975 | 13/18 | 0.906 | 12/21 | 0.782 |
| 20/21 | 0.997 | 5/6 | 0.966 | 15/21 | 0.901 | 5/9 | 0.766 |
| 17/18 | 0.996 | 17/21 | 0.956 | 17/24 | 0.897 | 13/24 | 0.752 |
| 14/15 | 0.995 | 12/15 | 0.951 | 2/3 | 0.866 | 8/15 | 0.743 |
| 11/12 | 0.991 | 19/24 | 0.947 | 15/24 | 0.831 | 11/21 | 0.733 |
| 19/21 | 0.989 | 7/9 | 0.940 | 13/21 | 0.826 | 3/6 | 0.707 |
| 8/9 | 0.985 | 16/21 | 0.931 | 11/18 | 0.819 | | |
| 21/24 | 0.981 | 9/12 | 0.924 | 9/15 | 0.809 | | |

If a phase belt consists of more than one coil, there is a *reduction* in the total induced voltage of the phase belt due to the space-phase relationships; this reduction is similar to the coil voltage reduction resulting from short pitching. Obviously, two adjacent coils in a phase belt are not in the same spatial location with respect to excitation-pole axes; therefore, the voltages in two adjacent coils will also have a time-phase relationship with each other, and the sum of the two coil voltages will be less than twice the voltage of one coil. If there are more than two coils connected in series in a phase belt, this "distributive effect" becomes progressively greater as the number of coils per phase belt is increased. The ratio of the phasor sum of the voltages of $S_{pm}$ coils per belt to $S_{pm}$ times the voltage of one coil is known as the *distribution factor* $K_d$. The distribution factor can be expressed analytically as

$$K_d = \frac{\sin(\alpha S_{pm}/2)}{S_{pm}(\sin \alpha/2)} \tag{3.12}$$

Table 3.2 lists the fundamental distribution factors for several phase belts in two- and three-phase machines.

***Skew Factor.*** In some distributed windings, the slots and the conductors laid in these slots are *skewed* with respect to parallelism with the axes of the machine's poles or the axis of the machine's shaft. This is also frequently done with damper windings on synchronous machines. The purpose behind skewing is the same as that for the use of short pitching and of distribution coils in a phase belt: improvement of machine voltage waveforms by reducing the harmonic content of induced voltages. However, skewing also reduces the magnitude of the voltage induced in a coil side or damper bar by varying the spatial phase relationship with respect to excitation-pole axes along the axial length of a coil side or damper bar. The angle of skew, $\lambda_{sk}$, is the angle at which the skewed coil or its projection crosses the shaft axis. The skew factor $K_{sk}$ is defined as

$$K_{sk} = \frac{\sin(\lambda_{sk}/2)}{\lambda_{sk}/2} \tag{3.13}$$

**TABLE 3.2** Distribution Factors for Fundamental Components in a Three-Phase Armature Winding

| Slots per pole per phase | Slots per pole | Factor $k_d$ | Slots per pole | Factor $k_d$ |
|---|---|---|---|---|
| 1 | 3 | 1.000 | 18 | 0.956 |
| 2 | 6 | 0.966 | 21 | 0.956 |
| 3 | 9 | 0.960 | 24 | 0.956 |
| 4 | 12 | 0.958 | $\infty$ | 0.955 |
| 5 | 15 | 0.957 | | |
| etc. | | | | |

***Winding Factor.*** The winding factor reflects the reduction in voltage due to short pitching, distributed phase belts, and skewing. It is defined as

$$K_w = K_p K_d K_{\text{sk}} \tag{3.14}$$

In this section, pitch, distribution, and skew factors have been introduced for the fundamental component of airgap magnetic field and winding induced voltages. When the effects of harmonic components of airgap flux must be included in the design or analysis of a machine, the machine's harmonic-pitch, distribution, and skew factors will be required. Table 3.3 lists pitch factors, and Table 3.4 lists distribution factors for a number of odd harmonics.

***Fractional-Slot Windings.*** A winding is termed a *fractional-slot winding* when the phase-belt number $S_{pm}$ is a noninteger. For example, consider placing a three-phase four-pole winding in a stator with 18 slots: $S_{pm}$ is seen to be 1.5. Note that the number of coils per phase must be an integral number in order for the induced phase voltages to be balanced. If the total number of slots divided by the number of phases is a noninteger, one or more slots is left vacant. If such a winding is to be placed on a rotating member, the vacant slot is filled with a "dummy coil" in order to maintain mechanical balance. In the present example, the number of slots per phase is an integer, 18/3 = 6, and a fractional-slot winding can be used to achieve balanced phase voltages.

Continuing with the 18-slot example above, a three-phase winding with several different values of $S_{pm}$ can be used. In this example, a relatively simple fit can be made by using two phase belts with one and two slots per pole per phase, respectively. In each polar group, there are 4.5 slots per pole, requiring three phase belts per pole. For two poles, the polar group would consist of two phase belts of two coils and one belt of one coil, giving five coils in these two polar groups. For the other two poles, there would be two phase belts of one coil and one of two coils, giving a total of four coils in these polar groups. The average of four polar groups is 4.5 coils, as required. Each phase winding would consist of two belts of two coils and two belts of one coil for a total of six coils per phase, as required. These would be selected in a symmetrical manner from the four polar groups in order to achieve balanced phase voltages. However, the flux density distribution of the four polar groups would be slightly different owing to the different number of coils per pole and to the different geometry of the slots for the two sets of poles. This would result in different distribution factors for the two sets of polar groups; however, the *average distribution factor* for the phase winding could be found in the

**TABLE 3.3** Pitch Factors*

| λ Pitch ratio | Slots per pole | | | | | | | | Pitch factor | | | |
|---|---|---|---|---|---|---|---|---|---|---|---|---|
| | 2 | 3 | 4 | 6 | 8 | 9 | 10 | 12 | $k_{p1}$ | $k_{p3}$ | $k_{p5}$ | $k_{p7}$ |
| | 1 | 1 | 1 | 1 | 1 | 1 | 1 | 1 | 1.000 | 1.000 | 1.000 | 1.000 |
| | | | | | | | | 11/12 | 0.991 | 0.924 | 0.793 | 0.609 |
| | | | | | | | 9/10 | | 0.988 | 0.891 | 0.707 | 0.454 |
| | | | | | | 8/9 | | | 0.985 | 0.866 | 0.643 | 0.342 |
| | | | | | 7/8 | | | | 0.981 | 0.831 | 0.556 | 0.195 |
| | | | | 5/6 | | | | 10/12 | 0.966 | 0.707 | 0.259 | −0.259 |
| | | | | | | | 8/10 | | 0.951 | 0.588 | 0.000 | −0.588 |
| | | | | | | 7/9 | | | 0.940 | 0.500 | −0.174 | −0.766 |
| | | | 3/4 | | 6/8 | | | | 0.924 | 0.383 | −0.383 | −0.924 |
| | | | | | | | 7/10 | | 0.891 | 0.156 | −0.707 | −0.988 |
| | | 2/3 | | 4/6 | | 6/9 | | 8/12 | 0.866 | 0.000 | −0.866 | −0.866 |
| | | | | | 5/8 | | | | 0.831 | −0.195 | −0.981 | −0.556 |
| | | | | | | | 6/10 | | 0.809 | −0.309 | −1.000 | −0.309 |
| | | | | | | | | 7/12 | 0.793 | −0.383 | −0.991 | −0.131 |
| | | | | | | 5/9 | | | 0.766 | −0.500 | −0.940 | 0.174 |
| | 1/2 | | 2/4 | 3/6 | 4/8 | | 5/10 | 6/12 | 0.707 | −0.707 | −0.707 | 0.707 |

* For curves illustrating the manner in which the pitch factors vary with the ratio of coil pitch to pole pitch, see Fig. 7.4, p. 649, *Standard Handbook for Electrical Engineers*, 8th ed., A. E. Knowlton (ed.), McGraw-Hill, New York, 1949.

**TABLE 3.4** Distribution Factors for Three-Phase Lap Windings, $K_{dn}$

| Slots per pole, $S_1/P$ | Harmonic, $n$ | Distribution factor, $K_{dn}$ |
|---|---|---|
| 3 | 1 | 1 |
| 3 | 3 | 1 |
| 3 | 5 | 1 |
| 3 | 7 | 1 |
| 3 | 9 | 1 |
| 3 | 11 | 1 |
| 6 | 1 | 0.9659258 |
| 6 | 3 | 0.7071073 |
| 6 | 5 | 0.2588201 |
| 6 | 7 | −0.2588176 |
| 6 | 9 | −0.7071054 |
| 6 | 11 | −0.9659253 |
| 9 | 1 | 0.9597952 |
| 9 | 3 | 0.6666673 |
| 9 | 5 | 0.2175688 |
| 9 | 7 | −0.1773622 |
| 9 | 9 | −0.3333333 |
| 9 | 11 | −0.1773643 |
| 12 | 1 | 0.9576622 |
| 12 | 3 | 0.6532821 |
| 12 | 5 | 0.2053359 |
| 12 | 7 | −0.1575584 |
| 12 | 9 | −0.2705982 |
| 12 | 11 | −0.1260796 |
| 15 | 1 | 0.9566772 |
| 15 | 3 | 0.6472141 |
| 15 | 5 | 0.2000009 |
| 15 | 7 | −0.149447 |
| 15 | 9 | −0.2472137 |
| 15 | 11 | −0.1094645 |
| 18 | 1 | 0.9561428 |
| 18 | 3 | 0.6439511 |
| 18 | 5 | 0.1971844 |
| 18 | 7 | −0.1452866 |
| 18 | 9 | −0.2357024 |
| 18 | 11 | −0.1017321 |
| 21 | 1 | 0.9558208 |
| 21 | 3 | 0.6419947 |
| 21 | 5 | 0.1955129 |
| 21 | 7 | −0.1428565 |
| 21 | 9 | −0.2291252 |
| 21 | 11 | −9.744083E-02 |
| 24 | 1 | 0.9556118 |
| 24 | 3 | 0.6407295 |
| 24 | 5 | 0.1944388 |
| 24 | 7 | −0.1413098 |
| 24 | 9 | −0.2249942 |
| 24 | 11 | −9.479173E-02 |

usual manner, using Eq. (3.12). In our example, the slot pitch would be, from Eq. (3.7), $180/4.5 = 40$. The distribution factor would be, from Eq. (3.12), $\sin(1.5 \times 40/2)/(1.5 \sin 40/2) = 0.5/0.513 = 0.975$.

## 3.4 CONCENTRIC WINDINGS

Concentric windings are very different from the double-layer lap windings discussed in Sec. 3.3. The coils making up a polar group form a concentric, or "spiral," pattern, as shown in Fig. 3.9, from whence arises their name. Concentric windings are used as field windings in many cylindrical-rotor synchronous machines. They are also used as armature windings in single-phase synchronous machines and, occasionally, in polyphase synchronous machines. Concentric windings are usually *single-layer windings*, in contrast to lap windings, although multilayer concentric windings are feasible. The coils of concentric windings may be constructed with more than one turn per coil and often with a different number of turns among the coils making up a polar group (as shown in Fig. 3.9), which is another difference from conventional lap windings.

It is seen from Fig. 3.9 that each of the concentric coils making up a polar group has a different pitch factor. Therefore, the pitch and distribution effects are combined into a concentric-winding factor $K_{wc}$, defined as[4]

$$K_{wc} = \frac{N_1 \sin n\beta_1 + N_2 \sin n\beta_2 + \cdots}{N_1 + N_2 + N_3 + \cdots} \tag{3.15}$$

where $N_1$, $N_2$, $N_3$, etc., are the number of turns in individual coils making up a polar group; $\beta_1$, $\beta_2$, $\beta_3$, etc., are the pitches in electrical degrees or radians of the individual coils; and $n$ is the order of the harmonic for which the winding factor is being calculated.

The complete winding factor for a concentric winding must also include the skew factor and is defined as

$$K_w = K_{wc} K_{sk} \tag{3.16}$$

in a manner similar to Eq. (3.14) for lap windings.

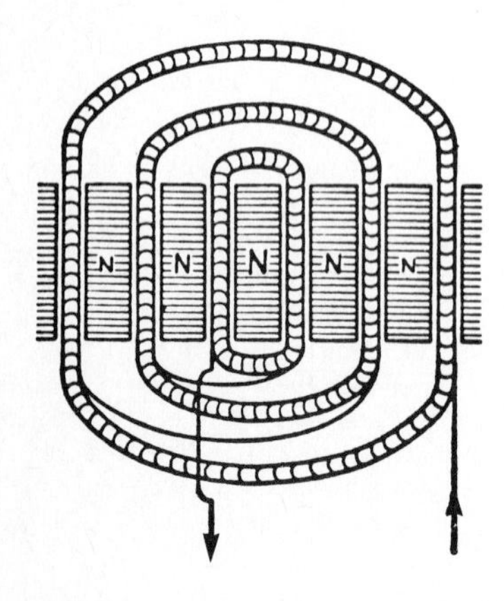


**FIG. 3.9** One pole group of three concentrically formed coils.

## 3.5 FIELD EXCITATION

The field windings of conventional polyphase synchronous machines are energized, or *excited*, from a dc source. Other types of synchronous machines are excited by permanent magnets. Also, certain synchronous machines have no external excitation and are called *singly excited* synchronous machines. Only the conventional, dc-excited class of synchronous machine excitation will be discussed here.

The source of energy for synchronous machine field windings may be any conventional dc power supply, including batteries, solar converters, dc generators, and electronic power supplies. Regardless of the source, there is always some measure of *control* of the energy source's voltage and/or current, which provides one of the most valuable features of synchronous machines. Excitation control provides a significant—and generally a relatively simple—means of controlling the synchronous machine's characteristics, such as terminal voltage, power factor, short-circuit current, torque, and transient response.

In large synchronous machines, the excitation source is often mounted on the same shaft as the synchronous machine itself. The exciter machine may be a dc or ac machine with appropriate control; if ac, the exciter is generally a conventional synchronous machine of the same configuration as the main machine. The alternating output of the exciter is rectified and applied to the field winding of the main machine. If the field of the main machine is on the rotor, which is the most common configuration, the excitation is supplied by means of a brush–slip-ring system. Excitation control can be provided by controlling the dc field of the exciter or by controlling the exciter armature voltage electronically, or by both methods. Except for battery energy sources, the typical excitation source may contain considerable ripple, which is generally undesirable since it greatly increases the harmonic content of the generated voltage; thus, an integral part of most excitation systems is a filter to reduce the ripple of the excitation output to an acceptable value. Since excitation ripple is of relatively low frequency, the cost of the filter may be a significant portion of the total excitation cost.

### Rotating-Rectifier Excitation

Except in large central station steam-generating plants, the use of slip rings to connect the exciter output to the synchronous machine field circuit may be undesirable. Slip rings present problems except where the environment is well controlled and where frequent inspection is performed, as in large central station plants. Therefore, in smaller sizes, synchronous machines are frequently constructed in a "brushless" configuration known as *rotating-rectifier machines*, especially when operation in harsh environments is required.

The rotating-rectifier configuration uses an exciter mounted on the same shaft as the main synchronous machine and is often combined structurally into a common housing with the main machine. The exciter has the reverse configuration of the conventional synchronous machine, with the field on the stator and the armature on the rotor. The output of the voltage generated in the exciter armature is rectified and applied directly to the main field circuit, eliminating the need for slip rings, since the exciter armature, rectifier, and main field are all mounted on the rotating shaft; hence, the name "rotating rectifier." Excitation control is achieved by controlling the stationary field of the exciter.

## Series Excitation

The exciter for the synchronous machine might be classified as "separate excitation" in the context of dc machine nomenclature. The exciter control may be derived from the main machine's output voltage, in which case the excitation is approximately analogous to shunt excitation in the dc case. Likewise, it is possible to provide the equivalent of dc *series* excitation; usually, this is accomplished in conjunction with shunt or separate excitation, providing the equivalent of dc "compound" excitation.

*Series excitation* refers to providing excitation derived from an element in series with the main machine's armature (or load) connections. Physically, it can be a current transformer, resistive current shunt, Hall device, or an actual series winding as used in dc machines. The first three of these schemes generally apply the output of the current-sensing device to the exciter control. Likewise, in the latter of the above schemes, the series winding is generally on the exciter rather than on the main field pole. An example of this scheme is shown in Fig. 3.10. This type of excitation is appropriate for isolated synchronous alternators that are used in motor-starting applications where high inrush current is required. Series excitation is seldom used in large central station alternators.

# 3.6 FIELD MMF AND FLUX IN AIRGAP

The purpose of machine excitation is to apply an mmf to the machine's airgap, where energy conversion occurs. This applied excitation mmf produces a magnetic flux density in the airgap. The flux density results from a combination of the applied mmf and the characteristics of the machine's magnetic circuit.

For a rectangular-shaped salient-pole structure, the airgap mmf due to the excitation windings on each pole presents a quasi-square-wave function around the

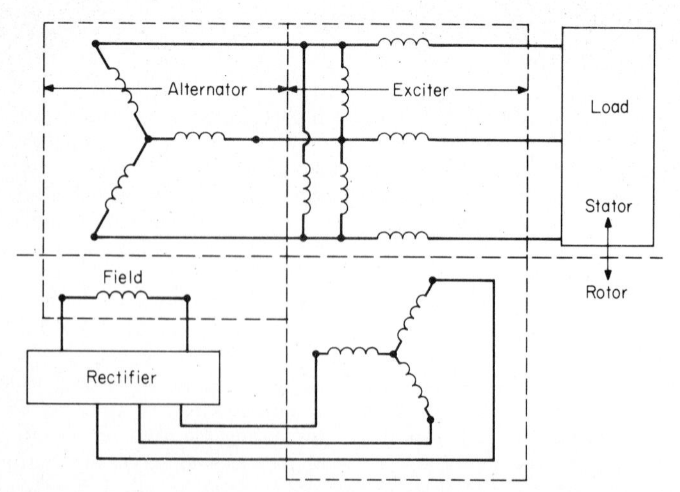


**FIG. 3.10** Schematic diagram of a self-excited rotating-rectifier alternator.

periphery of the airgap. Salient-pole structures are generally mushroom-shaped and may contain considerable "pole-shaping" along the peripheral edges of the poles in order to change the flux distribution along the airgap from a nearly square waveform to a more sinusoidal distribution. In representing mmf and flux distributions about the cylindrical airgap surface, a linear diagram is generally used. Such a diagram for a salient-pole field structure is shown in Fig. 3.11. A smooth magnetic surface results when "closed slot" construction is used, in which the slots are magnetically closed by axial steel inserts between adjacent teeth. However, this construction is seldom possible in large machines; usually, the armature airgap surface is the alternate slot-and-tooth structure described earlier. This circumferentially periodic variation in airgap reluctance will cause a ripple, or periodic variation, in airgap flux density resulting from the excitation mmf; the ripple is illustrated in Fig. 3.11*b*. The magnitude of the ripple depends upon the width of the slot opening; the period of the ripple depends, of course, on the number of slots in the stator laminations. This ripple in the airgap flux results in harmonics in the generated voltage, known as *slot harmonics*.

Cylindrical-rotor field structures result in roughly the same mmf and flux distributions about the airgap as do salient-pole structures. In general, cylindrical-pole structures with concentric windings result in a more sinusoidal flux distribution than is possible with salient-pole structures. However, the slots in cylindrical field structures introduce another ripple in airgap flux distribution and another set of slot harmonics. Certain combinations of stator and rotor slot numbers may result in some adverse operating conditions caused by slot harmonics.

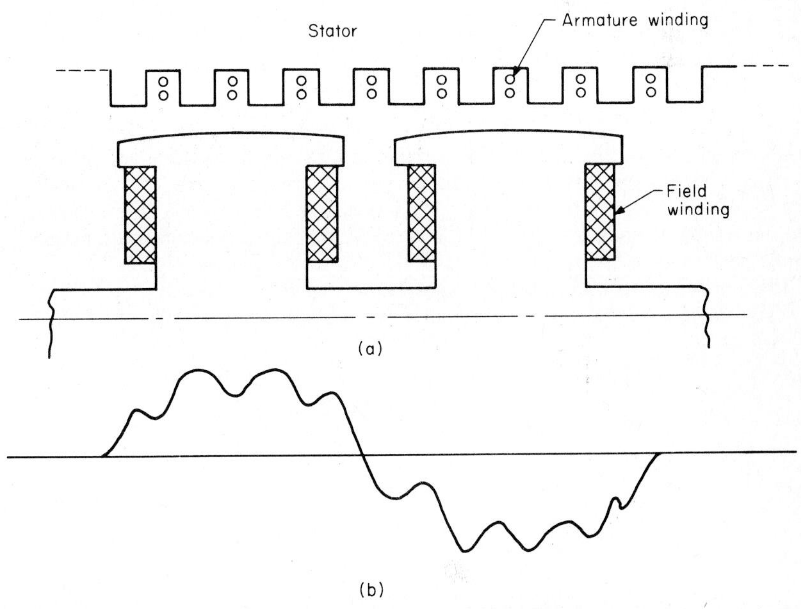


**FIG. 3.11** ***(a)*** Schematic layout of stator and salient rotor poles; ***(b)*** typical airgap flux waveform showing effects of stator slots.

Field excitation is generally defined on a per-pole basis as

$$M_f = N_f I_f \qquad \text{At per pole} \tag{3.17}$$

where $N_f$ = turns per pole
$I_f$ = current in the pole winding, A

Pole windings may be connected in series, parallel, or series-parallel combinations. Connecting all pole windings in series generally results in the most balanced mmf per pole but requires the highest voltage excitation source.

## 3.7 ARMATURE-WINDING MMF AND FLUX

The second source of excitation in synchronous machines is the armature winding. The source of excitation is the armature current. The armature mmf and/or flux are frequently termed *armature reaction.*

Armature-winding mmf can be described by means of straightforward magnetic circuit analysis. The mmf in the airgap of a synchronous machine due to one coil having a pitch $\lambda_c$ and carrying a steady current $I$ is shown in Fig. 3.12. The mmf is defined as

$$M = IT \qquad \text{At} \tag{3.18}$$

The mmf varies as a function of position $x$ along the airgap in the following manner:

$$\begin{aligned} m &= -\lambda_c M & -\pi < x < -\lambda_c\pi/2 \\ &= (2-\lambda_c)M & -\lambda_c\pi/2 < x < \lambda_c\pi/2 \\ &= -\lambda_c M & \lambda_c\pi/2 < x < \pi \end{aligned} \tag{3.19}$$

This elemental coil mmf is the basis for the 120° winding (as discussed in Sec. 3.3) in which a single coil sets up the two polarities, as shown in Fig. 3.12. However, the 60° winding is much more common in all types of rotating electric machines, and two coils such as shown in Fig. 3.12 are used, one for each north-south polarity of a pole pair.

It is convenient to resolve the rectangular waveshape of the coil mmf of Fig. 3.12 into sinusoidal components by means of Fourier series analysis. Thus

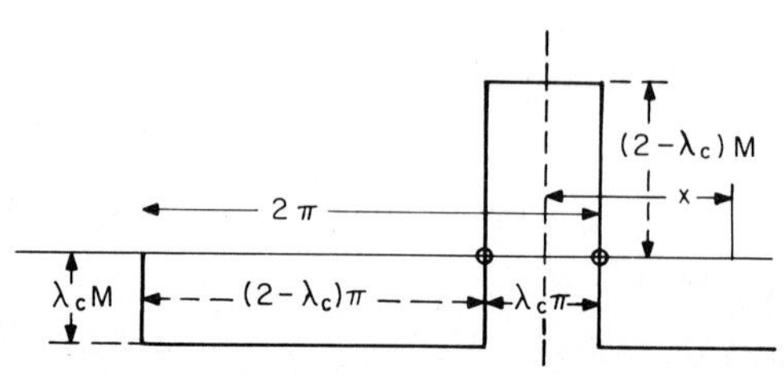


**FIG. 3.12** Airgap mmf of a single fractional-pitch coil.

$$m = \frac{4M}{\pi}\left(\sin\frac{\lambda_c\pi}{2}\cos x + \frac{1}{2}\sin\frac{2\lambda_c\pi}{2}\cos 2x\right.$$

$$\left. + \frac{1}{3}\sin\frac{3\lambda_c\pi}{2}\cos 3x + \cdots\frac{1}{k}\sin\frac{k\lambda_c\pi}{2}\cos kx + \cdots\right) \tag{3.20}$$

The components of Eq. (3.20) are known as *spatial harmonics*. The magnitude of the higher spatial harmonics generally decreases as the value of $S_{pm}$ increases; also, the pitch factor can greatly affect the magnitude of specific spatial harmonics, as shown in Table 3.3.

In ac machines the mmf function $M$ can be altered to indicate a sinusoidally varying function.

$$M = M_m \cos\omega t \tag{3.21}$$

which, when substituted into Eq. (3.20), gives

$$m = \frac{4M_m\cos\omega t}{\pi}\left(\sin\frac{\lambda_c\pi}{2}\cos x + \frac{1}{2}\sin\frac{2\lambda_c\pi}{2}\cos 2x\right.$$

$$\left. + \cdots\frac{1}{k}\sin\frac{k\lambda_c\pi}{2}\cos kx + \cdots\right) \tag{3.22}$$

where $M_m = \sqrt{2}(IT)$ = maximum mmf
$I$ = rms current in a series turn

## Rotating Magnetic Field

By means of trigonometric substitutions, a typical harmonic component of the form

$$m = \frac{4M_m}{\pi k}\sin\frac{k\lambda_c\pi}{2}\cos kx\cos 2\pi ft \tag{3.23}$$

can be resolved into

$$m = \frac{2M_m}{\pi k}\sin\frac{k\lambda_c\pi}{2}[\cos(kx - 2\pi ft)$$

$$+ \cos(kx + 2\pi ft)] \tag{3.24}$$

As seen from Eq. (3.24), the mmf of an ac-excited coil can be expressed as the sum of two equal traveling waves revolving about the airgap in two opposite directions at the speed

$$\frac{x}{t} = \frac{2\pi f}{k} \qquad \text{rad/s} \tag{3.25}$$

which, of course, is the harmonic synchronous speed.

## Polyphase-Winding MMFs

From earlier discussions it follows that the typical synchronous machine winding consists of a number of coil groupings, known as phase belts, arranged in a double-layer lap configuration. The mmf's of these coil and phase-belt groupings are modified as follows.[2]

***MMF of One Phase Winding.*** This can be expressed by combining terms of the form of Eq. (3.24) to include the effects of the distribution and pitch factors of the series coils making a complete phase winding:

$$m_\phi = \frac{2M_m S_{pm}}{\pi}\left[K_p K_d \cos(x-\omega t) + \frac{1}{3}K_{p3}K_{ds}\cos(3x-\omega t) + \frac{1}{5}K_{p5}K_{d5}\cos(5x-\omega t) + \cdots\right] + \frac{2M_m S_{pm}}{\pi}\left[K_p K_d \cos(x+\omega t) + \frac{1}{3}K_{p3}K_{d3}\cos(3x+\omega t) + \frac{1}{5}K_{p5}K_{d5}\cos(5x+\omega t) + \cdots\right] \tag{3.26}$$

It is seen that the mmf of one phase winding is *stationary* and cannot, of itself, develop a torque that results in machine rotation. It is the resultant of several phase windings that produces a rotating magnetic field.

***MMF of Polyphase Windings.*** Armature windings of synchronous machines have been wound for 1-, 2-, 3-, 4-, 5-, 6-, and 12-phase excitation. The most widely used polyphase winding for synchronous machines is the three-phase winding (as discussed in Sec. 3.3). The mmf of one phase winding of a polyphase winding has been given in terms of its Fourier series components by Eq. (3.26). In polyphase windings, $m$ such windings are displaced in both space and time by the characteristic angle of the polyphase system. The characteristic angles for polyphase systems are shown in Table 3.5.

In a three-phase system, the characteristic angle may take on the values of 0°, 120°, and −120°. Adding these three angles to both the time and space cosine functions in each harmonic term of Eq. (3.24) gives three new equations, one for each phase winding [the equation for 0°, of course, being identical to Eq. (3.24)]. Adding these three equations and introducing the pitch and distribution factors gives an expression for the resultant mmf of three identical windings displaced in space and time by 120 electrical degrees:

$$M_3 = \frac{4}{\pi}\left(\frac{3M_m S_{pm}}{2}\right)\left[K_p K_d \cos(x-\omega t) + \frac{1}{5}K_{p5}K_{d5} \cos(5x+\omega t) + \frac{1}{7}K_{p7}K_{d7}\cos(7x-\omega t) + \frac{1}{11}K_{p11}K_{d11}\cos(11x+\omega t) + \frac{1}{13}K_{p13}K_{d13}\cos(13x-\omega t) + \cdots\right] \tag{3.27}$$

The first term in Eq. (3.27) represents the fundamental mmf wave rotating at synchronous speed; the remaining terms including all odd harmonics that are not multiples of 3. Note that the direction of rotation of these harmonics is negative and then positive in alternate sequence. The third harmonic and its odd multiples, often called *triplen harmonics*, do not appear in the resultant mmf—and therefore do not appear in the line-line voltage of a three-phase machine. However, the triplen harmonics do appear in the phase mmf expression, Eq. (3.26).

**TABLE 3.5**

| Polyphase system | Characteristic angle (degree) |
|---|---|
| 2/4 | 90 |
| 3 | 120 |
| 5 | 72 |
| 6 | 60 |
| 12 | 30 |

## Single-Phase Alternator Windings

Single-phase alternators are used frequently for two major reasons: (1) when the alternator must be paralleled with an existing single-phase power system and (2) in an application of relatively low power demand that is expected to grow to a much higher demand.

Three types of winding schemes are used in single-phase alternators:

**1.** *Open delta:* This is a conventional polyphase double-layer lap winding with the three windings connected in "open delta"; i.e., the final two connections to "close" the delta winding are not performed, and these two terminals are brought out to form the two armature-winding terminals. This results in a line voltage (between the two armature terminals) equal to twice the voltage generated in one phase winding. The current in all three windings is identical in phase and magnitude, both of which result from the terminal voltage and the characteristics of the connected load. Thus, in two of the phase windings the generated voltage and current are inherently out of phase, resulting in a reduced power rating of these windings and of the total machine. The power rating of the single-phase open-delta armature winding is two-thirds that of the same machine operated as a three-phase alternator.

**2.** *Single-phase lap:* This is a conventional double-layer lap winding, but with *all* slots filled with the series-connected coils of one phase. Such a winding results in relatively low distribution and pitch factors, as compared to a three-phase winding placed in the same number of slots. However, the single-phase lap winding results in a generated voltage identical to that of the open-delta winding when wound in identical stator laminations.

**3.** *Concentric windings:* This single-layer winding is identical to that described in Sec. 3.4 for use in synchronous machine cylindrical-rotor field windings. For a given set of laminations, this winding likewise results in a generated voltage approximately equal to the voltages of the open-delta and single-phase lap windings.

It can be shown that all three single-phase windings result in approximately the same terminal voltage when wound in a given set of laminations for a given number of poles and power output. It can also be shown that the concentric winding results in the minimum weight of conductor required. And, finally, both the concentric and single-phase lap windings result in lower assembly costs than does the open-delta, given the smaller number of internal connections required.

### Airgap Flux

The fundamental mmf wave is given by the first term of Eq. (3.27). The magnitude portion of this term divided by the number of poles gives the mmf per pole, or

$$m_1 \text{ per pole} = \frac{6M_m S_{pm} K_p K_d}{p} \qquad \text{A per pole} \tag{3.28}$$

The magnetic field intensity across the gap is

$$H_1 = \frac{6M_m S_{pm} K_p K_d}{pg} \qquad \text{A per pole per m} \tag{3.29}$$

where $g$ = physical gap, m

The airgap flux density follows as

$$B_{gm} = \mu_0 H_1 = \frac{6\mu_0 M_m S_{pm} K_p K_d}{pg} \qquad \text{T} \tag{3.30}$$

where $\mu_0 = 4\pi \times 10^{-7}$, H/m

Equation (3.30) represents the peak value of the sinusoidally varying airgap flux.

## 3.8 HARMONICS

Synchronous machines are subject to the effects of harmonics in mmf, voltage, and current as a result both of spatial and of electrical harmonic sources.

### Spatial Harmonics

Spatial harmonics result basically from two phenomena: (1) Electrical windings in both the rotor and stator of synchronous machines are not "infinitely" or perfectly distributed about the airgap circumference but are grouped in discrete bundles known as phase belts, and (2) the magnetic surfaces of both the rotor and stator in synchronous machines are nonuniform and consist of alternate magnetic and nonmagnetic regions as a result of either the slot and/or tooth construction or the salient-pole construction. In general, both of these sources of spatial harmonics are inherent properties of practical machine construction. All tend to change the nature of the machine mmf, flux, voltage, and current parameters away from sinusoidal functions.

No useful power is generated from the cross coupling of components of different frequencies; therefore, harmonics—from any source—tend to increase heating effects and reduce efficiency, but produce no useful power output. Also, in many applications the harmonic currents can be a source of electromagnetic interference (emi), which may be intolerable for communications, computer, or medical equipment. In many motor applications, harmonics may present no major problems and only result in a small decrease in motor efficiency; thus, in some applications the relative effect of harmonics in synchronous machines is a trade-off between the machine cost and the purity of the machine's waveform.

Spatial harmonics can be decreased by several methods:

1. Increasing the number of slots in the armature lamination.
2. Increasing $S_{pm}$, that is, the number of slots per pole per phase, which generally follows from step 1.
3. Using coil pitch selectively to eliminate or reduce certain harmonics. Concentric windings are especially useful from this standpoint.
4. Closing the slot openings. This is generally an expensive solution and results in higher assembly costs.
5. Increasing the airgap length $g$. This generally lessens the permeance variations due to slots and/or teeth but also reduces the flux density in the airgap, which reduces the output of a given-size machine.
6. "Canning" the rotor (or stator). This method is seldom used to reduce harmonics but is often required for environmental considerations. It consists of placing a metal (usually aluminum) cylinder, or "can," around the rotor, or in some cases about the inside circumference of the stator, to reduce windage losses and prevent damage from chemicals, such as salt. Such a can has a generally smoothing effect upon spatial harmonics.

## Electrical Harmonics

Harmonics may appear in synchronous machine windings from electrical causes, usually external to the machine circuit. Some of the obvious sources of electrical harmonics are the nonlinear excitation currents of transformers, induction motors, and other inductive load devices. In the case of synchronous motors, the armature excitation source is often of an inherently nonlinear waveform, generally from power semiconductors, and thus harmonic voltages are applied to the motor from the source. Applied harmonic sources, such as typical semiconductor power supplies, can be resolved into Fourier series or harmonic components by methods similar to those used in the analysis of spatial harmonics. However, the application of nonsinusoidal sources to any type of electric machine results in a further proliferation of harmonics (given all the causes previously described in this chapter), and there comes a point beyond which any analytical analysis of the effects of machine parameters resulting from nonsinusoidal excitation becomes relatively meaningless. Experimental techniques are then of much more value.

## Harmonic Phase Relationships

The phase angle among harmonics in different phase windings varies according to the relationship

$$\theta_k = k \qquad \text{characteristic angle} \tag{3.31}$$

Thus, in a three-phase system the fifth harmonic ($k = 5$) of phase B is out of phase with the fifth harmonic in phase A by $5 \times 120$, or 600 electrical degrees, and that in phase C by $5 \times (-120)$, or $-600$. Note that this shift also results in opposite phase sequence, which is in agreement with Eq. (3.27). This phase change results in a unique situation for the triplen harmonic—the third and its odd multiples—where the phase angle in phases B and C are 360 and $-360$, respectively. the triplen harmonics are thus in phase with each other in the three phases. Therefore:

1. In wye-connected windings, triplen harmonics of the induced voltage cause *unsymmetric* phase voltages.
2. Triplen harmonics of the induced phase voltages, however, do not appear in the line voltages, since the latter are the differences of phase voltages.
3. In four-wire wye systems, with a wire or ground connecting the neutrals of the source and load, triplen harmonics in the phase windings will cause neutral or ground current to flow, often called "zero sequence" current.
4. In delta-connected windings, triplen harmonics in the phase voltages cause a single-phase circulating current to flow "around the delta." This may result in excessive heating of the windings and should be minimized.

## 3.9 SYNCHRONOUS MACHINE REACTANCES AND EQUIVALENT CIRCUITS

The previous sections of this chapter have defined the mmf's and flux densities resulting from the principal paths of the magnetic circuit of a synchronous machine. There are several other paths in the complete magnetic circuit of a synchronous machine that give rise to leakage reactance components of a synchronous machine. Leakage reactance components result from fluxes that exist in the various air paths. In general, leakage reactances result from magnetic flux components that do not enter into the airgap flux described in Eq. (3.30), i.e., the airgap flux density. Leakage flux generally exists in all nonmagnetic regions adjacent to armature conductors, as illustrated in Fig. 3.13. The three major components of armature-leakage flux are:

1. *End-turn leakage:* Flux linking the end turns of armature coils, beyond the magnetic core of the stator.
2. *Slot leakage:* Flux crossing the slot area from tooth to tooth, but not crossing the airgap.
3. *Zigzag leakage:* Flux "zigzagging" across the airgap from stator tooth to rotor tooth, but not entering into the principal airgap flux, described by Eq. (3.30).

The sum of these components of leakage flux is known as the armature-leakage flux of the machine and is generally expressed as a leakage reactance:

$$X_1 = 2\pi \times \text{flux linkage per ampere} \quad \Omega \tag{3.32}$$

### Synchronous Reactance

In an electric machine the airgap is the focus of the mutual magnetic circuit, which includes the rotor, stator, and yoke for a synchronous machine. The characteristics of this mutual path are likewise defined by a number of reactance terms, known collectively as synchronous reactance. Although the term *synchronous reactance* is often ascribed to just one specific quantity commonly used in the steady-state analysis of cylindrical-rotor machines, it is best to think of this quantity as "synchronous

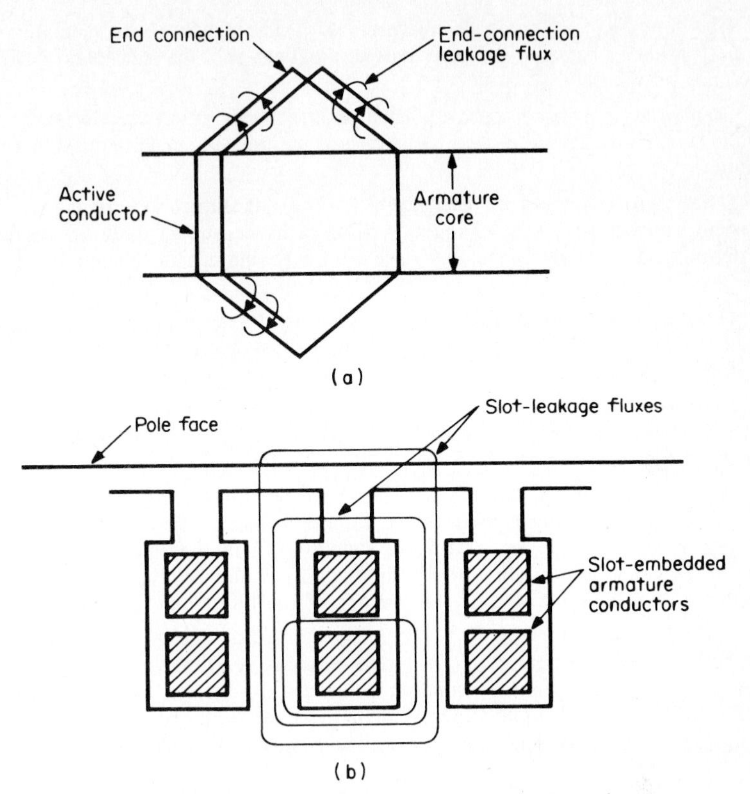


**FIG. 3.13** *(a)* End-connection leakage flux path; *(b)* slot-leakage flux paths.

reactance under one set of specific operating and structural conditions," as will be defined below. Other synchronous reactances describe the reluctance variations of salient-pole machines and various transient operating conditions. These several synchronous reactances result from variations of reluctance among flux paths in the various sections of the magnetic circuit and from changes in flux paths during transient conditions.

The principal reactance terms used in synchronous machine analysis are, expressed on a per-phase basis:

**1.** *Synchronous reactance* $X_s = X_{ad} + X_1$: This is the steady-state reactance used in the analysis of machines with cylindrical rotors. It is equal to the sum of a mutual reactance term based upon the fundamental flux produced by a balanced polyphase winding and the leakage reactance, defined in Eq. (3.32).

**2.** *Synchronous impedance* $Z_s = R_s + jX_s$: The phasor sum of synchronous reactance per phase and the per-phase armature-winding resistance.

**3.** *Direct-axis synchronous reactance* $X_d$: The synchronous reactance along the magnetic path through the pole axis in a salient-pole machine plus the leakage reactance per phase.

**4.** *Quadrature-axis synchronous reactance* $X_q$: The synchronous reactance along the path midway between the pole axes in a salient-pole machine plus the leakage reactance per phase.

**5.** *Direct-axis transient reactance* $x'_d$: The direct-axis reactance during transient conditions due to changes in flux paths caused by induced currents in the main field winding.

**6.** *Quadrature-axis transient reactance* $x'_q$: The quadrature-axis reactance during transient conditions due to changes in flux paths caused by induced currents in the main field winding. Since there are no field windings in the quadrature axis, $x'_q = X_q$.

**7.** *Direct-axis subtransient reactance* $x''_d$: The direct-axis reactance during transient conditions due to changes in flux paths caused by induced currents in the damper windings.

**8.** *Quadrature-axis subtransient reactance* $x''_q$: The quadrature-axis reactance during transient conditions due to changes in flux paths caused by induced currents in the damper windings.

**9.** *Negative-sequence reactance* $X_2$: The above reactances are "positive sequence" reactances, i.e., reactances based on the condition that the field and the armature's rotating magnetic field are rotating in synchronism in the same direction. When the two fields are rotating oppositely, the negative-sequence reactance is applicable, and it is generally calculated as the average of the direct-axis and quadrature-axis subtransient reactances.

**10.** *Zero-sequence reactance* $X_0$: The zero-sequence reactance is evaluated with the three phases of the armature winding connected in parallel, with a single-phase rated-frequency voltage applied across them. The zero-sequence reactance is used in evaluating synchronous machines during line-ground fault conditions.

## No-Load and Short-Circuit Values

With the armature terminals of a synchronous machine open-circuited and with the rotor driven at synchronous speed while the field winding is excited, a voltage will be generated in the armature windings; this is often termed the *no-load*, or *open-circuit*, voltage of the armature winding. Flux distribution over a pole pitch due to the field winding's excitation can generally be considered sinusoidal. With this assumption, the induced voltages in the three phases become

$$e_a = -E_m \cos \omega t \tag{3.33}$$

$$e_b = -E_m \cos (\omega t - 120°) \tag{3.34}$$

$$e_c = -E_m \cos (\omega t + 120°) \tag{3.35}$$

where $E_m = -p\omega S_{pm} T \phi$

Since the flux per pole, $\phi$, in the above equations is a function of the field current, the no-load characteristics of synchronous machines are plotted as an open-circuit saturation curve. A typical open-circuit saturation curve is shown as Fig. 3.14. The extension of the straight-line portion of the saturation curve is known as the airgap line.

A second characteristic of importance in synchronous machine analysis is the short-circuit saturation characteristic. If the armature terminals are shorted together

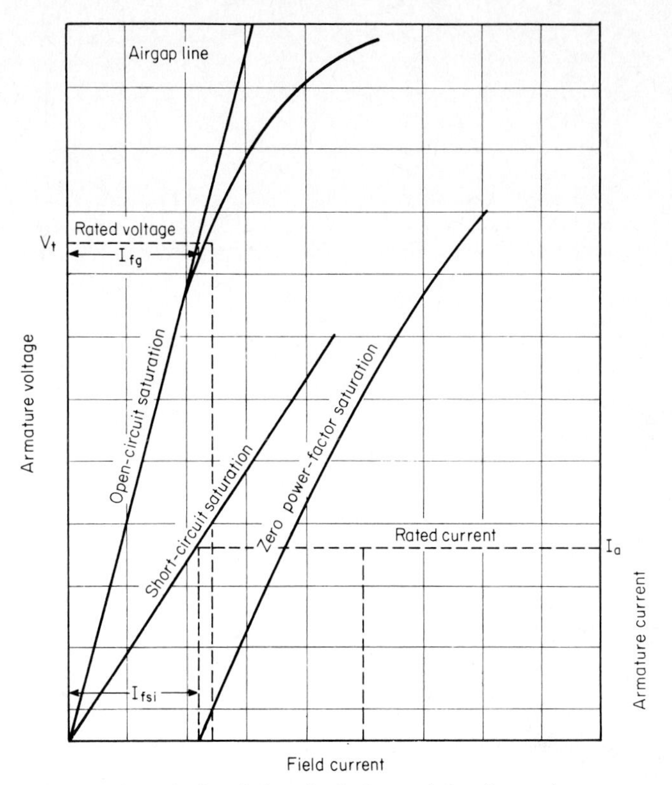


**FIG. 3.14** Open-circuit and short-circuit characteristics of a synchronous machine.

while the rotor is being driven at synchronous speed and the field current is varied from zero to some small value, current will flow in the shorted armature windings. The plot of this armature current versus the field current gives the short-circuit saturation curve, a typical example of which is shown in Fig. 3.14.

A third characteristic of value in synchronous motor analysis is the zero power-factor saturation characteristic. This is obtained by overexciting the machine being tested while it is connected to a load consisting of idle-running synchronous motors or to any other highly inductive load. By proper adjustment of the excitation of the machine being tested and of that of the load, the terminal voltage may be varied while the armature current of the machine being tested is held constant at the rated value. The curve desired consists of a plot of the terminal voltage versus the field current, as shown in Fig. 3.14.

## Determination of Synchronous Reactances from Test Data

Synchronous reactances can be determined from the experimentally obtained curves shown in Fig. 3.14. Following the standard IEEE procedure, reactances are obtained in per-unit.[5] From Fig. 3.14

$$x_d \text{ or } x_s = \frac{I_{fsi}}{I_{fg}} \tag{3.36}$$

where $I_{fsi}$ = field current corresponding to base current on short-circuit saturation curve

$I_{fg}$ = field current corresponding to base voltage on airgap line

To obtain the quadrature-axis synchronous reactance, the slip test is used.[5] It consists of driving the machine at slightly below synchronous speed with the field open-circuited and the armature windings excited by a three-phase, balanced, rated-frequency, positive-sequence source. The armature current, armature voltage, and voltage across the open-circuited field are measured, preferably by an oscillograph or storage oscilloscope. A pattern such as shown in Fig. 3.15 will be measured. The minimum and maximum armature voltages and armature currents are noted and converted to per-unit. The quadrature-axis synchronous reactance is the ratio of minimum voltage to maximum current (maximum rotor reluctance):

$$X_{qs} = \frac{E_{\min}}{I_{\max}} \tag{3.37}$$

Also, the direct-axis synchronous reactance can be obtained from the ratio of maximum voltage to minimum current:

$$X_{ds} = \frac{E_{\max}}{I_{\min}} \tag{3.38}$$

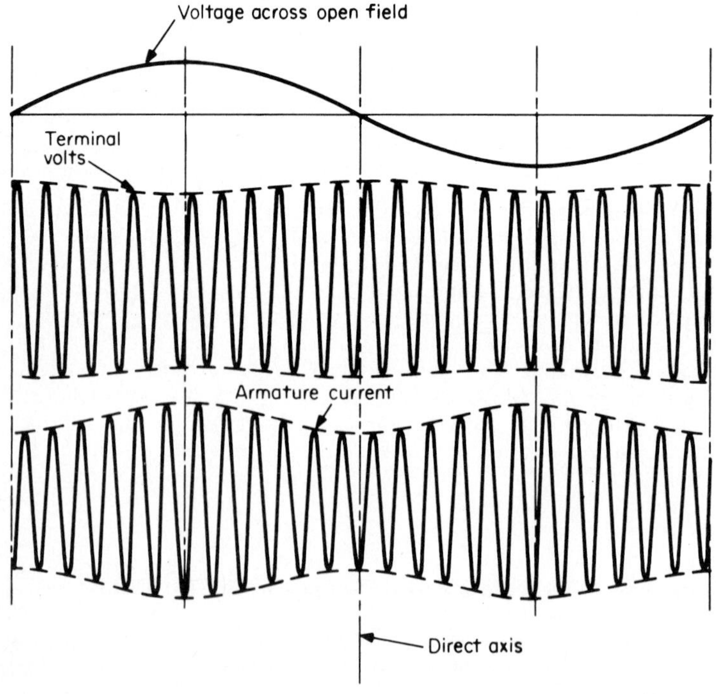


**FIG. 3.15** Slip method of obtaining quadrature-axis synchronous reactance.

The use of the subscript $s$ in these two equations is to indicate that these values are calculated by means of the slip test. In general, the slip test should not be used to obtain the direct-axis reactance; instead, the method described by Eq. (3.36) is preferred. A slight improvement of the accuracy of the quadrature reactance can then be made as follows:

$$X_q = X_d \frac{X_{qs}}{X_{ds}} \tag{3.39}$$

where $X_d$ is obtained from Eq. (3.36).

*The Potier reactance* $X_p$ is used to calculate excitation characteristics and voltage regulation as a function of alternator loading in large alternators where experimental loading is not feasible.[6] It is approximately equal to the leakage reactance and is a good means for obtaining a measure of leakage reactance from experimental no-load test data. The graphical procedure for obtaining the Potier reactance is shown in Fig. 3.16.[5] The intersection of the zero power-factor saturation curve with the rated terminal voltage ordinate locates point *d*. From *d*, the distance *ad* is laid off, with $ad = I_{fsi}$, the field current required for rated short-circuit current. From point *a*, a line parallel to the airgap line is drawn, which locates point *b* at the intersection with the no-load saturation curve. A line perpendicular to *ad* from *b* is drawn, locating point *c*. The distance *bc* is the Potier voltage to the scale on the ordinate, usually in per-unit. The Potier voltage divided by the rated current, in per-unit or rated amperes, equals the Potier reactance $x_p$ or $X_p$, depending upon the units used, which is also approximately equal to the machine leakage reactance $x_1$ or $X_1$.

The Potier reactance may also be obtained from load test data where experimental loading is practical.[5]

## Sudden Short-Circuit Test

Sudden short-circuit tests are performed both to evaluate the mechanical design of the synchronous machine and to determine the transient and subtransient reactances and time constants. Reference 5 should be consulted for the details pertaining to these tests.

## Torque Tests

A number of torque parameters are of interest in the design and application of synchronous machines, including the following:[6]

1. *Locked-rotor torque:* Torque developed at zero speed with the rotor locked and prevented from turning, with rated voltage and frequency applied.
2. *Pull-out torque:* The maximum possible sustained torque developed at synchronous speed for 1 min with rated voltage and frequency.
3. *Pull-in torque:* In a synchronous motor, the maximum constant torque under which the motor will pull into synchronism at rated voltage and frequency when its excitation is applied.
4. *Pull-up torque:* The minimum torque developed at rated voltage and frequency during start-up at subsynchronous speeds.
5. *Speed-torque characteristic:* The variation of developed torque with rated voltage

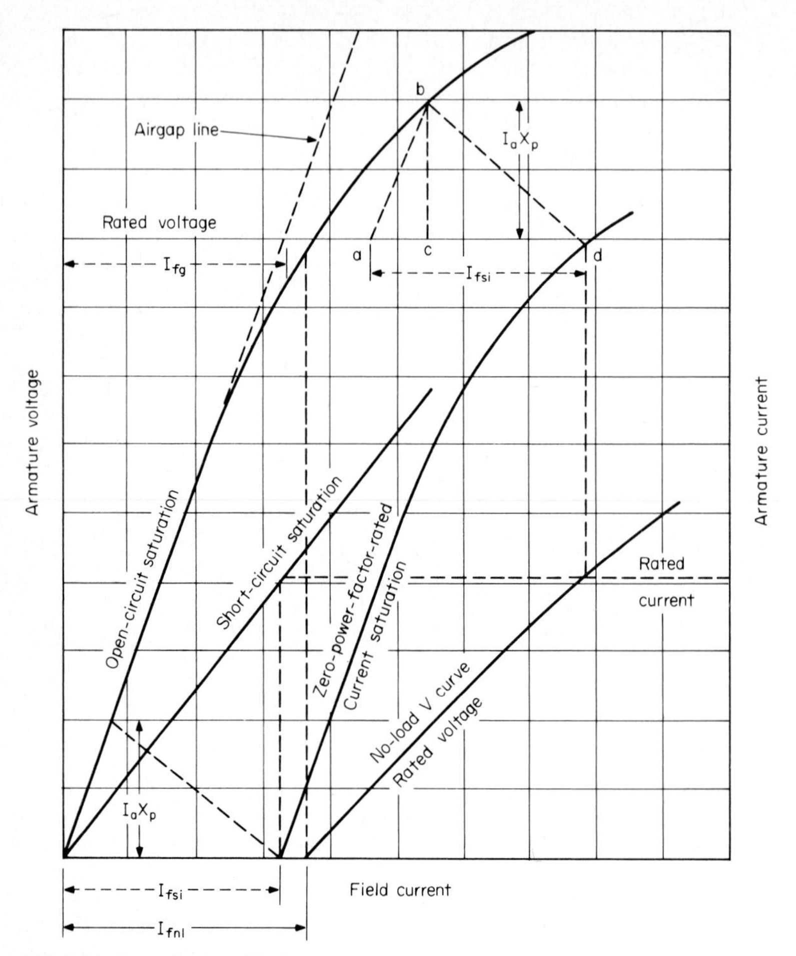


**FIG. 3.16** Determination of Potier reactance voltage.

and frequency applied to a synchronous motor as a function of speed, with zero excitation applied.

***Speed-Torque Characteristic.*** This characteristic is used to describe the operation of a synchronous motor when it is operating as an induction motor with no field excitation. In general, tests to obtain this characteristic in a synchronous machine are identical to those for induction motor speed-torque characteristics. Several of the torque parameters are readily obtained from the speed-torque characteristic of a synchronous machine. These are shown in the typical characteristic illustrated in Fig. 3.17. It should be noted, however, that there are some differences in winding impedances and in winding connections in a synchronous motor, as compared with an induction motor of similar horsepower: (1) The damper winding generally has a much higher resistance than a typical squirrel-cage winding of an induction motor of similar horsepower and frequency rating; thus, on the basis of the damper

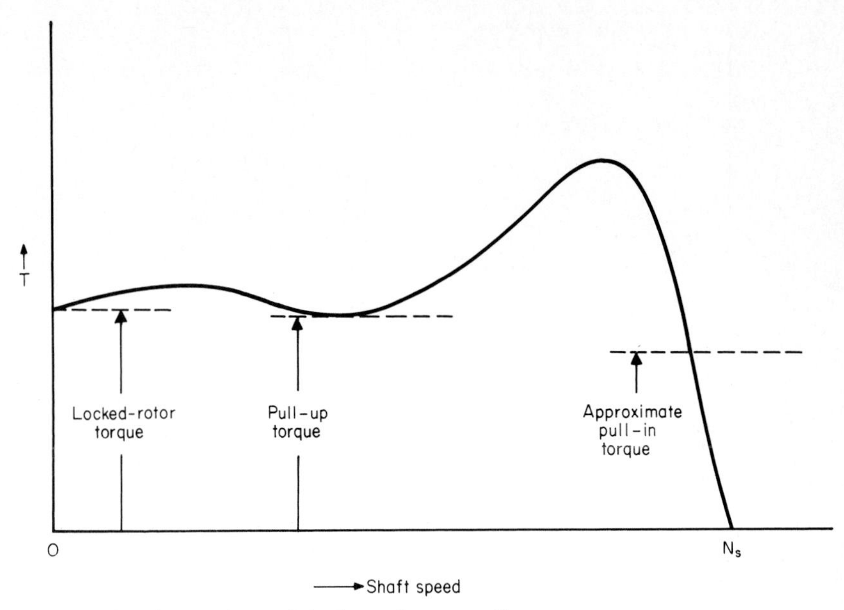


**FIG. 3.17** Speed-torque characteristic of a synchronous machine.

winding alone, the speed-torque characteristics of a typical synchronous motor appear more as those of a class C or D induction motor. (2) In addition to the damper winding, the speed-torque characteristic of a synchronous motor is significantly influenced by the field winding; most synchronous machines are designed for "closed field" starting, i.e., with the field winding short-circuited during start-up; the field winding, as "seen" from the armature winding, appears as a low-resistance winding and therefore—in itself—gives characteristics more like the class A or B induction motors. Thus, with the field short-circuited during start-up, the speed-torque characteristic of a synchronous machine is a composite of the characteristics resulting from the damper and field windings. Whether the field is shorted or open during starting is determined by the particular application of the synchronous machine. Figure 3.18 shows the differences in the speed-torque characteristic of a relatively small synchronous machine between the conditions of the field open and the field shorted.

***Pull-In Torque.*** This is a characteristic unique to synchronous machines. In general, it is a difficult parameter to measure or calculate. It is of most interest in the application of machines that are started and brought up to speed by means of the torque of the damper and field windings. *Pull-in torque* is the torque available to pull the rotor into synchronism with the armature rotating magnetic field.

There is no recognized method for determining pull-in torque.[7] Rather, the nominal pull-in torque is used to express this parameter. The *nominal pull-in torque* is arbitrarily defined as the torque developed by an induction motor at 95 percent of synchronous speed, as illustrated in Fig. 3.17.

***Pull-Out Torque.*** This is another characteristic unique to synchronous machines; it describes the maximum torque capabilities of a synchronous machine. When a synchronous machine is subjected to a torque greater than pull-out, whether as a

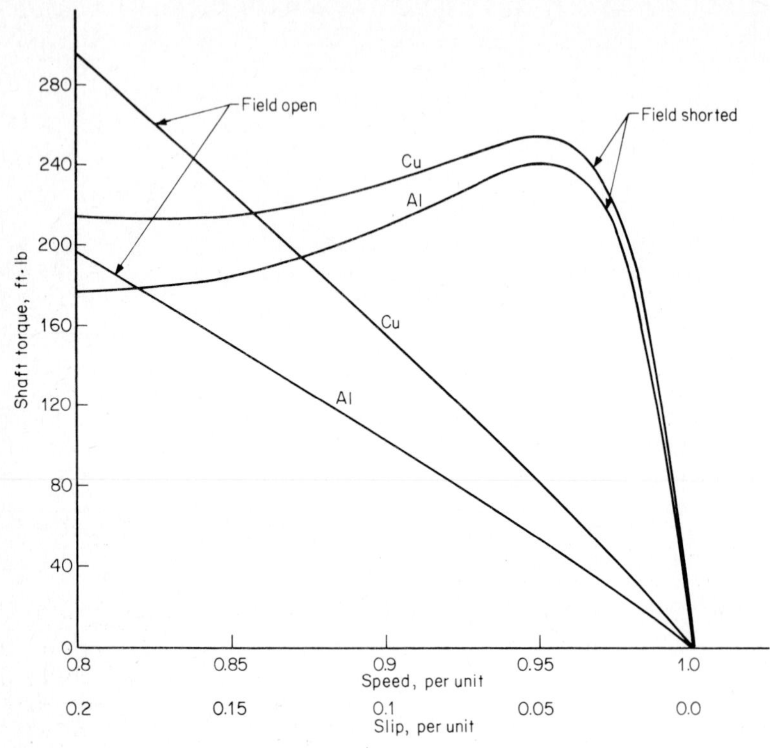


**FIG. 3.18** Synchronous motor speed-torque curves illustrating effect of field winding on motor torque. Effect of rotor-bar material also shown.

motor or a generator, the machine pulls out of synchronous speed, and a variety of conditions may result.

Pull-out torque may be obtained experimentally.[5] This is not feasible for machines of power ratings beyond the capacity of the power supplies generally available in the laboratory. Reference 5 gives a method for calculating an approximate value of pull-out torque from machine constants:

$$T_{\mathrm{po}} = \frac{K I_{f1} V}{I_{fsi} \eta \cos\theta} \tag{3.40}$$

where $T_{\mathrm{po}}$ = pull-out torque, per unit
$V$ = specified terminal voltage, per unit
$I_{f1}$ = specified per-unit field current
$I_{fi}$ = field current, per unit, corresponding to base armature current on the short-circuit saturation curve
$\cos\theta$ = rated power factor
$\eta$ = efficiency at rating, per unit

The factor $K$ in Eq. (3.40) is to allow for *reluctance torque* and for positive sequence $I^2R$ losses. This factor may be obtained from the machine manufacturer. It may also be estimated from the loading condition of the machine as follows:

$$K = \sin\delta + I_{fsi}V\frac{x_d - x_q}{2I_{f1}x_dx_q} \tag{3.41}$$

where $x_d$ = direct-axis synchronous reactance, per unit
$x_q$ = quadrature-axis synchronous reactance, per unit
$\delta$ = power angle, rad

## *3.10 STEADY-STATE CHARACTERISTICS—CYLINDRICAL-ROTOR MACHINES' EQUIVALENT CIRCUITS*

Since polyphase machines are, with few exceptions, perfectly balanced among the phases, equivalent circuits are drawn on a per-phase basis and all parameters are *phase* parameters unless otherwise noted. The equivalent circuit can be thought of as representing one phase of a balanced wye system, and the remaining phases can be represented by identical circuits and parameters except for the reference phase angle. When polyphase quantities are required from equivalent-circuit analysis, the phase parameters can be converted to polyphase parameters by standard methods for balanced polyphase circuits.

In steady-state, the cylindrical-rotor machine can be represented by the equivalent circuits shown in Figs. 3.19*a* and *b*. The equivalent circuit may be used with per-unit quantities or actual quantities in the units of volts, amperes, and ohms. The equivalent circuit is valid for either motor or generator (alternator) operation, the only difference in the equivalent circuit itself being the direction of the current. It is conventional to show this current arrow out of the terminals for generator operation (as shown in Fig. 3.19*a*) and into the terminals for motor operation.

The voltage source $E_f$ in Fig. 3.19 is the induced or generated voltage due to the field flux $\phi$ (often called the back emf in a motor). The impedance is defined in Sec. 3.9. The voltage $E_r$ is often called the voltage behind leakage, or Potier reactance. For many types of analysis, it is not necessary to introduce this voltage specifically, and the simplified circuit of Fig. 3.19*b* is then valid. As has been noted in Sec. 3.9, the reactance of the cylindrical-rotor machine is simply termed the *synchronous reactance*, and hence the symbol $X_s$ in Fig. 3.19*b*.

### Alternator Excitation

By varying the excitation, the terminal voltage of a synchronous generator on

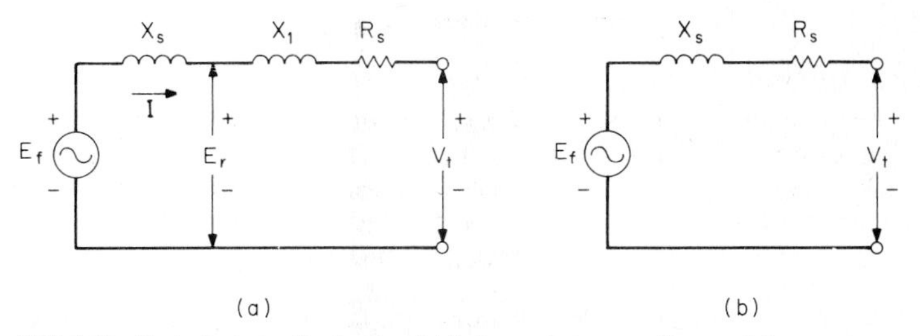


**FIG. 3.19** Equivalent circuits showing ***(a)*** leakage reactance and magnetizing reactance and ***(b)*** synchronous reactance.

no-load may be varied, and the no-load characteristic (Fig. 3.14) is thus obtained.

When the alternator is supplying load current at a power factor = cos $\theta$, there are two components of alternator load:

$$P = 3V_tI\cos\theta \tag{3.42}$$

$$\text{VAr} = 3V_tI\sin\theta \tag{3.43}$$

In these equations, $P$ is the real three-phase power and VAr is the reactive three-phase power (or, simply, the reactive voltamperes). The phasor sum of the two components is *apparent power* VA:

$$\text{VA} = \sqrt{P^2 + (\text{VAr})^2} \tag{3.44}$$

The units of voltage, current, and power are either per-unit or actual rms values. When an alternator is operated in an isolated condition (i.e., supplying only the load connected to its terminals), adequate field excitation must be supplied to maintain the terminal voltage $V_t$ for a given load current $I$ and power-factor angle $\theta$. As $I$ and $\theta$ are varied and $V_t$ is held invariant, the following changes in the phasor diagram can be noted:

1. For a constant $P$ (Eq. 3.42), it can be seen that the power angle $\delta$ remains approximately constant and that the magnitude of $E_f$ varies.
2. For a constant VAr (Eq. 3.43), it can be seen that $E_f$ remains approximately constant and that $\delta$ varies (Fig. 3.20).
3. If $I$ is constant and $\theta$ is varied, it is seen that the impedance triangle remains of constant magnitude but rotates about the end of the $V_t$ phasor, with the $IR$ phasor always parallel to the $I$ phasor (Fig. 3.20). This analysis of the alternator phasor diagram best illustrates the relationship between excitation and load power-factor angle $\theta$ or load VAr's (assuming $V_t$ constant):

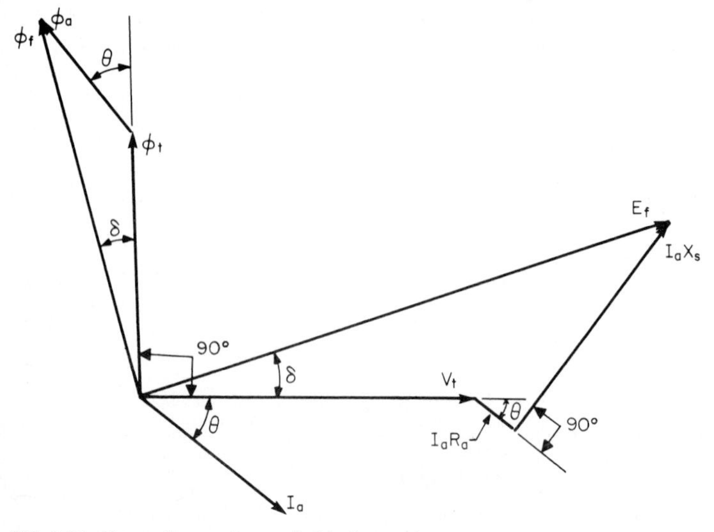


**FIG. 3.20** Phasor diagram for a cylindrical-rotor alternator.

**a.** *Underexcited:* When $\theta$ is approximately 90° leading (i.e., supplying a purely capacitive load), the excitation phasor $E_f$ is at a minimum and the alternator is said to be underexcited. As is seen from Fig. 3.21, the load current is in near quadrature with the terminal voltage, with a polarity that *aids* the field.

**b.** *Overexcited:* When $\theta$ is approximately 90° lagging (i.e., supplying a purely inductive load), the excitation phasor $E_f$ is at a maximum and the alternator is said to be overexcited. As can be seen from Fig. 3.22, the load current is in near quadrature with the terminal voltage, with a polarity that *opposes* the field.

Normal operation of an isolated alternator is, of course, in between these two extremes. One value of power factor, usually 0.8 lagging, is used as the power-factor rating of an alternator.

Another study using the phasor diagram (Fig. 3.20) is to observe the relationship between the terminal voltage $V_t$ and the excitation phasor $E_f$. In Fig. 3.20, if $E_f$ and $\theta$ are fixed and $I$ increases, $V_t$ will decrease somewhat, and vice versa. Now assume that $E_f$ and $I$ are fixed and that $\theta$ is permitted to vary. As $\theta$ decreases from the lagging value shown in Fig. 3.20 to zero and then to a leading value, $V_t$ will increase because of the aiding excitation supplied by the leading current. Likewise, as $\theta$ is made more lagging and approaches 90° lagging, $V_t$ approaches zero. This is the condition of the short-circuit test. Under short-circuit conditions, real power is very small (only alternator losses) and the power factor approaches zero lagging. Thus, in an isolated alternator the field current is the major means of controlling the terminal voltage. The mechanism for supplying this control is one of the earliest examples of feedback control, known as a voltage regulator.

The ability to regulate system voltage is a major reason for the use of synchronous alternators in almost all central station energy sources throughout the world. Voltage regulation is a matter of controlling the reactive-power flow into and out of an infinite bus. This characteristic of synchronous machines is often illustrated by means of a set of curves known as V-curves, which can be obtained experimentally on smaller machines as well as by means of the phasor diagram analysis described above; *V-curves* are plots of armature current versus either field current (if obtained experimentally) or excitation phasor $E_f$ at a condition of constant power and constant terminal voltage. Figure 3.23 illustrates a set of V-curves for several values of constant-power output. The minimum value of each V-curve is approximately at the condition of unity power-factor load—approximately, due to secondary effects of the machine losses. The V-curve characteristic is approximately the same whether the machine is operated as a motor or an alternator, and, again, the term *approximate* is used to account for the secondary effects of the machine losses.

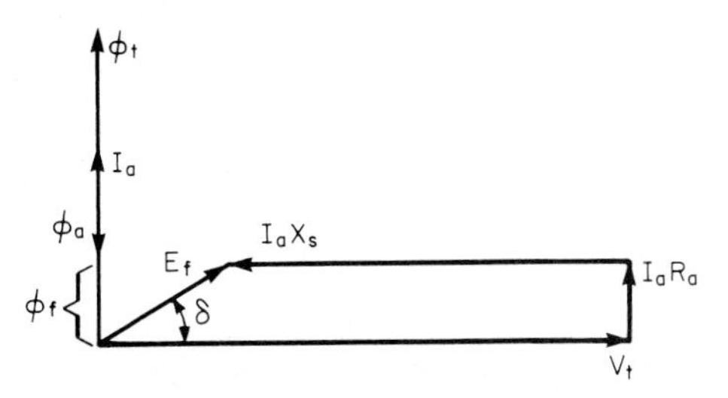


**FIG. 3.21** Phasor diagram for an alternator supplying a 90° leading (capacitive) load.

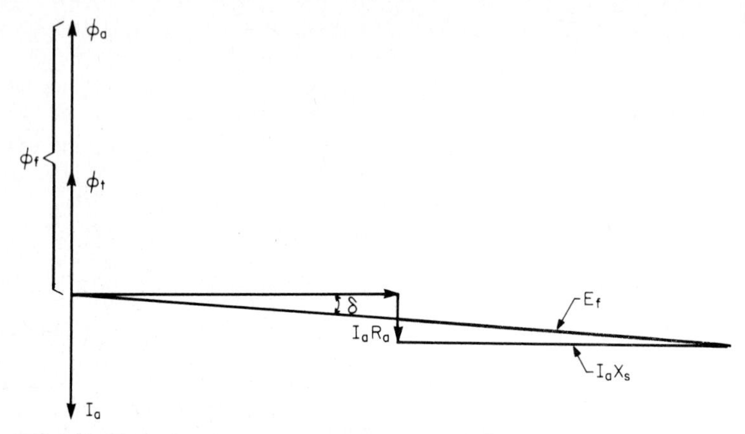


**FIG. 3.22** Phasor diagram for an alternator supplying a 90° lagging (inductive) load.

## Motor Excitation

Excitation characteristics of synchronous machines are identical whether the machine is operated as a motor or an alternator. Assuming a constant-voltage source, the motor excitation characteristics can be described in terms of the V-curves shown in Fig. 3.23. However, the "power factor" associated with the two regions of the V-curves (to the left and to the right of the minimum of the curves) is the inverse of that associated with that of the alternator. The power factor associated with a motor is that "looking into the motor" from the terminals. The motor appears as a lagging power-factor load (or inductive) when operated in the underexcited region, i.e., to the left of the minima of the V-curves; likewise, the motor appears as a leading power-factor load (or capacitive) when operated in the overexcited region, i.e., to the right of the minima of the V-curves. Thus, a synchronous motor can serve the dual role of an energy converter and "reactive-power converter." In fact, it should be apparent from this discussion (and from the previous descriptions of alternator excitation characteristics) that the basic synchronous machine can operate in all four quadrants of the real-power–reactive-power, or watt-VAr, plane with equal ease. This plane is illustrated in Fig. 3.24. However, except in a few rare applications, practical synchronous machines are seldom operated over the four quadrants shown in Fig. 3.24. Rather, motors operate in the right half of the watt-VAr plane, and alternators operate in the left half.

## Voltage Regulation

Voltage regulation of an alternator is a somewhat arbitrary definition of the excitation required to maintain constant terminal voltage as the load current and power factor are varied. (The adjective *arbitrary* is used because the definitions have changed through the years and the definition used in practice still varies among engineers and designers today.) Voltage regulation can be measured, of course, by actually loading the alternator and noting the change in field current as a function of load. However, this is not practical in large machines, and also it is usually desirable to determine excitation requirements during the design stages of the machine.

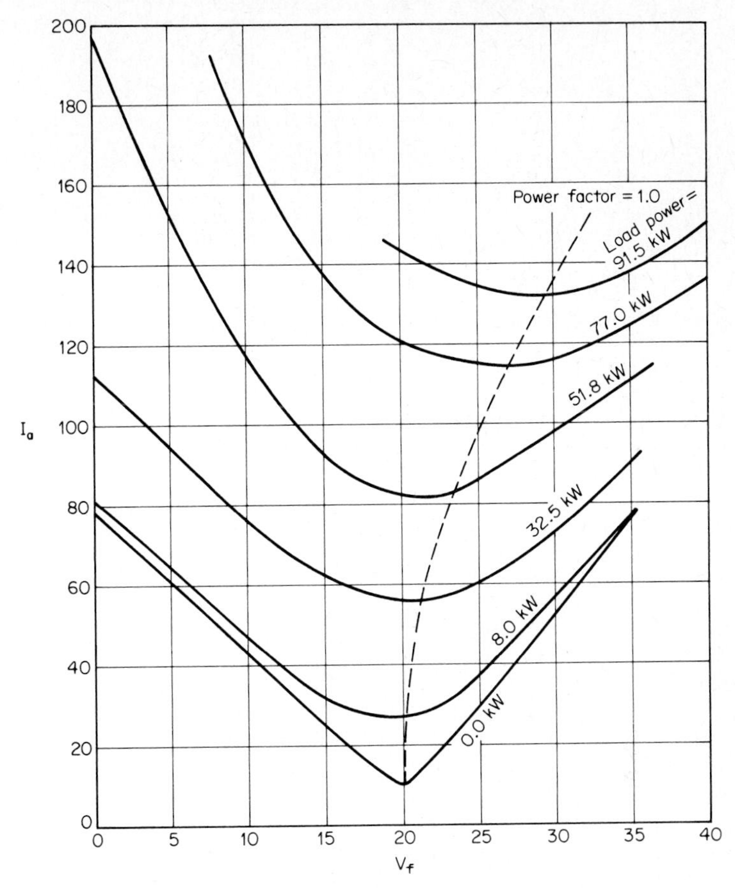


**FIG. 3.23** Synchronous motor V-curves.

As a first approximation, voltage regulation can be determined from the alternator phasor diagrams, as shown in Figs. 3.20, 3.21, and 3.22. From these phasor diagrams (and from the diagrams for salient-pole alternators to be presented in the next section), voltage regulation may be determined as

$$\text{Reg} \simeq \frac{E_f - V}{V} \qquad \text{in per-unit} \tag{3.45}$$

The value calculated by means of Eq. (3.45) depends upon whether the value $X_s$ used in the phasor diagrams is the saturated or unsaturated (airgap) value, and this choice is somewhat arbitrary and a matter of convention based upon past practices.

In order to make comparisons among alternator manufacturers and users, it is desirable to have a standard method of specifying voltage regulation. This has been provided by the Institute of Electrical and Electronics Engineers (IEEE). It should also be stated that the IEEE method is probably the most accurate method of calculating voltage regulations.[5]

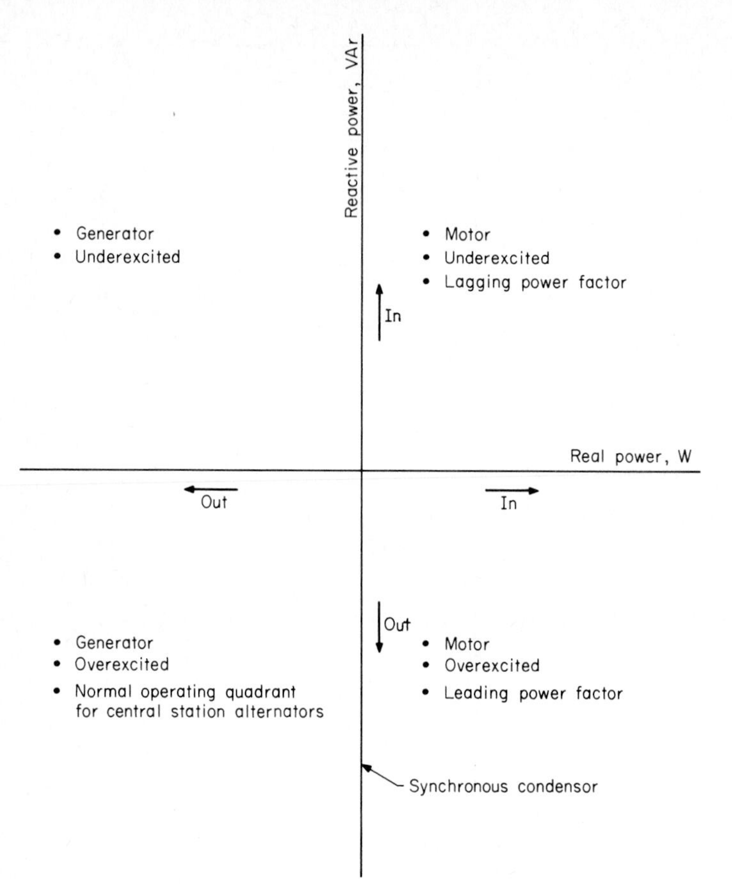


**FIG. 3.24** Four-quadrant synchronous machine operation.

## Power Characteristics

Synchronous machines can operate in the motoring or generating mode (Figs. 3.24 and 3.25) with equal ease—provided, of course, that the electrical and mechanical systems connected to the synchronous machine have similar bilateral capabilities. The electromagnetic *developed power* of a cylindrical-rotor machine can be shown to be[7–9]

$$P_d = \frac{E_f V_t}{X_s} \sin \delta \qquad \text{per unit} \tag{3.46}$$

based upon the equivalent circuit of Fig. 3.19*b*. The units of $P_d$ are in per-unit; if actual units are used on the right side of Eq. (3.46), the right side of this equation must be multiplied by the factor 3.0 for three-phase power in watts. The developed power is often termed *airgap power* since it is developed by the interaction of the excitation and armature flux linkages in the airgap. The *maximum developed power* of a cylindrical-rotor machine occurs when the power angle $\delta$ is 90°, and it equals (Fig. 3.25)

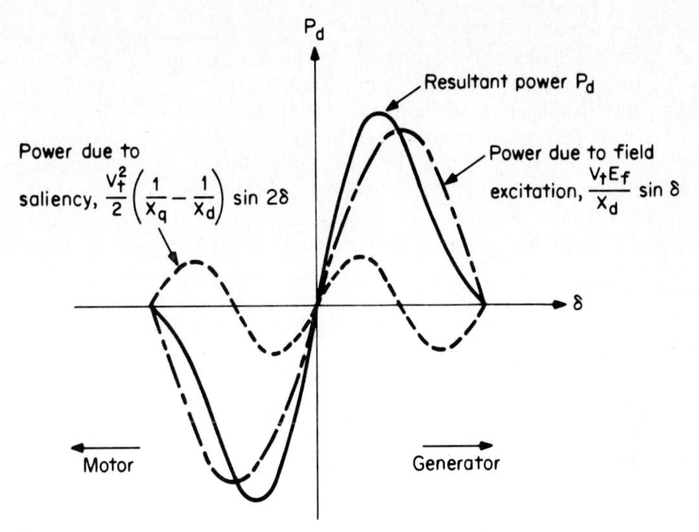


**FIG. 3.25** Power-angle characteristics of salient-pole machine.

$$P_{d,\max} = \frac{E_f V_t}{X_s} \tag{3.47}$$

The developed, or airgap, torque of a cylindrical-rotor synchronous machine is

$$T_d = \frac{P_d}{\omega_s} \tag{3.48}$$

where $\omega_s = 2\pi S/60$, rad/s
$S$ = synchronous speed, r/min

The units of $T_d$ are newton-meters or per-unit, depending upon the units used on the right side of Eq. (3.47). The electromagnetic developed power and torque are hypothetical quantities and cannot be measured. They describe the electromagnetic phenomena in the airgap, which is qualitatively identical for motor or generator action. The measurable power quantities in synchronous machines are input and output power. These are separated from the developed power by the internal losses of the machine.

## 3.11 MACHINE LOSSES

Synchronous machine losses are similar to losses in other types of rotating machines. Some losses associated with large synchronous machines include the following.

### Windage and Friction Loss

Windage loss is caused by air friction in the regions surrounding the rotor, primarily in the airgap. In large central station alternators, alternator efficiency is a significant factor in determining the cost of energy production of the central station. Therefore, many alternator design choices are made on the basis of operating efficiency rather than of initial cost. Most alternators rated 15,000 kW or larger used by utilities are housed in a sealed housing and the free spaces are filled with hydrogen rather than air in order to reduce windage losses. Hydrogen also serves as an improved cooling medium over air. Closed hydrogen systems are preferred, with the gas recirculated to the windings and air spaces after cooling in surface coolers. System pressures are in the neighborhood of 45 lb/in$^2$. The system is purged with $CO_2$ to avoid explosion hazards. Care must be exercised in the handling and storage of hydrogen, but modern practice has resulted in extremely high reliability of the hydrogen systems.

Friction losses are due to alternator bearing losses and slip-ring losses. An approximate combined friction and windage loss for air-cooled alternators is given by[7]

$$P_{FW} = K\left(\frac{U}{10{,}000}\right)^{2.5} D\sqrt{L} \qquad \text{kW} \tag{3.49}$$

where $U$ = rotor peripheral speed, ft/min
$D$ = rotor diameter, in
$L$ = rotor magnetic length, in
$K$ = 0.08 to 0.11 for slow-speed salient-pole machines
= 0.06 to 0.08 for higher-speed salient-pole machines
= 0.06 to 0.07 for cylindrical-rotor machines

## Core Loss

In synchronous machines, core loss is a very complex function of (1) the flux densities in the various portions of the magnetic circuit, (2) the fundamental frequency, (3) the magnitude and frequency of the various harmonics of airgap flux density, (4) the lamination thickness, (5) the temperature, (6) the type of magnetic material used in the magnetic circuit. Core loss is generally ascribed to the magnetic loss at the no-load condition of operation. An increase in core loss (and in certain other losses) as a function of load current is defined as a *stray-load loss*. Core loss is made up of two major components: (1) armature core loss due to time-varying magnetic flux in the armature and (2) pole-face loss on the surface of both the field structure and armature. Additional (usually minor) losses also exist in the shaft, housing, and other structural members owing to induced eddy currents.

Core loss can be reduced by constructing magnetic members of thinner laminations and by using low-loss types of magnetic materials, such as iron-nickel alloys, oriented silicon steel, or amorphous magnetic materials, although there is generally a cost penalty in all of these methods. Pole-face losses can be reduced by minimizing the slot openings, but this generally increases a winding's assembly cost.

## Stray-Load Loss

This loss has been defined just above as the increase in core loss as a function of load current. It results from losses induced by armature-leakage fluxes and other variations in airgap flux distribution. In very large machines, eddy-current losses induced in armature conductors of large cross section are also included in the stray-load loss. To minimize this component of stray-load loss, armature conductors are frequently constructed of laminated members, such as bundled conductors, strip conductors, square conductors, or Litz wire. Stray-load loss is a difficult loss component to measure accurately, and a value of 1 percent of the power output is often assumed as a typical value for stray-load loss.[8]

### Armature Conductor Loss

This loss is made up of the ohmic (or dc) loss and the "effective" (or ac) loss in the armature conductors. Effective loss is due to the nonuniform distribution of flux linkages over the conductor's cross section, often called *skin effect*, and is a function of the conductor's cross-sectional area and the frequency of the armature current. In large conductors, skin effect can result in a significant increase in armature copper loss even at 60 Hz. For this reason, armature conductors are frequently laminated or segmented.

The armature conductor loss varies as a function of conductor temperature in the typical manner. Conductor loss must be associated with a specific temperature when rating or specifying a machine. Typical temperatures used in synchronous machine ratings are temperature rises above ambient of 80, 105, and 130°C.

### Excitation Loss

This loss should generally include the entire loss of the excitation system, including the voltage regulator and the synchronous machine's field conductor loss. As a minimum, the field conductor loss must be included when specifying machine efficiency. The practice in this regard varies among manufacturers and users of synchronous machines. The field conductor loss is an ohmic loss and can be handled as the ohmic portion of the armature conductor loss.

## *3.12 STEADY-STATE CHARACTERISTICS—SALIENT-POLE MACHINES*

The magnetic structure of the field of salient poles is considerably different from that of cylindrical-rotor machines, as illustrated in Figs. 3.1 and 3.2. Because of this structural difference, the field permeance varies between a maximum, known as the direct-axis permeance, and a minimum, known as the quadrature-axis permeance. The reactances associated with these permeances have been defined in Sec. 3.9, which also described the means of determining these reactances experimentally.

### Excitation Characteristics

See Sec. 3.10 for cylindrical-rotor machines.

### Voltage Regulation

See Sec. 3.10 for cylindrical-rotor machines.

### Power Characteristics

The terminal power characteristics of salient-pole machines are identical to those of cylindrical-rotor machines at less than maximum loading. However, the maximum, or pull-out, power of a salient-pole machine differs from that of a cylindrical-rotor machine of similar size and steady-state rating.

It can be shown that the developed power of a salient-pole synchronous machine is[7–9]

$$P_d = \frac{E_f V_t}{X_d} \sin\delta + 0.5V_t\left(\frac{1}{X_q} - \frac{1}{X_d}\right)\sin 2\delta \tag{3.50}$$

$P_d$ is in per-unit if the terms on the right-hand side of Eq. (3.50) are in per-unit. If actual units are used for the terms on the right-hand side, Eq. (3.50) must be multiplied by a factor of 3.0 to give three-phase developed power. It is seen that the first term in Eq. (3.50) is the same as the developed power in a cylindrical-rotor machine; this term results from the electromagnetic torque developed by the interaction of the field and armature magnetic fields in the airgap. The second term in Eq. (3.50) results from the variable permeability of the field-pole structure; hence, this term defines a power term independent of field excitation that is called *reluctance power*, and the electromagnetic torque associated with this power quantity is called *reluctance torque*. Note that this term in the power equation varies with the second harmonic of the power angle $\delta$. The two components of a salient-rotor machine's developed power are illustrated in Fig. 3.25.

## 3.13 SERVICE CONDITIONS

Service conditions describe the environmental conditions in which a synchronous machine is to operate and are defined in American National Standards Institute (ANSI) C50.10-1977,[10] which is also an IEEE standard.

Usual service conditions include:

1. The cooling medium of air-cooled machines, except gas-turbine-driven alternators, does not exceed 40°C and is not less than 10°C.
2. The temperature of the cooling hydrogen in hydrogen-cooled machines does not depart, at rated pressure, from the values listed in ANSI C50.12-1977.[11]
3. The altitude, for air-cooled machines, does not exceed 1000 m (3300 ft).
4. The pressure of hydrogen-cooled machines when operating at altitudes above 1000 m is maintained at the same absolute internal pressure as that required for operation at sea level.

Unusual service conditions include:[10]

1. Contaminated air due to combustible dust, flammable gases, lint, nuclear radiation, chemical fumes, oil vapor, salt air, steam, etc.
2. Operation in pits or entirely enclosed volumes.
3. Operation at speeds other than rated, except during transient conditions.
4. Exposure to ambient conditions greater than 40°C or less than 10°C.
5. Exposure to cooling media other than those defined above.

6. Exposure to abnormal shock or vibration.
7. Departure from rated voltage or frequency, or both, exceeding the limits given in Ref. 11.
8. Unbalanced loading conditions in phase and/or in currents among the phases.
9. Operation where low audible noise levels are required.
10. Operation in unusual mechanical orientations, such as an inclined position, or with overhang, or an inverted position.
11. Operation above 100-m (3300-ft) altitude.

## 3.14 RATING

The *rating* of a synchronous machine implies the service conditions and loading conditions at which the machine can operate indefinitely. In general, the rating of a machine is also associated with warranties by the manufacturer for a certain period of time, although such warranties should be always obtained in written form. Synchronous machines are rated in terms of *output* capabilities, as are most other types of rotating machines. The principal parameters used in rating a machine are listed on the *nameplate* of the machine. These generally include:

1. Output (kilovoltamperes in an alternator; horsepower in a motor)
2. Terminal voltage, line to line
3. Frequency
4. Speed
5. Current
6. Power factor
7. Temperature rise at rated kilovoltampere (or horsepower) output
8. Service conditions

The rating of a machine on its nameplate is its *continuous*, or indefinite, rating.[10] For short periods of time, most rotating electric machines can operate at load conditions far exceeding these continuous, or steady-state, ratings. Ratings for shorter time periods, such as 1 h down to 1 min, are generally available from the manufacturer.

## 3.15 ELECTRICAL TRANSIENTS

Sudden changes from the conditions of steady-state operation result in synchronous machine voltages, currents, and power flows that are markedly different from the steady-state values. The period immediately following a sudden change from the steady-state is known as the *transient period*, and this period generally lasts for a short time period, usually measurable in seconds or even milliseconds. There are many causes of machine transients both within the machine and in the systems external to the machine—the electrical system connected to the electric terminals and the mechanical system connected to the shaft. Some of the more common causes of electrical transients in synchronous machines are:

1. Symmetrical short circuits in the external electrical system.
2. Unsymmetrical short circuits (one line to ground, line-to-line short, etc.) in the external electrical system.
3. Start-up of a relatively large induction motor in the external electrical system.
4. Synchronizing or paralleling synchronous alternators with the external electrical system.
5. Pullout exceeding the pull-out capability of the synchronous machine.
6. Sudden loss of external load (mechanical or electrical) on the synchronous machine

Electrical transients in synchronous machines are voltages and currents subsequent to the sudden changes listed above, or similar phenomena. The transients arise from the principle of the "conservation of flux linkages." In synchronous machines, this implies that the flux linkages among the five electric circuits of a synchronous machine (three armature windings, a field winding, and a damper winding) are the same before and immediately following sudden changes in system conditions. In many respects, synchronous machine electrical transients are similar to those of any resistive-inductive coupled circuit, as may be seen from Refs. 12 to 17.

## 3.16 EQUIVALENT CIRCUITS

Equivalent circuits to represent synchronous machine transients are shown in Figs. 3.26 and 3.27, respectively, for the *d* and *q* axes. Equivalent circuits are useful for computer analysis of synchronous machines and permit a machine to be treated as another circuit in many systems analysis programs. There are many variations to the circuits shown in Figs. 3.26 and 3.27, which depend upon the specific problem being studied; however, these figures are quite general and are applicable to most synchronous machines. Concordia gives much simpler circuits useful in the analysis of synchronous motor start-up;[18] these are shown as Figs. 3.28*a* and *b*. These circuits can be used to calculate the speed-torque curves, starting torque, pull-up torque, and pull-in torque of synchronous motors.

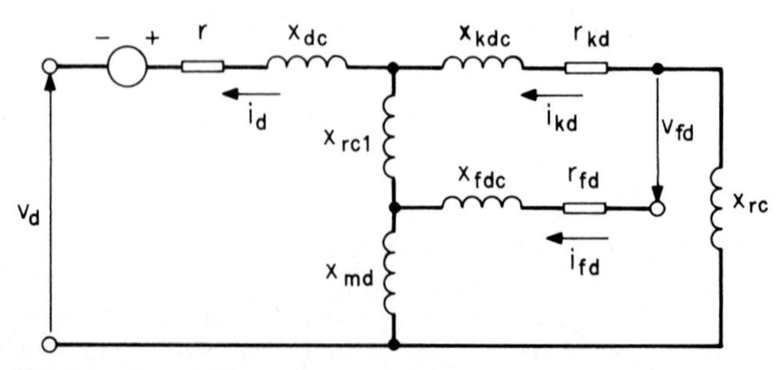


FIG. 3.26 General *d*-axis equivalent circuit.

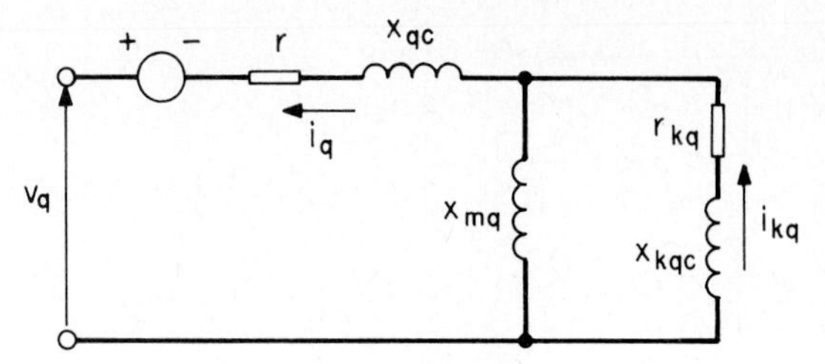


**FIG. 3.27** General $q$-axis equivalent circuit.

## 3.17 VOLTAGE DIP

*Voltage dip* is the instantaneous change in terminal voltage when there is a sudden change of alternator load or external impedance. Actually, this change can never occur instantaneously since there are some short-circuit time constants in the motor being started and in the transmission line between the motor and the alternator; however, these constants are normally relatively short compared to the alternator time constants that subsequently come into play, so it is usually assumed that this change occurs instantaneously. A second point of importance is that the power factor of the load affects the magnitude of the dip, and a purely resistive load may actually cause a sudden *increase* in terminal voltage; however, the principal concern here is that of the induction motor starting case, which is highly inductive.

Voltage dip refers only to the *initial* step change in voltage. Subsequent transient characteristics are a function of the control characteristics of the voltage regulator, the time constant in the exciter and the main alternator field, and the mechanical time constants of the alternator, engine, and engine governor. Voltage dip can be predicted from calculations by using the equivalent circuits of Figs. 3.26 and 3.27, with the sudden change in $i_d$ and $i_q$. Such an approach requires a full computer simulation of the machine, associated circuit, and load. Concordia has supplied a simplified approach to voltage dip that is widely used in the synchronous alternator industry. This is based upon the following assumptions:

1. Alternator resistance is zero.
2. Load (induction motor) resistance is zero.
3. Alternator mmf is sinusoidally distributed around the airgap.
4. Damper windings have negligible effect.
5. Alternator speed is constant.
6. Alternator is unloaded before motor starting load is applied.

The Concordia equation is[18]

$$v_t = \frac{X}{X + X_d'} e_f \tag{3.51}$$

where $X$ is the load (induction motor) reactance and the other parameters have been defined earlier. From Eq. (3.51) the voltage dip in volts is found from

$$e_f = v_t = \frac{x_d'}{x_d' + X} e_f \tag{3.52}$$

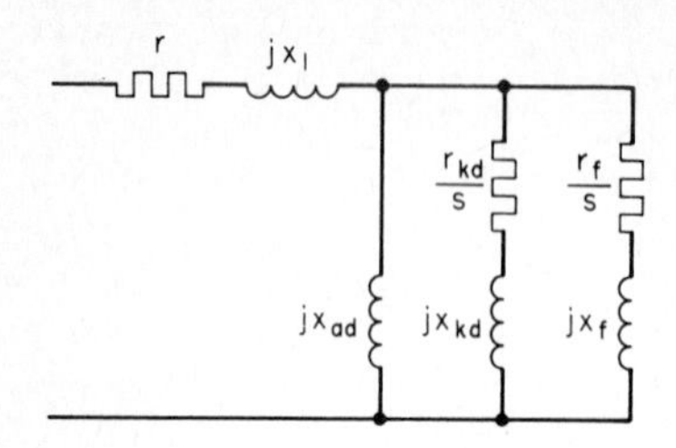


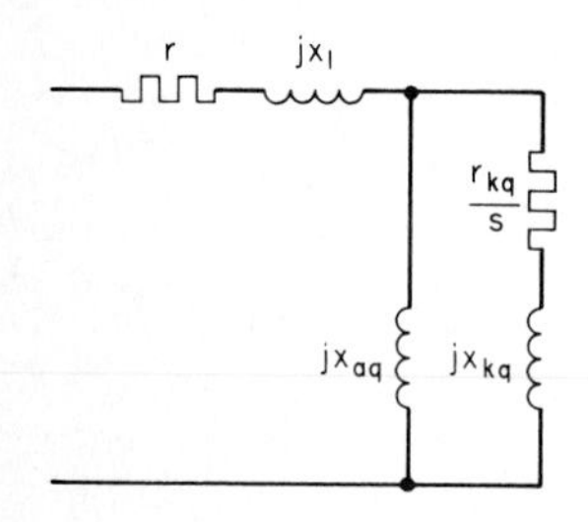


**FIG. 3.28** Direct- and quadrature-axis equivalent circuits.

Voltage dip in percent of no-load voltage $e_f$ is

$$\text{Percent dip} = \frac{100x'_d}{x'_d + \text{kVA}_b/\text{kVA}_m} \tag{3.53}$$

where $\text{kVA}_b$ = base $\text{kVA}_a$ of the alternator
$\text{kVA}_m$ = motor-starting kVA

This is the commonly used form of the voltage-dip equation. Voltage dip as a function of load is shown in Fig. 3.29.

## 3.18 SYNCHRONOUS MACHINE DESIGN

The design of a synchronous machine of any type and of any rating is a complex task. For many years, it was a most tedious task, requiring weeks of calculations of lengthy formulas. Synchronous machine design does not yield to a "synthesis" approach, given the many variables involved in the design process and given the fact that the machine output (power, voltage, torque, speed, etc.) is not related to the "input" of the design process (airgap length, rotor diameter, machine length, magnetic material coefficients, number of slots in the stator winding, airgap flux density, etc.) by any known analytical functions. Therefore, if the design obtained through this tedious process did not achieve the desired or "specified" performance, the entire process had to be repeated. With the advent of readily available mainframe computers, this process became computerized and was probably one of the first examples of "artificial intelligence," since the intelligence of the machine designer in using certain shortcuts or rules of thumb in the design process was put into the computer design program.

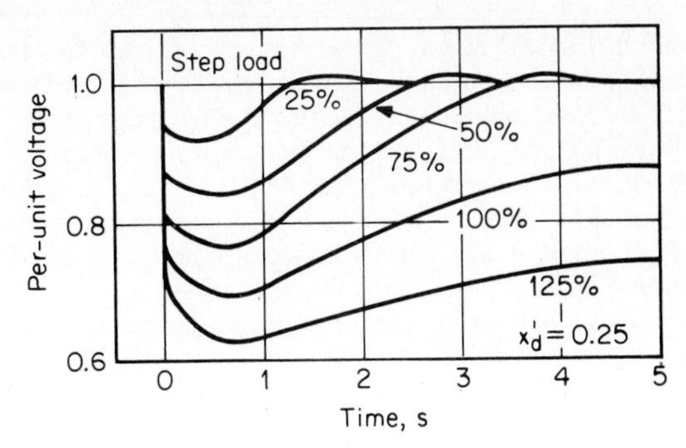


**FIG. 3.29** Typical terminal voltage dip during motor start. *(Courtesy of General Electric Company.)*

Synchronous machine design is a somewhat subjective process. In general, for any given set of performance specifications, there is an infinite number of designs capable of meeting the given set. If economic considerations are included along with performance specifications, the number of designs becomes considerably narrowed but is still substantial. If the operating lifetime, reliability, or mean time between failure is added to the specifications, the number of designs meeting performance, economic, and reliability specifications is further narrowed. In general, there are many machine designs that can satisfy given sets of specifications, no matter how detailed. The fact that many designs can meet performance, economic, and reliability specifications is the result of a number of factors: (1) the many magnetic, conductive, and insulation materials available to the designer today; (2) the many manufacturing techniques available to the designer; (3) the many types and layouts of armature windings; and most importantly (4) the multivariable nature of the design process itself.

Therefore, synchronous machine design is an *iterative* process. A set of dimensions, winding factors, material coefficients, etc., is chosen and the performance of the machine based upon these choices is calculated; the first calculation seldom realizes the required specifications, so the process is repeated with new "input parameters." This process is repeated until the required performance specifications are satisfied. Obviously, this iterative design process is well-suited to computer solution, and all contemporary synchronous machine design methods are computer-aided designs (CADs). There is still a significant role for the machine designer, however, not only in setting up the CAD process itself but also:

1. In making the initial choices of the design process, which are usually related to the winding layout, types of materials, lamination sizes, etc. (as will be discussed below).
2. In guiding the iterative design procedure to conserve computing time, required computer storage, etc., and in ensuring a convergence toward the desired specifications.
3. In studying various design trade-offs.
4. In ensuring that the resultant design is physically and economically realizable and manufacturable.

There is no "general" or "preferred" or "standardized" method for designing a synchronous machine, despite the almost complete computerization of the design process. There remain in the world today only a few manufacturers of the very large synchronous alternators used in central station power generation, and the design techniques of these manufacturers are highly proprietary. The design of small- and medium-size machines is somewhat more standardized, although the design procedures of individual manufacturers are still proprietary and generally not available to the purchaser or user. The standardization that has occurred is largely due to a design program developed for aerospace alternators by NASA; this, in turn, rests on the work of Ginsberg[19] and on Kuhlman's classic text on machine design.[20] Most of the following guidelines for synchronous machine design are based on these references.

### Design Specifications

The first step in the design of a synchronous machine is to specify its performance characteristics, or *output parameters*. These are generally based upon the machine's steady-state characteristics. Most synchronous machines are categorized by these steady-state ratings (listed in Sec. 3.14) and most machine catalogs describe their products on the basis of these ratings, known as *continuous* ratings. However, many other characteristics may be of importance in a specific application, such as:

**A.** Economic factors
- **1.** Initial cost
- **2.** Weight
- **3.** Mounting considerations, base, couplings, etc.
- **4.** Efficiency
  - *a.* At rated load
  - *b.* Over a certain duty cycle
  - *c.* Maximum
  - *d.* At a specific load
- **5.** Volume or space limitations
- **6.** Maintenance considerations; warranty

**B.** Environmental factors
- **1.** Ambient temperature of environment
- **2.** Vibration environment
  - *a.* Load-induced vibration in motors
  - *b.* Coupling to load or drive machine
  - *c.* Number of bearings (one or two)
- **3.** Corrosive influences
- **4.** Type of housing required
  - *a.* Open
  - *b.* Splash-proof; drip-proof
  - *c.* Totally enclosed or hermetically sealed
- **5.** Type and amount of cooling
  - *a.* Shaft-mounted fan

   - *b.* External blower
   - *c.* Liquid cooling
   - *d.* Forced hydrogen
6. Connected system voltage levels and phases
7. Impedance and other characteristics of connected electrical system
   - *a.* Permissible fault current into system
   - *b.* Relay and fault protection
8. Required machine protection, electrical and mechanical

C. Transient characteristics
1. Peak mechanical load magnitude and duration
2. Balanced three-phase short-circuit characteristics
3. Synchronizing torque and/or power
4. Voltage dip, when applicable
5. System stability considerations

D. Excitation characteristics
1. Excitation source
   - *a.* Physical configuration
   - *b.* Voltage and voltampere rating
   - *c.* Transient response
2. Voltage regulation: definition of expected load
3. Excitation protective circuitry

E. Means of starting or bringing up to speed

The above listings are given to serve as a guide in the design of a synchronous machine. The use of these ancillary specifications depends upon the particular application, and some are of more significance than others. However, in designing or even purchasing a machine for a given application, most of these factors should be considered in the initial stages of the design or purchase.

## Design Input Parameters

As stated in the previous paragraphs of this section, synchronous machine design cannot be synthesized. A number of assumptions must be made in order to initiate the design process. Further assumptions are required during the iterative steps of the design process.

Many of the assumptions used in synchronous machine designs are related to the physical configurations of the armature laminations to be used in the machine. Machine laminations are manufactured either by machine punching techniques or by chemical etching techniques. For synchronous machines, the former technique is almost universal, due to the relatively large physical size of synchronous machines used in power applications; the major cost of producing punched laminations is the initial tooling cost in producing the dies to punch out the slots, notches, and shaft diameter for the lamination. These same considerations generally apply to rotor (or field) laminations used in machines with damper windings. Therefore, the initial cost of a lamination is high, but the cost decreases rapidly with increasing lamination production. As a result of these manufacturing considerations, it is generally desirable to design a synchronous machine using existing lamination configurations, but if a large production is anticipated, obviously a new lamination configuration can be requested for the design. In either case, the lamination configuration is generally the first choice that must be made in designing a synchronous machine.

The second set of assumptions (or initial design choices) required to initiate the design process is the nature of the armature windings. Note that some of the choices affecting the winding ensue from the lamination configuration, such as the number of slots. Other obvious requirements are the type of winding (lap, concentric, etc.), winding connection (delta, wye, single-phase), wire and coil insulation, slot insulation, wire material (only copper and aluminum are used commercially, but nickel or silver alloy may be required for high-temperature alternators or alternators operating in unusual environments), and coil design. Wire current density is another important design parameter, since it affects the ohmic losses, weight, size, insulation, and possibly the lifetime of the armature winding.

In commercial machine design, many of the winding and lamination design parameters are dictated by economic and manufacturing considerations. Most manufacturers stock standard sizes of both laminations and preformed coils, and economic factors often favor modifying a design in order to use these stock parts. Economic factors may even favor the use of fractional-slot windings in order to use standard parts, despite the added design effort required for this class of windings. Automatic coil-winding machines or even manual coil-winding processes often favor certain coil designs and configurations. On the other hand, in the design of aerospace or experimental machines, new lamination and coil designs may be considered.

Some guidelines for choosing these design inputs are given at the end of this section under "Design Program" in connection with Table 3.6. At the same time, the following general restraints should be observed in initiating a synchronous machine design:

**1.** *Machine size versus machine rating:* A general relationship between synchronous machine size and its key parameters is given by

$$\text{kVA} \propto KSB_gJD^2L \tag{3.54}$$

where $D$ = stator bore (stator ID)
$L$ = armature stack length
$B_g$ = average airgap flux density
$J$ = armature conductor current density
$S$ = synchronous speed
$K$ = constant of proportionality

Equation (3.54) is an equation or proportionality and can seldom be used in an absolute sense. It is most useful in describing the relationship between the capacity of a synchronous machine and the parameters $D$, $L$, $B_g$, $J$, and $S$. Any consistent set of units can be used. In some cases, in which machines of reasonably similar shape and configuration are being designed, the proportionality constant $K$ can be given a numerical value based upon a known design and Eq. (3.54) can be used in a quasi-analytical sense.

**2.** *Airgap flux density:* The flux density of the magnetic materials in a synchronous machine is one of the physical limitations of the machine. If the magnetic field intensity is increased much beyond the level that results in saturation, magnetic materials exhibit diamagnetic characteristics and a relatively small increase in flux density results. Therefore, the magnetic materials in a synchronous machine should not, in general, be worked significantly beyond their saturation levels.

The degree of saturation in critical portions of the machine's magnetic circuit (mainly the armature teeth and the yoke) should be noted during the design process by observing the saturation constant $K_s$. Practice varies considerably among manufacturers in the amount of saturation permitted, and obviously there are design trade-offs between overall machine volume and the exciting current required to overcome the magnetic potential due to saturated magnetic materials. Related to this trade-off between the volume of magnetic material and the excitation required

**TABLE 3.6** Design Guide for Salient-Pole Alternator

| | Units | Symbol |
|---|---|---|
| I. Design Inputs: | | |
| A. *Alternator ratings*: | | |
| 1. Power output | kW | kW |
| 2. Phases | — | M |
| 3. Phase terminal voltage ($V_t$) | V | VPH |
| 4. Frequency | Hz | F |
| 5. Speed | r/min | S |
| 6. Poles | — | P |
| 7. Power factor | — | PF |
| 8. Ambient temperature | °C | TA |
| 9. Temperature rise | °C | ΔT |
| 10. Service factor | — | SF |
| 11. Frame size (if applicable) | | |
| 12. Permeability of free space | H/m | UO |
| 13. Core-loss factor | W/kg | KCL |
| B. *Armature laminations*: | | |
| 1. Size or designation | | |
| 2. Material | | |
| 3. Gauge or thickness | | |
| 4. Outer diameter | m | DS |
| 5. Inner diameter (bore) | m | D |
| 6. Stack length | m | L |
| 7. Stacking factor | — | SF1 |
| 8. Back-iron thickness | m | HC |
| 9. Slot style (Fig. 3.5) | — | |
| 10. Number of slots | — | S1 |
| 11. Slot dimensions; see Fig. 3.5 and Eq. 3.6 | mm | B1 |
| 12. | mm | B2 |
| 13. | mm | B3 |
| 14. | mm | BTM |
| 15. | mm | BTT |
| 16. | mm | D1 |
| 17. | mm | D2 |
| 18. | mm | D3 |
| 19. | mm | DT |
| 20. | mm | HS |
| 21. Lamination density | kg/m³ | RHOA |

**TABLE 3.6** Design Guide for Salient-Pole Alternator (*continued*)

| | Units | Symbol |
|---|---|---|
| C. *Armature winding*: | | |
| 1. Current density | $A/mm^2$ | J |
| 2. Type of wire | — | |
| 3. Insulation class | — | |
| 4. Winding type (lap, concentric, etc.) | | |
| 5. Winding connection (Y, Δ, 1$\phi$, etc.) | | |
| 6. Parallel paths | — | A |
| 7. Coils per slot | — | |
| 8. Series turns per coil | — | T |
| 9. Coil span (throw) | — | THRO |
| 10. Skew angle | | TS |
| 11. End-turn extension | mm | DE |
| 12. End-turn angle | rad | TE |
| 13. Conductor resistivity (20°C) | Ω·mm | RE |
| 14. Conductor temp. coef. of resistance | Ω/°C | ARA |
| 15. Conductor-material density | $kg/m^3$ | RHO1 |
| D. *Airgap*: | | |
| 1. Maximum gap | mm | GMX |
| 2. Minimum gap | mm | GMN |
| E. *Salient-pole design* (Fig. 3.30): | | |
| 1. Pole diameter (max.) | m | DP |
| 2. Pole span | deg | PS |
| 3. Pole length | m | PL |
| 4. Pole dimension | mm | PH1 |
| 5. Pole dimension | mm | PB1 |
| 6. Pole dimension | mm | PW |
| 7. Material | — | |
| 8. Gauge or thickness | — | |
| 9. Stacking factor | — | SF2 |
| 10. Turns per pole | — | T2 |
| 11. Conductor cross section | $mm^2$ | AW2 |
| 12. Conductor type | — | |
| 13. Coil connections | — | |
| 14. Conductor resistivity | Ω·mm | REF |
| 15. Coil-pole spacer thickness | mm | TP2 |
| 16. Other insulation | — | |
| 17. Conductor temp. coef. of resistance | Ω/°C | ARF |
| 18. Conductor-material density | $kg/m^3$ | RHO2 |
| 19. Field-lamination density | $kg/m^3$ | RHOF |
| F. *Damper winding* | | |
| 1. Bars per pole | — | S2 |
| 2. Bar diameter | mm | DB |
| 3. Bar skew | | SK2 |
| 4. Bar material | — | |
| 5. End-ring OD | m | DEO |
| 6. End-ring ID | m | DE1 |
| 7. End-ring width | mm | WER |
| 8. Damper-bar resistivity | Ω·mm | RED |
| 9. Damper-bar density | $kg/m^3$ | RHOD |

**TABLE 3.6** Design Guide for Salient-Pole Alternator (*continued*)

| | Units | Symbol | Equation |
|---|---|---|---|
| II. Design Outputs: | | | |
| A. *General*: | | | |
| 1. Angular shaft speed | rad/s | WM | $2\pi$(RPM)/60 |
| 2. Electrical angular speed | rad/s | WE | $2\pi$F |
| 3. Pole embrace | — | PSI | P(PS)/360 |
| 4. Phase current | A | I | kW/PF/M/VPH |
| 5. Power-factor angle ($\theta$) | rad | THT | THT=$\cos^{-1}$(PF) |
| B. *Airgap*: | | | |
| 1. Average gap | mm | LGA | (GMX+GMN)/2 |
| 2. Carter coefficient | — | KC | Eq. (3.6) |
| 3. Effective airgap | mm | LGE | Eq. (3.5) |
| 4. Airgap ratio | — | GR | GMX/GMN |
| C. *Armature winding*: | | | |
| 1. Pole pitch ($p$) | slots | LP | S1/P |
| 2. Coil pitch ($c$) | — | LC | (THRO)/(LP) |
| 3. Pitch factor | — | KP | Eq. (3.9) |
| 4. Distribution factor | — | KD | Eq. (3.12) |
| 5. Skew factor | — | KS | Eq. (3.13) |
| 6. Winding factor | — | KW | (KP)(KD)(KS) |
| 7. Connection factor | — | CS | 2(S1)(T)/(A)/(M) |
| 8. Slots per pole per phase | — | Q | S1/P/M |
| 9. Slot depth | mm | DT | D1+D2+D3 |
| D. *Armature resistance*: | | | |
| 1. Coil half length | mm | LA | $\frac{\pi(1000\text{D}+\text{DT})(\text{LC})}{\text{P}\cos(\text{TE})}$ +2(DE)+DT+L |
| 2. Conductor cross section | mm$^2$ | AW1 | I/J |
| 3. Armature resistance (20°C) | Ω/phase | RAA | (LA)(CS)(RE)/A/(AW1) |
| 4. Armature resistance ($\Delta T$) | Ω/phase | RAO | (RAA)[1+(ARA)($\Delta$T)] |
| E. *Armature reactances*: | | | |
| 1. Magnetizing reactance (unsaturated) | Ω/phase | XM | $\frac{2(\text{UO})\text{FLDM}(\text{CW})^2}{(\text{P})^2(\text{GE})}$ |
| 2. Gap flux density (max.) | T | BGM | P(VPH)/4.44/D/L/F/C1 |
| 3. Gap mmf (average) | A | MG | 2(BGM)(LGA)/UO/$\pi$ |
| 4. Core flux density (max.) | T | BCM | 1000D(BGM)/(SF1)/(HC)/F |
| 5. Core field intensity (max.) | A/m | HCM | Look up* |
| 6. Tooth flux density (max.) | T | BTM | $\pi$D(BGM)/(SF1)/(S1)/(BT1) |
| 7. Tooth field intensity (max.) | A/m | HTM | Look up* |
| 8. Tooth mmf (average) | A | MT | 2(DT)(HTM)/$\pi$/1000 |
| 9. Core mmf (average) | A | MC | $2\left[\text{D}+\frac{2(\text{DT})+(\text{HC})}{1000}\right]$ HCM |
| 10. Saturation factor | — | KI | 1+(MT+MC)/MG |
| 11. Magnetizing reactance (saturated) | Ω/phase | XMS | XM/KI |
| 12. Slot-leakage pitch factor | — | KL | Look up Fig. 3.31 |
| 13. Armature slot permeance factor | — | PS† | D1/B1+2(D2)/(B1+B2) +D3/3(B3) |

**TABLE 3.6** Design Guide for Salient-Pole Alternator (*continued*)

| | Units | Symbol | Equation |
|---|---|---|---|
| 14. Slot-leakage reactance | Ω/phase | XS | 2π(UO)FLM(CS)²(KS)(PS)/(S1) |
| 15. Fundamental field form (direct axis) | — | CD1 | Look up Fig. 3.32 |
| 16. Pole-embrace factor | — | PEF | $P \sin^{-1}(\frac{PS}{2})/180$ |
| 17. Demagnetizing factor | — | CM | 4(PEF)+ sin [π(PEF)]/4/ sin [π(PEF)/2] |
| 18. Fundamental field form (quad. axis) | — | CQ1 | [4(PSI)+1] 5−sin[π (PSI)]/π |
| 19. Direct-axis armature unsat. magnetizing reactance | Ω/phase | XAD | (XM)(CD1)(CM) |
| 20. Quad.-axis armature magnetizing reactance | Ω/phase | XAQ | (XM)(CQ1) |
| 21. End-coil leakage reactance | Ω/phase | XEC | FM(WE)(CS)²[0.3D (3(LC)−1)]/P² |
| 22. Zigzag-leakage reactance | Ω/phase | XZ | 0.675(P²)(XAD)/(S1) |
| 23. Belt-leakage reactance | | | |
| With cage | | | 0 |
| Without cage | Ω/phase | XB | 0.75(XAD)(KB) (KB: Look up Fig. 3.33) |
| 24. Leakage reactance | Ω/phase | XL | XS+XEC+XZ+XB |
| 25. Direct-axis synchronous reactance | Ω/phase | XD | XAD+XL |
| 26. Quad.-axis synchronous reactance | Ω/phase | XQ | XAQ+XL |
| F. *Alternator characteristics*: | | | |
| 1. Alternator kVA | kVA | kVA | kW/PF |
| 2. Power angle (δ) | rad | DELT | |

$$\text{DELT} = \tan^{-1}\left[\frac{\text{I(XQ)cos(THT)} - \text{I(RA)sin(THT)}}{\text{VPH+I(RA)cos(THT)+I(XQ)sin(THT)}}\right]$$

| | Units | Symbol | Equation |
|---|---|---|---|
| 3. Excitation phasor ($E_f$) | V | EF | |

$$\text{EF} = \text{(VPH)cos(DELT)} + \text{I(RA)cos(DELT + THT)} + \text{I(XD)sin(DELT + THT)}$$

| | Units | Symbol | Equation |
|---|---|---|---|
| 4. Approx. voltage regulation | (percent) | VR | 100(EV−VPH)/(VPH) |
| 5. Airgap flux per pole (max.) | Wb | PHI | DL(BGM)/P |
| 6. Pole average flux density (Fig. 3.30) | T | BAP | (PHI)/(PW)/(PL) |
| 7. Pole average field intensity | A/m | HAP | Look up (*B-H* curve) |
| 8. Approx. mmf drop per pole | A | MP | (HAP)(DP)/2 |
| 9. Required field excitation per pole | A | MF | MG+MP+MT+MC |
| 10. Field current per pole | A | IFP | MF/T2 |
| 11. Total field current (assume series) | A | IF=IFP | (See item I.E.13 in this table) |
| 12. Field mean turn length | m | LF | 2[1000(PL)+PW]+PB1 |
| 13. Field-winding resistance 20°C | Ω | RFA | P(REF)[LF+100]/AW2 |

**TABLE 3.6** Design Guide for Salient-Pole Alternator (*continued*)

| | Units | Symbol | Equation |
|---|---|---|---|
| 14. Field-winding resistance ($\Delta T$) | Ω | RFO | (RFA)[1+(ARF)(ΔT)] |
| 15. Field voltage | V | VF | (RFO)(IF) |
| 16. Field ohmic loss | W | P2 | (VF)(IF) |
| 17. Armature ohmic loss | W | P1 | M(I)$^2$(RAO) |
| 18. Armature magnetic volume | m$^3$ | UA | L(SF1)[π[(DS)$^2$−D$^2$]/4 −(S1)(DT)(B2)/10$^6$] |
| 19. Armature magnetic weight | kg | MA | (RHOA)(VA) |
| 20. Core loss | W | PCL | (MA)(KCL) |
| 21. Windage and friction loss | W | PWF | Eq. (3.45) or equivalent |
| 22. Stray-load loss | W | PSL | (10.)(kW) |
| 23. Alternator efficiency | — | EFF | kW(kW+P1+P2+PCL+PWL +PSL) |
| G. *Active-material weight*: | | | |
| 1. Armature-winding weight | kg | M1 | (LA)(CS)(AW)(RHO1)/10 |
| 2. Field-winding weight | kg | M2 | (LF)(AW2)(RHO2)/10$^9$ |
| 3. Field-pole volume | m$^3$ | UF | P(PL)(SF2)[(0.5)(DP/2)$^2$[PS −sin(PS)]+(DP/2)(PW)] |
| 4. Field-pole weight | kg | MF | (UF)(RHOF) |
| 5. End-ring weight | kg | MB | (RHOD)(WER/1000) [π(DEO)$^2$ −(DEI)$^2$]/4 |
| 6. Active weight | kg | MT | M1+M2+MA+MF+MB |

* Look up HCM and HTM from the magnetic materials characteristics (*B* versus *H*) used in the lamination. (See Chap. 2 for typical curves.)

† Add 0.125 for round bottom slots.

to overcome magnetic saturation are core loss and machine efficiency. In general, $K_s$ should not exceed 1.25. Magnetic saturation is an "absolute" limit—i.e., there are no short-time or intermittent limits above the saturation level.

**3.** *Conductor current density:* The second major physical limitation of the synchronous machine is the current density in field and armature conductors, *J*. Unlike magnetic flux density, there is no absolute limit to the conductor current density. Rather, current density involves a design trade-off. Increasing the current density decreases the wire size required to carry a given level of current. There is a "snowball" effect related to the choice of current density, for the conductor size affects the slot area required and hence affects the lamination design and perhaps even the lamination diameter, which is the major factor determining the overall machine size. Also, the volume of insulation required on the conductors and in the slot is affected. However, as the current density is increased, the ohmic losses increase by the simple relationship.

$$I^2R = J^2Al \tag{3.55}$$

where $J$ = current density, A/cm$^2$
$A$ = wire cross-sectional area, cm
$l$ = length of wire, cm

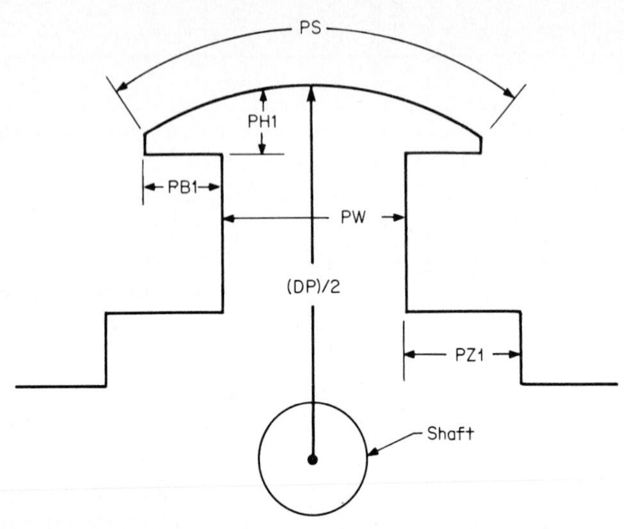


**FIG. 3.30** Simplified cross section of a rotor pole (four poles assumed) in a salient-pole machine.

Note that both $I$ and $J$ are rms values in Eq. (3.55). The current density should be based on the continuous rating of the machine. The ohmic loss affects the machine's efficiency, the grade of insulation material, and the nature and quantity of cooling required. Therefore, the choice of continuous current density is one of the most important choices in the machine design; it affects many other physical aspects of the machine, the initial machine cost, the operating efficiency, and the heat transfer

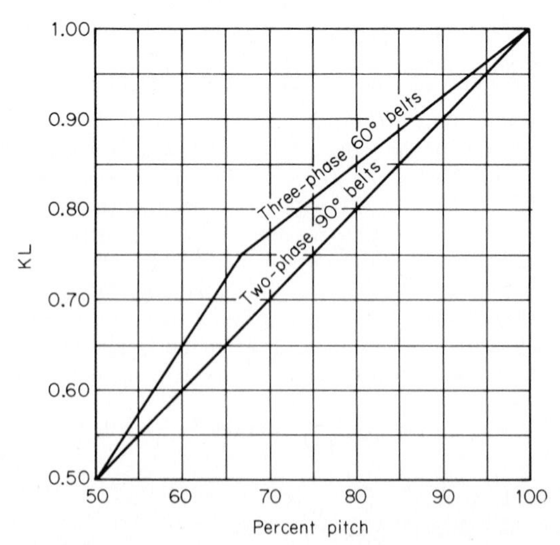


**FIG. 3.31** Slot-leakage pitch factor (item II.E.12, Table 3.6).

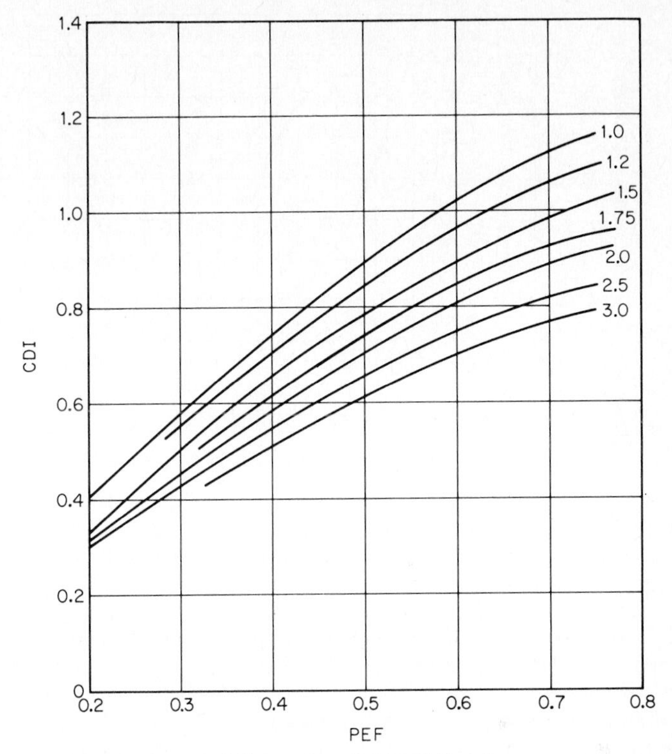


**FIG. 3.32** Fundamental field form (item II.E.15, Table 3.6).

required. Several guidelines for choosing current density are given below in the design program associated with Table 3.6.

**4.** *Thermal limit:* The basic, continuous rating of a synchronous machine is based upon its thermal characteristics. The temperature in a machine is determined largely by the choice of current density and, to a lesser extent, magnetic flux density and by the method of heat transfer used to cool the machine.

## Design Program

Section II of Table 3.6 is a basic design routine for calculating the performance of a synchronous alternator based upon the input parameters chosen from Sec. I. This design routine is based upon the theory and analytical expressions described in this chapter, and equations from this chapter are occasionally referred to directly in the design routine. The methods for determining synchronous reactance and other reactances are based upon Refs. 19 and 20, with some modifications.

An important difference in the manner of presenting the design equations in Sec. II of Table 3.6 is that specific parameters have been kept distinct—not combined into numerical constants. This combining of the constants is a major weakness of most previous design routines, including those of Refs. 19 and 20, for two reasons:

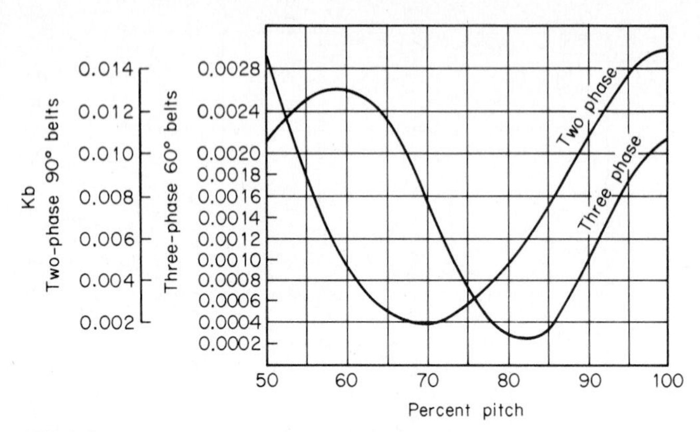


**FIG. 3.33** Belt-leakage factor, KB (item II.E.23, Table 3.6).

(1) It fixes the system of units to be used, and (2) the combining of constants and the wide use of "normalized" terms, factored terms, etc., make it impossible to identify the basic terms making up an equation (for reactance, say) and the equation becomes a meaningless collection of unknown terms that must be accepted on faith from the design routine's developer. It also adds an outrageous number of parameter definitions to the design routine. This combining approach was developed before large-scale computing facilities were available and at a time when per-unit notation was considered essential, neither of which reasons is valid at the present time; designers now prefer to use those equations whose terms can be identified and which can be related to phasor diagrams. Also, most early design routines were forced into the English system of units by combining permeability, resistivity, and other constants into one numerical term. The International System of Units (SI) is used worldwide today for engineering and scientific calculations, and SI units are specified by all standardizing agencies. However, when these constants are kept as parametric values, *any* system of units can be used in the design process.

The design routine is, of necessity, somewhat simplified, since many design calculations are based upon the detailed structural approaches taken by a particular design group. Also, a design group will tend to favor certain variations in winding layouts, field-pole structure and mounting technique, lamination design, etc., from the relatively basic approaches used in this routine. The field structure (Fig. 3.30) is particularly simplified and should be modified by the user as required. Also, note that the weight of the damper bars is included in the weight of the outer pole section. The design of the rotor damper winding is largely ignored in this design routine, except to call attention to the required input design parameters (Sec. I.F of Table 3.6) and to include its weight in the active weight summary.

The design routine is based upon the steady-state characteristics of the alternator, which is why the damper winding plays so small a role in the design. The transient performance of the machine is not calculated in the processes of Table 3.6 since these parameters are based upon the synchronous parameters and have almost no influence upon the physical and structural design of the machine. However, before a final set of design dimensions is chosen, the transient performance should be calculated in order to ensure proper transient operation.

This specific design routine is for a salient-pole alternator. The design approach would be almost identical for a salient-pole motor except for a different set of input rating specifications. The design approach for cylindrical-rotor machines would require, obviously, a different pole-design specification and considerable modification of the reactance and field-resistance equations, the field-excitation calculations, and the active-weight calculations. In Sec. II of Table 3.6, several parameters are obtained from a "lookup" method; these parameters are generally the nonlinear magnetic field intensities for a specific magnetic material or certain design parameters, such as CD1, the curve for which is supplied. Graphical characteristics can be handled in a computer design routine in a variety of ways, such as deriving analytical approximations, curve-fitting routines, or table storage, and the user should use the method most convenient and most readily available. Reference 8 gives several methods for handling magnetic *B-H* characteristics.

## REFERENCES

1. F. W. Carter, "The Magnetic Field of the Dynamo-Electric Machine," *Journal of the IEE*, vol. 64, 1926, p. 1115.
2. P. L. Alger, *The Nature of Polyphase Induction Machines*, Wiley, New York, 1951.
3. C. S. Suskind, *Alternating-Current Armature Winding*, McGraw-Hill, New York, 1951.
4. C. G. Veinott, *Theory and Design of Small Induction Motors*, McGraw-Hill, New York, 1957.
5. *Test Procedures for Synchronous Machines*, IEEE Standard 115-1983.
6. *IEEE Standard Dictionary of Electrical and Electronics Terms*, ANSI/IEEE Standard 100-1984.
7. D. G. Fink and H. W. Beaty (eds.), *Standard Handbook for Electrical Engineers*, 11th ed., McGraw-Hill, New York, 1977.
8. S. A. Nasar and L. E. Unnewehr, *Electromechanics and Electric Machines*, 2d ed., Wiley, New York, 1983.
9. A. E. Fitzgerald and C. Kingsley, *Electrical Machinery*, 3d ed., McGraw-Hill, New York, 1968.
10. *General Requirements for Synchronous Machines*, ANSI Standard C50.10-1977.
11. *Cooling Requirements for Synchronous Machines*, ANSI Standard C50.13-1977.
12. R. H. Park, "Two-Reaction Theory of Synchronous Machines—I: Generalized Method of Analysis," *AIEE Transactions*, vol. 48, 1929, pp. 716–727.
13. Y. H. Ku, *Electric Energy Conversion*, Ronald Press, New York, 1959.
14. L. Salvatore and M. Savino, "Exact Relationships Between Parameters and Test Data for Models of Synchronous Machines," *Electric Machines and Power Systems*, vol. 8, no. 3, May 1983, pp. 169–184.
15. I. M. Canay, H. J. Rohrer, and K. E. Schnirel, "Effect of Differential Electrical Faults and Switching Operations on Maximization of Mechanical Torques in Large Turbogenerator Sets," *Electric Machines and Electromechanics Quarterly*, vol. 4, no. 2, Sept. 1979, pp. 183–206.
16. F. Harashima, H. Naitoh, and T. Haneyoshi, "Dynamic Performance of Self-Controlled Synchronous Motors Fed by Current-Source Inverters," *IEEE Transactions on Industry Applications*, vol. 1A-15, no. 1, Jan. 1979, pp. 36–46.
17. I. M. Canay, "A Unified Theory of Torsional Oscillations in Electrical Machines and Electromechanical Systems," *Electric Machines and Power Systems*, vol. 8, no. 5, July 1983, pp. 341–356.
18. C. Concordia, *Synchronous Machines*, Wiley, New York, 1951.
19. D. Ginsberg, "Design Calculations for AC Generators," *AIEE Transactions*, vol. 69, 1950.
20. J. H. Kuhlman, *Design of Electrical Apparatus*, Wiley, New York, 1946.

# CHAPTER 4
# INDUCTION MACHINES

**H. E. Jordan**

## *4.1 PRINCIPLE OF OPERATION*

The basic principle of operation of an induction machine is illustrated by the revolving-horseshoe-magnet and copper-disk experiment pictured in Fig. 4.1. When the horseshoe magnet is rotated, the moving magnetic field passing across the copper disk induces eddy currents in the disk. These eddy currents are in such a direction as to cause the disk to follow the rotation of the horseshoe magnet. With the direction of rotation shown in the figure, the eddy currents will be as displayed in Fig. 4.1 according to Fleming's right-hand rule.* By applying Fleming's left-hand rule,† the force on the copper disk is determined to be in the direction of rotation of the magnet.

Whereas the copper disk will rotate in the same direction as the rotating magnetic field, it will never reach the same speed as the rotating magnet, because if it did, there would be no relative motion between the two and therefore no current induced in the copper disk. The difference in speed between the rotating magnetic field and the copper disk is known as *slip*, which is essential to the operation of an induction motor. In induction motors the rotating magnetic field is set up by windings in the stator, and the induced currents are carried by conductors in the rotor. The rotating horseshoe magnet and copper disk are considerably different in structure from today's induction motor, but the basic principles of operation are the same.

**Fleming's right-hand rule:* Place the thumb and the first and second fingers of the right hand so that all three are mutually perpendicular. With the hand in this position, the first finger is pointed in the direction of the field, the thumb is in the direction of motion of the conductor, and the second finger is in the direction of the induced voltage. Note that the relative motion of the conductor is opposite to the direction of rotation of the magnetic field.

†*Fleming's left-hand rule:* Place the thumb and the first and second fingers of the left hand so that all three are mutually perpendicular to each other. With the first finger in the direction of the field and the second finger in the direction of the current, the thumb indicates the direction of the force.

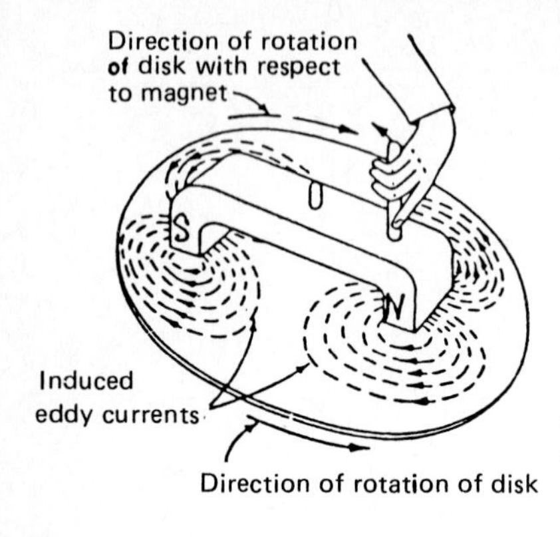


**FIG. 4.1** Rotation of a copper disk following the rotation of a permanent magnet. *(From C. G. Veinott, Theory and Design of Small Induction Motors, McGraw-Hill, New York, 1959.)*

## 4.2 ROTATING MAGNETIC FIELD

The rotating magnetic field is essential to the functioning of an induction motor. In practical machines this rotating magnetic field is achieved by a combination of a space displacement of the windings and a time-phase displacement of the exciting voltages.

This concept will now be explained by using a two-phase machine as an example, although the basic principle is applicable to any multiphase ac machine, including the popular three-phase induction motor.

Figure 4.2 displays a magnitude-versus-time relationship of current in a two-phase electrical supply system providing power to a two-phase motor. The abscissa is plotted as $\omega t$ (where $\omega = 2\pi f$, $f$ = the frequency of the ac supply, and $t$ = time), and the ordinate is the magnitude of the current. If this set of two-phase currents is used to supply the A- and B-phase coils of a two-phase winding, as shown in Fig. 4.3, the result will be a uniformly revolving magnetic field in the airgap of the machine.

The essential elements for the production of this rotating field are the *time* displacement of the exciting currents and the *space* displacement of the coils in the winding. The phase A and B coils are shown displaced by $\pi/2$ rad, 90°, in space in Fig. 4.3, and $i_A$ and $i_B$ are displaced $\pi/2$ electrical radians in time. Assuming that each of the coils individually produces a sinusoidal magnetomotive force (mmf) in the airgap, then vector addition can be used to calculate the resulting mmf of the two coils.

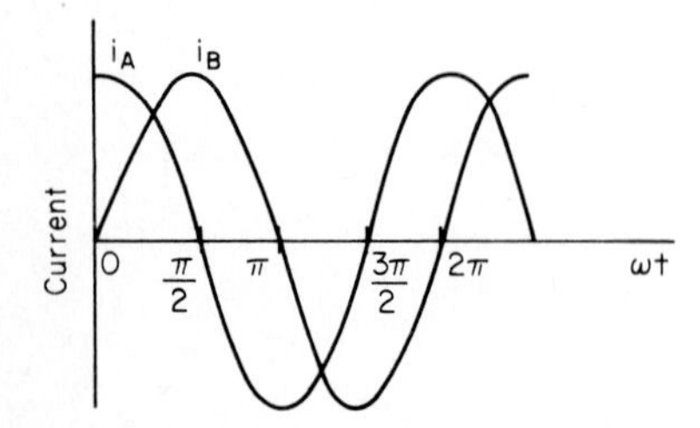


**FIG. 4.2** Two-phase electrical supply.

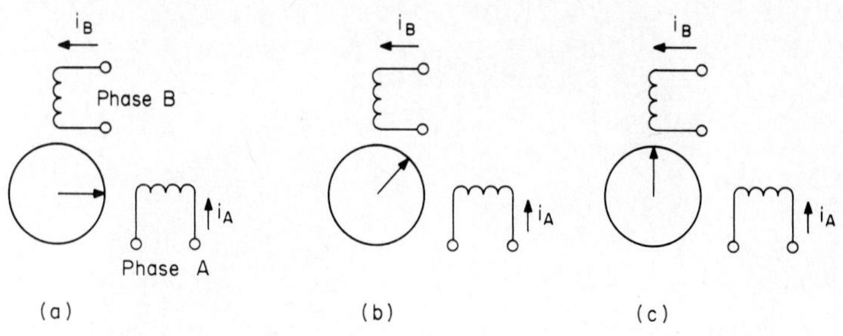


**FIG. 4.3** Rotating magnetic field at three different instants: (*a*) $\omega t = 0$; (*b*) $\omega t = \pi/4$; (*c*) $\omega t = \pi/2$.

Figure 4.3 shows the two phases occupying one-half of the airgap's circumference, which makes it a two-pole winding; in a four-pole winding, each set of two-phase pole groups would occupy one-fourth of the airgap circumference; etc. The rotating magnetic field travels past a pair of poles for each complete time cycle of excitation. Therefore, the rotating field travels around the airgap at a speed of

$$\text{r/s} = \frac{f}{p/2} \tag{4.1}$$

where r/s = revolutions per second of the rotating magnetic field
$f$ = frequency of excitation, Hz
$p$ = number of poles

The speed of rotation is more commonly expressed in revolutions per minute and is called synchronous r/min, where

$$\text{Synchronous r/min} = \frac{120f}{p} \tag{4.2}$$

Although the concept has been explained in terms of a two-phase machine for the sake of simplicity, it can be extended to any polyphase machine to show that a uniformly rotating magnetic field will be produced if the coils are properly distributed in space and excited with a balanced set of polyphase voltages. For instance, a balanced three-phase winding will produce a uniformly rotating mmf (or magnetic field) as given by

$$F = {}^3\!/_2 NI \cos(\theta - \omega t) \tag{4.3}$$

## 4.3 CONSTRUCTION AND GENERAL FEATURES

Figure 4.4 shows the cutaway view of a fan-cooled induction motor with the various parts labeled. The stator winding is similar to that of a synchronous machine (see Chap. 3).

### Rotors

The rotor is formed from laminated electrical steel punchings, and the rotor winding consists of bars contained in slots punched in the laminations. These bars are

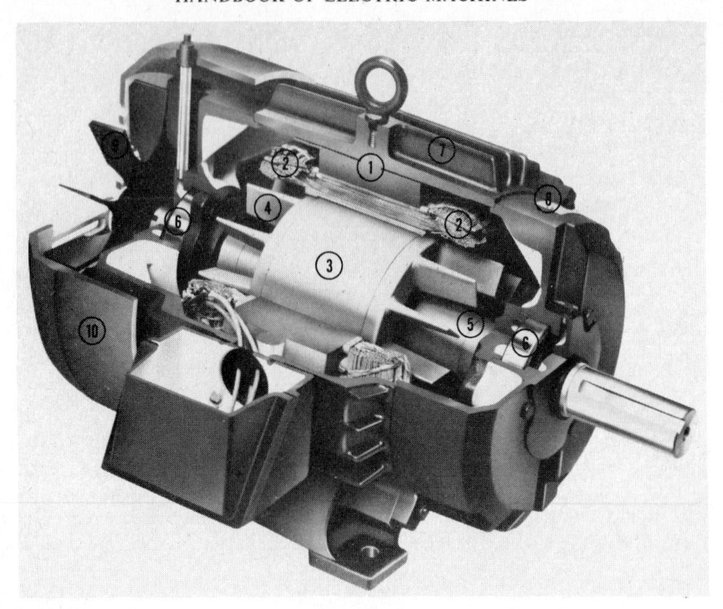

**FIG. 4.4** Cutaway of a fan-cooled induction motor. *(Courtesy of Reliance Electric Company.)*

short-circuited at both ends by a short-circuiting ring. A bar-end ring structure, without the laminated core, is called a squirrel cage. In small- and medium-horsepower sizes, rotors are made by casting aluminum into the rotor core. In the larger sizes of ac motors, cast-aluminum rotors are not practical, and copper bars are inserted into the rotor slots. These copper bars are short-circuited at both ends by a copper end ring, and the end ring is brazed or soldered onto the bars. Sometimes bronze or other alloys are used to replace copper in making the cage and end ring. The size at which the transition between cast-aluminum and copper rotors takes place varies among motor manufacturers, but virtually all rotors in motor sizes of several thousand horsepower and above are built with bar-type rotors.

## Coil Windings

Another construction feature dependent on motor size is the type of coil winding used. In small- and medium-size ac motors, most coils are random-wound. These coils are made with round wire, which is wound into the stator slots and assumes a diamond shape in the end turns; however, the wires are randomly located within a given coil, and hence the name "random-wound." For larger ac motors and particularly for high-voltage motors, 2300 V and above, form-wound coils are used. These coils are constructed from rectangular wire, which is bent into shape around forms and then taped. The coil is formed to the proper size so that the complete coil can be inserted into the stator slots at the time the stator is wound. Form-wound coils are used for high-voltage windings because it is relatively easy to add extra insulation on the individual coils before inserting them into the stator.

## Enclosures

Enclosures are an important feature in the motor design, and the type of enclosure is usually dictated by the application requirements. General-purpose ac motors are built in a *drip-proof enclosure*, which is defined by the National Electrical Manufacturers Association (NEMA) as an enclosure that will not permit damage to the windings or internal parts of the motor due to water entering the motor from an angle not exceeding 15° from the vertical.[1] This type of motor is used in applications in which excessive water or corrosive or hazardous environments are not present. Motors that are to be exposed to more severe environments than the drip-proof motor can withstand are built with more restrictive enclosures. Typical enclosures used for severe environments are: totally enclosed, nonventilated (TENV); totally enclosed, fan-cooled (TEFC); totally enclosed, separately ventilated; and motors built with heat exchangers.

Totally enclosed motors are characterized by having no interchange of air between the inside and the outside of the motor; thus, contamination in the atmosphere cannot enter the motor. The TENV construction has no openings in the frame or brackets and no external fan for providing a flow of air across the motor frame. This construction depends entirely upon convection and radiation to dissipate heat generated internally within the motor; consequently, it is usually limited to small horsepower sizes, 5 hp or less. Some larger horsepower ratings are built as TENV motors, but at the expense of housing the rating in a much larger frame size than the same rating that a general-purpose motor would require.

A TEFC motor is pictured in Fig. 4.4. The external fan forces air across the motor frame, enhancing the motor's heat-dissipating ability. Many TEFC motors are constructed with cooling fins to increase the frame's surface area and therefore its heat-dissipating ability. Heat generated inside the motor must travel by conduction through the stator core, through the frame, and by convection from the frame to the external air. Some fan-cooled motor designs circulate air internally and transfer heat by convection from the internal air to the motor frame, where it can eventually pass to the external air. The heat-transfer process in a TEFC motor is less efficient than in a drip-proof motor, where the outside air comes into direct contact with the active electrical components of the machine; therefore, larger frame sizes are often required for a given horsepower rating using TEFC construction than for a drip-proof construction. This is particularly true for the larger horsepower ratings. The NEMA Standards recognize this and assign different frame sizes for drip-proof and TEFC motors in many ratings.[2]

The separately ventilated machines (i.e., having a separate source of ventilating air) can be of either a drip-proof or a TENV construction. If a drip-proof or open construction is used, the ventilating air is normally piped into one or more of the openings in the machine and discharged through other openings. If a TENV construction is used, the ventilating air is forced across the outside of the frame from a source other than the motor's integrally mounted fan. Ratings in a given frame size can often be upgraded by providing a separate source of ventilation, as the air velocity can be increased over that which would be available from the motor's own fan.

For large horsepower ratings, fan-cooled constructions become impractical. Basically, this is due to the fact that while the surface area for ventilation increases only as the first power of the diameter, the area inside the machine for active electrical and magnetic materials increases as the square of the diameter; thus, the capability for producing electrical ratings increases more rapidly than does the surface area for dissipating the losses. Machines with heat exchangers are used in the larger horsepower sizes. One commonly used type of heat exchanger is a tube-cooled machine in which air-to-air or air-to-water heat exchangers are used to dissipate the motor's internal heat.

Numerous other special constructions are used. In the mining industry and in other severely hazardous environments, explosion-proof motors are used.[3] These motors are of TENV or TEFC construction—but, in addition, must be designed so that any spark occurring inside the motor for whatever reason (not normal operation) cannot escape outside of the motor enclosure. Air-conditioning compressors, pumps for nuclear reactors, vehicle traction drives, furnace fans, gasoline pumps, and many other unique applications utilize motors of special mechanical construction. In some of these applications the motor is so designed that it can be incorporated as an integral part of the unit it is driving. One example of this is a motor used to drive hermetic compressors for air-conditioning equipment. These motors are incorporated into the air-conditioning unit so that the refrigeration gas can be used to cool the motor windings and significantly upgrade the horsepower contained in a particular enclosure.

## 4.4 PERFORMANCE CHARACTERISTICS: SPEED-TORQUE AND SPEED-CURRENT CURVES

Figure 4.5 displays the speed-torque and speed-current curves for a polyphase induction motor and shows for these curves the three regions of major interest: motoring, plugging, and generating.

*Synchronous speed*, defined in Eq. (4.2) and designated as $N_s$ in Fig. 4.5, is the speed at which the rotor is revolving in synchronism with the rotating magnetic field generated by the stator windings, and there is therefore no rotor current and no electrically generated torque.

*Slip* is the difference in speed between the rotor and the airgap rotating magnetic field and is defined by

$$\text{Slip} = \frac{\text{synchronous r/min} - \text{rotor r/min}}{\text{synchronous r/min}} \tag{4.4}$$

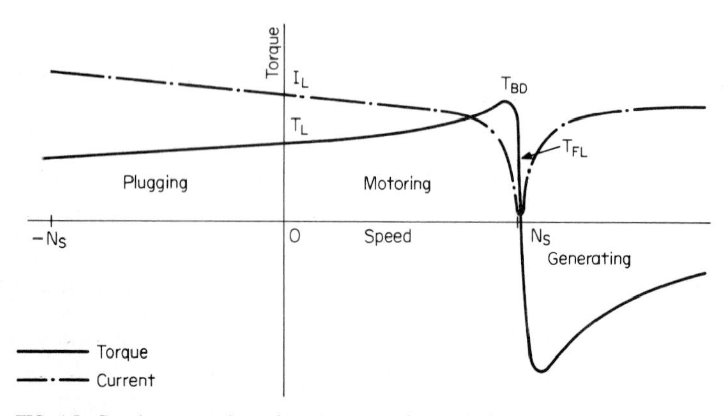


**FIG. 4.5** Speed-torque and speed-current curves for a polyphase induction motor.

Equation 4.4 yields slip as a per-unit value, although it is often expressed as a percentage.

*Breakdown torque* is the maximum torque that the motor generates in the motoring region; it is shown at the point $T_{BD}$ in Fig. 4.5. The locked-rotor torque $T_L$ in Fig. 4.5 is the torque produced by the motor at zero speed and is important because this is the torque that must overcome any breakaway forces imposed by the load. Normally, a motor operates in the region between $T_{BD}$ and synchronous speed, at the rated full-load torque indicated by $T_{FL}$ on the curve. Slips in the region between zero speed and synchronous speed $N_s$ range between 1.0 at zero speed and 0 at synchronous speed.

At speeds above synchronous speed the machine operates as an induction generator, and this region of the speed-torque curve is labeled "Generating" in Fig. 4.5. This region is also characterized by a maximum torque point, also called breakdown torque, and in general the breakdown torque in the generating region exceeds $T_{BD}$ in the motoring region. Slip values for speeds above synchronous speed are negative, as is evident from Eq. (4.4). Operation in the generating region results from overhauling loads or from the motor being driven by a prime mover. For induction generating action to occur, the machine must have a source of leading exciting current. This leading excitation can be supplied from the power system if the motor is connected to such a system, or it can be provided by capacitors of appropriate value connected across the motor terminals. In recent years, induction generators have found application as wind-driven generators; they are connected to a power line and deliver power to the system whenever wind velocities reach a certain minimum value.

The third region of the speed-torque curve displayed in Fig. 4.5 is the plugging region. A motor traverses this region when it is operating in one direction and then the direction of the rotating magnetic field is suddenly reversed. In Fig. 4.5 the plugging region extends from minus $N_s$ (slip = 2.0) to zero speed. Plugging is frequently used to accomplish fast speed reversal but is accompanied by high motor losses and large inrush currents.

Motor designs are classified by the type of speed-torque characteristic that they produce. NEMA has defined several design classifications for induction machines:

1. *Design A:* The locked-rotor (starting inrush) current exceeds the value specified for design B motors. The torque values, namely, locked-rotor and breakdown torques, will also usually exceed those specified for design B. Design A motors are used where greater-than-normal locked-rotor torques are required, even at the expense of increased inrush current. Also, a design A motor can be designed with better efficiency than a comparable design B machine, and this characteristic is sometimes used to improve motor efficiency.

2. *Design B:* Design B motors meet specified maximum values of locked-rotor current and minimum values of breakdown, locked-rotor, and pull-up torques.[1] General-purpose ac motors are design B machines. The speed-torque characteristics of design A, B, C, and D machines are displayed in Fig. 4.6. The slip for design B motors at rated load is usually 0.5 to 3.0 percent, and always less than 5 percent.

3. *Design C:* This type of design has a higher locked-rotor torque than design B. The locked-rotor current specifications are the same as for design B, and design C motors are used for applications where higher breakaway torques are encountered in getting the load started; conveyors are a typical application.

4. *Design D:* Design D motors are high-slip machines and have speed-torque characteristics distinctly different from the other design categories, as is evident from

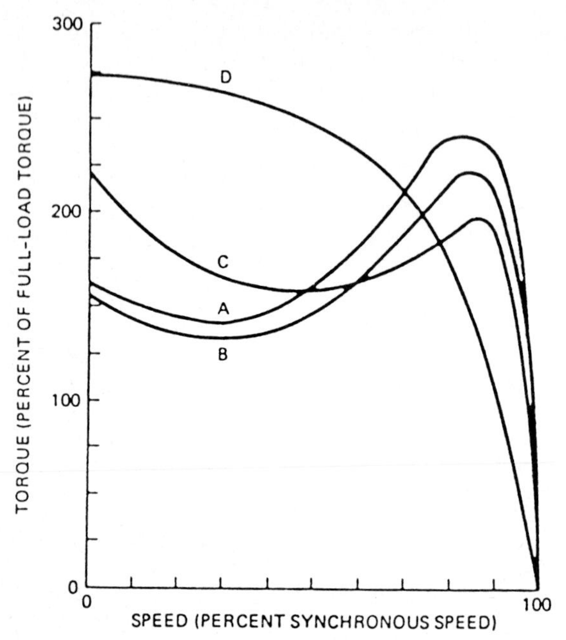


**FIG. 4.6** Speed-torque curves for motors with NEMA design A, B, C, and D characteristics. *(From NEMA Standards Publication MG10, Energy Management Guide for Selection and Use of Polyphase Motors, ©1983 by NEMA. Reprinted by permission of the National Electrical Manufacturers Association.)*

Fig. 4.6. Design D motors have higher locked-rotor torque values than design B machines, and these values are specified by the NEMA Standards.[1] They are used for applications where starting or accelerating the load is a more demanding feature of the application than is running under full load. Slips for design D motors at rated load equal or exceed 5 percent. Rotor losses for design D exceed those for design B motors; consequently, running efficiency is reduced. Design D motors are therefore usually not selected for loads to be driven continuously near the rated horsepower output of a machine. Typical applications for design D motors are punch presses, cranes, and hoists.

## 4.5 NEMA SPECIFICATIONS FOR MOTOR PERFORMANCE CHARACTERISTICS

NEMA Standards Publication MG1, "Motors and Generators," specifies many performance characteristics that facilitate the application of motors.[1] These standards specify performance levels that the users can expect so long as they apply a motor that meets NEMA standards.

Locked-rotor current is important inasmuch as it defines the loading that will be impressed upon the power system during the starting period. Table 4.1 specifies

**TABLE 4.1** Locked-Rotor Current of Three-Phase 60-Hz Integral-Horsepower Squirrel-Cage Induction Motors Rated at 230 V*

| Hp | Locked-rotor current, A | Design letters |
|---|---|---|
| 1/2 | 20 | B, D |
| 3/4 | 25 | B, D |
| 1 | 30 | B, D |
| 1 1/2 | 40 | B, D |
| 2 | 50 | B, D |
| 3 | 64 | B, C, D |
| 5 | 92 | B, C, D |
| 7 1/2 | 127 | B, C, D |
| 10 | 162 | B, C, D |
| 15 | 232 | B, C, D |
| 20 | 290 | B, C, D |
| 25 | 365 | B, C, D |
| 30 | 435 | B, C, D |
| 40 | 580 | B, C, D |
| 50 | 725 | B, C, D |
| 60 | 870 | B, C, D |
| 75 | 1085 | B, C, D |
| 100 | 1450 | B, C, D |
| 125 | 1815 | B, C, D |
| 150 | 2170 | B, C, D |
| 200 | 2900 | B, C |
| 250 | 3650 | B |
| 300 | 4400 | B |
| 350 | 5100 | B |
| 400 | 5800 | B |
| 450 | 6500 | B |
| 500 | 7250 | B |

* The locked-rotor current of single-speed three-phase constant-speed induction motors rated at 230 V, when measured with rated voltage and frequency impressed and with rotor locked, shall not exceed the following values.

***Source:*** NEMA Standards Publication MG1, *Motors and Generators*, ©1982 by NEMA. Reprinted by permission of the National Electrical Manufacturers Association.

maximum values of locked-rotor current as a function of horsepower for design B motors. For many of the ratings, design C and D motors must meet the same locked-rotor current as design B. The table gives the current in amperes at 230 V; locked-rotor current limits for other voltages can be calculated by

$$I_{L1} = \frac{230}{V_1} I_{L2} \tag{4.5}$$

where $V_1$ = voltage for which the locked-rotor current is desired
$I_{L1}$ = locked-rotor current corresponding to voltage $V_1$
$I_{L2}$ = locked-rotor current at 230 V

Table 4.2 specifies locked-rotor torque as a percentage of full-load torque for horsepowers ranging from ½ to 500 hp. Locked-rotor torque values are also a function of the motor's synchronous speed, as indicated in the table. The motor's locked-rotor torque must exceed the breakaway torque of the load for successful application.

**TABLE 4.2** Locked-Rotor Torque of Single-Speed Polyphase Squirrel-Cage Integral-Horsepower Motors with Continuous Ratings*

| | Synchronous speed, r/min | | | | | | | |
|---|---|---|---|---|---|---|---|---|
| | 60 Hz | 3600 | 1800 | 1200 | 900 | 720 | 600 | 514 |
| Hp | 50 Hz | 3000 | 1500 | 1000 | 750 | | | |
| ½ | | | | | 140 | 140 | 115 | 110 |
| ¾ | | | | 175 | 135 | 135 | 115 | 110 |
| 1 | | | 275 | 170 | 135 | 135 | 115 | 110 |
| 1½ | | 175 | 250 | 165 | 130 | 130 | 115 | 110 |
| 2 | | 170 | 235 | 160 | 130 | 125 | 115 | 110 |
| 3 | | 160 | 215 | 155 | 130 | 125 | 115 | 110 |
| 5 | | 150 | 185 | 150 | 130 | 125 | 115 | 110 |
| 7½ | | 140 | 175 | 150 | 125 | 120 | 115 | 110 |
| 10 | | 135 | 165 | 150 | 125 | 120 | 115 | 110 |
| 15 | | 130 | 160 | 140 | 125 | 120 | 115 | 110 |
| 20 | | 130 | 150 | 135 | 125 | 120 | 115 | 110 |
| 25 | | 130 | 150 | 135 | 125 | 120 | 115 | 110 |
| 30 | | 130 | 150 | 135 | 125 | 120 | 115 | 110 |
| 40 | | 125 | 140 | 135 | 125 | 120 | 115 | 110 |
| 50 | | 120 | 140 | 135 | 125 | 120 | 115 | 110 |
| 60 | | 120 | 140 | 135 | 125 | 120 | 115 | 110 |
| 75 | | 105 | 140 | 135 | 125 | 120 | 115 | 110 |
| 100 | | 105 | 125 | 125 | 125 | 120 | 115 | 110 |
| 125 | | 100 | 110 | 125 | 120 | 115 | 115 | 110 |
| 150 | | 100 | 110 | 120 | 120 | 115 | 115 | |
| 200 | | 100 | 100 | 120 | 120 | 115 | | |
| 250 | | 70 | 80 | 100 | 100 | | | |
| 300 | | 70 | 80 | 100 | | | | |
| 350 | | 70 | 80 | 100 | | | | |
| 400 | | 70 | 80 | | | | | |
| 450 | | 70 | 80 | | | | | |
| 500 | | 70 | 80 | | | | | |

* The locked-rotor torque of design A and B 60- and 50-Hz single-speed polyphase squirrel-cage motors, with rated voltage and frequency applied, shall be not less than the following values, which are expressed as a percentage of full-load torque. For applications involving higher torque requirements, see the locked-rotor torque values for design C and D motors.

*Source:* NEMA Standards Publication MG1, *Motors and Generators*, ©1982 by NEMA. Reprinted by permission of the National Electrical Manufacturers Association.

**TABLE 4.3** Breakdown Torque of Single-Speed Polyphase Squirrel-Cage Integral-Horsepower Motors with Continuous Ratings*

| | Synchronous speed, r/min | | | | | | | |
|---|---|---|---|---|---|---|---|---|
| | 60 Hz | 3600 | 1800 | 1200 | 900 | 720 | 600 | 514 |
| Hp | 50 Hz | 3000 | 1500 | 1000 | 750 | | | |
| 1/2 | | | | | 225 | 200 | 200 | 200 |
| 3/4 | | | | 275 | 220 | 200 | 200 | 200 |
| 1 | | | 300 | 265 | 215 | 200 | 200 | 200 |
| 1 1/2 | | 250 | 280 | 250 | 210 | 200 | 200 | 200 |
| 2 | | 240 | 270 | 240 | 210 | 200 | 200 | 200 |
| 3 | | 230 | 250 | 230 | 205 | 200 | 200 | 200 |
| 5 | | 215 | 225 | 215 | 205 | 200 | 200 | 200 |
| 7 1/2 | | 200 | 215 | 205 | 200 | 200 | 200 | 200 |
| 10–125, inclusive | | 200 | 200 | 200 | 200 | 200 | 200 | 200 |
| 150 | | 200 | 200 | 200 | 200 | 200 | 200 | |
| 200 | | 200 | 200 | 200 | 200 | 200 | | |
| 250 | | 175 | 175 | 175 | 175 | | | |
| 300–350 | | 175 | 175 | 175 | | | | |
| 400–500, inclusive | | 175 | 175 | | | | | |

* The breakdown torque of design A and B 60- and 50-Hz single-speed polyphase squirrel-cage motors, with rated voltage and frequency applied, shall be not less than the following values, which are expressed as a percentage of full-load torque.

***Source:*** NEMA Standards Publication MG1, *Motors and Generators*, ©1982 by NEMA. Reprinted by permission of the National Electrical Manufacturers Association.

Table 4.3 specifies the breakdown torque in a format similar to that of Table 4.2. Because the breakdown torque is the maximum torque that the motor can sustain, it is necessary to compare this value against the peak expected load torque for loads that are subject to periodic overload conditions. Of course, a motor cannot operate continuously at or near breakdown torque, and if this condition is anticipated, the motor temperature rise may exceed acceptable values and should be calculated or measured. Perhaps a higher-horsepower motor rating will be necessary.

## 4.6 EQUIVALENT CIRCUIT

The induction machine under steady-state conditions can be represented by an electrical network, and since electrical network solutions are readily handled by computers, this method affords an easy technique for analyzing machine performance.

Two versions of the equivalent circuit for a polyphase induction machine are given in Fig. 4.7. Figure 4.7*a* shows the equivalent circuit with a parallel branch representing the magnetizing reactance and iron-loss resistor. This implies that the iron loss varies as the square of the voltage impressed across $r_{FE}$, which, although not a completely accurate representation of the iron-loss phenomenon, is a sufficiently close approximation for use in most practical design calculations. Figure 4.7*b* is the same equivalent circuit shown in Fig. 4.7*a* except that the parallel branches $r_{FE}$

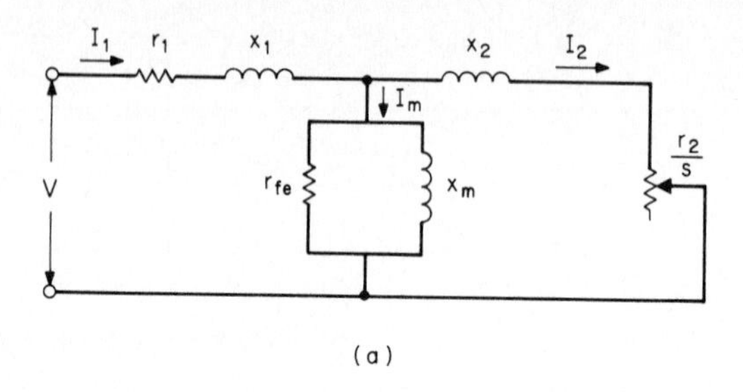


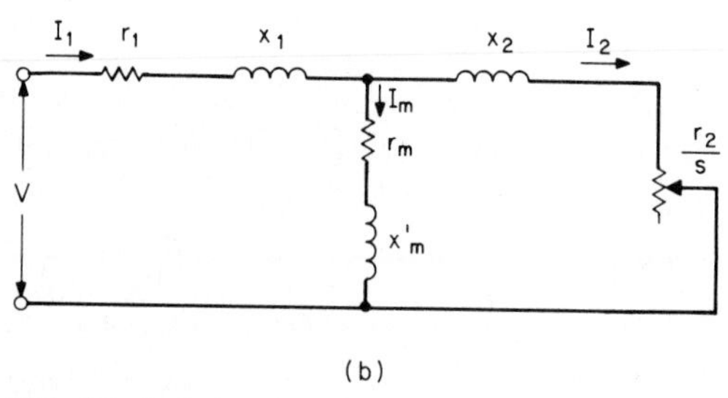


**FIG. 4.7** Equivalent circuits representing a polyphase induction motor: (*a*) equivalent circuit with a parallel branch for impedances representing iron loss and magnetizing reactance; (*b*) equivalent circuit with iron-loss and magnetizing impedances in series.

and $x_m$ have been converted to an equivalent series branch containing $r_m$ and $x'_m$. The conversion between the iron-loss and magnetizing-reactance parameters from a parallel to an equivalent series branch have been derived by Veinott and are given in the following equations:[4]

$$r_m = \frac{x_m^2}{r_{\text{FE}}} \frac{1}{1 + (x_m/r_{\text{FE}})^2} \tag{4.6}$$

$$x'_m = x_m \frac{1}{1 + (x_m/r_{\text{FE}})^2} \tag{4.7}$$

All of the constants in the equivalent circuit are line-to-neutral quantities based on a wye-connected machine. If the machine is delta-connected, then the phase quantities must be transformed as follows to an equivalent wye circuit in order to be used in the equivalent-circuit representation:

$$\mathbf{Z}_Y = \frac{\mathbf{Z}_\Delta}{3} \tag{4.8}$$

where $\mathbf{Z}_\Delta$ = complex impedance $(r + jx)$ in one phase of a delta-connected machine
$\mathbf{Z}_Y$ = impedance to use in equivalent-circuit representation of Fig. 4.7, which is based on a wye-connected machine

The circuit parameters appearing in the equivalent circuit are defined as follows:

$r_1$ = stator-winding resistance.

$r_2$ = rotor resistance referred to stator winding.

$x_1$, $x_2$ = leakage reactances of stator and rotor, respectively. The rotor leakage reactance is referred to the stator winding. Several leakage flux paths contribute to the total leakage reactance comprising $x_1$ and $x_2$. Veinott and Alger give equations for detailed calculations of these reactances.[4,5]

$r_{FE}$ = iron-loss resistance in parallel with the magnetizing reactance.

$r_m$ = iron-loss resistance in series with magnetizing reactance. Veinott gives the equation for calculating $r_m$ as[4]

$$r_m = \frac{\text{core loss}}{mI_o^2} \tag{4.9}$$

where $m$ = number of phases
$I_o$ = magnetizing current

$x_m$ = magnetizing reactance in the parallel branch circuit shown in Fig. 4.7*a*.

$x'_m$ = magnetizing reactance in the series magnetizing circuit of Fig. 4.7*b*.

Several operating conditions of interest can be analyzed by inserting the proper value of slip $s$ in the equivalent circuit:

**1.** *No-load:* Slip = 0, and $r_2/s$ is infinite. The secondary circuit is open and no rotor current flows. Therefore, exciting current flows only in the stator, and the losses are the primary $I^2R$ core losses, plus friction and windage.

**2.** *Locked-rotor:* Slip = 1.0. Under this condition the branch circuit comprising $x_2$ and $r_2$ has an impedance that is relatively low compared to the magnetizing reactance, and $I_1 \simeq I_2$.

**3.** *Full load:* For most induction machines, the full-load operating range is between 0.005 and 0.03; therefore, $r_2/s$ ranges between 33 and 200 times the value that this resistor assumes during locked-rotor conditions. The secondary circuit is primarily resistive, and most of the power entering the machine is dissipated in the $r_2/s$ resistor.

**4.** *Breakdown torque:* The slip at breakdown torque is given by

$$S_m \simeq \frac{r_2}{\sqrt{r_1^2 + (x_1 + x_2)^2}} \tag{4.10}$$

This relationship follows from the maximum-power transfer theorem, which states that maximum power will be transferred to the load resistor when it is equal to the impedance of the circuit looking back from the load resistors with the input terminal short-circuited. Equation (4.10) is an approximation but gives a value acceptably close to the actual slip at maximum torque for most calculations.

**5.** *Torque and power output:* The loss in the $r_2/s$ resistor of the equivalent circuit is equal to the output torque expressed in synchronous watts.[4,6] The conversion between synchronous watts and lb·ft or oz·ft is given by

$$T = \frac{k}{\text{syn. r/min}} \frac{mI_2^2 r_2}{s} \tag{4.11}$$

where $T$ = torque
$m$ = number of phases
$k$ = 7.04 for torque in lb/ft, or 112.7 for torque in in/lb

and where all other quantities are defined in the equivalent circuit of Fig. 4.7.

The secondary $I^2R$ losses can be calculated from the total losses in the $r_2/s$ resistor by Eqs. (4.12) and (4.13):

$$\text{Secondary } I^2R \text{ loss} = \text{secondary input} \times \text{slip} \tag{4.12}$$

$$\text{Secondary } I^2R \text{ loss} = mI_2^2r_2 \tag{4.13}$$

## 4.7 CALCULATING METHOD FOR SOLVING THE EQUIVALENT CIRCUIT

We now consider an example of determining motor performance from the equivalent circuit. Table 4.4 gives the parameters for a typical 10-hp four-pole three-phase motor. The circuit resistances and reactances listed in the table correspond to the circuit elements in Fig. 4.7*b*. Additionally, two components of loss needed for the efficiency calculation—namely, friction and windage and stray-load loss—are included in the table since they do not result directly from the equivalent-circuit solution.

Using equations given by Veinott, the currents in the three branches of the equivalent circuit can be calculated.[4] The secondary-circuit impedance $Z_2$ depends upon slip $s$. Solving the equivalent circuit is an iterative process: A value of $s$ is assumed and the circuit solution performed; if the desired output is not obtained, a new value of $s$ is assumed and the circuit solution repeated until the desired output

**TABLE 4.4** 10-Hp Motor Parameters for Example Calculation

| Item | Value |
|---|---|
| Horsepower | 10 |
| Voltage, V | 460 |
| Phases | 3 |
| Frequency, Hz | 60 |
| Per-unit base power, kW | 7.460 |
| Per-unit base voltage, V | 265.581 |
| Per-unit base current, A | 9.363 |
| Per-unit base impedance, Ω | 28.365 |
| $r_{1,hot}$, pu | 0.02633 |
| $r_{2,hot}$, pu | 0.01862 |
| $x_1$, pu | 0.08406 |
| $x_2$, pu | 0.10936 |
| $x'_m$, pu | 2.46737 |
| $r_m$, pu | 0.16195 |
| Friction and windage, kW | 0.074 |
| Stray-load loss, kW | 0.097 |

is obtained. For the example calculation below, a value of $s$ that yields the rated output of the machine, 10 hp, was used.

The following equivalent-circuit calculations are on a per-unit basis, and the per-unit (pu) base quantities are given in Table 4.4:

$$\mathbf{I}_1 = \frac{\mathbf{V}(\mathbf{Z}_m + \mathbf{Z}_2)}{\mathbf{Z}_1\mathbf{Z}_m + \mathbf{Z}_1\mathbf{Z}_2 + \mathbf{Z}_2\mathbf{Z}_m} \tag{4.14}$$

$$\mathbf{I}_2 = \frac{\mathbf{V}\mathbf{Z}_m}{\mathbf{Z}_1\mathbf{Z}_m + \mathbf{Z}_1\mathbf{Z}_2 + \mathbf{Z}_2\mathbf{Z}_m} \tag{4.15}$$

$$\mathbf{I}_m = \frac{\mathbf{V}\mathbf{Z}_2}{\mathbf{Z}_1\mathbf{Z}_m + \mathbf{Z}_1\mathbf{Z}_2 + \mathbf{Z}_2\mathbf{Z}_m} \tag{4.16}$$

where $s = 0.02339$
$\mathbf{V} = 1.0$
$\mathbf{Z}_1 = 0.02633 + j0.08406$
$\mathbf{Z}_2 = 0.01862/0.02339 + j0.10936$
$\mathbf{Z}_m = 0.16195 + j2.46737$

Substituting these numerical values into Eqs. (4.14) to (4.16) yields the following branch-circuit currents:

$$\mathbf{I}_1 \text{ pu} = 1.275\underline{/-29.261^\circ}$$

or

$$\mathbf{I}_1 = 11.94 \text{ A}$$

$$\mathbf{I}_2 \text{ pu} = 1.147\underline{/-12.620^\circ}$$

or

$$\mathbf{I}_2 = 10.74 \text{ A}$$

$$\mathbf{I}_m \text{ pu} = 0.373\underline{/-91.043^\circ}$$

or

$$\mathbf{I}_m = 3.49 \text{ A}$$

The calculations of input power, output power, torque, and losses are performed using current, voltage, and impedances expressed as amperes, volts, and ohms where these quantities can be obtained from the per-unit values by multiplying by the appropriate per-unit base. For example,

$$\begin{aligned} I_1 &= I_1 \text{ pu} \times \text{per-unit base current} \\ &= 1.275 \times 9.363 \\ &= 11.94\ A \end{aligned} \tag{4.17}$$

$$\begin{aligned} \text{Secondary input} &= mI_2^2 \frac{r_2}{s} \\ &= \frac{3 \times 10.74^2}{1000} \frac{0.01862 \times 28.365}{0.02339} \\ &= 7.814 \text{ kW} \end{aligned} \tag{4.18}$$

where $m$ = number of phases

$$\text{Secondary } I^2R \text{ loss} = mI_2^2 r_2$$

$$= \frac{3 \times 10.74^2 \times 0.01862 \times 28.365}{1000}$$

$$= 0.183 \text{ kW} \tag{4.19}$$

$$\text{Output} = \text{secondary input} - (\text{secondary } I^2R \text{ loss}$$

$$+ \text{friction and windage} + \text{stray-load loss})$$

$$= 7.814 - (0.183 + 0.074 + 0.097)$$

$$= 7.460 \text{ kW} \tag{4.20}$$

$$\text{Output power} = \frac{7.460 \times 1000}{746}$$

$$= 10 \text{ hp}$$

$$\text{Primary } I^2R \text{ loss} = mI_1^2 r_1$$

$$= \frac{3 \times 11.94^2 \times 0.02633 \times 28.365}{1000}$$

$$= 0.319 \text{ kW} \tag{4.21}$$

$$\text{Iron loss} = mI_m^2 r_m$$

$$= \frac{3 \times 3.49^2 \times 0.16195 \times 28.365}{1000}$$

$$= 0.168 \text{ kW} \tag{4.22}$$

$$\text{Total loss} = \text{primary } I^2R + \text{secondary } I^2R + \text{iron}$$

$$+ \text{friction and windage} + \text{stray-load loss}$$

$$= 0.319 + 0.183 + 0.168 + 0.074 + 0.097$$

$$= 0.841 \text{ kW} \tag{4.23}$$

$$\text{Input power} = \text{output power} + \text{total losses}$$

$$= 7.460 + 0.841$$

$$= 8.301 \text{ kW} \tag{4.24}$$

$$\text{Efficiency} = \frac{\text{output power}}{\text{input power}}$$

$$= \frac{7.460}{8.301} \times 100$$

$$= 89.9\% \tag{4.25}$$

$$\text{Power factor} = \text{cosine of angle of } I_1$$

$$= \cos\,(-29.261^\circ)$$

$$= 87.2\% \tag{4.26}$$

$$\text{Speed} = \text{synchronous r/min } (1 - s)$$
$$= 1800(1 - 0.02339)$$
$$= 1757.9 \text{ r/min} \tag{4.27}$$

$$\text{Torque} = \frac{\text{horsepower output} \times 5252}{\text{speed}}$$
$$= \frac{10 \times 5252}{1757.9}$$
$$= 29.88 \text{ lb} \cdot \text{ft} \tag{4.28}$$

## 4.8 TEST PROCEDURES FOR INDUCTION MACHINES

The U.S. testing standard for induction machines is IEEE 112.[6a] This standard generally describes how tests are conducted but does not provide values that the machine must meet. The purpose of the IEEE test standard is to describe testing *procedures*; NEMA and other industry standards specify the test *results* that must be met if a motor is to qualify to meet a specific standard. For example, NEMA specifies the maximum current for a design B motor under locked-rotor conditions (as tabulated in Table 4.1), and IEEE 1126a describes how to conduct the locked-rotor test. Many types of tests are described in IEEE 112, a few of which are:

- Performance tests to determine efficiency and power factor.
- Tests for measuring speed-torque and speed-current characteristics.
- No-load test.
- Locked-rotor test.
- Temperature tests to determine the temperature rise of the machine operating under load.
- Insulation resistance.
- High-potential test.
- Shaft current and bearing insulation test.
- Vibration tests.
- Noise tests: IEEE 112 refers to IEEE Standard 85 for noise measurements.[6a]

There are numerous international testing standards for rotating machinery, but two particular ones are of some interest in the United States, both from the standpoint of international trade and because one of the international standards, namely JEC37, has found some limited use in the United States. These two international standards are the International Electrotechnical Commission Standard IEC 34-2 and the Japanese Standard JEC37.[7,8] The References at the end of this chapter list the publishers of these standards, and several authors have given a comparison of test results obtained by IEEE, IEC, and JEC test methods.[9,10]

## Load Tests and Efficiency

IEEE 112 provides five different methods for measuring motor efficiency:[6a]

**1.** *Method A—brake:* A mechanical brake is used as the means of loading the motor. The brake must have a means for adjusting the torque load to the desired value and also a means of measuring the torque. This test is generally used for fractional-horsepower motors because of the heat generated by the brake.

**2.** *Method B—dynamometer:* The output of the motor is loaded into a dynamometer that is capable of being adjusted to the desired load torque. The output torque is measured by a scale, load cell, or strain gage.

**3.** *Method C—duplicate machines:* This method of measuring efficiency may be used only when duplicate machines are available. The two machines are mechanically coupled together and electrically connected to two separate sources of power, one of which must be an adjustable-frequency power source. By controlling the adjustable-frequency source to operate at frequencies above and below the fixed-frequency source, each machine is caused to operate as both a motor and a generator during the test. All the power measurements are electrical, and the losses of the two machines can be determined by the difference between input and output power from the system. Segregated loss calculations can be made as described under method C of IEEE 112, and the efficiencies of the machines operating both as a motor and a generator can then be calculated.

**4.** *Method E—input measurement:* To conduct this test, the machine must be coupled to a variable load, but the load output power does not need to be measured. Operating the test equipment at varying loads provides sufficient test information for the individual losses to be segregated and the motor efficiency calculated. Since the total losses cannot be determined by this method, a separate test must be conducted to measure one component of loss, namely, stray-load loss. A reverse-rotation test and a rotor-removed test are necessary to obtain the stray-load loss measurement. As an alternative method, NEMA permits an assumed value of stray-load loss to be used.[1] This assumed value is based on horsepower size; it is 1.2 percent of rated output for motor sizes of 1500 hp and smaller and 0.9 percent of rated output for 1500 hp and larger.

**5.** *Method F—equivalent circuit:* When tests under load are not made, motor efficiency can be determined by constructing the equivalent circuit based on no-load, locked-rotor, and impedance tests. Analysis of data from these three tests provides the equivalent-circuit constants. The efficiency and power factor can then be calculated from the equivalent circuit as described in Sec. 4.7.

## Tests at No-load

The motor is operated at rated voltage and frequency without any load on the shaft, and readings of power, voltage, and current are obtained. The no-load input readings may vary until the friction losses in the bearings have stabilized; a run-in period may be required to stabilize these losses.

The iron loss and the friction and windage loss can be obtained by conducting no-load tests at a series of voltages varying from 125 percent of rated voltage down to a voltage at which the current increases with further voltage reduction. A typical curve of power and current versus voltage is displayed in Fig. 4.8. The friction and

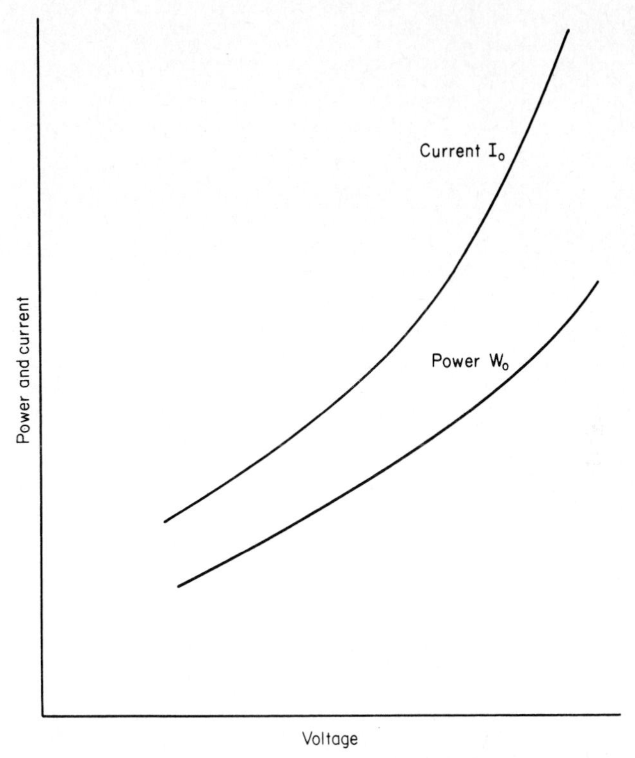


**FIG. 4.8** No-load power and current curves for an induction motor.

windage loss can be obtained by plotting the power input readings from this test and extending the curve back to 0 V. Greater accuracy is achieved by plotting the power input against the voltage squared for the lower voltage values, as is shown in Fig. 4.9.

The iron loss can be calculated from the no-load test by

$$W_{\mathrm{FE}} = W_o - KI_o^2 r_1 - W_{\mathrm{FW}} \tag{4.29}$$

where $W_o$ = input power
$I_o$ = current
$r_1$ = stator-winding resistance
$K$ = conversion factor to obtain $I_o^2 r_1$ for all three phases, 1.5 if $r_1$ is line-to-line resistance and winding is wye- or delta-connected
$W_{\mathrm{FW}}$ = friction and windage loss

## Segregation of Losses

Now that $W_{\mathrm{FW}}$ and $W_{\mathrm{FE}}$ have been determined from the no-load test, the remaining three components of loss can be calculated by using data taken from the efficiency tests described earlier in this section. The detailed steps involved in calculating

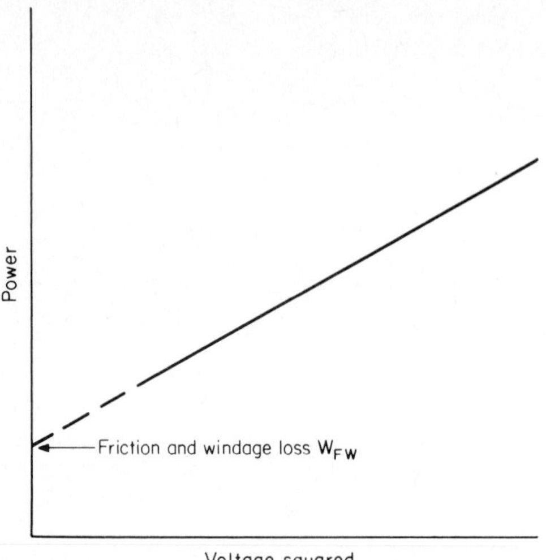


**FIG. 4.9** Determination of friction and windage loss from no-load saturation test. *(From H. E. Jordan, Energy Efficient Electric Motors and Their Applications, Van Nostrand Reinhold, New York, 1983.)*

the individual losses differ somewhat between the different test methods (B, C, etc.), but conceptually the methods are quite similar. Using method B as the example, a general description of the technique is given in the following paragraphs. Method B, dynamometer, including a regression-curve fit for correcting stray-load losses, is recommended by NEMA for efficiency measurements in the medium-size ac motors, 1 to 125 hp.[1]

The stator copper loss is calculated by

$$W_{CU} = KI_1^2 r_1 \tag{4.30}$$

where $I_1$ = stator current at the load point
$r_1$ = stator-winding resistance
$K$ = conversion factor to obtain $I_1^2 r_1$ for all three phases, 1.5 if $r_1$ is line-to-line resistance and winding is wye- or delta-connected

The secondary $I^2R$ loss is calculated by

$$W_{SEC} = (W_1 - W_{FE} - W_{CU})s \tag{4.31}$$

where $W_1$ = total watts input at the load point
$s$ = slip in per-unit

The stray-load loss is the remaining portion of the total losses after the other four losses ($W_{FW}$, $W_{FE}$, $W_{CU}$, and $W_{SEC}$) have been accounted for. The equation for calculating stray-load loss is

$$W'_{LL} = (W_1 - P_o) - (W_{FE} + W_{FW} + W_{CU} + W_{SEC}) \tag{4.32}$$

where $P_o$ = mechanical output power of the motor

Now $W'_{LL}$ contains not only the true stray-load loss but measurement errors as well. A technique for minimizing the effect of the random measurement errors on the stray-load-loss value is described in NEMA Standard MG1. This stray-load-loss correction technique requires that test data for several load points be available; therefore, load tests are taken from 150 percent load down to 25 percent load in six equal increments. The stray-load losses are plotted as shown in Fig. 4.10, and a straight line is then mathematically fit to these test-data points by using a linear regression-curve fit. The equation of the straight line is

$$W_{LL} = AT^2 + B \tag{4.33}$$

where $A$ = slope of straight line resulting from linear regression-curve fit
$B$ = intercept on $Y$ axis
$T$ = torque

Since the stray-load loss should be zero at zero load, the line is shifted to intercept the origin, and the corrected stray-load loss is calculated from

$$W_{LLC} = AT^2 \tag{4.34}$$

Combining the stray-load loss calculated from Eq. (4.34) with the other losses, the motor efficiency can be calculated by

$$\text{Efficiency} = \frac{W_1 - (W_{LLC} + W_{CU} + W_{FE} + W_{FW} + W_{SEC})}{W_1} \tag{4.35}$$

Temperature and dynamometer corrections need to be included in the efficiency calculations to obtain accurate results. The detailed steps and equations for performing the calculation are given in NEMA Standard MG1.[1] A sample of a completed test analysis is displayed in Fig. 4.11.

## Locked-Rotor and Speed-Torque Tests

Tests are taken with the rotor locked to determine the starting torque and current. On small machines, the torque and current can be measured directly with

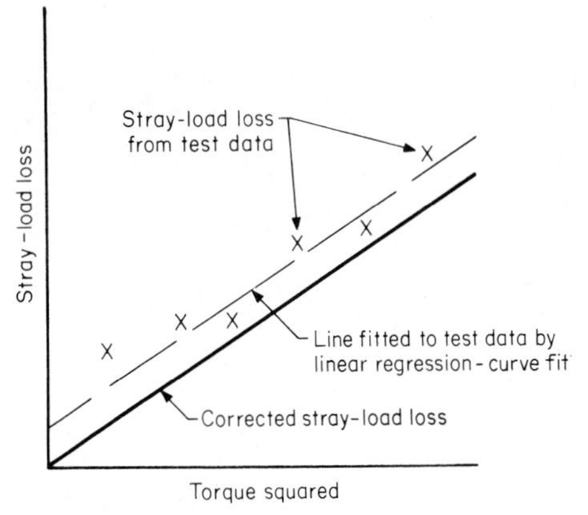


**FIG. 4.10** Stray-load loss correction by linear regression-curve fit.

Type P Design B Frame 215T Hp 10 Phase 3
Frequency 60 Volts 460 Synchronous r/min 3600 Serial No. P21G33
Degrees C Temperature Rise 80 Time Rating Continuous Model No.

Stator Winding resistance Between Terminals 1.376 Ohms @ 23 °C
Specified Temperature For Resistance Correction (ts) = 83.2 (See MG 1-12.53.a)

| Item | Description | 1 | 2 | 3 | 4 | 5 | 6 |
|---|---|---|---|---|---|---|---|
| 1 | Ambient Temperature, °C | 25.2 | 25.2 | 25.2 | 25.2 | 25.2 | 25.2 |
| 2 | ($t_t$) Stator Winding Temperature, °C | 71 | 71 | 70 | 70 | 70 | 67 |
| 3 | Slip, r/min | 154 | 125 | 96 | 72 | 47 | 23 |
| 4 | Speed, r/min | 3440 | 3470 | 3500 | 3525 | 3551 | 3576 |
| 5 | Line Voltage, volts | 460 | 460 | 460 | 460 | 460 | 460 |
| 6 | Line Current, amperes | 17.13 | 14.17 | 11.37 | 8.68 | 6.17 | 3.88 |
| 7 | Power Input, watts | 12770 | 10574 | 8350 | 6270 | 4250 | 2277 |
| 8 | Core Loss, watts | 61 | 63 | 65 | 67 | 68 | 70 |
| 9 | Stator $I^2R$ Loss, watts, at ($t_t$) °C | 718 | 492 | 315 | 184 | 93 | 36 |
| 10 | Power Input to Rotor, watts | 11991 | 10019 | 7970 | 6019 | 4089 | 2171 |
| 11 | Rotor $I^2R$ Loss, watts | 513 | 348 | 212 | 120 | 53 | 14 |
| 12 | Friction and Windage Loss, watts | 174 | 174 | 174 | 174 | 174 | 174 |
| 13 | Total Conventional Loss, watts | 1466 | 1077 | 766 | 545 | 388 | 294 |
| 14 | Torque, lb.ft * | 22.45 | 18.72 | 14.98 | 11.24 | 7.49 | 3.75 |
| 15 | Dynamometer Correction lb.ft * | 0 | 0 | 0 | 0 | 0 | 0 |
| 16 | Corrected Torque lb.ft * | 22.45 | 18.72 | 14.98 | 11.24 | 7.49 | 3.75 |
| 17 | Power Output, watts | 10915 | 9223 | 7444 | 5626 | 3776 | 1904 |
| 18 | Apparent Total Loss, watts | 1805 | 1351 | 906 | 644 | 474 | 373 |
| 19 | Stray-Load Loss, watts | 339 | 274 | 140 | 99 | 86 | 79 |
| Intercept 47.63369 | Slope 0.57333 Corr. Factor 0.97 | Point Deleted 0 | | | | | |
| 20 | Stator $I^2R$ Loss, watts, at ($t_s$) °C | 747 | 511 | 329 | 192 | 97 | 38 |
| 21 | Corrected Power Input to Rotor, watts | 11962 | 10000 | 7956 | 6011 | 4085 | 2169 |
| 22 | Corrected Slip, r/min | 160 | 130 | 100 | 75 | 49 | 24 |
| 23 | Corrected Speed, r/min | 3440 | 3470 | 3500 | 3525 | 3551 | 3576 |
| 24 | Rotor $I^2R$ Loss, watts, at ($t_s$) °C | 532 | 361 | 221 | 125 | 56 | 14 |
| 25 | Corrected Stray-Load Loss, watts | 289 | 201 | 129 | 72 | 32 | 8 |
| 26 | Corrected Total Loss, watts | 1803 | 1310 | 918 | 630 | 427 | 304 |
| 27 | Corrected Power Output, watts | 10967 | 9264 | 7432 | 5640 | 3823 | 1973 |
| 28 | Power Output, hp | 14.7 | 12.4 | 9.96 | 7.56 | 5.12 | 2.64 |
| 29 | Efficiency, percent | 85.9 | 87.6 | 89.0 | 90.0 | 90.0 | 86.6 |
| 30 | Power Factor, percent | 93.6 | 93.7 | 92.2 | 90.7 | 86.5 | 73.7 |

$t_t$ = Temperature of stator winding as determined from stator resistance or temperature detector during test.
* Indicate Torque Units as N·m or lb·ft.

**FIG. 4.11** Calculation form for input-output test of induction machine, with segregation of losses and smoothing of stray-load loss. *(The form used is from NEMA Standards Publication MG1, Motors and Generators, ©1982 by NEMA. Reprinted by permission of the National Electrical Manufacturers Association.)*

rated voltage and frequency applied. On large machines, the current and torque values obtained under locked-rotor conditions are frequently too large either for the system supplying the power to the motor or for the measuring equipment; in these situations, it is necessary to take readings at a reduced voltage and to calculate the locked-rotor torque and current to be expected at rated voltage. If saturation is neglected, the locked-rotor torque varies as the square of the voltage and the current varies as the first power of the voltage; however, many machines are operated in a sufficiently saturated region that the theoretical square and first-power laws for torque and current require modification. One technique for obtaining a more accurate relationship is to use a ratio of the measured torque, current, and voltage values from the highest two voltage readings in order to calculate the correct exponent. This is illustrated in Eqs. (4.36) and (4.38) for calculating the exponent in the torque relationship. A similar set of equations for obtaining the exponent applicable to current measurements is given in Eqs. (4.37) and (4.39).

$$\frac{T_1}{T_2} = \left(\frac{V_1}{V_2}\right)^x \tag{4.36}$$

$$\frac{I_1}{I_2} = \left(\frac{V_1}{V_2}\right)^y \tag{4.37}$$

where $T_1$, $T_2$ = torques measured at voltages $V_1$ and $V_2$, respectively; subscript 1 denotes values obtained from the higher of the two voltage readings
$I_1$, $I_2$ = currents measured at voltages $V_1$ and $V_2$
$x$, $y$ = exponents

Solving Eqs. (4.36) and (4.37) for $x$ and $y$, respectively, yields

$$x = \frac{\log (T_1/T_2)}{\log (V_1/V_2)} \tag{4.38}$$

$$y = \frac{\log (I_1/I_2)}{\log (V_1/V_2)} \tag{4.39}$$

The exponents from (4.38) and (4.39) are used to extend the values of torque and current measured at the highest test voltage to the expected torque and current at rated voltage:

$$T_R = T_1 \left(\frac{V_R}{V_1}\right)^x \tag{4.40}$$

$$I_R = I_1 \left(\frac{V_R}{V_1}\right)^y \tag{4.41}$$

where $T_R$, $I_R$ = expected values of torque and current at rated voltage
$V_R$ = rated voltage

Other methods of extrapolating the test data are acceptable, such as plotting curves of torque and current versus voltage and extrapolating the curves.

Since locked-rotor torque and current are a function of temperature, it is desirable to obtain the test at the lowest possible temperature, which means that the readings must be taken quickly after the application of power. IEEE 112 requires that the readings for any point shall be taken within 5 seconds (s) after power is applied for motors that are rated 10 hp and below and within 10 s for all motors above a 10-hp rating.

The *speed-torque curve* of a motor is the relationship between torque and speed and is usually plotted in the range from zero to synchronous speed. Similarly, the *speed-current* relationship is the curve of current versus speed for the same speed region.

Three methods are described in IEEE 112 for obtaining speed-torque and speed-current curves: (1) measured output, (2) acceleration, and (3) input. The measured-output method utilizes a dc generator or dynamometer to measure the torque output of the motor at various speeds. The acceleration method determines torque by measuring rate of acceleration and applying the equation

$$T = J\left(\frac{dn}{dt}\right) \tag{4.42}$$

where $T$ = torque
$J$ = polar moment in the inertia of the rotating parts
$dn/dt$ = acceleration

The input method calculates the output torque from measurements of input power, current, and speed. Losses at each speed are calculated in a manner similar to the techniques described earlier in this section. These losses are subtracted from the input power to determine the net output power. A specific procedure for determining the losses is described in IEEE 112.

The speed-torque and speed-current tests are subject to the same power and temperature limitations described under the locked-rotor test and are therefore frequently run at reduced voltage. The method described under the locked-rotor test for extending the data to rated voltage is also applicable to the speed-torque and speed-current tests.

## Determination of Equivalent-Circuit Constants from Tests

Test verification of the circuit constants is useful in helping the designer improve his or her calculating procedure. In addition, one of the accepted testing methods for determining efficiency—namely, method F, equivalent circuit—depends upon experimental determination of the equivalent-circuit constants.

Figure 4.12 is the circuit configuration used when circuit constants are to be determined from tests. This figure is the same as Fig. 4.7, except that the iron-loss resistor has been omitted.

Veinott gives a set of simple relationships for obtaining the equivalent-circuit constants based on the circuit of Fig. 4.12.[4] Only a no-load test and a locked-rotor test are needed. The required test data are:

$V_o$ = voltage per phase, V, no-load
$I_o$ = line current, A, no-load
$V_L$ = voltage per phase, V, locked-rotor
$I_L$ = line current, A, locked-rotor
$W_L$ = power per phase, W, locked-rotor

Using these measured quantities, the equivalent-circuit constants can be calculated:

$$X_o = \frac{V_o}{I_o} \tag{4.43}$$

$$X_M = X_o - x_1 \tag{4.44}$$

$$r_2 = \frac{W_L}{I_L^2} - r_1 \tag{4.45}$$

$$x_1 + x_2 = \sqrt{\left(\frac{V_L}{I_L}\right)^2 - \left(\frac{W_L}{I_L}\right)^2} \tag{4.46}$$

Equation (4.44) requires that the value of $x_1$ be known in order to determine $X_M$. However, only the quantity $x_1 + x_2$ can be determined from the locked-rotor test

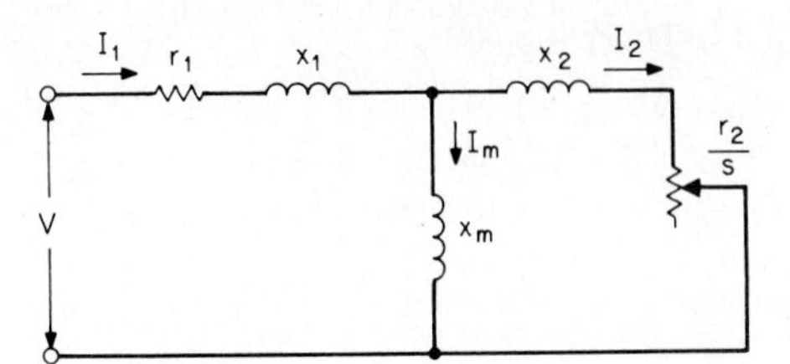


**FIG. 4.12** Induction motor equivalent circuit for calculating circuit constants from test data.

data when using Eq. (4.46). IEEE 112 recommends the following:[6a] When design details are available, use the calculated $x_1/x_2$; otherwise, use $x_1/x_2 = 1.0$ for design A and D motors and wound-rotor motors, $x_1/x_2 = 0.67$ for design B motors, and $x_1/x_2 = 0.43$ for design C motors.

Equations (4.43) to (4.46) are easy to apply and provide a quick method for determining the equivalent-circuit constants once test data are available. However, this method suffers from accuracy limitations arising primarily from three sources:

1. Equations (4.45) and (4.46) assume that $X_M >> x_1 + x_2$ and therefore that the current flowing in the magnetizing branch during a locked-rotor test is negligible.
2. Since $x_1$ and $x_2$ are affected by saturation in the leakage flux paths and since the current during a locked-rotor test is usually many times the full-load current, the values of $x_1 + x_2$ obtained from Eq. (4.46) will be too low for equivalent-circuit calculations in the region of full load.
3. Machines of 1 hp and above will usually exhibit a nonuniform current density in the rotor bars under locked-rotor test. Therefore, a higher value of $r_2$ will be obtained from Eq. (4.45) than would be experienced during normal running conditions when the rotor slip frequency is relatively low.

To compensate for these phenomena, IEEE 112 specifies a more extensive method for evaluating the equivalent-circuit constants than using Eqs. (4.43) to (4.46) when these constants are to be used in method F for efficiency determination. The technique requires two additional tests:

1. *Low-frequency test:* A locked-rotor test is taken at a reduced frequency, which is specified as a maximum of 25 percent of rated frequency at rated current. The use of reduced frequency and rated current provides a better approximation for the values of $x_1$ and $x_2$ in terms of saturation and current-density distribution.
2. *Load test with measured values for slip, current, and input power:* A procedure is specified in IEEE 112 whereby the equivalent circuit is solved iteratively until a value of $r_2/s$ is found such that the measured input power equals the calculated power. Then, by multiplying $r_2/s$ by the measured slip, the value of $r_2$ is determined.

When efficiency measurements are being made by method F, the procedure in IEEE 112 should be followed rigorously. For example, if a guaranteed efficiency is being verified using method F, then IEEE 112 should be used in its entirety. On the other hand, the simplified method using Eqs. (4.43) to (4.46) is useful for a rapid determination of the equivalent-circuit constants for making design checks and problem diagnosis.

## 4.9 MOTOR APPLICATIONS

Many motors are applied in situations in which either the speed or the load varies. This section discusses calculating methods that are useful in applying motors under duty cycle conditions. The term *duty cycle operation* is generally used to refer to an operating condition under which the motor is either accelerating or decelerating or the load is varying during a significant part of the operating period.

### Acceleration

In applications where the motor must accelerate or decelerate frequently, selection of the motor rating must be made taking into account the temperature rise, which has to be within safe limits for the motor's class of insulation.

One of the most commonly encountered conditions is a requirement to accelerate a high-inertia load. The motor's acceleration losses and their contribution to the motor's temperature rise must be determined in order to ensure a successful application.

Considering the case when acceleration takes place with no appreciable retarding torque exerted by the load, the energy dissipated in the rotor is given by

$$W_R = 0.231 Wk^2 N_s^2 (S_i^2 - S_f^2) \times 10^{-6} \tag{4.47}$$

where $W_R$ = secondary $I^2R$ loss, kWs
$Wk^2$ = load + motor polar moment of inertia, lb·ft$^2$
$N_s$ = synchronous speed, r/min
$S_i$ = initial slip, per unit
$S_f$ = final slip, per unit

It is interesting to note that none of the parameters in Eq. (4.47) is motor-design-dependent. The equation is applicable regardless of the particular motor being used, so long as it is a polyphase induction motor. The concept is illustrated in a diagram published by Cook and displayed in Fig. 4.13.[11] If the motor accelerates from zero to synchronous speed, the rotor loss (the area below the diagonal *OB*) is equal to the stored energy in the rotating mass (the area above *OB*). For accelerations from zero up to any other speed less than synchronous speed, this equality is not true. In

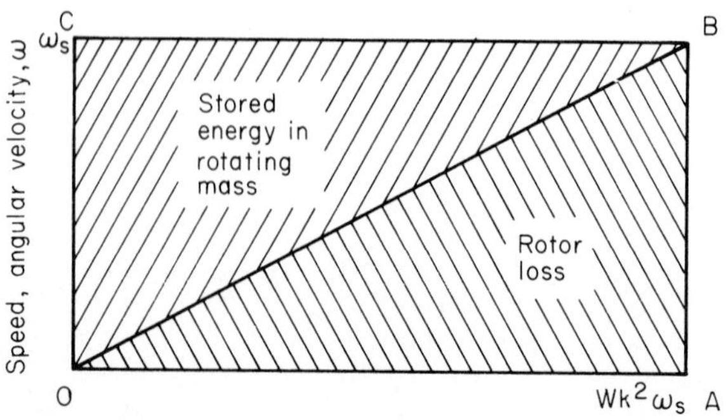


**FIG. 4.13** Relation of rotor loss to stored energy in load for acceleration.

fact, the ratio of rotor energy loss to energy stored in the rotating mass is greater than unity.

For the frequently occurring case when acceleration is from standstill to nearly synchronous speed,

$$S_i^2 - S_f^2 \simeq 1.0 \tag{4.48}$$

and Eq. (6.1) simplifies to

$$W_R = 0.231 W k^2 N_s^2 \times 10^{-6} \tag{4.49}$$

## Deceleration

Many motor applications require frequent starting, stopping, and reversing. Equation (4.47) can be modified to cover all three cases, namely, acceleration, plug stop, and plug reversal:

For acceleration: $S_i = 1.0$ and $S_f = 0$

$$S_i^2 - S_f^2 = 1.0$$

For a plug stop: $S_i = 2.0$ and $S_f = 1.0$

$$S_i^2 - S_f^2 = 3.0$$

For a plug reversal: $S_i = 2.0$ and $S_f = 0$

$$S_i^2 - S_f^2 = 4.0$$

From the tabulation above, it is evident that a plug stop generates 3 times as many losses as an acceleration and that a plug reversal generates 4 times as many losses. Clearly, either plug stopping or plug reversing produces much greater thermal stress on the motor than does accelerating from standstill.

## Stator Losses

The stator $I^2R$ losses are also a significant component of the losses during duty cycle operation. Consider the circuit in Fig. 4.12 and recognize that during the speed-changing conditions occurring in duty cycle operation, the stator and rotor $I^2R$ losses greatly exceed the other components of loss. Therefore, with little loss in accuracy, the total losses can be calculated simply as the sum of the stator $I^2R$ and rotor $I^2R$ losses. Since $I_1 \simeq I_2$, the stator $I^2R$ loss is

$$W_S = W_R \frac{r_1}{r_2} \tag{4.50}$$

Equation (4.50) suggests a possibility for reducing the total losses under duty cycle conditions. The ratio $r_1/r_2$ can be reduced substantially by using a design D motor that has a relatively high value of $r_2$. This technique can reduce $W_S$ by a factor of 4:1 or 5:1 compared to a design B motor. Of course, if the motor must run under load for a considerable length of time, the extra secondary $I^2R$ losses in a design D motor may offset the beneficial effects of the $W_S$ reduction during speed changing. The application, therefore, must be evaluated based upon the expected duty cycle.

## Temperature Rise during Duty Cycle Operation

Most speed changes occur during a relatively short period of time, whereby the heat generated by the $I^2R$ losses is absorbed in the stator and rotor windings. The assumption often used is that all of the stator and rotor losses calculated by Eqs. (4.47), (4.49), and (4.50) are stored in the windings in which they are generated. This assumption does not allow for heat loss by conduction or radiation, so it is pessimistic; for speed-changing periods of only a few seconds, however, the assumption is accurate enough for most application calculations. Based on an all-heat-storage assumption, the temperature rise is given by

$$\theta = \frac{562 \times 10^{-6} W_x}{w_c c_w} \tag{4.51}$$

where $\theta$ = temperature rise, °C
$W_x$ = losses ($W_S$ or $W_R$), Ws
$w_c$ = weight of conductor material (stator winding if $W_S$ is used, or rotor bars and end ring if $W_R$ is used), lb
$c_w$ = specific heat, J/(°C·lb)

A tabulation of specific heats for materials commonly used in motor stators and rotors is given in Table 4.5.

**TABLE 4.5** Specific Heats of Materials

| Material | $c_w$, J/(°C·lb) |
|---|---|
| Copper | 0.094 |
| Aluminum | 0.222 |
| Brass | 0.092 |
| Phosphor bronze | 0.093 |

## Acceleration Time

The time to accelerate from one speed to another can be calculated by using an equation taken from basic mechanics:

$$T = J\frac{d\omega}{dt} \tag{4.52}$$

where $T$ = torque
$J$ = polar moment of inertia
$d\omega/dt$ = acceleration

Solving Eq. (4.52) for time $t$ and converting to engineering units results in

$$t = \frac{0.00352 Wk^2 (N_2 - N_1)}{T} \tag{4.53}$$

where $t$ = time, s
$T$ = torque, ft·lb

$N_1, N_2$ = rotational speeds at beginning and end of the period being considered, r/min
$Wk^2$ = load + motor polar moment of inertia, lb·ft$^2$

Equation (4.53) assumes a constant value of torque over the accelerating interval. This is rarely the case, of course, since the motor's speed-torque curve varies with speed. For most application calculations, it is satisfactory to assume some average value of torque obtained by averaging the torque values from the speed-torque curve over the speed interval of interest. For greater accuracy, the averaging should be done for the reciprocal of torque, $1/T$. The effective torque will be the average of the $1/T$ values, which is not the same as averaging torque $T$ and taking the reciprocal of the average torque.

## Calculation of Losses and Acceleration Time when Retarding Load Torque Is Present

Equations (4.47), (4.49), and (4.53) are all based on the assumption that there is no appreciable load torque present during the speed-changing period. It is therefore assumed that all the torque produced by the motor is available to accelerate the $Wk^2$ of the motor rotor and the connected load. However, there are many applications which do impose a retarding load torque and for which the equations mentioned above are not strictly applicable. Fans are a typical example of this type of load, since their torque varies as the square of the speed.

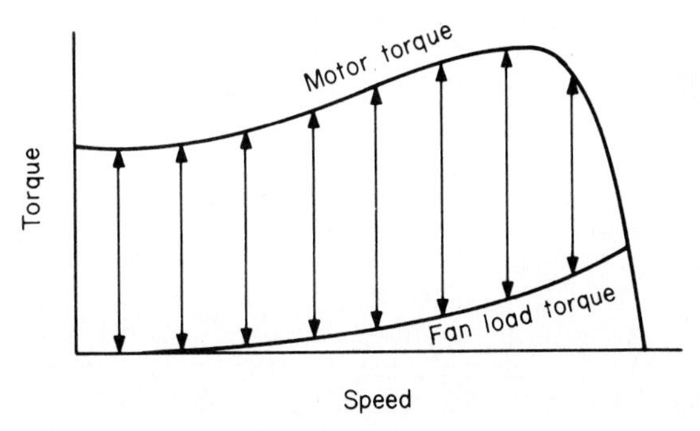


**FIG. 4.14** Motor and fan load speed-torque curves.

Figure 4.14 displays a motor speed-torque curve and a fan load-torque versus speed curve. The torque available for acceleration is the difference between these two curves, and it is this difference that must be used to calculate both the losses and the accelerating time. The procedure for calculation of losses and accelerating time is to divide the speed-torque curve into segments small enough that the torque values over a small increment of speed can be assumed constant. The division of the curves into segments is shown in Fig. 4.14. The presence of a retarding load torque both complicates and lengthens the calculating procedure considerably, and a detailed description of the technique for performing these calculations is given in Refs. 11 and 12.

## 4.10 ENERGY-EFFICIENT MOTORS

Energy conservation has become increasingly important since the Arab oil embargo of the early 1970s. Product lines of energy-efficient motors are readily available. These motors not only lead to the conservation of valuable energy resources but also provide attractive financial rewards to their users.

The technology by which energy savings are accomplished through motor design and the methods for calculating the economic paybacks resulting from motor loss reductions are the subjects of the following sections.

The technology involved in making major reductions in motor losses can best be understood by examining the loss components individually and determining what can be done to reduce them.

1. *Primary $I^2R$ loss:* Primary $I^2R$ losses occur in the stator and are ohmic losses due to current in the stator conductors. Increasing the cross-sectional area of the conductors reduces the primary resistance and therefore the primary $I^2R$ losses. In motors where aluminum conductors are used, usually in small horsepowers, a change to copper conductors will also have a major impact on reducing primary $I^2R$ losses.
2. *Iron loss:* The iron loss is mainly in the stator magnetic sections. Although the flux densities in the rotor are approximately the same magnitude as in the stator, the rotor flux is at very low frequency (slip frequency), and therefore the slip-frequency iron loss in the rotor is small. Iron loss is reduced by increasing the stack length, thereby reducing the flux densities and improving the grade of electrical steel used in the motor construction. Generally, both techniques are used. Most energy-efficient motors contain silicon steel and are made of thin-gauge laminations; taken in combination, these reduce both the hysteresis and eddy-current losses.
3. *Secondary $I^2R$ loss:* The secondary $I^2R$ loss is the ohmic loss in the rotor conductors. An increase in the conductor area will, of course, reduce the secondary $I^2R$ losses, just as it does for the primary $I^2R$ losses. However, there is a limit to the amount of reduction that can be achieved in this manner, since lower rotor resistance also results in lower starting torque. Since most energy-efficient motors meet the same starting-torque and inrush-current requirements as a standard motor, the reduction of secondary $I^2R$ losses is somewhat limited.
4. *Friction and windage loss:* Friction losses occur in the bearings, and windage losses occur as fans move air through and over the motor for cooling purposes. In fan-cooled motors the predominant component of this loss is due to the external fan, which drives cooling air across the motor frame. An improved aerodynamic fan design will reduce the friction and windage loss. Since energy-efficient motors have less loss than their standard counterparts, less cooling air is required and smaller fans can frequently be used.
5. *Stray-load loss:* Stray-load loss is defined as the difference between the total motor losses and the sum of the other four losses referred to above. Stray-load loss is caused by many factors. Major contributors are the imperfections of the airgap flux density waveform due to slotting, saturation, and mechanical imperfections in the airgap. Careful attention to the design and manufacture of the motor can significantly reduce the stray-load loss.

Taken in total, many energy-efficient motors have loss reductions of 40 percent or more, as compared with the equivalent standard motors.

## 4.11 CALCULATION OF MOTOR LOSSES

The equation for calculating motor losses at any particular load point is

$$L = \text{hp} \times 0.746\left(\frac{1.0}{\eta} - 1.0\right) \tag{4.54}$$

where $L$ = motor losses, kW
hp = power delivered to load, hp
$\eta$ = efficiency, per unit

Note that the losses are proportional to the reciprocal of efficiency. This sometimes results in confusion in evaluating savings to be realized when comparing two motors. For example, a 3 percentage point change in efficiency for a 10-hp motor, raising the efficiency from 80 to 83 percent, results in a loss reduction of 0.337 kW:

$$L_1 = 10 \times 0.746\left(\frac{1}{0.8} - 1.0\right)$$

$$= 1.865 \text{ kW}$$

$$L_2 = 10 \times 0.746\left(\frac{1}{0.83} - 1.0\right)$$

$$= 1.528 \text{ kW}$$

$$\text{Difference in losses} = 1.865 - 1.528$$

$$= 0.337$$

$$\text{Percent loss reduction} = \frac{0.337}{1.865} \times 100$$

$$= 18.1\%$$

The same 3 percent change in efficiency from 92 to 95 percent for a 100-hp motor, which is a typical rating that would have efficiencies of this magnitude, produces a savings of 2.561 kW, or a 39.5 percent loss improvement. In evaluating energy savings opportunities, the total loss reduction is important, not just the percentage points of efficiency improvement.

An additional equation useful in comparing the savings in losses between two motors of different efficiencies is

$$L_S = \text{hp} \times 0.746\left(\frac{1}{\eta_1} - \frac{1}{\eta_2}\right) \tag{4.55}$$

where $L_S$ = savings in motor losses, kW
hp = power delivered to load, hp
$\eta_1$ = efficiency expressed in per-unit for the lower value of the two motors being considered
$\eta_2$ = efficiency, higher value

## 4.12 ECONOMIC PAYBACK CALCULATIONS

The savings in motor losses can be equated to dollar savings. Two additional factors need to be considered: the hours of operation and the cost of electric energy. The general equation for dollar savings is

$$S = L_S \times H \times C \tag{4.56}$$

where $S$ = savings per year, dollars
$L_S$ = savings in motor losses, from Eq. (4.55)
$H$ = operating time per year, hours
$C$ = cost of electric energy, dollars per kWh

Equations (4.55) and (4.56) can be combined into a more convenient form to calculate savings:

$$S = \text{hp} \times 0.746\left(\frac{1}{\eta_1} - \frac{1}{\eta_2}\right) \times H \times C \tag{4.57}$$

The payback period for the extra cost of the energy-efficient motor is calculated by

$$\text{PBP} = \frac{\text{CD}}{S} \tag{4.58}$$

where PBP = payback period, years
CD = cost difference of two motors, dollars
$S$ = savings from Eqs. (4.56) or (4.57)

Equation (4.58) is a simple payback calculation that neglects the effect of interest on the money invested in the more expensive motor during the payback period. Frequently, payback periods are so short that interest costs can be neglected. However, if this is not the case, Ref. 12 discusses the calculation of payback periods, using present-worth factors to account for the effect of the time value of money.

## 4.13 DESIGN CALCULATION PROCEDURE FOR INDUCTION MACHINES

### The Design Process

The design process for an induction machine involves several steps. The sequence for designing an induction motor is displayed in the diagram of Fig. 4.15.

In this section, equations are given for step 3, "Calculate performance." It is beyond the scope of this discussion to provide the experience, background, and design rules required in step 2. In the usual case, a similar design is available to serve as a starting point, and the experience and skill of the designer determine how many iterations must be carried out before a successful design is achieved.

The design calculations are based on the induction motor equivalent circuit displayed in Fig. 4.12. Section 4.6 described how the equivalent circuit can be used to predict the motor performance, once values have been assigned to the equivalent-circuit resistances and inductances. Here in Sec. 4.13, a set of equations is given to calculate the equivalent-circuit constants, resistances, and inductances from the physical dimensions of the machine parts and the properties of the materials used in the machine construction.

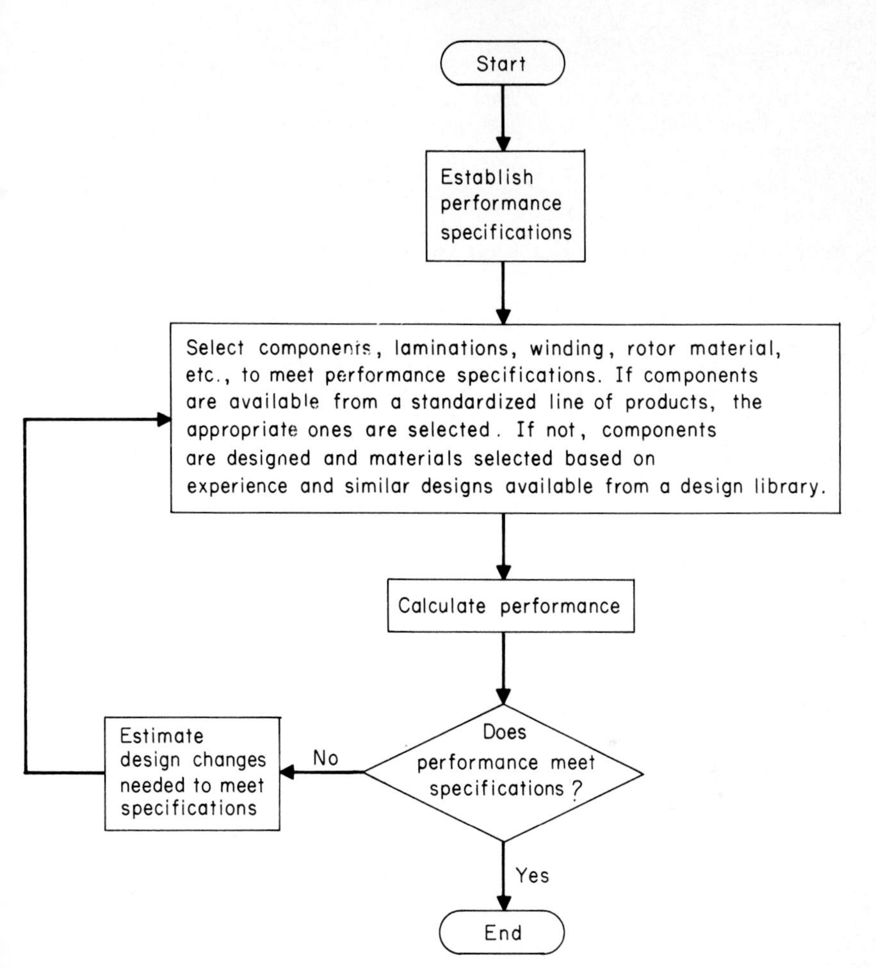


**FIG. 4.15** Design process for induction machine.

An induction motor is similar in many respects to a transformer. It has both a primary and secondary, the secondary being the rotor of an induction motor, and, to simplify the calculation process, calculations are performed with all voltages, currents, and impedances referred to either the primary or secondary side. For an induction motor, all the quantities are usually referred to the primary windings, and this is the convention that will be used here.

The basic equations for calculating each of the resistances and reactances used in the equivalent circuits are given in the following subsections. However, some of the terms used in the equations (such as slot constants, saturation factor, and Carter coefficients) are common to many types of electric machines and have been analyzed extensively in the literature. Literature references are cited for these items, and these references or other equivalent calculating procedures will have to be added to the material in this chapter in order to build a complete design procedure.

## Resistances

***Primary Resistance*** $r_1$**.** The *primary resistance* is the ohmic value of one phase of the stator winding. The resistance value is calculated from the basic equation for resistance:

$$R = \frac{\rho l}{A} \tag{4.59}$$

Specifically, for calculating $r_1$, the mean length of conductor $l$ and the effective cross-sectional area $A$ of the conductor must be evaluated. Tho formula for calculating $r_1$ is

$$r_1 = \rho \times C \times \text{MLC} \tag{4.60}$$

where $\rho$ = (resistance of wire per inch of length)/(number of parallel circuits × correction factor), where the correction factor = 1.0 for Y connection, or $\sqrt{3}$ for Δ connection; if more than one strand per conductor is used, $\rho$ must be the equivalent resistance of the parallel strands per inch of length
$C$ = series conductor per phase (see "Definition of Symbols" at end of section)
MLC = mean length of conductor; this is one-half the average length of a single coil turn and must be calculated from the geometry of the winding

***Secondary Resistance*** $r_2$ ***Referred to Primary.*** The secondary resistance is comprised of two parts: the bar and end-ring resistances.

Bar:
$$r_2' = (Ck_w)^2 m \frac{L_b \rho_b}{S_2 \times (\text{bar area, in}^2)} \tag{4.61}$$

where $L_b$ = length of rotor bar, including increased length due to skew (if applicable), in
$\rho_b$ = resistivity of bar material
= $0.693 \times 10^{-6}$ for copper at 25°C, Ω·in
= $1.260 \times 10^{-6}$ for aluminum at 25°C, Ω·in

End ring:
$$r_{\text{ER}} = (Ck_w)^2 m \frac{2D_r \rho_r}{\pi p^2 A_r} \tag{4.62}$$

where $\rho_r$ = resistivity of end-ring material; for copper and aluminum, refer to values for $\rho_b$ given under Eq. (4.61)
$D_r$ = mean diameter of end ring, in
$A_r$ = cross-sectional area of end ring, in²

If the end ring is wide in the radial direction compared to a pole pitch, the current must travel radially as well as circumferentially to achieve nearly uniform current-density distribution. This has the effect of altering the end-ring resistance from the value calculated by Eq. (4.62). This effect can be accounted for by multiplying the right-hand side of Eq. (4.62) by a $K_{\text{ring}}$ factor as described by Trickey.[13] The total secondary resistance $r_2$ is obtained from:

$$r_2 = r_2' + r_{\text{ER}} \tag{4.63}$$

Equations (4.61), (4.62), and (4.63) are for single-cage rotors. Calculations for double-cage rotors have been discussed by Alger and Wray, Jordan, and others.[14,15]

The values of resistivity $\rho_r$ are given with Eq. (4.61) for a temperature of 25°C. The final value of $r_2$ in Eq. (4.63) must be corrected to the expected operating temperature of the machine.

Equation (4.61) is applicable to all regions of the speed-torque curve so long as the appropriate value of "bar area" is used. In the region of full load, the current in the rotor bar is at a very low (slip) frequency and the current density is distributed uniformly over the bar area. However, for slip >> 0, particularly at standstill, where the frequency of the rotor current equals the stator applied frequency, the current density may not be uniform. This phenomenon is known as *deep-bar effect* and must be considered for motors 1 hp and above, since the bar depths are sufficient for motors of this size to cause appreciable change in $r_2$. Alger discusses the deep-bar rotor and provides curves for the correction factors that can be applied to $r'_2$.[5]

## Magnetizing Reactance $x_m$

The *magnetizing reactance* is the voltage induced by the fundamental component of the airgap flux density divided by the magnetizing current that establishes the airgap flux. An equation for magnetizing reactance $x_m$ is

$$x_m = 2\pi f(Ck_w)^2 m \times 10^{-8}\left(\frac{0.3234 \times A_g \times C_{\mathrm{SK}}}{p \times g \times \mathrm{SF}}\right) \tag{4.64}$$

where $A_g$ = airgap area spanned by one pole pitch (pole pitch $\times$ $L_1$)
$C_{\mathrm{SK}}$ = skew coupling constant [refer to Eq. (4.69)]
$g$ = effective airgap (actual airgap $\times$ Carter coefficients[16,17])
SF = saturation factor[4] (airgap At + iron AT)/airgap At

The methods for calculating Carter coefficients and magnetic saturation factors are available in Refs. 4, 16, and 17.

## Leakage Reactances $x_1$ and $x_2$

The leakage flux paths in an induction machine are numerous and difficult to analyze. *Leakage flux* can generally be defined as flux that does not link both the primary and secondary windings and thus does not contribute to the mutual flux field through which energy conversion is achieved. Leakage reactances are important, for they are major parameters limiting the machine's breakdown torque.

Designers and authors are not in complete agreement as to which of the leakage flux components are important in calculating induction motor performance. In actual practice, each design procedure utilizes a set of components unique to the particular procedure for calculating the total leakage reactance. The components are selected and equations developed for calculating each component such that the final values of $x_1$ and $x_2$ used with the equivalent circuit give results that closely agree with tests. There is no one "best way" that everyone agrees on. Veinott, Alger, Kuhlman, and others have all given sets of equations that yield acceptable performance calculations, but there are many differences between the various methods.[4,5,18] The equations used here are based primarily on Veinott, with some modifications.

***Primary Leakage Reactance*** $x_1$. Four components are included in the calculations of the primary leakage reactance: slot, zigzag, skew, and end leakage. The general format for reactance calculations is to divide the calculations into two parts:

1. Calculate the term common to all components, called the reactance constant.
2. Calculate a term proportional to the permeance of the leakage path, called the slot constant, zigzag constant, etc.

Reactance constant: $$K_x = 2\pi f(Ck_w)^2 m \times 10^{-8} \tag{4.65}$$

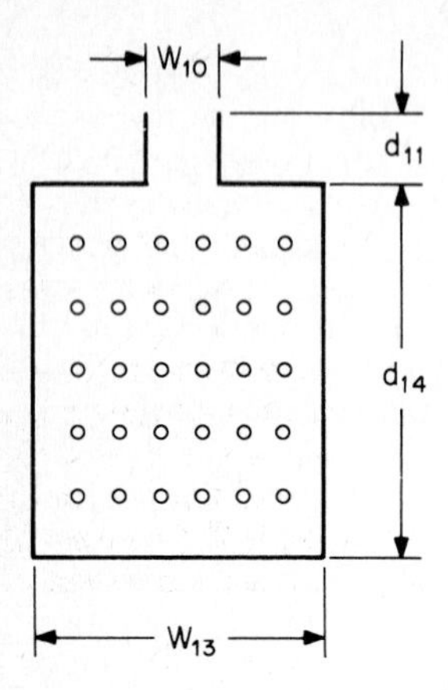


**FIG. 4.16** Rectangular slot uniformly filled with current-carrying conductors. *(From C. G. Veinott, Theory and Design of Small Induction Motors, McGraw-Hill, New York, 1959.)*

**1.** *Slot leakage:* The slot constant for a rectangular slot (Fig. 4.16) is given by

$$K_{ss} = \frac{3.19L_1(d_{14}/3W_{13} + d_{11}/W_{10})}{S_1 k_w^2} \tag{4.66}$$

A method for calculating the slot constant for shapes other than the rectangular one shown in Fig. 4.16 has been given by Waldschmidt.[19]

**2.** *Zigzag leakage:* The zigzag-leakage reactance accounts for the flux that encircles a slot by way of an overlapping tooth on the opposite side of the airgap. Figure 4.17 illustrates this concept. Alger gives an equation for this reactance, which, after conversion to be consistent with the notation used here, is[5]

Zigzag slot constant:

$$K_Z = \frac{\pi^2 K_M}{24}\left(\frac{p^2}{S_1^2} + \frac{p^2}{S_2^2}\right) \tag{4.67}$$

where $K_m = 0.3234A_gC_{SK}/pg\text{SF}$

**3.** *Skew leakage:* Denote the skew of a slot in electrical degrees, $\alpha$. Then the skew slot constant is given by[4]

$$K_{SK} = (1 - C_{SK}^2)\frac{K_m}{2} \tag{4.68}$$

where

$$C_{SK} = \frac{\sin(\alpha/2)}{(\pi\alpha)/360} \tag{4.69}$$

**4.** *End leakage:* The paths for the leakage flux that link the end turns of an induction motor are perhaps the most varied and complex of the many leakage paths. To complicate matters further, many motor constructions locate metal baffles in the region of the end turns, and these can materially affect the total end-turn reactance. The problem has been addressed experimentally by Barnes and analytically by Honsinger.[20,21] The equation for the end-leakage constant developed by Barnes is given in Eq. (4.70) after modification for consistency with other reactance equations:

$$K_E = \frac{1}{p}\frac{\phi_e L_e}{2n} \tag{4.70}$$

where $\phi_e L_e/2n$ is obtained from the curves in Fig. 4.18.

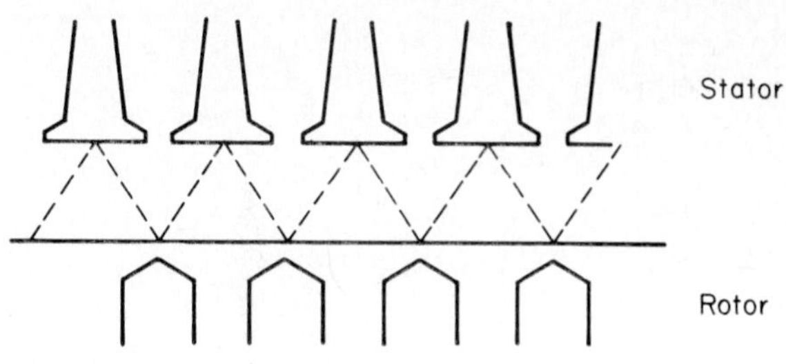


**FIG. 4.17** Zigzag flux path.

The total primary leakage reactance in ohms per phase is

$$x_1 = K_x(K_{ss} + K_Z + K_{SK} + K_E) \tag{4.71}$$

***Secondary Leakage Reactance*** $x_2$. The secondary leakage reactance is comprised of three of the four components that make up $x_1$: slot, zigzag, and skew leakage.

**1.** *Slot leakage:* The slot leakage is calculated in the same manner as explained for $x_1$. Equation (4.72) is used for a rectangular slot with a closed top and a slot bridge of magnetic material; appropriate modifications of Eq. (4.72) are used if the slot differs from a rectangular shape.[19] The $d_{24}/3d_{23} + 2d_{21}/d_{23}$ terms in Eq. (4.72) account for the leakage flux crossing the slot. Many induction motors are made with closed rotor slots. Leakage flux passes through the magnetic steel bridge above the slot, and this flux also contributes to the slot-leakage reactance. Calculation of the reactance due to the bridge portion of the slot-leakage flux cannot be characterized by means of a simple dimensional ratio, as can be done for flux that crosses the slot. The difficulty in calculating this reactance arises from the fact that most slot bridges are sufficiently narrow that the leakage flux passing through the bridge causes magnetic saturation, and the permeance of the path depends upon the level of saturation. A rigorous analysis of this phenomenon is a difficult undertaking. The alternative approach, which is the one used here, is to assign an empirical multiplying constant to account for the increase in $x_2$ due to bridge leakage. The slot constant for a rectangular slot with closed top (see Fig. 4.19) is

$$K_{SR} = \frac{C_B \times 3.19L_2(d_{24}/3d_{23} + 2d_{21}/d_{23})}{S_2} \tag{4.72}$$

where $C_B$ = bridge-leakage multiplying constant

Typical values of $C_B$ for medium-size induction motors range from 1.5 to 3.0.

**2.** *Zigzag and skew leakage:* Zigzag and skew leakage are shared between the stator and rotor. Equations (4.67) and (4.68) calculate constants based on half of the leakage flux in these paths, and these constants are used in Eq. (4.71) to calculate $x_1$. Thus, half of the leakage reactance due to zigzag and skew leakage is assigned to $x_1$. The other half is assigned to $x_2$, and therefore Eqs. (4.67) and (4.68) are used again for the zigzag- and skew-leakage constants to be used in the calculation of $x_2$. The equation for $x_2$ is

$$x_2 = K_x(K_{SR} + K_Z + K_{SK}) \tag{4.73}$$

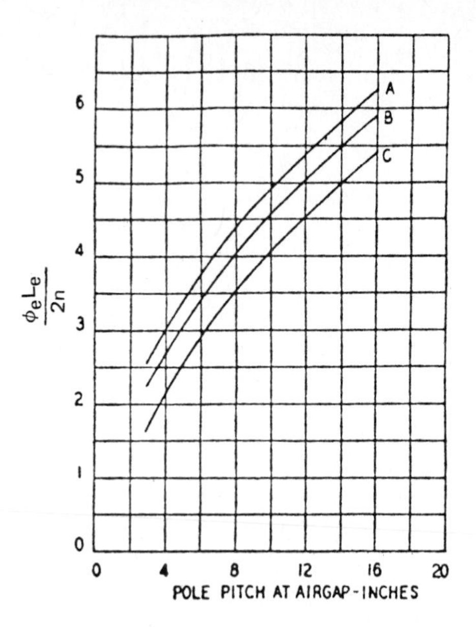


Curve A—Coils having 100 per cent pole pitch (also Gray's curve)
Curve B—Coils having 66 per cent pole pitch
Curve C—Coils having 50 per cent pole pitch

**FIG. 4.18** Empirical end-winding constant versus pole pitch. *(From E. C. Barnes, "An Experimental Study of Induction Machine End-Turn Leakage Reactance," AIEE Transactions, vol. 70, 1951.)*

## Iron-Loss Resistance $r_{\mathrm{FE}}$ and $r_{\mathrm{m}}$

The iron-loss-resistance calculation is based on the calculated value of iron loss, which consists of the fundamental-frequency losses in the stator teeth and core, plus the high-frequency and surface losses. These calculations are common to many types of electric machines and are described in Refs. 4 and 5. Curves published by the electrical steel manufacturers must be used to determine the watts per pound of fundamental-frequency core loss at the calculated value of flux density.

The high-frequency and surface losses associated with the magnetizing airgap field should also be included as part of the iron loss. Veinott gives an equation and table for calculating these losses based on a classic paper by Spooner and Kinnard.[4,22]

An equation for calculating the iron-loss resistance $r_{\mathrm{FE}}$ is

$$r_{\mathrm{FE}} = \frac{V_1^2 k_p^2}{(\text{iron loss in watts})/3} \tag{4.74}$$

where $k_p = x_m/(x_1 + x_m)$

The conversion of the parallel branch of $x_m$ and $r_{\mathrm{FE}}$ to an equivalent series $x_m'$ and $r_m$ was described in Sec. 4.6.

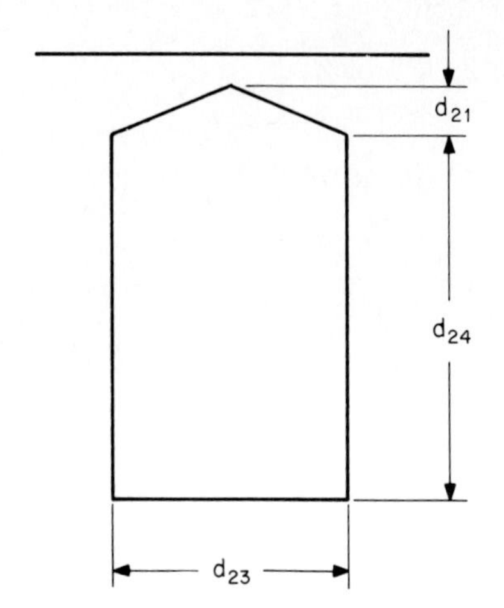


**FIG. 4.19** Rectangular rotor slot with closed top.

## Magnetic Saturation Effect on Equivalent-Circuit Constants

The equations given in the preceding sections are suitable for use in the normal range of operation for an induction motor, namely, in the region of rated load. However, in the region of the speed-torque curve extending from locked-rotor to breakdown torque, magnetic saturation of the leakage flux paths plays an important role in defining the motor's performance. There is no single well-agreed-upon method for incorporating saturation effects into the calculation of the equivalent-circuit parameters. Chang and Lloyd have described one technique for including saturation in the parameter calculations.[23] In the method described here, an empirical multiplying factor is used; $x_1$ and $x_2$ are the parameters most significantly affected.

$$x_{1\mathrm{LR}} = C_{\mathrm{LR}} x_1 \tag{4.75}$$

$$x_{2\mathrm{LR}} = C_{\mathrm{LR}} x_2 \tag{4.76}$$

$$x_{1\mathrm{BD}} = C_{\mathrm{BD}} x_1 \tag{4.77}$$

$$x_{2\mathrm{BD}} = C_{\mathrm{BD}} x_2 \tag{4.78}$$

where $C_{\mathrm{LR}}$ = saturation constant at locked-rotor
$C_{\mathrm{BD}}$ = saturation constant at breakdown torque
$x_{1\mathrm{LR}}, x_{2\mathrm{LR}}$ = leakage reactances at locked-rotor
$x_{1\mathrm{BD}}, x_{2\mathrm{BD}}$ = leakage reactances at breakdown torque

The reactances calculated by Eqs. (4.75) to (4.78), used together with the deep-bar corrections discussed earlier, yield satisfactory results for the principal regions of interest in the motor's speed-torque curve. When calculating the entire speed-torque curve, the saturation factors $C_{\mathrm{LR}}$ and $C_{\mathrm{BD}}$ can be prorated based on speed. Typical values for $C_{\mathrm{LR}}$ range from 0.50 to 0.75, and for $C_{\mathrm{BD}}$ from 0.60 to 0.85.

### Definition of Symbols Used in Sec. 4.13

$C$ = effective series conductors per phase = $(2\mathrm{TPC} \times S_1 \times \text{equiv } Y)/m$

Equiv $Y$ = 1/(number of parallel circuits) for Y connection, and $1/(\sqrt{3}\times$ number of parallel circuits) for Δ connection

$f$ = frequency

$k_w$ = winding factor = pitch factor × distribution factor

$L_1$ = length of stator core

$L_2$ = length of rotor core

$m$ = number of phases

$p$ = number of poles

$S_1$ = number of stator slots

$S_2$ = number of rotor slots

TPC = turns per coil

$V_1$ = terminal voltage per phase, (line-to-line volts)/$\sqrt{3}$ for a three-phase motor, V

## REFERENCES


1. NEMA Standards Publication MG1, *Motors and Generators*, National Electrical Manufacturers Association, Washington, DC, 1978.
2. NEMA Standards Publication MG13, *Frame Assignments for Alternating-Current Integral-Horsepower Induction Motors*, National Electrical Manufacturers Association, Washington, DC, 1974.
3. P. J. Tsivitse, *Motor Application and Maintenance Handbook*, pt. 5: "Mining Motors," McGraw-Hill, New York, 1969.
4. C. G. Veinott, *Theory and Design of Small Induction Motors*, McGraw-Hill, New York, 1959.
5. P. L. Alger, *The Nature of Polyphase Induction Machines*, Wiley, New York, 1951.
6. A. F. Puchstein and T. C. Lloyd, *Alternating Current Machines*, Wiley, New York, 1942.

6a. *IEEE Standard Test Procedure for Polyphase Induction Motors and Generators*, IEEE Standard 112-1984.

7. *Rotating Electrical Machines*, pt. 2: "Methods for Determining Losses and Efficiency of Rotating Electrical Machinery from Test," Publication 34-2, International Electrotechnical Commission, Geneva, Switzerland, 1972.
8. *Standard of the Japanese Electrotechnical Committee, Induction Machines*, JEC37, Denki Shoin, Tokyo.
9. H. E. Jordan and A. Gattozzi, *Efficiency Testing of Induction Machines*, IEEE Industry and Applications Society Conference Record, New York, 1979.
10. P. G. Cummings, W. D. Bowers, and W. J. Martiny, *Induction Motor Efficiency Test Methods*, IEEE Industry and Applications Society Conference Record, New York, 1979.
11. J. W. Cook, "Squirrel-Cage Induction Motors Under Duty Cycle Conditions," *Electrical Manufacturing*, February, 1956.
12. Howard E. Jordan, *Energy Efficient Electric Motors and Their Applications*, Van Nostrand Reinhold, New York, 1983.
13. P. H. Trickey, "Induction Motor Resistance-ring Width," *AIEE Transactions*, 1936.
14. P. L. Alger and J. H. Wray, "Double and Triple Squirrel Cages for Polyphase Induction Motors," *AIEE Transactions*, vol. 72, pt. III, 1953.

15. H. E. Jordan, "Synthesis of Double-Cage Induction Motor Design," *AIEE Transactions*, vol. 78, pt. III, 1959.

16. F. W. Carter, "Air Gap Induction," *Electrical World*, vol. 38, Nov. 30, 1901.

17. C. A. Adams, "The Design of Induction Motors," *Proc. AIEE*, June 1905.

18. J. H. Kuhlman, *Design of Electrical Apparatus*, Wiley, New York, 1940.

19. K. J. Waldschmidt, "A General Method for Slot Constant Calculation," *AIEE Transactions*, vol. 77, pt. III, 1958.

20. E. C. Barnes, "An Experimental Study of Induction Machine End-Turn Leakage Reactance," *AIEE Transactions*, vol. 70, 1951.

21. V. B. Honsinger, "Theory of End-Winding Leakage Reactance," *AIEE Transactions*, vol. 78, pt. III-A, 1959.

22. T. Spooner and I. F. Kinnard, "Surface Iron Losses with Reference to Laminated Materials," *AIEE Transactions*, 1924.

23. S. S. L. Chang and T. C. Lloyd, "Saturation Effects on Leakage Reactance," *AIEE Transactions*, vol. 68, 1949.

# CHAPTER 5
# DIRECT-CURRENT MACHINES

**M. G. Say***

This chapter is concerned with basic "conventional" rotating heteropolar dc machines of cylindrical geometry, excluding homopolar, permanent magnet, brushless, linear, and control machines.* Conventional dc generators are now uncommon. Motors of low rating are employed with pure dc sources (e.g., secondary batteries and solar-cell arrays). Most industrial dc motors are fed from ac supplies through rectifiers and are built with due regard to the consequent waveform distortion.

## 5.1 CONSTRUCTION

The essential relationship between the stator and rotor magnetomotive force (mmf) axes for a heteropolar dc machine is that shown for a two-pole case in Fig. 5.4*b*. The constructional forms to realize this relationship are shown in Figs. 5.1 to 5.3.

### Small Motor

Figure 5.1 is typical of two-pole machines in ratings up to about 1 kW. For production economy, and where the supply is rectified ac, the stator and rotor cores are both laminated. The stator core plates are clamped between end brackets by through bolts; the rotor plates are pressed onto the shaft. The field and armature coils are wound with resin-insulated circular copper wires. To avoid unduly weak teeth, the rotor (armature) slots are tapered. Adequate commutation can be achieved without commutating poles.

*By the same author: M. G. Say and E. O. Taylor, *D.C. Machines*, 2d ed., Pitman, London, 1985.

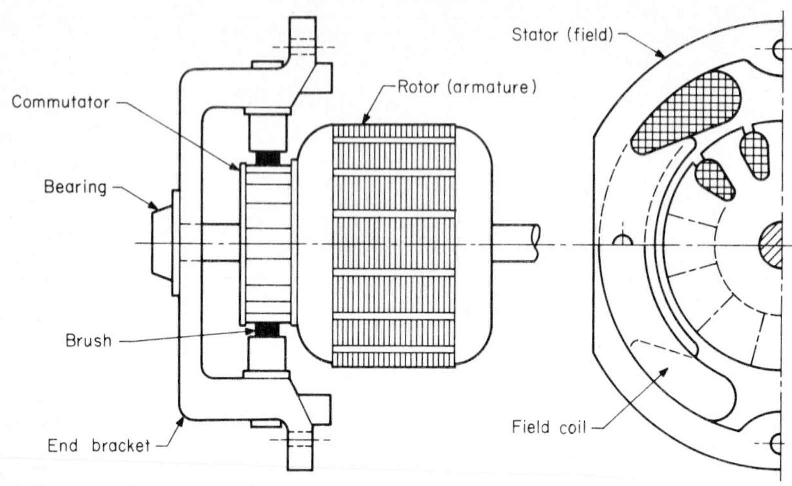


**FIG. 5.1** Small dc motor.

## Medium-Rated Motor

Industrial machines (Fig. 5.2) in the range of 50 to 500 kW have a conventional frame (serving also as a yoke) formed from thick plate into a circular shape (left) or a square shape (right); alternatively, both may be laminated. In either case, the poles are bolted to the yoke, and interpoles are essential. The axial half-section at the bottom of Fig. 5.2 shows the armature laminations compressed between pressure castings, and the axial cooling ducts. The commutator sector assembly exposes the inner surface to axial cooling air through an annular duct. A typical slot is illustrated; the tape-insulated conductors in two layers are retained by wedges or by banding.

## Large Generator

In Fig. 5.3 the main and commutating poles are assemblies of thick plates bolted to the solid yoke. The main-field windings shown include series, shunt, and compensating coils. The armature core is carried on a "spider" and has radial cooling ducts. The commutator has the conventional V-ring clamping, with through bolts for tightening. The brush arms are mounted in a rigid structure, with facility for slight adjustment of the brush position.

## General Details

***Stator Frame.*** Formerly, solid mild-steel frames, solid poles, and laminated pole shoes (to reduce the effects of armature-tooth pulsations) were common. Fully laminated stators are necessary for machines fed through rectifiers, or subject to impulsive loads, or for which rapid response and accurate speed control are required. The time constant of a solid stator is about 0.5 second (s) and of a laminated one is a few milliseconds. Segmental plates are used for the larger machines for economy

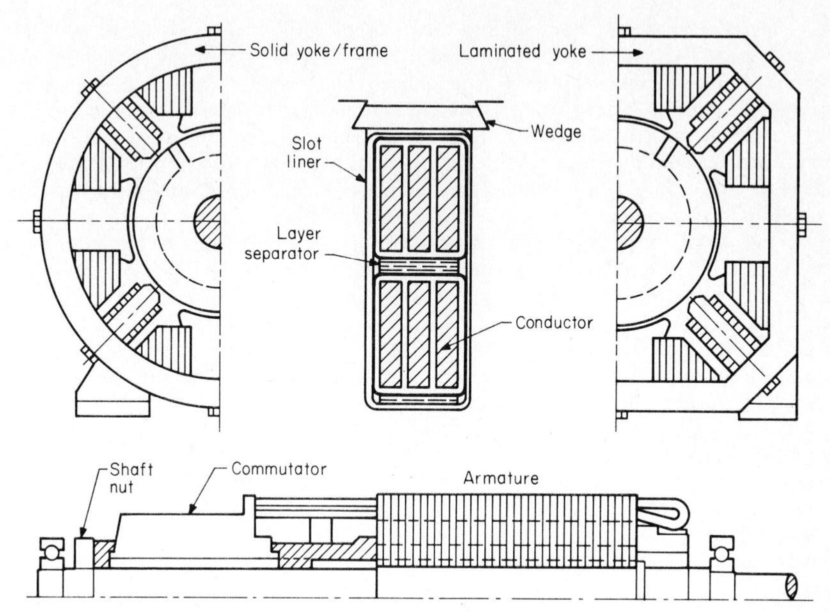

**FIG. 5.2** Medium-rated dc motor.

in plate material. Laminae are compressed between end plates, which provide fixed locations for the end brackets.

***Field Windings.*** Shunt coils are wound from circular- or rectangular-section conductors, insulated and mounted in a bobbin, taped and impregnated, slipped onto the pole body, and held by the pole shoe. Series and commutating windings are generally of copper strip wound on edge.

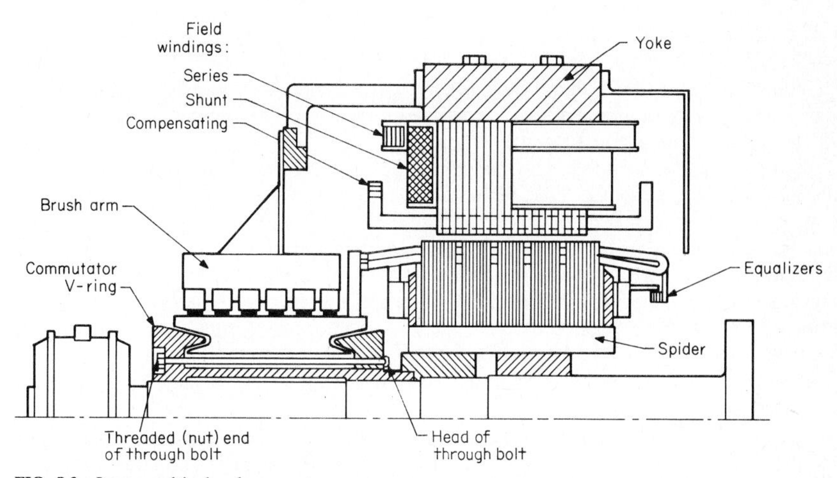

**FIG. 5.3** Large multipolar dc generator.

***Armature Winding.*** Many motors have a simple two-circuit wave winding. Wire-wound coils of conductors up to an area of about 1 $mm^2$ are enamel-insulated; for larger ratings, conductors are of rectangular-section copper, insulated, assembled, taped half-lap and placed in the slots with insulating liners, and retained by slot wedges or by glass-cord banding. Banding is applied in the semicured state under controlled tension; during the subsequent baking, the resin flows to form a strong homogeneous band. Class F polyester enamels, sheets, tapes, and varnishes are widely used in windings to raise the output/mass ratio.

***Commutator.*** The "standard" commutator is assembled from wedge-section tapered sectors of hard copper, separated by micanite plates 0.5 to 0.8 mm thick, to form a cylindrical structure coaxial with the armature. The coil ends of the armature winding are hard-soldered to the sectors, which are slotted to receive them or are provided with lugs, "risers," and clips for this purpose. The conglomerate of copper sectors and separators is rigidly held by clamping or molding. The traditional method is the use of steel V-rings, insulated from the sectors by micanite carefully molded to shape, and clamped by through bolts, as illustrated in Fig. 5.3; a simpler but effective structure is the molded version in Fig. 5.2. In service, the commutator must withstand high operating temperature without distortion and withstand peripheral speeds of 50 m/s without bursting. It must present a closely cylindrical surface with a very small bar-to-bar tolerance, and the micanite separators must be recessed by about 1 mm to prevent brush "bounce" and excessive wear.

***Brushgear.*** An axial line of carbon brushes (or, in a small machine, a single brush) is carried on a brush arm to bear with optimum pressure on the commutator surface. The effective (but not the physical) brush axis is normally aligned with the interpole axis; for nonreversing machines, it may be shifted by a small angle. The brush width circumferentially is typically between two and three sector widths (10 to 20 mm) and about 30 to 40 mm axially. The brush material comprises a grade of carbon-graphite, of which several grades are available. The brush tops are insulated, and the flexible copper "pigtails" are sleeved to avoid (1) thermal damage to the constant-pressure springs and (2) unequal current sharing.

***End Shields.*** These have spigot-and-bearing housings machined in one setting operation to ensure concentricity and the maintenance of airgap-length uniformity.

***Bearings.*** End-shield bearings are usual in ratings up to about 500 kW. Sealed roller or ball bearings eliminate greasing and reduce the number of parts held in stock. Journal bearings may be used where they offer advantage.

***Shaft.*** Conventional shafts have several variations of diameter along the axis to locate the several components. Modern designs seek to limit the variations in order to reduce machining and stresses.

***Ventilation.*** To up-rate the output of a machine of given frame size, it is necessary to achieve a high electric loading; consequently, a higher $I^2R$ loss must be dissipated. Heat from the rotor is dissipated mainly by convection from the airgap surface, augmented in larger machines by rotor-mounted fans, directly or through ducts or screens. The air passes over the rotor cooling surface and between the stator poles and is then ejected. Where ambient conditions are difficult (as in hazardous atmospheres) a totally enclosed construction is employed. Motors with a wide-range speed control may have to run at low speeds, for which rotor-mounted fans cannot give adequate ventilation; it is then necessary to mount a constant-speed fan unit on the frame to provide forced cooling.

***Noise.*** Magnetically generated noise, arising from flux variations as the rotor teeth pass a pole shoe, may be reduced by slot skewing. Aerodynamic noise depends on the speed and the number of rotor teeth and is more difficult to suppress.

## 5.2 MAGNETIC CIRCUIT

In Fig. 5.4, showing part of the magnetic circuit of a four-pole machine, the current direction is indicated by black (outward) and crosshatched (inward) blocks, and the magnetic flux distribution by dotted lines and polarity arrows. In Fig. 5.4*a*, the *no-load* condition, only the main-field windings are excited. The field mmf of one pole sets up a flux through the yoke, pole body, pole shoe, airgap, armature teeth, and core; this is the useful (*working*) flux. But there is also a nonuseful (*leakage*) flux from the pole body to the yoke, and between the shoes of neighboring main poles via the unexcited interpoles. The flux distribution is symmetrical about the centerline of the pole.

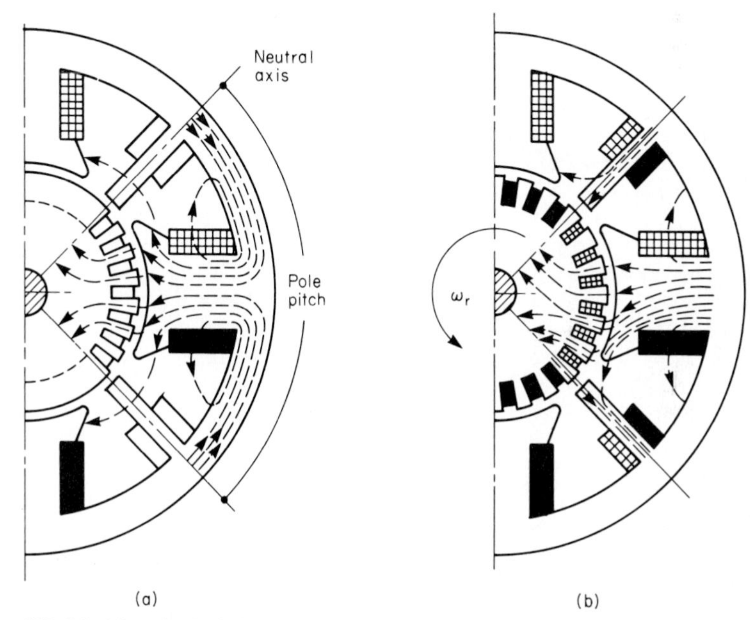


**FIG. 5.4** Magnetic circuit (motor): ***(a)*** no-load and ***(b)*** on-load conditions.

In Fig. 5.4*b*, for the machine *on load* as a motor, current flows also in the armature and interpole windings, developing two additional mmf's acting on the magnetic circuit. The armature mmf distorts the distribution of the gap flux density (the *armature reaction* effect), reducing it on one pole flank and increasing it on the other, the total pole flux (ignoring saturation) remaining unchanged. On the interpolar axis, the armature and interpole mmf's are in opposition. The gap flux "distortion" in Fig. 5.4*b* that results from armature reaction is the basic means by which a rotational torque is developed in accordance with the force-on-a-current principle.

The complex on-load flux distribution is conventionally treated by separating the working flux from the leakage. The former is based on the configuration in

Fig. 5.4*a*, with adjustment for armature reaction effects. Leakage fluxes, which have substantial parts in air, are estimated from dimensional design data.

## Airgap MMFs

The distribution of the working flux in the airgap is a major feature in commutation. Consider the several mmf components that act to drive flux across the gap.

***Main Field.*** For the pole in Fig. 5.5*a*, where the circumferential pole arc is $b$ and the pole pitch is $Y$, the ratio $b/Y$ at the armature surface is usually about 2:3.

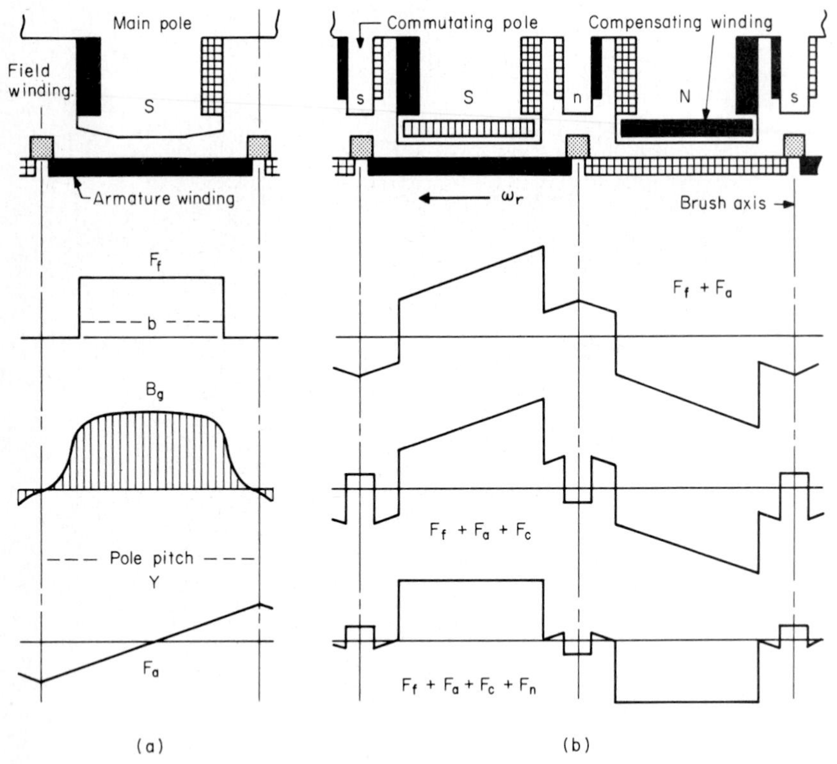


**FIG. 5.5** Airgap mmf's: ***(a)*** main field; ***(b)*** the combination of the main field and armature field.

The field mmf impressed on the gap is $F_f$, represented by the rectangle so marked. The resulting flux density distribution (neglecting rotor slotting) is $B_g$, which at a given point varies with the gap length there.

***Armature.*** Assuming the armature conductors to be uniformly distributed, their mmf per pole has a maximum $F_a$ at the interpolar axis and a triangular wave distribution.

## Resultant Gap MMF

The combination of the main-field and armature mmf's $F_f + F_a$ is shown for a motor in Fig. 5.5*b*, together with the modifications introduced by the interpole (commutating) and compensating (neutralizing) mmf's $F_c$ and $F_n$.

$F_f + F_a$. The armature reaction mmf (1) distorts the gap mmf distribution and (2) develops a flux in the neutral (commutating) zone. This distortion results in high motional electromotive forces (emf's) in the armature coils (and therefore between their associated commutator sectors) near the leading edge of the pole. The flux in the commutating zone produces, in coils undergoing commutation, a rotational emf that opposes current reversal, thereby impairing commutation.

$F_f + F_a + F_c$. With the interpoles excited in series with the armature, $F_a$ is neutralized in the commutating zone and a reversed flux polarity now assists commutation.

$F_f + F_a + F_c + F_n$. A neutralizing, or compensating, winding which has an appropriate number of turns per pole and which is located in the pole shoes, uniformly distributed, and connected in series with the armature opposes $F_a$ to restore the gap mmf distribution to symmetry, giving the gap flux density $B_g$ in Fig. 5.5*a*. Neutralization is applied to machines subject to impulsive load or to those which run at high speed with a weak main-field excitation. Interpoles are fitted to all industrial machines of rating more than a few kilowatts; in small unidirectional machines, reliance is placed on "resistance commutation" and brush shift.

## Magnetic Circuit Calculation

The excitation required to develop the working gap flux $\phi_g$, by which the rotational armature emf is generated, is usually based on the *no-load* conditions in Fig. 5.4*a*. The method is a summation of the mmf's required by each series element of flux path (yoke, pole, gap, armature teeth, and armature core) in a single pole pitch. Table 5.1 shows a typical assessment for a machine with a useful gap flux $\phi_g$ = 80 mWb per pole (corresponding, with pole leakage, to a mean pole flux of 92 mWb and a yoke flux of 96 mWb, Sec. 5.4), with a gap length $l_g$ = 5.0 mm (extended to $l'_g$ = 5.75 mm as a result of the armature slotting), and with the net (i.e., "iron") areas of laminated ferromagnetic parts. Tabulated are the areas $a$, the flux $\phi = Ba$, the flux density $B$, the mmf $H$ per meter of path length $l$, and the total mmf $F$

**TABLE 5.1** Magnetic Circuit Calculation for No-Load

| Part | $a$, m$^2$ | $\phi$, mWb | $B$, T | $H$, kA·t/m | $l$, m | $F$, kA·t |
|---|---|---|---|---|---|---|
| Airgap | 0.084 | 80 | 0.95 | 796 | 0.00575 | 4.58 |
| Teeth | 0.038 | 80 | 2.10 | 40 | 0.048 | 1.90 |
| Core* | 0.033 | 40 | 1.20 | 1.0 | 0.16 | 0.16 |
| Pole | 0.058 | 92 | 1.60 | 3.0 | 0.20 | 0.60 |
| Yoke* | 0.040 | 48 | 1.20 | 0.7 | 0.30 | 0.21 |
| Total | | | | | | 7.45 |

* Each branch.

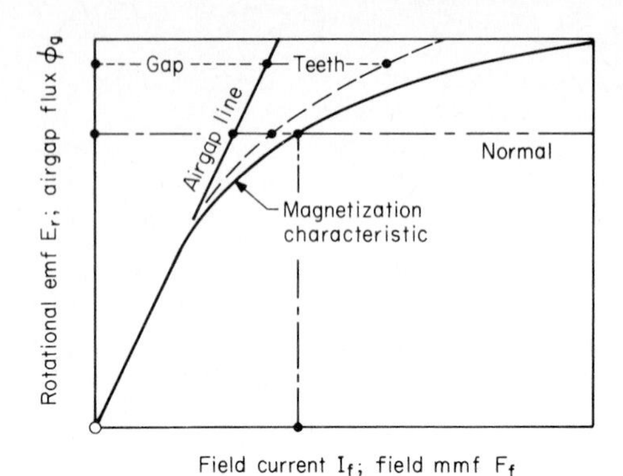


**FIG. 5.6** Magnetization characteristic.

expended on each part's path. Note that the yoke and the armature core have two paths in parallel.

A number of calculations for a range of useful gap fluxes enables one to draw the magnetization characteristic (Fig. 5.6). At low flux levels the airgap mmf (the *airgap line*) predominates, at normal flux the airgap takes typically 0.8 of the total field mmf, and for higher flux the teeth demand a substantial proportion of the total field mmf as a consequence of rising saturation.

The magnetization characteristic is checked by driving the machine at constant speed with the armature terminals connected to a voltmeter to measure the rotational emf over a range of field current. The gap flux is calculated from the winding details. Such a characteristic shows a small emf at *zero* field current because of remanent flux in the magnetic circuit.

## Interpoles

On load, the interpoles are excited but have no direct effect on the main flux circuit. The resulting flux distribution in the yoke and armature core can be approximated by superposing the main and interpole fluxes. The two components, $\phi$ in the main circuit and $\phi_c$ in the interpole circuit, are shown in Fig. 5.7. Consider that part of the yoke between the main pole $S$ and interpole $n$: The flux is the summation of $\frac{1}{2}\phi$ and $\frac{1}{2}\phi_c$, giving $\frac{1}{2}(\phi - \phi_c)$, and there is a similar effect in the associated armature core. In the yoke and armature core between the main pole $N$ and the interpole $n$, however, the two flux directions are opposed, giving $\frac{1}{2}(\phi - \phi_c)$. Because the interpole flux on load may be comparable with the main flux, additive fluxes lead to conditions of high saturation level in parts of the yoke and of low flux density (or even of reversal) in adjacent parts, with similar variations in the armature core and teeth.

## Load Conditions

When the armature, interpole, compensating, and main-field windings are all developing mmf's, the flux distribution is complex. The magnetic saturation nonlinearity

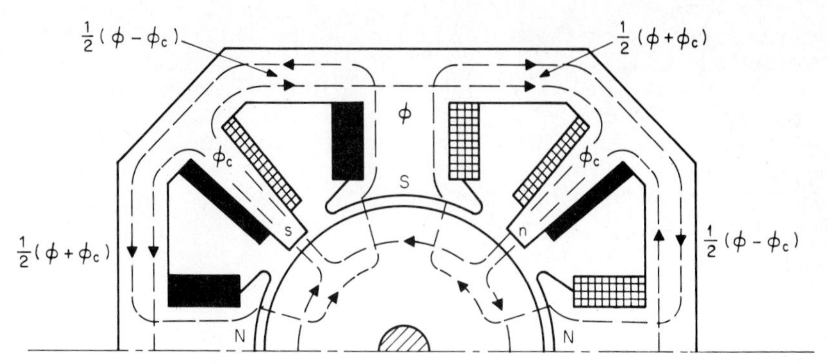


**FIG. 5.7** Main and interpole magnetic circuits.

makes it invalid to superpose flux contributions in terms of their individual mmf's. For large machines, in which precise flux patterns are important in their influence on performance, new techniques have been developed. These include numerical integration and energy-based methods. Computerized solutions using finite elements in the magnetic circuit can cope with magnetostatic problems of this kind in two dimensions, with iterative allowance for magnetic saturation. Transient behavior is much more difficult to analyze; at present, its analysis relies on iteration, with short steps of time and with appropriate *B-H* relations adjusted progressively.

## 5.3 WINDINGS

The stator windings excite the main field, the interpole field, and, where required, the compensating field. The main-field windings may be connected in parallel (shunt) or in series with the armature or may be separately excited. The interpole and compensating windings must produce mmf's proportional to the current in the armature; consequently, they are connected in series with it.

The rotor winding has a basic function. Its object is to set up a conductor-current pattern fixed with respect to the main poles so that optimum interaction torque is developed at rest and at any speed. As a given conductor passes from the region of a N-pole field into that of the successive S-pole field, the current must reverse. The winding must therefore be closed and be connected through fixed brushes to a commutator.

### Field Windings

Shunt and separately excited field windings are designed on the basis of their applied voltage, and series windings are designed for a given current.

***Shunt.*** A winding supplied at voltage $V$ to produce an mmf $F = NI$ in $N$ turns, each of cross section $s$, has a resistance $r = \rho L_{mt}N/s = V/I$; hence,

$$s = \frac{\rho L_{mt}F}{V}$$

where $\rho$ is the conductor resistivity at working temperature and $L_{mt}$ is the mean length of turn. The field current is then $I = Js$, the current density $J$ being chosen so that the $I^2R$ loss $P = VI$ can be dissipated from the winding without the specified temperature rise being exceeded. The thermal dissipation depends on the exposed coil surface, which in turn depends on the space factor $k_s$ (the ratio of the total conductor section $Ns$ to the total coil cross section $Ns/k_s$). The field time constant $\tau = L/r$ is proportional to $N^2$; expressing the inductance as $L = qN^2$, the time constant is

$$\tau = \frac{(Ns/k_s)qk_s}{\rho L_{mt}}$$

*Series.* For a current $I$, an mmf $F = NI$, and an $I^2R$ loss $P$, the winding resistance is $r = P/I^2 = \rho L_{mt}N/s$; hence,

$$s = \frac{\rho L_{mt}NI^2}{P} = \frac{\rho L_{mt}IF}{P}$$

## Armature Winding

This is an arrangement of identical coils, each spanning approximately a pole pitch and so formed as to produce a winding in two layers. Typical coils are shown in Fig. 5.8. For a single-turn coil, a copper bar is cut to length, pressed into a diamond shape, with an upper and lower coil side, and finally taped. A two- or three-turn coil is formed from preinsulated copper strap. For small machines, coils are wire-wound on jigs and subsequently formed and taped.

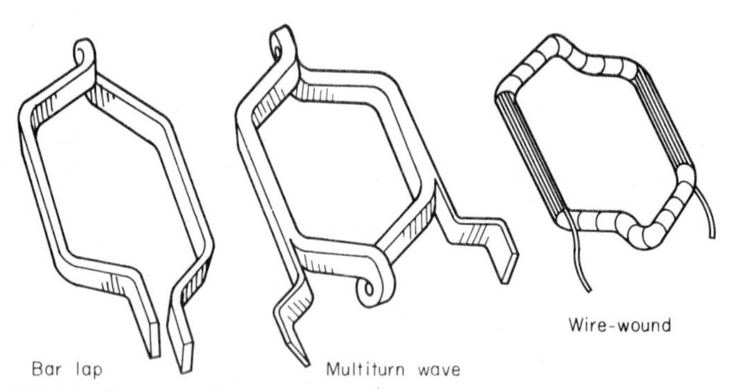


**FIG. 5.8** Armature coils.

The armature core is slotted to receive the coils. The top and bottom coil sides are given odd and even numbers, respectively, in sequence (Fig. 5.9). Slot 1 has four coil sides (with coil sides 1 and 3 above, and 2 and 4 below). Coil sides 1 and 3 are joined by the coil overhang (or end winding) to coil sides 22 and 24 at the bottom of slot 6; the lower coil sides (2 and 4) of slot 1 are part of another coil. The pitch, or span, of a coil in terms of coil sides must be an odd number; thus, in Fig. 5.9, $(22 - 1) = (24 - 3) = 21$. The following nomenclature is employed:

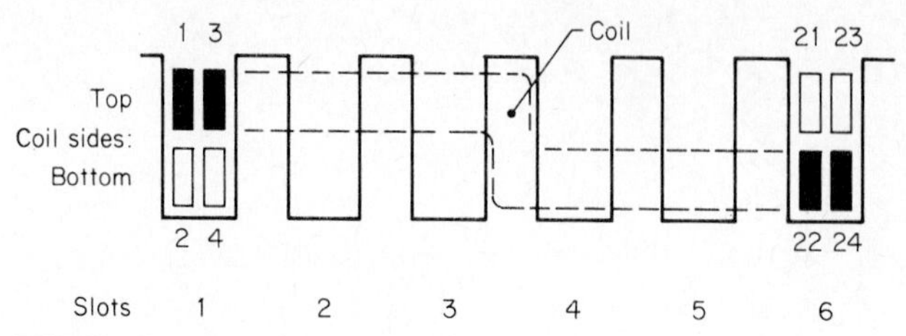

**FIG. 5.9** Slot and coil-side nomenclature.

$S$ = number of slots
$m$ = number of coil sides per slot
$C$ = number of coils or commutator sectors ($C = \frac{1}{2}mS$)
$N$ = number of armature turns ($N = Cz$)
$z$ = number of conductors per coil side ($z = N/S$)
$p$ = number of pole pairs
$a$ = number of pairs of armature circuits in parallel
$y_b$ = coil span or back pitch in coil sides
$y_f$ = front pitch (commutator end) in coil sides
$y_r$ = resultant pitch in coil sides ($y_r = y_b \pm y_f = 2y_c$)
$y_c$ = commutator pitch (in coils or commutator sectors between starts of electrically successive coils)

To preserve symmetry, the $2a$ parallel circuits into which the winding is divided by the brushes must all be identical. For this, $S/a$ and $p/a$ must both be integers.

Figure 5.10 shows diagrammatically the coils, pitches, and appearance of two successive coils; Fig. 5.10*a* is a simple *lap* winding and Fig. 5.10*b* is a simple *wave* winding, the terms being self-explanatory. The top coil sides are indicated by a solid line, the bottom ones by a dotted line. Each coil, whether single-turn or multiturn, starts at one commutator sector and ends on another. In the following examples, the number of coils is restricted for clarity.

***Simple Lap Winding.*** The winding rules are

$$y_b - y_f = y_r = 2 \qquad y_c = 1 \qquad a = p$$

Figure 5.11 is the *layout* diagram for a four-pole winding having 22 coils and two coil sides per slot; thus, $C = S = 22$. The coil span must approximate a pole pitch; since in coil sides this is $44/4 = 11$ (an odd number), we take $y_b = 11$ and $y_f = 11 - 2 = 9$. Consider coil 10 (shown in heavy line): It starts at sector 10, has a top coil side 19 and a bottom coil-side $(19 + y_b) = (19 + 11) = 30$, and ends at sector $(10 + 1) = 11$, where coil 11 starts. The brush spacing in commutator sectors is $22/4 = 5\frac{1}{2}$, and they are placed with this spacing as indicated by $a$, $b$, $c$, and $d$, with $a$ and $c$ positive and $b$ and $d$ negative. The corresponding *sequence* diagram in Fig. 5.12 gives the same information more clearly. The brushes divide the closed winding into four circuits ($2a = 4$) in accordance with the rule $a = p = 2$.

If magnetic asymmetry (due to an eccentric shaft, for example) causes the emf's in the $2a$ supposedly identical circuits to differ, currents will circulate between the circuits via the brush-contact surfaces. To mitigate this effect, *equalizer*

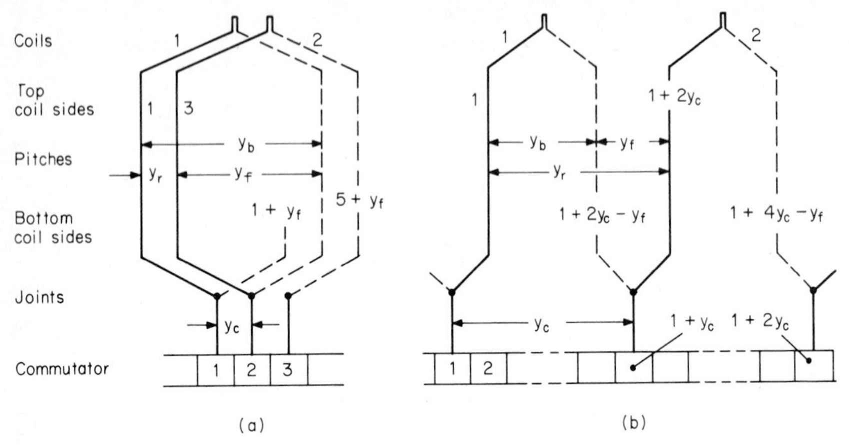


**FIG. 5.10** Basic coil connections: ***(a)*** lap; ***(b)*** wave.

interconnections are made at the back of the winding to join points whose emf's should at every instant be the same. Such points are spaced at intervals $C/p$, and a number of equalizer rings, each with $p$ joints so spaced, are fitted. Two rings are shown in the layout and sequence diagrams of Figs. 5.11 and 5.12. Circulating equalizer currents flow in the armature coils, but not in the commutator connections or brushes.

***Simple Wave Winding.*** The winding rules are

$$y_c = \frac{C \pm 1}{p} \qquad y_r = 2y_c = y_b + y_f \qquad a = 1$$

Successive coils are located in successive double pole pitches. The commutator pitch $y_c$ is such that, after one tour around the armature, the last coil side is one sector ahead or astern of sector 1, at which the first coil started. The *layout* diagram (Fig. 5.13) is for a four-pole armature with 21 coils. Applying the above winding rules, $y_c = (21 \pm 1)/2 = 10$ or 11 (say 10), and $y_r = 2y_c = y_b + y_f = 20$ (say $y_b = 11$ and $y_f = 9$). The corresponding *sequence* diagram (Fig. 5.14) shows how the brushes, despite their uniform spacing of $21/4 = 5\,^1/_4$ sector pitches, are associated with only two regions of the winding and give only one pair of parallel circuits ($a = 1$). It would be possible to dispense with one positive and one negative brush, but the current per brush would be doubled and there would be a minor asymmetry. Nevertheless, it is common in traction motors to adopt the two-brush form for easier access and maintenance.

***Limitations.*** A *lap* winding can be devised for any number $S$ of slots (although this is not necessarily practical or economic), but if equalizers are to be fitted, $S$ must be an integral multiple of $p$. Generally, $S$ is made a multiple of $2p$ so that $y_b$ can be equal to a pole pitch. In a *wave* winding, $C = py_c \pm 1$, so $C$ must be 1 more or 1 less than a multiple of the number $p$ of pole pairs. Any odd number will serve for four- and eight-pole machines; for six poles, $C$ must not be a multiple of 3.

***Duplex Windings.*** These are lap or wave windings with twice as many parallel circuits as the corresponding simple lap or wave windings. If the resultant-pitch rule for a lap winding were changed to $y_r = 4$, the result would be two separate lap windings connected to alternate commutator sectors. By using brushes wide enough

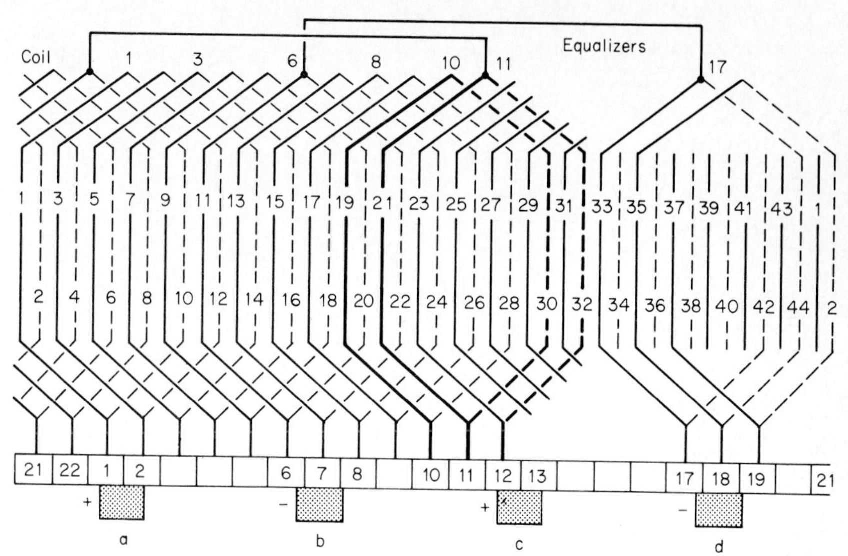


FIG. 5.11 Four-pole lap winding: layout.

to span at least two adjacent commutator sectors, the two windings would operate in parallel with, effectively, $a = 2p$. There are several versions of duplex and multiplex windings, but their use is rare.

## Armature EMF

The rotational emf generated between the positive and negative brushes in a machine that has $2p$ poles, $2a$ parallel circuits, $N$ total armature turns (or $Z = 2N$ active conductors), and a coil span approximating a pole pitch and is running at angular speed $\omega_r$ (rad/s) or rotational speed $n$ (r/s) is

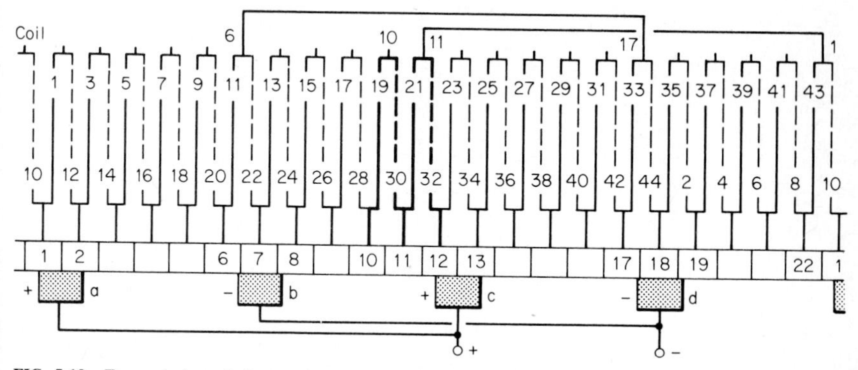


FIG. 5.12 Four-pole lap winding: sequence.

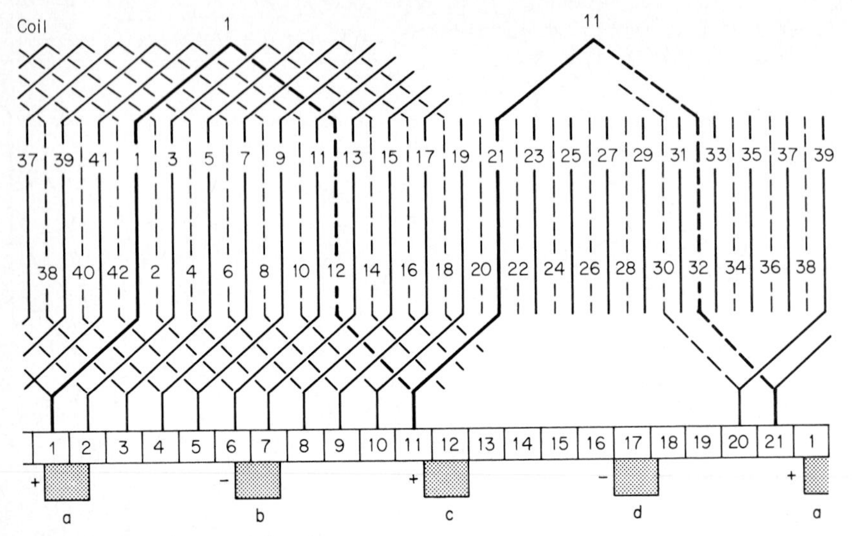


**FIG. 5.13** Four-pole wave winding: layout.

$$E_r = \frac{1}{\pi}\frac{p}{a}N\omega_r\phi_g = K\omega_r\phi_g = \frac{p}{a}Zn\phi_g \tag{5.1}$$

where $\phi_g$ (Wb per pole) is the useful gap flux linking the armature winding. For simple lap and wave windings,

Lap ($a = p$): $\qquad E_r = nZ\phi_g$

Wave ($a = 1$): $\qquad E_r = pnZ\phi_g$

## Armature Resistance

With $2a$ armature circuits and therefore $N/2a$ turns per circuit, the resistance per path for turns of cross section $s$ and mean length $L_{mt}$ will be $(\rho L_{mt}/s)(N/2a)$, where

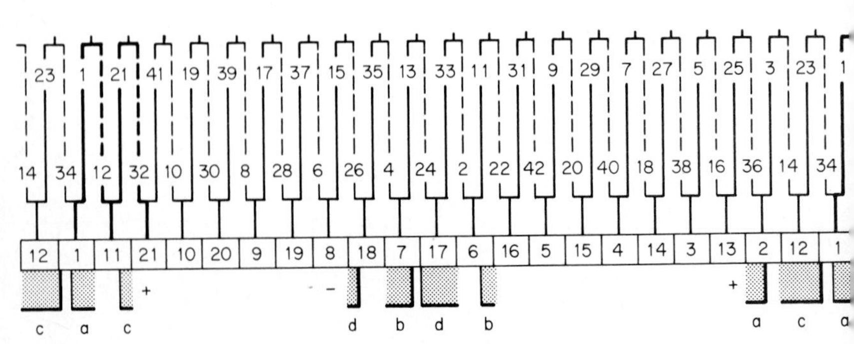


**FIG. 5.14** Four-pole wave winding: sequence.

$\rho$ is the resistivity of the conductor material at working temperature. The armature resistance between brushes is then

$$r_a = \frac{\rho L_{mt}}{s}\frac{N}{4a^2}$$

The armature terminal resistance has, in addition, the resistance of the brushes and of the brush-commutator contact surfaces. The former can be assessed from the resistivity of the brush material. The latter is nonlinear and is usually regarded as equivalent to a constant voltage drop (e.g., 1 V per brush set), substantially independent of the current.

### Airgap Flux MMFs

It is now possible to give values to the mmf components in Fig. 5.5.

***Field MMF.*** This is provided by one or more exciting windings to give $F_f = \Sigma N_f I_f$, assessed as in Table 5.1.

***Armature MMF.*** For a machine with $2p$ poles, $2a$ parallel armature paths, a total of $N$ turns, and an input current $I$, the current per path is $I/2a$ and the total mmf is $NI/2a$. The mmf is distributed as in Fig. 5.5*a*, its maximum value per pole being $F_a = NI/4ap$ on the brush axis. The distribution is triangular for a hypothetical uniformly distributed winding, but in fact the distribution is "stepped" as a result of the accommodation of the armature winding in slots; the stepping becoming more prominent with a reduction of the number of slots per pole and of the airgap length.

***Compensating MMF.*** To neutralize $F_a$ over the pole arc $b$, the mmf $F_n = F_a(b/Y)$ is required.

***Commutating-Pole MMF.*** Interpole windings are required in order to offset uncompensated armature mmf in the commutating zone and to set up a flux density $B_c$ to assist rapid reversal of the armature current. The mmf required is given by $F_c = F_a - F_n + kB_c$, where $k$ is a function of the interpole magnetic circuit. The commutating-pole flux magnetic circuit shares the yokes with the main-pole flux, increasing and reducing the flux respectively in successive yokes (Fig. 5.7).

## 5.4 LEAKAGE AND INDUCTANCE

Figure 5.4 shows the magnetic field patterns set up for no-load and for on-load conditions. The distribution of the gap mmf in Fig. 5.5 enables the useful gap flux distribution to be obtained, on which the armature emf and the commutating conditions depend. There remains the nonuseful *leakage* flux that follows paths between the main poles and interpoles and around the armature slots and end windings; these affect several aspects of performance, introducing inductances that delay commutation and the response to speed control.

The flux $\phi$ in a magnetic path across which an mmf $F$ is expended is $\phi = F\Lambda = F/S$, where $\Lambda$ is the permeance and $S$ is the reluctance of the path. The complex geometry of leakage flux paths makes estimation difficult. Normally it is assumed that leakage flux paths extend between ferromagnetic surfaces of infinite permeability that have some predetermined magnetic potential difference; then attention can be concentrated on the "air" parts of each path. The flux pattern is first established (e.g., by a flux plot) and divided into a number of parallel sections of area $a$ and length $l$. For $x$ sections, the leakage permeance is $\Lambda = \mu_0 \Sigma a_x/l_x = \mu_0 \Sigma \lambda_x$ and the problem reduces to the estimation of the *permeance coefficient* $\lambda = a/l$ for each section. Quite complex fields can be solved by digital computation.

## Main Pole

The pole-leakage flux is roughly mapped in Fig. 5.15. Four regions can be considered: (1) $\phi_1$ and $\phi_2$ between the facing edges and outer sides, respectively, of the pole-shoes and (2) $\phi_3$ and $\phi_4$ for the pole bodies. Then the total leakage flux leaving the two facing sides and the two end surfaces of one pole structure is

(1): $$\phi_{sl} = 2\phi_1 + 4\phi_2$$

(2): $$\phi_{pl} = 2\phi_3 + 4\phi_4$$

Between adjacent shoes the magnetic potential difference is the mmf required to drive the useful flux $\phi_g$ across two airgaps and the armature teeth and core. Between adjacent pole bodies, however, the magnetic potential difference falls as the leakage path considered nears the yoke, becoming equal to the yoke mmf where the pole bodies are attached to the yoke. The permeance coefficients are derived from the several dimensions $b$, $c$, and $l$.

For a useful gap flux $\phi_g$ produced by a field current $I_f$ in a pole winding of $N_f$ turns, the mean flux per pole approximates $\phi_p = \phi_g + \phi_{sl} + {}^1/_2\phi_{pl}$, and the field-coil inductance is $L_f = N_f\phi_p/I_f$. A rough approximation that neglects leakage, slotting, and saturation is

$$L_f = \frac{1}{2}\pi N_f^2 Dl \frac{b}{Y}\frac{\mu_0}{l_g}$$

This is a direct application of the relationship $\phi = F\Lambda$, using the dimensions $D$ and $l$ of the rotor, the pole-arc/pole-pitch ratio $b/Y$, and the gap length $l_g$, based on the

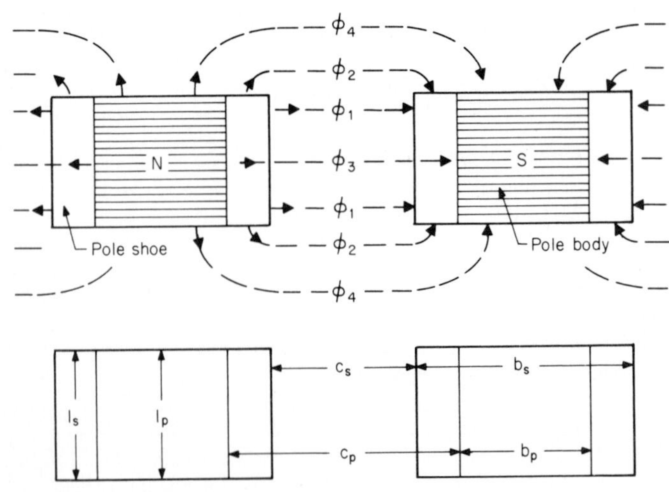


**FIG. 5.15** Pole-leakage flux.

useful flux $\phi_g$. The result may be multiplied by a factor (e.g., 1.2) to take rough account of the leakage flux.

## Armature

Provided that the brush axis is not shifted from the $q$ axis, the armature mmf develops a leakage flux disposed symmetrically about the interpolar $q$ axis. For a four-pole machine, Fig. 5.16*a* shows a leakage flux pattern for the armature due to its own current, with the main poles unexcited. Most of the flux takes a path through the armature core and teeth and is completed circumferentially through the pole shoe; the remainder (where the mmf is greatest) penetrates the interpolar zone. The flux pattern is built up by a superposition of components, in particular the slots (Fig. 5.16*b*) and the end-winding overhang (Fig. 5.16*c*).

***Slot Leakage.*** Consider a slot in which all conductors carry the same current in the same direction (Fig. 5.17*a*). The flux set up by the slot current crosses the slot width and completes its closed path around the core and teeth. Then the flux density at any level $x$ in the slot is the current linked multiplied by the slot permeance, which is that of the air path of length $w_s$ in Fig. 5.17*b*. With the dimensions in Fig. 5.17*b*, the overall slot permeance per unit of axial length is

$$\lambda_s = \frac{h_1}{3w_s} + \frac{h_2}{w_s} + \frac{2h_3}{w_s + w_w} + \frac{l_g}{y_s}$$

The second and third terms on the right-hand side are simple area/length ratios, using in the third term the mean of the slot and wedge widths $w_s$ and $w_w$; the mmf is the total slot current. The first term for the region occupied by the conductors

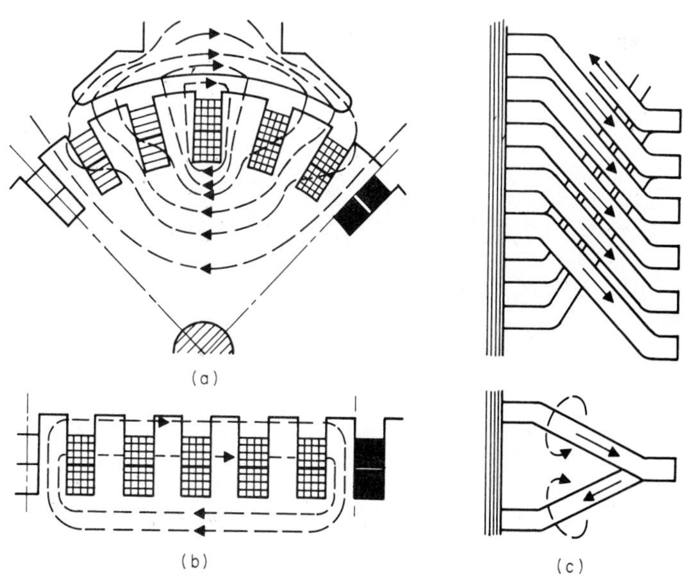


**FIG. 5.16** Armature-leakage flux: ***(a)*** a leakage flux pattern for the armature due to its own current, with the main poles unexcited; ***(b)*** slot-leakage flux; ***(c)*** end-connection flux.

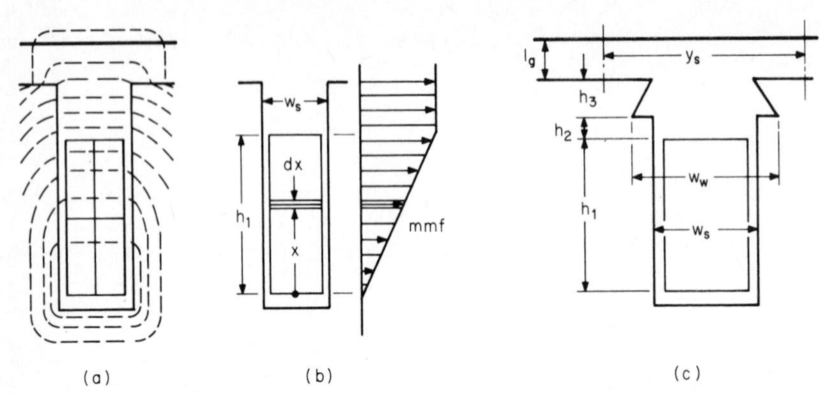


**FIG. 5.17** Slot-leakage flux: ***(a)*** a slot in which all conductors carry the same current in the same direction; ***(b)*** slot mmf; ***(c)*** slot dimensions.

has an mmf and an effective conductor-linked area proportional to $x/h_1$. The fourth term approximates the permeance presented to slot flux that completes its path in the pole shoe and crosses the airgap twice. For an axial slot length $l_s$, the total slot permeance is $\mu_0 l_s \lambda_s$.

***Overhang Leakage.*** The end-winding configuration is complex in respect to its conductors, current, and field. The following empirical formulas are in terms of $l_o$ (the length of a coil in the overhang) and the equivalent radius $r$ or perimeter $q$ of a coil:

$$\lambda_o = \frac{1}{2\pi} \ln \frac{l_o}{2r} \qquad \lambda_o = 0.18\left(\ln \frac{l_o}{q} + 0.07\right)$$

***Overhang and Slot Leakage.*** It is usual to express the overhang and slot leakage in terms of the axial length $l_s$ of a slot. Then an armature coil of $z$ conductors in series, carrying a current $I$ and of slot length $l_s$, has a leakage flux

$$\phi_s = 2\mu_0 I z l_s \lambda_a$$

where $\lambda_a = \lambda_s + (l_o/l_s)\lambda_o$

***External Leakage.*** In Fig. 5.16 it can be seen that, besides the slot and overhang components, there is a considerable armature flux that passes through the armature core and teeth, crosses the airgap twice, and completes its path through the pole shoe. Figure 5.18*a* shows the conditions for a noncompensated machine. Disregard the slotting (i.e., treat the armature conductors as producing an equivalent, uniformly distributed current sheet) and assume that the armature-produced flux extends only over the pole arc $b$. Then at any point within this region, the flux density can be inferred from the armature mmf there. At the pole edge the flux density is $B_b = \frac{1}{4} NI(b/Y)(\mu_0/l_g)$, and the corresponding field energy density is $\frac{1}{2} B_b^2/\mu_0$. Summing the field energy in the region and equating it to $\frac{1}{2} L_a I^2$ gives

$$L_a = \frac{1}{48}\pi N^2 Dl \left(\frac{b}{Y}\right)^3 \left(\frac{\mu_0}{l_g}\right)$$

When, as in Fig. 5.18*b*, there is a pole-face compensating winding, the external armature flux pattern is radically changed. If the armature and compensating current

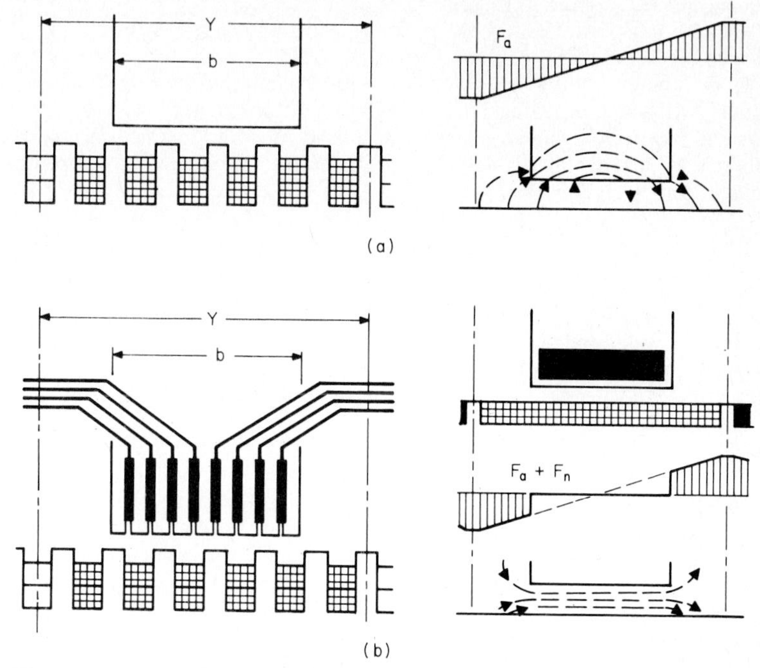


**FIG. 5.18** Armature external-leakage flux: ***(a)*** the conditions for a noncompensated machine; ***(b)*** a pole-face compensating winding.

sheets within the pole arc $b$ are exactly matched, the flux of their combination is circumferential. For a two-pole machine with $N$ armature turns and an input current $I$, the field-energy density summed in the gap space gives the inductance

$$L_a = \frac{1}{2} N^2 l l_g \frac{b}{Y} \frac{\mu_0}{\pi D}$$

This inductance is due to the armature and compensating windings in combination, not to the armature alone, and therefore applies to the armature *circuit*. The total armature-circuit inductance requires the addition of the effects of armature- and compensating-winding slots, overhang, and interpoles.

## 5.5 COMMUTATION

The commutation process provides the essential switching and position sensing by means of which the current directions in successive regions of the rotor winding are reversed in order to secure for any speed a direct armature emf and a sustained direction of torque. The conventional method involves passing a current through the sliding contact between a carbon brush and a circular assembly of copper bars (or commutator sectors), each bar being connected to a pair of coil ends. As a commutator sector passes under a brush, the current in the coil to which it is connected is reversed. The brush-commutator assembly is an electromechanical switching system that operates by mechanically making and breaking electric contact between the brush and the sectors, a process that inherently occurs at any speed.

Proper commutation is crucial. Because the commutator is carried on the rotor, it must withstand centrifugal forces without distortion and must be limited in temperature rise to avoid softening. The brushes must be retained in holders with a sliding fit in order to enable optimum pressure to be applied for proper surface contact. The voltage between sectors has to be low to prevent sparkover.

## Current Reversal

Figure 5.19 refers to an idealized case. The armature coils are of full pitch, with one turn per coil, and the brush width is equal to that of the commutator sectors. Consider a particular coil *C*, connected to sectors 1 and 2, as it moves past a brush. The normal coil current is $I_a$, and the brush current $I_b = 2I_a$ is assumed to be kept constant. In condition (i), the brush current is provided from coils *B* and *C*. In (ii), the coils having moved by half a sector pitch, coil *C* is short-circuited and the brush current is furnished by the contributions $I_a$ from coils *B* and *D*. For the further movement in (iii), coil *C* has been incorporated into the group of reversed current: thus, its current is $I_a$, which, with a like current in *D*, forms the brush current $I_b$.

If the coil inductance were negligible, the division of the brush current $I_b$ between adjacent sectors [as in condition (iv) of Fig. 5.19] would be governed only by the combined resistance of the coil and the brush-contact surface. The contact surfaces $a_1$ and $a_2$ can at any instant be regarded simplistically as providing equivalent resistances $r_1$ and $r_2$. As the areas change with the motion of the commutator, $r_1$ rises and $r_2$ falls to aid current reversal by *resistance commutation.* This effect is relied upon for acceptable commutation in small machines without interpoles.

At operating speed, the time $t_c$ during which a coil is short-circuited by a brush during current reversal is typically 2 ms. For a brush current $I_b = 50$ A, the mean rate of change of coil current is therefore of the order of 50 kA/s. The effect of coil inductance can therefore be considerable.

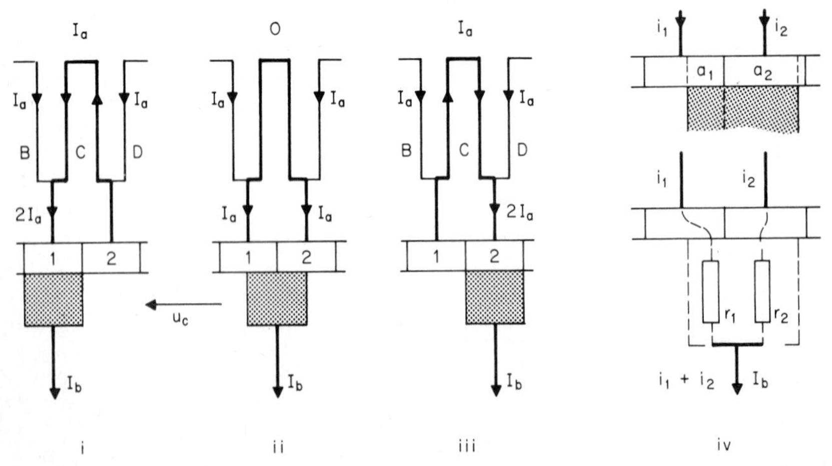


**FIG. 5.19** Current reversal.

## Commutation EMFs

Practical commutation is a process complicated by the effects of inductance. Features contributing to the inductance of a coil undergoing commutation are:

1. There may be more than two coil sides per slot. Besides the self-induced emf in a commutating coil, mutual-inductive effects occur as the result of current change in other coil sides in the same slot.
2. The coils may be short-pitched. The commutation of top coil sides is then not contemporaneous with that of bottom coil sides.
3. The brushes normally span more than one commutator sector and may cover between two and four sectors.
4. Sudden current changes set up eddy currents in the armature conductors and teeth, introducing additional $I^2R$ loss.

In general terms, there is in each commutating coil a pulsational (self- and mutual-inductive) emf $e_{pc}$ that opposes current change. To counter $e_{pc}$ and accelerate current reversal, it is necessary to induce in the coil a rotational emf $e_{rc}$ by means of interpoles in the commutating zone. The magnetic polarity of an interpole is the same as that of the main pole ahead (in the direction of rotation) for a generator, and opposite for a motor. Precise neutralizing of $e_{pc}$ by $e_{rc}$ is not possible, for the former depends on switching instants, rates of change, and inductances, and the latter depends on the motion of conductors across a constant magnetic field.

## Inductive EMFs

If the brushes are set in the $q$ axis (the neutral, or interpolar, axis), only the changes in slot- and overhang-leakage fields contribute to the inductive emf in a coil undergoing commutation. Both the self-flux and the mutual flux (i.e., that set up by adjacent coils being simultaneously commutated) must be taken into account. Writing $p$ for $d/dt$, the inductive emf in coil 1, which carries an instantaneous current $i_1$ and has a self-inductance $L_{11}$, is then

$$e_{pc} = L_{11}pi_1 + \Sigma L_{1x}pi_x$$

where $L_{1x}$ is the mutual inductance between coil 1 and coil $x$.

Evaluation of $e_{pc}$ requires the rate of change of current during reversal to be known. The conditions are so complex that some assumption must be made. The simplest is the *straight-line* commutation in Fig. 5.20*a*; here, $pi$ is given by $2I/t_c$ and the inductive emf is constant, as indicated by the hatched curve. The *sinusoidal* commutation in Fig. 5.20*b* leads to a cosine wave of $e_{pc}$. Where interpoles are used and are either too strongly magnetized (Fig. 5.20*c*) or too weak (Fig. 5.20*d*), the rate of change of current is too fast or too slow, respectively. In Fig. 5.20*d*, reversal has to be completed by a spark between the brush and the associated commutator sector as they separate; the dotted lines shown in the sparking zone are purely hypothetical. Sparking leads to a higher $I^2R$ loss, a higher temperature rise, and damage to the commutator surface, which intensifies brush wear.

The self- and mutual-inductance parameters $L_{11}$ and $L_{1x}$ are derived from the permeance coefficients in Sec. 5.4, with appropriate modifications.

***Self-Inductance.*** For equal currents in the same direction in all coil sides in a slot, the bottom coil sides are linked by the flux produced by the top coil sides as

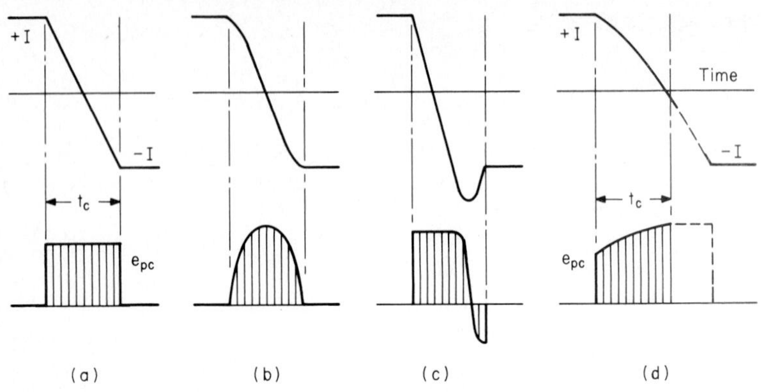


**FIG. 5.20** Current reversal and inductive emf: ***(a)*** straight-line commutation; ***(b)*** sinusoidal commutation; ***(c)*** overcommutation; ***(d)*** undercommutation.

well as by their self-flux. The dimensions concerned are given in Fig. 5.21*a*, where $h_1$ is the depth of *each* layer. Then the *slot* permeance coefficients are

Top:
$$\lambda_{st} = \frac{h_1}{3w_s} + \frac{h_2}{w_s} + \frac{w_t}{2l_{gc}}$$

Bottom:
$$\lambda_{sb} = \lambda_{st} + \frac{h_1}{3w_s}$$

The term $w_t/2l_{gc}$ is an approximation for the permeance to the leakage flux that twice crosses the gap $l_{gc}$ at the interpole shoe. If the spacing $h_3$ between the layers is significant, the term $h_3/w_s$ can be added to $\lambda_{sb}$. The *overhang* (Fig. 5.21*b*), expressed in terms of per-unit axial length $l$ of the slots, is

Overhang:
$$\lambda_o = \frac{l_o}{l} \ln \frac{l_o}{2r}$$

where $r$ is the equivalent radius of a coil side.

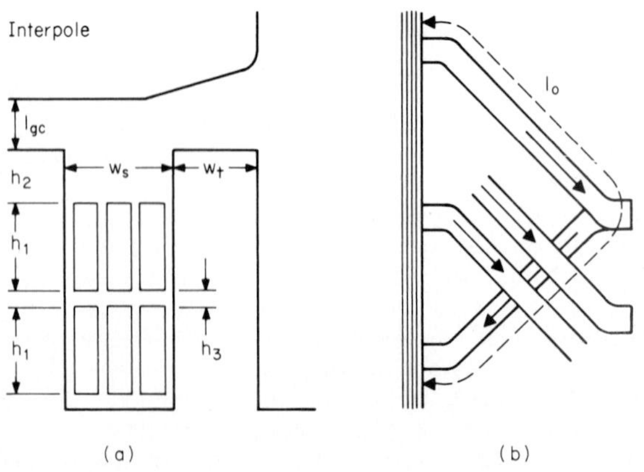


**FIG. 5.21** ***(a)*** Slot dimensions; ***(b)*** overhang dimensions.

The overall permeance per unit of slot length for self-flux is, for coil 1, $\lambda_{11} = \lambda_{st} + \lambda_{sb} + 2\lambda_o$. The self-inductance for $z$ turns per coil is

$$L_{11} = L_{st} + L_{sb} + 2L_o = \mu_0 l \lambda_{11} z^2$$

***Mutual Inductance.*** Between a given coil and the others in the same pair of slots, the mutual inductance can be determined as for the self-inductance, the values in some cases being identical. The several inductances comprise $L_{tt}$ between coil sides in the top layer, $I_{bb}$ between those in the bottom layer, and $L_{tb}$ and $L_{bt}$ between the top and bottom coil sides, with the values for the slot portion given by

Top: $\qquad L_{tt} = L_{st}$

Bottom: $\qquad L_{bb} = L_{sb}$

Top to bottom: $\qquad L_{tb} = L_{bt} = \mu_0 l \lambda_{st} z^2$

In the overhang (Fig. 5.21*b*), with the top and bottom coil sides crossing at an angle not very different from 90°, the mutual inductance between layers is negligible, and only coils in the same layer contribute to mutual flux; hence,

Overhang: $\qquad L_{oo} = \mu_0 l \lambda_o z^2$

***Total Inductance.*** For coil 1, the induced emf of self-inductance in the slots and overhang is $L_{11} p i_1$. The emf of mutual inductance is the summation of contributions $L_{1x} p i_x$ for current changes $p i_x$ therein. These concern the various inductances $L_{tt}$, $L_{bb}$, $L_{tb}$, $L_{bt}$, and $L_{oo}$. All the current rates of change $pi$ are assumed identical, but they do not occur simultaneously.

***Commutation Time.*** Let the commutator have sectors of width $w_c$ and a surface peripheral speed $u_c$, and let the brush width be $w_b$. Then the "ideal" commutation time for a given coil is $t_c = w_b/u_c$, during which the current is reversed. But the *effective* commutation time is the greater duration $T_c$, during which mutual-inductive emf's occur in the coil as a result of current changes in magnetically associated coils, and account is taken of the effect of short pitching.

Consider the simplified two-pole lap winding in Fig. 5.22. It has $C = 30$ coils and commutator sectors, $m = 6$ coil sides per slot, and therefore $S = 10$ slots. The coils are short-pitched by one slot, so that coil 1 spans from slot 1 to slot 5. The inductive emf is therefore given by

$$\begin{aligned} e_{p1} = {} & (L_{st} + L_{sb} + L_o) p i_1 \\ & + L_{tt}(p i_2 + p i_3) + L_{bb}(p i_2 + p i_3) \\ & + L_{tb}(p i_{19} + p i_{20} + p i_{21}) + L_{bt}(p i_{13} + p i_{14} + p i_{15}) \\ & + L_{oo}(p i_2 + p i_3) \end{aligned}$$

The first term is for the self-inductance of the slot and the overhang of coil 1. The remaining terms are for the mutual inductance, respectively, between the top and bottom coil sides in the slot, the top to bottom coils and bottom to top coils in the slot, and finally the mutual inductance in the overhang.

Assuming *straight-line* commutation, the various emf's have the waveform indicated in Fig. 5.20*a*. When sector 2 just reaches brush $a(+)$, the current $i_1$ in coil 1 begins to reverse along the current-time line 1*t* in Fig. 5.23, completing reversal in time $t_c$. After a delay of $t_d = w_c/u_c$ for sector 2 to pass the leading edge of the brush, the current $i_2$ in coil 2 begins to reverse along the line 2*t*, and similarly for

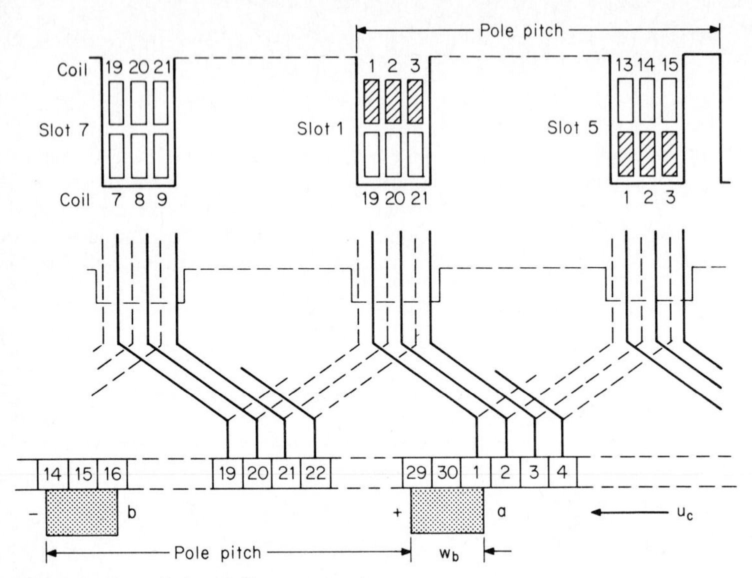


**FIG. 5.22** Two-pole lap winding: commutation.

coil 3. The total time of commutation for the top coil sides of the coils in slot 1 is thus $t_c + 2t_d$.

Were the coils of full pitch, the bottom coil sides of the coils considered would commutate at brush $b(-)$ simultaneously. But the first bottom coil in slot 1 is part of coil 19, which does not start to reverse until sector 20 reaches brush $b(-)$, i.e., $3t_d$ later, as indicated by the lines 19*b*, 20*b*, and 21*b* in Fig. 5.23.

The bottom coil of coil 1 in slot 5 has emf's induced by the current changes in coils 13, 14, and 15. Coil 13 is seen to have completed commutation one-half of a sector pitch $w_c$ prior to $t = 0$, and the current-change lines for coils 13, 14, and 15 precede those for 1, 2, and 3 by a time corresponding to the short pitching. Thus, $T_d = t_c + 8t_d$, giving the total time for the commutation of coils 1, 2, and 3 in slot 1.

Rearranging the expression above for $e_{p1}$ gives

$$\begin{aligned} e_{p1} = {} & (L_{st} + L_{sb} + L_o)pi_1 \\ & + (L_{tt} + L_{bb} + L_{oo})pi_2 \\ & + (L_{tt} + L_{bb} + L_{oo})pi_3 \\ & + (L_{bt})pi_{13} + (L_{bt})pi_{14} + (L_{bt})pi_{15} \\ & + (L_{tb})pi_{19} + (L_{tb})pi_{20} + (L_{tb})pi_{21} \end{aligned}$$

These terms are plotted in Fig. 5.23 in accordance with their origin (i.e., 1 for the self-inductance emf of coil 1; 2 and 3 for the mutual-inductance emf in coil 1 due to the currents in 2 and 3; etc.). The component emf's are summed to show the waveform of the inductive emf $e_{p1}$ in coil 1. Since an interpole cannot generate an equal and opposing emf $e_{rc}$, its effective emf is based on the mean value of $e_{p1}$.

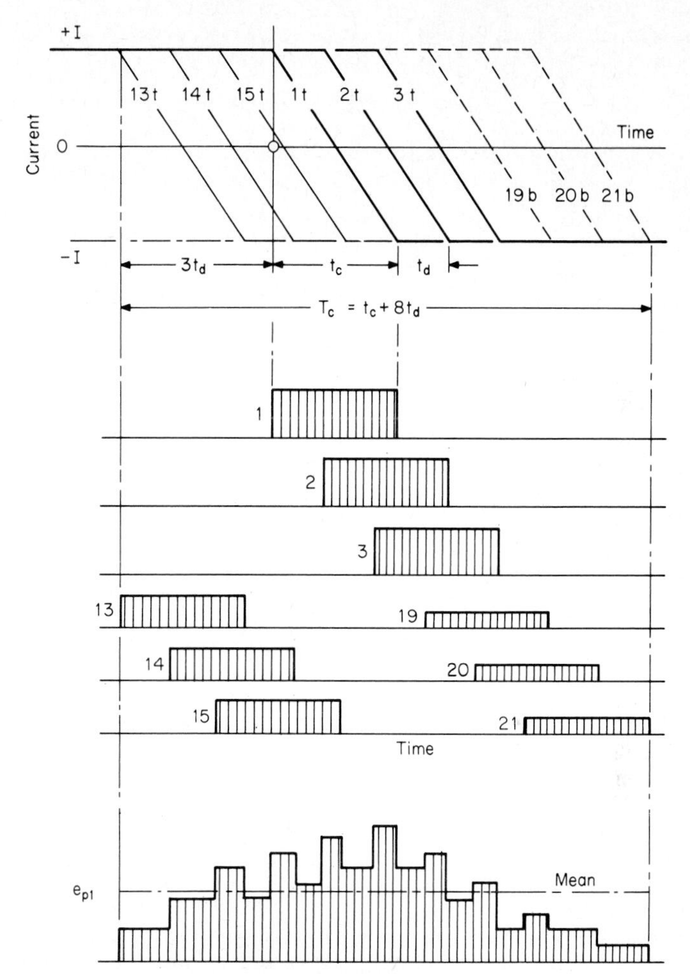


**FIG. 5.23** Commutation time and inductive emf.

In general, the commutation time $T_c$ for the coils in one slot—for lap windings with coils short-pitched by $q$ slots and for wave windings with $2a$ armature paths and $2p$ poles—is given by

Lap: $$T_c = t_c + [m(q + 1/2) - 1]t_d$$

Wave: $$T_c = t_c + \left(1/2m - \frac{a}{p}\right) t_d$$

***Approximate Mean Inductive EMF.*** The permeance coefficient of a slot of any physical dimensions depends only on its shape. For a rectangular slot which has a depth/width ratio of 3.5 and which houses $1/2m$ coil sides per layer, each with $z$ conductors carrying a current $I$, the slot-leakage flux approximates to $\phi_s = 4mzI$

$\mu$Wb/m of axial slot length $l$, and the overhang-leakage flux $\phi_o$ is about $0.1\phi_s$ per meter of overhang length $l_o$. The total for a coil side is

$$\phi = \phi_s + \phi_o = 4mzI(l + 0.1l_o)$$

and twice this for a complete coil. In the commutation process, $\phi$ is reversed in time $T_c$, and the mean inductive emf is

$$e_{pc} = \frac{16mzI(l + 0.1l_o)}{T_c}$$

which gives $e_{pc}$ in millivolts if $T_c$ is in milliseconds, serving as a preliminary approximation.

## Rotational EMF

To counterbalance the inductive emf, interpoles are located between the main poles and excited by the armature current to generate a motional emf $e_{rc}$ in coils undergoing commutation. The effect on the radial distribution of the airgap mmf is shown (idealized) in Fig. 5.5*b*. In the commutating zone the useful flux can be regarded as the superposition of three components:

1. That due to the *armature* mmf $F_a = N_a I_a / 4ap$, where $N_a$ is the *total* number of armature turns and $I_a$ is the *total* armature current. This peak value is located on the neutral, or $q$, axis and applies where a compensating winding is not fitted to the main poles.
2. That due to the *main poles*. Although the main-pole flux density in the $q$ axis is ideally zero, the coils undergoing commutation extend over slots on either side of the axis and therefore lie in the fringing fields.
3. That due to the *interpoles*. The peak of the useful flux density $B_{cm}$ should generate a rotational emf equal to the peak inductive emf in the commutating coils. Thus, $B_{cm} = e_{pc}/l_c u$ for an interpole of axial length $l_c$ (which usually differs from the armature axial core length $l$). The useful interpole flux is $\phi_c = B_{cm} l_c w_z / k_c$, where $w_z$ is the width of the commutating zone as measured circumferentially at the armature surface, and $k_c$ is the form factor of the flux distribution in the airgap.

***Interpole Flux.*** In Fig. 5.7 the superposition of main-pole and interpole fluxes is shown in general terms, with the consequential effect on the flux in the yoke and armature core. On full load, the mmf developed by the interpole winding has the same order of magnitude as that developed by a main-pole winding. Between a main pole and the adjacent interpole of opposite polarity, there is consequently a considerable leakage flux. The plot in Fig. 5.24*a*, derived from a computer solution using finite elements, shows the large leakage field of the interpole in a case where the main poles carry a compensating winding. The interpole flux at the root (i.e., the yoke end) may be 3 or more times the useful flux in the airgap at the shoe end that produces, across the interpole airgap $l_{gc}$, the emf $e_{rc}$. It is thus necessary in large machines to make a careful estimate of the leakage flux and to avoid excessive saturation at the root by tapering the interpole width (Fig. 5.24*b*). The leakage flux is reduced by making the axial length $l_c$ of the interpoles shorter than the armature axial length $l$.

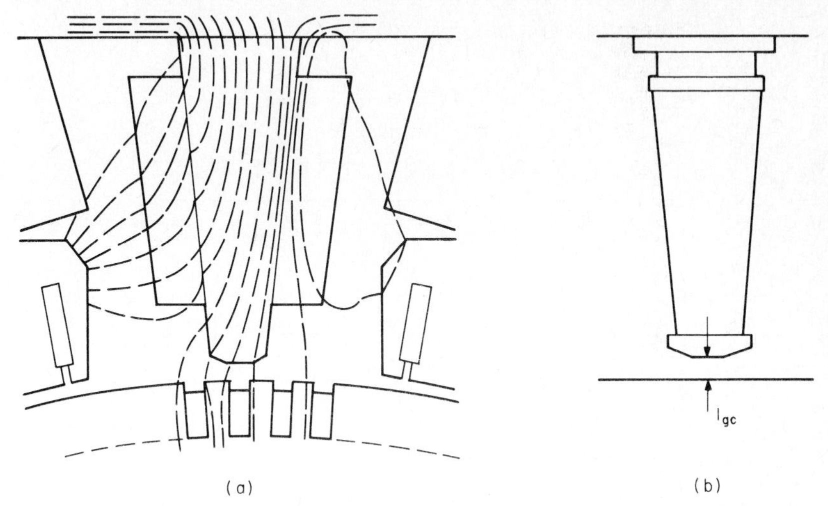


**FIG. 5.24** Interpole leakage flux: ***(a)*** plot derived from a computer solution using finite elements, showing the large leakage field; ***(b)*** tapered interpole width.

***Interpole MMF.*** With an armature mmf $F_a$ to be overcome, and a further mmf to produce the commutating field, the interpole mmf approximates

$$F_c = \frac{F_a + KB_{cm}l_{gc}}{\mu_0}$$

if the ferromagnetic parts of the magnetic circuit are not unduly saturated; the empirical factor $K$ is typically between 1.1 and 1.3. Since the number of teeth under an interpole shoe usually varies as the armature rotates, causing the gap permeance to fluctuate, the airgap length $l_{gc}$ is made larger than the main-pole gap $l_g$ by a factor normally between 1.2 and 1.7. And since good commutation is vital to the proper performance of a machine, shims may be provided at the root of an interpole (as in Fig. 5.24*b*). They may be magnetic or nonmagnetic. They enable the gap length $l_{gc}$ and the path reluctance to be modified in the light of test results.

Where the main poles have a neutralizing winding, the total mmf on the axis of the interpole is the resultant of the interpole and compensating winding mmf's in combination. The number of compensating turns is limited, and the appropriate number is chosen, the interpole carrying the remainder.

The use of interpoles does not entirely solve the problem of commutation. Magnetic saturation causes an interpole to become less effective as the armature current is increased, to the point where it is not capable of producing "black" (i.e., sparkless) commutation. Although brush-contact voltage drop will aid sparkless commutation with the interpole slightly overexcited or underexcited, proper commutating conditions are obtained only when the inductive emf's are counterbalanced by the rotational emf's in the commutating zone. The effect is shown by the "black band" tolerance diagram in Fig. 5.25, obtained by test. It indicates the band of tolerance within which the interpole as set is capable of suppressing sparking when its mmf diverges from the set value. On light load, the tolerance band is wide; on full load, a well-designed and properly set interpole may provide a ±0.05 per-unit (5 percent) tolerance, a condition common in industrial machines. (A lower tolerance

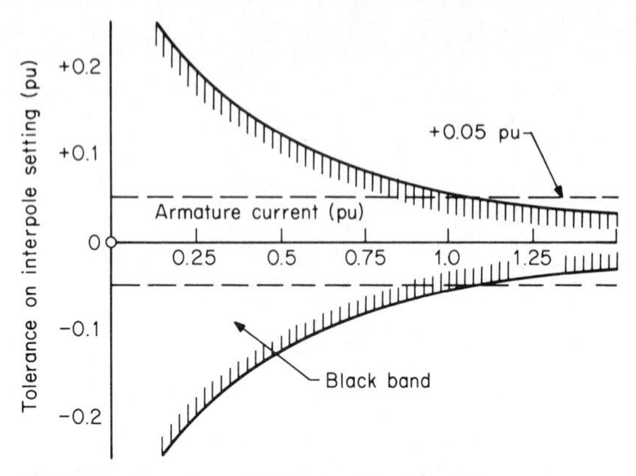


**FIG. 5.25** Interpole mmf tolerance: "black band."

would be required in a high-speed machine.) But greater current narrows the tolerance band, and if the rating of a given frame is to be increased to exploit better cooling and high-temperature armature-winding insulation, it is necessary to give detailed attention to pole-shoe contours, slot shapes, gap lengths, and saturation conditions.

## 5.6 COMMUTATOR AND BRUSHGEAR

Commutator sectors are made of hard-drawn or silver-bearing copper with good thermal and electrical conductivity and a high softening temperature. In conventional machines the sectors, separated by hard micanite plates 0.5 to 0.8 mm thick, are assembled between steel V-rings and compressed axially by through bolts, using the *arch-bound* or the *wedge-bound* constructions in Figs. 5.26*a* and *b*. An alternative is to dovetail the sectors into a steel hub, the copper being bonded to and insulated from the hub by fused glass. Many manufacturers have turned to the *molded* construction (Fig. 5.26*c*), in which the sector assembly, bonded internally to insulated steel rings, is consolidated by an insulating compound injected at high pressure and temperature. This method allows for ready cooling of the internal surface of the commutator and avoids much of the machining and thermal cycling that is necessary

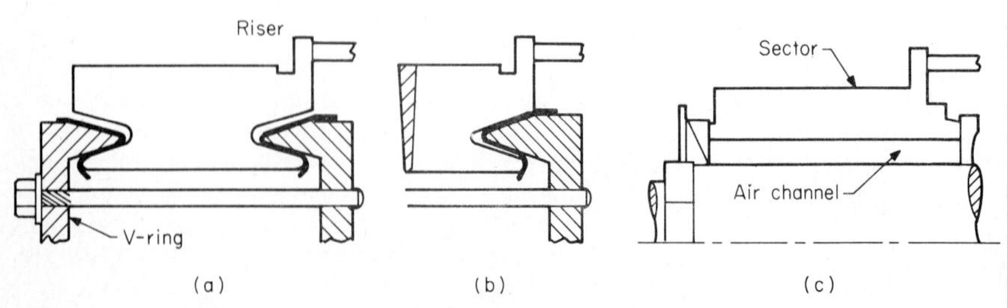


**FIG. 5.26** Commutators: *(a)* arch-bound; *(b)* wedge-bound; *(c)* molded.

with the conventional commutator to obtain a truly circular and concentric form. Circularity is essential if rapid brush and commutator-surface wear, brush "bounce," and sparking are to be prevented, and a bar-to-bar tolerance on radius may have to be very small (e.g., 5 μm). The final machining may demand the use of a diamond tool.

The many sectors of industrial commutators for higher voltages require the outside diameter of a commutator to be about 0.8 of the armature diameter lest the sectors be too thin. A long high-speed commutator may have insulated steel rings shrunk onto the outer surface to counter the centrifugal force.

## Brushgear

Brushes are made from carbon, graphite, and carbon-copper mixtures; the material is finely ground, mixed with a binder, and formed into blocks, which are then heat-treated. Many varieties of types and characteristics are available from specialist manufacturers. The general characteristics are:

*Natural graphite:* Has good lubricating properties and a low noise level. With a nonmetallic binder, the higher-contact voltage drop aids commutation in small machines without interpoles.

*Hard carbon:* Has robust mechanical properties and uniform wear characteristics. Is suitable for small machines and larger low-speed machines where the commutation conditions are easy.

*Electrographite:* Has a low rate of wear. Is widely used for industrial machines.

*Metal graphite:* Has a low contact drop and a high current density suitable for low-voltage high-current machines.

The brush type chosen depends on the armature voltage and the commutator peripheral speed. Generally, electrographitic brushes are chosen for industrial machines. Typical properties of brush types are given in Table 5.2.

***Dimensions.*** The required contact area is found from the current per brush set and from the current density. The brush dimensions (Fig. 5.27*a*) are the peripheral thickness $t$, the axial width $w$, and the height $h$; $t$ is determined from the number of sectors to be spanned, $w$ is limited to about $3t$ (the area required being furnished by several individual brushes in line if necessary), and $h$ must be several times $t$ to give stability and provision for wear.

**TABLE 5.2** Typical Brush Characteristics

| Type | Voltage drop per brush $v$, V | Current density $J$, A/cm² | Brush pressure $p$, N/cm² | Commutator peripheral speed $u_c$, m/s | Coefficient of friction $\mu_f$ |
|---|---|---|---|---|---|
| Natural graphite | 0.7–1.2 | 10 | 14 | 20–60 | 0.15–0.25 |
| Hard carbon | 0.7–1.9 | 6.5–8.5 | 1.4–2.0 | 20–30 | 0.10–0.20 |
| Electrographite | 0.7–1.8 | 8.5–11 | 1.8–2.1 | 30–60 | 0.10–0.20 |
| Metal graphite | 0.4–0.7 | 10–20 | 1.8–2.1 | 20–30 | 0.10–0.20 |

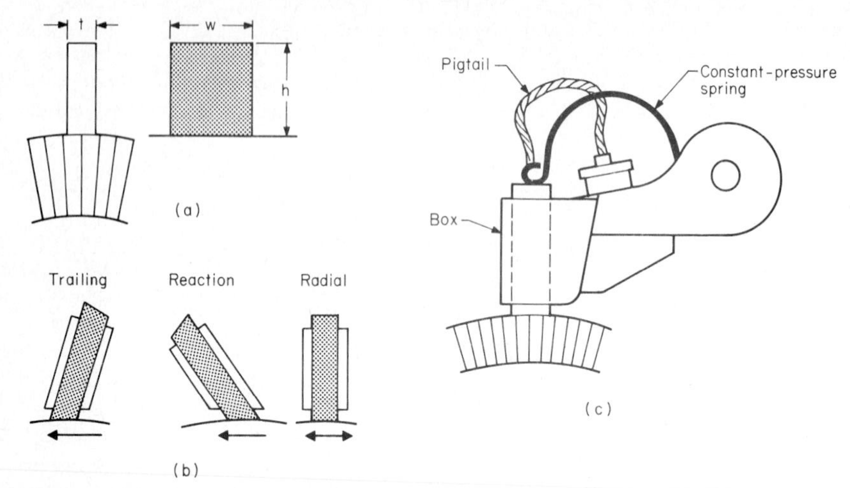


**FIG. 5.27** Brush details: ***(a)*** dimensions; ***(b)*** orientation; ***(c)*** holder.

***Orientation.*** In nonreversing machines the brushes may be given either a *trailing* or a *reaction* rake, with angles from the radial direction limited to 15° for the former and 30 to 40° for the latter (Fig. 5.27*b*). The reaction type is preferred if the brushes are narrow. For reversible machines, the *radial* orientation is common because it gives the same stability in each direction. The tendency of the brush to lie diagonally in its box is countered by a greater $h$ dimension and a close-fitting box.

***Holder.*** Figure 5.27*c* illustrates a typical brush holder, showing the flexible pigtail conductor (which is incorporated with the brush in manufacture) and the constant-pressure spiral spring.

The conductivity of a brush to currents circulating in coils undergoing commutation may be reduced by splitting it into two wafers, each $\frac{1}{2}t$ thick. In large machines the effect is enhanced by the use of double brush holders.

***Contact Phenomena.*** The process of current transfer between the brush and commutator surfaces is complex. The technological effects of voltage drop, friction, temperature rise, and wear are functions of the brush material, commutator peripheral speed, brush pressure, magnitude and direction of current, and ambient conditions. The actual contact between the brush and commutator surfaces occurs at relatively few points, the remaining area being separated by very small gaps on the order of 1 $\mu$m, across which conduction takes place by an electronic field-emission process. Combined with friction, the effect is to form a "skin," or patina, on the commutator surface. The skin is considered to be beneficial. Its formation is inhibited by low ambient temperature and pressure (as in aircraft service) and by atmospheric dryness and contamination.

***Loss.*** Brush loss comprises the effects of voltage drop and friction. The contact loss $vI$ falls with increasing brush pressure, but the friction loss rises. Their sum has a minimum corresponding to the condition of optimum wear. In general, the contact voltage drop at a positive brush differs from that at the negative, markedly for hard-carbon brushes but only marginally for electrographitic types. The friction coefficient varies with the brush current and with speed, but not with polarity.

### Flashover

An arc between successive brush arms or from a brush to the frame can be damaging. It may occur in steady-state or transient conditions and at any load. In the *steady state*, flashover may be initiated by the recessed gaps between the commutator sectors becoming filled with carbon-copper dust, forming paths between those sectors unable to withstand the voltage between them. Large and sudden *transient* load currents cannot be matched by the interpoles, the flux of which is delayed by eddy currents or limited by saturation. The resulting heavy sparking may develop flashover. Similar effects may result from the sudden interruption and reestablishment of the supply to a motor. Protection against flashover damage may require the use of high-speed circuit breakers. If the type of duty demands, flash barriers of arc-resisting insulating sheets, set close to the commutator surface, may have to be provided to prevent flashover.

## 5.7 PERFORMANCE EQUATIONS

The basic equation is that of the electromechanical power conversion in the armature, expressed symbolically in terms of

$E_r$ = armature rotational emf, V

$I_a$ = armature output or input current, A

$M_e$ = electromagnetic torque developed, N·m

$\omega_r$ = armature angular speed, rad/s

$n$ = armature rotational speed, r/s

Then $E_r I_a = M_e \omega_r = M_e 2\pi n$, applying to all conditions, whether steady or transient.

The practical realization of this equation involves much more detail. A suitable analytic model is required, and its relative parameters are found by test or from design data. Because a conventional dc machine can operate as a motor or a generator and (except where there is a series-field winding) can pass from one mode to the other automatically in response to load conditions, the same model applies to both. The complexity of the model and the equations that describe it depend on the use to which the equations are to be put. Four such models are:

1. *Simplified steady state:* The gap flux is a linear function of the main-field mmf alone (i.e., saturation and armature reaction effects are ignored), commutation is perfect, the field- and armature-circuit resistances are constant, and the inductances play no part.
2. *Practical steady state:* The resistances and brush-contact drop are assumed constant, saturation and armature reaction are included, commutation is perfect, and the operating conditions are determined by the load or the prime-mover characteristics.
3. *Linear transient state:* This is the extension of the simplified steady-state model to include the effect of constant inductance parameters.
4. *Practical transient state:* Rapid changes of speed and load result in complex variation in saturation level; eddy currents are induced; the interpole and compensating windings do not achieve their intended aims because of satura-

tion, and commutation is impaired; and the precise positioning of the effective brush axis is crucial. Coils undergoing commutation represent short-circuited loops on the $d$ axis and are therefore affected by changes of the main-pole flux. A few of these effects can be incorporated into a linear analysis if the saturation conditions can be linearized. Representative parameters are not readily evaluated from design data and may have to be inferred from test oscillograms.

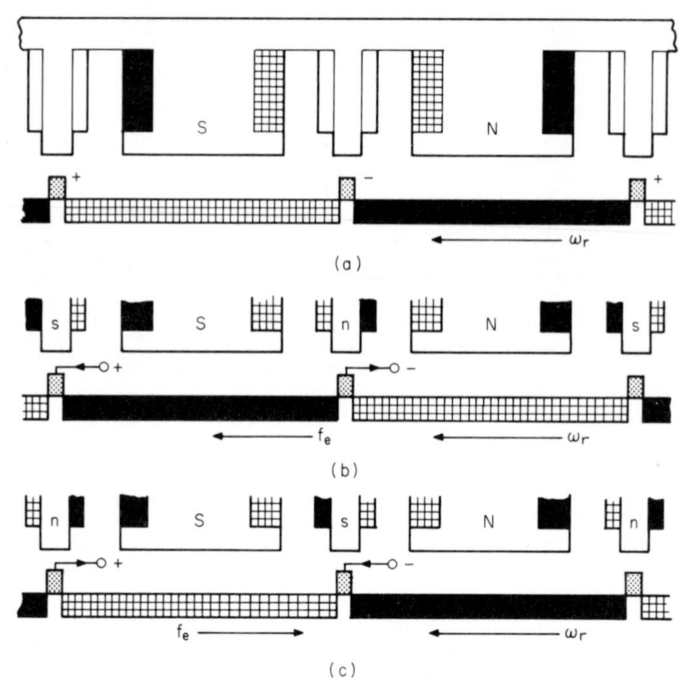


**FIG. 5.28** Motor and generator action: ***(a)*** armature emf; ***(b)*** motor; ***(c)*** generator.

## Generator and Motor Action

Figure 5.28 shows developed diagrams of a machine to illustrate steady-state action in the motor and generator modes; for both, the angular speed $\omega_r$ is a rotor movement from right to left. Figure 5.28*a* relates the armature rotational emf $E_r$ to the main-pole polarities N and S. For the motor (Fig. 5.28*b*), the armature rotational emf $E_r$ opposes the armature current, being maintained by an external voltage $V_a$; the interaction force $f_e$ has the direction of $\omega_r$, implying the production of motor torque. For the generator (Fig. 5.28*c*), $E_r$ drives the armature current out of the positive terminal, reversing the torque and making it necessary to drive the machine mechanically against $f_e$. Figures 5.28*b* and *c* also show the interpole excitation, giving the polarities *s* and *n*, which automatically reverse with the armature current.

## Connections

A dc machine has its main-field windings on the stator, their magnetic axis being the direct, or $d$, axis. The armature (rotor) winding has a magnetic axis fixed by the position of the brushes, normally in the quadrature, or $q$, axis. The two members can be excited by the same or by different direct currents, giving the three basic connections shown in Fig. 5.29 for the motor mode: $A$ is the armature winding, $F$ the shunt-field (or separate $d$-axis) winding, and $S$ the $d$-axis series-field winding. In addition, the stator carries on the $q$ axis an interpole winding $C$ (and, where required, a compensating winding $N$). The three connections are:

1. *Separate or shunt excitation:* The *separate* field current $I_f$ in $F$ is derived from a source of voltage $V_f$, and the current $I_a$ in $A$ is derived from a source $V_a$; the connection becomes one of *shunt* excitation if both the field and the armature currents are obtained from the same source of voltage $V = V_a = V_f$ (Fig. 5.29*a*).
2. *Series excitation:* Here the armature and series-field windings $A$ and $S$ are in series (Fig. 5.29*b*).
3. *Compound excitation:* In this case the gap flux is developed by the combination [either cumulative (assisting) or differential (opposing)] of the mmf's of windings $S$ and $F$ (Fig. 5.29*c*).

The connections apply also to dc generators.

## Steady-State Performance

Applying Eq. (5.1) and assuming fixed conditions of voltage, current, speed, and torque, the basic relations for electromagnetic torque $M_e$, angular speed $\omega_r$, and electromechanical power conversion $P$ are

$$E_r = \omega_r(K\phi_g) \qquad V_a = E_r \pm I_a r_a$$
$$M_e = I_a(K\phi_g) \qquad P = E_r I_a = M_e \omega_r \tag{5.2}$$

where $r_a$ is the armature-circuit resistance, and the + and − signs in $V_a$ refer to the motor and generator modes, respectively.

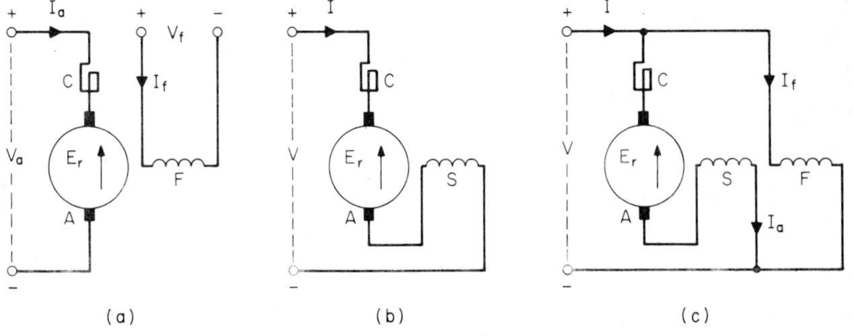


**FIG. 5.29** Field and armature connections: ***(a)*** separate or shunt excitation; ***(b)*** series excitation; ***(c)*** compound excitation.

Simplified performance characteristics that embody the basic motor torque-speed and generator voltage-current relations can be obtained by disregarding armature reaction and magnetic saturation effects, the working gap flux $\phi_g$ being assumed directly proportional to the resultant field excitation. Then $K\phi_g$ in Eq. (5.2) can be expressed as $GI_f$ for separate or shunt connections, $GI_s$ for series connections, or $G(I_f + xI_s)$ for compound connections, $x$ being the ratio of series to shunt turns per pole in the field windings $S$ and $F$. The factor $G$ relates the armature rotational emf per unit of speed to the resultant field current and is termed the *motional inductance.* The performance equations now become

$$E_r = \omega_r(GI') \qquad V_a = E_r + I_a r_a$$
$$M_e = I_a(GI') \qquad P = E_r I_a = M_e \omega_r \tag{5.3}$$

where $I'$ is the effective field current. Idealized characteristics of motors and generators are summarized in Fig. 5.30.

***Motor: Separate or Shunt Excitation.*** The field current $I_f = V_f/r_f$ is constant, giving a constant gap flux corresponding to $GI_f$; the torque $M_e = I_a(GI_f)$ is directly proportional to the armature current. Because $E_r = \omega_r(GI_f) = V_a - I_a r_a$, the speed is $\omega_r = (V - I_a r_a)/(GI_f)$, drooping from its no-load value $V_a/(GI_f)$ as the torque demand is increased. The armature resistance is usually small, and the speed-torque characteristic is then described as "constant speed."

***Motor: Series Excitation.*** The current is $I$ in both the armature and the field windings; hence, $M_e = I(GI) = GI^2$, a parabolic torque-current characteristic. The speed is $\omega_r = E_r/(GI)$, and ignoring the voltage drop $I(r_a + r_s)$, it approximates $V/(GI)$, a hyperbola, with the description "varying (or inverse) speed." At low loads the speed may be very high; consequently, series motors are often coupled permanently to their mechanical loads to prevent overspeeding.

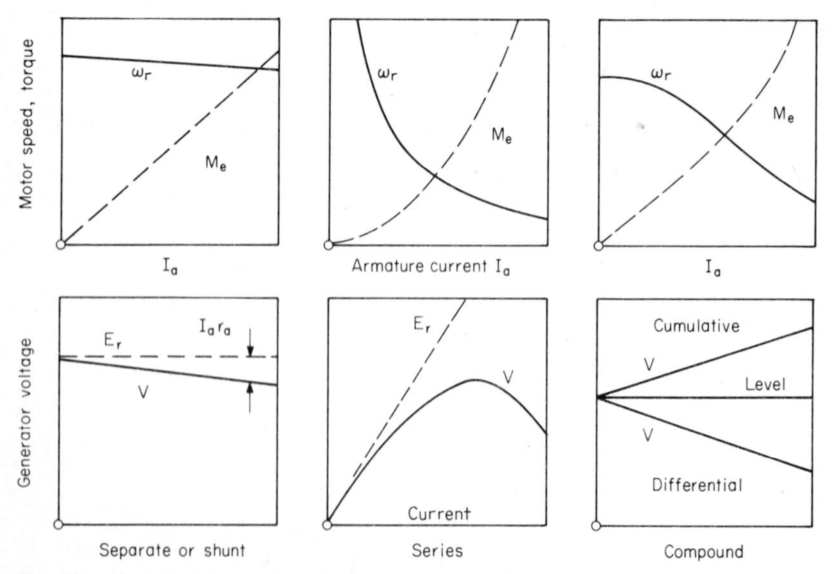


**FIG. 5.30** Idealized performance characteristics.

***Motor: Compound Excitation.*** The curves in Fig. 5.30 are for cumulative compounding, the mmf's of the field windings $S$ and $F$ being additive. The torque-current and speed-current relations resemble those of a shunt motor for low loads and of a series motor for higher loads. There is no risk of overspeeding, for there is an upper speed limit determined by the shunt field $F$.

***Generator: Separate or Shunt Excitation.*** With a constant-speed drive and a constant field current, the rotational emf $E_r$ is constant. The output voltage is $V_a = E_r - I_a r_a$, resulting in a slightly drooping output voltage-current relation.

***Generator: Series Excitation.*** The machine cannot generate on no-load (i.e., with the output terminals on open circuit) because there is no current for field excitation. The circuit is not completed until the terminals are closed through a load resistance $R$. For short-circuit conditions ($R = 0$), the current is at maximum and the output voltage is zero.

***Generator: Compound Excitation.*** The voltage-current characteristic is that of a shunt machine with an auxiliary series-field winding assisting (cumulative) or opposing (differential) the shunt-field mmf. The output voltage is thus increased or decreased as the load current rises. A small additive series excitation can provide a constant output voltage, the resistance voltage drops being compensated to give *level compounding.*

## Transient Performance

This is far more complex (and more important) than operation in the steady state. The electromagnetic part of the machine system is represented by electric circuits with self-, mutual, and rotational inductance as well as resistance; the mechanical part must take account of frictional, elastic, damping, and inertial torques as well as load torque. Linear analysis must neglect the effects of armature slotting, eddy currents, commutation phenomena, constructional asymmetry, vibration, noise, and such mechanical features as have to be accounted for in a real machine. In particular, magnetic saturation cannot be readily included; neither can remanence and brush-contact nonlinearities. But *numerical* solutions can deal with several nonlinearities, provided that the relevant parameters can be evaluated.

Retaining the assumption that saturation does not occur, the expressions in Eq. (5.3) can be modified for transient conditions. Consider a separately excited machine with a field winding $F$ and an armature winding $A$, with respective mmf's on the $d$ and $q$ axes and with all variables represented by their instantaneous values. The instantaneous power $p_f = v_f i_f$ into $F$ supplies the excitation loss $i_f^2 r_f$ and the rate of change $dw_f/dt$ of the stored field energy. The power into $A$ is $p_a = i_a^2 r_a + dw_a/dt$ for the $I^2R$ loss and the rate of change of armature magnetic energy, together with the instantaneous conversion power $e_r i_a = M_e \omega_r$. Expressing magnetic linkage in terms of constant inductances $L_f$ and $L_a$ and of the rotational inductance $G$, the system equations are

$$v_f = (r_f + L_f p) i_f$$

$$v_a = \omega_r (G i_f) + (r_a + L_a p) i_a \tag{5.4}$$

$$M_e = (G i_f) i_a = M_l + M_a + M_f + M_d$$

where the operator $p$ is $d/dt$ and the torque components are the load torque $M_l$, the accelerating torque $M_a = J p \omega_r$ for a total system inertia $J$, the friction torque $M_f$, and the damping torque $M_d$. Interpoles add a small inductance into the armature circuit, as do compensating windings.

Series motors and generators have high saturation effects; thus a linear analysis is not feasible. Where the machine field is compounded, the system equations become

$$v_f = (r_f + L_f p)i_f + L_{fs} p i_a$$

$$v_a = G(i_f + x i_a)\omega_r + (r_a + L_a p)i_a + L_{fs} p i_f \tag{5.5}$$

$$M_e = G(i_f + x i_a)i_a = M_l + M_a + M_f + M_d$$

where $L_{fs}$ is the mutual inductance between the field windings $F$ and $S$, $r_a$ and $L_a$ refer to the whole armature circuit, and $x$ is the ratio of turns of $S$ to $F$. The rotational emf and torque terms include components contributed by the fluxes (taken separately) of $F$ and $S$, which, on the linear assumption, are additive.

Equations (5.4) and (5.5) can give an insight into transient performance. They are stated in terms of the motor mode but can readily be adjusted for the generator mode. However, more accurate results are obtainable by digital computation, using saturation and other nonlinearity data assessed from testing. Saturation is further complicated in transient conditions by the induction of eddy currents, an effect mitigated by complete lamination.

### Armature Reaction

The no-load magnetization characteristic (Fig. 5.6) does not hold for load conditions, since distortion of the gap flux distribution (Fig. 5.4) increases the saturation in the high-density regions of poles and teeth. For a given field current $I_f$, the flux $\phi_g$ is reduced on load as if $I_f$ were reduced to $(I_f - I'_f)$. An alternative is to write $V_a = (E_r \pm I_a r'_a)$, where $r'_a$ is an equivalent armature-circuit resistance increased to take account of armature reaction. But both $I'_f$ and $r'_a$ depend on the load conditions and the degree of saturation and require a point-by-point evaluation.

## 5.8 MOTOR PERFORMANCE

The idealized characteristics for the motor in Fig. 5.30 are modified in a practical machine by saturation, armature reaction, and loss torque. The electromagnetic conversion torque $M_e$ is reduced to an output torque $M_l$ at the shaft by armature core loss, by friction (bearings and commutator), and by windage (air disturbance by the rotating armature augmented by rotor-mounted fans for cooling-air circulation). The practical speed and torque characteristics are shown by the thick lines in Fig. 5.31, to a base of total input current at a constant supply voltage; they relate to the steady-state condition.

### Speed Control

From Eq. (5.1), $E_r = K\omega_r\phi_g = V_a - I_a r_a$; hence,

$$\omega_r = \frac{V_a - I_a r_a}{K\phi_g} \quad \text{rad/s} \qquad \text{or} \qquad n = \frac{V_a - I_a r_a}{2\pi K\phi_g} \quad \text{r/s}$$

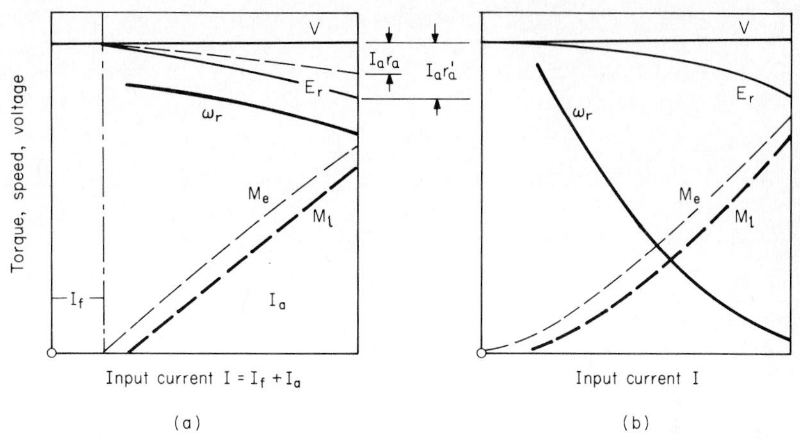


**FIG. 5.31** Motor steady-state characteristics: ***(a)*** separate or shunt; ***(b)*** series.

The steady speed of a dc motor can therefore be controlled by variation of (1) the airgap flux $\phi_g$, (2) the armature applied voltage $V_a$, and (3) the armature-circuit resistance by the addition of an external resistor $R$. Figures 5.32 and 5.33 summarize the speed-torque characteristics obtained with shunt and series motors. The thick lines refer to normal operation.

## Field Control

This is obtained by changing the field current.

***Shunt Motor.*** A rheostat $R$ is included in the field circuit, reducing the field current and consequently raising the speed, but retaining the constant-speed shunt characteristic. For a given armature current, however, the torque is reduced, so that a roughly constant output power is obtained. With *separate* excitation, a speed lower than normal can be achieved, as indicated by the dotted speed characteristic.

***Series Motor.*** Here the higher speed (and lower torque for a given current) is obtained either by (1) shunting the field winding as a whole by a "diverter" resistor $R$ or (2) subdividing the winding on each pole and excluding a part of it from the motor circuit.

## Armature-Voltage Control

If a voltage higher or lower than the rated value is economically obtainable, the speed is raised or lowered without change of torque for a given armature current. In Figs. 5.32 and 5.33 voltage 2 is about one-half of voltage 1, giving a 2:1 speed ratio.

***Shunt Motor.*** The constant-speed characteristic is retained, but at a speed lower than rated value.

***Series Motor.*** The varying-speed characteristic is retained. For series motors in traction and haulage, where motors share the same load equally, economic voltage control is obtainable by a series-parallel connection (Fig. 5.34). It is important to

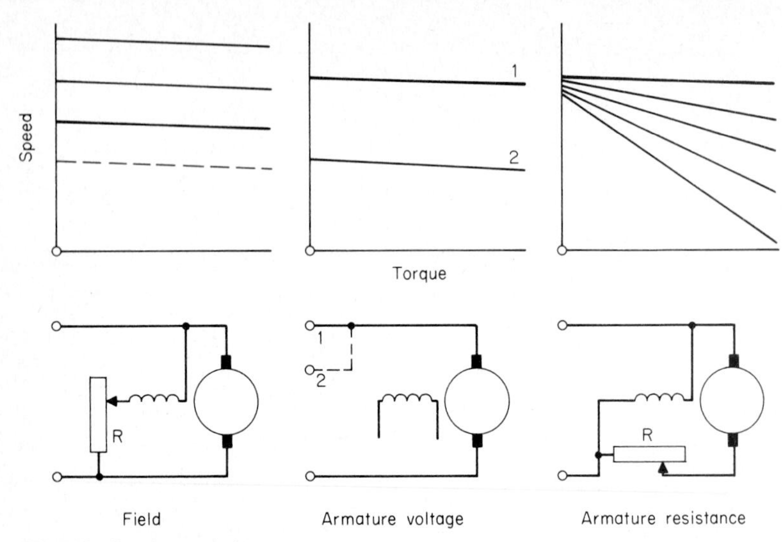


FIG. 5.32 Speed control: shunt motor.

switch from the series to the parallel connection (and vice versa) without interrupting the current. One method of transition, the "bridge," is to shunt each motor in series connection by a resistor, then to open-circuit the bridge, and finally to cut out the resistors. Another method, often adopted with battery-driven vehicles (such as golf carts and forklift trucks), is to operate two batteries with series or parallel connection, the former being employed in accelerating the vehicle from rest; further speed control is obtained by series resistors.

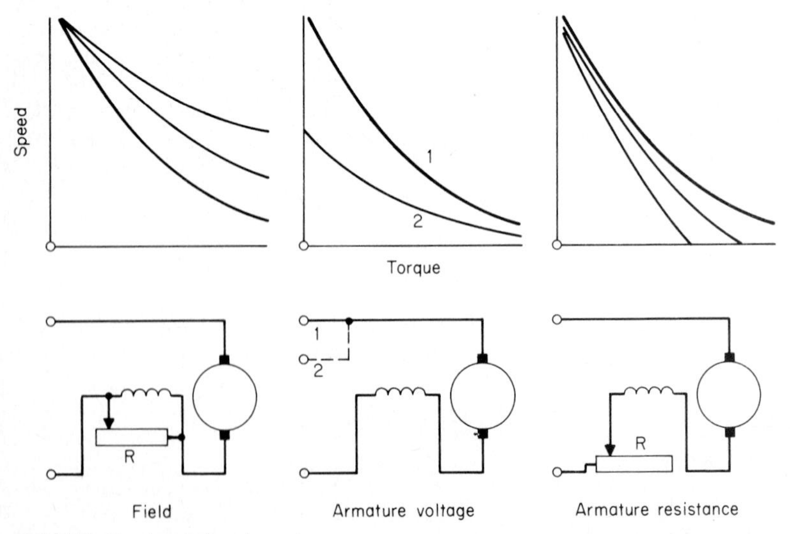


FIG. 5.33 Speed control: series motor.

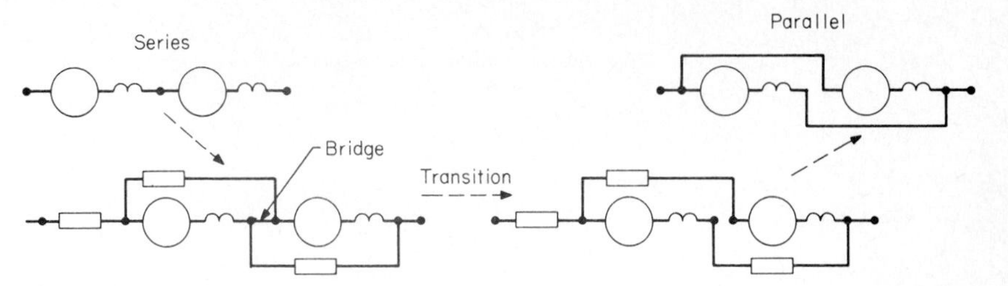


**FIG. 5.34** Series-parallel control.

***Separately Excited Motor.*** Most dc motors in industry are fed from ac supplies through controlled rectifiers. The field is fed through semiconductor diodes, or thyristors, or (for small machines) power transistors. The armature is supplied with a fluctuating but unidirectional voltage by means of a thyristor bridge assembly, the voltage being varied by adjustment of the firing angle.

## Armature Resistance Control

This method, although sometimes unavoidable, is wasteful because of the high $I^2R$ loss entailed. The speed-torque characteristic droops from the no-load to the load condition.

***Shunt Motor.*** The characteristic for a small value of $R$ may be applied to motor flywheel sets for load equalization. In general, the method is used for starting a machine on load by cutting out sections of $R$ as the speed rises.

***Series Motor.*** The use of armature resistance for run-up, and of series-parallel switching with field weakening for variable-speed running, is common for railroad traction systems.

## Starting

This is a common transient condition. Consider a motor run up from rest against a load torque $M_l$ (usually a speed-dependent quantity) and accelerated to a final angular speed $\omega_1$ in time $t_1$. The electromagnetic torque $M_e$ supplies the components $M_l$, $M_a$, $M_f$, and $M_d$ in accordance with Eqs. (5.4) or (5.5). The total $I^2R$ loss depends on the current-time relationship, but one component can be shown to be equal to that of the stored energy $\frac{1}{2}J\omega_1^2$ in the system inertia $J$. Thus, the armature input energy for acceleration is $J\omega_1^2$, and the frequent starting of high-inertia loads may result in excessive armature temperature rise and electric loading. The latter can be reduced by current limitation, but the acceleration time is lengthened.

Conventional starting is accomplished by means of a rheostat in the armature circuit. Ignoring inductance, the prospective initial armature current in a motor switched directly on line is $V_a/r_a$, because at rest there is no counter-emf of rotation. This peak may be many times the rated current and is permissible only for small machines. A rheostat limits the initial current typically to $1\frac{1}{2}$ to 2 times full-load current, thereby limiting mechanical shock to the shaft and load. The rheostat is cut out in sections, the process generally being automatic, e.g., by current-sensitive relays that operate between upper and lower current levels. In sophisticated control,

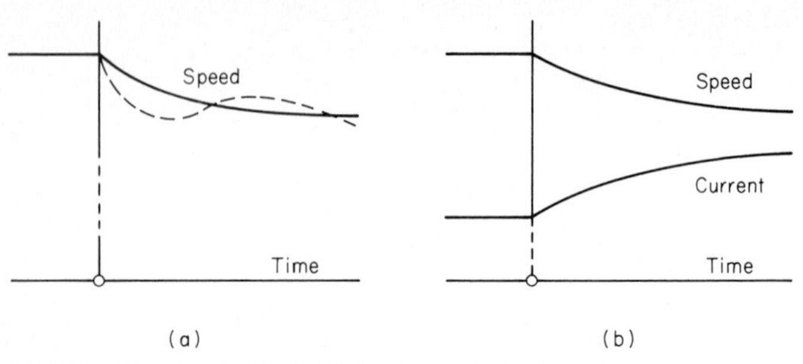


**FIG. 5.35** Step change of load: ***(a)*** shunt motor; ***(b)*** series motor.

particularly with separately excited machines fed from ac supplies through thyristors, a "soft" start that gives optimum run-up conditions is readily obtained.

## Loading

When changes of supply voltage, field current, or mechanical torque are impressed on a motor in operation, the machine adjusts to the new condition through a combination of electrical and mechanical transients. Typical cases are illustrated in Fig. 5.35*a* for a shunt motor and Fig. 5.35*b* for a series motor, in each case subjected to a step rise in load torque. In Fig. 5.35*a* the solid line gives the exponential speed drop in a machine with negligible armature inductance, while the dotted line is for a case of considerable inductance, leading to a damped oscillation.

In some drives the conditions vary cyclically. For instance, the load torque on a ship's propeller shaft fluctuates because of the action of the screw blades, load fluctuation occurs in rolling-mill drives, and air compressors impose rhythmic torque variation. These are typical of small changes around a steady mean operating condition, represented by the quantities $V_f$, $I_f$, $V_a$, $I_a$, and $M_e$. Then if the fluctuations are small, Eq. (5.4) can be written as changes

$$\Delta v_f = (r_f + L_f p)\,\Delta i_f$$

$$\Delta v_a = (r_a + L_a p)\,\Delta i_a + G\omega_r\,\Delta i_f + GI_f\,\Delta\omega_r$$

$$\Delta M_e = GI_f\,\Delta i_a + GI_a\,\Delta I_f = Jp\,\Delta\omega_r$$

superimposed on the steady mean values. If the variations can further be assumed to be sinusoidal and of frequency $\omega$, the operator $p$ can be replaced by $j\omega$ and treated as a problem in terms of phasors. The expressions above omit the effects of friction and damping, but these can be included by suitably reducing $\Delta M_e$.

## Braking

Mechanical brakes to hold a motor at rest against a load torque are sometimes necessary for safety in elevators and hoists. A more predictable braking effect is obtained by electric braking. Three common methods are shown by the connection diagrams for shunt and series motors in Fig. 5.36.

***Regenerative Braking.*** If the rotational emf $E_r$ of a motor exceeds the terminal voltage $V_a$, the current direction is reversed and the machine becomes a generator. Regeneration is obtainable with shunt and separately excited machines, and with compound-excited machines if the series compounding is relatively small. Series machines require the reversal of either the armature or the field connections and are more stable if provided with a stabilizing shunt winding. The method is possible only for the condition that $E_r$ is greater than $V_a$, which usually means an overspeed regime with the armature driven, such as by a hoist load being lowered or by a load of considerable inertia. Regeneration down to standstill is not possible, but the useful range of regeneration can be widened by increasing the field current or reducing the armature voltage, as in series-parallel control of motors in pairs.

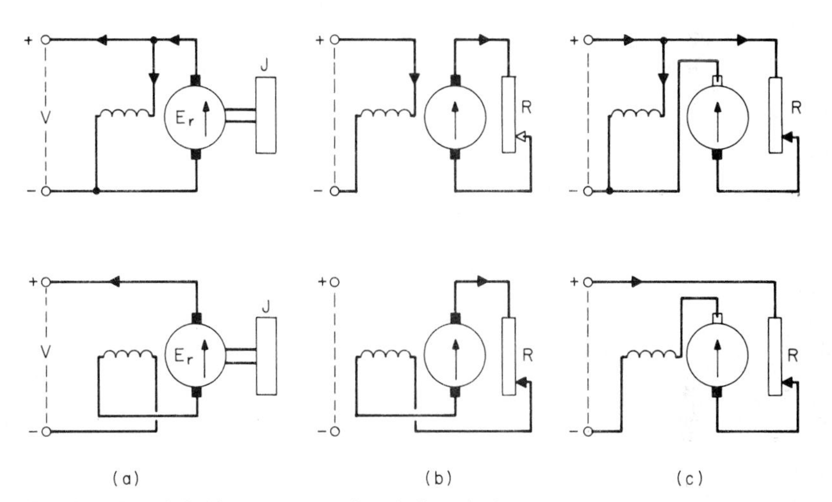


**FIG. 5.36** Electric braking: ***(a)*** regenerative; ***(b)*** dynamic; ***(c)*** countercurrent.

***Dynamic Braking.*** In a shunt machine the armature circuit is disconnected from the supply and closed onto a resistor $R$, but the field winding remains connected to the supply. Then energy extracted from the load inertia is dissipated in $R$, which can be adjustable in order to regulate the retardation rate; fast braking is obtained with a low value of $R$. For a series motor, it is necessary to reverse the connections of either the field or the armature, and the braking torque collapses if $R$ is too great, inhibiting self-excitation.

***Countercurrent Braking.*** This method (often termed *plugging*) gives rapid braking down to standstill, for the connections are such that in the armature circuit $E_r$ aids $V_a$, imposing a large current and braking torque. The machine may run up in reversed direction unless disconnected at zero speed. A series resistance $R$ is included in the armature circuit to limit the current, which, on connection for countercurrent breaking, would jump to $(V_a + E_r)/r_a$ if not so limited. The method is mechanically severe, and it is wasteful in that the whole of the stored kinetic energy is dissipated and the *total* energy lost is 3 times as much.

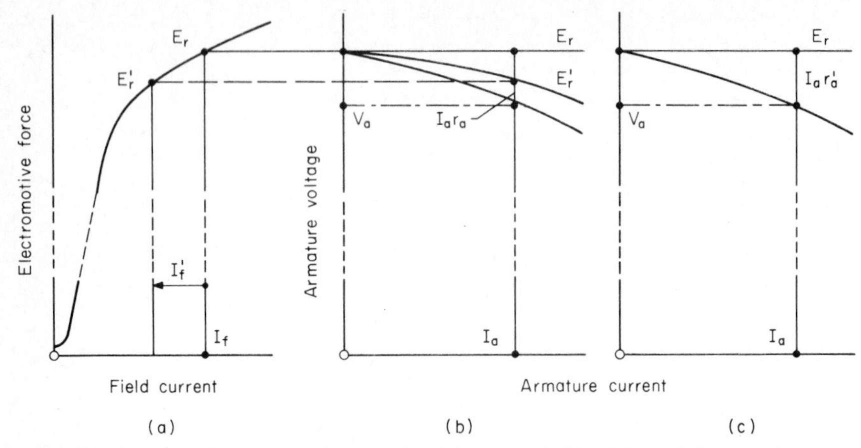


**FIG. 5.37** Generator voltage-current characteristics: ***(a)*** on no-load; ***(b)*** and ***(c)*** load characteristics.

## 5.9 GENERATOR PERFORMANCE

In several of its traditional applications, the dc generator has been superseded by solid-state electronic devices. However, it is still to be found in: diesel-electric locomotives; rolling-mill drives, in which motors are required to give special output characteristics; battery charging for standby or emergency plants; auxiliaries in ships, trains, and light aircraft; exciters for synchronous machines; and some control and servo systems.

Given the open-circuit magnetization characteristic for constant speed (Fig. 5.6), the steady-state load characteristic can be derived. The effect of armature reaction where the generator is not provided with a pole-shoe compensating winding is dealt with as for a motor (Fig. 5.31) by a putative reduction in the field current by $I'_f$ or an increase in the effective armature resistance to $r'_a$. Figure 5.37*a* shows the magnetization characteristic for the appropriate speed, giving the open-circuit rotational emf $E_r$ for a field current $I_f$. Figure 5.37*b* shows the reduction of $E_r$ to $E'_r$ by reduction of $I_f$ by $I'_f$, where $I'_f$ is a function of the armature current $I_a$; then $V_a = E'_r - I_a r_a$. In Fig. 5.37*c* the same load characteristic is obtained by use of the effective armature resistance, whereby $V_a = E_r - I_a r'_a$. In most constant-speed cases the speed in practice falls slightly with load because of the governing of the prime mover.

### Separate Excitation

The foregoing applies, with a gap flux and rotational emf depending on the excitation and speed.

### Shunt Excitation

The machine is excited by connecting the field circuit across the armature terminals. The output voltage $V$ and current $I$ are given by

$$V = V_a = V_f \qquad I = I_a - I_f \qquad I_f = \frac{V}{r_f}$$

$$V_a = E_r - I_a r'_a = K\omega_r \phi_g - (I + I_f)r'_a$$

Particular operating features concern the self-excitation process and the self-limiting load characteristic.

***Self-Excitation.*** Let a shunt machine be driven up to a fixed speed with both the armature and the field circuits open. Since there is no field current, there is no rotational armature emf $E_r$; and if the field circuit is closed, there is still no field current. Self-excitation depends on the presence in the magnetic circuit of *remanence*, a residual magnetic flux capable of generating a small emf, so that when the field circuit is closed, a small field current will flow. Provided that this is in such a direction as to augment the remanent (or residual) gap flux, the first condition for self-excitation is fulfilled. The second condition is that the field resistance shall not exceed an upper limit. Figure 5.38 shows the open-circuit magnetization characteristic for the specified speed, indicating the residual emf. The line *OG* is the voltage-current relation of the field current $I_f$ flowing through the resistance $r = r_f + r_a$; for self-excitation, it must lie to the right of the magnetization characteristic, the limit being the airgap line. Thus, for a field current *OX*, the voltage *XR* is absorbed in the field-circuit resistance, leaving an emf *RS* for increasing the current through the inductance of the field winding. The change of slope of the magnetization characteristic determines the stable point *G* of the field current, with the value *OF*. Because $E_r$ is speed-dependent, self-excitation at a lower speed will occur only if the field-circuit resistance *r* can be reduced to compensate for it.

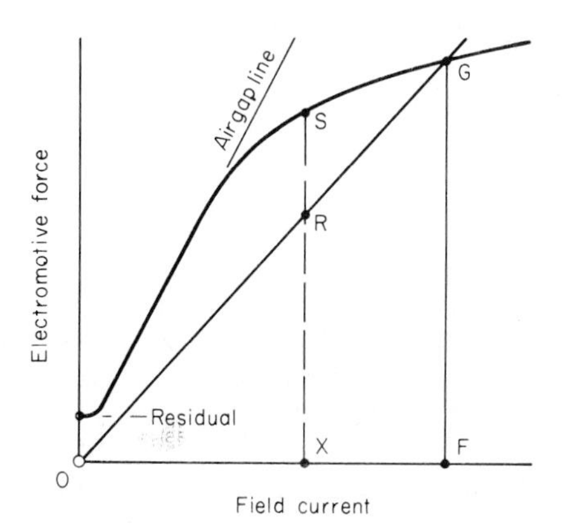


**FIG. 5.38** Shunt generator: self-excitation.

***Load Characteristics.*** Typical characteristics relating output voltage *V* and current *I* for a machine driven at constant speed are shown in Fig. 5.39. On no-load, the armature-voltage drop $I_f r_a$ is usually negligible, so that $V = E_{r0}$, the emf being determined by the magnetization characteristic and the field-resistance line *OG* in Fig. 5.39*a*. On load, as in Fig. 5.39*b*, with a current *I* supplied to a variable resistive load *R*, the terminal voltage *V* falls as a result of two effects: (1) the armature-voltage drop $I_a r'_a = (I + I_f)r'_a$, and (2) the concomitant reduction in

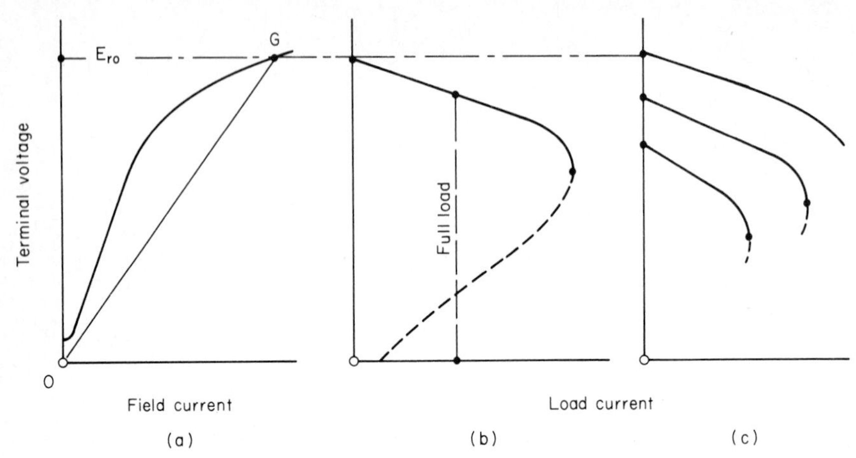


**FIG. 5.39** Shunt generator: ***(a)*** no-load characteristics; ***(b)*** load characteristics; ***(c)*** load characteristics with varying field current.

field current $V/r_f$. These cumulative effects increase as more load current is taken (by reducing $R$); and at an output current 2 or 3 times the rated value, a limit is reached, followed by a collapse of $V$ and $I$. For $R = 0$, $V$ is almost zero and $I$ is generated by the residual gap flux. The $V/I$ curves in Fig. 5.39*c* show the effect of lowering $I_f$ by including a field rheostat. Only a small increase in field resistance is feasible, for the field line *OG* in Fig. 5.38 approaches the slope of the airgap line and the machine becomes unstable. However, it can run stably if provided with a second field system carrying a control current.

## Series Excitation

This is an uncommon case. No current can flow unless the output terminals are closed through a load resistance *R*, and self-excitation is possible only if the field line $(r_s + r_a + R)$ has a slope less than that of the airgap line of the magnetization characteristic for a given speed of drive.

## Compound Excitation

A shunt machine can be (and often is) provided with a series winding *S*. The two methods of connection in Fig. 5.40 are the long shunt and the short shunt; there is little practical difference between them. The mmf $F_s$ of the series winding on normal load is usually not more than about 10 percent of the shunt-field mmf $F_f$. In cumulative compounding, $F_s$ acts in the same sense as $F_f$ to raise the terminal voltage on load or to keep $V$ substantially constant (level compounding). In differential compounding, the resultant field excitation is $F_f - F_s$, and the terminal voltage falls considerably with load, a drooping $V/I$ characteristic that can be used for generator protection when liable to heavy impulsive overloads or short circuits, or to release the kinetic energy in a flywheel in load-equalization schemes.

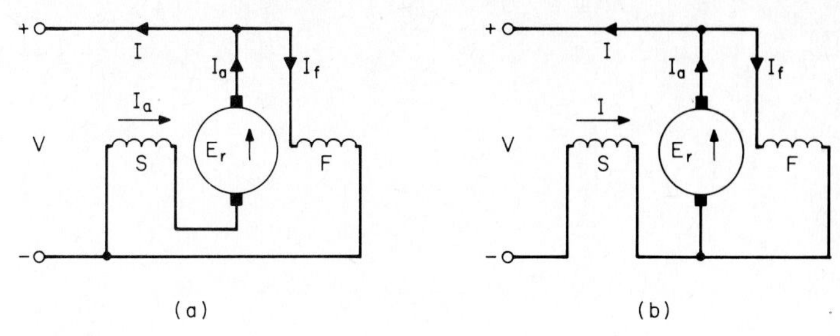


**FIG. 5.40** Compound generator: ***(a)*** long-shunt and ***(b)*** short-shunt connections.

## Transient Performance

The effects of sudden or cyclic variation in the load on a generator can be analyzed as for motors, using Eqs. (5.4) and (5.5) with the appropriate modifications. A particular case is that of a generator short circuit. This is a severe condition, especially in a large generator. Some of the effects are: high saturation levels; the induction of eddy currents in unlaminated conductor and core material; magnetic coupling of interpoles and compensating windings to the field system; severe armature reaction effects, producing abnormally large voltages between commutator sectors, with possible flashover; interpole saturation, inhibiting the development of the high level of commutating flux required. Representative parameters are difficult to evaluate because they vary with current and saturation level (though they can be linearized piecemeal).

Based on linear inductance parameters, Fig. 5.41 shows the transient currents $i_a$ and $i_f$ in the armature and field circuits of a separately excited generator, initially excited but not loaded. The peak and decay of $i_a$ are reflected in $i_f$, in which the rise of armature flux is opposed by the field current in accordance with the constant-linkage theorem.

# 5.10 LOSSES, EFFICIENCY, AND RATING

The sources of loss in a dc machine are as follows:

*Conduction:* $I^2R$ loss in windings and their associated resistance regulators. In industrial machines the conduction loss is referred to a standard winding temperature.

*Exciter:* Applicable to separately excited machines.

*Core:* The cyclic reversal of the magnetic flux in the armature body and teeth as they pass the successive N and S field poles has the basic frequency $f = np$ in a $2p$-pole machine running at a rotational speed $n$ (r/s). It is customary to consider the armature core loss as comprising the components of hysteresis proportional to $f$ and of eddy current proportional to $f^2$.

*Pole face:* Local high-frequency flux pulsations in the pole shoes as the teeth run past them.

*Brush contact:* Normally taken as the product of the brush current and a constant-voltage drop, depending on the brush grade.

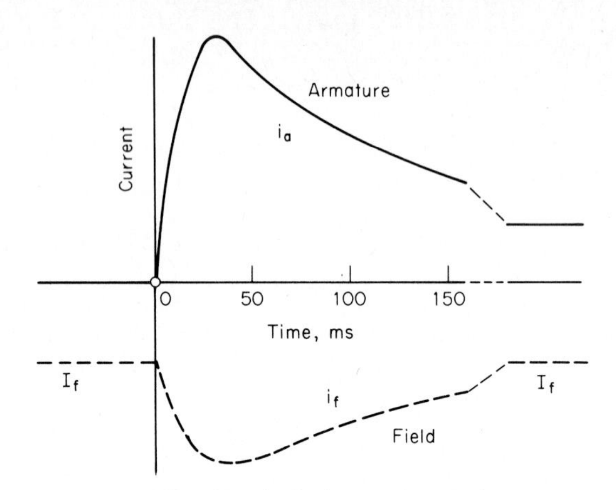


**FIG. 5.41** Sudden short circuit of separately excited generator.

*Brush friction:* The friction loss depends on the brush grade and pressure and on the commutator peripheral speed.

*Bearing friction:* Calculable empirically from the contact area of the bearing and the rubbing speed. The loss is reduced if ball and/or roller bearings can be used.

*Windage:* The air drag on the armature surface is significant in large high-speed machines. The loss is increased if rotor-mounted fans are employed for circulating cooling air.

*Load (or stray):* Several sources of loss, not included in the above, may appear on load: core loss due to gap flux distortion, eddy currents in windings, and loss in armature coils when short-circuited by a brush during commutation.

## Efficiency

The per-unit power efficiency of a motor with an output $P$, a loss $p$, and an input $P_i = P + p$ is

$$\eta = \frac{P}{P_i} = 1 - \frac{p}{P_i}$$

and in specifications is normally quoted for rated output. The losses may be classified as *load*- or *speed*-dependent or as the sum of *fixed* (constant) and *varying* (load) components. They are obtained in design by calculated estimates and on load by measurement.

In general terms, the main features of the efficiency/input-current characteristic $\eta/I$ in the steady state can be shown by using simple approximations for the loss elements. Core loss is assumed to be proportional to (flux $\times$ frequency)$^2$ and therefore to $(nI_f)^2$. Brush-contact loss $vI$ is proportional to $I$. Mechanical (friction and windage) loss is assumed to be proportional to $n^2$.

***Shunt Motor.*** Assuming the field flux and the speed to be fixed, the loss $p$ is represented by the three components:

*Fixed:* Field and regulator $I^2R$; no-load core; mechanical.

*Varying:* Brush contact, proportional to $I$; armature $I^2R$ and stray, proportional to $I^2$.

The total loss can therefore be written $p = k_0 + k_1 I + k_2 I^2$, the input power is $P_i = VI$, and the efficiency is

$$\eta = 1 - \frac{p}{P_i} = 1 - \frac{p}{VI} = 1 - \left(\frac{K_0}{I} + K_1 + K_2 I\right)$$

where $K_0 = k_0/V$, etc. The power-efficiency/input-current relation for the shunt motor has the characteristic shape shown in Fig. 5.42. Here $I_0$ is the no-load current, with $VI_0$ approximating the "fixed" loss $K_1$. Maximum efficiency occurs for $K_1 = K_0/I$ (provided that the loss due to brush-voltage drop is small), i.e., for the fixed loss equal to the variable armature $I^2R$ and stray loss.

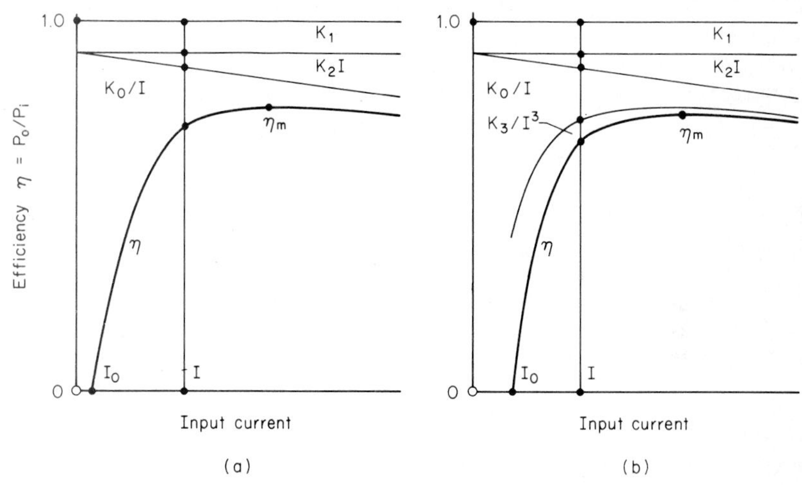


**FIG. 5.42** Steady-state power efficiency: ***(a)*** shunt; ***(b)*** series.

***Series Motor.*** Both the current $I$ and the speed $n$ vary with load, and the loss distribution is more complex. Using the idealized expressions in Eq. (5.3), the product $nI$ is assumed to be constant. Mechanical loss is taken as proportional to $n^2$ and therefore to $1/I^2$. Core loss is proportional to the squares of flux and frequency, i.e., to $(nI)^2$, which is constant. The loss components are therefore

*Fixed:* Armature core.

*Varying:* Brush contact, proportional to $I$; armature-field $I^2R$ and stray, proportional to $I^2$; mechanical, proportional to $1/I^2$.

The total loss is $p = k_0 + k_1 I + k_2 I^2 + k_3/I^2$, and the efficiency is

$$\eta = 1 - \left(\frac{K_0}{I} + K_1 + K_2 I + \frac{K_3}{I^3}\right)$$

with the result shown in Fig. 5.42. The no-load current corresponds to maximum speed, which is self-limiting in a small motor but requires a large machine to remain connected to its load.

***Generator.*** The power efficiency of a generator driven at an approximately constant speed and supplying a steady load can be found as for a motor, but here the losses are added to obtain the electrical equivalent of the prime-mover drive power.

### Energy Efficiency

A modern dc motor is rarely operated in the steady state, for its prime advantage is its flexible adaptation to wide-ranging speed control, braking, reversal, and automatic interchange between the motor and generator modes. Assessment of energy loss therefore has to be made on a basis of its duty cycle and the consequent fluctuation of its temperature rise. The energy efficiency of a machine throughout a duty cycle is the ratio $W_o/W_i$ of the output and input energies. Optimizing the energy efficiency is likely to be of considerably greater importance than raising the steady-state power efficiency, for the motor-load system is electromechanical, often with inertial stored kinetic energy as a dominant characteristic and with temperature rise as an operating limitation.

### Thermal Rating

All losses result in the production of heat, raising the temperature in the region of origin above that of its immediate environment in accordance with (1) the rate of heat production, (2) the thermal capacity, and (3) the rate of heat transfer to the cooling medium (normally the ambient air). The temperature of a region generating loss in service must not exceed that above which the material is endangered or its life unduly shortened; this is particularly true of commutators and winding insulating materials. The problem of ventilating a machine is discussed in Chap. 12, and the temperature limits for various classes of insulant are discussed in Chap. 13. These determine the *thermal rating.*

The duty of a particular machine is individual to it. For assessing thermal performance, however, international standards have been set up for commonly required duties: namely, continuous, short-time, and periodic intermittent operation (with or without extended starting and electric braking); inertia and regenerative loading; and duty-cycle duration.

## *5.11 DESIGN*

In industry, the dc generator is obsolescent except for such applications as rolling-mill drives, standby sets, and small windmills for space or water heating. But dc motors remain competitive for speed control, especially with controlled-rectifier supplies.

The main design interest lies in the economic batch production of motors in a range of ratings (e.g., 10 to 40 or 50 to 500 kW) provided by a coherent series of frame sizes defined by their shaft-center heights. In a given frame (Fig. 5.43), the armature diameter is fixed, but its active core length may be varied to cover outputs within the frame's capability. The *rating* is the set of quantities (voltage, current,

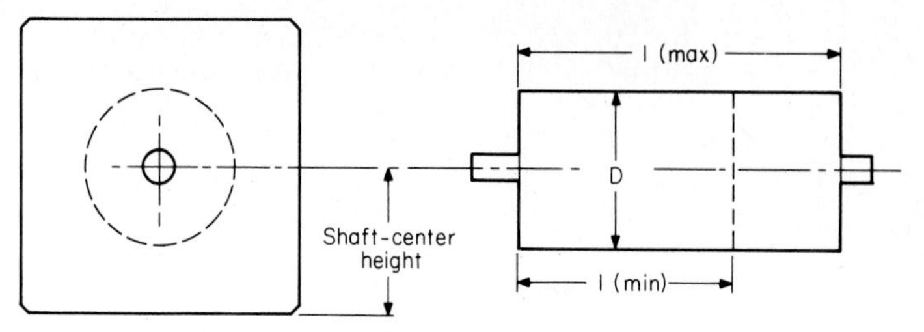


**FIG. 5.43** Frame size.

speed, torque, etc.) conforming to the specified performance; it is not unique, but depends on the conditions to be met and the duty to be performed. In competitive design, each parameter has to be optimized with due regard to constraints imposed by materials, specific loadings, commutation, temperature rise, power/mass ratio, energy efficiency, and manufacture. The design process makes extensive use of sophisticated computer programs.

In Sec. 1.2, the conversion power $P_e$ is given in terms of specific magnetic and electric loadings $B$ and $A$, the diameter $D$ and active core-length $l$ of the armature, and the speed $n$, in the form $P_e = \pi^2 BAD^2 ln$. For a dc machine with $2p$ stator poles, $N$ armature turns in $2a$ parallel paths, and a rated armature current $I_a$, the gap flux per pole and the product $I_a N$ are

$$\phi_g = B\frac{\pi D}{2p} \qquad \text{and} \qquad I_a N = A(\pi D a)$$

The rotational emf (which differs from the armature applied voltage $V_a$ only by the usually small resistance voltage drop $I_a r_a$) is $E_r = 2(\phi_g n)(p/a)N$, from which the turns and paths of a practical armature winding can be determined.

The following notes refer mainly to four-pole motors in a unified series of related frame sizes.

## Specific Loadings

With materials presently available, there is little scope for $B$ to be appreciably above 0.7 to 0.8 T for large and 0.4 to 0.5 T for small machines because slotting makes the density in the armature teeth 2 to 2½ times $B$, for which the saturation level is high, demanding increased field mmf and $I^2R$ loss. The saturation level in yokes and interpoles can, however, be low to stabilize commutation. Normally the whole magnetic circuit is laminated to limit core loss, to reduce the effects of rectifier-supply harmonics, and (in motors for rapid response to speed control) to shorten time constants. Low-loss electrical sheet steel is used in order to minimize core and pole-face losses. The former loss depends on the frequency $f = np$ of flux reversal, a common upper limit being 50 Hz (i.e., 1500 r/min for a four-pole machine); the latter is a function of speed and the number of teeth covered by the pole arc.

The main-pole flux and the speed determine the number of armature turns in series. The electric loading $A$ can be raised only by increasing the conductor current, with dissipation of the greater $I^2R$ loss by more effective ventilation.

Typical values of $B$ and $A$ are given for a range of diameters $D$ in Fig. 5.44. The upper limit of $A$ can be reached for machines having an optimized cooling system.

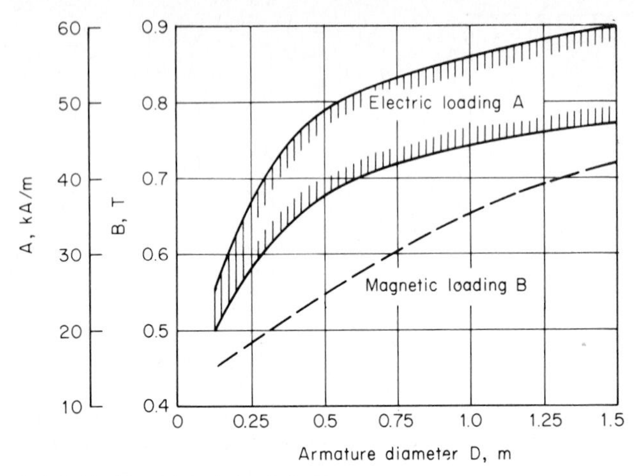


**FIG. 5.44** Magnetic and electric loadings.

## Armature Dimensions

The shaft-center height determines the outside dimensions of the frame. The square frame for a four-pole machine permits the diameter $D$ to be greater than for a circular frame. The conversion power is proportional to $D^2$; it is also proportional to $l$, and to realize a high power/mass ratio, the axial lengths of the commutator and winding overhang must be minimized. Where the motor inertia is limited by specification (it is proportional to $D^2l$) the diameter-length relation may have to be modified.

## Commutator and Brushgear

Commutation imposes constraints on (1) $E_{pc}$, the inductive emf in a commutating coil; (2) $E_c$, the permissible mean voltage between adjacent sectors; and (3) the commutator construction, peripheral speed $u_c$, and number of sectors $C$. (1) and (2) necessitate single-turn armature coils in large machines. Typical values are: $E_{pc} =$ 10 V, $E_c = 20$ V, and $u_c = 40$ m/s. The commutator diameter may approach that of the armature core in order to accommodate the required number of sectors of width not less than 4 mm and to increase the cooling surface.

## Ventilation

Heat from the armature is dissipated mainly by convection, and an effective system of axial air ducts to convey air through the commutator and armature core is essential for a high specific output. The core duct's diameter and positioning with respect to the slots must optimize heat transfer without unduly deflecting flux paths or causing saturation. The use of class F or H for winding insulation affords a high conductor current rating. Cooling air is provided by a frame-mounted motor-driven fan, operating independently of the armature speed; it may be designed to produce a turbulent airflow to give a "scrubbing" action for better thermal pickup.

## Mechanical Design

This must concentrate on cost-effective production in terms of materials and labor. Castings and injection moldings are employed for their ready availability and ease of shaping. Sealed bearings do not need greasing, simplifying maintenance. End brackets can have the spigot and the bearing housing machined in one operation to ensure armature concentricity in the frame, stabilization of the radial airgap length, and avoidance of unbalanced magnetic pull.

## Future Possibilities

The increasing concern with energy cost and conservation is likely to change a policy of "purchase on price" to one of "purchase on energy efficiency," for the greater cost of a low-loss machine might well be recoverable in a remarkably short time. Metallurgical advances may develop (1) core steels of substantially higher saturation levels, to save core material and reduce excitation mmf's, and (2) cheaper and more tractable permanent magnets to extend their application into the medium-rated industrial range, eliminate field loss, and increase the armature diameter of a given frame. Some headway has been made with solid-state commutation; a practical electronic scheme that would achieve commutation without interpoles would solve one of the designer's major problems.

# CHAPTER 6
# SMALL ELECTRIC MOTORS

C. G. Veinott

## 6.1 INTRODUCTION

As compared with large motors, small electric motors involve more types of machines, more units, and more dollars. This chapter is devoted to a discussion of what kinds of small electric motors are available, their principles of operation, where and how to use each type, and a peek at their design considerations. The many types and varieties will first be cataloged and then discussed in detail.

Most small electric motors have to operate on single-phase alternating current because that is the type of energy most readily and most economically available. It used to be that many large cities used dc in the downtown areas and ac elsewhere, so this gave rise to a need for universal motors, which operate on either source; another use for universal motors is where speeds in excess of 3600 r/min are needed. If the motor must operate on battery power, dc motors or inverter-driven ac motors are required.

## 6.2 TYPES OF SINGLE-PHASE AND SMALL MOTORS

Single-phase motors will run on a single winding, usually called the *main winding*. However, they are not self-starting, so a second winding, usually called the *auxiliary winding*, is needed. Single-phase motors, operating upon only the main winding, inherently have a low power factor, develop a double-frequency pulsating torque greater than the load torque, and have high rotor $I^2R$ losses due to the backward field; these problems can be greatly helped by using one or more capacitors. The above considerations lead to several types:

**1.** Split-phase motors

   **a.** Standard split-phase motors

   **b.** Special-service split-phase motors

2. Capacitor motors
   a. Capacitor-start motors
   b. Two-value capacitor motors
   c. Permanent-split capacitor motors
   d. Split-phase capacitor motors
3. Shaded-pole motors

Other types of small motors are:

4. Synchronous motors
5. Universal motors
6. DC motors
7. Polyphase motors

## 6.3 SPLIT-PHASE MOTORS

A *split-phase motor* is a single-phase induction motor that has a main and an auxiliary (starting) winding; the two windings are mutually displaced by 90 electrical degrees. The auxiliary winding has a higher ratio of resistance to reactance than the main winding in order to achieve a phase-splitting effect, and a starting switch cuts it out of the circuit as the motor approaches operating speed.

### Principles of Operation

The essential parts of a split-phase motor are schematically represented in Fig. 6.1*a*, and the phase-splitting effect of the higher ratio of $R/X$ in the auxiliary winding is illustrated in Fig. 6.1*b*. It is the latter effect that causes the motor to start and come up to speed (if the load is not too great).

Once the motor is up to speed, the auxiliary winding must be cut out of the circuit in order to avoid excessive losses and low efficiency and to prevent burnout

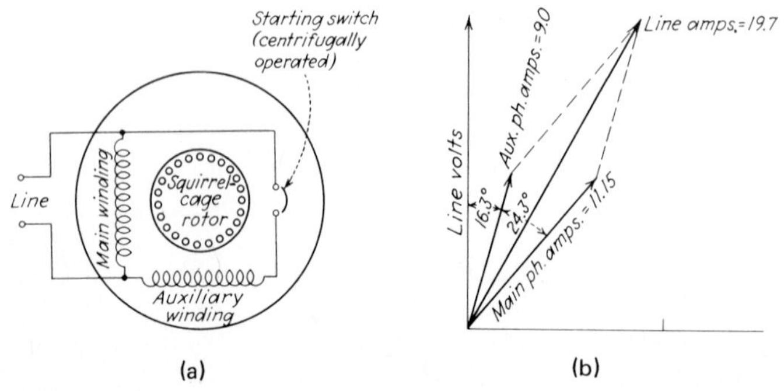


**FIG. 6.1** The split-phase motor: (*a*) schematic representation; (*b*) phasor diagram of locked-rotor currents.

of the winding itself. Figure 6.1*a* illustrates the use of a centrifugal switch, which is the most commonly used device for this purpose; however, magnetic relays are also often used.

## Standard and Special-Service Motors

Applications for split-phase motors fall into two broad general classes:

1. Those which require frequent starting and a relatively large total running time per year, such as oil burners and furnace blowers. (Standard split-phase motors.)
2. Those which require infrequent starting and a relatively small total running time, such as home-laundry equipment, home workshops, and cellar drainers. (Special-service motors.)

*Standard split-phase motors* are built in a wide variety of horsepower and speed ratings for the first class of applications. Some manufacturers call these general-purpose motors because of their wide usage, but this is not strictly correct since such motors do not meet the torque requirements specified by NEMA for general-purpose motors. A typical speed-torque curve is illustrated in Fig. 6.2*a*. These motors have moderate starting torques and moderate locked-rotor currents.

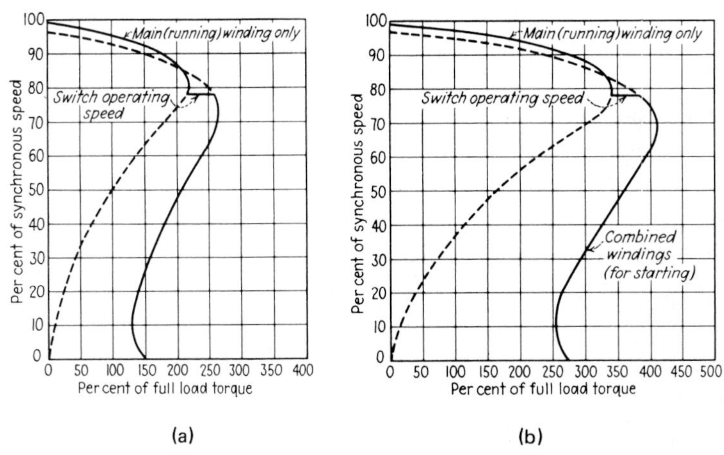


**FIG. 6.2** Speed-torque characteristics of split-phase motors: ***(a)*** a "standard" motor; ***(b)*** special-service motor.

*Special-service motors* (sometimes called high-torque motors) are used for the second class of applications. A typical speed-torque curve is given in Fig. 6.2*b*. Compared with standard motors, special-service motors have higher torques, lower efficiencies, and higher locked-rotor currents. Because they are built in high-volume production and *in few models*, they are generally lower in first cost. However, they should not generally be used on lighting circuits because of their high locked-rotor currents.

Many two-speed split phase motors, usually of the pole-changing variety, are built for home-laundry service.

## 6.4 CAPACITOR MOTORS

Four different types of capacitor motors are on the market today. The characteristics differ so much from one another that the term *capacitor motor* should never be used by itself. All have a main winding, usually wound in two sections (which can be connected externally in series or in parallel) for dual-voltage operation. Most types use a single-section auxiliary winding, which is spatially displaced from the main winding by 90 electrical degrees; split-phase capacitor motors often use a second auxiliary winding in space phase with the other. Tapped-winding capacitor motors have one or more "extra," or "booster," windings to achieve speed reduction; these are wound in space phase with the main winding.

### Capacitor-Start Motors

This is by far the most popular type of ac motor for heavy-duty general-purpose applications requiring high starting torques. It is used for applications such as compressors, tire-changing tools, jet pumps, swimming-pool pumps, farm and home workshop tools, and conveyors. It is generally available in ratings from $^1/_8$ hp and up, wound for two, four, or six poles, and usually arranged for dual-voltage operation.

A single-voltage capacitor-start motor is represented schematically in Fig. 6.3*a*, which shows the two windings in quadrature and the customary centrifugal switch to cut out the auxiliary winding when the motor is up to speed. A switch of some kind is needed to prevent burning out the auxiliary winding and/or the capacitor when the motor is running continuously. Magnetic switches are used for some applications. Figure 6.3*b* is a typical phasor diagram of the motor under locked-rotor conditions. It illustrates why this motor develops so much more torque than the split-phase motor:

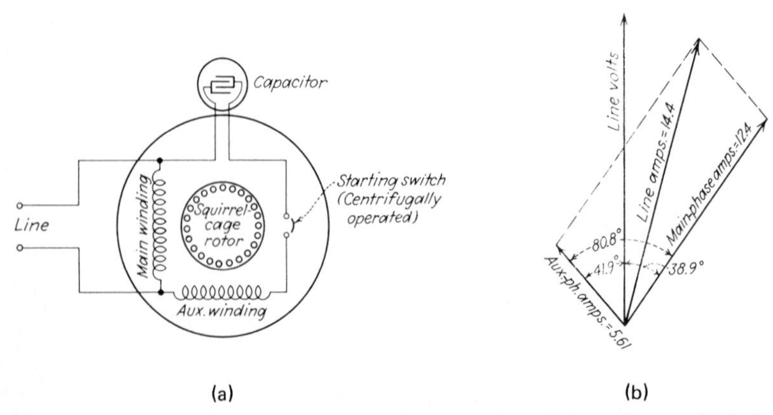


**FIG. 6.3** The capacitor-start motor: ***(a)*** schematic representation; ***(b)*** phasor diagram of locked-rotor currents.

The time-phase displacement of the currents is much greater, and each winding can carry more current for the same line current because of the greater phase displacement between the currents in the respective windings. Figure 6.4 shows a typical speed-torque curve; note how much higher the torques are, especially at low speeds, as compared to the split-phase motors of Fig. 6.2, and note also how rapidly the capacitor voltage rises at speeds above 80 percent of synchronous.

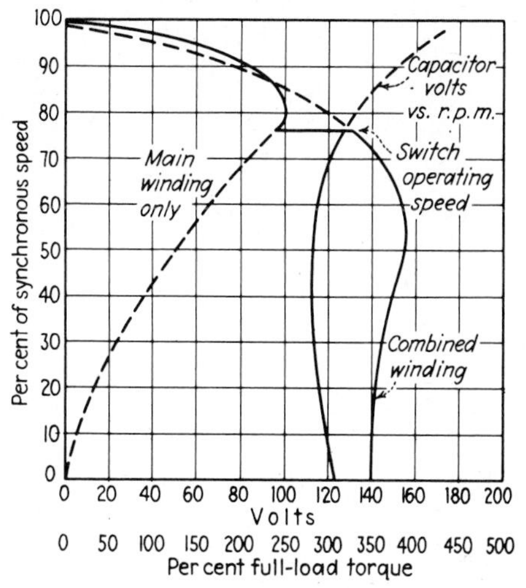


**FIG. 6.4** Speed-torque characteristics of a capacitor-start motor.

Most commercial capacitor-start motors are arranged for operation on either 115- or 230-V circuits. How this is achieved is illustrated in Fig. 6.5, which shows a dual-voltage capacitor-start motor. In Fig. 6.5*a* the motor has six line leads; in Fig. 6.5*b* the four main-winding leads are permanently connected to the *back* of a five-post terminal board, while the two leads from the auxiliary phase are connected to the *front* of the terminal board to permit the user to connect the motor for either direction of rotation. In Fig. 6.5*c* the same motor is connected for 230 V merely by changing the position of the links. The fifth post, not really needed in Fig. 6.5, is provided for thermally protected motors; the protector is connected between posts 1 and 2 at the back of the board. Most commercial capacitor-start motors are supplied with (1) a cast-integral conduit box in the front end shield, (2) a terminal board to which the stator member of the starting switch is affixed, and, most often, (3) a built-in thermal protector. Quick-connect tabs to facilitate user connections to the line are also commonly used.

## Two-Value Capacitor Motors

The two-value capacitor motor, sometimes referred to by the ambiguous term *capacitor-start, capacitor-run*, sprang to life in the early 1930s for applications requiring high locked-rotor and breakdown torques. It pretty much superseded

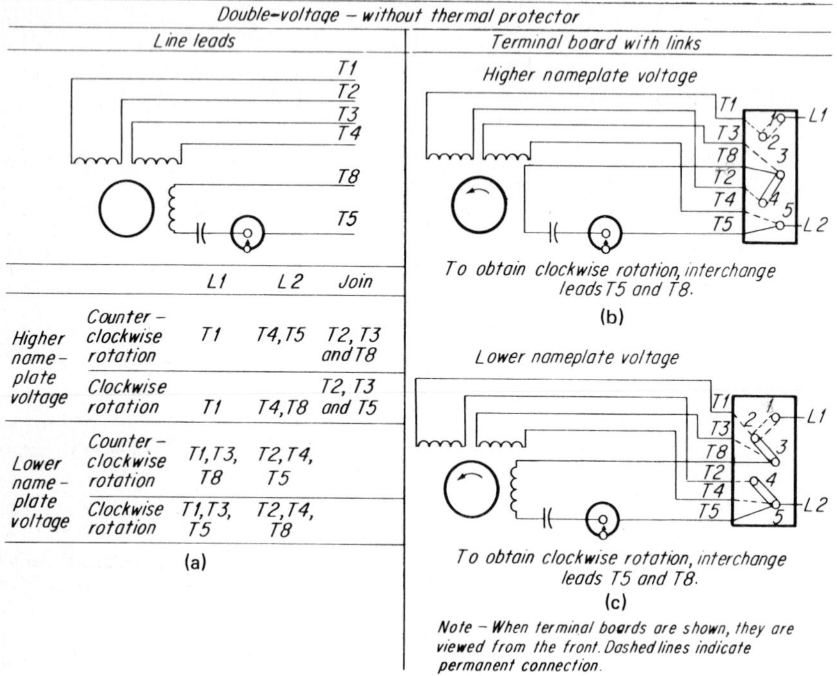


**FIG. 6.5** Wiring and line-connection diagrams for dual-voltage reversible capacitor-start motors: ***(a)*** with six external line leads; ***(b)*** with terminal board, connected for higher voltage; ***(c)*** with terminal board, connected for lower voltage.

repulsion-type motors and was, in turn, largely replaced by the capacitor-start motor. It regained some popularity recently because of the surging interest in higher efficiencies, for it has 5 to 10 points more efficiency and has less double-frequency pulsating torque, hence tending to produce less noise than the capacitor-start motor. Also, it is often used to squeeze an additional rating or two in a given frame size.

A two-value capacitor motor is represented schematically in Fig. 6.6. Note that it is similar to the capacitor-start motor of Fig. 6.3, except that a running capacitor has been connected permanently in series with the auxiliary winding. A starting switch is required for the same reasons as it is in a capacitor-start motor (items 2, 3, and 4 below). The speed-torque characteristics of this motor are substantially as shown in Fig. 6.4. Considering any given motor, the effects of adding the running capacitor are to:

**1.** Increase the breakdown torque from 5 to 30 percent.

**2.** Improve the efficiency by 5 to 10 points.

**3.** Raise the power factor up above 90 percent.

**4.** Reduce noise under full-load operating conditions.

**5.** Increase the locked-rotor torque 5 to 20 percent.

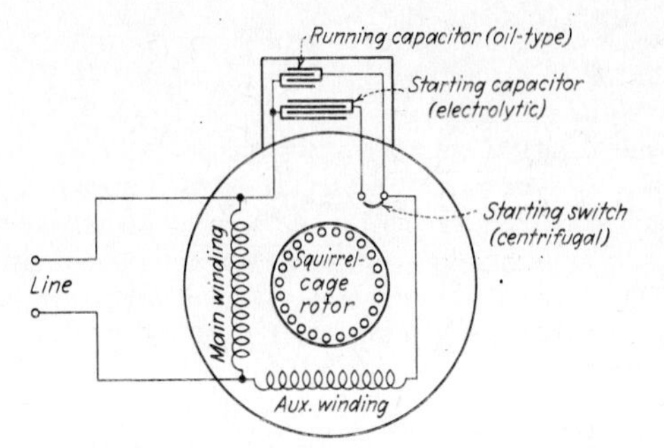


**FIG. 6.6** Schematic representation of a two-value capacitor motor.

The windings for the two-value capacitor motor are very similar to those for capacitor-start motors and are also usually dual-voltage. Sometimes, however, as illustrated in Fig. 6.7, this motor uses a two-section auxiliary winding, and the two sections are not generally identical. On the starting connection a single section is used, but on the running connection both sections are connected in series with the running capacitor. This arrangement is used because starting capacitors usually carry a much lower voltage rating than running capacitors do, so this permits using a smaller running capacitor, because it is subjected to a higher voltage than it would be if only a single-section auxiliary winding were used.

## Permanent-Split Capacitor Motors

Once called single-value motors, permanent-split motors are usually used for special-purpose applications, such as shaft-mounted fans and blowers, instruments (often using synchronous rotors), and servomotors. In the smaller sizes, they often compete with shaded-pole motors. They are more efficient, have better power factor and more output per pound, but are more costly than shaded-pole motors. With shaded-pole motors, they share the important advantage that they require no starting switch or relay. Generally speaking, they are not suited for belted applications or for any other continuous-duty application requiring substantial locked-rotor torque. However, they

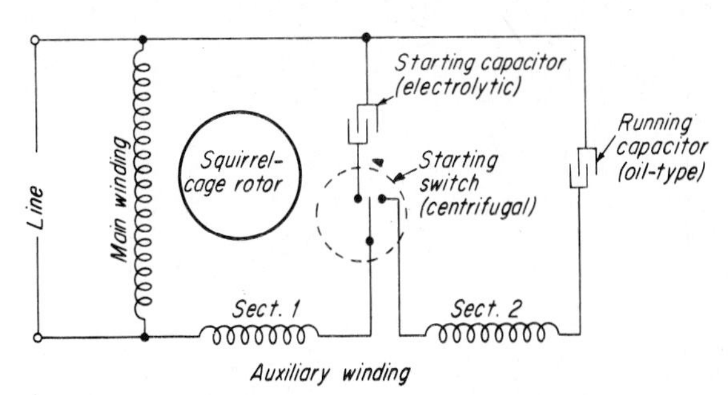


**FIG. 6.7** Two-value capacitor motor with a two-section auxiliary winding.

do lend themselves well to high-torque, reversing, intermittent-duty service. They are generally available in ratings from 1 millihorsepower (mhp) to $^1/_3$ hp, and, for shaft-mounted fans and blowers, up to $^3/_4$ hp.

A *permanent-split* capacitor motor uses but a single capacitor, which is in the circuit all the time. A schematic representation of it would be Fig. 6.3 with the starting switch omitted, or Fig. 6.6 with the starting switch and starting capacitor omitted. The main winding may have one or two sections, according to whether the motor is to be single- or dual-voltage. Generally, all permanent-split motors have a higher rotor resistance than other comparable single-phase motors, and more slip at full load. Within the past few years, full-load speeds of permanent-split capacitor motors have slowly fallen, primarily to obtain better starting torques and more stable characteristics at reduced speeds, for this motor type is better suited to speed control than are many others.

Permanent-split motors for shaft-mounted fans and blowers in heating, air-conditioning, and similar applications are predominantly six-pole, largely to reduce air noise. Motors with outputs of 2 to 5 mhp are used in industrial instruments; they are generally rated on the basis of full-load torque, while the breakdown may be about 175 percent of full-load torque. They can be wound for either single- or dual-voltage.

***Tapped-Winding Capacitor Motors.*** These are very popular for driving fans, because they can offer two or more speeds. They achieve speed reduction by reducing the airgap flux by means of increasing the number of effective turns in the main winding. A tapped-winding motor is schematically represented in Fig. 6.8. There is an *intermediate* winding, in space phase with the main winding, and the customary auxiliary phase, consisting of a capacitor and auxiliary winding that is spatially displaced from the main winding by 90 electrical degrees. For high-speed operation, the main winding is connected directly across the line, while for low speed the main and intermediate windings are connected in series across the line. In the figure, one side of the auxiliary phase is connected to the high-speed tap (T connection); alternatively, it may be connected to the low-speed tap (L connection) or to the other side of the line. Thus, there are three ways to connect the tapped-winding motor. The L connection (not shown) is most useful for 115-V motors and requires less capacitance than the T connection. However, the T connection is more suited to 230-V applications than the L, for the latter is likely to produce excessive voltages across the capacitor. The third arrangement (auxiliary phase across the line) requires an extra pole on the control switch. Three speeds are obtained by using two intermediate windings, using one or two in series to obtain the second and third speeds; as many as five speeds have been reported.

***Motor Limitations, when Used for Multispeed Applications.*** Of all types of single-phase motors, permanent-split motors are best suited to multispeed operation of fans, but they do have some inherent limitations that need to be mentioned. Figure 6.9 can assist the reader in following this discussion:

**1.** *Speed depends upon the load:* At no-load, the motor will operate at practically the same speed on the *high*, *medium*, and *low* connections. If a *light fan*, requiring one-half the power of the *normal* fan, is used, the motor will operate at speeds shown by points *d*, *e*, and *f*. Note that the speed reduction on the low-speed connection is about 25 percent instead of the 50 percent obtained with the normal fan. And with a heavy fan, even more speed reduction is obtained; moreover, at low speeds the operating speed is apt to be unstable, especially if there should happen to be a harmonic cusp in the speed-torque curve. In any case, there would be danger of serious overheating.

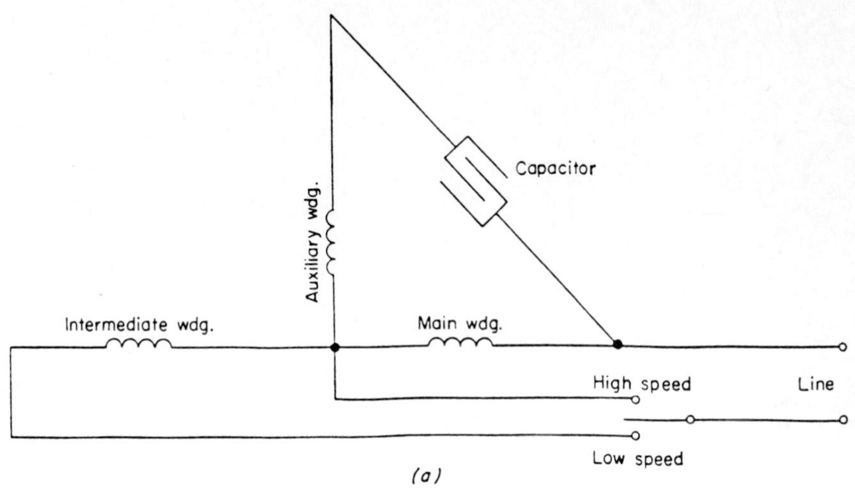


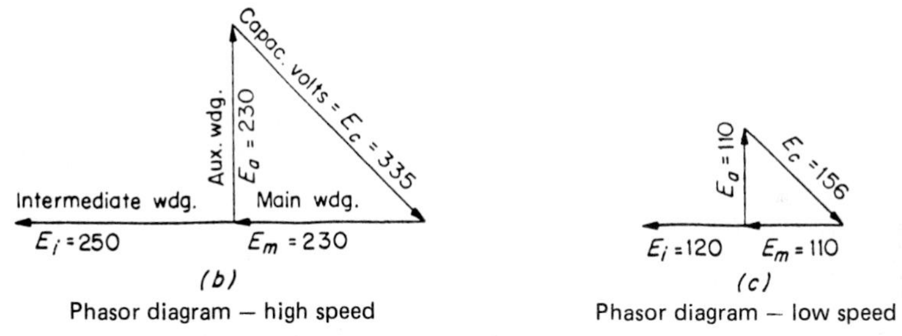


**FIG. 6.8** Tapped-winding capacitor motor, T connection.

**2.** *Locked-rotor torque necessarily low:* Clearly, the locked-rotor torque on low speed has to be less than the fan torque at the desired low speed. At half speed, the fan torque is 25 percent of full-load torque, and so the locked-rotor torque has to be less than 25 percent.

**3.** *Not suited to belted drives:* Because of the inherently low locked-rotor torque, and because there is apt to be a considerable variation in belt friction, particularly if a V-belt is used, and because the low speeds would therefore be uncertain, this type of speed control is not suited to belted applications.

**4.** *Unstable low-speed connection:* Because on the low-speed connection the fan torque and motor torque curves cross at a very small angle, the speed characteristics tend to be unstable; the speed is sensitive to changes in voltage and also to restrictions in the inlet and outlet ducts of the fan or blower it is driving. As noted before, if there are any harmonic cusps in the motor speed-torque curve, this difficulty is exacerbated.

*External impedance*, inserted in series with a permanent-split motor, can often be used to vary the speed of the motor if it is driving a fan.

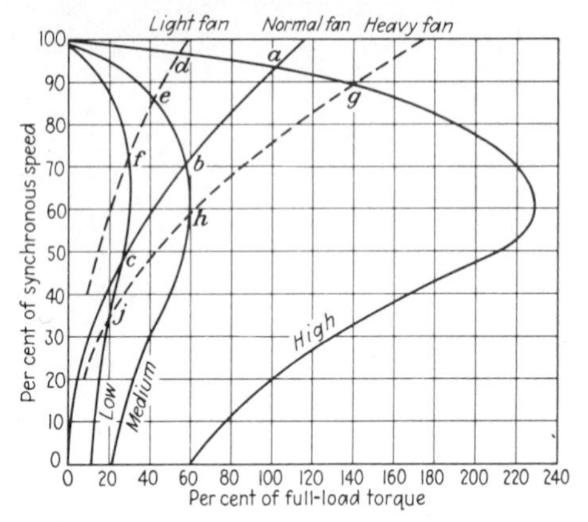


**FIG. 6.9** How speed control of a fan is obtained with a tapped-winding capacitor motor.

***High-Torque, Reversing, Intermittent-Duty Motor.*** If the starting switch of a capacitor-start motor is omitted, the motor becomes a permanent-split capacitor motor with the high torques of the capacitor-start motor. In this case, however, the motor generally has to be intermittently rated, even though a more durable capacitor may be substituted. Motors of this sort are widely used for operating valves, induction regulators, rheostats, arc-welding control, etc. In order to achieve simplicity of reversing control, the novel arrangement shown in Fig. 6.10 is often used. When the reversing switch is in the UP position, winding $A$ becomes the main winding and $B$ the auxiliary winding. When the switch is in the DOWN position, the roles of the windings become reversed. Interchanging the roles of the two windings causes the motor to reverse. The single-pole double-throw switch with an OFF position is all that is required for complete control of the motor. A simpler control system can scarcely be imagined! If windings $A$ and $B$ are identical, the same torque will be developed in both directions of rotation. If they are not identical, different torques are obtained in the two directions of rotation. These motors are often provided with a mechanical brake.

## Split-Phase Capacitor Motors

These motors start as split-phase motors and run as permanent-split capacitor motors. They have replaced split-phase motors where high efficiency for continuous operation is of primary importance. Because the split-phase starting arrangement permits the use of lower rotor resistance, the split-phase capacitor motor usually operates at a higher efficiency and higher speed than a comparable permanent-split capacitor motor. One arrangement is represented schematically in Fig. 6.11. It is to be noted that this arrangement is similar to that of Fig. 6.1*a*, except that the starting switch is double-throw and that a capacitor has been added; hence, the stator windings are the same in form and arrangement as those of a split-phase motor. The motor starts as a split-phase motor and runs as a permanent-split motor.

A disadvantage of the simple arrangement shown is that, in order to obtain locked-rotor torque similar to a split-phase motor, the auxiliary winding is normally

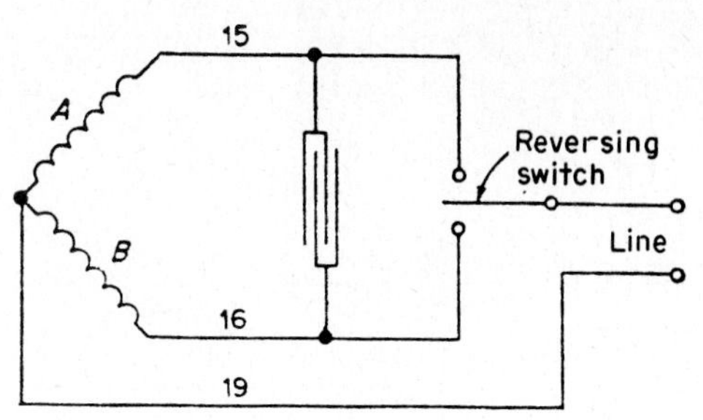


**FIG. 6.10** Schematic representation of a high-torque, reversing, permanent-split capacitor motor.

wound with fewer turns than the main winding. This results in a relatively low voltage across the capacitor, with resultant less effect than if a higher voltage were impressed across it. The double-section auxiliary winding of Fig. 6.12 overcomes both of these difficulties. By proper design of the turns in section 2, the voltage on the capacitor is increased to its normal operating value, and the same auxiliary-winding current produces more effect because of the added turns in section 2. Thus, this latter arrangement enables the designer to obtain more of the beneficial effects of a permanent-split motor. Most split-phase capacitor motors have this dual-section arrangement.

## 6.5 SHADED-POLE MOTORS

Shaded-pole induction motors are used in a wide variety of applications requiring a motor of $^1/_4$ hp or less, even down to 1 mhp. In the subfractional range (horsepower ratings below $^1/_{20}$), it is the standard general-purpose constant-speed ac motor. It is simple in construction, low in cost, and extremely rugged and reliable—like a polyphase induction motor—because it needs no commutator, switch, collector rings, brushes, governor, or contacts of any sort. Its torque characteristics and applications are similar to those of a permanent-split capacitor motor, except that it has a lower efficiency and a lower power factor. But efficiency and power factor are seldom of real importance in these sizes. It is used in an extremely

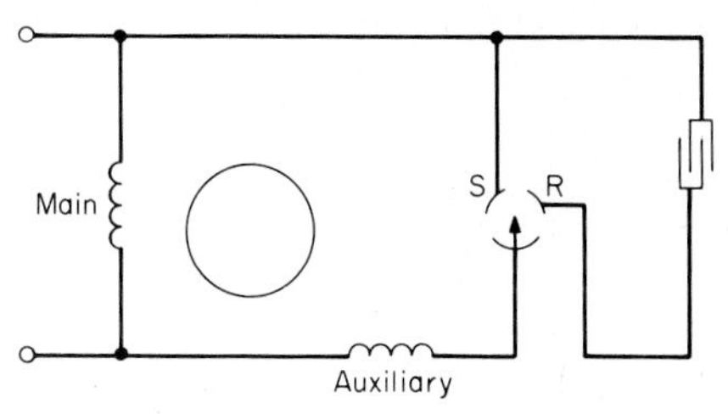


**FIG. 6.11** Split-phase capacitor motor with single-section auxiliary winding. (*Courtesy of Emerson Motor Division.*)

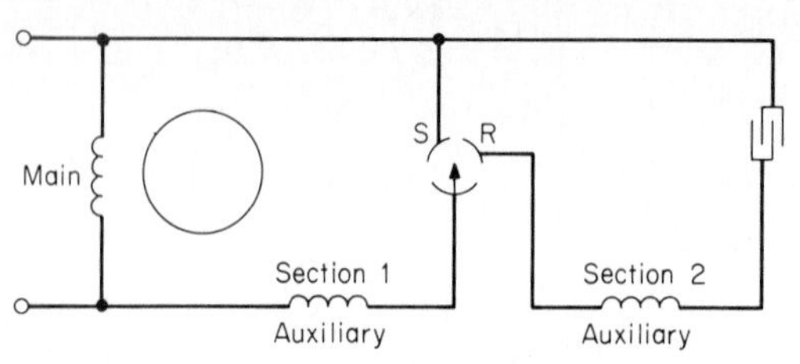


**FIG. 6.12** Split-phase capacitor motor with two-section auxiliary winding. (*Courtesy of Emerson Motor Division.*)

wide variety of applications: home appliances, such as rotisseries, fans of all kinds, humidifiers, and slide projectors; small business machines, such as photocopiers; vending machines; advertising displays; etc. They are available with integral gear reducers to obtain almost any speed, even down to less than one revolution per month. Some manufacturers offer integral clutches and/or brakes. Although the simple shaded-pole motor is inherently nonreversible, reversible motors are built in different forms.

A *shaded-pole motor* may be defined as a single-phase induction motor provided with an auxiliary short-circuited winding or windings displaced in magnetic position from the main winding.

## Essential Parts and Construction

Figure 6.13 gives a schematic representation of a shaded-pole motor. Figure 6.14 shows a form of construction long popular in the larger sizes: The stator has salient poles, one coil on each pole; the stator coils are held in place by wedges, usually of a magnetic material (which appears to improve performance); there is a short-circuited "shading" coil around a portion of each main pole; and the rotor is of squirrel-cage construction. In smaller ratings, a very popular and low-cost arrangement is the C-frame construction illustrated in Fig. 6.15; this figure also illustrates the use of more than one shading coil. For some time now, most shaded-pole motors have had a notch cut in the leading pole tip to increase its reluctance. One way of doing this is illustrated in Fig. 6.16; another way is to punch a hole in the leading tip so that it saturates, giving somewhat the same effect as the increased airgap.

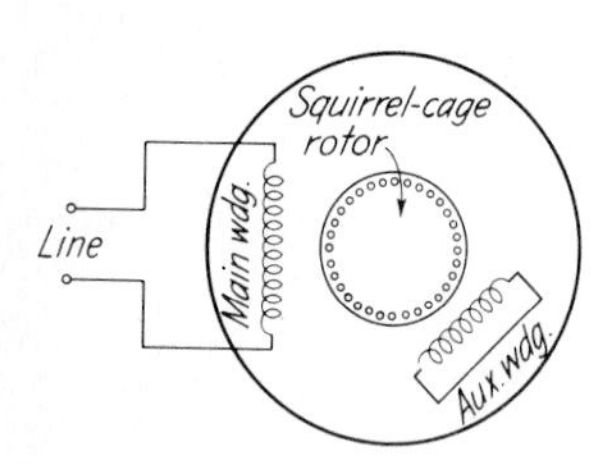


**FIG. 6.13** Schematic representation of a shaded-pole motor.

## Principle of the Shading Coil

Like any other induction motor, the shaded-pole motor is caused to run by the action of a revolving magnetic field—however imperfect—set up by the stator. It is the shading coil that delays the flux in its portion of the pole, and this delay creates a sort of rotating field.

AC contactors use shading coils to minimize noise and chattering of the contacts; watthour meters also use them for the light-load adjustment. In order to understand how a shading coil functions, let us first consider a simple ac circuit containing resistance only: the current in this circuit is in phase with the impressed voltage. Suppose we insert an inductive reactance into this circuit: now *the current will lag behind the impressed voltage and will be reduced in value.* Referring to the motor of Fig. 6.14, assume that neither the shading coil nor the rotor conductors have yet been introduced. Let an alternating mmf be set up by passing a current through the stator winding: then an alternating flux will be set up across the airgap that is proportional to, and in phase with, the alternating mmf. Now let us surround a part of each pole with a shading coil having resistance and no reactance: then *the flux through the shading coil will lag behind the mmf produced by the stator coil and will be reduced in value from what it was before the shading coil was added.* Since the flux in the unshaded portion of the pole will still be in phase with the stator mmf, it follows that the flux threading the shading coil will lag behind the flux in the unshaded portion.

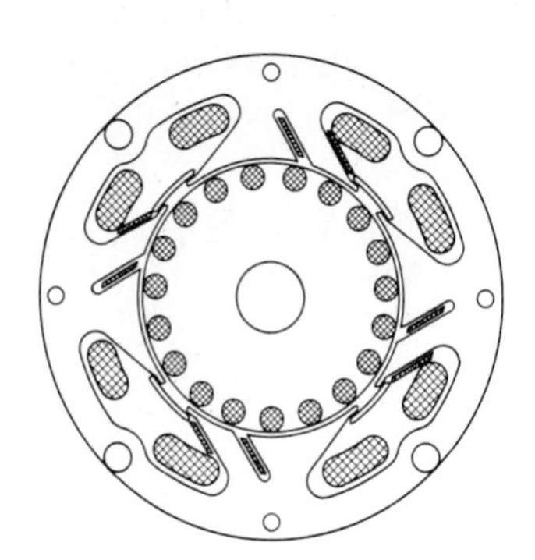

**FIG. 6.14** A shaded-pole motor with tapered poles and magnetic wedges.

Let us now consider what we did in each case. In the electric circuit, by linking with it a magnetic circuit, we introduced *reactance* in which was generated a back voltage proportional to the first time derivative of the current. In the magnetic circuit, by linking with it an electric circuit, we introduced a back mmf (due to

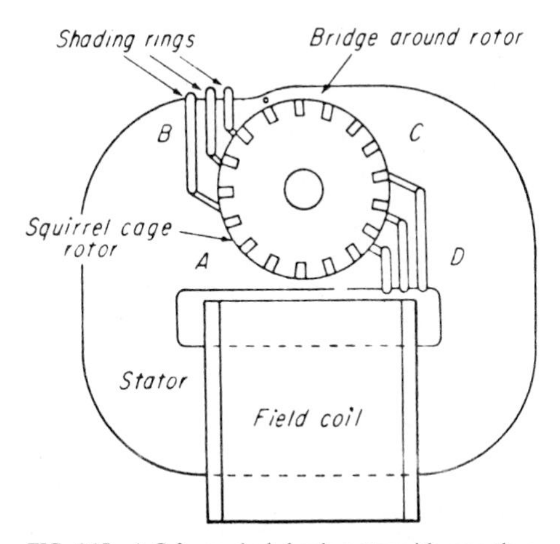


**FIG. 6.15** A C-frame shaded-pole motor with more than one shading coil. (*Courtesy of Barber-Colman Company.*)

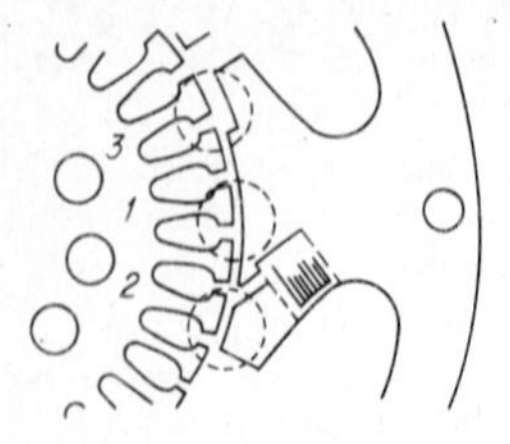


**FIG. 6.16** Shaded-pole motor with notched leading pole tip. (*Courtesy of Controls Company of America.*)

the induced current) proportional to the first time derivative of the flux; this effect caused the shaded flux to lag behind its mmf *and to be reduced in magnitude.* Hence, the flux in the shaded portion of the pole was caused to lag behind the flux in the unshaded portion, for the latter still remained in phase with its mmf.

## Ratings, Performance, and Torque Characteristics

Shaded-pole motors are built in a very wide variety of ratings and performance characteristics.

Motors rated from $^1/_4$ down to $^1/_{20}$ hp (fractional-hp range) are generally built in a round-frame construction, such as that illustrated in Fig. 6.14, and may have four, six, or even eight poles. Motors rated 40 mhp or less (subfractional-hp range) are often built on a C frame, such as that illustrated in Fig. 6.15. The full-load speeds of these small motors, for 60 Hz, may range from 2000 to 3000 r/min.

The efficiencies may be as high as 30 percent in fractional-hp sizes, and as low as 2 percent in subfractionals.

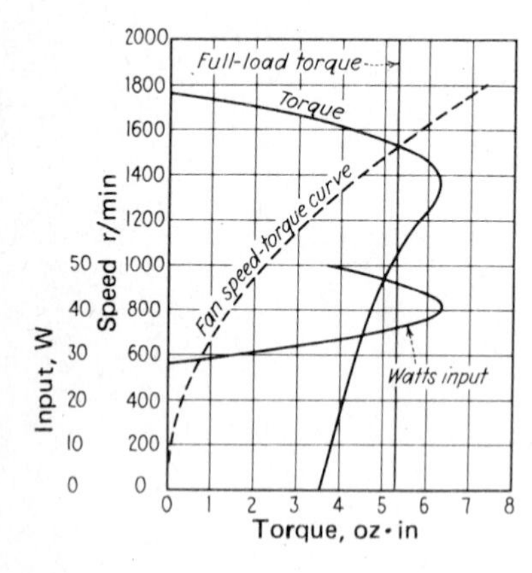


**FIG. 6.17** Performance curves of a shaded-pole motor.

A "typical" speed-torque curve, if there is such a thing, could be Fig. 6.17. This curve is generally similar in shape to the speed-torque curve of a typical polyphase motor, except that the locked-rotor and breakdown torques are considerably lower, in terms of full-load torque, than are those of a polyphase motor. Also, the change in watts input from no-load to locked-rotor is much smaller than is usual for other types of motors. This is not so much due to low locked-rotor watts as it is to high no-load watts.

### Reversibility

The shaded-pole motors described above can operate in only one direction. Some are made reversible by using two shading coils, one on each side of the pole; others achieve reversibility by mechanically coupling two identical motors of opposite rotation.

Two, three, or more speeds are fairly easy to obtain by simply tapping the winding to put more or less turns across the line. This type of speed control is similar in principle to the tapped-winding capacitor motor previously discussed and is subject to the same limitations.

## 6.6 SYNCHRONOUS MOTORS

Exactness of speed is what has made synchronous motors so popular in fractional- and subfractional-horsepower sizes. The speed of a synchronous motor is exactly proportional to the line frequency, which is regulated on all large 60-Hz power systems so closely that clocks can be driven by synchronous motors; the frequency is rarely in error by more than a small fraction of 1 percent, and these small errors are averaged out over a period of time. Hence, synchronous motors are widely used for clocks and timing motors.

Synchronous motors are built over the widest horsepower and speed range of any type of motor. The horsepower ratings range from the thousands down to the millionths; the speeds range from 24,000 r/min down to as low as one revolution per month (with reduction gearing). In integral-horsepower sizes, they are dc-excited, but dc excitation is not used in fractional- and subfractional-horsepower sizes. In sizes from $^1/_8$ hp down to about 1 mhp, these motors are used for such applications as teleprinters, facsimile picture transmission, instruments of all kinds, and sound-recording or sound-reproducing apparatus. They are also often used in textile applications, in which several motors may be driven from a common variable-frequency source. In ratings of 1 mhp and less, they are generally self-starting, but some clock motors are intentionally made so that manual starting is required.

Synchronous motors are built in many types and constructions. Most of them have a feature in common with induction motors: a stator with windings that, when properly energized, set up a magnetic field that rotates at a speed—known as synchronous speed—exactly proportional to the line frequency. Synchronous motors are so built that they lock into step with this rotating field; it is the rotor construction that determines whether or not the rotor rotates in step with the field. Hence, synchronous motors are usually classified by (1) the type of rotor construction and (2) the stator winding arrangement. Integral-horsepower synchronous motors use a salient-pole rotor with a dc-excited field coil on each pole, but synchronous operation is obtained in fractional-horsepower sizes by means other than dc excitation. Popular constructions are: *reluctance*, *hysteresis*, *permanent magnet*, and *inductor*. The reluctance, hysteresis, and permanent magnet types use stators like those of induction motors, but the inductor type uses an altogether different construction for both the stator and the rotor.

### Reluctance Motors

A *reluctance motor* has a stator similar in construction to that of an induction motor, but the rotor has salient poles, without dc excitation. It starts as an induction motor.

As implied in this definition, the reluctance motor is really a special case of the familiar dc-excited synchronous motor *with an open field circuit*. The small reluctance motor is usually built from induction-motor parts, except that some teeth are cut out of the squirrel-cage rotor; in addition, flux barriers are often added in the yoke section to accentuate the effect of saliency. The squirrel cage serves both to start the motor—as an induction motor—and to provide damping against hunting during synchronous operation.

One form of rotor punching, for a six-pole motor, is shown in Fig. 6.18; in each of six places, teeth have been removed. If (as in this example) the sizes and locations of these cutouts are slightly unsymmetrical, better locked-rotor torque is obtained than if the cutouts were uniform in size and spacing, because cogging effects are thereby reduced. Note that although the teeth were removed, the end rings and bars were left intact. The projecting groups of teeth that remain provide something that can be "grabbed" by the rotating field and "held" so that the rotor stays in step with the rotating field. Two rotor laminations with *barrier slots* are illustrated in Fig. 6.19. These slots are positioned to increase the reluctance to the quadrature-axis field without affecting the direct-axis reluctance.

*Pull-in torque* is defined as the maximum constant torque against which the motor will pull its connected load into synchronism. This torque is greatly affected by the inertia of the connected load. When the motor is driving a load at a steady speed, all of the internal torque (torque developed by electromagnetic action) is available to carry the load; however, when pulling into step, some of the internal torque is consumed in overcoming inertia because of the need for a sudden increase in speed to pull the motor into step. Consideration of how a synchronous motor pulls into step will clarify this point.

Starting from rest, the motor accelerates as an induction motor. As it approaches synchronous speed, the salient poles of the rotor slip by the poles of the rotating magnetic field, but the nearer the rotor gets to synchronous speed, the slower the

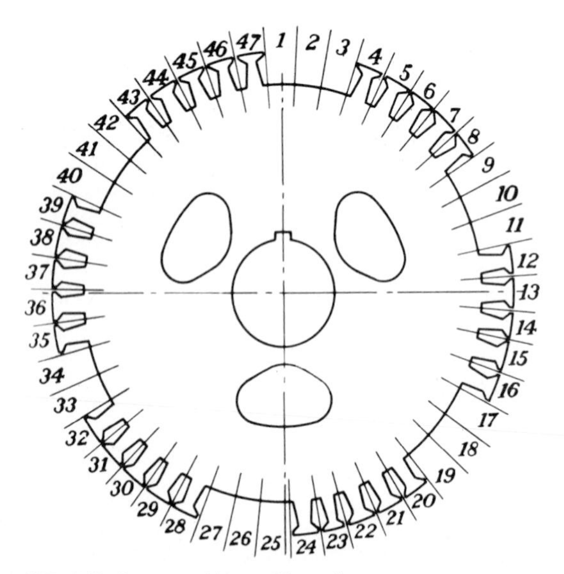


**FIG. 6.18** Rotor punching used in a reluctance motor.

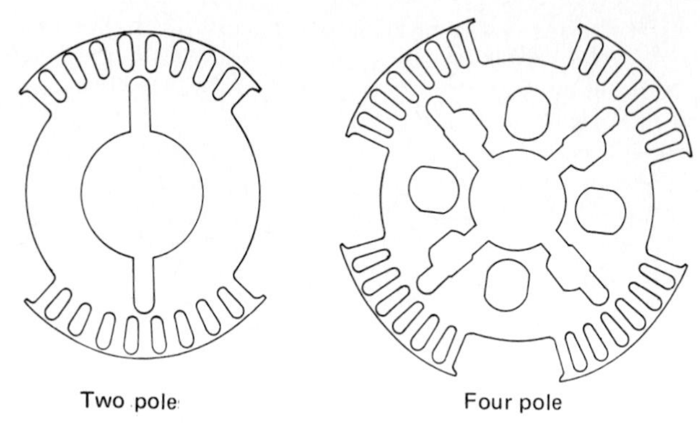


**FIG. 6.19** Rotor laminations for reluctance motors, showing flux-barrier slots. (*Courtesy of Reliance Electric Company.*)

poles slip. As a rotor pole passes and begins to lag behind a pole of the rotating field, a forward torque is created, tending to accelerate the rotor. If the rotor has not pulled into step by the time its salient pole has slipped back half a pole pitch, a backward torque will try to decelerate the rotor. The average effect of these successive pulls in opposite directions is zero; therefore, unless the rotor can accelerate from the induction speed to synchronous speed while a rotor pole is falling behind the rotating field by half a pole pitch, the motor will not synchronize at all.

The inertia of the connected load affects the pull-in torque because the change in speed—from maximum as an induction motor to synchronous—has to be effected while the rotor slips less than half a pole pitch with respect to the revolving field. The accelerating torque needed to effect a speed change in a certain time is directly proportional to the inertia of the connected load. Because accelerating torque subtracts directly from internal torque, it follows that the greater the inertia, the less the effective pull-in torque.

Rotor resistance also affects the pull-in torque, because the higher the rotor resistance, the greater the slip at which the motor ceases to accelerate on the starting connection; therefore, the greater speed change needed for synchronization increases the accelerating torque required, leaving less pull-in torque.

Popular types of reluctance motors are: (1) split-phase, (2) capacitor, (3) shaded-pole, and (4) polyphase.

## Hysteresis Motors

Unlike reluctance motors, hysteresis motors pull into or fall out of step smoothly, for there are no physical poles or projections on the rotor. Hence, hysteresis motors can pull into step just about any load they can carry, regardless of its inertia. They do not pull into step in any predetermined position with respect to the applied voltage wave; in this respect they differ from reluctance motors, which always pull into step with the rotor in some definite position. With ordinary rotors there is one position per pole, but with "polarized" synchronous motors there is one position per pair of poles. Although hysteresis motors synchronize smoothly, they may produce a "flutter" (variation in speed during a revolution), which is objectionable in certain applications.

A *hysteresis motor* is a synchronous motor without salient poles and without dc excitation that starts by virtue of the hysteresis losses induced in its hardened-steel member by the revolving field of the primary and operates normally at synchronous speed because of the retentivity of the secondary core.

***Construction.*** The rotor construction for a typical hysteresis motor is illustrated in Fig. 6.20. Hysteresis rings are made of a special magnetic material, such as a steel alloy (containing 3.5 percent chromium or 3 to 36 percent cobalt) or alnico, and are carried on a supporting arbor made of a nonmagnetic material, such as brass; the assembly is carried on the shaft. Hysteresis rings are usually from thin stock, and several of them are assembled to give a built-up laminated rotor. In the smallest sizes, the rotor may consist of a single solid ring, or cylinder.

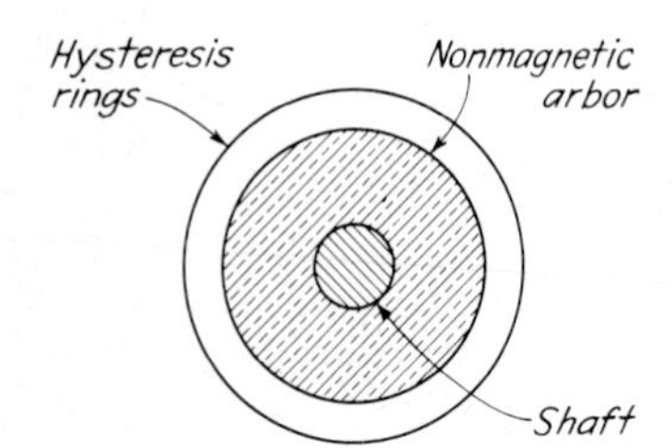


**FIG. 6.20** Rotor construction for a hysteresis motor.

In the *inverted*, or "umbrella," construction the hysteresis rings are pressed into a cup of a nonmagnetic material secured to the shaft. This assembly revolves around the outside of the stator, like an umbrella. This design gives more rotor inertia, providing a more uniform angular velocity; it is ideal for gyroscopes, since the hysteresis rings can be pressed directly into the flywheel. The smoothness of operation makes it ideal for driving tape recorders and video recorders.

***Polyphase Hysteresis Motors.*** Polyphase motors are discussed here because their theory is easier to explain than is the theory of other types of hysteresis motors. Suppose that the rotor of Fig. 6.20 is placed inside a polyphase stator that is producing a four-pole magnetic field of constant strength, rotating at synchronous speed. When we sketch in the paths of the four-pole magnetic flux, it at once becomes evident that:

1. The flux will enter the rotor radially at two points and leave it radially at two points.
2. The flux will flow circumferentially in a clockwise direction at two points and counterclockwise at two other points.
3. When the field rotates faster than the rotor, each and every point in the rotor will be subjected to an alternating flux, although at any instant the density and direction of flux will vary spatially around the rings.

Hence, at slips other than zero, there is alternating flux and, concomitantly, hysteresis loss in the rings. *The hysteresis loss in the rings at standstill is equal to the synchronous torque developed, expressed in synchronous watts.* The synchronous torque developed will be constant all the way up to synchronous speed. Thus, the speed-torque curve of an "ideal" polyphase hysteresis motor will have the shape of a rectangle, as shown in Fig. 6.21.

Many single-phase-wound stators will also produce synchronous torque, but since other types of stators do not produce uniformly rotating fields of constant strength,

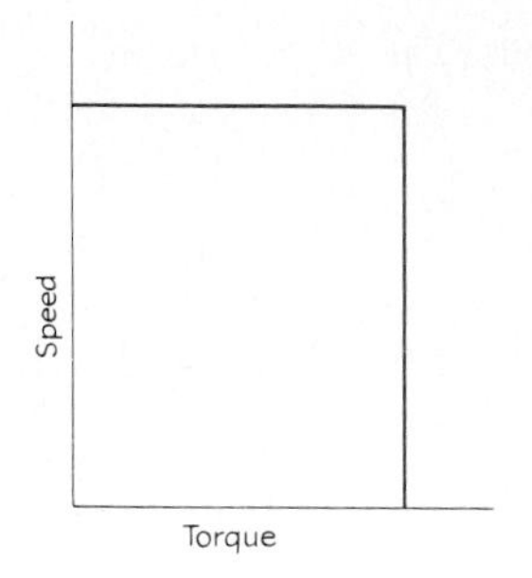


**FIG. 6.21** Speed-torque curve of an "ideal" polyphase hysteresis motor.

the resultant speed-torque curves will depart from the "ideal" that is shape illustrated in Fig. 6.21.

## Permanent Magnet Motors

The ceramic magnet materials developed over the last few years have made possible the development of small permanent magnet synchronous motors that have magnetic salient poles but no physical poles. It is the extremely high coercivity of the ceramic materials that has made this development possible. A typical rotor is illustrated in Fig. 6.22.

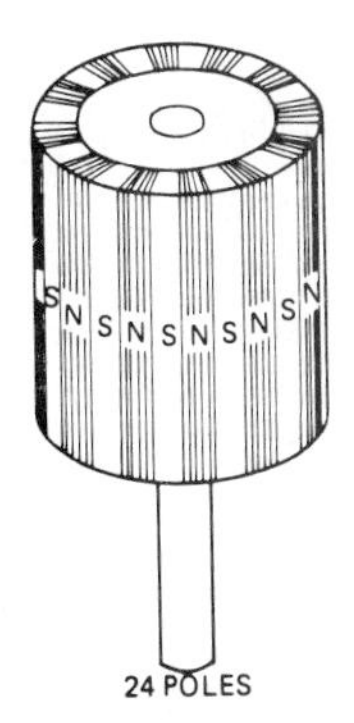


**FIG. 6.22** Rotor for a permanent magnet synchronous motor. (*Courtesy of Hurst Manufacturing Corporation.*)

This construction provides no torque at subsynchronous speeds. In order to operate at all, the rotor and its load have to accelerate from standstill to synchronous speed in less than a half-cycle of the supply frequency. This makes the permanent magnet motor very sensitive to inertial loads and limits the amount of shaft-connected inertia that can be brought up to synchronous speed. Sometimes this problem can be mitigated by using a coupling with some torsional resiliency.

*Starting in the wrong direction of rotation* is often a problem with this type of motor.

## Inductor Motors

Another widely used type of small synchronous motor is known as the inductor type, although there are different variations of the same principle. One form of construction is illustrated in Fig. 6.23, which shows one of the twin structures needed. A permanent magnet (indicated by the shaded area) causes all of the rotor teeth to be set up as south poles in the structure shown, and as north poles in the twin member. The stator has 8 teeth, each with a coil on it, while the rotor has 10. As the rotor revolves, the airgap opposite each stator tooth goes through a cyclic variation, causing an ac voltage to be induced in each coil. (This is the principle of the inductor alternator.) In a variation of this, the rotor is made of two stamped spiders and thus has less inertia, which is an advantage.

The inductor construction is widely used for stepper motors.

## 6.7 UNIVERSAL MOTORS

*Universal motors* are series motors designed to operate either on direct current or on alternating current up to 60 Hz. They develop more horsepower per pound than do other types of ac motors, principally because of their high operating speeds. The power ratings vary from 10 mhp to 1 hp for continuous-rated motors, and even higher for intermittent-rated motors. Full-load operating speeds may range from 4000 to 16,000 r/min in the larger sizes, and up to 20,000 r/min in the smaller sizes.

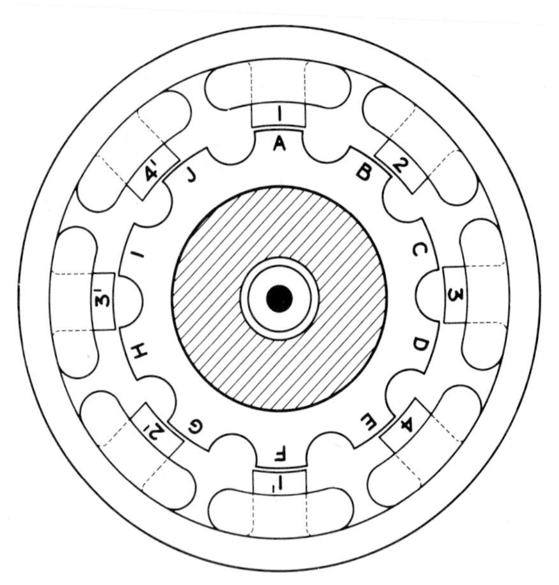


**FIG. 6.23** Schematic sectional view of inductor motor with one stator tooth for each stator coil. Actual motor has several teeth per coil, but principles of operation are the same. (*Courtesy of General Electric Company.*)

Very popular applications are portable drills, saws, routers, vacuum cleaners, sewing machines, food mixers, blenders, etc. Most motors today are of salient-pole construction, noncompensated, and are usually supplied as parts that the ultimate users build directly into their appliances.

Figure 6.24 shows the speed-torque characteristics for a $^1/_4$-hp 6000 r/min universal motor. Note that the torque characteristics are not identical for ac and dc operation, nor even for 25 and 60 Hz.

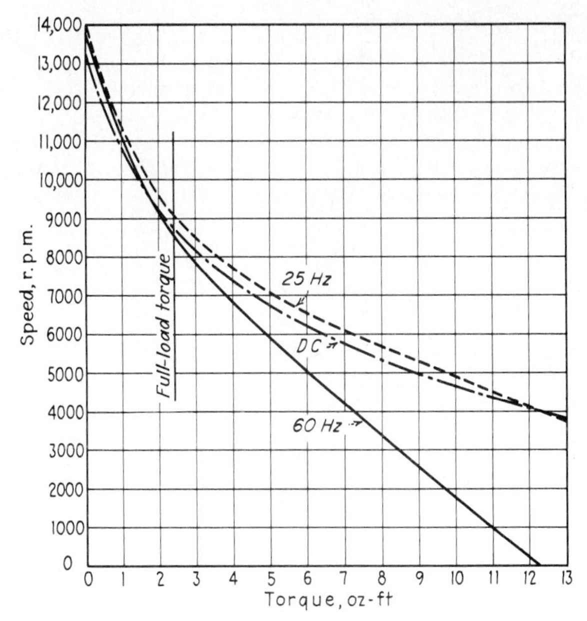


**FIG. 6.24** Speed-torque curves of a noncompensated concentrated-pole universal motor rated at $^1/_4$ hp at 8000 r/min.

## 6.8 DC MOTORS

Originally, fractional-horsepower dc motors were mostly used only where ac power was not available, and they represented but a small part of the small-motor industry. Recently, the use of small dc motors has been growing, thanks in part to the growing popularity of adjustable-speed drives and in part to the availability of low-cost controllable silicon rectifiers. Today, most fractional-horsepower dc motors operate on rectified power; in subfractional-horsepower sizes, many of them use permanent magnets and operate directly from batteries.

DC motors are available in the sizes and ratings common to fractional-horsepower motors; one popular application is for driving golf carts. Permanent magnet motors require no power for excitation and are therefore much more efficient than motors with wound fields; this makes them especially useful where they are battery-driven. DC motors using electronic commutation are now offered by a few manufacturers and are usually called brushless dc motors. They are not to be confused with a battery-powered inverter driving an ac motor.

DC motors cannot be depended upon to operate as closely to their rated full-load speed as do ac motors, unless they are externally regulated. At full load, fractional-horsepower single-phase induction motors will generally operate within 2 percent of their rated speed, whereas similar dc motors may vary as much as $7^1/_2$ percent above or below their rated speeds. (See also Chap. 5, "Direct-Current Machines.")

## 6.9 POLYPHASE MOTORS

Polyphase induction motors are available in the larger fractional-horsepower ratings. They are designed for across-the-line starting and use single-cage rotors. To offset their natural tendency to have higher slips than corresponding single-phase motors, fractional-horsepower polyphase motors are designed for 40 percent more breakdown torque than comparable single-phase motors. (See Chap. 4, "Induction Machines.")

## 6.10 DESIGN OF A SINGLE-PHASE MOTOR

To design a small induction motor requires making such an enormous amount of tedious error-free calculations that it is neither economical nor practical to do it by hand. Nevertheless, the following will attempt to show how it might be done manually, based on equations and procedures developed and published by the author.[1,2] It will take an illustrative problem, starting first by designing the punchings. The actual calculations, however, will be made on a personal computer, using a set of interactive programs developed by the author and generally following the author's procedures.[3] The algorithm behind each step will be briefly explained.

The method of attack will be first to put together some sort of preliminary design specification and to make a preliminary design analysis. The next steps will be to review the results of the design analysis, change some parameters, make a new design analysis, review it, and repeat the process until a satisfactory design is obtained. In executing this plan, the procedure will be (1) to note first only the most important performance items needed to assess each current design and then (2) to examine more and more performance features as we zero in on the final design. What features are important may well vary with each problem, for this approach is so flexible that a wide range of choices of parameters can be made.

The features of interest for the illustrative problem are those listed and monitored in Table 6.1. The Target column of this table shows that we wish to produce a design having the breakdown and locked-rotor torques of the $^1/_4$-hp general-purpose motor of Table 6.2 without exceeding the locked-rotor amperes or full-load slip of Table 6.1. The successive columns of Table 6.1 record the progress at each step.

**TABLE 6.1** Steps in Arriving at a Design

| Feature* | Target | TRIAL-1 | TRIAL-2 | TRIAL-3 | TRIAL-4 | TRIAL-5 |
|---|---|---|---|---|---|---|
| % Full | < =65 | 104.9 | 60.5 | 57.1 | 55.4 | 64.4 |
| BT1 | < =100 | 64.5 | 84.6 | 94.9 | 98.2 | 98.2 |
| BDT | = >34 | 19.2 | 30.7 | 33.5 | 33.7 | 35.5 |
| LRT | = >46 | | | 32.9 | 54.1 | 56.4 |
| LRA | < =26 | | | 17.2 | 20.9 | 21.2 |
| FL RPM | = >1725 | | 1744.8 | 1750.4 | 1740.3 | 1741.3 |
| STG MFD | | 100 | 100 | 100 | 189 | 189 |
| FL EFF | | | | | 0.6396 | 0.6506 |

* Where % Full = percentage of slot fullness; BT1 = stator tooth density; BDT = breakdown torque; LRT = locked-rotor torque; LRA = locked-rotor amperes; FL RPM = full-load revolutions per minute; STG MFD = starting microfarads; FL EFF = full-load efficiency.

**TABLE 6.2** Performance Characteristics of General-Purpose 60-Hz Motors

| | | Single and Polyphase | | | Single-phase | | | | Three-phase | |
|---|---|---|---|---|---|---|---|---|---|---|
| | | Full-load | | | Torque, oz·ft | | Locked-amp | | Torque, oz·ft | |
| Horsepower rating | Poles | Watts | r/min | Torque, oz·ft | Break-down | Lock | 115 V | 230 V | Break-down | Lock |
| 1/20 | 2 | 37.3 | 3450 | 1.22 | 3.7 | | 20 | 12 | 5.2 | |
| | 4 | 37.3 | 1725 | 2.44 | 7.1 | | 20 | 12 | 9.9 | |
| | 6 | 37.3 | 1140 | 3.69 | 10.4 | | 20 | 12 | 14.6 | |
| | 8 | 37.3 | 850 | 4.95 | 13.5 | | 20 | 12 | 18.9 | |
| 1/12 | 2 | 62.2 | 3450 | 2.03 | 6.0 | | 20 | 12 | 8.4 | |
| | 4 | 62.2 | 1725 | 4.06 | 11.5 | | 20 | 12 | 16.1 | |
| | 6 | 62.2 | 1140 | 6.15 | 16.5 | | 20 | 12 | 23.1 | |
| | 8 | 62.2 | 850 | 8.24 | 21.5 | | 20 | 12 | 30.1 | |
| 1/8 | 2 | 93.3 | 3450 | 3.05 | 8.7 | | 20 | 12 | 12.2 | |
| | 4 | 93.3 | 1725 | 6.09 | 16.5 | 24 | 20 | 12 | 23.1 | |
| | 6 | 93.3 | 1140 | 9.22 | 24.1 | 32 | 20 | 12 | 33.7 | |
| | 8 | 93.3 | 850 | 12.36 | 31.5 | | 20 | 12 | 44.1 | |
| 1/6 | 2 | 124.4 | 3450 | 4.06 | 11.5 | 15 | 20 | 12 | 16.1 | 13 |
| | 4 | 124.4 | 1725 | 8.12 | 21.5 | 33 | 20 | 12 | 30.1 | 24 |
| | 6 | 124.4 | 1140 | 12.29 | 31.5 | 43 | 20 | 12 | 44.1 | 35 |
| | 8 | 124.4 | 850 | 16.48 | 40.5 | | 20 | 12 | 56.7 | 46 |
| 1/4 | 2 | 186.5 | 3450 | 6.09 | 16.5 | 21 | 26 | 15 | 23.1 | 19 |
| | 4 | 186.5 | 1725 | 12.18 | 31.5 | 46 | 26 | 15 | 44.1 | 35 |
| | 6 | 186.5 | 1140 | 18.44 | 44.0 | 59 | 26 | 15 | 61.6 | 49 |
| | 8 | 186.5 | 850 | 24.7 | 58.0 | | 26 | 15 | 81.2 | 65 |
| 1/3 | 2 | 248.7 | 3450 | 8.12 | 21.5 | 26 | 31 | 18 | 30.1 | 24 |
| | 4 | 248.7 | 1725 | 16.24 | 40.5 | 57 | 31 | 18 | 56.7 | 45 |
| | 6 | 248.7 | 1140 | 24.6 | 58.0 | 73 | 31 | 18 | 81.2 | 65 |
| | 8 | 248.7 | 850 | 33.0 | 77.0 | | 31 | 18 | 107.8 | 86 |
| 1/2 | 2 | 373 | 3450 | 12.18 | 31.5 | 37 | 45 | 25 | 44.1 | 35 |
| | 4 | 373 | 1725 | 24.4 | 58.0 | 85 | 45 | 25 | 81.2 | 65 |
| | 6 | 373 | 1140 | 36.9 | 82.5 | 100 | 45 | 25 | 115.5 | 92 |
| 3/4 | 2 | 559.5 | 3450 | 18.28 | 44.0 | 50 | 61 | 35 | 61.6 | 50 |
| | 4 | 559.5 | 1725 | 36.6 | 82.5 | 110 | 61 | 35 | 115.5 | 92 |
| 1 | 1 | 746 | 3450 | 24.4 | 58.0 | 61 | | | | |

***Source:*** Compiled from NEMA and original sources.

## Motor to Be Designed

The illustrative problem mentioned above is to design a capacitor-start motor rated at 1/4 hp, 115 V, 60 Hz, four poles, with performance characteristics like those in Table 6.2. Both punchings are to be designed.

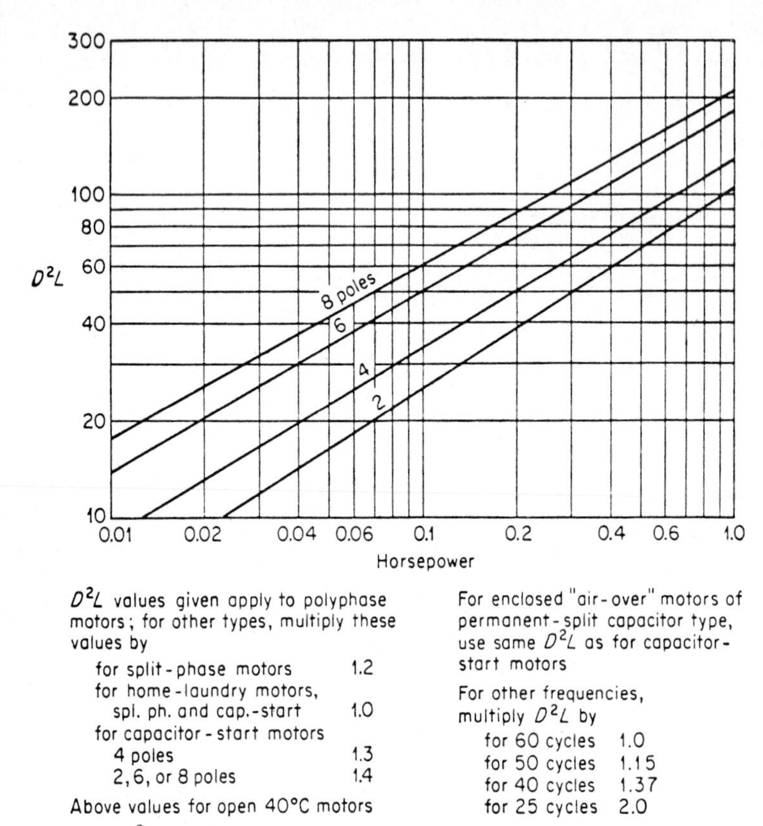


**FIG. 6.25** $D^2L$ values for fractional-horsepower motors.

## Design of Stator and Rotor Cores

Find $D^2L$ from Fig. 6.25, and use 80 percent of it.

$$D^2L = 57 * 1.3 * 0.8 = 59.28 \tag{6.1}$$

If $D^2L < 75$, use the 48 frame; if $75 < D^2L < 170$, use the 56 frame. Let's put this on the 48 frame: the outside diameter (OD) of the stator punching is 5.375. Then $L = 59.28/5.375^2 = 2.05$, say 2.0.

***Stator Punching Design.*** Figures 6.26 and 6.27 give a method for designing a stator punching. A computer program for doing this is illustrated in Fig. 6.45. For the first try in Fig. 6.45, only the OD, number of slots, and number of poles were specified. The computer selected a bore dimension of 3.227; this inside-diameter (ID) figure was rounded to 3.25. We then let the computer compute the dimensions of the punching, plus the *net slot-winding area* calculated by the methods of Fig. 6.29 and plus the stator slot constants (for calculating slot-leakage reactance) calculated according to Fig. 6.35. The results are illustrated in Fig. 6.45. A punching identification number "DEM-S1" was assigned, and the data on the line beginning with "DEM-S1" were stored on disk for later use.

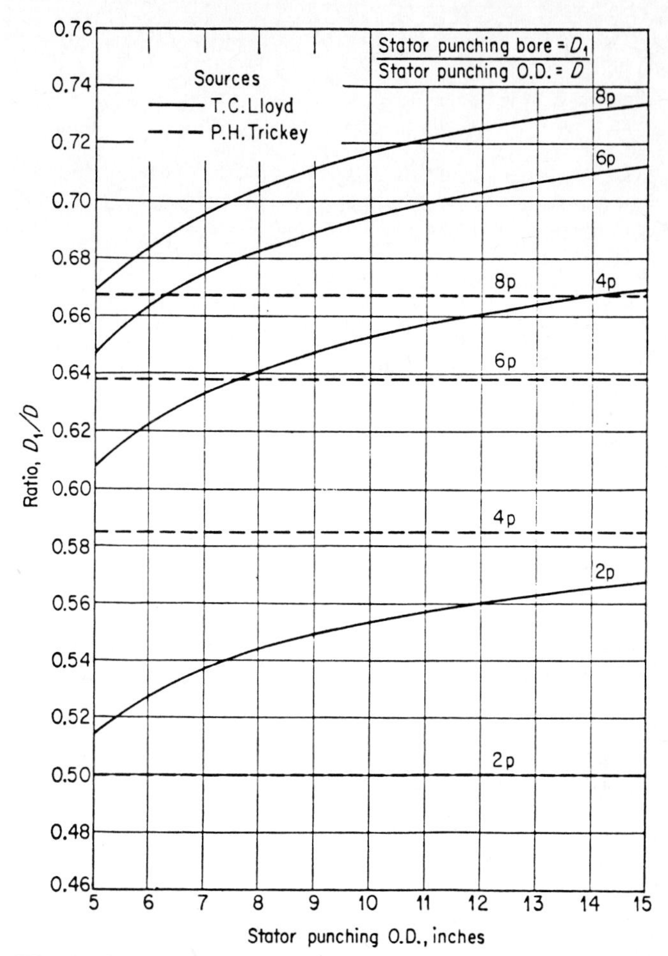


**FIG. 6.26** Stator punching proportions.

***Rotor Punching Design.*** To establish the OD of the rotor, it is first necessary to determine the airgap; unless better information is available, we can use the equation at the top of Fig. 6.28, which yields a value of 0.012, so

$$\text{Rotor OD} = 3.25 - 2 * 0.012 = 3.226 \tag{6.2}$$

Now Fig. 6.28 shows how to design a lamination with open slots, but the available computer program (Fig. 6.45) designs a closed rotor slot with rounded tops and bottoms. First we need to find the rotor tooth width needed to carry the flux in the stator teeth. The width of all 36 stator teeth, shown in Fig. 6.45, is S1*T1 = 5.0271. If we use 48 rotor slots and assume that the rotor tooth area needs to be only 90 percent of the stator teeth area, then

$$\text{Rotor tooth width T2} = 0.9 * 5.0271/48 = 0.0943 \text{ in} \tag{6.3}$$

### *Design of a Stator Punching*

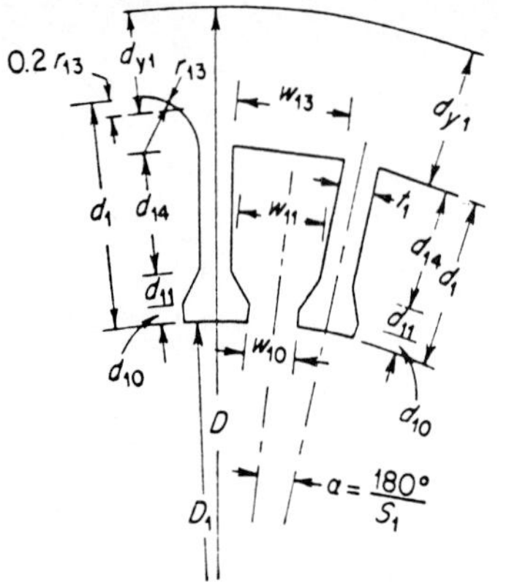

1. $D$ = O.D.
2. $D_1$ = bore See Fig. 6.26
3. $S_1$ = no. of slots
4. $w_{10}$ = slot opening
   - if $S_1 = 24$, $w_{10} = 0.027 + 0.0175D_1$
   - if $S_1 = 36$, $w_{10} = 0.015 + 0.0175D_1$
   - if $S_1 = 48$, $w_{10} = 0.0175D_1$
5. $d_{10}$ = depth of tip = 0.03
6. $d_{11}$ = depth of mouth = 1.0 to $1.5d_{10}$
7. $t_1$ = tooth width
   - for 2 poles $t_1 = (1.10 + 0.08D_1)D_1/S_1$
   - for 4+ poles $t_1 = (1.27 + 0.09D_1)D_1/S_1$
8. $d_{y1}$ = yoke depth $= \left(\dfrac{B_t}{B_y}\right)\left(\dfrac{S_1 t_1}{\pi P}\right)$

   where $B_t$ = tooth density, $B_y$ = yoke density, $P$ = no. of poles
   A good value for $B_t/B_y$ is 1.15
9. $\alpha$ = half slot angle = $180°/S_1$

*Flat-bottom slots*

10. $w_{11} = \dfrac{\pi[D_1 + 2(d_{10} + d_{11})]}{S_1} - t_1$
11. $d_{14} = 0.5(D - D_1) - (d_{10} + d_{11} + d_{y1})$
12. $w_{13} = w_{11} + 2d_{14} \tan \alpha$

*Round-bottom slots*

13. $A = (D_1 \sin \alpha - t_1)/2$
14. $B = 0.5(D - D_1) - d_{y1}$
15. $d_1 = \dfrac{B(1 + \sin \alpha) + 0.2A}{1 + 0.8 \sin \alpha}$
16. $r_{13} = \dfrac{A + B \sin \alpha}{1 + 0.8 \sin \alpha}$
17. $d_{14} = d_1 - (r_{13} + d_{10} + d_{11})$
18. $w_{11} = \dfrac{2(r_{13} - d_{14} \sin \alpha)}{\cos \alpha}$
19. $w_{13} = 2r_{13}$ approximately

Functions of Half-angle $\alpha = 180/S_1$

| $S_1$ | $\alpha$ | sin | cos | tan |
|---|---|---|---|---|
| 24 | 7°30′ | .13053 | .99144 | .13165 |
| 30 | 6°00′ | .10453 | .99452 | .10510 |
| 36 | 5°00′ | .08716 | .99619 | .08749 |
| 42 | 4°17.1′ | .07472 | .99721 | .07493 |
| 48 | 3°45′ | .06540 | .99786 | .06554 |
| 54 | 3°20′ | .05814 | .99831 | .05824 |
| 60 | 3°00′ | .05234 | .99863 | .05241 |
| 66 | 2°43.6′ | .04757 | .99887 | .04762 |
| 72 | 2°30′ | .04362 | .99905 | .04366 |
| 90 | 2°00′ | .03490 | .99939 | .03492 |
| 96 | 1°52.5′ | .03271 | .99946 | .03273 |
| 108 | 1°40′ | .02908 | .99958 | .02910 |
| 120 | 1°30′ | .02618 | .99966 | .02619 |
| 144 | 1°15′ | .02181 | .99976 | .02182 |

**FIG. 6.27** Design of a stator punching for a small induction motor for parallel-sided teeth, flat- or round-bottom slots.

*Design of a Rotor Punching*

1. Airgap $= g = .005 + \dfrac{.0042D_1}{\sqrt{P}}$
2. $D_2 = D_1 - 2g$
3. $t_2 = 0.95S_1t_1/S_2$
4. $w_{20} \geqq$ lamination tkns., to .03
5. $d_{20} \geqq w_{20}$

*Round slots*

6. $r_{21} = \dfrac{\pi(D_2 - 2d_{20}) - S_2t_2}{2(S_2 + \pi)}$
7. $d_c = 2r_{21} - 0.015$
   Adjust $d_c$ to nearest size and recompute $r_{21}$. Then
8. $t_2 = \dfrac{\pi(D_2 - 2d_{20} - 2r_{21})}{S_2} - 2r_{21}$

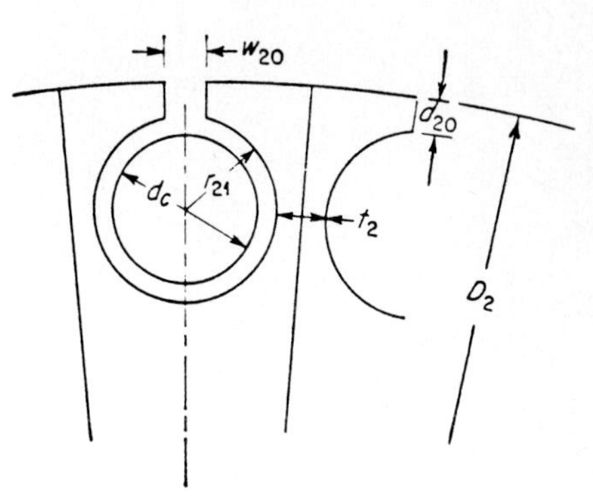


*Trapezoidal slots*

9. Radial clearance $= C_r = d_{24} - d_b = .010 - .015$
10. Tangential clearance $= C_t = w_{23} - w_b = .010$
11. $d_{21} = (1 \text{ to } 1.5)d_{20}$
12. $d_{24} = 0.10 + 0.03D_1$
    Above equation is for usual slot proportions; for max. cond. area, use
    12a. $d_{24} = \dfrac{D_2 - 2(d_{20} + d_{21} + C_r)}{4} - \dfrac{S_2(t_2 + C_t)}{4\pi}$
13. $d_b = d_{24} - C_r$
14. Adjust $d_b$ to nearest size
15. $d_{24} = d_b + C_r$
16. $d_2 = d_{24} + d_{21} + d_{20}$
17. $w_{23} = \dfrac{\pi(D_2 - 2d_2)}{S_2} - t_2$
18. $w_b = w_{23} - C_t$
19. Adjust $w_b$ to nearest size
20. $w_{23} = w_b + C_t$
21. $w_{21} = w_{23} + 2d_{24} \tan\left(\dfrac{180}{S_2}\right)$
22. $t_2 = \dfrac{\pi(D_2 - 2d_2)}{S_2} - w_{23}$

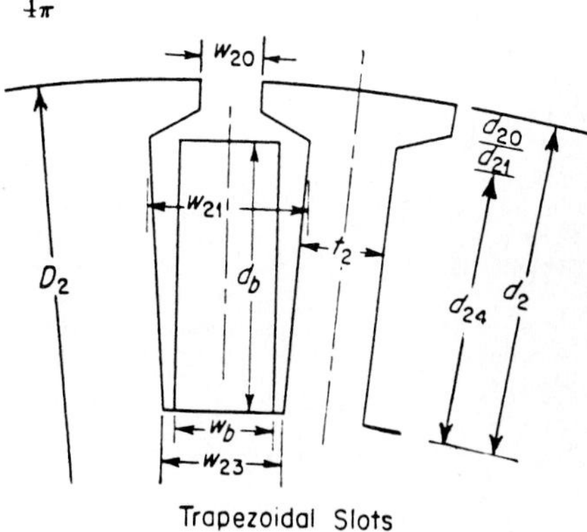


FIG. 6.28 Design of a rotor punching for a small induction motor for round or for rectangular conductors.

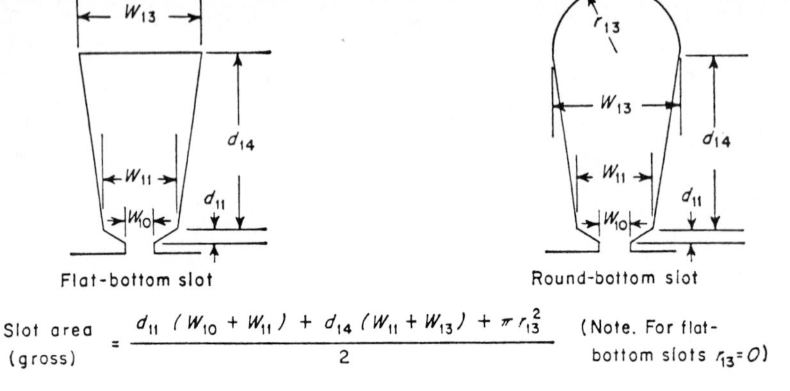


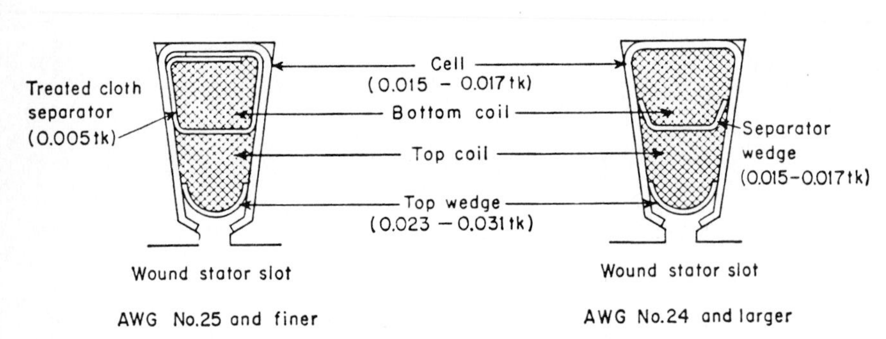


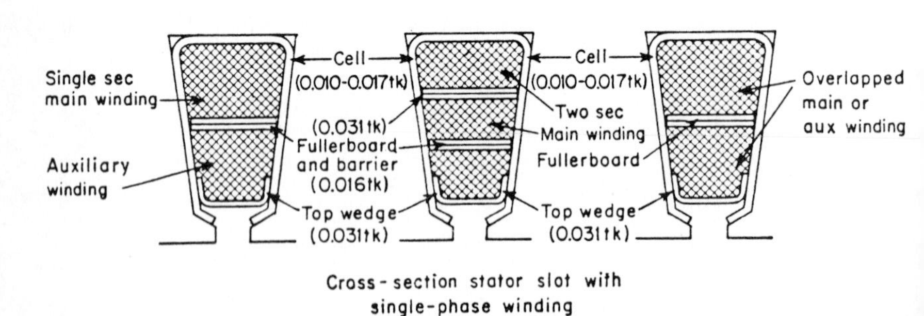


Net slot wdg. area = gross slot area — gross-sectional area stator insulation

**FIG. 6.29** Typical methods of insulating stator slots and method for figuring net slot-winding area.

*Primary Resistance and Weight of Wire*

| Poles | $\gamma$ |
|---|---|
| 2 | 1.3 |
| 4 | 1.5 |
| 6+ | 1.7 |

$LMC$ = length of mean conductor
$= L_1 + \gamma(\text{ACT})\pi D_e/S_1$

$$r_1 = \left(\frac{LMC}{12}\right)\left(\frac{C}{q}\right)\left(\frac{\text{ohms}/1000'}{1000}\right)$$

$$\text{Weight} = \left(\frac{LMC}{12}\right) Cqm \left(\frac{\text{lb}/1000'}{1000}\right)$$ (Use $m$ = 1 for single-phase)

*Secondary Resistance*

$$\text{“Bar”} = \frac{\sqrt{L_2^2 + SK^2} \times 100}{S_2 A_b \times \%\ \text{conductivity}}$$

$$\text{“Ring”} = \frac{63.7 D_r \times K_{\text{ring}}}{A_r P^2 \times \%\ \text{conductivity}}$$ (For $K_{\text{ring}}$, see Fig. 6.39 )

Resistance constant $= 0.693m(Ck_w)^2 \times 10^{-6}$ (Use $m$ = 2 for single-phase)
$r_2$ = Resistance constant × (Bar + Ring)

*Iron Losses*

| | Density | Volume |
|---|---|---|
| Sta. core | Same as in Fig. 6.32 | Area × $P$ × length |
| Sta. teeth | Same as in Fig. 6.32 | Area × $P$ × length |
| Surface | See Eq. (6.8) | |

*Friction and Windage*

For sleeve-bearing fractional-horsepower motors

$$\text{Friction} = 1.25\ (\text{journal OD})^3 \times \left(\frac{\text{rpm}}{100}\right)$$

**FIG. 6.30** Resistance and loss calculations in small induction motors.

The first four inputs for the rotor tooth design are now fixed. A good bridge depth is 0.008. As for the rotor slot area, let's start with the biggest one possible. To do this, simply enter 1.0 for the slot area. Now from these inputs, and using procedures similar to those shown in Fig. 6.28, the program comes up with the CALCULATED SLOT DIMENSIONS, a statement that SLOT IS LARGEST POSSIBLE, BUT SMALLER THAN SPECIFIED, and the PUNCHING DATA

## Single-phase-motor Design Summary Sheet

| Slot No. | 1 | 2 | 3 | 4 | 5 | 6 | 7 | 8 | 9 | 10 | 11 | 12 | 13 | 14 | 15 | 16 | 17 | 18 | Bare diam. | Diam. over ins. |
|---|---|---|---|---|---|---|---|---|---|---|---|---|---|---|---|---|---|---|---|---|
| Mn. wdg. (Total Strands) | 18<br>18 | 36 | 36 | 36 | 18 | × | × | × | 18 | 36 | 36 | 36 | 18<br>18 | each section | | | | | 2 − .0360 | .0421/.0386 |
| Aux. wdg. (Total Strands) | | | 12 | 12 | 36 | 24 | 12<br>12 | 24 | 36 | 12 | 12 | | | | | | | | | |
| % Full | | | | | | | | | | | | | | | | | | | | |

| | Main wdg. | Aux. wdg. | | |
|---|---|---|---|---|
| $ACT$ | 8 | 8.25 | | |
| $\gamma$ | 1.3 | 1.3 | | |
| $LMC$ | 9.03 | 9.24 | | |
| Weight | 3.52 | 0.94 | $D_e$ | 4.80 |
| Ohms/sect. | .890 | 3.53 | $C_{xm}$ | 0.971 |
| Res. @ 25C | $r_1$ = .445 | $r_{1a}$ = 3.53 | $C_{xa}$ | |
| $C$ | 288 | 384 | $a$ | 1.363 |
| $k_w$ | .822 | .842 | Skew, in inches at gap | .523 |
| $Ck_w$ | 237 | 323 | Elect. angle of skew, $\alpha$ | 15° |
| $V$ | 115 | | Rot. cond. size | 0.162 diam. Copper |
| $\phi$ = 364 × .97 = 353 | | | Res. ring. dimensions | 2.25 I.D. × ⅛ tk. Copper |

STATOR SLOT: 0.03, 0.100, 0.04, 0.73, 0.305, 0.66, 0.4788, R; INSUL. S.W.A. 0.204

ROTOR SLOT: 0.03, 0.03, 0177, R

| | Stator | | Rotor | | $K_S$ | 2.24 | Slot | 1.49 | Summary of Performance | | | |
|---|---|---|---|---|---|---|---|---|---|---|---|---|
| | | | | | | | | | | Goal | Calc. | Test |
| OD | $D$ | 7.500 | $D_2$ | 3.966 | $K_{ZZ}$ | .1574 | $ZZ$ | 2.05 | | | | |
| ID | $D_1$ | 4.000 | $D_b$ | | $N_{SP}$ | 15.0 | End | 2.51 | No-load amps. | | 5.28 | |
| Stack length | $L_1$ | 2.50 | $L_2$ | 2.50 | $K_B$ | 1.28 | Belt | .91 | No-load watts | | 135 | |
| No. of slots | $S_1$ | 24 | $S_2$ | 36 | $Q$ | .0056 | Skew | 1.05 | F.L. amps. | | 10.2 | |
| Tooth pitch | $\lambda_1$ | .524 | $\lambda_2$ | .346 | $C_{SK}$ | .9972 | $P_L$ | 8.01 | F.L. watts | | 972 | |
| Tooth face | $t_{10}$ | .424 | $t_{20}$ | .316 | $K_m$ | 300 | $P_L/2$ | 4.00 | F.L. Eff. | | 76.8 | |
| Tooth width | $t_1$ | .236 | $t_2$ | .149 | $K_x$ | .212 | $P_m$ | 193.5 | F.L. P.F. | | 82.6 | |
| Yoke depth | $d_{y1}$ | 1.02 | $d_{y2}$ | .671 | Bar | 3.41 | $P_0$ | 197.5 | F.L. App. Eff. | | 63.4 | |
| Carter fact. | $K_1$ | 1.115 | $K_2$ | 1.023 | Ring | 4.72 | $X$ | 1.698 | F.L. Rpm | | 3492 | |
| Slot const. | $K_{S1}$ | 1.20 | $K_{S2}$ | 1.623 | $B + R$ | 8.13 | $X_0$ | 41.87 | F.L. Losses | | 226 | |
| | $K_1K_2$ | 1.141 | $D_r$ | 3.73 | Res. const. | .0778 | $K_p$ | .97/.979 | Breakdown T. | 58 | 55–65 | |
| | $g$ | .017 | $A_b$ | .02061 | $r_2$ | 0.633 | $K_r$ | .94/.959 | Lock-rotor T. | 61 | 63.2 | |
| | $\lambda_p$ | 6.26 | $A_r$ | .1072 | $r_2/x$ | .429 (hot) | $K_c$ | .92/ | Lock-rotor amp. | 63 | 61.1 | |

| $F$ = .93 | Area | Density | Length | $AT$/in. | $AT$ | Watts/in.³ | Watts | Remarks |
|---|---|---|---|---|---|---|---|---|
| Sta. core | 4.74 | 74.5 | 5.09 | 4.4 | 22.4 | 0.58 | 28.0 | $r_1$ hot = 0.512 |
| Sta. teeth | 6.58 | 84.2 | .73 | 9.2 | 6.7 | 0.74 | 7.1 | $r_2$ hot = 0.728 |
| Rot. core | 3.12 | 104.0 | 2.29 | 17.5 | 40.1 | Surface | 20.6 | $r_2/X_0$ = 0.01739 |
| Rot. teeth | 6.24 | 81.7 | .06 | 8.5 | 0.5 | Core loss | 55.7 | |
| Air gap | 15.65 | 33.4 | .0194 | | 202.8 | F & W = 19.0 | | |
| Saturation Factor $SF_m$ = 1.344 | | | | Total | 272.5 | | | |

Type Cap.-start Frame _____ Service _____

1 Hp 115/230 Volts 60 Cycles 2 Poles 3450 Rpm Date 11-14-59

**FIG. 6.31** A design summary sheet.

FOR DESIGN-ANALYSIS PROGRAM, in which is included a slot area of 0.0374; all these are shown in the bottom half of Fig. 6.45. We assign the name "DEM-R1" to this punching and file the punching data on disk, to be used later.

### Design of Stator and Rotor Windings

The first step is to call up the DESIGN-ANALYSIS program, which successively asks for the six lines of inputs shown in Fig. 6.46. (This figure illustrates the format of the design-analysis program used, but the data are for the fourth trial.) The first

## Magnetic-circuit Calculations for Induction Motors

(all dimensions in inches)

| | Stator | | Rotor | |
|---|---|---|---|---|
| O.D. | $D$ | outside diameter | $D_2$ | finished O.D. |
| I.D. | $D_1$ | bore diameter | $D_b$ | bore diameter |
| Stack length | $L_1$ | measured axially | $L_2$ | axial length |
| Slots | $S_1$ | total no. of slots | $S_2$ | total no. of slots |
| Tooth pitch | $\lambda_1$ | $\pi D_1/S_1$ | $\lambda_2$ | $\pi D_2/S_2$ |
| Tooth face | $t_{10}$ | $\lambda_1 - w_{10}$ | $t_{20}$ | $\lambda_2 - w_{20}$ |
| Tooth width | $t_1$ | effective tooth width | $t_2$ | effective tooth width |
| Depth of yoke | $d_{y1}$ | | $d_{y2}$ | |
| Carter factor | $K_1$ | $\dfrac{\lambda_1(5g + w_{10})}{\lambda_1(5g + w_{10}) - w_{10}^2}$ | $K_2$ | $\dfrac{\lambda_2(5g + w_{20})}{\lambda_2(5g + w_{20}) - w_{20}^2}$ |

| | | | |
|---|---|---|---|
| Gap constant | $K$ | $K_1 \times K_2$ | Stator slot opening $= w_{10}$ |
| Air gap | $g$ | $(D_1 - D_2)/2$ | Rotor slot opening $= w_{20}$ |
| Pole pitch | $\lambda_p$ | $\pi(D_2 + g)/P$ | Number of poles $= P$ |

$V$ = impressed volts per phase

$$\phi = \frac{45{,}000V}{fCk_w} \times K_\phi \text{ (kilolines)}$$

$f$ = frequency in cps

$Ck_w$ = effective series conds. per ph.

$K_c = K_p/(2 - K_r)$ [from est. of $K_p$ and $K_r$]

$$K_\phi = \frac{V - I_1r_1\text{P.F.}}{V} \text{ or } \frac{OE}{ZE} \text{ (circle diag.) [usually estimated]}$$

$F$ = stacking factor

$L = L_1$ or $L_2$, whichever is lesser

| | Area | Density: Single-Ph. | Density: Polyphase | Length of Magnetic Path | $C = AT$/in. |
|---|---|---|---|---|---|
| Stator core | $2FL_1d_{y1}$ | $\dfrac{\phi}{\text{area}}$ | $\dfrac{\phi}{\text{area}}$ | $\dfrac{\pi(D - d_{y1})}{2P}$ | from curve |
| Stator teeth | $\dfrac{FL_1S_1t_1}{P}$ | $\dfrac{\phi}{.637 \times \text{area}}$ | $\dfrac{\phi}{.637 \times \text{area}}$ | actual | from curve |
| Rotor core | $2FL_2d_{y2}$ | $\dfrac{\phi K_c}{\text{area}}$ | $\dfrac{\phi K_p}{\text{area}}$ | $\dfrac{\pi(D_b + d_{y2})}{2P}$ | from curve |
| Rotor teeth | $\dfrac{FL_2S_2t_2}{P}$ | $\dfrac{\phi K_c}{.637 \times \text{area}}$ | $\dfrac{\phi K_p}{.637 \times \text{area}}$ | actual | from curve |
| Air gap | $L\lambda_p$ | $\dfrac{\phi K_r}{.637 \times \text{area}}$ | $\dfrac{\phi K_p}{.637 \times \text{area}}$ | $gK = g_e$ | $313g_e \times$ gap dens. |

Saturation factor = total $AT$/Gap $AT$

**FIG. 6.32** Magnetic circuit calculations for induction motors.

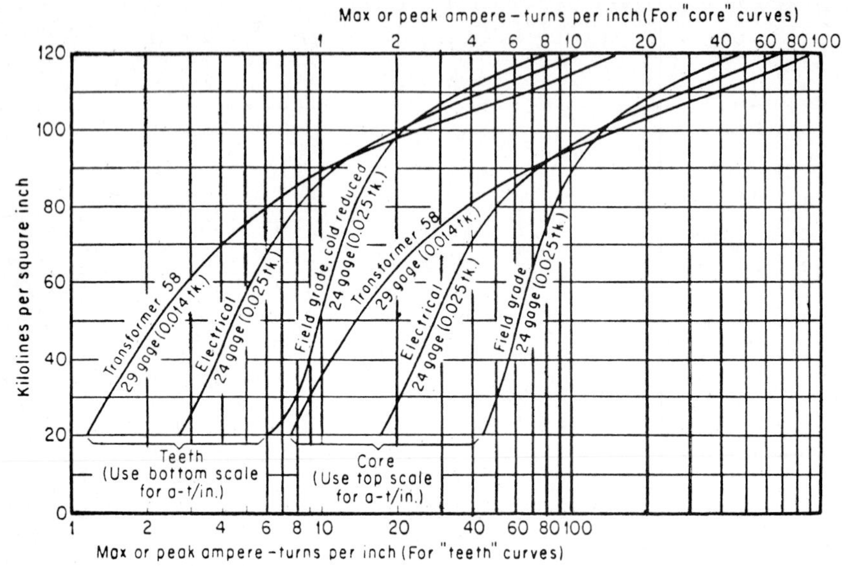


**FIG. 6.33** Saturation curves for induction motors.

line is self-explanatory (except for the 3 under 12345; this 3 permits the punching design data to be read from the disk). The value of the starting microfarads (SMFD) was originally guesstimated at 100 and later adjusted to 189.

***Ring Area.*** The rotor description is entered on the second line. The minimum volume of the rotor cage, for a given rotor resistance, has been shown to occur when

$$\mathrm{AR} = \mathrm{AB} * \mathrm{S2}/(\mathrm{P} * \mathrm{PI}) \quad \text{or} \quad \mathrm{AR} = 0.0374 * 48/(4 * \mathrm{PI}) = 0.143 \text{ in} \tag{6.4}$$

***Conductor CD.*** This item (CD = diameter at centers of rotor conductors) is supplied by the punching design program.

***Ring ID.*** If the ring section is square, each side would be 0.378; thus, the ring ID = 3.226 −2 * 0.378 = 2.47.

***Skew.*** A good rule is to skew one tooth pitch of the member having the least number of slots; this suggests for our case that we skew one stator tooth pitch, *measured at the rotor surface*. At the rotor surface, one stator tooth pitch is 0.282.

***Percent Conductivity.*** Assuming aluminum, use 50.

***Friction and Windage.*** Lacking better information, use the equation at the bottom of Fig. 6.30, obtaining

$$\mathrm{F\&W} = 1.25 * 0.5 * 0.5 * 1800/100 = 5.6 \text{ W} \tag{6.5}$$

This fills the second line. Now go to the third line.

***Main Winding.*** Select a good distribution (such as 2 2 1 x x x 1 2 2) for the first try, and pick an arbitrary number of turns (such as 50 50 25) and an arbitrary size (say gauge no. 18). Consult Table 6.3 to obtain the bare and insulated diameters of 0.0403 and 0.0437, respectively. Use section 1 only. Assume copper wire (conductivity = 100 percent) and QM = 1 for a series connection, where QM is the number of main-winding circuits.

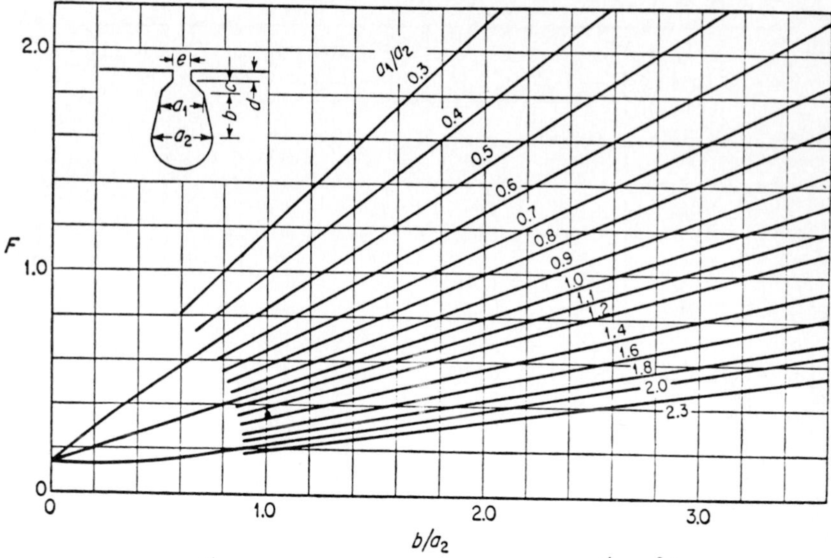


**FIG. 6.34** Slot constants for round-bottom slots.

***Auxiliary Winding.*** A distribution that goes well with the main-winding distribution used is 1 2 1 x x x x 1 2 1; the outer element of this coil spans a full pitch, so TOA = 10. Since the auxiliary winding usually has 25 to 50 percent more turns, TPC = 1.4 × (50 + 50 + 25)/(1 + 2 + 1) = 43.75, say 44, so

$$\text{Aux. wdg.} = 44\ 88\ 44\ \text{x x x x}\ 44\ 88\ 44 \tag{6.6}$$

For wire, use three gauges smaller than the main winding uses; in this case, from Table 6.3, the diameters are 0.0285 and 0.0314.

The data for the punchings just designed are then read from the disk, thus completing all the parameters needed to define the first trial design. The next step is to make a design analysis.

## Analysis of the First Trial Design

Analysis of a set of design parameters must be made in order to determine if this particular design suits the intended purposes or, if not, to serve as a basis for modifying parameters in order to obtain a satisfactory design. Such an analysis involves several steps. Important features used to monitor progress in this particular problem are listed in Table 6.1. The table also lists how closely each trial design meets the objectives. If this problem is being worked out manually, it is convenient to summarize the design-analysis results on a form such as Fig. 6.31. The steps in making a design analysis are outlined below. The format of computer-calculated results is illustrated in Figs. 6.46 and 6.47, but lest the reader become confused, the data there shown are for TRIAL-4.

**TABLE 6.3** Wire Table

| Gauge no. | Diameter bare (nominal), in | Area | | Diameter over insulation, in* | | Heavy, lb/1000 ft | Resistance, Ω/1000 ft at 25°C |
|---|---|---|---|---|---|---|---|
| | | Circular mils | mm² | Single | Heavy | | |
| 14 | 0.0641 | 4109 | 2.08 | 0.0666 | 0.0682 | 15.9 | 2.575 |
| 15 | 0.0571 | 3260 | 1.652 | 0.0594 | 0.0609 | 12.6 | 3.247 |
| 16 | 0.0508 | 2581 | 1.308 | 0.0531 | 0.0545 | 10.05 | 4.094 |
| 17 | 0.0453 | 2052 | 1.040 | 0.0475 | 0.0488 | 6.34 | 5.163 |
| 18 | 0.0403 | 1624 | 0.823 | 0.0424 | 0.0437 | 5.02 | 6.51 |
| 19 | 0.0359 | 1289 | 0.653 | 0.0379 | 0.0391 | 4.00 | 8.21 |
| 20 | 0.0320 | 1024 | 0.519 | 0.0339 | 0.0351 | 3.17 | 10.35 |
| 21 | 0.0285 | 812 | 0.411 | 0.0303 | 0.0314 | 2.51 | 13.05 |
| 22 | 0.0253 | 640 | 0.324 | 0.0270 | 0.0281 | 1.99 | 16.46 |
| 23 | 0.0226 | 511 | 0.259 | 0.0243 | 0.0253 | 1.58 | 20.76 |
| 24 | 0.0201 | 404.0 | 0.2047 | 0.0217 | 0.0227 | 1.26 | 26.17 |
| 25 | 0.0179 | 320.4 | 0.1621 | 0.0194 | 0.0203 | 0.998 | 33.00 |
| 26 | 0.0159 | 252.8 | 0.1282 | 0.0173 | 0.0182 | 0.793 | 41.62 |
| 27 | 0.0142 | 201.6 | 0.1024 | 0.0156 | 0.0164 | 0.630 | 52.48 |
| 28 | 0.0126 | 158.8 | 0.0806 | 0.0140 | 0.0147 | 0.501 | 66.17 |
| 29 | 0.0113 | 127.7 | 0.0649 | 0.0126 | 0.0133 | 0.396 | 83.4 |
| 30 | 0.0100 | 100.0 | 0.0507 | 0.0112 | 0.0119 | 0.316 | 105.2 |
| 31 | 0.0089 | 79.2 | 0.0400 | 0.0100 | 0.0108 | 0.251 | 132.7 |
| 32 | 0.0080 | 64.2 | 0.0324 | 0.0091 | 0.0098 | 0.198 | 167.3 |
| 33 | 0.0071 | 50.4 | 0.0253 | 0.0081 | 0.0088 | 0.158 | 211.0 |
| 34 | 0.0063 | 39.7 | 0.0203 | 0.0072 | 0.0078 | 0.126 | 266.0 |
| 35 | 0.0056 | 31.4 | 0.0157 | 0.0064 | 0.0070 | 0.0966 | 335.5 |
| 36 | 0.0050 | 25.0 | 0.0127 | 0.0058 | 0.0063 | 0.0791 | 423.0 |
| 37 | 0.0045 | 20.3 | 0.0101 | 0.0052 | 0.0057 | 0.0628 | 533.4 |
| 38 | 0.0040 | 16.0 | 0.0081 | 0.0047 | 0.0051 | 0.0498 | 672.6 |

* Insulations include Formvar, Nyform, Nyleze, polyester, Polythermaleze, and ML.

***Source:*** C. G. Veinott and J. E. Martin, *Fractional- and Subfractional-Horsepower Electric Motors*, 4th ed., McGraw-Hill, New York, 1986. From *NEMA Motor and Generator Standards*, NEMA Publ. MW 1000-1981, National Electrical Manufacturers Association, Washington, 1981, and data from Essex, Fort Wayne, IN 46804.

1. *Resistances and weights of wire:* The resistances, weights of wire, and rotor resistance can be calculated from Fig. 6.30.
2. *Slot fullnesses: Slot fullness* is defined as

$$100 * ND^2/\text{Net SWA} \tag{6.7}$$

where $N$ = total number of wires in slot
$D$ = diameter over insulation of each wire
SWA = net slot-winding area

3. *Magnetic circuit calculations:* How to make these is summarized in Fig. 6.32; the saturation curves are given in Fig. 6.33, and the curves for fundamental

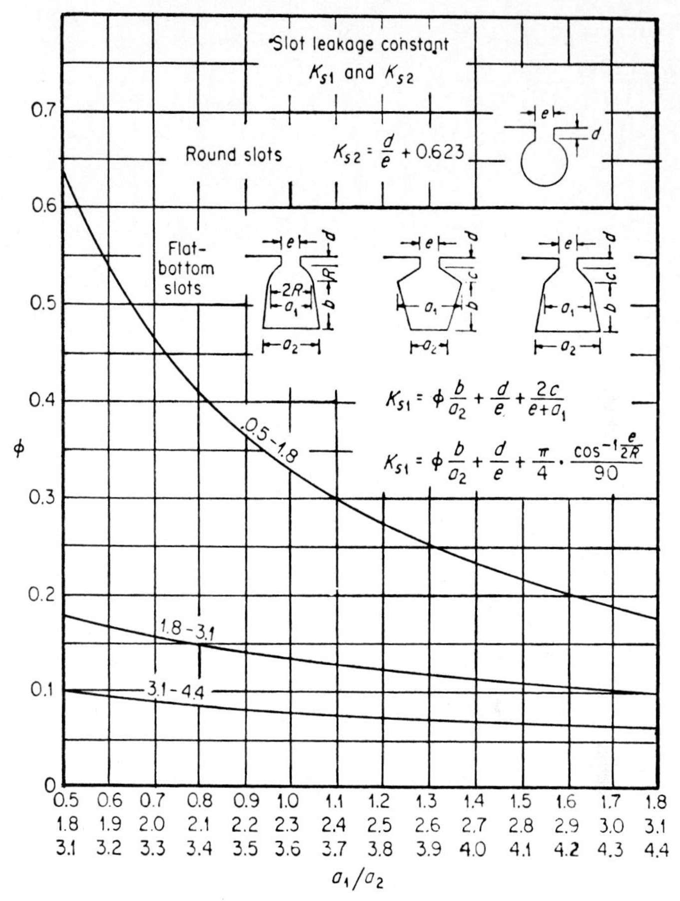


**FIG. 6.35** Slot constants for flat-bottom trapezoidal slots.

frequency losses are given in Fig. 6.40. To calculate the surface losses, one can use the following equation: *

$$\text{Watts loss} = C_{sf} \times 10^{-8} \times B_g{}^{2.3} \times \left(\frac{f}{P}\right)^{1.55} \times D_1{}^{2.05} \times \sqrt{S_1} \times \left(\frac{w_{10}}{g}\right)^{1.22} \times L \tag{6.8}$$

4. *Reactance calculations:* The calculations of leakage and magnetizing reactances are summarized in Fig. 6.38. Supporting curves are given in Figs. 6.34 to 6.37.
5. *Performance calculations:* A method for making calculations on the running winding only is detailed in Fig. 6.41. These calculations need to be made for full-load and breakdown-torque conditions. Locked-rotor calculations are illustrated

*$C_{sf}$: for electric grade, $1.85 \times 10^{-8}$; for field grade, $2.3 \times 10^{-8}$.

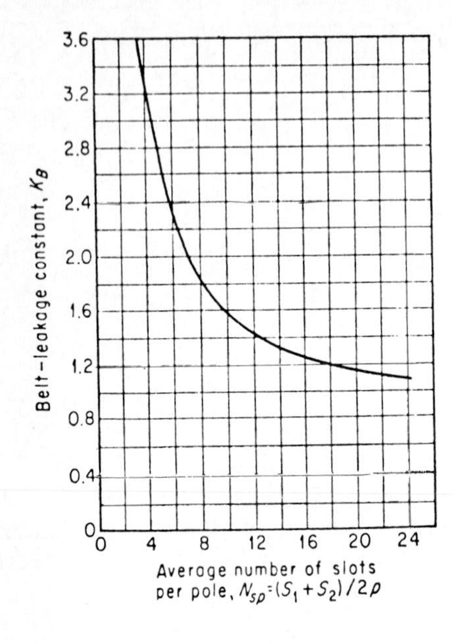


**FIG. 6.36** Belt-leakage constant.

in Fig. 6.42. Running performance calculations on both windings are detailed in Figs. 6.43 and 6.44.

## Additional Trial Designs

The major items to notice about TRIAL-1 are: The breakdown torque falls far short, the stator tooth density is quite low, and the slot fullness of 104.9 is obviously impossible. These facts all call for decreasing the turns. Based on the breakdown torque only, the new turns would be

$$50 * \mathrm{SQRT}(19.2/34) = 37.5 \tag{6.9}$$

If we kept the same amount of copper in the main winding as before, 37.5 turns would theoretically be correct, but since we obviously have to reduce the slot fullness from 104.9, we will have to use fewer than 37.5 turns, so let's try 36 turns.

Now if we used the same size wire as before, the slot fullness would go down to

$$104.5 * 36/50 = 75.2 \tag{6.10}$$

which is still too full to be practicable, so we must use a wire one gauge smaller for the next try. Obtaining the diameters from the wire table (Table 6.3), the main winding for TRIAL-2 becomes

$$36\ 36\ 18\ \mathrm{x}\ \mathrm{x}\ \mathrm{x}\ 18\ 36\ 36\ 0.0359\ 0.0391 \tag{6.11}$$

For the *auxiliary winding*, let us use approximately 40 percent more conductors than for the main winding and reduce the wire gauge by one, for the auxiliary winding also contributes to slot fullness. Then the auxiliary winding for TRIAL-2 becomes

$$32\ 64\ 32\ \mathrm{x}\ \mathrm{x}\ \mathrm{x}\ \mathrm{x}\ 32\ 64\ 32\ 0.0253\ 0.0281 \tag{6.12}$$

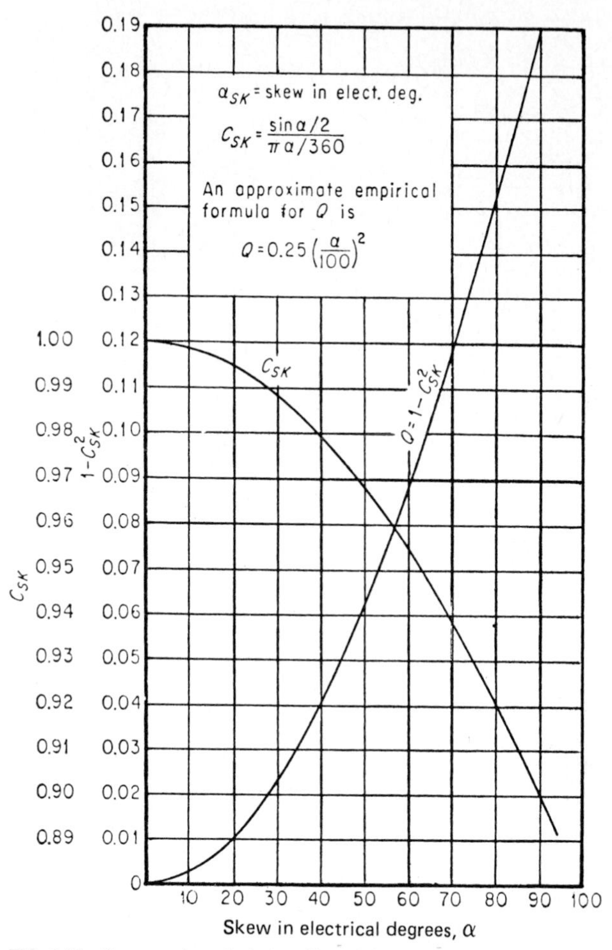


**FIG. 6.37** Constants for calculating effect of skew upon magnetizing and leakage reactances.

These two new windings are entered as TRIAL-2 and a design analysis is made, the significant results from which are noted in Table 6.1. We note:

1. Breakdown torque 30.7, still low
2. Slot fullness OK
3. Tooth density OK

Let's first boost the breakdown torque by reducing the main-winding turns to

$$36 * \mathrm{SQRT}(30.7/34) = 34.2, \text{ say } 34 \tag{6.13}$$

and the auxiliary turns by the same ratio to 30 60 30, with the same wire gauge in both as before, and make a design analysis for TRIAL-3; results of interest are recorded in Table 6.1.

| Reactance Constants | Reactance Equations |
|---|---|
| $C_x$ for single-phase motors: $= \dfrac{(C_1^2 + C_2^2 + \cdots)}{(C_1 + C_2 + \cdots)^2} \times \dfrac{1}{k_w^2} \times \dfrac{S_1}{4P}$ | (Note: for single-phase motors, use $m = 2$ below.) ACT = weighted ave. coil throw, in slots |
| $C_x$ for polyphase motors: | $D_e$ = dia. at centers of sta. slots |

| $m$ | $\psi$ | $p$ = p.u. pitch | $C_x$ |
|---|---|---|---|
| 3 | 60° | .33– .67 | $.25(6p - 1)$ |
| " | " | .67–1.00 | $.25(3p + 1)$ |
| " | " | 1.00–1.33 | $.25(7 - 3p)$ |
| " | 120° | .67–1.33 | 0.75 |
| " | " | 1.33–2.00 | $.375p + .25$ |
| 2 | 90° | 0 –1.00 | $p$ |
| " | " | 1.00–2.00 | $2 - p$ |
| " | 180° | 0 –2.00 | $0.5p$ |

Reactance equations:

$$\text{Slot} = \frac{3.19 m K_S L_1}{S_1}$$

$$\text{Zig-zag} = \frac{1.065 m L_1}{S_1 g} K_{zz}$$

$$\text{End} = \frac{1.57 m D_e(\text{ACT})}{S_1 P}$$

$$\text{Belt} = .00118 m K_m K_B$$

$$\text{Skew} = .3234 m K_m K_p Q$$

Reactance constants:

$K_{S1}$ = stator slot constant

$K_{S2}$ = rotor slot constant

$K_S = K_{S1} C_x^* + (S_1/S_2) K_{S2}$

$$K_{zz} = \frac{(t_{10} + t_{20})^2}{4(\lambda_1 + \lambda_2)}$$

$$N_{Sp} = \frac{S_1 + S_2}{2P}$$

$K_B$ See Fig. 6.36

$Q$ See Fig. 6.37

$C_{sk}$ See Fig. 6.37

$$K_m = \frac{A_g}{g_e SF_m P}$$

$K_x = 2\pi f(CK_w)^2\, 10^{-8}$

$P_L$ = Slot + ZZ + End + Belt + Skew

$P_m = .3234 m K_m C_{sk}$

$P_o = P_m + P_L/2$

$X = K_x P_L$

$X_0 = K_x P_o$

* For greater accuracy, the following term should be added:

$$0.25 K_{S1w}(1 - C_x)$$

where $K_{S1w}$ = slot constant of wound portion of stator slot.

**FIG. 6.38** Reactance calculations on induction motors.

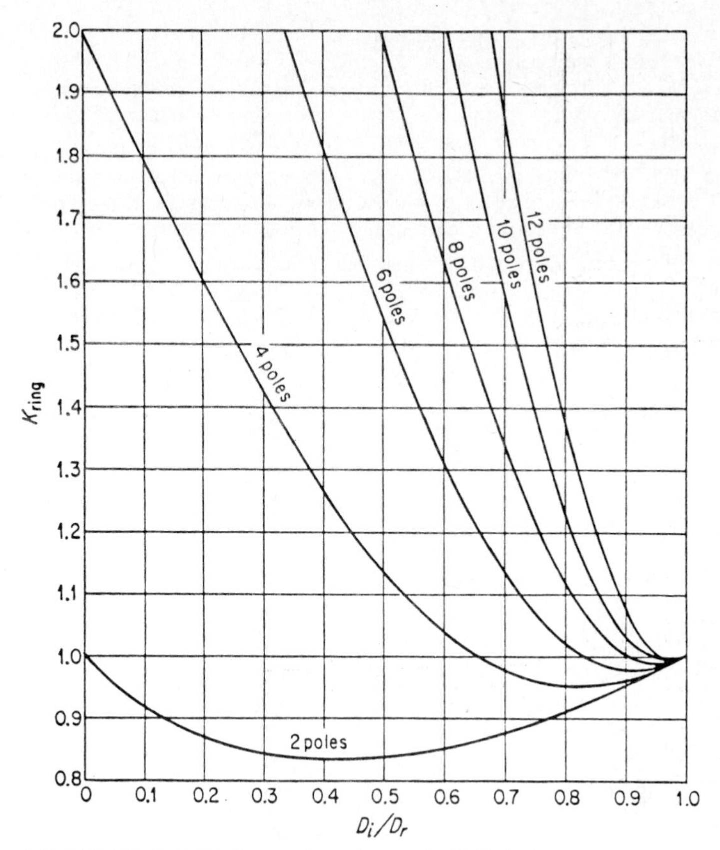


**FIG. 6.39** Effect of wide rings on ring resistance. (*P. H. Trickey.*)

TRIAL-3 is much closer to our objective; we need to study more results of this analysis before proceeding further. Some comments are in order:

1. Slot fullness = 57.1. This is lower than need be; efficiency could be improved by increasing it.
2. The stator tooth density of 94.9 is now OK but should probably not be increased much if quietness is of importance.
3. The breakdown torque of 33.5 is still marginally lower than the desired 34.
4. The locked-rotor torque of 32.9 is way below the objective of 46.
5. The locked-rotor amperes of 17.2 allows a substantial margin below the 26 A specified.
6. The full-load r/min of 1750.4 is high, permitting us to increase the rotor resistance in order to increase the starting torque.

The locked-rotor torque (LRT) now needs our main attention. How can it be increased?

1. It can be increased by reducing the main-winding turns, but it is obvious that if we reduced the turns enough to bring the LRT up to 46 (that is, by a ratio of 46/32.9), the densities would go up in the same ratio and be excessive, and the breakdown torque would be much more than necessary. (Generally, it is uneconomic to give more breakdown torque than needed.)
2. The LRT can be increased by increasing the rotor resistance; for a small increase in R2, the LRT would go up by almost the same percentage. However, increasing R2 reduces the locked-rotor current *and the breakdown torque.*
3. The LRT can be increased by attention to the auxiliary phase. Help in how to do this can be obtained from the LOCKED-ROTOR CALCS. WITH OTHER MFD. VALUES at the bottom of Fig. 6.47, which also illustrates the format of these results. Similar results for TRIAL-3, as read from the computer screen, were

| | MFD | LRT | LRA |
|---|---|---|---|
| Actual | 100 | 32.85 | 17.19 |
| For maximum LRT | 158.5 | 38.77 | 20.41 |
| For maximum LRT/LRA | 134.7 | 38.64 | 19.27 |

This table of results, compiled from TRIAL-3, shows that the LRT could be increased, but not enough to meet the requirements, simply by increasing the MFD.

The manual calculation of such results is illustrated (for a different problem) in Fig. 6.42. The calculation is not particularly difficult or time-consuming to do by hand.

In sum, then, let us do the following for TRIAL-4:

1. Increase R2 about 25 percent to boost the LRT. For the next trial, the simplest way to do this is to reduce the conductivity of the rotor from 50 to 40 percent. For the final design, one should reduce the size of the bar or of the ring, or both, since a material of 40 percent conductivity may not be available. We reduce the conductivity here as a faster way to find out the rotor resistance needed; then we can achieve it later by redesigning the conductors and bars.
2. Reduce the main-winding turns to compensate for the effect of increased rotor resistance, which decreases the breakdown torque (BDT), and because the BDT is already marginally weak. For an educated guess, let's reduce the turns from 34 34 17 to 33 33 16.
3. Auxiliary winding: Use one size larger wire, with the turns reduced in order to maintain about the same slot fullness. New auxiliary winding:

$$24\ 48\ 24\ 0.0285\ 0.0314 \tag{6.14}$$

4. Starting microfarads (STG MFD): For TRIAL-3, the MFD for maximum starting torque per ampere was 134.7. Now since the auxiliary turns have been reduced in the ratio 30/24 = 1.25, it follows that for the same ampere-turns, the auxiliary-winding current (and hence the capacitance) must be increased at least by the same factor, or up to 169. From a capacitor table, it can be found that the next standard MFD rating above 169 is 189, so let's use 189.

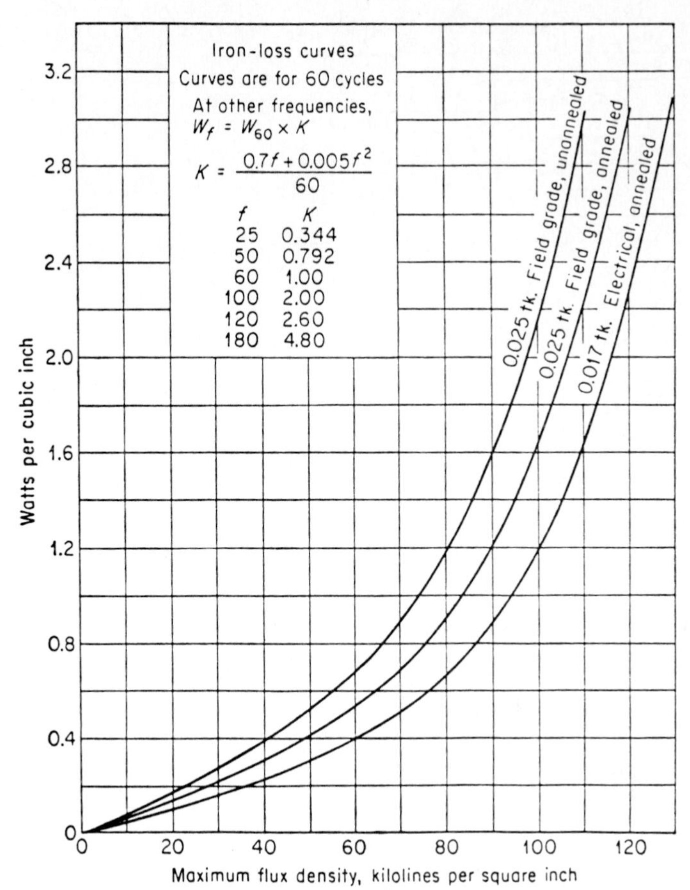


**FIG. 6.40** Fundamental-frequency iron losses in induction motors.

Figures 6.46 and 6.47 show the winding just developed as TRIAL-4, together with its design-analysis and performance calculations. For purposes of comparison, some of the results are summarized in Table 6.1.

Starting from scratch, and having to design laminations, we have achieved our objectives on only the *fourth trial design.* We may wish to make further refinements, but how easy it is to do on a microcomputer, for it requires only 26 seconds to make and display the design-analysis and performance calculations needed to evaluate each change.

If we wish to go further, we might increase the size of the main winding by one-half a wire gauge, for example, which would fill the slots fuller and increase the BDT and efficiency. Just to illustrate the effect, the author did just this in TRIAL-5, in which the only changes made (from TRIAL-4) were that the main winding was wound in two sections in parallel, that one section used wire two gauges smaller than the wire used in TRIAL-4, and that the other used wire three gauges smaller. Significant results are included in Table 6.1; note the increases in % Full, BDT, LRT, LRA, FL RPM, and FL EFF.

## Single-phase Motor Performance by Revolving-field Theory

| Given: | |
|---|---|
| $r_1$ hot | 3.80 |
| $r_2$ hot | 4.65 |
| $X$ (s.c. react.) | 8.3 |
| $X_0$ (o.c. react.) | 110 |
| $K_p$ | .964 |
| $K_r$ | .929 |
| Fe loss (ff) | 10.0 |
| Fe loss (hf) | 9.0 |
| Fe loss (total) | 19.0 |
| Fr. & Wind. | 10.0 |
| Hp rating | ⅙ |
| Volts | 110 |
| Cycles | 60 |
| Poles | 4 |

| Calculate: | | | |
|---|---|---|---|
| $x_1 = X/(1 + K_p)$ | 4.23 | $M_3 = 0.5K_pX_0$ | 53.0 |
| $M_1 = 0.5K_rr_2$ | 2.16 | $M_4 = 0.5K_px_1$ | 2.04 |
| $M_2 = r_2/X_0$ | .0423 | | |

| | | | | | | | No load | Break-down torque |
|---|---|---|---|---|---|---|---|---|
| 1 | $s$ | .044 | | | | | | .19 |
| 2 | $2 - s$ | 1.956 | | | | | | 1.81 |
| 3 | $M_1/s$ | 49.1 | | | | | | 11.37 |
| 4 | $M_1/(2 - s)$ | 1.104 | | | | | | 1.193 |
| 5 | $(M_2/s)^2$ | .924 | | | | | | .0496 |
| 6 | $[M_2/(2 - s)]^2$ | .000467 | | | | | | .00055 |
| 7 | $1.0 + (5)$ | 1.924 | | | | | | 1.0496 |
| 8 | $1.0 + (6)$ | 1.0005 | | | | | | 1.00055 |
| 9 | $R_f = (3)/(7)$ | 25.5 | | | | | | 10.83 |
| 10 | $R_b = (4)/(8)$ | 1.103 | | | | | | 1.192 |
| 11 | $r_1 =$ | 3.8 | | | | | | 3.80 |
| 12 | $R = R_f + R_b + r_1$ | 30.40 | | | | | | 15.82 |
| 13 | $M_3 \times (5)$ | 49.0 | | | | | | 2.62 |
| 14 | $M_3 \times (6)$ | .025 | | | | | | .0292 |
| 15 | $M_4$ | 2.04 | | | | | | 2.04 |
| 16 | $(13) + (15)$ | 51.04 | | | | | | 4.66 |
| 17 | $(14) + (15)$ | 2.065 | | | | | | 2.07 |
| 18 | $X_f = (16)/(7)$ | 26.5 | | | | | | 4.44 |
| 19 | $X_b = (17)/(8)$ | 2.06 | | | | | | 2.07 |
| 20 | $x_1$ | 4.23 | | | | | | 4.23 |
| 21 | $X = X_f + X_b + x_1$ | 32.8 | | | | | | 10.74 |
| 22 | $Z = \sqrt{(12)^2 + (21)^2}$ | 44.7 | | | | | | 19.12 |
| 23 | $I = V/(22)$ | 2.46 | | | | | | 5.75 |
| 24 | $I$ (corrected) * | 2.52 | | | | | 1.860 | |
| 25 | $R_f - R_b$ | 24.4 | | | | | | 9.638 |
| 26 | Pri. loss $= (24)^2 \times r_1$ | 24.1 | | | | | 13.1 | |
| 27 | Sec. loss (f) $= (23)^2 sR_f$ | 6.8 | 19.9 | | | | 0.0 | |
| 28 | " " (b) $= (23)^2(2 - s)R_b$ | 13.1 | | | | | 7.5 | |
| 29 | Fe loss (ff) | 10.0 | | | | | 10.0 | |
| 30 | $(23)^2 \times (25) \times (1 - s)$ | 141.2 | | | | | 19.0 | 258.1 |
| 31 | Input = 26 to 30 inc. | 195.2 | | | | | 49.6 | |
| 32 | Fe loss (hf) + F & W | 19.0 | | | | | | 19.0 |
| 33 | Output = (30) − (32) | 122.1 | | | | | | 239.1 |
| 34 | Rpm = $(1 - s)$ syn. rpm | 1721 | | | | | | 1458 |
| 35 | Torq. = 112.7(33)/(34) | 8.00 | | | | | | 18.5 |
| 36 | Eff. = (33)/(31) | 62.6 | | | | | | |
| 37 | P.F. = $(31)/(24)V$ | 70.4 | | | | | | |
| 38 | App. Eff. = (36) × (37) | 44.1 | | | | | | |
| 39 | % Full load | 98.2 | | | | | | |

* $I$ (corrected)

$$= (23) + \frac{\text{Fe (ff)}}{V} \times \frac{(12)}{(22)}$$

| | No-load Calcs. |
|---|---|
| 24 | $I_0 = 2V/(X_0 + X)$ |
| 27 | Sec. loss (f) = 0 |
| 28 | Sec. loss (b) $= I_0^2 M_1$ |

**FIG. 6.41** Performance calculations on a single-phase motor running on main winding, using the double-revolving field theory.

## Starting Performance for Single-phase Motors

| Calculation | | | 1 | 2 | 3 | 4 | 5 | 6 | 7 |
|---|---|---|---|---|---|---|---|---|---|
| (20) = either sin (16) or $\frac{(19)}{(5) \times (13)}$ | | | Mfd. for max. $T_S$ aux. wdg. 12 12 36 24 12 12 No. 20 Dbl. formvar | Mfd. for max. $T_S/I_L$ Same wdgs. as No. 1 | | | | | |
| | Resistance | Capacitor | | | | | | | |
| 1 | $r_1$ | $r_1$ | 0.512 | ↓ | | | | | |
| 2 | $r_2' = r_2C_r$ | $r_2' = r_2C_r$ | 0.698 | | | | | | |
| 3 | $R_m$ | $R_m$ | 1.210 | | | | | | |
| 4 | $X_m' = X_mC_R$ | $X_m' = X_mC_R$ | 1.710 | | | | | | |
| 5 | $Z_m$ | $Z_m$ | 2.09 | | | | | | |
| 6 | — | $R_c$ | .47 | | | | | | |
| 7 | $r_{1a}$ | $r_{1a}$ | 3.53 | | | | | | |
| 8 | $r_{2a}'$ | $r_{2a}'$ | 1.297 | | | | | | |
| 9 | $R_a$ | $R_{ac}$ | 5.297 | ↓ | | | | | |
| 10 | — | $-X_{ac}$ | 1.687 | 2.64 | | | | | |
| 11 | $X_a'$ | $X_a'$ | 3.17 | 3.17 | | | | | |
| 12 | — | $-X_c$ | 4.857 | 5.81 | | | | | |
| 13 | $Z_a$ | $Z_{ac}$ | 5.559 | 5.918 | | | | | |
| 14 | $\theta_m$ | $\theta_m$ | | | | | | | |
| 15 | $\theta_a$ | $-\theta_{ac}$ | | | | | | | |
| 16 | (14) − (15) | (14) + (15) | | | | | | | |
| 17 | $X_mR_a$ | $X_mR_{ac}$ | 9.058 | 9.058 | | | | | |
| 18 | $R_mX_a$ | $-R_mX_{ac}$ | 2.041 | 3.212 | | | | | |
| 19 | (17) + (18) | (17) + (18) | 11.099 | 12.27 | | | | | |
| 20 | sin $(\theta_m - \theta_a)$ | sin $(\theta_m - \theta_{ac})$ | 0.955 | | | | | | |
| 21 | $\times a$ | $\times a$ | 1.363 | | | | | | |
| 22 | $\times 1.88P/f$ | $\times 1.88P/f$ | .0627 | 3.281 | | | | | |
| 23 | $\times r_2'$ | $\times r_2'$ | 0.698 | | | | | | |
| 24 | $\times I_m$ | $\times I_m$ | 55.0 | ↓ | | | | | |
| 25 | $\times I_a$ | $\times I_{ac}$ | 20.7 | 19.43 | | | | | |
| 26 | $= T_S$ (oz-ft) | $= T_S$ (oz-ft) | 64.9 | 63.2 | | | | | |
| 27 | $R_{ma}$ | $R_{mac}$ | 6.507 | 6.507 | | | | | |
| 28 | $X_{ma}$ | $X_{mac}$ | .023 | −.930 | | | | | |
| 29 | $Z_{ma}$ | $Z_{mac}$ | 6.507 | 6.573 | | | | | |
| 30 | $I_L = I_mZ_{ma}/Z_a$ | $I_L = I_mZ_{mac}/Z_{ac}$ | 64.4 | 61.1 | | | | | |
| 31 | C.M./$I_a$ | $E_c = I_{ac}X_c$ | 100.5 | 112.9 | | | | | |
| 32 | | $VA_c = I_{ac}^2X_c$ | 2081 | 2193 | | | | | |
| 33 | | Mfd. | 545 | 455 | | | | | |
| 34 | | | | | | | | | |
| 35 | | | | | | | | | |
| 36 | | | | | | | | | |
| 37 | | | | | | | | | |

FIG. 6.42 Locked-rotor calculations on capacitor-start auxiliary windings.

## Performance Calculations on Capacitor Motors

| | |
|---|---|
| $V_m$ | 115 |
| $X_0$ | 20.55 |
| $X_m$ | 1.52 |
| Hp | 75 |
| $K_r$ | .926 |
| $K_p$ | .960 |
| $X_a$ | 2.37 |
| $r_2$ | .854 |

| | |
|---|---|
| $V_a$ | 115 |
| $K_\epsilon = V_a/V_m$ | 1.00 |
| $a$ | 1.25 |
| $a^2$ | 1.563 |
| Syn. rpm | 1800 |
| Fe loss | 94.7 |
| F & W | 13.3 |
| Fe + F & W | 108.0 |

| | | | |
|---|---|---|---|
| $Z_1$ | $r_1$ (hot) = .695 | $x_1 = \dfrac{X_m}{1 + K_p} = .775$ | |
| $Z_{1a}$ | $r_{1a}$ (hot) = 2.70 | $x_{1a} = \dfrac{X_a}{1 + K_p} = 1.21$ | |
| $Z_c$ | $R_c$ (hot) = 0.46 | $X_c = -5.82$ | |
| $Z_{1a} + Z_c$ | $r_{1a} + R_c = 3.16$ | $x_{1a} + X_c = -4.61$ | |
| $M_1 = 0.5K_r r_2 =$ | .395 | $M_3 = 0.5K_p X_0 =$ | 9.86 |
| $M_2 = r_2/X_0 =$ | .0415 | $M_4 = 0.5K_p x_1 =$ | .372 |

| | | | r-term | j-term | r-term | j-term | r-term | j-term | r-term | j-term |
|---|---|---|---|---|---|---|---|---|---|---|
| 1 | $s$ = slip | $M_4$ | .20 | .372 | | | | | | |
| 2 | $(M_2/s)^2$ | $M_3(M_2/s)^2$ | .04306 | .4245 | | | | | | |
| 3 | $M_1/s$ | (1) + (2) | 1.9750 | .7965 | | | | | | |
| 4 | $(M_2/s)^2 + 1$ | | 1.04306 | | | | | | | |
| 5 | $2 - s$ | $M_4$ | 1.800 | .3720 | | | | | | |
| 6 | $(M_2/2-s)^2$ | $M_3(M_2/2-s)^2$ | .00053 | .0052 | | | | | | |
| 7 | $M_1/2 - s$ | (5) + (6) | .2194 | .3772 | | | | | | |
| 8 | $(M_2/2 - s)^2 + 1$ | | 1.00053 | | | | | | | |
| 9 | $R_f$ = (3)/(4) | $X_f$ = (3)/(4) | 1.8935 | .7636 | | | | | | |
| 10 | $R_b$ = (7)/(8) | $X_b$ = (7)/(8) | .2193 | .3770 | | | | | | |
| 11 | (9) − (10) | | 1.6742 | .3866 | | | | | | |
| 12 | $a \times$ (11) | | 2.093 | .4833 | | | | | | |
| 13 | $Z_1$ | | .695 | .775 | | | | | | |
| 14 | (9) + (10) | | 2.1128 | 1.1406 | | | | | | |
| 15 | $a^2 \times$ (14) | | 3.3023 | 1.7828 | | | | | | |
| 16 | $Z_{1a} + Z_c$ | | 3.160 | −4.61 | | | | | | |
| 17 | $Z_T$ = (13) + (14) | | 2.8078 | 1.9156 | | | | | | |
| 18 | $Z_{Ta}$ = (15) + (16) | | 6.4623 | −2.8272 | | | | | | |
| 19 | $Z_T \times Z_{Ta}$ (Vector) | | 23.561 | 4.441 | | | | | | |
| 20 | [Vector (12)]$^2$ | | 4.147 | 2.023 | | | | | | |
| 21 | (19) − (20) | | 19.414 | 2.417 | | | | | | |
| 22 | $V_m$/(21) | | 5.833 | −.7262 | | | | | | |
| 23 | (18) | | 6.4623 | −2.8272 | | | | | | |
| 24 | $jK_\epsilon \times$ (12) (Vector) | | −.4833 | 2.093 | | | | | | |
| 25 | (23) + (24) | | 5.9790 | −.7342 | | | | | | |
| 26 | $K_\epsilon \times$ (17) | | 2.8078 | 1.9156 | | | | | | |
| 27 | $-j \times$ (12) (Vector) | | .4833 | −2.0930 | | | | | | |
| 28 | (26) + (27) | | 3.2911 | −.1774 | | | | | | |
| 29 | $\bar{I}_m$ = (25)(22) (Vector) | | $A$ = 34.342 | $B$ = −8.625 | $A$ = | $B$ = | $A$ = | $B$ = | $A$ = | $B$ = |
| 30 | $\bar{I}_a$ = (28)(22) (Vector) | | $g$ = 19.068 | $h$ = −3.425 | $g$ = | $h$ = | $g$ = | $h$ = | $g$ = | $h$ = |
| 31 | $\bar{I}_L = \bar{I}_m + \bar{I}_a$ | | 53.410 | −12.050 | | | | | | |
| | | | 382.745 | | | | | | | |

**FIG. 6.43** Combined-winding calculations on a capacitor-start motor. Part I: calculation of impedances and currents.

Performance Calculations on Capacitor Motors

| 32 $Ah$ | $-117.62$ | | | | | | | |
|---|---|---|---|---|---|---|---|---|
| 33 $Bg$ | $-164.46$ | | | | | | | |
| 34 $Ah - Bg$ | $+46.84$ | | | | | | | |
| 35 $I_m = \sqrt{A^2 + B^2}$ | 35.41 | | | | | | | |
| 36 $I_a = \sqrt{g^2 + h^2}$ | 19.37 | | | | | | | |
| 37 Line Amps | 54.75 | | | | | | | |
| 38 Watts, Mn. Ph. $= V_m A$ | 3946.8 | | | | | | | |
| 39 Watts, Aux. Ph. $= V_a g$ | 2192.8 | | | | | | | |
| 40 Line Watts = (38) + (39) | 6139.6 | | | | | | | |
| 41 $I_m^2$ | 1253.76 | | | | | | | |
| 42 $(aI_a)^2$ | 586.63 | | | | | | | |
| 43 (41) + (42) | 1840.39 | | | | | | | |
| 44 $2a \times (34)$ | 117.10 | | | | | | | |
| 45 (43) + (44) | 1957.49 | | | | | | | |
| 46 (43) − (44) | 1723.29 | | | | | | | |
| 47 (45) $\times R_f$ | 3706.5 | | | | | | | |
| 48 (46) $\times R_b$ | 377.9 | | | | | | | |
| 49 (47) − (48) | 3328.6 | | | | | | | |
| 50 $1 - s$ | .800 | | | | | | | |
| 51 (49) × (50) | 2662.9 | | | | | | | |
| 52 (Fe + F & W)$(1 - s)$ | 86.4 | | | | | | | |
| 53 Output Watts = (51) − (52) | 2576.5 | | | | | | | |
| 54 Efficiency = (53)/(40) | .4197 | | | | | | | |
| 55 P.F. = (40)/(37) $\times V_m$ | .9755 | | | | | | | |
| 56 Rpm = (50) × syn. rpm | 1440.0 | | | | | | | |
| 57 Torque, oz-ft | 201.6 | | | | | | | |
| 58 Sec. $I^2R(f)$ = (47) $s$ | 741.3 | | | | | | | |
| 59 Sec. $I^2R(b)$ = (48)$(2 - s)$ | 680.2 | | | | | | | |
| 60 Mn. Wdg. Cu. loss $= r_{1m} I_m^2$ | 871.4 | | | | | | | |
| 61 Aux. Wdg. Cu. loss $= r_{1a} I_a^2$ | 1013.4 | | | | | | | |
| 62 Capacitor loss $= R_c I_a^2$ | 172.6 | | | | | | | |
| 63 (Fe + F & W)$(1 - s)$ | 86.4 | | | | | | | |
| 64 Total losses | 3565.3 | | | | | | | |
| 65 Output + losses | 6141.8 | | | | | | | |
| 66 Cap. Volts $= I_a \sqrt{R_c^2 + X_c^2}$ | 113.02 | | | | | | | |
| 67 Aux. wdg. volts $= V -$ (30)$Z_c$ | 169.07 | | | | | | | |
| 68 $T_p = \sqrt{[(11r)^2 + (11j)^2]\,(45)(46)}$ | 3158 | | | | | | | |
| 69 | | | | | | | | |
| 70 | | | | | | | | |
| 71 | | | | | | | | |

**FIG. 6.44** Combined-winding calculations continued. Part II: calculation of losses, output, efficiency, etc.

```
        VICA-07: STATOR PUNCHING DESIGN - WITH ROUND-BOTTOM SLOTS
                INPUTS, INCLUDING COMPUTER-DEVELOPED INPUTS

    IDENT-NO    **O. D.   ***SLOTS  ***POLES  ---I. D.   ----TYPE   ----B
    DEM-S1       5.3750        36.        4.    3.2500         1.     1.15

     ---CELL   ---MID-S   --EFF-TW --NET-SWA TIP-DEPTH MOUTH-DPH SLOT-OP
       .0120      .0030      .1396     .0000     .0300     .0400     .07

             CALCULATED SLOT DIMENSIONS
RM = RADIUS OF THE BOTTOM OF THE SLOT =     .1162
BS = DEPTH OF TRAPEZOIDAL PART        =     .4395
FS = DEPTH OF TOOTH TIP               =     .0300
DS = DEPTH OF SLOT MOUTH              =     .0400
PS = TOTAL SLOT DEPTH                 =     .6257
A1S= WIDTH OF SLOT AT THE TOP         =     .1564
A2S= WIDTH OF SLOT AT THE BOTTOM      =     .2324
S1*T1 = WIDTH OF ALL THE TEETH        =    5.0271

              PUNCHING DATA FOR DESIGN-ANALYSIS PROGRAM
*STA-PCG SLS SLT STA-P STA-P SLOT *SLOT EFF. EFF. YOKE **NET SLOT-CONS
**IDENT# S1  TYP *O.D. *I.D. OPNG *C.D. **TW **TL DPTH **SWA FKS1A FKS
DEM-S1   36.  1. 5.375 3.250 .071 3.946 .140 .568 .460 .0910  .852 1.0


                VICA-07: CLOSED-SLOT ROTOR PUNCHING DESIGN
                         ROUND TOPS AND BOTTOMS

                                 INPUTS
                      ("EFF. ID" IS AN OPTIONAL INPUT)
 **PCHG-# ***O. D. NO.SLOTS -EFF.-TW --BRIDGE BAR-AREA ---POLES -EFF.-
 DEM-R1     3.2260      48.    .0943    .0080   1.0000        4.    .75

                      CALCULATED SLOT DIMENSIONS
R1 = TOP RADIUS              =      .0543
R2 = BOTTOM RADIUS           =      .0300
BR=DEPTH, STRAIGHT PART      =      .3711
FR = BRIDGE THICKNESS        =      .0080
PR = TOTAL SLOT DEPTH        =      .4634
DBEF=MAX. EFF. USABLE ID     =     1.5191
S2*T2=WIDTH OF ALL TEETH     =     4.5264
  SLOT IS LARGEST POSSIBLE, BUT SMALLER THAN SPECIFIED

              PUNCHING DATA FOR DESIGN-ANALYSIS PROGRAM
*ROT-PCG SLS SLT ROT-P ROT-P SLOT **BAR EFF. EFF. YOKE  SLOT SLOT-CONS
**IDENT# *S2 TYP *O.D. *I.D. OPNG DEPTH **TW **TL DPTH *AREA FKS2A FKS
DEM-R1   48. 13. 3.226  .750 .000  .455 .094 .439 .781 .0374 1.000 1.9
   DOR=  2.755 = "COND-CD" INPUT FOR VICA-02, 2d LINE
```

**FIG. 6.45** Laminations designed for illustrative problem.

```
IAL-4              VICA-02A: CAPACITOR-START MOTOR                          PAGE  1
                                  INPUTS

UN-NO. ****DATE 12345 IRON TYPE H. P. VOLTS HERTZ **P RISE RMFD. SMFD.
IAL-4   5/7/84    3       2.   1.  .250  115.    60.  4.  50.   .00  189.

 CORE------- R O T O R    D E S C R I P T I O N ------ %-STRAY F-AND-W
STACK BAR-AREA RNG-AREA  COND-CD RING-ID **SKEW %-COND *L-LOSS *AT-SYN
2.000   .03740   .14300    2.755   2.470   .282     40.    .000     5.6

--------- M A I N    W I N D I N G    D E S C R I P T I O N -----------
OM C1 C2 C3 C4 C5 C6 STD1 DBARE D-INS STD2 DBARE D-INS EXTNS %COND *QM
 9 33 33 16  0  0  0    1. .0359 .0391   0. .0000 .0000  .000  100.   1.

--------- A U X .    W I N D I N G    D E S C R I P T I O N -----------
OA C1 C2 C3 C4 C5 C6 STD1 DBARE D-INS STD2 DBARE D-INS EXTNS %COND *QA
10 24 48 24  0  0  0    1. .0285 .0314   0. .0000 .0000  .000  100.   1.

TA-PCG SLS SLT STA-P STA-P SLOT *SLOT EFF. EFF. YOKE **NET SLOT-CONSTS
IDENT# *S1 TYP *O.D. *I.D. OPNG *C.D. **TW **TL DPTH **SWA FKS1A FKS1W
M-S1    36.  1. 5.375 3.250 .071 3.946 .140 .568 .460 .0910  .852 1.083

OT-PCG SLS SLT ROT-P ROT-P SLOT **BAR EFF. EFF. YOKE *SLOT SLOT-CONSTS
IDENT# *S2 TYP *O.D. *I.D. OPNG DEPTH **TW **TL DPTH *AREA FKS2A FKS2W
M-R1    48. 13. 3.226  .750 .000  .455 .094 .439 .781 .0374 1.000 1.957

                  RESULTS OF DESIGN ANALYSIS CALCULATIONS
             ACT       LMC   COLD R    WT-#1     WT-#2 C=CONDS       KW      CKW
IN WDG     6.415     5.313    2.389    1.190      .000    656.    .8703    570.9
X. WDG     7.000     5.616    4.690     .928      .000    768.    .9114    699.9

OT NO.         1         2        3        4         5        6        7        8
C.FULL      55.4      55.4     52.9     52.0      52.0       .0       .0       .0

GNETIC CIRCUIT       AREA DENSITY   LENGTH        AT VOLUME   W/CU IN    WATTS
ATOR CORE           1.711   85.64     1.93      7.45   13.21     1.07    14.19
ATOR TEETH          2.344   98.22      .57      7.63    5.32     1.58     8.40
TOR CORE            2.905   44.90      .62       .59 SURFACE LOSS=        9.59
TOR TEETH           2.098   97.67      .44      5.70 TOTAL CORE LOSS     32.18
R GAP               5.086   41.71   .01388    181.25 FR. & WINDAGE        5.60
SATURATION FACTOR=    1.1179   TOTAL AT= 202.62 FLUX PER POLE =  146.55

   BAR RES   RING RES    R2 COLD   AIR-GAP SKEW ANGL BASE OHMS    A-RATIO
    1.2706      .8336     2.1042    .01200   20.0340   70.9115     1.2260

   P SLOT X  S SLOT X      END X    SKEW X ZIZ-ZAG X    BELT X    TOTAL X
      .9062     .9658      .6785     .6624    1.0994     .3688     4.6812

T CONS  R1         X1         R2        X2         XM       R1A        X1A
    .04017     .03737     .03561    .02864     .91351    .07888     .06160
    2.8488     2.6500     2.5250    2.0311    64.7787    5.5932     4.3682
```

**Fig. 6.46** Inputs and design-analysis calculations for the illustrative design problem.

TRIAL-4 VICA-02A: CAPACITOR-START MOTOR PAGE

LOAD PERFORMANCE

| PER-CENT LOAD | .0 | 25.0 | 50.0 | 75.0 | 100.0 | 125.0 | 150. |
|---|---|---|---|---|---|---|---|
| SLIP, PER-UNIT | .00532 | .01138 | .01796 | .02516 | .03317 | .04219 | .0527 |
| RPM | 1790.4 | 1779.5 | 1767.7 | 1754.7 | 1740.3 | 1724.1 | 1705. |
| LINE WATTS | 78.2 | 126.7 | 178.3 | 233.0 | 291.7 | 355.2 | 425. |
| LINE AMPS | 3.171 | 3.235 | 3.378 | 3.604 | 3.916 | 4.316 | 4.81 |
| TORQUE, OZ-FT | .00 | 2.95 | 5.95 | 8.99 | 12.08 | 15.23 | 18.4 |
| HORSEPOWER OUT | .0000 | .0624 | .1251 | .1876 | .2501 | .3124 | .374 |
| EFFICIENCY | .0002 | .3674 | .5233 | .6006 | .6396 | .6562 | .657 |
| POWER FACTOR | .2145 | .3407 | .4589 | .5621 | .6477 | .7155 | .767 |
| PRI I2R LOSS | 28.6 | 29.8 | 32.5 | 37.0 | 43.7 | 53.1 | 66. |
| A-PHASE I2R LOSS | .0 | .0 | .0 | .0 | .0 | .0 | . |
| SEC I2R (F) | .2 | 1.0 | 2.5 | 4.7 | 7.9 | 12.3 | 18. |
| SEC I2R (B) | 11.9 | 12.4 | 13.5 | 15.4 | 18.2 | 22.1 | 27. |
| IRON LOSSES | 31.8 | 31.5 | 31.0 | 30.6 | 30.1 | 29.5 | 28. |
| FR. & WINDAGE | 5.5 | 5.5 | 5.4 | 5.3 | 5.2 | 5.1 | 5. |
| STRAY-LOAD LOSS | .0 | .0 | .0 | .0 | .0 | .0 | . |
| TOTAL LOSSES | 78.2 | 80.2 | 85.0 | 93.0 | 105.1 | 122.1 | 145. |
| PULSATING TORQUE | 19.46 | 19.56 | 20.07 | 20.98 | 22.27 | 23.86 | 25.7 |

SPEED-TORQUE DATA

RUNNING CONNECTION

| SLIP | R. P. M. | TORQUE | WATTS | LINE A | MAIN A | AUX. A | AUX. V | CAP |
|---|---|---|---|---|---|---|---|---|
| .00000 | 1800.0 | -2.74 | 34.7 | 3.177 | | | | |
| .05000 | 1710.0 | 17.69 | 407.7 | 4.686 | | | | |
| .10000 | 1620.0 | 28.34 | 690.9 | 7.060 | | | | |
| .15000 | 1530.0 | 32.78 | 893.3 | 9.040 | | | | |
| .20000 | 1440.0 | 33.69 | 1035.2 | 10.586 | | | | |
| .25000 | 1350.0 | 32.70 | 1134.8 | 11.787 | | | | |
| .30000 | 1260.0 | 30.77 | 1205.4 | 12.727 | | | | |
| .40000 | 1080.0 | 25.84 | 1292.8 | 14.069 | | | | |

STARTING CONNECTION

| SLIP | R. P. M. | TORQUE | WATTS | LINE A | MAIN A | AUX. A | AUX. V | CAP |
|---|---|---|---|---|---|---|---|---|
| .00000 | 1800.0 | -33.43 | 2165.3 | 18.847 | 12.273 | 13.588 | 277.8 | 19 |
| .05000 | 1710.0 | 13.50 | 2214.8 | 19.344 | 11.350 | 11.334 | 241.5 | 15 |
| .10000 | 1620.0 | 40.00 | 2245.4 | 19.695 | 11.210 | 9.848 | 214.5 | 13 |
| .15000 | 1530.0 | 55.19 | 2265.3 | 19.953 | 11.456 | 8.898 | 193.9 | 12 |
| .20000 | 1440.0 | 63.83 | 2278.5 | 20.150 | 11.865 | 8.323 | 178.0 | 11 |
| .25000 | 1350.0 | 68.55 | 2287.5 | 20.303 | 12.325 | 8.006 | 165.5 | 11 |
| .30000 | 1260.0 | 70.84 | 2293.9 | 20.425 | 12.785 | 7.863 | 155.5 | 11 |
| .40000 | 1080.0 | 71.44 | 2301.6 | 20.603 | 13.626 | 7.879 | 140.9 | 11 |
| .50000 | 900.0 | 69.49 | 2305.7 | 20.725 | 14.331 | 8.085 | 131.0 | 11 |
| .60000 | 720.0 | 66.55 | 2307.9 | 20.808 | 14.911 | 8.357 | 124.2 | 11 |
| .70000 | 540.0 | 63.32 | 2309.1 | 20.866 | 15.387 | 8.642 | 119.3 | 12 |
| .80000 | 360.0 | 60.09 | 2309.7 | 20.903 | 15.778 | 8.918 | 115.8 | 12 |
| .90000 | 180.0 | 57.01 | 2310.0 | 20.924 | 16.098 | 9.180 | 113.3 | 12 |
| 1.00000 | .0 | 54.12 | 2310.1 | 20.931 | 16.359 | 9.426 | 111.3 | 13 |

| | MFDS. | LOCKED-ROTOR CALCS. WITH OTHER MFD. VALUES. | | | | | | |
|---|---|---|---|---|---|---|---|---|
| MAX T | 235.9 | 55.33 | 2530.0 | 23.190 | 16.359 | 10.500 | 124.0 | 11 |
| MAX T/I | 188.0 | 54.01 | 2303.9 | 20.872 | 16.359 | 9.394 | 111.0 | 13 |

**FIG. 6.47** Running performance and speed-torque calculations for the illustrative design problem.

## Some General Guidelines

The above design was worked out by successive trials. In each case, we had to make an intelligent change in design parameters in order to come closer to the objective. The following general rules should be helpful to the uninitiated in deciding which way to vary the design parameters in order to move in the desired direction. Usually, following the particular rule applicable, these generalizations should materially help the designer zero in on the desired objectives, if they are possible of attainment, or at least help one get as close as practicable. Each of these rules assumes that the only feature changed is the one discussed and that all other features are held constant.

1. The BDT and LRT are inversely proportional to the square of the number of effective main-winding conductors.
2. The flux densities are inversely proportional to the first power of the number of effective main-winding conductors.
3. For changes in R1, R2, X1, and X2, the BDT is inversely proportional to the simple sum of these four parameters.
4. For small changes in main-winding conductors, the full-load slip is nearly inversely proportional to the square of the number of effective main-winding conductors.
5. For small changes in R2, LRT is proportional to R2; but for larger changes in R2, LRT does not go up as fast as R2, because increasing R2 reduces the locked-rotor current.
6. For a given set of main and auxiliary windings, there is a value of MFD that gives a maximum value of LRT, and another value of MFD that makes the ratio of LRT/LRA a maximum. Usually, the optimum value falls between these two values.
7. The maximum obtainable LRT increases as the number of effective auxiliary-winding conductors is decreased.
8. If the main winding is unchanged but the stack length is varied, the densities vary inversely with the stack length, but the torques vary inversely as the square root of the stack length.
9. To change the stack length without changing the torques, make the effective conductors in the main winding vary inversely as the square root of the stack length.

## *REFERENCES*


1. C. G. Veinott and J. E. Martin, *Fractional- and Subfractional-Horsepower Electric Motors*, 4th ed., McGraw-Hill, New York, 1986.
2. C. G. Veinott, *Theory and Design of Small Induction Motors*, McGraw-Hill, New York, 1959; now available from University Microfilms, Inc., Box 1647, Ann Arbor, MI 48106.
3. ———, "VICA: A System for Single-Phase Motor Design by Personal Computer," *Electric Machines and Power Systems*, vol. 9, no. 2, Mar.–Apr. 1984.

# CHAPTER 7
# LINEAR ELECTRIC MACHINES

**I. Boldea**
**S. A. Nasar**

Just as an electromagnetic torque produces rotary motion in a rotating electric machine, electromagnetic forces may be used to produce linear motion—resulting in linear motion electric machines (LEMs). In principle, there is a LEM counterpart for every rotary machine. This chapter presents a study of linear electric motors—their analysis, design, testing, and representative applications. In order to keep the analysis rather simple but still reasonably accurate, approximate quasi-one-dimensional analytical theory is used throughout this chapter. This approach enables the determination of the performance and design parameters of LEMs in a simplified manner.

## *7.1 DIFFERENCES BETWEEN LEM'S AND ROTATING MACHINES*

For the sake of pointing out the main differences between a LEM and its rotary counterpart, let us consider the now classical example of a three-phase rotary induction motor. By an imaginary process, this device may be transformed into a linear induction motor (LIM) if the stator and the rotor of the rotary motor are cut by a radial plane and unrolled (Fig. 7.1). For the LIM, we designate the "stator" as the primary and the "rotor" as the secondary. Either of the two may be stationary while the other is moving.

Notice that the primary (core and windings) now has finite lengths called the active lengths of the LIM, having a beginning and an end. The presence of these two ends leads to the phenomenon of end effects, which is unique to LEMs and does not occur in conventional rotating machines, whose primary is cylindrical and closed in the direction of motion in most common machines.

The end effects depend on the speed of the LEM, and such a speed-dependent phenomenon has no counterpart in a rotary motion machine. Also, for LEMs with one primary and one secondary placed face to face, as in Fig. 7.1, a nonzero net

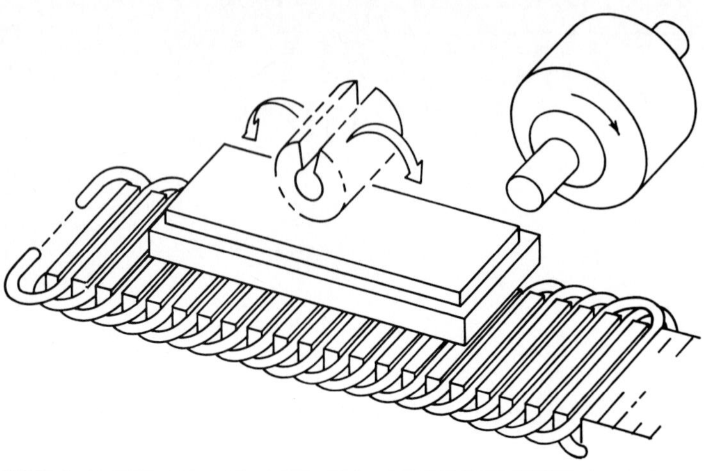

**FIG. 7.1** Obtaining a linear induction motor.

normal force (attraction or repulsion) occurs (Fig. 7.2). The presence of a normal force, besides thrust, in a LEM is another feature distinguishing the LEM from a rotary motion machine, in which such a force is considered abnormal.

## *7.2 CLASSIFICATION OF LEM'S*

On the basis of operating principles, there are three main types of LEMs:

1. Linear induction motors (LIMs) (Fig. 7.2)
2. Linear synchronous motors (LSMs) (Fig. 7.3)
3. DC commutator linear motors (DC LEMs) (Fig. 7.4)

### Linear Induction Motors (LIMs)

The operating principle of a LIM is identical to that of its rotary counterpart. The polyphase winding of the primary produces a traveling field. This field induces

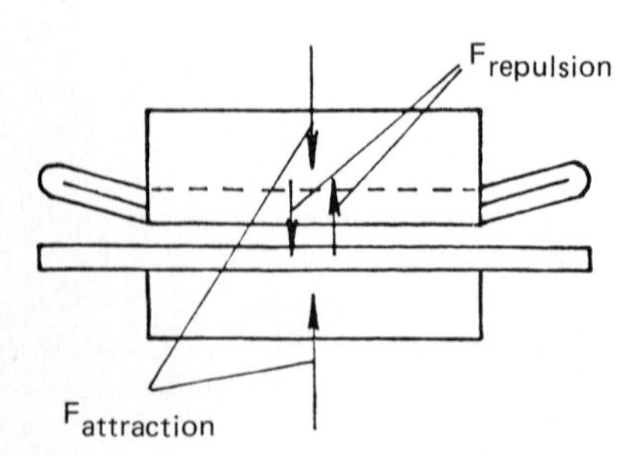


**FIG. 7.2** Normal forces in a LIM.

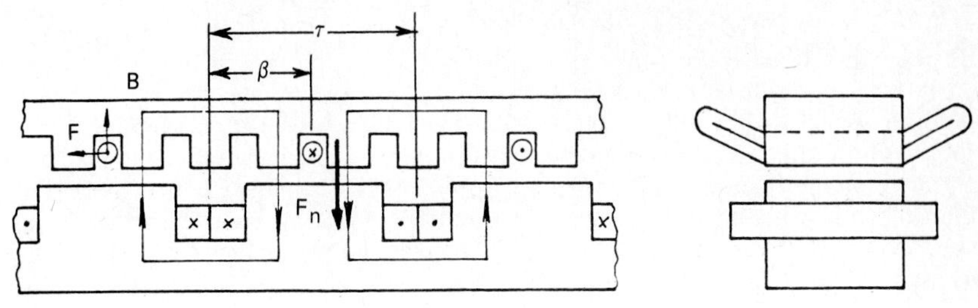


FIG. 7.3 Linear synchronous heteropolar motor (LSM).

voltages and, consequently, currents in the secondary. A thrust is thus developed by the interaction of the induced currents and the traveling field. At the entry and exit ends of the primary, additional currents are induced in the secondary because of the rather steep variation of the primary magnetic fields in these zones. These additional currents then decay in the active zone (under the primary) and behind the motor primary stack, with the time constants dependent primarily on the speed, pole pitch, secondary surface conductivity, airgap, and primary frequency. These currents produce the so-called end effects.

The consequences of end effects are: additional longitudinal drag force, additional joule losses, distortion of the longitudinal distribution of the flux density, and a reduction in power factor and efficiency. In order to evaluate the performance of the LIM realistically, it is of utmost interest to know when the end effects are negligible. When they are negligible, the theory gets simplified and most of the approaches to rotary induction machines may be used.

The secondary structures of LIMs may be chosen from a wide variety of configurations. Secondaries made of conducting sheets backed by solid iron lead to low secondary costs at the expense of some penalty in performance. An attraction-type normal force occurs between the primary and the secondary iron stacks, and a

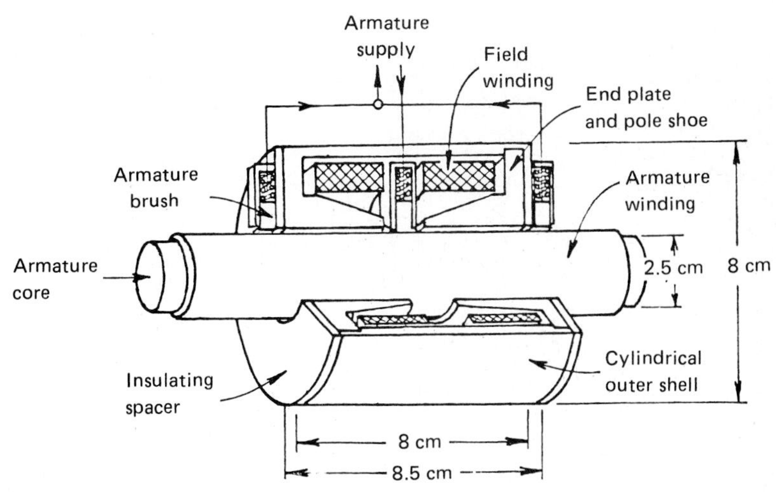


FIG. 7.4 DC linear motor.

repulsion-type normal force exists between their mmf's. The net normal force may be attractive, repulsive, or even zero under certain conditions.

It is now evident that the limited length of the primary core and winding—combined with a solid secondary and the occurrence of a nonzero net normal force—constitutes a major difference between linear and rotary induction machines. Finally, for mechanical reasons the airgap of LIMs is about 5 to 10 times larger than in corresponding rotary induction machines. In comparison with their rotary counterparts, LIMs have slightly lower efficiencies and power-factor because of their end effects and larger airgaps.

## Linear Synchronous Motors (LSMs)

There is a linear counterpart for every rotary synchronous motor. Thus, there are heteropolar LSMs (Fig. 7.3) and homopolar LSMs (Fig. 7.5). Whereas the principle of operation is the same for rotary and linear synchronous machines, there are some differences as well. For economic reasons, only a few topologies are considered practical: the active-guideway LSM (Fig. 7.3), with a conventional or a superconducting field winding on the movable body, and the passive-guideway LSM (Fig. 7.5).

The passive-guideway LSM results in lower overall investment costs, especially if the guideway is manufactured of solid iron. In this latter case, however, the primary core and the limited length of the field winding cause, through motion, eddy currents in the secondary solid-iron structure both at the entry and exit ends of the primary. These eddy currents die out much the same way as in a LIM, and their consequences are also called end effects. The end effects consist of additional secondary joule losses, a drag force, a distortion of airgap magnetic field distribution, and a reduction of normal attractive force. A rather small end effect also occurs at the entry and exit ends of the active-guideway LSM, but in general the additional currents induced in the armature winding are negligibly small for practical cases.

A normal attractive force occurs between the primary and secondary of iron-core LSMs. However, in air-core LSMs the net normal force could be attractive or repulsive in nature, depending on the time lag between the armature and field-winding mmf's. Thus, the normal force is attractive when the phase lag between the two mmf's is less than 90° and is repulsive when this angle is more than 90°.

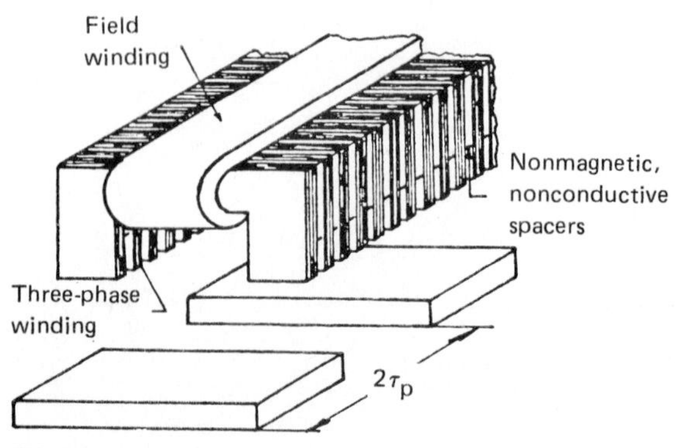


**FIG. 7.5** Linear synchronous homopolar motor.

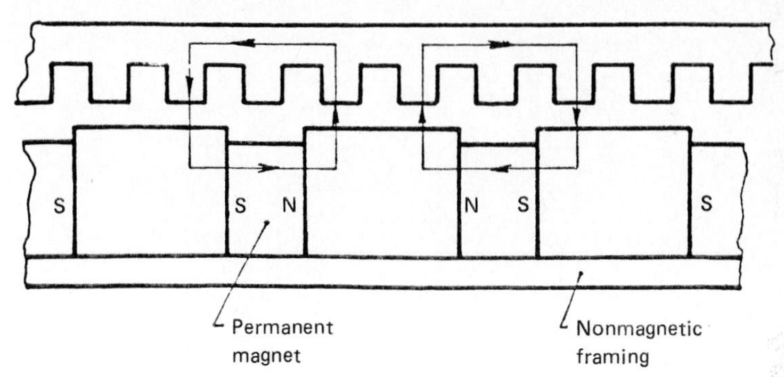

FIG. 7.6 Permanent magnet LSM.

The airgap of an LSM is about the same as in its rotary motor's topological counterparts. In principle, leading power factors may be obtained. In practice, however, at least for the active-guideway LSM, a lagging power factor of only about 0.7 to 0.8 may be obtained because the energized track section length should not be reduced below a certain level if reasonable investment costs in the power stations are to be obtained.

In passive-guideway LSMs, a leading power factor may be obtained in practice, but it should be kept in mind that in this case the longitudinal component of the armature field has a demagnetizing effect on the field-winding field, lowering the normal attractive force of the motor, the angle between the two mmf's being over 90°. Thus, a leading power factor is obtained at the expense of a lower normal (attraction) force in an iron-core LSM. In general, better energy conversion levels may be obtained with LSMs than with LIMs, but at higher costs per motor. The capability of iron-core LSMs to produce much higher normal forces than LIMs for the same thrust makes the LSM more suitable for applications in vehicle suspension and propulsion.

Besides these average- and large-power LSMs, there are permanent magnet LSMs (Fig. 7.6) and reluctance or permanent magnet LSM steppers (Fig. 7.7)

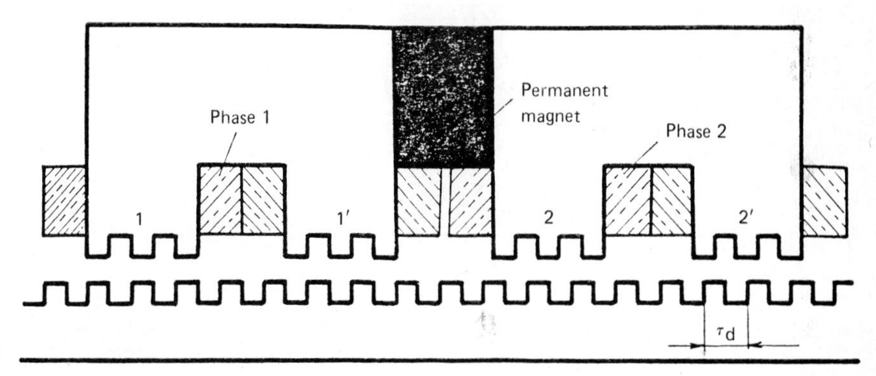


FIG. 7.7 Linear synchronous stepper.

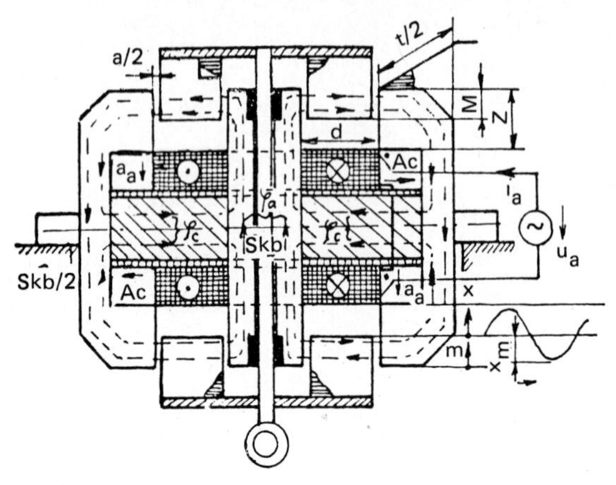

FIG. 7.8 Linear synchronous oscillator.

corresponding to their rotary counterparts. There are also synchronous oscillators used for small compressors (Fig. 7.8).

## DC Linear Motors

Heteropolar (Fig. 7.9) and homopolar (Fig. 7.10) dc linear motors are used for short-stroke low-thrust applications. The tubular structure is preferred for a better usage of the armature copper and for balancing the normal forces. The tubular structure may be obtained from the flat linear structure by an additional rolling along the direction of motion. The penalty is the limited length for which the tubular structure could be used.

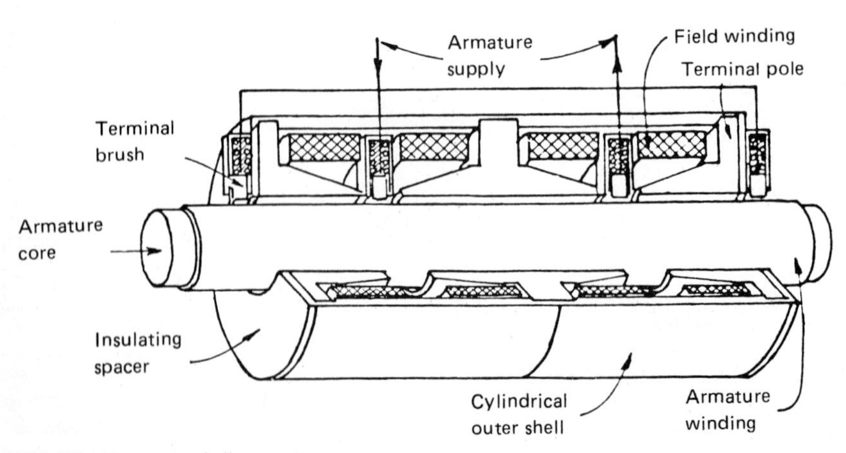


FIG. 7.9 Heteropolar dc linear motor.

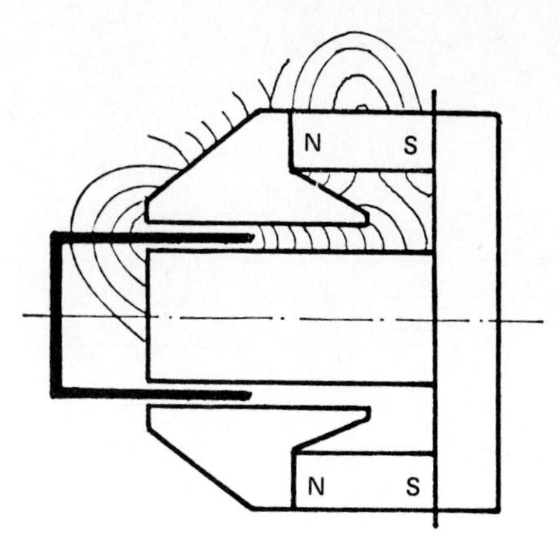


FIG. 7.10 Homopolar dc linear motor.

A flat heteropolar dc linear motor with electronic commutation (Fig. 7.11) has been proposed for the high-power long excursions typically needed for transportation purposes. It is, again, an active-guideway structure with the field winding on the movable body. The electronic commutation renders this motor similar to the inverter-fed position-controlled LSM.

## 7.3 LINEAR INDUCTION MOTORS

Linear induction motors have found numerous applications, such as in propulsion systems for urban and interurban people movers, sliding doors, aluminum-strip tension, bottle transfer, and car-crash test facilities. In order to obtain economical solutions, many topological arrangements have been proposed.

The most common of LIMs is the three-phase LIM, but two-phase LIMs or shaded-pole LIMs for low speeds (below 3 m/s) and low thrusts (less than 50 N) are also feasible when only a single-phase source is available. In contrast to its conventional rotary counterpart, a LIM may have a moving *primary* (with a fixed secondary) or a moving *secondary* (the primary being stationary). Depending on the relative lengths of the primary and secondary, a LIM may be a short-primary, short-secondary, or short-primary short-secondary LIM (Fig. 7.12).

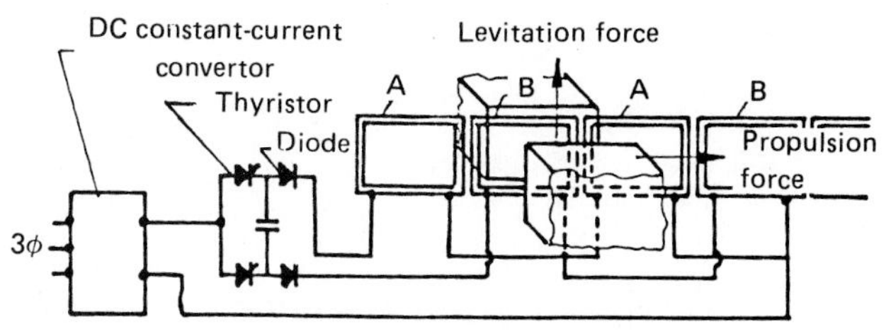


FIG. 7.11 Heteropolar dc linear motor with electronic commutation.

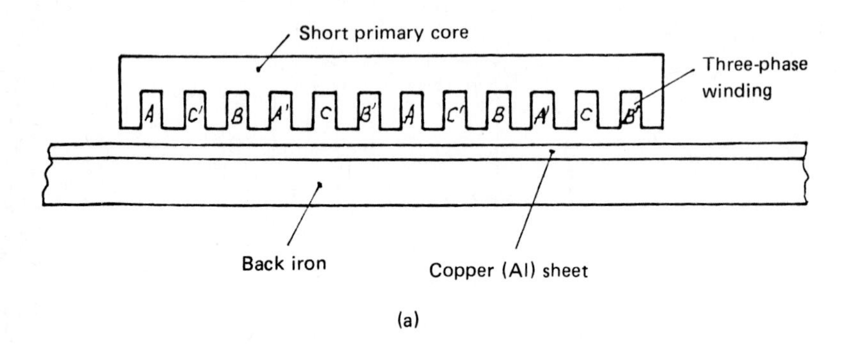


(a)

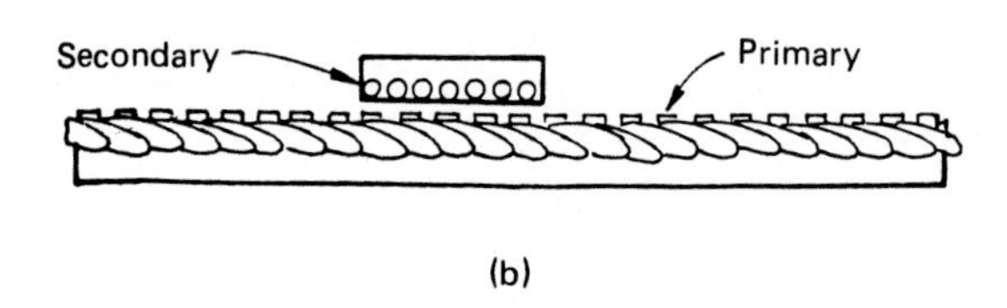


(b)

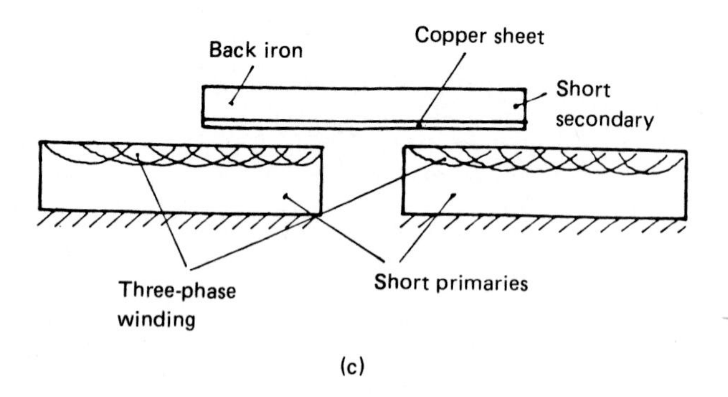


(c)

**FIG. 7.12** Various single-sided LIM configurations: ***(a)*** short-primary LIM; ***(b)*** short-secondary LIM; ***(c)*** short-primary short-secondary LIM.

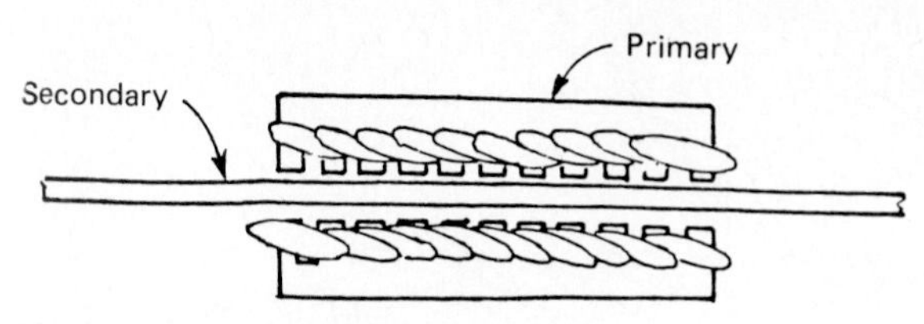


**FIG. 7.13** Double-sided LIM.

Because of its poor power factor and efficiency, the short-secondary LIM is hardly practical, and the short-primary short-secondary LIM is used in movable-secondary applications, with the primaries energized through proximity transducers only in the presence of the moving secondary. Also, the LIM may have two primaries face to face to form a double-sided LIM (DSLIM), as in Fig. 7.13. If the LIM has only one primary (Fig. 7.12), as in most applications, it is called a single-sided LIM (SLIM).

The secondary of a LIM (Fig. 7.14) may consist of a conducting plate or sheet (in a SLIM) backed by a solid (or laminated) ferromagnetic material or be made of the usual ladder (the counterpart of squirrel cage) in a SLIM. A wound-type secondary is rather uncommon.

The SLIM with a sheet on a solid-iron secondary is considered a very practical solution because of its ruggedness and the low costs of the secondary, although it has a slightly lower efficiency and power factor in comparison with the more expensive solution of the ladder-type secondary. The ladder-type secondary reduces the magnetic airgap by the sheet thickness, thus reducing the magnetizing current for the same flux density in the airgap. However, by placing the secondary conductors in slots, the repulsive normal force between the primary and secondary mmf's is considerably reduced. Thus, for the same airgap flux density, the normal attractive force for a given thrust is sensibly higher in ladder-type secondaries. In wheel-suspension transportation applications, this high normal force may be undesirable. The ladder-type secondary should be used cautiously.

So far, we have considered only flat LIMs. If the flat LIM is rerolled about an axis parallel to the direction of traveling motion, as in Fig. 7.15, a tubular LIM is obtained. One obvious advantage of the tubular over the flat LIM is that it does not have any end connections. A better usage of primary copper and secondary

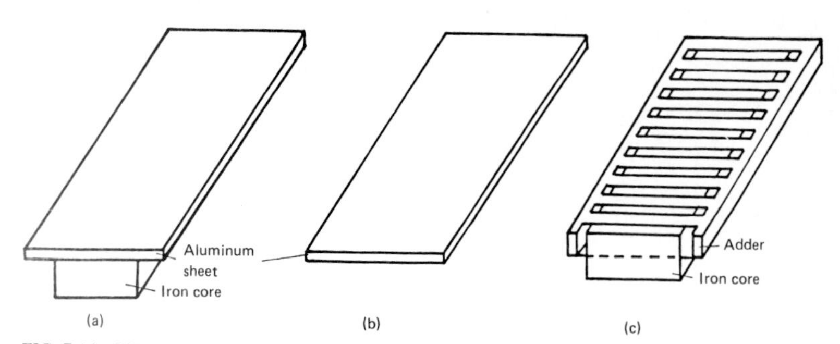


**FIG. 7.14** LIM secondaries: ***(a)*** conduction sheet on solid iron; ***(b)*** conducting plate; ***(c)*** ladder-type secondary.

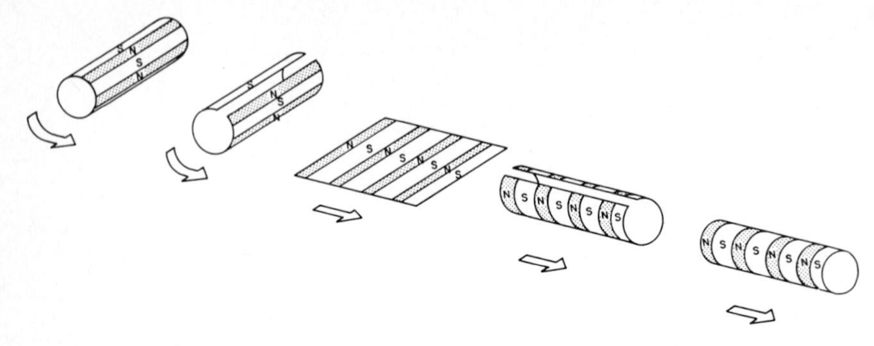

FIG. 7.15 Tubular LIM.

conducting sheet material is obtained. In a tubular motor, as mentioned earlier, the net normal force is zero.

The cylindrical structure of the primary and secondary limits the length of LIM excursion to less than 1 m in the case of horizontal motion and to a few meters in a vertical-motion application. Because of zero normal force and short excursions, the airgap should be as small as mechanically feasible, and the secondary circular coils (or rings) should be embedded in the circular grooves of the solid-iron secondary body (Fig. 7.16). Special cooling measures should be taken for both the primary and the secondary in a heavy-duty application.

Flat and tubular LIMs belong to the class of motors in which the magnetic flux lies in the plane of motion, as in all rotary induction machines. Such motors may

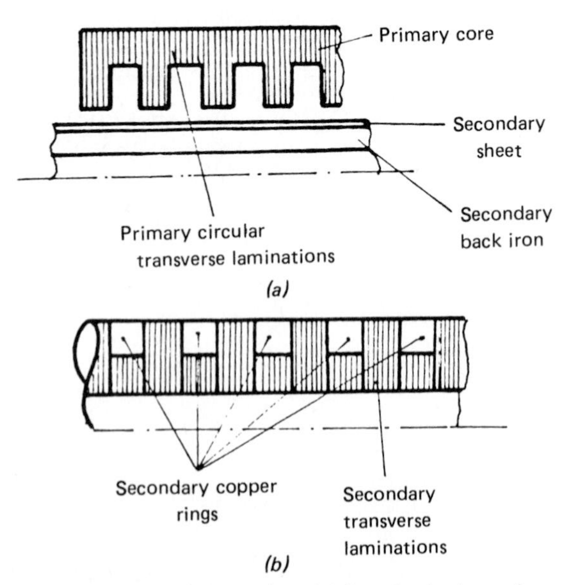


FIG. 7.16 Tubular LIM secondary: *(a)* sleeve (or sheet) secondary; *(b)* composite secondary.

be called longitudinal (or axial) flux motors. The construction of LIMs for high speeds (100 m/s) at industrial frequency (50 or 60 Hz) has led to high pole pitches. On the other hand, the secondary (and consequently the primary) width should be restricted in order to keep the secondary costs within reasonable limits. The small stack-width/pole-pitch ratio leads to long end connections, whereas the large pole pitch (over 0.6 m) leads to thick primary and secondary back irons.

To circumvent part of these disadvantages, a transverse flux topology has been proposed (Fig. 7.17). With the development of silicon controlled rectifiers (SCRs), a variable voltage and frequency supply for LIMs is available for high-speed applications. Frequencies up to 250 Hz are easily achievable, and thus the pole pitch may be lowered to optimal values, rendering the TFLIM hardly practical.

In what follows, only the flat and tubular heteropolar longitudinal flux LIMs will be treated in detail.

## Winding Layouts

Recall from Sec. 7.1 that the LIM may be developed from a rotary induction motor by the "cutting" and "splitting" procedure. This approach, however, requires certain changes in the primary windings. In principle there are two main arrangements (Figs. 7.18*a* and *b*). The layout in Fig. 7.18*a* can be obtained without any change in the coils at the ends in a double-layer winding with full-pitched coils. This winding effectively has five poles, although the poles at the ends are weaker. This type of winding (with $2p + 1$ poles) is characterized by the fact that the airgap flux density in the central zone [$(2p - 1)\tau$ in length] in the absence of secondary is a pure traveling wave, which is the distribution common in rotary induction machines.

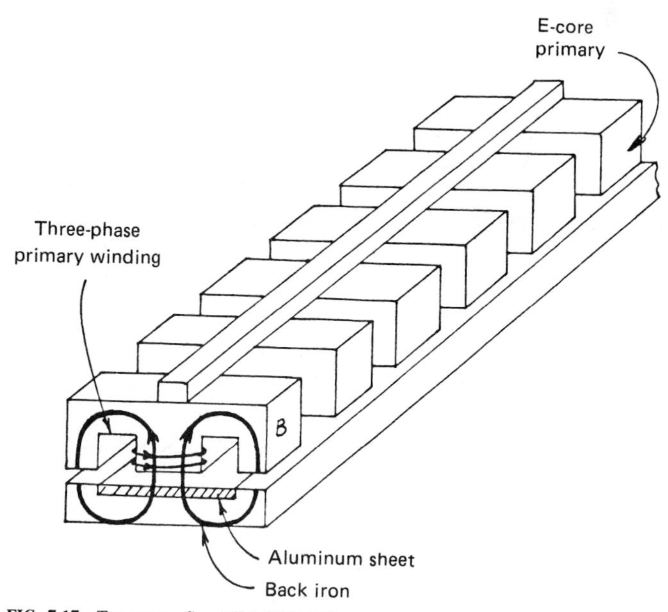


**FIG. 7.17** Transverse flux LIM (TFLIM).

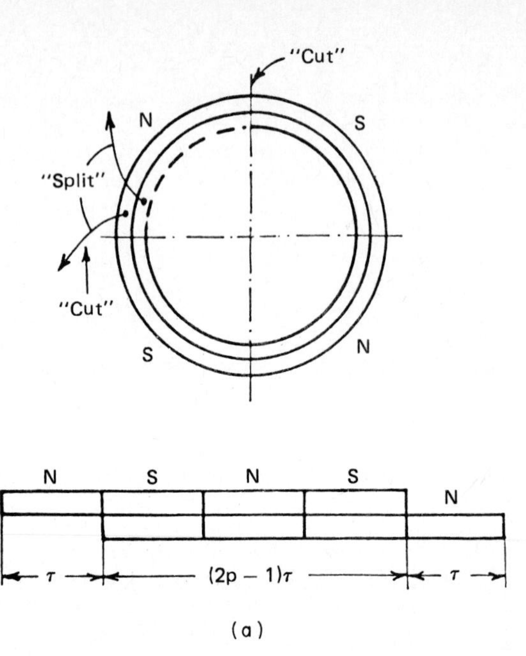


(a)

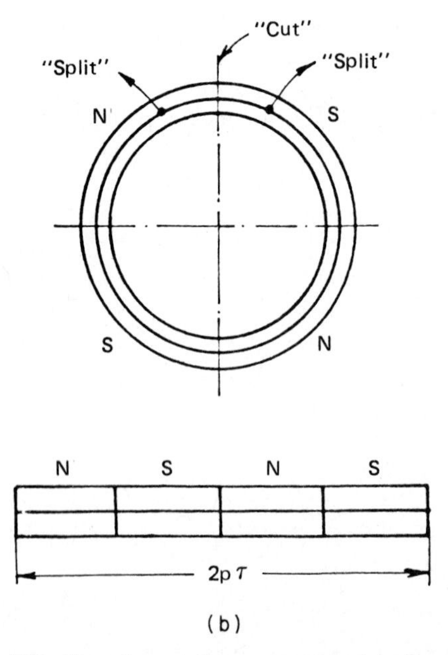


(b)

**FIG. 7.18** Three-phase winding arrangements: ***(a)*** with an odd number of poles; ***(b)*** with an even number of poles ($2p$).

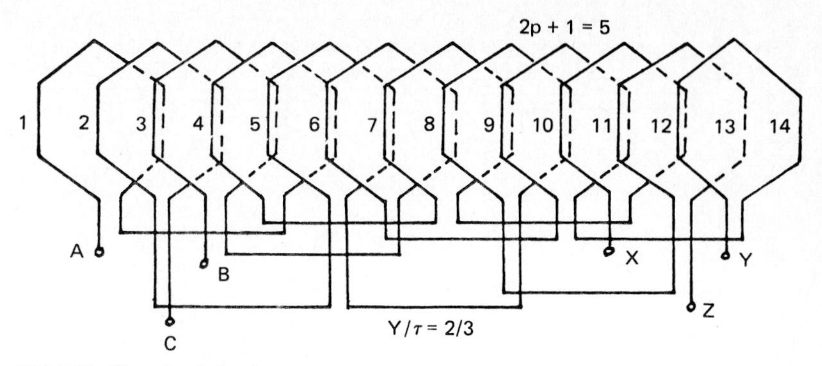


**FIG. 7.19** Five-pole winding layout.

Chorded coils may be used to reduce both the primary mmf harmonics level and the length of the end connections. An example of such a winding is given in Fig. 7.19. The procedure for designing such a winding (with $2p + 1$ poles) is identical to the case of the rotary motor from which it was obtained (that is, a $2p$-pole machine with the same number of slots per pole per phase and the same chording ratio). The only difference is that, when making the winding layout, the last $3q$ coils' returning branches occupy one layer of the $2p + 1$th pole. Whereas the end poles are half-filled and thus weaker, this type of a winding is so far considered the best for average and high-power and high-speed LIMs.

The arrangement in Fig. 7.18*b* has an even number of poles and is applicable for single-layer windings. In the absence of a secondary, the airgap flux density exhibits a pure traveling wave all along the $2p$ poles of the machine. The presence of an additional alternative flux component in the back-iron core of the primary and secondary (in SLIMs) becomes negligible when the solid back iron of the secondary gets highly saturated. A $2p$-pole motor has the disadvantage of long end connections in comparison with the $(2p + 1)$-pole machine. However, in a four- or two-pole LIM the $2p$-pole winding is preferred since it makes a better use of the primary core (all slots being filled). An example of a four-pole LIM winding is given in Fig. 7.20. The procedure for designing it is the same as for its rotary counterpart.

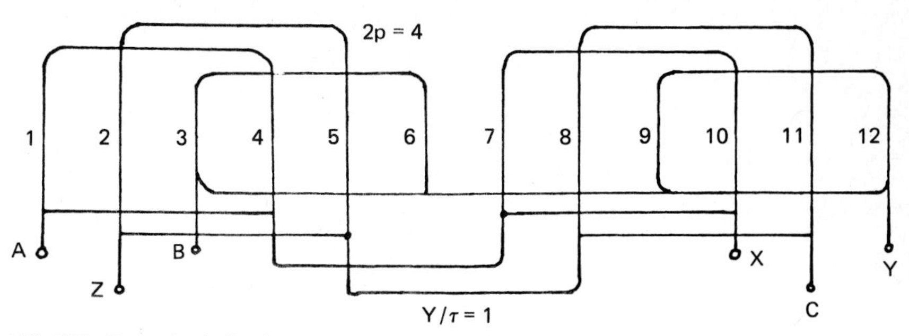


**FIG. 7.20** Four-pole winding layout.

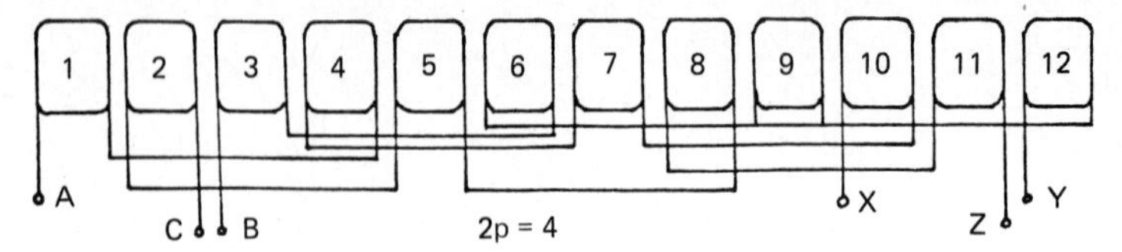


FIG. 7.21 Inexpensive winding for micro-LIMs.

For small thrusts (less than 20 N) and small speeds (less than 3.6 m/s), significant reductions in end-connection length and winding manufacture costs are obtained in special one-layer winding (Fig. 7.21), at the expense of an increased harmonic content in primary mmf. This winding is suitable for flat LIMs. For tubular LIMs, concentric coils are used to form a typical one-layer winding (Fig. 7.22).

## Specific Phenomena in LIMs

The finite length of the LIM primary or secondary (or both) iron cores and windings along the direction of motion, together with the large airgap and the solid body structure of the secondary, leads to specific phenomena not present in rotary machines. The specific phenomena introduced by the finite length of the primary core and winding along the direction of motion are known as end effects. The solid structure of the secondary, combined with the finite width of the primary stack and the secondary, causes the edge effects. The large airgap/pole-pitch ratio introduces a kind of airgap leakage, leading to an apparent increase in the airgap length. The secondary sheet on solid iron experiences significant skin effects even at rated speed since the secondary frequency hardly runs lower than 5 Hz. Finally, in single-sided LIMs a normal force occurs between the primary and secondary mmf's and their magnetic cores.

In double-sided LIMs the normal force between the primary and aluminum-sheet secondary is a pure repulsive force, and an attractive force occurs between the two sides of the primary core. A superposition of these effects leads to the possibility of using a quasi-one-dimensional approach, which is adequate for design purposes.[1] Thus, in the following, all specific phenomena except end effects are accounted for first by correction coefficients obtained from one-dimensional approaches to emphasize only one specific phenomenon at a time.

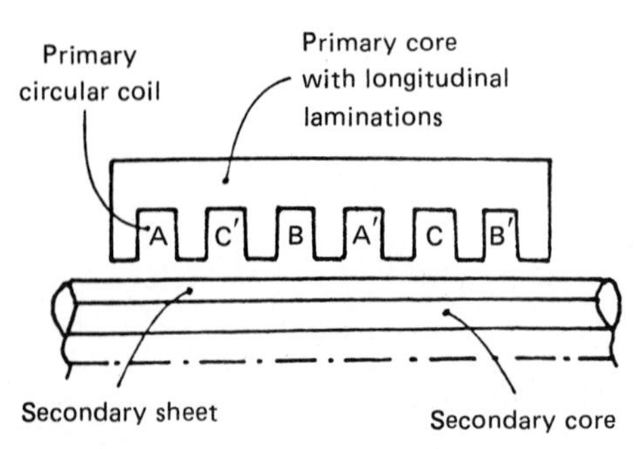


FIG. 7.22 Circular coil winding for tubular LIMs.

***Airgap Leakage.*** In most LIMs the ratio of the mechanical airgap ($g_m = 2$ to 15 mm) to the pole pitch is greater than 0.04. The secondary sheet ($d = 2$ to 6 mm) adds to the mechanical airgap to form the magnetic airgap $g_o$:

$$g_o = g_m + d \tag{7.1}$$

A large airgap leads to airgap leakage. In other words, part of the airgap flux does not cross the airgap. The airgap leakage factor is the ratio $K_1$, given by

$$K_1 = \frac{\sinh(\pi g_o/\tau)}{\pi g_o/\tau} \qquad g_{o1} = K_1 g_o \tag{7.2}$$

The airgap $g_o$ is thus modified by the factor $K_1$. For a SLIM, $g_o$ must be replaced by $g_o/2$.

***Skin Effect.*** Skin effect occurs both in the aluminum sheet and in the secondary solid back iron of SLIMs. To limit the skin effect, the secondary frequency should not exceed 10 to 15 Hz for all speeds; this requires the use of static power converters for most LIM applications. For the aluminum sheet, the correction coefficient $K_{\text{sk}}$ (known from rotary induction machines) may be applied:

$$K_{\text{sk}} = \frac{2d}{d_s}\left[\frac{\sinh(2d/d_s) + \sin(2d/d_s)}{\cosh(2d/d_s) - \cos(2d/d_s)}\right] \tag{7.3}$$

where $d_s$ is the field penetration depth in the secondary sheet:

$$d_s \simeq \frac{1}{[f_1 \mu_o s \sigma_{\text{Al}} + {}^1\!/_2(^\pi\!/_\tau)^2]^{1/2}} \tag{7.4}$$

Thus, the equivalent conductivity $\sigma_{\text{Ale}}$ of aluminum sheet is apparently decreased by the skin effect by $K_{\text{sk}}$:

$$\sigma_{\text{Ale}} = \frac{\sigma_{\text{Al}}}{K_{\text{sk}}} \tag{7.5}$$

For the double-sided LIM, $d$ in Eq. (7.3) is replaced by $d/2$ because the field penetrates the aluminum sheet from both sides.

***Edge Effects.*** The limited width of the primary stack $2a$ and of the solid-structure secondary $2c$ results in a longitudinal component of secondary current density $J_x$ besides the transverse one $J_y$ within the active zone $|y| \leq a$. As observed from Fig. 7.23, the reaction field of the longitudinal component decreases the primary field toward the stack center ($y = 0$) while increasing it toward the stack edges ($y = \pm a$). The resultant transverse distribution of the airgap field becomes nonuniform. The presence of the longitudinal component of the induced currents $J_x$ in the active zone ($|y| \leq a$) causes an apparent increase of secondary resistance by $K_{\text{tr}}$ and an apparent reduction of the magnetizing reactance by $K_{\text{tm}}$ (or an apparent increase of the airgap length).

When we consider the case of a SLIM with aluminum sheet on the solid back iron in the secondary, the double-sided LIM then evolves as a special, simplified case of a SLIM. The presence of solid back iron in the secondary leads to high levels of saturation and to eddy currents in the solid iron. For a given motor, the level of saturation depends on the primary mmf and the secondary frequency.

Based on two-dimensional theory, the edge effect in the secondary aluminum sheet is accounted for. The two correction coefficients mentioned above yield the expressions[2]

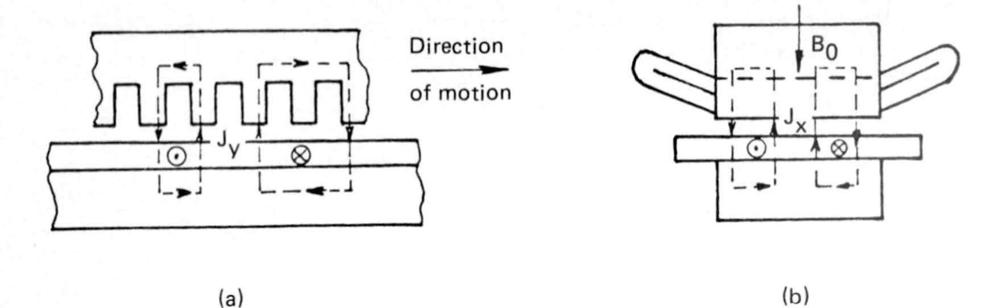


**FIG. 7.23** Reaction-field components: *(a)* longitudinal view; *(b)* transverse view.

$$K_{\mathrm{tr}} = \frac{K_x^2}{K_R}\left[\frac{1+(sG_{ep}K_RK_x)^2}{1+s^2G_{ep}^2}\right] \geq 1 \tag{7.6}$$

$$K_{\mathrm{tm}} = \frac{K_R}{K_x}K_{\mathrm{tr}} \leq 1 \tag{7.7}$$

with

$$K_R = 1 - \mathrm{Re}\left[(1-jsG_{ep})\frac{\lambda}{\alpha a_e}\tanh \alpha a_e\right] \tag{7.8}$$

$$K_x = 1 + \mathrm{Re}\left[(sG_{ep}+j)\frac{sG_{ep}\lambda}{\alpha a_e}\tanh \alpha a_e\right] \tag{7.9}$$

$$\lambda = \frac{1}{(1+\sqrt{1+jsG_{ep}})}\tanh \alpha a_e \tanh(\pi/\tau)(c-a_e) \tag{7.10}$$

$$\alpha = \frac{\pi}{\tau}\sqrt{1+jsG_{ep}} \qquad G_{ep} = \frac{2\mu_o f_1 \sigma_{\mathrm{Al}e}\tau^2 d}{\mu g_o K_1 K_{\mathrm{sk}} K_c (1+K_p)} \qquad g_o = g_m + d \tag{7.11}$$

where $f_1$ = primary frequency
$\tau$ = pole pitch
$g_m$ = mechanical airgap
$d$ = aluminum-sheet thickness
$K_c$ = Carter coefficient
$K_p$ = correction coefficient (still to be determined) accounting for secondary back-iron magnetic reluctance
$G_{ep}$ = goodness factor
$2a_e$ = equivalent stack width accounting for lateral fringing ($2a_e \simeq 2a + g_o$)

It should be noted that the skin effect and airgap leakages have already been accounted for by their correction coefficients, but the back-iron saturation and eddy currents are still to be considered. We assume an exponential decay of the airgap field into the solid back iron, with a penetration depth of $\delta_i$.

Edge effect also occurs in the solid back iron, but because the iron's conductivity is much smaller than that of aluminum and the width of the solid iron equals that of the primary iron stack, the edge-effect correction coefficients [Eqs. (7.06) to (7.11)] become

$$K_{\mathrm{tri}} \simeq \frac{1}{1-\tau/\pi a_e \tanh(\pi/\tau)a_e} \qquad K_{\mathrm{tmi}} \simeq 1 \tag{7.12}$$

Thus, the depth of penetration $\delta_i$ is

$$\delta_i = \mathrm{Re}\left\{\frac{1}{[(\pi/\tau)^2 + 2\pi j\mu_i f_1 s(\sigma_i/K_{\mathrm{tri}})]^{1/2}}\right\} \tag{7.13}$$

The depth of penetration may be significantly increased by making the back iron of a few laminations. In this case, $K_{\mathrm{tri}}$ increases by replacing $a_e$ by $a_e/i$, where $i$ = number of laminations. The coefficient $K_p$ in Eq. (7.11) that accounts for the back-iron reluctance may be simply calculated by allowing for a $\delta_i$ thickness of the back iron and a $\tau/\pi$ mean length of field path between adjacent poles in the back iron; that is,

$$K_p \simeq \frac{\tau^2}{\pi^2}\frac{\mu_o}{\mu_i \delta_i g_o K_c} \tag{7.14}$$

The value of the back-iron equivalent permeability $\mu_i$ is still to be determined, but first let us introduce the edge effects and the presence of back iron in the expressions of equivalent airgap $g_e$ and conductivity $\sigma_e$. In principle the real motor is replaced by an ideal equivalent motor (without edge effects, back-iron saturation, or iron eddy currents), whose equivalent airgap $g_e$ and secondary conductivity $\sigma_e$ account for all these phenomena such that

$$g_e = \frac{g_o K_c (1 + K_p) K_e}{K_{\mathrm{tm}}} \tag{7.15}$$

$$\sigma_e = \frac{\sigma_{\mathrm{Al}}}{K_{\mathrm{sk}} K_{\mathrm{tr}}} + \frac{\sigma_i \delta_i}{K_{\mathrm{tri}} d} \tag{7.16}$$

Thus, the equivalent goodness factor $G_e$ of the LIM becomes

$$G_e = \frac{2 f_1 \mu_o \tau^2 \sigma_e d}{\pi g_e} \tag{7.17}$$

Provided that $\mu_i$ is known, the problem is solved. An iterative procedure to determine $\mu_i$ is now presented. The value of $\mu_i$ is dependent on the primary mmf $\mathcal{F}_{m1}$ and the slip frequency $f_2 = sf_1$ for a given motor. The airgap flux density $B_g$ is given by

$$B_g = \frac{j \mathcal{F}_{m1} \mu_o}{g_e (1 + jsG_e)} \qquad \mathcal{F}_{m1} = \frac{3\sqrt{2} W_1 I_1 K_{w1}}{\pi p} \tag{7.18}$$

where $W_1$ = turns per phase
$K_{w1}$ = winding factor
$p$ = pole pairs

Since the decay of flux density in the back iron is considered exponential, the longitudinal (tangential) component of flux density on the back-iron surface $B_{xi}$ is

$$B_{xi} \simeq \frac{B_g}{\delta_i}\frac{\tau}{\pi} \tag{7.19}$$

From the real magnetization curve we get the equivalent permeability $\mu_e$ corresponding to 0.9 $B_{xi}$. It should be mentioned that in order to calculate $B_{xi}$ we have to assign an initial value to $\mu_{ii}$ and then calculate $K_p, \delta_i, g_e, \sigma_e, G_e$, and $B_g$.

Evidently $\mu_e \neq \mu_{ii}$, and thus the process should be repeated until satisfactory (imposed) convergence is obtained. Experience has shown that a slight underrelaxation factor leads to fast convergence:

$$\mu' = \mu_e + 0.1(\mu_{ii} - \mu_e) \tag{7.20}$$

Finally, we conclude that for each value of the primary mmf $\mathcal{F}_{m1}$ and secondary frequency $f_2$, the real LIM for a given motor could be replaced by an ideal LIM characterized by a different airgap $g_e$ and goodness factor $G_e$. For the double-sided LIM, in most cases there is no solid iron in the secondary, and thus $K_p = 0$ and $\sigma_i = 0$ in the above expressions.

***End Effects.*** The end effects are divided into static and dynamic end effects. Static end effects are due to the open character of the magnetic circuit in LIMs; consequently, the self- and mutual inductances of the primary phases are not symmetric. However, the differences are negligibly small when the number of poles is larger than 6, and in most well-designed LIMs this condition is met. The dynamic end effects are caused by the relative motion of the short primary with respect to the secondary. Short movable-primary LIMs are commonly used in diverse applications, whereas fixed-primary movable-secondary LIMs are less frequently applied.

The reaction field of the induced currents due to end effects causes a distortion of the airgap flux density's longitudinal distribution (along the direction of motion). The distribution of specific thrust along the direction of motion also becomes nonuniform. Finally, in general, a significant reduction of thrust, efficiency, and power factor at small slips for a given primary mmf is the main consequence of end effects.

The following presents a quasi-one-dimensional approach to account for the chief consequences of end effects. To keep the mathematical expressions as simple as possible, the following assumptions are made:

- The primary mmf is sinusoidally distributed along the length of $2p$ poles.
- The edge effects, skin effect, saturation, and airgap leakage are accounted for by correction coefficients already defined; thus, the secondary-induced currents exhibit only the transverse thrust-producing component.
- The length of the primary core is still considered infinite.

Under these conditions, from Maxwell's equations the following expression for **H** is obtained (assuming sinusoidal time variations):

$$\frac{\partial^2 \mathbf{H}}{\partial x^2} - \frac{\pi}{\tau} G_e(1-s)\frac{\partial \mathbf{H}}{\partial x} - j\left(\frac{\pi}{\tau}\right)^2 G_e \mathbf{H} = -j\frac{\pi}{\tau g_e} J_m e^{-j(\pi x/\tau)} \tag{7.21}$$

The solution is of the form

$$\mathbf{H} = \mathbf{A} e^{\gamma_1 (x - 2p\tau)} + \mathbf{B} e^{\gamma_2 x} + \mathbf{B}_n e^{-j(\pi x/\tau)} \tag{7.22}$$

where
$$\mathbf{B}_n = \frac{jJ_m}{g_e(\pi/\tau)(1 + jsG_e)} \tag{7.23}$$

The third term in Eq. (7.22) is the conventional one, whereas the first two account for the end effects. The coefficients $\gamma_1$ and $\gamma_2$ are given by

$$\gamma_1 = \frac{a_1}{2}\left(\sqrt{\frac{b_1+1}{2}} + 1 + j\sqrt{\frac{b_1-1}{2}}\right) = \gamma_{1r} + j\gamma_i \tag{7.24}$$

$$\gamma_2 = -\frac{a_1}{2}\left(\sqrt{\frac{b_1+1}{2}} - 1 + j\sqrt{\frac{b_1-1}{2}}\right) = \gamma_{2r} - j\gamma_i \tag{7.25}$$

with
$$a_1 = \frac{\pi}{\tau} G_e(1-s) \qquad b_1 = \sqrt{1 + \frac{16}{G_e^2(1-s)^4}} \tag{7.26}$$

Since the primary stack is infinitely long, the field equations for the entry and exit zones are similar to Eq. (7.21), but with the right-hand-side term considered to be zero. Thus

$$\mathbf{H}'_e = \mathbf{A}'_e e^{\gamma_1 x} \qquad x \le 0 \tag{7.27}$$

$$\mathbf{H}'_o = \mathbf{A}'_o e^{\gamma_2 (x - 2p\tau)} \qquad x \ge 2p\tau \tag{7.28}$$

The boundary conditions require the continuity of the airgap flux density and secondary current density at $x = 0$ and at $x = 2p\tau$.

Finally, we get

$$\mathbf{A} = \frac{-jJ_m(\tau\gamma_2/\pi + sG_e)}{(\tau/\pi)(\gamma_2 - \gamma_1)(1 + jsG_e)} \tag{7.29}$$

$$\mathbf{B} = \frac{jJ_m[(\tau/\pi)\gamma_1 + sG_e]}{(\tau/\pi)(\gamma_2 - \gamma_1)(1 + jsG_e)} \tag{7.30}$$

$$\begin{aligned} A'_e &= \mathbf{A}e^{-2p\tau\gamma_1} + \mathbf{B} + \mathbf{B}_n \\ A'_o &= \mathbf{A} + \mathbf{B}e^{2p\tau\gamma_2} + \mathbf{B}_n \end{aligned} \tag{7.31}$$

The secondary current density $\mathbf{J}_2$ is

$$\begin{aligned} \mathbf{J}_2 &= \frac{g_e}{d}\left[\gamma_1 \mathbf{A}_e e^{\gamma_1(x-2p\tau)} + \gamma_2 \mathbf{B} e^{\gamma_2 x} \right. \\ &\qquad \left. - \frac{jsG_e J_m}{g_e(1 + jsG_e)} e^{-j(\pi/\tau)x}\right] \qquad 0 \le x \le 2p\tau \\ \mathbf{J}_2 &= \frac{g_e}{d}\gamma_1 \mathbf{A}'_e e^{\gamma_1 x} \qquad x \le 0 \\ \mathbf{J}_2 &= \frac{g_e}{d}\gamma_2 \mathbf{A}'_o e^{\gamma_2(x-2p\tau)} \qquad x \ge 2p\tau \end{aligned} \tag{7.32}$$

## Power and Forces

The secondary $I^2R$ loss $P_2$ is obtained from

$$P_2 = \frac{a_e d}{\sigma_e} \int_{-\infty}^{+\infty} |\mathbf{J}_2|^2 dx \tag{7.33}$$

Similarly, the reactive power leaving the primary core, $Q_2$, is given by

$$Q_2 = a_e \omega_1 \mu_o g_e \int_{-\infty}^{+\infty} |\mathbf{H}|^2 dx \tag{7.34}$$

The thrust $F_x$ is found from

$$F_x = \mu_o a_e J_m \operatorname{Re}\left( \int_o^{2p\tau} \mathbf{H}^* e^{-j\pi x/\tau} dx \right)$$

The first term in $\mathbf{H}$ in Eq. (7.22) attenuates rapidly (within a few millimeters from the entry end); thus, $F_x$ becomes

$$F_x = \mu_o a_e J_m \operatorname{Re}\left( \mathbf{B}_n^* 2p\tau + \mathbf{B}^* \frac{\{\exp 2p\tau[\gamma_2^* - j(\pi/\tau)] - 1\}}{\gamma_2^* - j(\pi/\tau)} \right) \tag{7.35}$$

The first term is the conventional one and the second term is responsible for the end effects. The first term is zero at synchronism ($s = 0$), but the second term may be either positive (propulsion) or negative (drag). The normal force $F_n$ has two components: an attractive component $F_{na}$ (between the primary and secondary magnetic cores) and a repulsive component $F_{nr}$ (between the primary and secondary mmf's). These are given by

$$F_{na} \simeq \frac{a_e \mu_o}{2} \int_o^{2p\tau} |\mathbf{H}|^2 dx \tag{7.36}$$

$$F_{nr} = \mu_o J_m a_e d \operatorname{Re}\left( \int_o^{2p\tau} \mathbf{J}_2 e^{-j\pi x/\tau} dx \right) \tag{7.37}$$

The net normal force $F_n$ is

$$F_n = F_{na} + F_{nr} \tag{7.38}$$

In a double-sided LIM the expressions of both components still hold when the secondary is placed symmetrically between the two primaries. However, the attractive force acts upon the two primary stacks and the repulsive force acts only upon the aluminum-sheet secondary.

## Efficiency and Power Factor

Efficiency and power factor may be defined both with respect to the secondary ($\eta_2$ and $\cos \phi_2$) and to the primary ($\eta_1$ and $\cos \phi_1$). These are expressed as

$$\eta_2 = \frac{F_x U}{F_x U + P_2} \qquad \cos \phi_2 = \frac{F_x U + P_2}{Q_2^2 + (F_x U + P_2)^2} \tag{7.39}$$

$$\eta_1 = \frac{F_x U}{F_x U + P_2 + 3R_1 I_1^2 + P_{i1}} \qquad \cos \phi_1 = \frac{F_x U/\eta_1}{(F_x U/\eta_1)^2 + (Q_2 + 3X_{1\sigma} I_1^2)^2} \tag{7.40}$$

where $R_1$ and $X_{1\sigma}$ are the primary phase resistance and leakage reactance, respectively, and $P_{i1}$ is the primary core loss.

## Low-Speed and High-Speed Operation of LIMs

The airgap flux density, Eq. (7.22), has three components. One is the conventional unattenuated forward-traveling wave, having the pole pitch $\tau$ and the speed $U_s = 2\tau f_1$. The other two, the attenuated traveling waves responsible for the end effects, are moving forward and backward, having the pole pitch $\tau_e = \pi/\gamma_i$ and the speed $U_e = 2\tau_e f_1$. The backward end-effect wave, the first term in Eq. (7.22), attenuates much faster than the forward end-effect wave since $\gamma_{1r} >> |\gamma_{2r}|$. Also, $\tau_e \geq \tau$ for all conditions. Thus, only the forward end-effect wave, in fact, matters. Here, we assume that end effects could be neglected if the end-effect forward wave attenuates $e$ times within 10 percent of motor length; that is, if (Fig. 7.24)

$$\frac{1}{|\gamma_{2r}|} \leq \frac{2p\tau}{10} \tag{7.41}$$

Thus, for a given goodness factor $G_e$ and slip, the high-speed operating condition depends on the number of poles. Furthermore, at low speeds, low primary frequency is used for starting medium- and high-speed LIMs. Thus, the goodness factor decreases in the low-speed region and the low-speed conditions are met.

## Design Criteria

Since applications of LIMs range from standstill to high speeds, unified design criteria are difficult to formulate. For average- and high-speed applications, two basic criteria are considered:

1. Maximum mechanical power per unit input (kVA).
2. Maximum mechanical power per unit weight (kg).

For standstill and low-speed (less than 3 m/s) applications, appropriate criteria are considered to be:

3. Maximum thrust per unit input.
4. Maximum thrust per unit weight.

The above criteria are conflicting, and a compromise between them should be made.

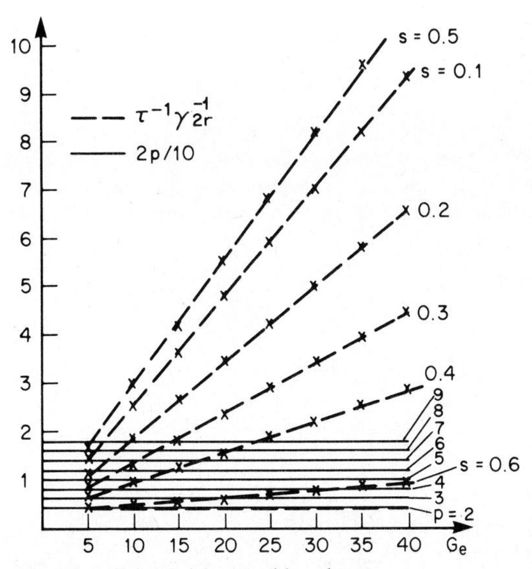


FIG. 7.24 High-speed–low-speed boundary.

Moreover, the influence of the second criterion, (2) or (4), is strongly dependent on the application since the costs of the secondary should also be considered. Thus, in fact the first criterion will be used first and the second will be applied subsequently.

### LIM Design Guidelines for Low Speeds

At low speeds, end effects may be neglected. Other specific phenomena are accounted for by correction coefficients. Both flat and tubular LIMs are now considered.

***Flat LIMs.*** The known equivalent circuit may be used for design purposes (Fig. 7.25). The magnetizing reactance $X_m$ and secondary resistance $R_2'$ contain the pertinent equivalent correction coefficients:

$$X_m = \frac{12\mu_o\omega_1 a_e K_{w1} W_1^2 \tau}{\pi^2 p g_e} = K_m W_1^2 \tag{7.42}$$

$$R_2' = \frac{X_m}{G_e} = \frac{12 a_e K_{w1} W_1^2}{d\tau p \sigma_e} = K_{R'2} W_1^2 \tag{7.43}$$

The primary phase resistance $R_1$ and leakage reactance $X_{1\sigma}$ are given by the following expressions:[1]

$$R_1 = \frac{1}{\sigma_{co}} \frac{(4a + 2l_{ec})}{W_1 I_1} j_{co} W_1^2 = K_{R1} W_1^2 \tag{7.44}$$

$$X_1 = \frac{2\mu_o\omega_1}{p}\left\{\left[\lambda_s\left(1+\frac{3}{2p}\right)+\lambda_d\right]\frac{2a}{q}+\lambda_e l_{ec}\right\}W_1^2 = K_{1\sigma}W_1^2 \tag{7.45}$$

where $l_{ec}$ = coil end-connection length; $\sigma_{co}$ = copper electrical conductivity; $j_{co}$ = design current density; and $\lambda_s$, $\lambda_d$, $\lambda_e$ are the slot, differential, and end-connection specific permeances, respectively, and are given by

$$\lambda_s \simeq \frac{1}{12}\frac{h_s}{W_s}(1+3\beta_1) \qquad \lambda_d = \frac{5(\tau/qW_s)}{5+4(\tau/qW_s)} \tag{7.46}$$

$$\lambda_e = 0.3(3\beta_1 - 1) \tag{7.47}$$

$\beta_1$ being the chording factor (coil span per pole pitch).

All specific phenomena are incorporated in $g_e$ and $\sigma_e$, which are functions of the primary current $I_1$ and the slip frequency $s\omega_1$. Further, for low-speed LIMs, the expressions of thrust and normal force become simplified. Thus, the thrust $F_x$ may be written as

$$F_x = \frac{3I_2'^2 R_2'}{s2\tau f_1} = \frac{3I_1^2 R_2'}{s2\tau f_1[(1/sG_e)^2 + 1]} \tag{7.48}$$

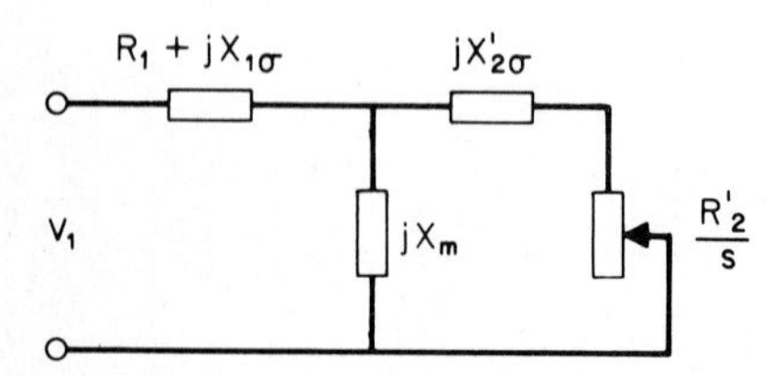


**FIG. 7.25** Equivalent circuit (for low-speed LIMs).

Neglecting the iron losses, the efficiency $\eta_1$ and power factor $\cos\phi_1$ are

$$\eta_1 = \frac{F_x 2\tau f_1(1-s)}{F_x 2\tau f_1 + 3R_1 I_1^2} \qquad \cos\phi_1 = \frac{F_x 2\tau f_1 + 3R_1 I_1^2}{3V_1 I_1} \tag{7.49}$$

The normal force $F_n$ is composed of an attractive component and a repulsive component. The final expression is

$$F_n = 2a_e \frac{p\tau^3}{\pi^2} \frac{\mu_o J_m^2}{g_e^2(1+s^2G_e^2)}\left(1 - g_e s\omega_1\sigma_e \frac{\tau}{\pi}\mu_o s G_e\right) \tag{7.50}$$

As expected in the low-slip region, the normal force is attractive. For high slips, however, $F_n$ may become repulsive. In general, in low-speed LIMs the normal force is always attractive.

***Design Algorithm and an Example (Flat LIM).*** According to the criterion of maximum thrust at rated slip $s_{rk}$, from Eq. (7.48) it follows that

$$s_{rk}G_e = 1 \tag{7.51}$$

The initial data are:

Rated voltage $V_{1r} = 220$ V
Rated thrust $F_{xr} = 1500$ N
Rated frequency $f_1 = 50$ Hz
Rated speed $U_r = 6$ m/s
Rated airgap flux density $B_{gr} = 0.15$ to $0.35$ T
Mechanical airgap $g_m = 2$ to $6$ mm
Slot depth per slot width, $h_s/W_s = 4$ to $6$
Secondary overhang $c - a = \tau/\pi$
Design current density $j_{co} = 4$ A/mm$^2$

The major unknowns are: stack halfwidth per pole pitch; $a_e/\tau$; pole pitch $\tau$; number of poles at $2p + 1$; rated slip $s_{rk}$; aluminum-plate thickness $d_a$; number of turns per phase $W_1$. The primary frequency could also be a variable. To find these unknowns, the above equations are reformulated. Thus, from Eqs. (7.48) with (7.51) and (7.43) combined with (7.18) we obtain

$$F_{xr} = 6C_s B_{gr}\tau^3 \frac{a_e}{\tau} k_{w1}(p - 1/2) \tag{7.52}$$

with

$$C_s = \frac{1}{9q}\frac{h_s}{W_s}\left(\frac{3W_s q}{\tau}\right)^2 K_{\text{fil}} j_{co} \tag{7.53}$$

$$\frac{3W_s q}{\tau} = 1 - \frac{B_{gr}}{B_{\text{tr}}} \tag{7.54}$$

where $B_{\text{tr}}$ = tooth-design flux density
$K_{\text{fil}}$ = slot fill factor

Also, by choosing $a_e/\tau$ as a parameter, from Eqs. (7.12) and (7.13) we get

$$\tau^2\omega_1 s_{rk} \simeq \frac{K_{ti}\pi^2}{\mu_i\sigma_i}\sqrt{\left(\frac{B_{xir}}{B_{gr}}\right)^4 - 4} = C_\omega \tag{7.55}$$

An iterative procedure for the magnetic permeability is used to find $K_{ti}$ from the definition of slip:

$$\omega_1 s_{rk} \tau = \omega_1 \tau - U_r \pi \tag{7.56}$$

Thus, Eq. (7.55) becomes

$$\tau(\omega_1 \tau - U_r \pi) = C_\omega \tag{7.57}$$

Provided that $a_e/\tau$ is known, from Eq. (7.57) the pole pitch $\tau$ is found. From Eq. (7.52) $p$ is obtained. If $p$ is not an integer, $a_e/\tau$ is slightly modified in Eq. (7.52) until, from Eqs. (7.55) to (7.57), an adequate $\tau$ is obtained, leading to a $p$ that is an integer.

Next, the depth of penetration $\delta_i$ from Eq. (7.13) is calculated. From the definition of $G_e$ [Eqs. (7.15) to (7.17), (7.48), and (7.52)], with $s_{rk}G_e = 1$ we obtain

$$\sigma'_a d_a = \left( \frac{3\tau C_s \pi}{\omega_1 s_{rk} B_{gr}} - \frac{\sigma_i \delta_i}{K_{ti}} \right) K_{\text{tr}} \tag{7.58}$$

Unfortunately, the unknown $d_a$ enters also in the expression for $K_{\text{tr}}$ [Eqs. (7.6) to (7.11)]. Thus, only by an iterative procedure could the aluminum-plate thickness $d_a$ be determined. The equivalent airgap $g_e$ [(Eq. (7.15)] and conductivity $\sigma_\epsilon$ [Eq. (7.16)] can now be calculated. Finally, from Eq. (7.17) $G_e$ is found. The ampere-turns per phase $W_1 I_1 r$, from Eq. (7.18), becomes

$$I_{1r} W_1 = \frac{B_{gr} g_e \pi p}{3 k_{\omega 1} \mu_o} \tag{7.59}$$

Finally, the number of turns per phase $W_1$ is given by

$$W_1 = \frac{V_{1r}}{W_1 I_{1r} \left\{ \left[ K_{R1} + \dfrac{K'_{R2}}{s_{rk}(1 + 1/s_{rk}^2 G_e^2)} \right]^2 + \left( K_1 + \dfrac{K_m}{1 + s_{rk}^2 G_e^2} \right)^2 \right\}^{1/2}} \tag{7.60}$$

where $s_{rk}G_e = 1$

Now, for our example with $a_e/\tau = 0.9$, we obtain $s_{rk}\omega_1 = 93s^{-1}$; $\tau = 0.085$ m; $d_a = 1.8 \times 10^{-3}$ m; $W_1 = 384$ turns per phase; $I_{1r} = 32.757$ A; the apparent input power $S_1 = 21.62$ kVA; and $2p + 1 = 11$.

The mechanical output per apparent power input is

$$\eta_1 \cos \phi_1 = \frac{F_{xr} U_r}{S_1} = \frac{1500 \times 6}{21{,}620} = 0.416 \tag{7.61}$$

which is acceptable. Also, the thrust per unit stack area is approximately 1 N/cm$^2$. For a double-sided LIM, the procedure is the same, but simplified due to the absence of secondary back iron.

***Tubular LIMs.*** The tubular LIM is used, in general, for short-stroke applications. The design criteria should remain the same as above, and $s_{rk}G_e = 1$. But in very short stroke applications (less than 0.5 m) the rated slip is taken as $s_{rk} = 1$ since the maximum thrust should occur at standstill. The design procedure remains the same, but the expressions slightly change to deal with the cylindrical structure and the absence of transverse edge effects and primary coil end connections. These expressions are

$$R_{1t} = \rho_{co} j_{co} \frac{\pi D_{\text{avp}} W_1^2}{I_{1r} W_1} \qquad X_{1\sigma t} = 2\mu_o \omega_1 \frac{\pi D_{\text{avp}}}{pq} (\lambda_s + \lambda_d) W_1^2 \tag{7.62}$$

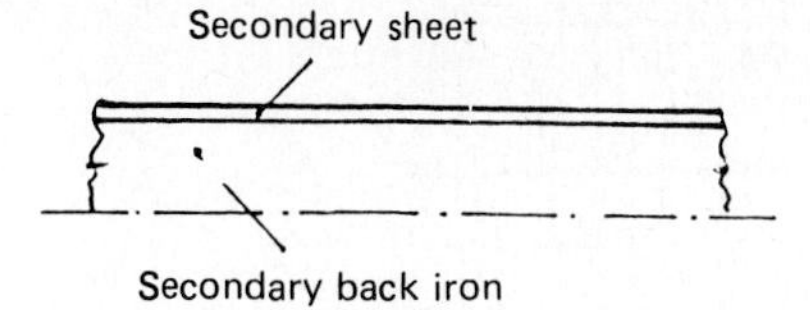


**FIG. 7.26** Sheet-on-iron tubular secondary.

$$X_m = \frac{6\mu_o\omega_1}{\pi^2}\frac{\pi D_o\tau}{pg_e}K_{w1}^2W_1^2 \qquad R_2' = \frac{6\pi D_{\text{avs}}}{p\tau d_a\sigma}K_{w1}^2W_1^2 \tag{7.63}$$

where $D_{\text{avp}}$ = average diameter of primary coils
$D_o$ = primary bore diameter
$D_{\text{avs}}$ = secondary sheet average diameter

Equations (7.62) and (7.63) are valid for sheet-on-iron secondary and longitudinal-lamination primary tubular LIMs (Figs. 7.26 and 7.27). Tubular LIMs without a primary core have also been built. They are characterized by a poorer power factor but better cooling conditions. Besides, a slotted solid-iron secondary with circular conducting rings in slots represents a solution with higher energy conversion performance. The airgap is reduced, and the cooling conditions are superior. For high-thrust low-frequency applications, such a secondary should be preferred (Fig. 7.28). In this case the secondary resistance and leakage reactance are

$$R_2' = 12\rho_{co}\frac{\pi D_{\text{avs}}}{A_{\text{ring}}N_{s2}}K_{w1}^2W_1^2 \tag{7.64}$$

$$X_2'\sigma = 24\mu_o\omega_1(\pi D_{\text{avs}})\frac{(\lambda_{s2}+\lambda_{d2})}{N_{s2}}K_{w1}^2W_1^2 \tag{7.65}$$

According to these conditions, the maximum thrust is obtained when

$$s_{rk}G_e = \frac{1}{\sqrt{1+(X_2'\sigma/X_m)^2}} \tag{7.66}$$

***An Example of the Design of a TLIM.*** A tubular LIM with transverse laminations

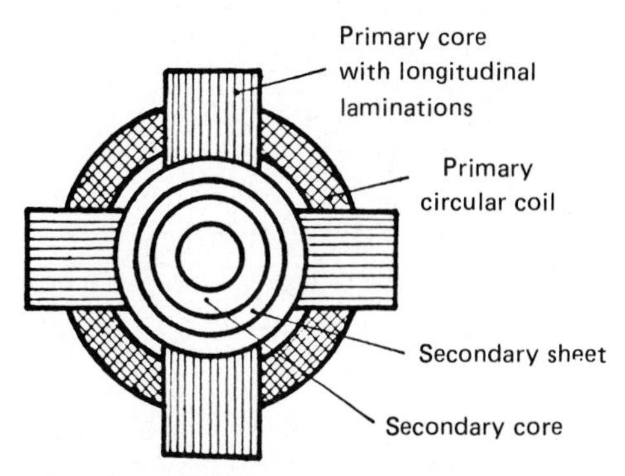


**FIG. 7.27** Longitudinal lamination tubular LIM.

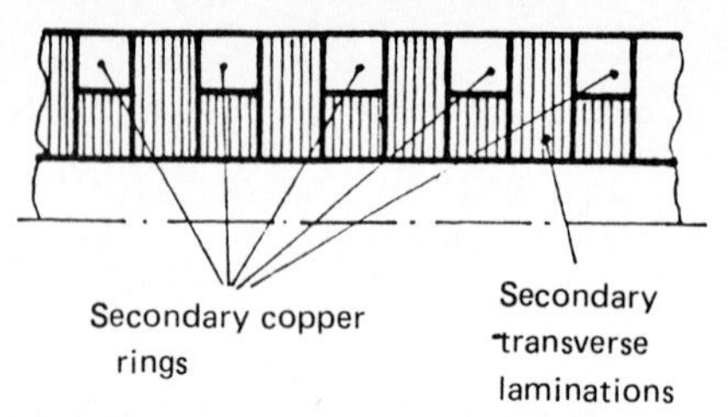


**FIG. 7.28** Tubular secondary with conducting rings in slots.

in the primary and secondary and with copper secondary rings has the following data:

Primary internal diameter $D_i = 0.15$m
Primary external diameter $D_e = 0.27$m
Primary slot depth $h_s = 0.06$m
Pole pitch $\tau = 0.06$m
Number of poles, $2p = 8$
Slots per pole and phase, $q = 1$
Secondary slots per primary length, $N_{s2} = 48$
Airgap $g_o = 3 \times 10^{-3}$ m
Primary slot width $W_{s1} = 1.25 \times 10^{-2}$ m
Secondary slot pitch $\tau_{s2} = 10^{-2}$ m
Secondary slot width $W_{s2} = 6 \times 10^{-3}$ m
Secondary slot depth $h_{s2} = 6 \times 10^{-3}$ m
Design current density $J_{co} = 3 \times 10^6$ A/m$^2$
Slot fill factor $K_{\text{fil}} = 0.6$

Let us calculate the primary frequency at which the maximum thrust is obtained at standstill. First we calculate the secondary resistance and leakage reactance from Eqs. (7.64) and (7.65), respectively:

$$R_2' = 1.922 \times 10^{-4} K_{w1}^2 W_1^2 \tag{7.67}$$

$$X_2 = 0.985 \times 10^{-6} f_1 K_{w1}^2 W_1^2 \tag{7.68}$$

Also, Eq. (7.63) yields

$$X_m = 10^{-5} f_1 K_{w1}^2 W_1^2 \tag{7.69}$$

Thus, the maximum thrust condition for standstill leads to the equation

$$G_e = \frac{X_m}{R_2'} = \frac{1}{[1 + (X_{2\sigma}'/X_m)^2]^{1/2}} \tag{7.70}$$

Making use of Eqs. (7.67) to (7.69) in (7.70), we obtain the primary frequency $f_1 = 10.22$ Hz. The primary ampere-turns per phase $I_{1r}W_1$ may be calculated from the slot dimensions ($h_s$, $W_s$) and the design current density $J_{co}$;

thus

$$W_1 I_{1r} = pqh_s W_s J_{co} K_{\text{fil}} = 4 \times 1 \times 6 \times 10^{-2} \times 1.25 \times 10^{-2} \times 3 \times 10^6 \times 0.6$$

$$= 5400 \text{ At/phase} \tag{7.71}$$

The primary phase resistance $R_{1t}$ [Eq. (7.62)] and the leakage reactance $X_{1\sigma t}$ [Eq. (7.63)] are

$$R_{1t} = \rho_{co} J_{co} \frac{\pi D_{\text{avp}} W_1^2}{I_{1r} W_1}$$

$$= \frac{2.1 \times 10^{-8} \times \pi 2.1 \times 10^{-1}}{5400} \times 3 \times 10^6 W_1^2$$

$$= 1.7693 \times 10^{-5} W_1^2 \tag{7.72}$$

$$X_{1\sigma t} = \frac{2\mu_o \omega_1 \pi D_{\text{avp}}}{pq} (\lambda_{s1} + \lambda_{d1}) W_1^2$$

$$= \frac{2 \times 1.25 \times 10^{-6} \times 2\pi \times 10.22 \times \pi \times 0.21}{4 \times 1} \times 2.041 \times W_1^2$$

$$= 0.5398 \times 10^{-4} W_1^2 \tag{7.73}$$

As all the parameters in the equivalent circuit contain only $W_1$ as an unknown, once the rated voltage is given, $W_1$ can be computed. Because this choice does not affect the performance, we proceed to compute the apparent power $S_1$ as follows:

$$S_1 = 3I_1^2 \left| \left[ R_{1t} + jX_{1\sigma t} + \frac{jX_m(R_2' + jX_2')}{jX_m + R_2' + jX_2'} \right] \right| \tag{7.74}$$

Eliminating $R_{1t}$, $X_{1\sigma t}$, $X_m$, $R_2'$, and $X_2'$ from Eqs. (7.72) and (7.73) and substituting Eqs. (7.67) to (7.69) in Eq. (7.74), with $W_1 I_{1r}$ from Eq. (7.71), we obtain

$$S_1 = 3(I_{1r} W_1)^2 |0.7693 + j5.398 + 4.533 + j5.118| \times 10^{-5} = 10.317 \text{ kVA} \tag{7.75}$$

The electromagnetic power $P_e$ is represented by the third term in (7.75):

$$P_e = F_{xr} 2\tau f_1 = 3.964 \text{ kW} \tag{7.76}$$

with thrust $F_{xr} = 3.232$ kN. The corresponding power factor $\cos\phi_1 = 0.4497$.

The thrust per input kVA and the thrust per unit bore area are 0.313 N/VA and 1.429 N/cm$^2$, respectively. This high performance has been obtained by having deep slots in the primary, so it should be considered close to the attainable maximum. The heating-cooling problems should be treated to complete the design. By changing the pole pitch, the performance change and the design algorithm proposed for the flat LIM apply here also.

## Design Guidelines for Medium- and High-Speed LIMs

In designing high-speed LIMs, we apply a global design criterion, namely, the optimum goodness factor. The conventional performance does improve with increasing $G_e$, but so do the deteriorating end effects. The realistic goodness factor $G_e$ for which the total thrust is zero at zero slip is called the optimum goodness factor $G_o$.

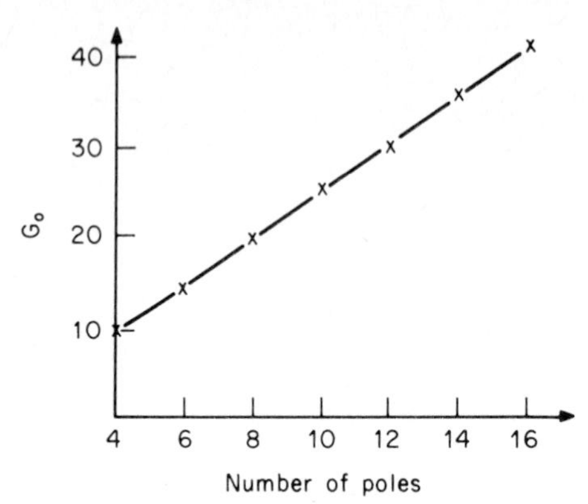


**FIG. 7.29** Optimum goodness factor.

As determined from Eq. (7.35), $G_o$ (Fig. 7.29) depends only on the number of pole pairs $p$.

The initial data are similar to those pertinent to a low-speed LIM, and so are the additional data on slot geometry, slot depth, airgap flux density, etc. The pole pitch should be less than 0.35 m for a double-sided LIM and less than 0.3 m for a SLIM. The design procedure is the same as for a low-speed LIM, maintaining the optimum goodness factor and choosing the rated slip $s$ such that the secondary (internal)power factor $\cos \phi_2$ is between 0.80 and 0.85. A reasonable resultant power factor of 0.5 to 0.6 at rated speed slip is thus achieved. The condition is

$$s_r G_o = 1.635 \text{ to } 2.06 \tag{7.77}$$

In single-sided LIMs the goodness factor depends on the secondary frequency and primary mmf (or airgap flux density). The optimum goodness factor found for zero slip is also valid for rated speed. Besides the optimum goodness factor (Fig. 7.29), other criteria, such as maximum thrust or maximum efficiency at rated speed, have also been proposed. The power factor has not been considered. So we suggest using the above proposed design algorithm and checking the maximum mechanical power (input kVA and mechanical power per weight). In some applications (transportation), the secondary costs may prevail.

## *7.4 LINEAR SYNCHRONOUS MOTORS (LSM'S)*

In principle there is a linear counterpart for every rotary synchronous machine. Thus, several LSM types have been proposed. The range of applications serves as an appropriate basis of classification. The two types of LSMs considered here are transportation LSMs (average and high speeds) and low-speed LSMs.

Transportation linear synchronous motors are heteropolar conventional (Fig. 7.30), permanent magnet (Fig. 7.31), and superconducting (Fig. 7.32) LSMs, as well as transverse flux (Fig. 7.33) and longitudinal flux (Fig. 7.34) linear inductor synchronous motors (LISMs); variable reluctance LSMs (Fig. 7.35) have also been proposed for transportation applications. For low-speed low-thrust applications, linear motion synchronous steppers (Fig. 7.36) and oscillators have been considered.

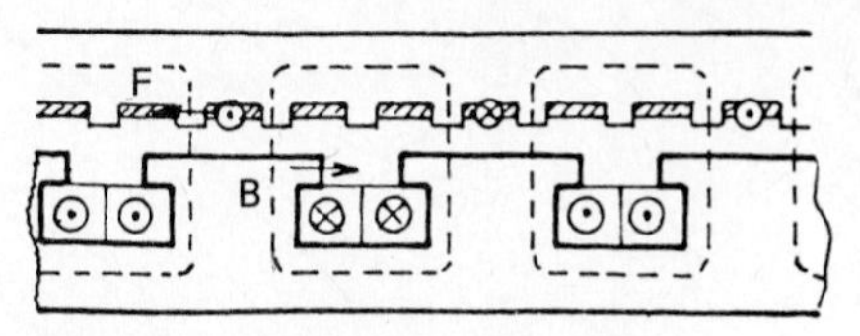


FIG. 7.30 Heteropolar conventional LSM.

## Heteropolar Transportation LSMs

The three main types of heteropolar LSMs are characterized by some common aspects, such as an active three-phase stator placed along a track energized in sections of 1 to 5 km in length. Practically, the same equations are valid as for their rotary counterparts since end effects are negligible. Also, in single-sided LSMs a nonzero normal force occurs.

## Superconducting LSM

In the superconducting (SC) field-winding LSM, the armature reaction is negligible. Thus, only the armature energized-section phase-leakage reactance and resistance voltage drops occur. The steady-state phase equation becomes

$$\mathbf{V}_{1sc} = \mathbf{I}_1(R_1 + jX_{1\sigma}) - \mathbf{E}_{1sc} \tag{7.78}$$

with the phasor diagram and equivalent circuit shown in Fig. 7.37. Only the fundamental of the induced voltage $E_1$ and of the primary current $I_1$ have been considered in Eq. (7.78).

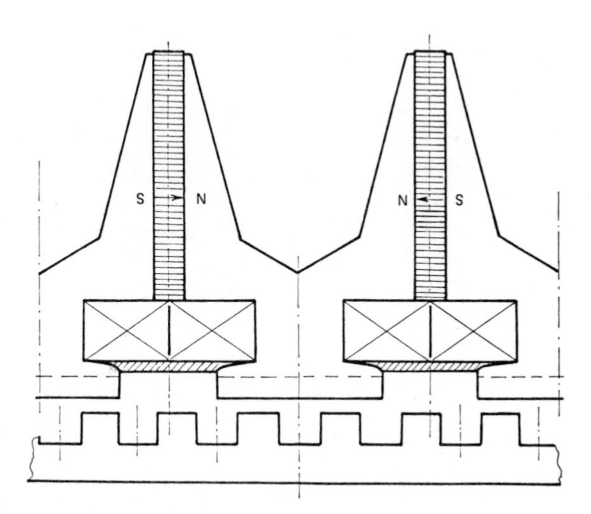


FIG. 7.31 Permanent magnet LSM.

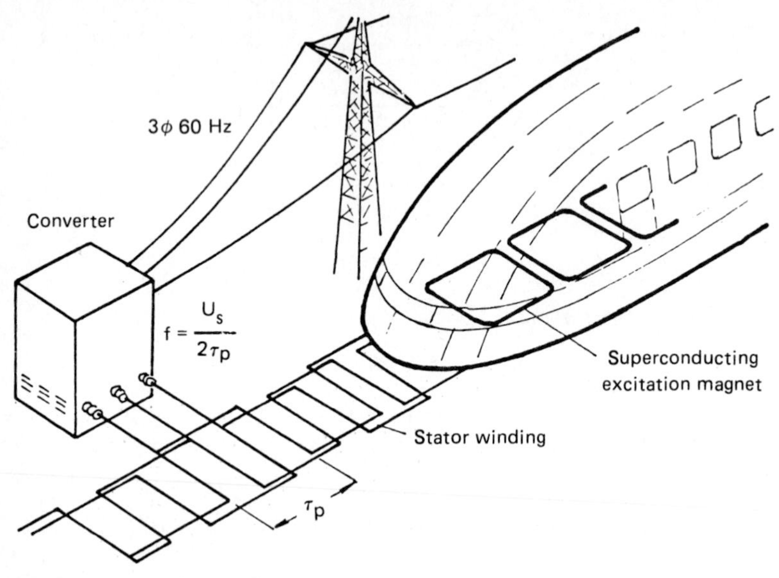


FIG. 7.32 Superconducting LSM.

Since the airgap is between 0.2 and 0.3 m, the geometry of the SC coils may be chosen such that harmonics are kept negligible. The thrust $F_x$ is obtained from the phasor diagram, such that

$$F_x = \frac{3E_1 I_1 \cos \gamma_o}{U_s} \tag{7.79}$$

where the synchronous speed $U_s$ is calculated as in rotary machines. Hence,

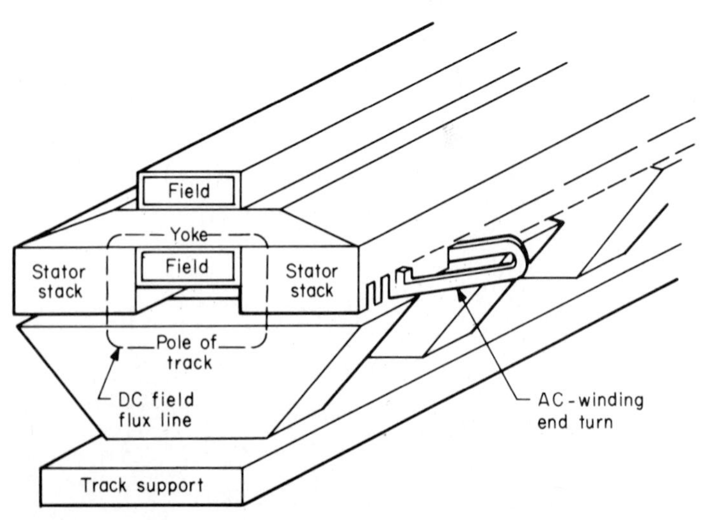


FIG. 7.33 Transverse flux LISM.

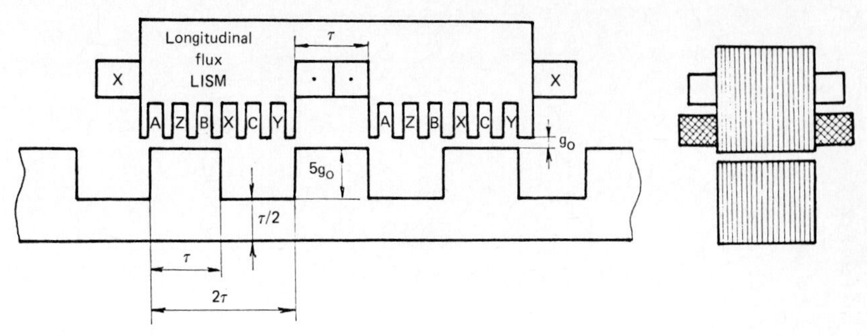


**FIG. 7.34** Longitudinal flux LISM.

$$U_s = \frac{\omega_1 \tau}{\pi} = 2\tau f_1 \tag{7.80}$$

Expressions for $R_1$, $X_{1\sigma}$, and $E_1$ depend on the type of armature winding. Usually a single-layer single conductor per pole and phase leads to a continuous wire per section winding, which is easy to manufacture (Fig. 7.38). For this type of winding,

$$R_1 = \frac{2\rho_{Al}}{A_{Al}}(L + \tau)p' \tag{7.81}$$

where $p'$ = poles for energized section
$A_{Al}$ = conductor cross section

$$X_{1\sigma} = \frac{\mu_o \omega_1}{2\pi}\left[2p'L \ln\left(2\frac{\tau}{d}\right) + p'\tau \ln \frac{2l}{d}\right]$$

$$E_1 \simeq Lp'B_1U_s\sqrt{2} \tag{7.82}$$

where $B_1$ = fundamental of SC winding field at armature surface
$L$ = active coil span

Due to the high currents involved, the large armature skin effect is important at high speeds; consequently, stranded (and twisted) aluminum cables made of many elementary thin wires are used.

The reactive power $Q_1$ is related to the induced voltage by

$$Q_1 = 3E_1 I_1 \sin \gamma_o \tag{7.83}$$

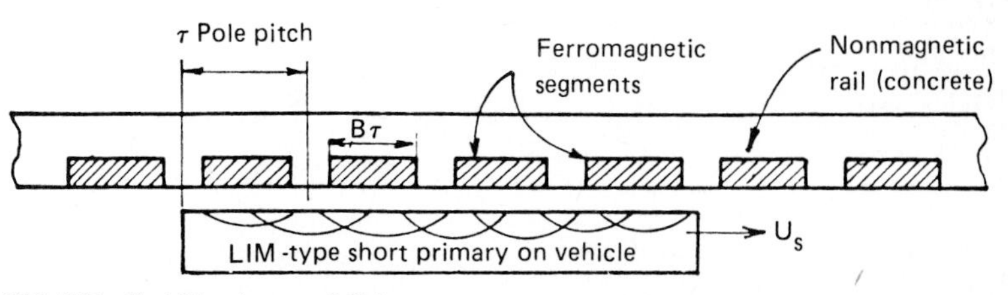


**FIG. 7.35** Variable reluctance LSM.

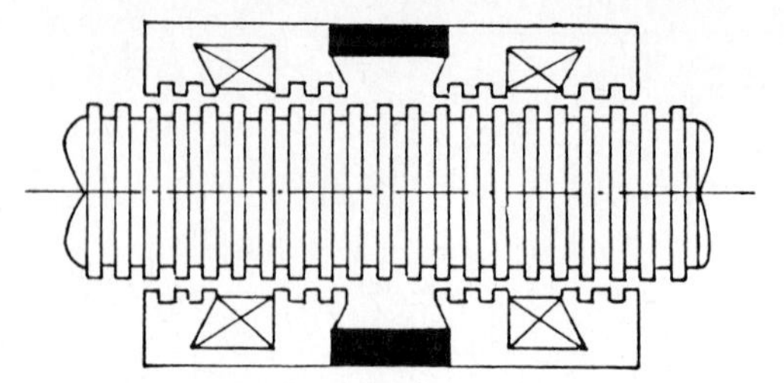

**FIG. 7.36** Linear stepper.

Thus, the magnetic energy $W_{m1}$ is related to $Q_1$ as

$$W_{m1} = \frac{Q_1}{2\omega_1} \tag{7.84}$$

By the small-displacement approach, the normal force $F_n$ at constant $I_1$ and $\gamma_o$ is given by

$$F_n = \frac{3I_1 \sin\gamma_o}{2\omega_1} \frac{\delta E_1}{\delta g_o} \tag{7.85}$$

The induced voltage depends on the airgap through the dependence of $B_1$ (flux density) on $g_o$. For a given SC coil (for constant mmf), the derivative of $B_1$ with respect to $g_o$ is known. Using Ampère's law, the longitudinal and transverse distribution of the SC coil's flux density in the plane of the armature winding may be calculated. The pole pitch $\tau$ and the coil's relative length $l_e/\tau$ should be considered in an optimization process aimed at producing an almost sinusoidal longitudinal flux density distribution.

Because the variable supply frequency is imperative, a current-controlled inverter system is likely to be used. We should thus calculate the performance for a given current and a load angle $\gamma_o$.

***A Numerical Example.*** Consider the following initial data:

SC coils' mmf $W_f I_f = 1$ MA

SC coil length $L = 0.9$ m

SC width $l_w = 0.75$ m

Pole pitch $\tau = 1.5$ m

Rated airgap $g_o = 0.25$ m

Rated speed $U_s = 120$ m/s

Motor number of poles, $2p = 20$

Poles per section, $2p' = 1000$ (1.5 km)

Rated load angle $\gamma_o = 20°$

Fundamental of flux density $B_1 = 0.4$ T

Thrust required from motor = 40 kN

Let us calculate the motor performance at rated thrust. First we calculate the induced voltage $E_1$ from Eq. (7.82):

$$E_1 = LpB_1U_s\sqrt{2}$$

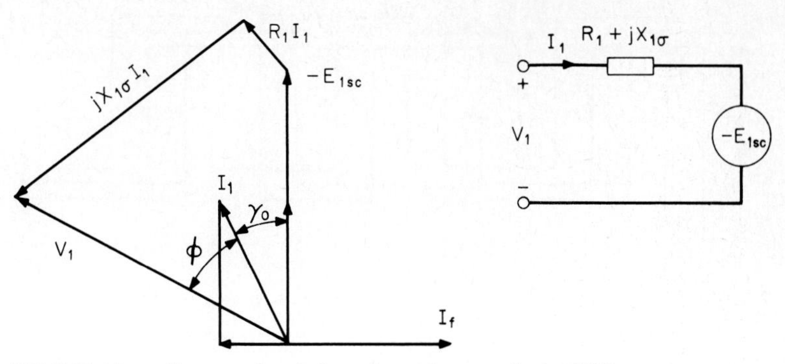


**FIG. 7.37** Phasor diagram and equivalent scheme of superconducting LSM.

$$= 0.9 \times 10 \times 9.4 \times 120\sqrt{2}$$

$$= 609.12 \text{ V}$$

Now the rated current $I_r$ is, from Eq. (7.79),

$$I_r = \frac{F_x U_s}{3E_1 \cos \gamma_o}$$

$$= \frac{4 \times 10^4 \times 120}{3 \times 609 \times 12 \times \cos 20°}$$

$$= 2795 \text{ A}$$

Considering a design current density $j_{Al} = 3 \text{ A/mm}^2$, the phase resistance $R_1$ and phase reactance $X_{1\sigma}$ per section length become, from Eqs. (7.81) and (7.82), respectively,

$$R_1 = \frac{\rho A}{A_{Al}}(L + \tau)2p'$$

$$= \frac{3 \times 5 \times 10^{-8}}{2795/(3 \times 10^6)}(0.9 + 1.5) \times 1000$$

$$= 9.016 \times 10^{-2} \ \Omega$$

$$X_{1\sigma} = 0.345 \ \Omega$$

The total active power $P_{1r}$ is

$$P_{1r} = 3RI^2 + F_x U_s$$

$$= 3 \times 9.016 \times 10^{-2} \times 2u9t^2 + (40{,}000 \times 120)$$

$$= 6.909 \text{ MW}$$

Similarly, the reactive power $Q_{1r}$ becomes

$$Q_{1r} = 3X_{1\sigma}I_r^2 + 3E_1 I_r \sin \gamma_o$$

$$= 3 \times 345 \times 2.793^2 + 3 \times 0.609 \times 12 \times 2{,}795 \times \sin 20°$$

$$= 11.19 \text{ MVAr}$$

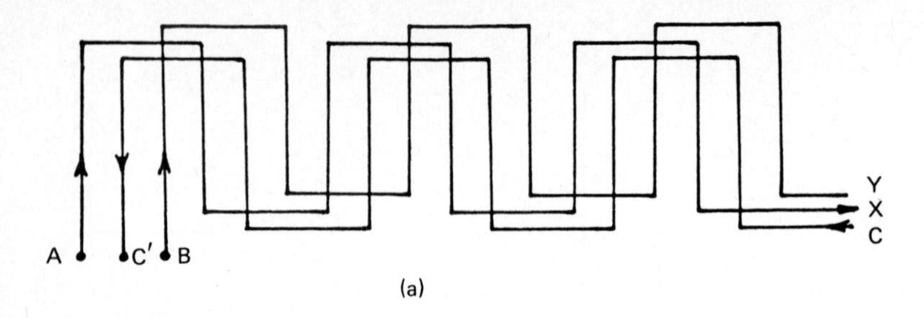


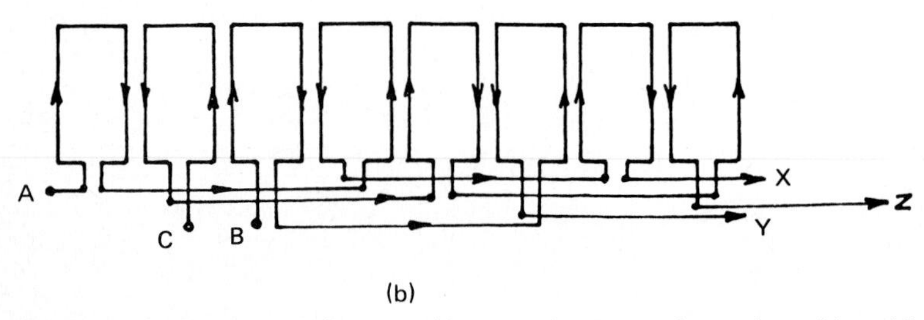


**FIG. 7.38** LSM armature windings: ***(a)*** with one conductor per pole per phase; ***(b)*** special winding easy to fabricate and install.

Thus, the rated power factor is

$$\cos \phi_1 = \frac{P_{1r}}{(P_{1r}^2 + Q_{1r}^2)^{1/2}}$$

$$= \frac{6.909}{(6.909^2 + 11.19^2)^{1/2}}$$

$$= 0.535$$

The required phase voltage $V_r$ is

$$V_r = \frac{(P_{1r}^2 + Q_{1r}^2)^{1/2}}{3I_r} = 1526 \text{ V}$$

Finally, the rated efficiency $\eta_r$ is

$$\eta_r = \frac{F_x U_s}{P_{1r}}$$

$$= \frac{4 \times 10^4 \times 120}{6.909 \times 10^6}$$

$$= 0.6947$$

The corresponding normal force $F_n$ is, from Eq. (7.85),

$$F_n \simeq \frac{3I_1 \sin \gamma_o}{2\omega_1} \left( \frac{-1}{g_o} E_1 \right)$$

$$= \frac{-3 \times 2795 \times \sin 20^\circ \times 609.12}{2 \times 2\pi \times 40 \times 0.25}$$

$$= -1.390 \times 10^4 \text{ N}$$

The normal force is of a repulsive character since $\gamma_o > 0$. For $\gamma_o < 0$, the normal force is of an attractive type. Also, the value of $F_n$ is small (smaller than the propulsion force).

The performance is satisfactory, considering the section length, the large airgap, and the high thrust developed.

For a detailed presentation of the SCLSM, the interested reader should consult Refs. 3 and 4.

## Conventional Field-Winding Heteropolar LSM

In this case a salient inductor structure is considered. The stator winding is placed in open slots and made of stranded (twisted) aluminum (or copper) cable with one conductor per slot per pole per phase. The equations (for steady state) and the phasor diagram are typical of a synchronous machine. The inductor (movable) does not generally contain a damper winding but has in its slots a linear generator, trapping the power contained in the stator slot harmonic field.

The voltage equation for steady state is

$$U_1 = \mathbf{I}_1 R_1 + j(X_{1\sigma} + X_a)\mathbf{I}_1 + jX_{dm}\mathbf{I}_d + jX_{qm}\mathbf{I}_q - \mathbf{E}_1 \tag{7.86}$$

where the $X_{1\sigma}$ is known as the leakage reactance corresponding to the slot and end connections of an energized section; the additional leakage reactance $X_a$ is related to the magnetic field spread above the active surface of the stator core corresponding to the portion of the section not "covered" by the inductor.

It is well known that the armature reaction cannot be neglected in this case. The rated airgap is 10 to 15 mm, and the parameters in Eq. (7.86) are given by

$$R_1 = \frac{2\rho_{Al}}{A_{Al}}(L + \tau + 0.04)p' \tag{7.87}$$

$$X_{1\sigma} = \omega_1\mu_o\left[(\lambda_s + \frac{p\lambda_d}{p'})L + \lambda_e(\tau + 0.02)\right]2p' \qquad \lambda_e \simeq 6 \tag{7.88}$$

where $\lambda_s$, $\lambda_d$, and $\lambda_e$ are the slot, differential, and end-connection specific permeances, respectively, as known from Eqs. (7.46) and (7.47).

An approximate value of $X_a$ may be obtained by considering that the open stator is in fact characterized by an equivalent airgap of $\tau/\pi$; hence,

$$X_a = \frac{6\mu_o\omega_1}{\pi^2}\frac{\tau L}{(\tau/\pi)}(p' - p) \tag{7.89}$$

Also, the magnetizing reactances $X_{dm}$ and $X_{qm}$ are

$$X_{dm} = \frac{6\mu_o\omega_1}{\pi^2}\frac{\tau L}{g_o K_c} p K_{dm} \tag{7.90}$$

$$X_{qm} = \frac{X_{dm}K_{qm}}{K_{dm}} \tag{7.91}$$

$$K_{dm} \simeq 4 \int_0^{1/3} \cos^2\left(\frac{\pi}{\tau}x\right) d(\frac{x}{\tau})r = 0.948 \tag{7.92}$$

$$K_{qm} \simeq 4 \int_0^{1/3} \sin^2\left(\frac{\pi}{\tau}x\right) d(\frac{x}{\tau}) = 0.3845 \tag{7.93}$$

The stator slot area $h_sW_s$ is given by

$$h_sW_s = \frac{I_r}{j_{Al}K_{\text{fil}}} \tag{7.94}$$

The slot width $W_s$ is given by

$$W_s = \frac{\tau}{3}\left(1 - \frac{B_1}{B_t}\right) \tag{7.95}$$

where $B_1$ is the fundamental of airgap flux density and $B_t$ is the tooth-design flux density. The thrust, efficiency, and power factor expressions are the same as for the SCLSM, but the iron losses have to be included.

The attractive normal force may be determined by using the resultant airgap flux density distribution $B_{r1}$:

$$B_{r1}(x) = B_{r1}\cos\frac{\pi}{\tau}x + \frac{3\mu_o w_1 I_1\sqrt{2}k w_1}{\pi p g_o}\sin\left(\frac{\pi}{\tau}x - \gamma_o\right) \tag{7.96}$$

Thus, the normal force $F_n$ is

$$F_n = \frac{1}{2}\frac{L2p}{\mu_o}\int_{-\tau/3}^{-\tau/3} B_{r1}^2(x)\,dx \tag{7.97}$$

***A Numerical Example.*** Consider an LSM with the following data:

Airgap flux density fundamental (produced by the field winding) $B_1 = 0.7$ T
Rated airgap $g_o = 1$ cm
Pole pitch $\tau = 0.2$ m
Rated speed $U_s = 120$ m/s
Stack width $L = 0.25$ m
Rated thrust $F_{xr} = 20$ kN
Rated efficiency is imposed to be $\eta_r = 0.8$
Energized section length = 1.2 km
Current load angle $\gamma_o = 20°$

It is required to determine the rated current, number of poles, slot dimensions, rated voltage and power factor, and the normal force.

First, since the efficiency $\eta_r$ is given, we may calculate the losses $P_{co}$ in the energized section:

$$P_{co} = 3R_1I_r^2 = 3.2\frac{\rho_{Al}}{I_r}j_{Al}(L + \tau + 0.04)p'I_r^2 = F_xU_s\left(\frac{1}{R_r} - 1\right) \tag{7.98}$$

The stator slot area is obtained from

$$h_sW_s = \frac{I_r}{j_{Al}K_{\text{fil}}}$$

$$= \frac{647}{3 \times 0.6}$$

$$= 360 \text{ mm}^2 \tag{7.99}$$

with

$$B_t = 1.5 \text{ T}$$

$$W_s = \frac{\tau}{3}\left(1 - \frac{B_1}{B_t}\right)$$

$$= \frac{200}{3}\left(1 - \frac{0.7}{1.5}\right)$$

$$= 35.55 \text{ mm} \tag{7.100}$$

So the slot active depth is $h_s = 360/35.55 = 10.23$ mm. This value is quite acceptable, leading to a low slot leakage. Now the number of poles per motor (per field winding) $2p$ is found from the thrust, as given by

$$F_x \simeq \frac{3E_1 I_1 \cos\gamma_o}{U_s} = \frac{3LpB_1U_s\sqrt{2}}{U_s} I_r \cos\gamma_o \tag{7.101}$$

$$p = \frac{F_x}{3L\sqrt{2}B_1 I_r \cos\gamma_o}$$

$$= \frac{20 \times 10^3}{3 \times 0.18 \times 1.41 \times 0.7 \times 647 \cos 20^\circ}$$

$$= 63.46 \tag{7.102}$$

The value $p = 64$ is chosen.

The total length of such a motor would be $2p\tau = 2 \times 64 \times 0.4 = 25.6$ m. Because such a motor is also generally used as a levitation system, the inductor length should be divided into separate units of less than 2 m in length. In this case, 16 units of 8 poles for each unit (1.6 m in length) would be an appropriate choice. Now the induced voltage is

$$E_1 = LpB_1U_s\sqrt{2}$$

$$= 0.25 \times 64 \times 0.7 \times 120 \times \sqrt{2}$$

$$= 1895 \text{ V} \tag{7.103}$$

To determine the rated phase voltage and power factor, we should first calculate the machine reactances:

$$X_{dm} = \frac{6\mu_o\omega_1}{\pi^2}\frac{\tau L}{g_o K_c}\frac{p}{2}K_{dm}$$

$$= \frac{6 \times 1.25 \times 10^{-6}}{\pi^2} \times \frac{2\pi \times 300 \times 0.2 \times 0.25 \times 32 \times 0.948}{1 \times 10^{-2} \times 1.25}$$

$$= 1.739 \ \Omega$$

$$X_{qm} = \frac{K_{qm}}{K_{dm}} X_{dm}$$

$$= \frac{0.3845}{0.948} \times 1.737$$

$$= 0.7053 \ \Omega$$

The leakage reactances, from Eqs. (7.87) to (7.89), become

$$X_{1\sigma} = 1.77 \ \Omega \qquad X_a \simeq 3.3 \ \Omega \qquad \text{also} \qquad R_1 = 0.159 \ \Omega$$

In the expression for the thrust, we have neglected the so-called reluctance thrust due to the fact that $X_{dm} \neq X_{qm}$.

Since $I_r$ and $\gamma_d$ are given, the current components $I_d$ and $I_q$ are

$$I_d = I_r \sin \gamma_o$$

$$= 647 \times 0.342$$

$$= 221.28 \text{ A}$$

$$I_q = I_r \cos \gamma_o$$

$$= 647 \times 0.9396$$

$$= 607.98 \text{ A}$$

Now from the phasor diagram we obtain

$$V_r \cos (\gamma_o + \phi) = E_1 - (X_{dm} + X_{1\sigma} + X_a)I_d + R_1 I_q$$

$$= 1895 - (1.739 + 1.77 + 3.3) \times 221.28 + (0.159 \times 607.98)$$

$$= 486 \text{ V}$$

$$V_r \sin (\gamma_o + \phi) = (X_{qm} + X_{1\sigma} + X_a)I_q + R_1 I_d$$

$$= (0.7053 + 1.77 + 3.3) \times 607.98 + (0.159 \times 221.28)$$

$$= 3546 \text{ V}$$

Thus, $V_r = 3586$ V, $(\gamma_o + \phi) = 82.2°$, $\phi = 62.2°$, and $\cos \phi_r = 0.466$. The power factor is rather poor due to an appreciable influence of section leakage, which is considerably higher than in an ironless armature core. If a row of such motors is installed on a train, both the efficiency and the power factor can be relatively improved.

The normal force $F_n$ is, from Eqs. (7.96) and (7.97), 80.0 kN. The influence of the reaction field both on the phase voltage and on the normal force is rather small ($X_{dm} I_d << V_r$, and $X_{qm} I_q << V_r$).

## Permanent Magnet LSM

The conventional field winding in an LSM may be replaced by high-energy permanent magnets. This solution is proposed for low speeds: up to 20 m/s.

Rare earth magnets would be ideal for the purpose. But less expensive permanent magnets may be used if adequate flux concentration schemes are applied in order to obtain high flux densities (0.5 to 0.7 T) in the airgap (Fig. 7.39).

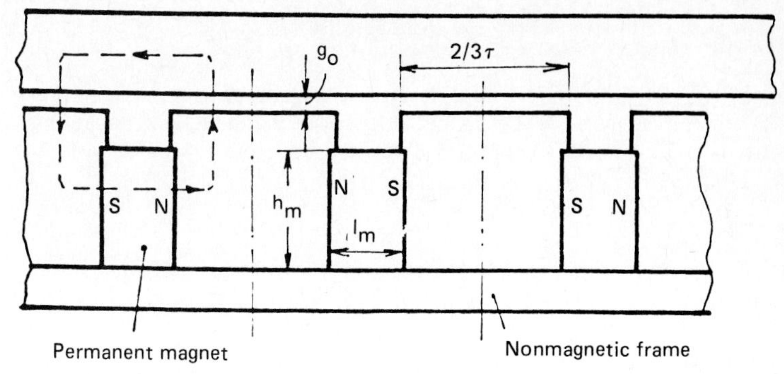


**FIG. 7.39** Permanent magnet inductor.

The approach for performance assessment is, in principle, the same as above, except for the inductor design. The inductor must be designed considering the permanent magnet characteristics, as shown by the following example.

***A Numerical Example.*** The initial data are: pole pitch $\tau = 0.2$ m; pole shoe $\tau_p = 0.66\ \tau$; stack width $L = 0.25$ m; number of poles per motor $= 4 + 1/2 + 1/2$; airgap $g_o = 1 \times 10^{-2}$ m; airgap flux density $B_o = 0.5$ T; permanent magnet material—ferrite $F_2$, with $B_r = 0.395$; $H_o = 200$ KA/m and $\mu_r = 4.5\ \mu_o$ (Fig. 7.40). The conservation of flux and Ampère's law for the inductor yield the following (the iron magnetic voltage drops being neglected due to the large airgap and low permeability of the permanent magnet):

$$\frac{B_m A_m}{K_\sigma} = B_o A_o \qquad A_m = L h_m \qquad A_o = \frac{\tau}{3} L \tag{7.104}$$

$$H_m l_m + \frac{2}{\mu_o} B_o g_o = 0 \tag{7.105}$$

From these two equations we obtain the linear dependence between $H_m$ and $B_m$ (the straight line in Fig. 7.40):

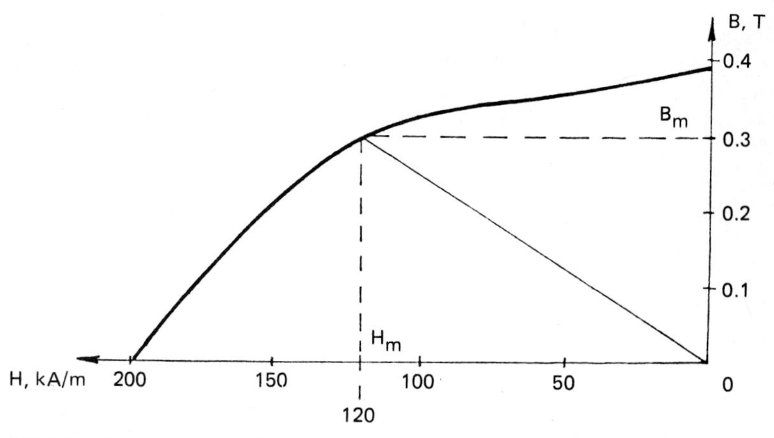


**FIG. 7.40** Permanent magnet characteristic.

$$H_m = \frac{-1}{K_\sigma} \frac{2g_o}{l_m} \frac{3h_m}{\tau} \frac{B_m}{\mu_o} \tag{7.106}$$

In Eq. (7.104), $K_\sigma = (1.1 \text{ to } 1.4)$ is a leakage coefficient accounting for the leakage flux of the permanent magnet lost between the inductor's neighboring poles. Considering that $l_m = \tau/3$ (Fig. 7.39), we find $H_m$ from Eq. (7.105):

$$\begin{aligned} H_m &= \frac{-2B_o g_o}{\mu_o l_m} \\ &= \frac{-2 \times 0.5 \times 1 \times 10^{-2}}{1.25 \times 10^{-6} \times 6.66 \times 10^{-2}} \\ &= 120.12 \text{ kA/m} \end{aligned}$$

From the hysteresis cycle, for $H_m = 120.12$ kA/m, $B_m = 0.3$ T. Thus, $h_m$ from Eq. (7.104), with $K_\sigma = 1.4$, is given by

$$\begin{aligned} h_m &= \frac{B_o}{B_m} \times \frac{\tau}{3} K_\sigma \\ &= \frac{0.5}{0.3} \frac{2 \times 10^{-1}}{3} \times 1.4 \\ &= 0.155 \text{ m} \end{aligned}$$

The permanent magnet height $h_m$ may be about 2 times smaller for an $SmCo_5$ magnet.

There is one more aspect of this motor that must be considered: namely, the longitudinal magnetizing reactance $X_{dm}$ is considerably smaller than the transverse reactance $X_{qm}$. This is due to the fact that the permanent magnet permeability $\mu_p$ is very small ($1.05\mu_o$ to $4.5\mu_o$), providing a high magnetic reluctance path along the *d* axis. Thus, the expression for $X_{qm}$ still holds, but $X_{dm}$ is slightly modified as

$$X_{dm} \simeq \frac{6\mu_o}{\pi^2} \omega_1 p K_{dm} \frac{\tau L}{g_o \left[1 + (\tau l_m \mu_o / 6 g_o h_m \mu_p)\right]} \tag{7.107}$$

In our case the bracketed value in the denominator of Eq. (7.107) is 5.3 instead of 1, for the case of an LSM with a field winding, implying a 5.3:1 reduction in $X_{dm}$.

From now on, the performance and preliminary design assessment are the same as in the two previous examples. It is to be noted that the development of LSMs (with conventional, SC, or permanent magnet excitation) is in progress and that the above data serve only for preliminary update design purposes.

## REFERENCES


1. S. A. Nasar and I. Boldea, *Linear Motion Electric Machines*, Wiley, New York, 1976.
2. H. Bolton, "Transverse Edge Effects in Sheet Rotor Induction Motors," *Proc. IEE*, vol. 116, 1969, pp. 725–739.
3. I. Takano, Y. Sato, and N. Ogimara, "End Effect of a Magnetically Suspended Ultra High Speed Train," *El. Eng. in Japan*, vol. 95, no. 1, 1975, pp. 59–66.
4. T. Obstuka and Y. Kyotani, "Superconducting MAGLEV Test," *IEEE Trans.*, vol. MAG-15, no. 6, 1979, pp. 1416–1421.

## *BIBLIOGRAPHY*

Amemiya, Y., and S. Aiba: "Linear Synchronous Motor Using Superconducting Propulsion Coils for Both Propulsion and Levitation," *El. Eng. in Japan*, vol. 99, no. 2, 1979, pp. 59–67.

Boldea, I.: *Vehicles on Magnetic Cushion* (in Romanian), Romanian Academy Publ. House, Bucharest, 1981.

——— and S. A. Nasar: "Field Winding Drag Force in Linear Synchronous Homopolar Motors," *EME*, vol. 2, no. 3, 1978, pp. 253–268.

——— ———: *Linear Motion Electromagnetic Systems*, Wiley, New York, 1985.

——— ———: "Linear Synchronous Homopolar Motor—A Design Procedure for Propulsion and Levitation," *EME*, vol. 4, nos. 2–3, 1979, pp. 125–136.

——— ———: "Simulation of High-Speed Linear Induction Motor End Effects in Low-Speed Tests," *Proc. IEE*, vol. 121, no. 9, 1974, pp. 961–964.

Budig, K. P.: *Alternating Current Linear Induction Machinery*, Oxford University Press, New York, 1981.

Laithwaite, E. R.: *Induction Machines for Special Purposes*, George Newnes, London, 1968.

Lorenzen, H. W., and W. Wild: "The Synchronous Linear Motor" (in German), Internal Report, Technical University of Munich, 1976.

Nondahl, T. A., and D. W. Novotny: "Three-Phase Pole-by-Pole Model of a Linear Induction Machine," *Proc. IEE*, vol. 127, pt. B, no. 2, 1980, pp. 68–82.

Oberretl, K.: "Three-Dimensional Analysis of the Linear Motor Taking Account of Edge Effects and the Distribution of the Windings" (in German), *Arch. Electrotech.*, vol. 55, 1973, pp. 181–190.

Papageorgiou, C., and E. M. Freeman: "Criteria and Conditions for the Best Linear Induction Motor," *Proc. ICEM*, Budapest, Sept. 9–10, 1982, pp. 976–979.

Poloujadoff, M.: *The Theory of Linear Induction Machinery*, Oxford University Press, New York, 1981.

Weh, H.: "The Integration of Magnetic Levitation and Electric Propulsion Functions" (in German), *ETZ-A*, vol. 96, no. 3, 1975, pp. 131–135.

# CHAPTER 8
# ELECTRONIC CONTROL OF MOTORS

**S. B. Dewan**
**G. R. Slemon**
**A. Straughen**

## *8.1 SOLID-STATE DEVICES*

The family of solid-state devices that may be employed in converters designed for motor control is small but growing. Consisting originally only of the silicon diode, or rectifier, and the SCR structure usually called by the generic name "thyristor," but also called the silicon controlled rectifier, it has now grown to include bipolar and field-effect power transistors, the bidirectional thyristor often called the triac, and the gate-turnoff thyristor. In the design of motor control, as opposed to the design of converter circuits, the devices may usually be considered to be ideal switches. Moreover, none of these ideal switches can perform operations that cannot also be performed by one or more thyristors, and indeed in high-power drives only thyristors and power diodes are employed in the power circuits.

The difference between a system controlled by, say, power transistors and a similar system controlled by thyristors lies chiefly in the nature of the control signals applied to the devices, that is, the method of turn-on, or gating, and turn-off, or commutation. While the mere removal of the gating signal from a transistor results in turn-off, a thyristor must often be commutated by auxiliary circuitry after the gating signal has been removed. Under steady state, the analysis of the operation of the power circuit is independent of the type of solid-state devices employed. For this reason, all converters are considered to include only power diodes and thyristors. Any waveforms of gating signals shown in the diagrams thus apply to thyristors only. If the power circuit of a converter is anywhere given, the thyristors are shown without auxiliary circuitry for turn-on, commutation (turn-off), and protection, while the electronic units controlling the converters are regarded simply as "black boxes"

---

*Note:* The source of the greater part of the material in this chapter is the text *Power Semiconductor Drives* by S. B. Dewan, G. R. Slemon, and A. Straughen, published by Wiley-Interscience, New York, 1984.

that provide the gating signals. Where thyristors must be forced-commutated by auxiliary circuitry, the thyristor symbols in the circuit are enclosed in circles, as shown in Fig. 8.1.

## 8.2 SYSTEM BEHAVIOR

The diagrams and formulas in this chapter apply to steady-state operation. It is assumed that any driven mechanical system may be represented by a characteristic of speed at the motor coupling $n$ r/min versus torque $T_L$ at the coupling. These characteristics may, of course, be quite varied. Similarly, the characteristic of any motor driving or regenerating under constant excitation can be represented by a curve of speed at the coupling versus either airgap or shaft torque. From the intersection of these two characteristics, one can readily arrive at a prediction of steady-state system performance. Some losses in the motor and all losses in the converters may often be ignored to produce a result sufficiently accurate for the great majority of engineering applications.

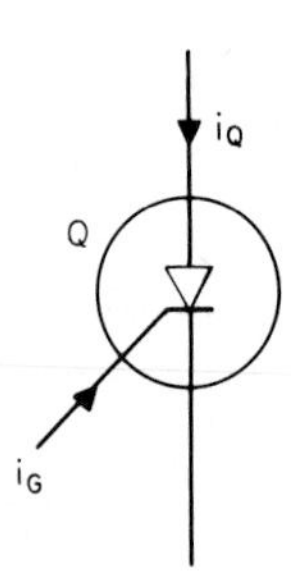


**FIG. 8.1** Forced-commutated thyristor.

The general problem may be represented by Fig. 8.2, in which the two speed-torque characteristics for the motor ($M$) and the load ($L$) are represented over part of their range. The equilibrium condition giving steady-state operation is represented by the point of intersection $p$. While the load characteristic is given and unalterable (although there may be a family of such curves), the motor characteristic depends upon the type of motor, the excitation applied to it, the converter(s) supplying that excitation, and possibly the power system supplying the converters. The practical

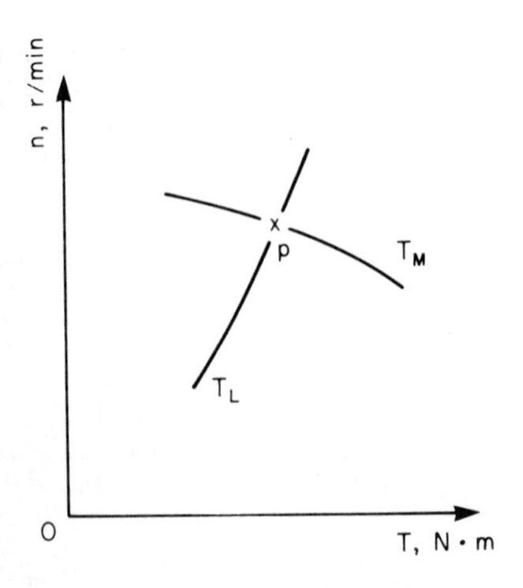


**FIG. 8.2** Load and drive characteristics.

problem to be solved may take one of two forms and consist of answering one or both of two questions:

1. What kind of drive and power supply are required to achieve operation at all required points on the mechanism characteristic?
2. For a given drive system, what are the conditions of motor excitation and operation of the converter(s) to give equilibrium at a particular point on the load characteristic?

To answer (1), a comparison of the results of analysis of a number of possible drive systems must be made. To answer (2), an analysis of the chosen drive system is required.

## 8.3 RATING AND DERATING OF MOTORS

The direct voltages and currents provided by rectifiers and choppers (dc-to-dc converters), as also the alternating voltages and currents provided by ac power controllers, inverters, and cycloconverters, may have a large harmonic content over much of their operating range. This means that a dc motor must operate on "mixed" current (a combination of dc and multifrequency ac), while an ac motor must operate on current having a waveform that may be far from a pure sinusoid. The effect of this kind of excitation on both types of motor is an increase in both resistive losses and core losses over those which would occur with true direct or sinusoidal sources. Motors are manufactured that are specially designed to tolerate these operating conditions. If standard motors must be used, then they must be derated by restricting the permissible root-mean-square (rms) line current to less than the nameplate value to avoid overheating due to resistive losses and core losses. The result of this procedure may be a reduction of as much as 20 percent in the permissible continuous output power.

With derated standard motors or motors designed for excitation from converters, it is then possible to carry out performance calculations which assume either that the values of direct variables are equal to those of the average values actually occurring or that the values of alternating variables are equal to those of the fundamental components of the actual waveforms.

## 8.4 POWER SUPPLY

The choice of the power converter(s) and motor cannot be made arbitrarily. It depends upon the nature of the available power supply.

### AC Sources

In general, for a system developing up to 2 kW of drive power, a single-phase ac source is adequate. For higher powers, a three-phase source is desirable, and it is usually essential for more than 5 kW of drive power. There may be restrictions on the harmonic content of current drawn from the source and on the power factor at which the drive system may operate.

### DC Sources

It is unusual for a dc source to be available for general purposes in an industrial plant; thus, if dc is particularly desirable, it must normally be provided by conversion from the ac power system. If some form of rectification is employed, then the possibility of regeneration from the drive system must be considered. An ac motor driving a dc generator is acceptable, particularly if a large synchronous motor operating at a leading power factor is employed. Such a machine set is capable of accepting regenerated energy. In traction systems, large-scale conversion equipment capable of regeneration may be employed in substations. The advantage of dc for power distribution in a rail-guided system is great, since the guiding rails and a third rail or overhead wire may be employed as conductors, whereas the high self-inductance of steel rails prohibits their use with ac of standard power frequency.

## 8.5 SEPARATELY EXCITED DC MOTOR*

In a separately excited dc motor, if $I_f$ is the field current, then the flux per pole, $\phi$, linking the armature winding can be expressed by

$$\phi = F(I_f) = F(V_f/R_f) \qquad \text{Wb} \tag{8.1}$$

In the armature circuit

$$E_a = k\Omega_m\phi \qquad \text{V} \tag{8.2}$$

where $k$ is the machine constant for this value of $\phi$, and

$$V_t = E_a + R_a I_a \qquad \text{V} \tag{8.3}$$

In the model of Fig. 8.3, the friction factor $B$ accounts for viscous friction of both motor and mechanism. If the motor possesses appreciable coulomb friction, it may be included in the working torque $T_w$. Then for a given field current, the steady-state airgap torque of the motor is

$$T = k\phi I_a = B\Omega_m + T_w \qquad \text{N}\cdot\text{m} \tag{8.4}$$

Figure 8.4 shows a curve of speed versus viscous friction torque marked $B\Omega_m$ and two curves of speed versus airgap torque marked $T_w + B\Omega_m$, where in the case shown $T_w$ is positive or negative and is constant at all speeds. From Eqs. (8.2) to (8.4)

$$\Omega_m = \frac{V_t}{k\phi} - \frac{R_a T}{(k\phi)^2} = \frac{k\phi V_t - R_a T_w}{(k\phi)^2 + R_a B} \qquad \text{rad/s} \tag{8.5}$$

This equation describes the motor characteristic of speed versus airgap torque for the given conditions of motor excitation specified by $V_t$ and $\phi$. Where the lines described by Eqs. (8.4) and (8.5) intersect is the point of steady-state operation. Curves for two values of $V_t$ are shown. For regeneration in the second quadrant of Fig. 8.4, armature current $I_a$ reverses, so that the flow of energy at the motor terminals reverses.

*See also Chap. 5, "Direct-Current Machines."

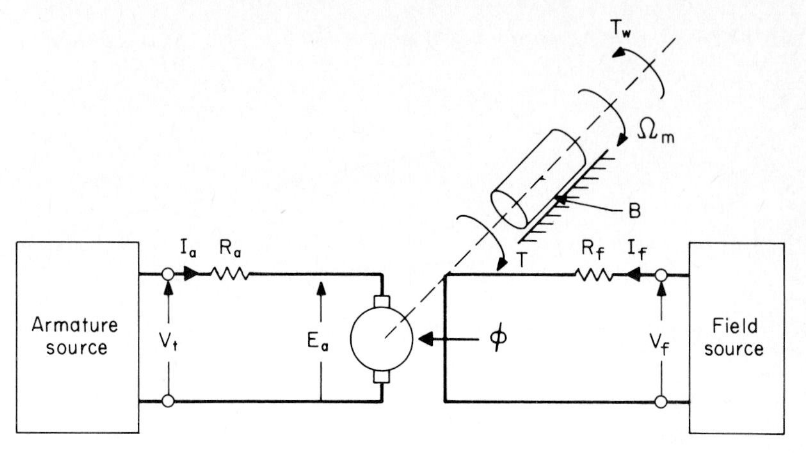


**FIG. 8.3** Separately excited dc motor.

The maximum permissible value of $V_t$ may be set either by the acceptable value for the motor or by the limit of the voltage of the dc source. Once this maximum value is reached, further speed increase is obtained by field weakening, that is, reduction of $\phi$ in Eq. (8.5). The effect of field weakening is shown by the lines for two values of $\phi$ in Fig. 8.5. With field weakening, the armature current must normally be limited to the rated value, and since $V_t$ is constant, the power input to the motor is constant. The maximum power output of the motor will therefore be approximately constant, so that

$$P = T\Omega_m = \text{const.} \qquad \text{W} \tag{8.6}$$

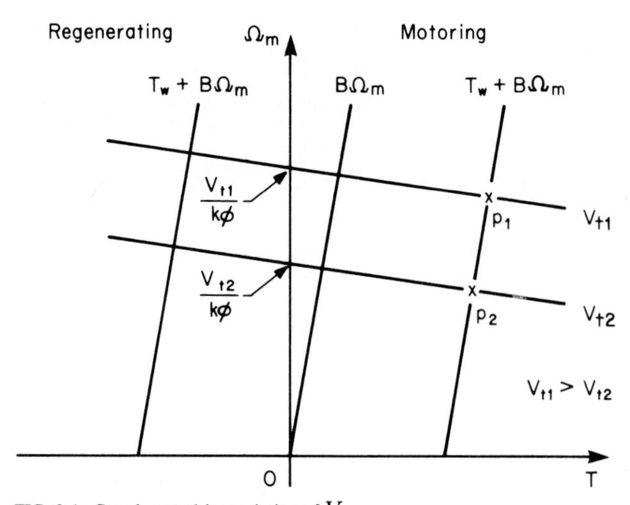


**FIG. 8.4** Speed control by variation of $V_t$.

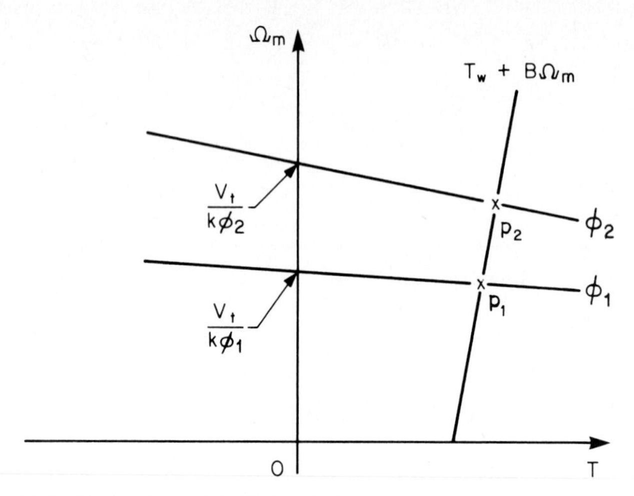

**FIG. 8.5** Speed control by field weakening.

This means that, including regeneration and reverse running by reversal of $V_t$, the motor can operate continuously at any point within the area shown in the diagram of Fig. 8.6. For short periods of time, the torque may exceed that defined by this diagram.

The steady-state operation of the motor and load can be represented by the block diagram of Fig. 8.7, which is constructed from Eqs. (8.2) to (8.4). For constant motor excitation and variable working torque $T_w$, Eq. (8.5) shows that this drive has speed regulation, as indicated in Fig. 8.4. The speed regulation may be reduced by the addition to the system of a speed transducer and a regulator, as shown in Fig. 8.8. The transfer function of the controlling system, comprising the regulator, converter, and converter logic, is then

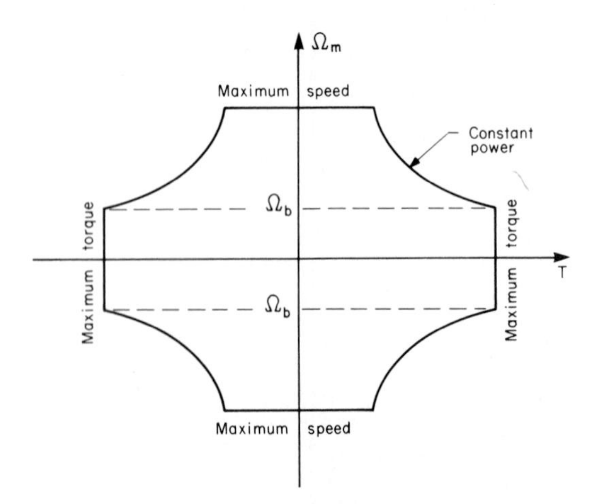

**FIG. 8.6** Continuous torque and speed limits.

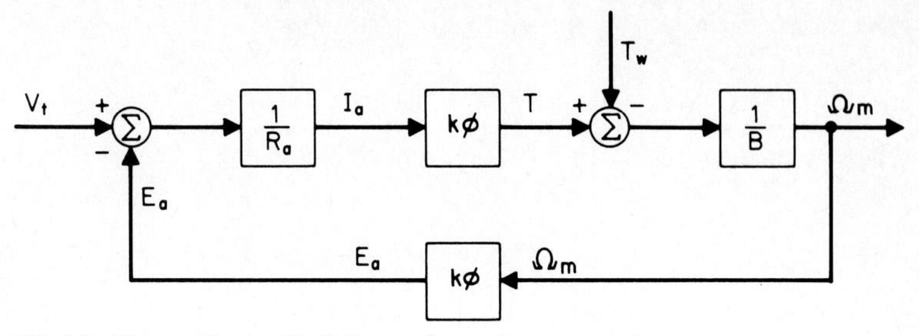


FIG. 8.7 Motor-mechanism block diagram for steady-state operation.

$$k_1 = \frac{V_t}{V - V_t} \tag{8.7}$$

where $V$ is the speed command in the form of a voltage. The transfer function of the speed transducer is

$$k_2 = \frac{V_T}{\Omega_m} \qquad \mathrm{V/(rad \cdot s)} \tag{8.8}$$

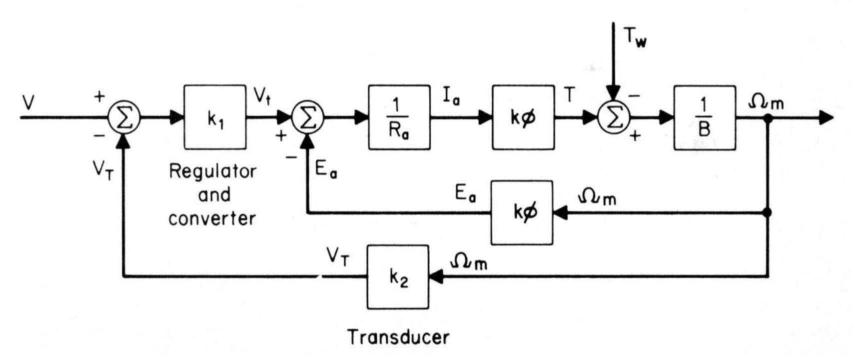


FIG. 8.8 Steady-state operation of a speed control system.

There are two excitations to the system of Fig. 8.8, namely, $V$ and $T_w$. The steady-state response to any pair of these two values may be obtained by setting each in turn to zero and obtaining the response to the other. The two responses may then be combined by superposition to give the speed of the drive. Thus, setting $T_w$ to zero yields for the motor-mechanism combination of Fig. 8.7

$$\frac{\Omega_m}{V_t} = \frac{k\phi/R_aB}{1 + (k\phi)^2/R_aB} \qquad \mathrm{rad/s/V} \tag{8.9}$$

Normally

$$\frac{(k\phi)^2}{R_aB} >> 1 \tag{8.10}$$

in which case

$$\frac{\Omega_m}{V_t} = \frac{1}{k\phi} \tag{8.11}$$

The closed-loop steady-state transfer function for the entire system of Fig. 8.8 is then

$$\frac{\Omega_m}{V} = \frac{k_1}{k\phi + k_1 k_2} \tag{8.12}$$

Setting $V$ to zero (when a positive value of $T_w$ will drive the motor backward) yields

$$\frac{\Omega_m}{-T_w} = \frac{R_a}{R_a B + k\phi(k_1 k_2 + k\phi)} \tag{8.13}$$

The two responses from Eqs. (8.12) and (8.13) may then be combined to give the steady-state speed.

## 8.6 SINGLE-PHASE RECTIFIER DRIVES

Rectifiers may be classified by the number of phases of the ac source supplying them or by the number of pulses of current that they supply to the load circuit during one cycle of the ac source. In the following, the only systems considered will be those drawing from the source an alternating current with a symmetrical waveform. Such a waveform contains no direct component or harmonics. A single-phase rectifier is therefore a two-pulse rectifier.

### Fully Controlled Rectifier Drives

The main power circuit elements of two commonly used rectifier configurations are shown in Figs. 8.9*a* and *b*. The method of controlling the field current is not indicated. Provided that the rated motor armature terminal voltage and that of the ac source are compatible, the transformer may be omitted from Fig. 8.9*a* if not required for isolation. In this bridge circuit, thyristors $Q_1$ and $Q_2$ are turned on simultaneously at the chosen delay angle $\omega t = \alpha$, where

$$v_s = \sqrt{2}V \sin \omega t \qquad \text{V} \tag{8.14}$$

Thyristors $Q_3$ and $Q_4$ are turned on one-half cycle later, when the gating signals are already removed from $Q_1$ and $Q_2$. In the circuit of Fig. 8.9*b*, $Q_1$ and $Q_2$ are turned on alternately.

For low-potential motors operating down to low speeds, the circuit of Fig. 8.9*b* is preferable, since there is only one thyristor in series with the armature to reduce the armature terminal voltage and cause resistive losses. (The forward voltage across the terminals of a conducting thyristor may be of the order of 2 V.) However, the transformer of Fig. 8.9*b* is larger than that of Fig. 8.9*a*, since only one-half of the secondary winding carries current at any instant. Each of the configurations of Fig. 8.9 can be represented by the equivalent circuit of Fig. 8.10, which includes the resistance, inductance, and electromotive force (emf) of the motor armature.

Figure 8.11 shows steady-state waveforms for the equivalent circuit of Fig. 8.10. Equation (8.2) gives the value of armature emf $E_a$. These waveforms are typical

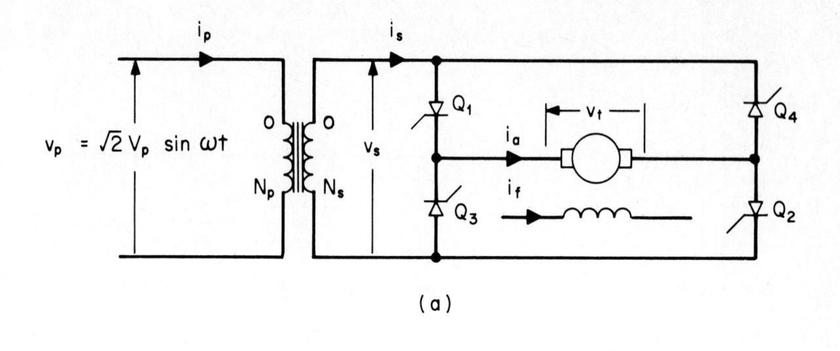


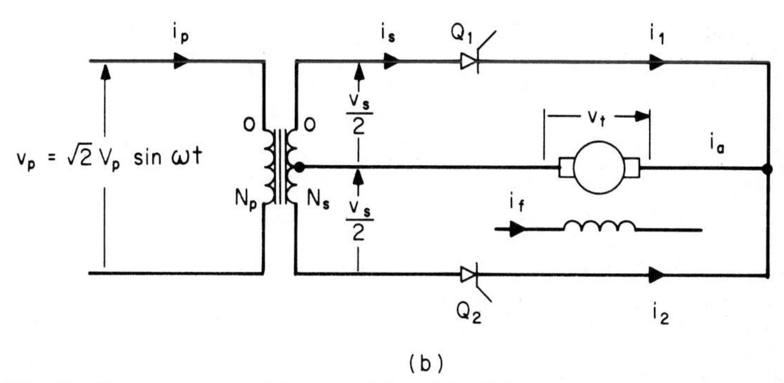


**FIG. 8.9** Single-phase controlled rectifier drives: ***(a)*** and ***(b)*** are two possible configurations.

of the behavior of a single-phase rectifier and small motor in which armature inductance is normally low, so that the current is discontinuous, although continuous current may occur when motor inductance is significant and the load is close to the rated value. In Fig. 8.11 the instant at which the armature current falls to zero is designated $\omega t = \beta$, and at that instant

$$\frac{\sqrt{2}V}{Z}\sin(\beta-\varphi) - \frac{k\phi\Omega_m}{R_a} + \left[\frac{k\phi\Omega_m}{R_a} - \frac{\sqrt{2}V}{Z}\sin(\alpha-\varphi)e^{(\alpha-\beta)/\tan\varphi}\right] = 0 \quad (8.15)$$

in which

$$Z = \left[(\omega L_a)^2 + R_a^2\right]^{1/2} \quad \Omega \quad (8.16)$$

$$\varphi = \tan^{-1}\frac{\omega L_a}{R_a} \quad \text{rad} \quad (8.17)$$

Provided that

$$\alpha > \eta = \sin^{-1}\frac{k\phi\Omega_m}{\sqrt{2}V} \quad \text{rad} \quad (8.18)$$

(where angle $\eta$ is defined in Fig. 8.11), for a given set of values of $V$, $L_a$, $R_a$, $k\phi$, $\Omega_m$, and $\alpha$, a $\beta$ may be obtained numerically from Eq. (8.15). The average armature current is then

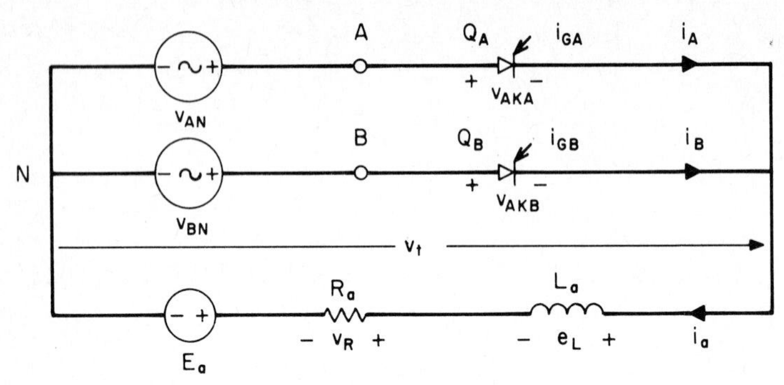


**FIG. 8.10** Equivalent circuit of the systems in Fig. 8.9.

$$\bar{i}_a = \frac{\bar{v}_t - k\phi\Omega_m}{R_a} \qquad \text{A} \tag{8.19}$$

where $\bar{v}_t$, the average value of $v_t$, may be obtained from the waveform of Fig. 8.11 as

$$\bar{v}_t = \frac{1}{\pi}\left[k\phi\Omega_m(\pi + \alpha - \beta) - \sqrt{2}V(\cos\beta - \cos\alpha)\right] \qquad \text{V} \tag{8.20}$$

The average internal torque of the motor is then

$$\bar{T} = k\phi\bar{i}_a \qquad \text{N}\cdot\text{m} \tag{8.21}$$

This gives a point on the curve of $\Omega_m$ versus $\bar{T}$.

For the range

$$\eta > \alpha > (\beta - \pi) \qquad \text{rad} \tag{8.22}$$

the waveform of $v_t$ is that shown in Fig. 8.12*a*, and $\bar{T}$ may be obtained by setting $\alpha = \eta$ in Eqs. (8.15) and (8.20). For the range

$$\eta > (\beta - \pi) > \alpha > 0 \qquad \text{rad} \tag{8.23}$$

the waveform of $v_t$ is that shown in Fig. 8.12*b*, and Eqs. (8.15) and (8.20) no longer apply. However, since systems are not normally designed to employ such low values of $\alpha$, that is not important. For *any* value of $\alpha$ at $\bar{T} = 0$, the speed of the motor is given by

$$\Omega_m = \frac{\hat{v}_t}{k\phi} \qquad \text{rad/s} \tag{8.24}$$

Figure 8.13 shows waveforms for continuous-current operation for $\alpha > \eta$. For the range

$$\eta > \alpha > 0 \qquad \text{rad} \tag{8.25}$$

the sharp trough of the curve of $i_a$ at $\omega t = \alpha$ is replaced by a smooth curve. The waveforms of Fig. 8.13 are approximated in that *overlap* is ignored. This term indicates an interval during which both thyristors of Fig. 8.10 are conducting simultaneously and is due to a combination of transformer-leakage inductance and protective *di/dt* inductors in series with the thyristors. The overlap interval, during

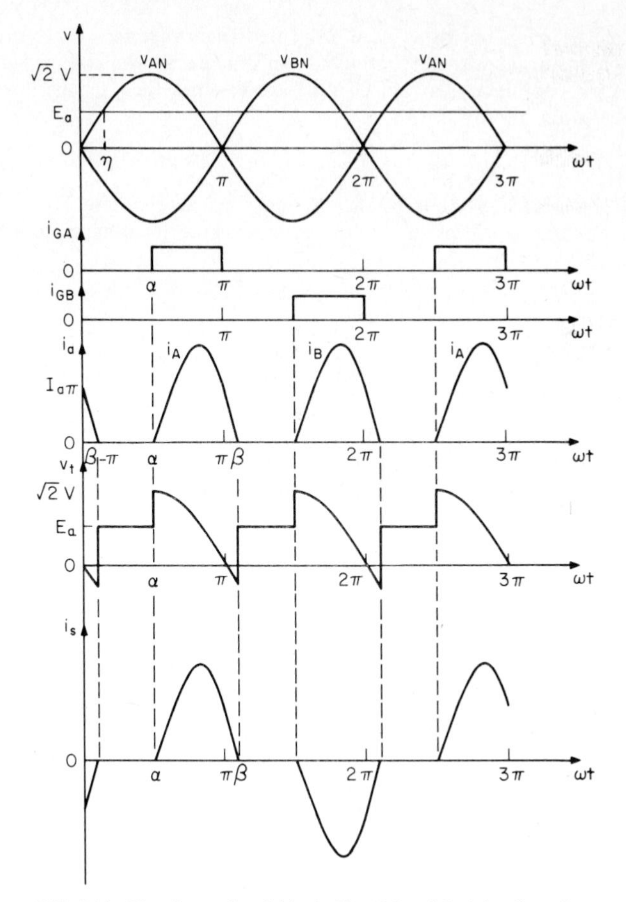


**FIG. 8.11** Waveforms of variables in Figs. 8.9 and 8.10 for discontinuous-current operation.

which $v_t \simeq 0$, is designated by angle $\mu$ and results in a small reduction in $\bar{v}_t$ that is neglected here.

For continuous-current operation,

$$\bar{v}_t = \frac{2\sqrt{2}V}{\pi} \cos\alpha \qquad \text{V} \tag{8.26}$$

Equations (8.19) and (8.21) apply, and

$$\Omega_m = \frac{2\sqrt{2}V}{\pi k\phi} \cos\alpha - \frac{R_a\bar{T}}{(k\phi)^2} \qquad \text{rad/s} \tag{8.27}$$

A family of curves of per-unit speed versus per-unit average internal torque for a 1-hp (746-W) 230-V dc motor is shown in Fig. 8.14. The rectifier was supplied with 270 V ac at 60 Hz.

In the fourth quadrant of Fig. 8.14 the torque is positive, but the speed is negative. The machine is regenerating and supplying energy to the single-phase source via the rectifier, which is operating as a fixed-frequency inverter. Waveforms for this condition of operation with discontinuous current are shown in Fig. 8.15. Inverter operation with continuous current is also possible, but then the permissible upper limit to $\alpha$ is given by

$$\alpha_{\max} = \pi - \omega t_{\text{off}} - \mu \qquad \text{rad} \tag{8.28}$$

where $t_{\text{off}}$ is the turn-off time of the thyristors.[1] The characteristic marked 180° in Fig. 8.14 is thus an idealization. In practice, values of $\alpha$ close to 180° are not necessary.

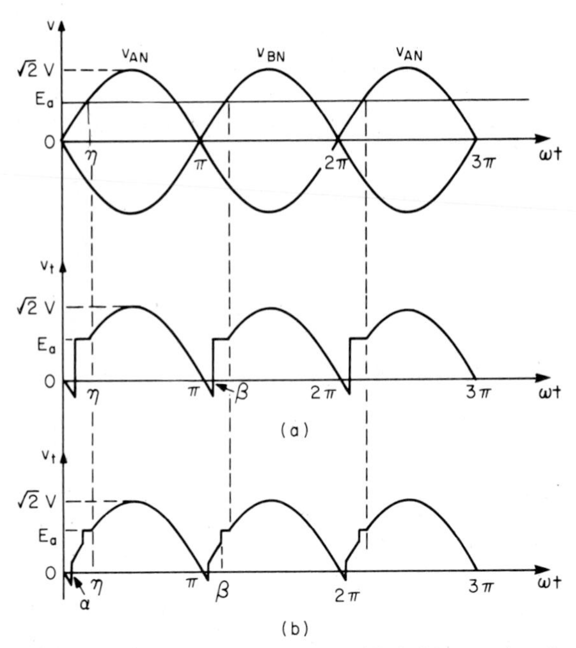


**FIG. 8.12** Waveforms of $v_t$ in Figs. 8.9 and 8.10: ***(a)*** $\eta > \alpha > (\beta - \pi)$; ***(b)*** $(\beta - \pi) > \alpha > 0$.

The rectifier-motor systems of Fig. 8.9 are capable of operating in the two quadrants of Fig. 8.14. Four-quadrant operation with a smooth transition from one quadrant to another is obtained by employing two rectifiers to form a dual converter, as shown in Fig. 8.16. The gating signals to the two rectifiers are such that at all settings $\bar{v}_{\text{op}} = -\bar{v}_{\text{on}}$. For this, it is necessary that $\alpha_p = \pi - \alpha_n$ rad. Harmonic currents in the mesh formed by the lines connecting the dc terminals of the two rectifiers are reduced or eliminated either by fitting an inductor in the mesh or blanking out the gating signals in the rectifier that is not required in the desired quadrant. The four-quadrant speed-torque diagram may be considered to be that of Fig. 8.14, with the second and third quadrants containing the same family of curves rotated through 180° about the origin.

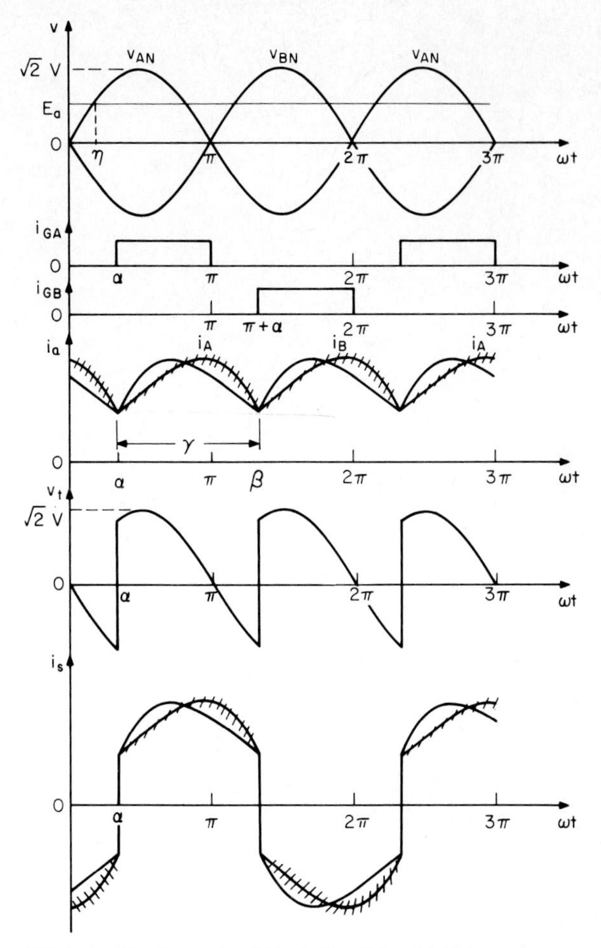


**FIG. 8.13** Waveforms of variables in Figs. 8.9 and 8.10 for continuous-current operation.

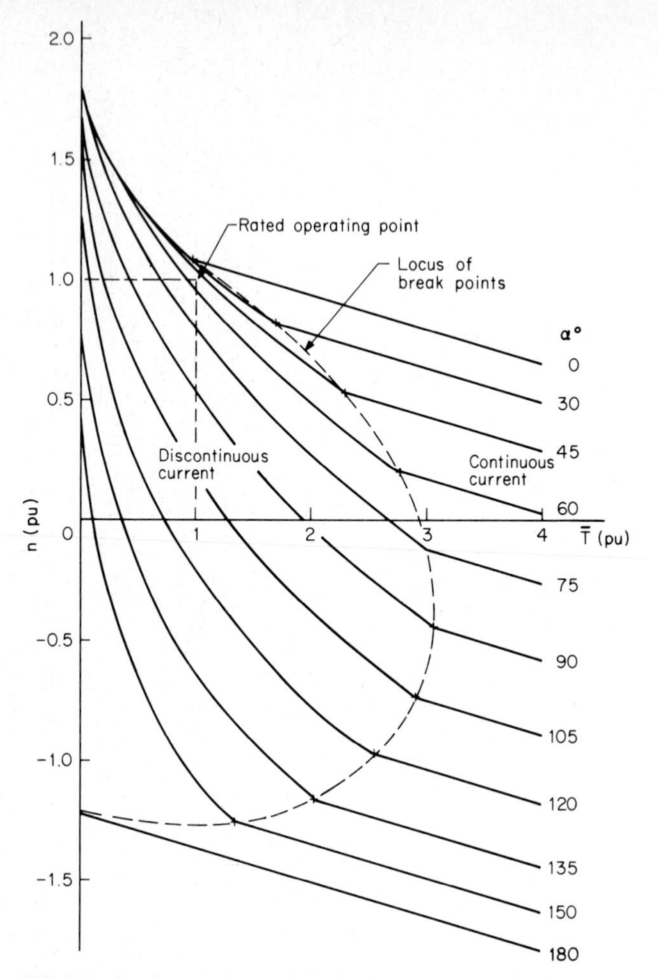


**FIG. 8.14** Speed-torque characteristics for a single-phase rectifier drive.

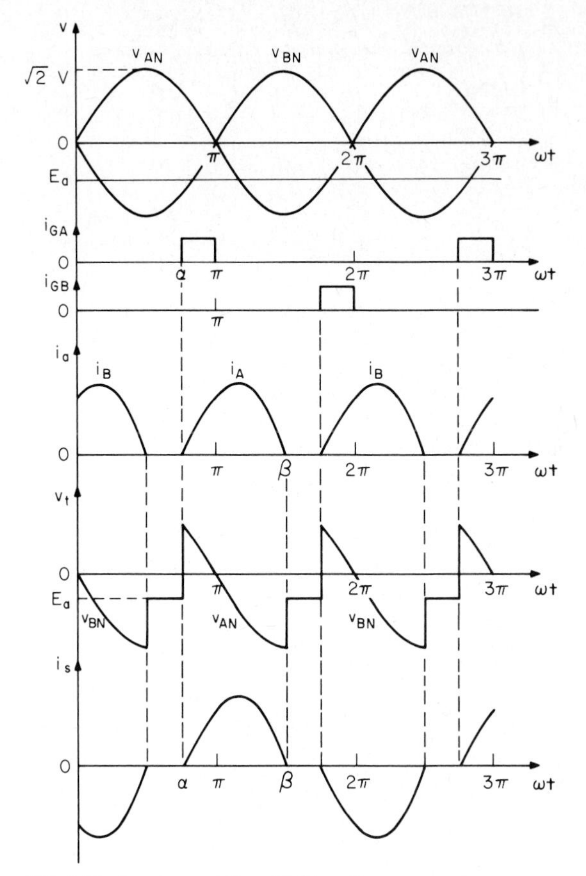


**FIG. 8.15** Inverter operation with discontinuous current.

## Rectifier Drives with Freewheeling

If thyristors $Q_2$ and $Q_3$ of Fig. 8.9*a* are replaced by diodes, giving a half-controlled rectifier, they provide a path for $i_a$ when that current is prolonged by armature-circuit inductance, a path through which $i_a$ can flow unopposed by voltage $v_s$. This process is called *freewheeling.* In this way continuous-current operation may be obtained under some conditions that would produce discontinuous current in the fully controlled rectifier. The same result could be achieved by connecting a diode in the required direction across the motor armature in either of the connections of Fig. 8.9. Waveforms for the boundary condition between discontinuous- and continuous-current operation for a rectifier with freewheeling are illustrated in Fig. 8.17. During freewheeling, the current $i_D$ through the diodes decays exponentially. For a larger value of $\alpha$, there would be a gap between the two current pulses, and for that interval, $v_t$ would be equal to $E_a$. For a smaller value of $\alpha$, $i_D$ would not fall to zero.

If the pulse of $i_a$ starts from zero, as in Fig. 8.11, then at $\omega t = \pi$, $i_a$ will have a value given by

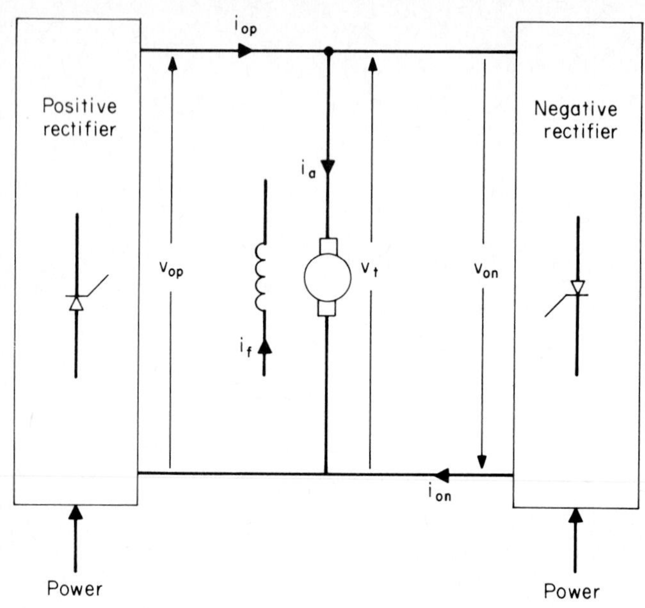


**FIG. 8.16** Dual converter.

$$I_{a\pi} = \frac{\sqrt{2}V}{Z}\sin\varphi - \frac{k\phi\Omega_m}{R_a} + \left[\frac{k\phi\Omega_m}{R_a} - \frac{\sqrt{2}V}{Z}\sin(\alpha-\varphi)\right]e^{(\alpha-\pi)/\tan\varphi} \quad \text{A} \tag{8.29}$$

For the boundary condition of Fig. 8.17,

$$I_{a\pi b} = \frac{k\phi\Omega_m}{R_a}\left[e^{\alpha/\tan\varphi} - 1\right] \quad \text{A} \tag{8.30}$$

If, for any set of operating conditions, $I_{a\pi} \geq I_{a\pi b}$, then the current is continuous. If the expressions in Eqs. (8.29) and (8.30) are equated, then for a given speed they may be solved to give the required value of $\alpha$ for boundary-condition operation. However, if $\alpha$ is found to be negative, then continuous-current operation is not possible.

If continuous-current operation is achieved, then

$$\bar{v}_t = \frac{\sqrt{2}V}{\pi}(1+\cos\alpha) \quad \text{V} \tag{8.31}$$

and

$$\Omega_m = \frac{\sqrt{2}V}{\pi k\phi}(1+\cos\alpha) - \frac{R_a\bar{T}}{(k\phi)^2} \quad \text{rad/s} \tag{8.32}$$

This is a linear relationship similar to that in Eq. (8.27).

The speed-torque curves for discontinuous current may be obtained by a procedure similar to that yielding those for the fully controlled rectifier in Fig. 8.14. However, while the harmonic content of the current is much reduced by freewheeling, the range of torque over which discontinuous current occurs still more than covers the normal operating range of the drive, except at very low speeds. This situation can be improved by including an inductor in the motor armature circuit.

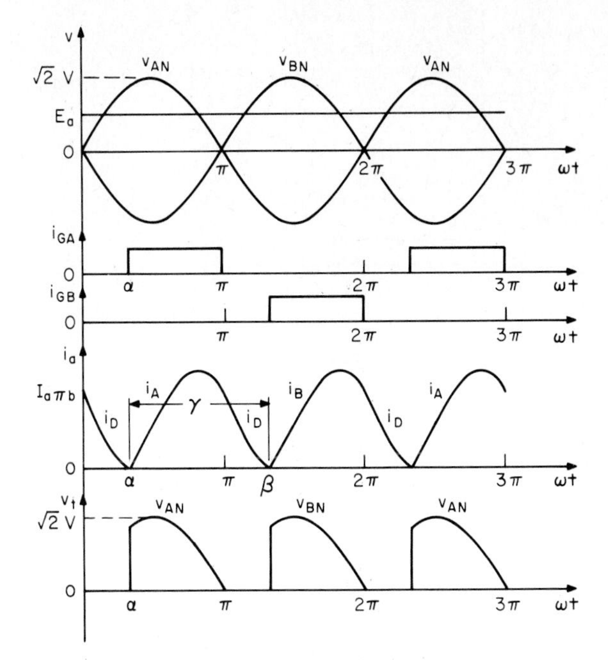


**FIG. 8.17** Single-phase rectifier with freewheeling—boundary condition.

A disadvantage of the freewheeling rectifiers so far described is that regeneration, as in the fourth quadrant of Fig. 8.14, is not possible, since $v_t$ cannot go negative. Regeneration with freewheeling in the fourth quadrant can be obtained, however, by controlling the thyristors of the fully controlled rectifier of Fig. 8.9*a* individually instead of in pairs. The waveforms of the gating signals and other circuit variables for operation in the first quadrant of Fig. 8.6 are shown in Fig. 8.18. All of the gating signals consist of 180° trains of pulses. Those for thyristors $Q_2$ and $Q_4$ remain fixed in the position shown in Fig. 8.18. Those for thyristors $Q_1$ and $Q_3$ are moved simultaneously along the $\omega t$ axis. Thyristor $Q_1$ is the reference device turned on at $\omega t = \alpha$. The gating signals for $Q_1$ and $Q_3$ can be moved to the right until $\alpha = \pi - \omega t_{\text{off}} - \mu$ rad.

Figure 8.19 shows the waveforms for operation in the fourth quadrant of Fig. 8.6. In theory the gating signals for $Q_2$ and $Q_4$ can be moved to the right until $\alpha = 2\pi$. In practice this degree of regeneration is not necessary. For fourth-quadrant operation,

$$\bar{v}_t = \frac{-\sqrt{2}V}{\pi}(1 + \cos\alpha) \qquad \text{V} \tag{8.33}$$

in which expression the small positive part of the waveform of $v_t$ is ignored.

The power circuit conditions giving rise to continuous-current operation are the same as those for the other rectifier with freewheeling. However, the operation of this circuit is liable to discontinuities that do not arise when freewheeling is not employed, so computer simulation of any projected system over the whole range of anticipated operating conditions is advisable.[2]

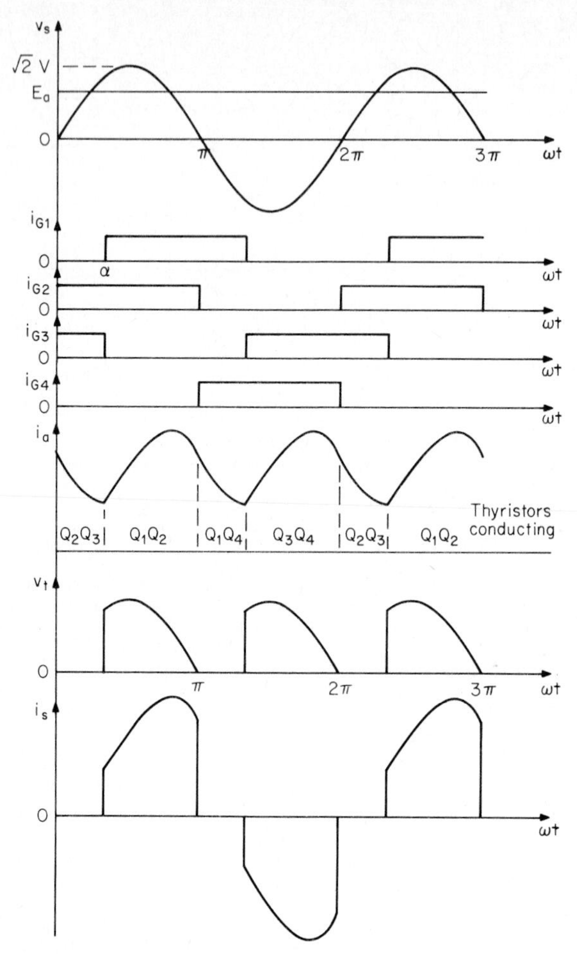


**FIG. 8.18** Waveforms of variables in a system with freewheeling: $\bar{v}_t > 0$.

## Power in Load and Source Circuits

If a dc motor is driven with continuous current, some useful approximate figures showing the driving performance of the system can be calculated. The following assumptions are made:

1. Only second harmonics of current need be considered when calculating rms currents.
2. $2\omega L_a >> R_a$.
3. Armature current is continuous and of average magnitude $\bar{i}_a$, so that the waveforms of $i_s$ in Figs. 8.13 and 8.18 may be assumed to be rectangular and of amplitude $\bar{i}_a$.
4. The rectifiers are ideal.

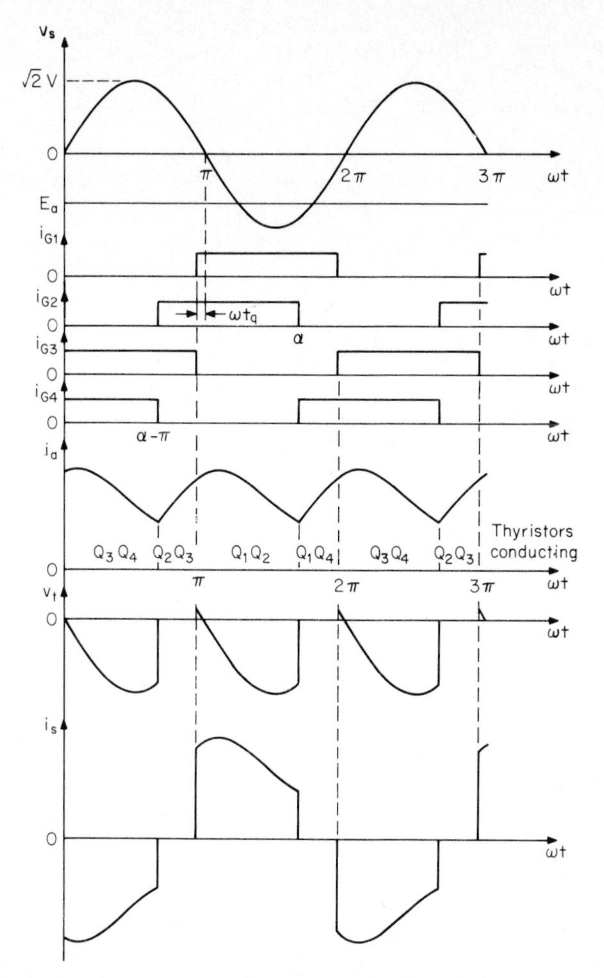


**FIG. 8.19** Waveforms of variables in a system with freewheeling: $\bar{v}_t < 0$.

By Fourier analysis, it can then be shown that $c_2$, the amplitude of the second harmonic of $v_t$, varies with $\alpha$ as in Fig. 8.20.

The rms second-harmonic armature current is

$$I_{R2} = \frac{c_2}{2\sqrt{2}\omega L_a} \quad \text{A} \tag{8.34}$$

and the rms armature current is

$$I_R = \left[(\bar{i}_a)^2 + I_{R2}^2\right]^{1/2} \quad \text{A} \tag{8.35}$$

where $\bar{i}_a$ is obtained from Eq. (8.19).

For rectifiers without freewheeling (Fig. 8.13), the rms value of the source current is

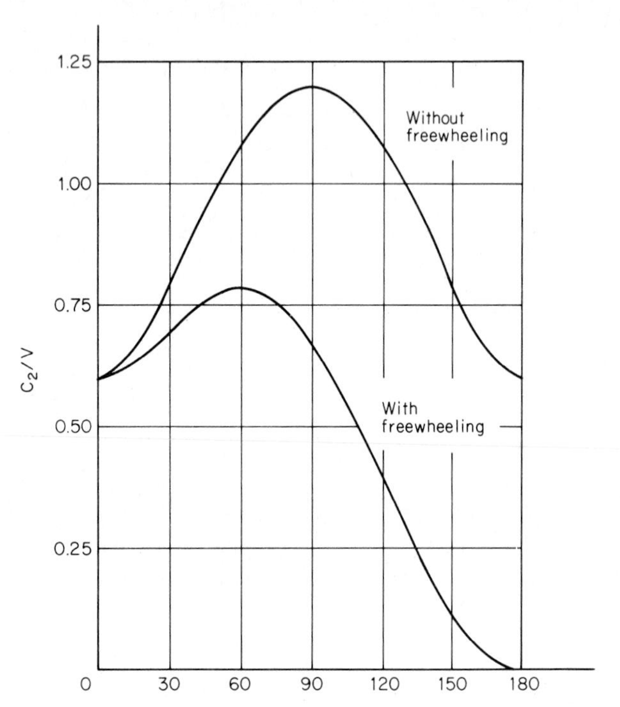


**FIG. 8.20** Variation of $c_2/V$ as a function of $\alpha$ for single-phase rectifier drives.

$$I_p = \frac{N_s}{N_p}\bar{i}_a \qquad \text{A} \tag{8.36}$$

The power input to the rectifier is

$$P_{\text{in}} = \frac{2\sqrt{2}}{\pi}\frac{N_s}{N_p}\bar{i}_a V_p \cos\alpha = R_a I_R^2 + k\phi\Omega_m \bar{i}_a \qquad \text{W} \tag{8.37}$$

and the apparent power factor (pf) at the rectifier ac terminals is

$$\text{pf} = 0.9\cos\alpha \tag{8.38}$$

For rectifiers with freewheeling (Fig. 8.18),

$$I_p = \frac{N_s}{N_p}\left[\frac{\pi-\alpha}{\pi}\right]^{1/2}\bar{i}_a \qquad \text{A} \tag{8.39}$$

$$P_{\text{in}} = \frac{2\sqrt{2}}{\pi}\frac{N_s}{N_p}\bar{i}_a V_p \cos^2\frac{\alpha}{2} = R_a I_R^2 + k\phi\Omega_m \bar{i}_a \qquad \text{W} \tag{8.40}$$

$$\text{pf} = \left[\frac{2}{\pi(\pi-\alpha)}\right]^{1/2}(1+\cos\alpha) \tag{8.41}$$

## 8.7 THREE-PHASE RECTIFIER DRIVES

The circuit of the three-phase bridge rectifier, the most frequently employed in motor control systems, is shown in Fig. 8.21*a*. A half-controlled configuration, with three thyristors replaced by diodes, is possible, but it has the disadvantage of introducing even harmonics into the line-current waveforms and is therefore unsuitable for large-power applications. The three ac sources whose voltage waveforms are shown in Fig. 8.21*b* may represent the terminal voltages of the secondary windings of a three-phase transformer. If the terminal voltage of the ac supply and the rated voltage of the motor are compatible, no transformer need be installed, unless isolation is required.

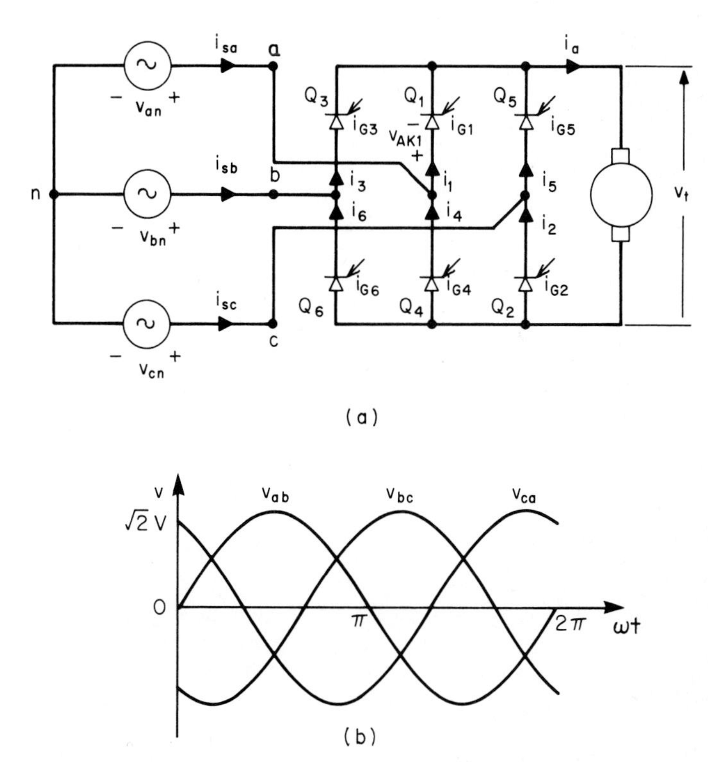


**FIG. 8.21** Three-phase fully controlled bridge rectifier drive: ***(a)*** circuit and ***(b)*** three-phase voltages.

### Fully Controlled Rectifier Drives

The thyristors of Fig. 8.21*a* are turned on in the sequence in which they are numbered, two thyristors necessarily conducting at any instant. Since the gating of each thyristor initiates a pulse of load current, this is a six-pulse rectifier. The waveforms of the circuit variables for continuous-current operation in the first quadrant of Fig. 8.6 are shown in Fig. 8.22. This system can also operate with discontinuous current, but only over a small range of low torque. With

$$v_{ab} = \sqrt{2}V \sin \omega t \qquad \text{V} \tag{8.42}$$

the delay angle for the reference thyristor $Q_1$ is now measured from the point $\omega t = \pi/3$ rad in accordance with the following definition: *The delay angle is zero when the rectifier, supplying a purely resistive load circuit without emf, delivers the maximum load current.* This means that $\alpha = 0$ when the controlled rectifier operates as if it were a diode rectifier.

The system of Fig. 8.21*a* can also operate in the fourth quadrant of Fig. 8.6, with the rectifier operating as a fixed-frequency inverter, as shown in Fig. 8.23. Thus, two of these rectifiers, connected as a dual converter, provide a four-quadrant drive with regeneration in the second and fourth quadrants. Once again, the permissible upper limit to $\alpha$ is given by Eq. (8.28).

For continuous-current operation,

$$\bar{v}_t = \frac{3\sqrt{2}V}{\pi} \cos \alpha \qquad \text{V} \tag{8.43}$$

Eqs. (8.19) and (8.21) apply, and

$$\Omega_m = \frac{3\sqrt{2}V}{\pi k\phi} \cos \alpha - \frac{R_a \bar{T}}{(k\phi)^2} \qquad \text{rad/s} \tag{8.44}$$

Equation (8.15) may be applied to discontinuous-current operation provided that $\alpha$ is replaced by $\alpha + \pi/3$ and $\beta$ by $\alpha + 2\pi/3$. This gives

$$\frac{\sqrt{2}V}{2} \sin(\beta - \varphi) - \frac{k\phi\Omega_m}{R_a} + \left[\frac{k\phi\Omega_m}{R_a} - \frac{\sqrt{2}V}{Z} \sin\left(\alpha + \frac{\pi}{3} - \varphi\right)\right] e^{(\alpha + \frac{\pi}{3} - \beta)/\tan\varphi} = 0 \tag{8.45}$$

For a given set of values of $V$, $L_a$, $R_a$, $k\phi$, $\Omega_m$, and $\alpha$, angle $\beta$ may be obtained numerically from Eq. (8.44), and Eq. (8.19) applies, where

$$\bar{v}_t = \frac{3}{\pi}\left\{k\phi\Omega_m\left(\frac{2\pi}{3} + \alpha - \beta\right) - \sqrt{2}V\left[\cos\beta - \cos\left(\alpha + \frac{\pi}{3}\right)\right]\right\} \qquad \text{V} \tag{8.46}$$

Equations (8.19) and (8.21) then yield the corresponding torque.

A family of curves of per-unit speed versus per-unit average internal torque for a 25-hp (18.65-kW) 230-V dc motor is shown in Fig. 8.24. The rectifier was supplied with 208 V line to line at 60 Hz.

## Rectifier Drives with Freewheeling

If a diode is connected across the motor armature of Fig. 8.21 in such a direction as to permit the armature inductance to prolong the flow of current in the armature, then freewheeling is obtained. In this circuit, $v_t$ cannot have a negative value, and the operation of the system is confined to the first quadrant of Fig. 8.6.

If the armature current is continuous and $v_t > 0$ throughout the cycle, no diode current flows and the operation of the system is the same as it would be without

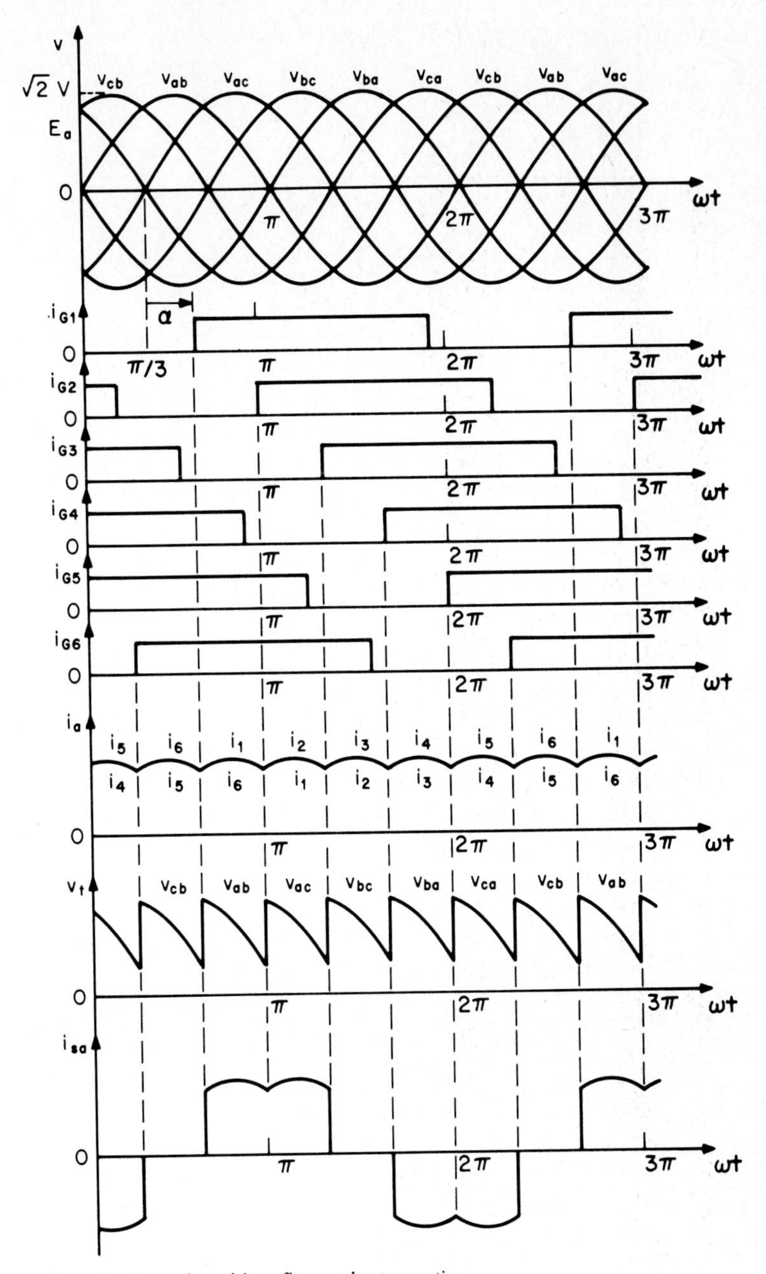


**FIG. 8.22** Three-phase drive—first-quadrant operation.

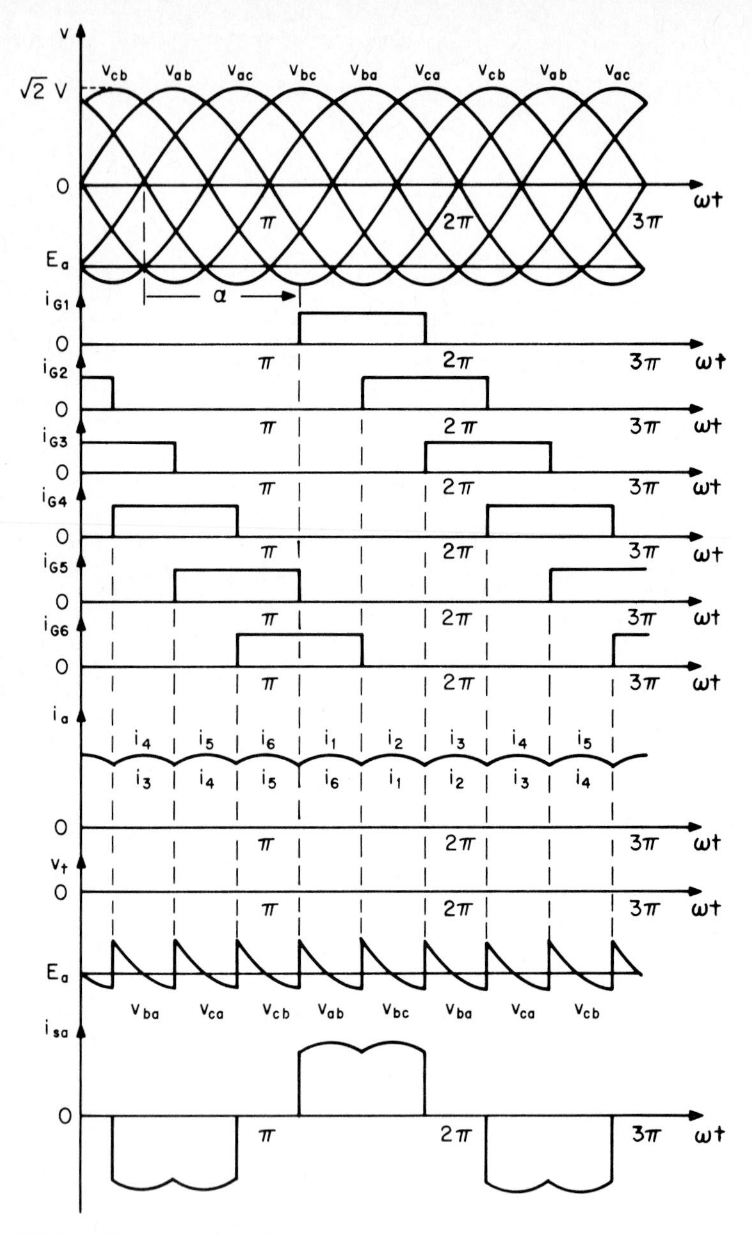


**FIG. 8.23** Three-phase drive—fourth-quadrant operation.

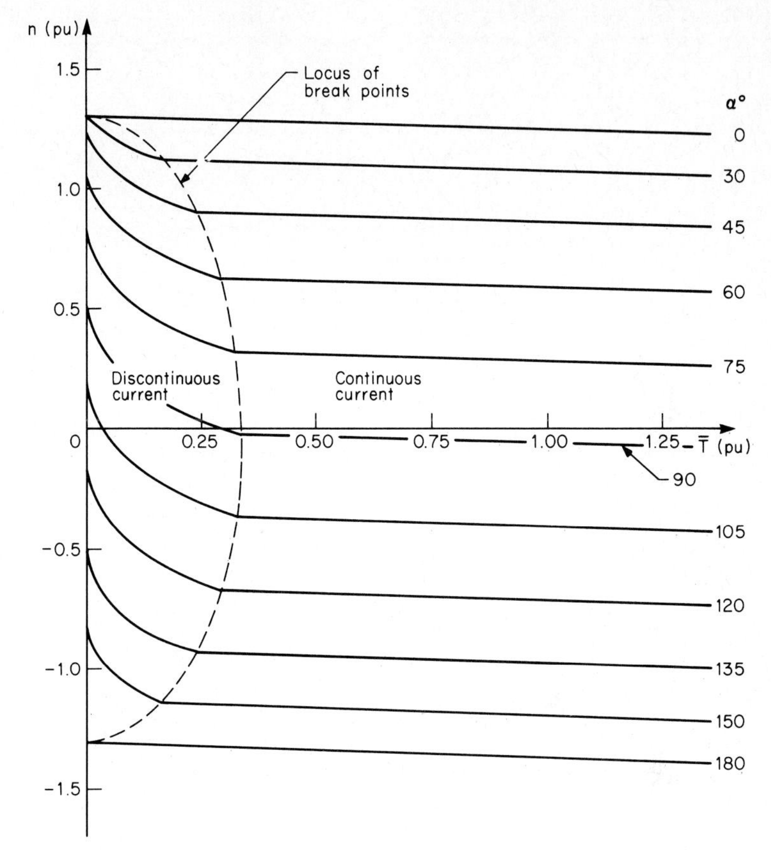


**FIG. 8.24** Speed-torque characteristics for a three-phase rectifier drive.

the diode. These conditions occur for $\alpha < \pi/3$. However, if $\alpha > \pi/3$, then $v_t$ and $v_D$ become zero for interval $\pi < \omega t < \alpha + 2\pi/3$ and diode current flows, as shown in Fig. 8.25. It can also be seen from Fig. 8.25 that $\bar{v}_t = 0$ when $\alpha = 2\pi/3$. Thus, for operation with freewheeling,

$$\bar{v}_t = \frac{3\sqrt{2}V}{\pi}\left[1 + \cos\left(\alpha + \frac{\pi}{3}\right)\right] \quad \text{V} \tag{8.47}$$

Equations (8.19) and (8.21) apply to this system also, so that

$$\Omega_m = \frac{3\sqrt{2}V}{\pi k\phi}\left[1 + \cos\left(\alpha + \frac{\pi}{3}\right)\right] - \frac{R_a\bar{T}}{(k\phi)^2} \quad \text{rad/s} \tag{8.48}$$

The speed-torque curves for discontinuous current may be obtained by a procedure similar to that yielding those for the fully controlled rectifier in Fig. 8.24. This has been done for the system of Fig. 8.21 with a freewheeling diode added to the rectifier circuit. The resulting curves are shown in Fig. 8.26. The reduction in breakpoint

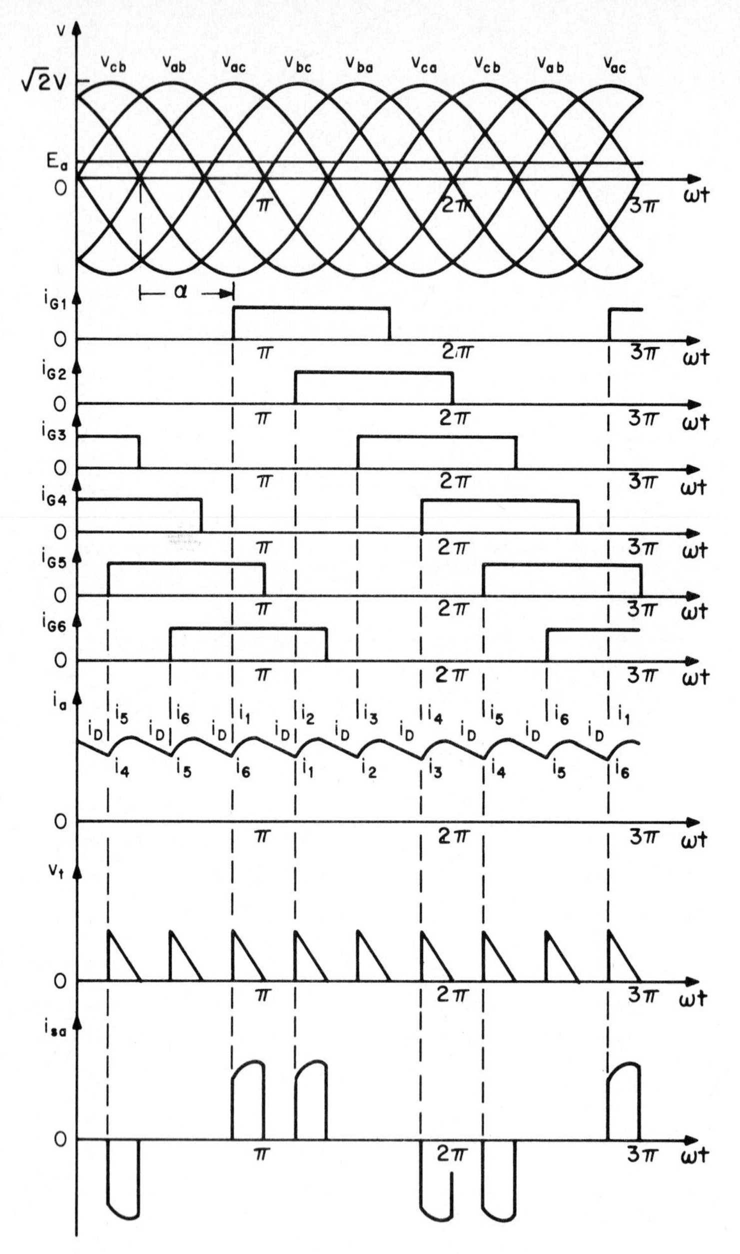


FIG. 8.25 Waveforms of the variables in a three-phase system with freewheeling.

torque at low speed may suit many load speed-torque characteristics and result in continuous-current operation over the entire speed range.

The effect of the prolonged gating signals shown in Fig. 8.27 is to provide freewheeling paths through two thyristors, so that current no longer flows in the source line when $v_t$ tends to go negative, and the negative parts of the $v_t$ waveform are eliminated. The waveform of $i_{sa}$ is identical with that shown in Fig. 8.25. As for the rectifier with a freewheeling diode, freewheeling does not occur for $0 < \alpha < \pi/3$. Thus, for first-quadrant operation, $\bar{v}_t$ is expressed as a function of $\alpha$ by Eqs. (8.43) and (8.47), except that for the latter the range of $\alpha$ must be slightly reduced to $\pi/3 < \alpha < (2\pi/3 - \omega t_{\text{off}} - \mu)$.

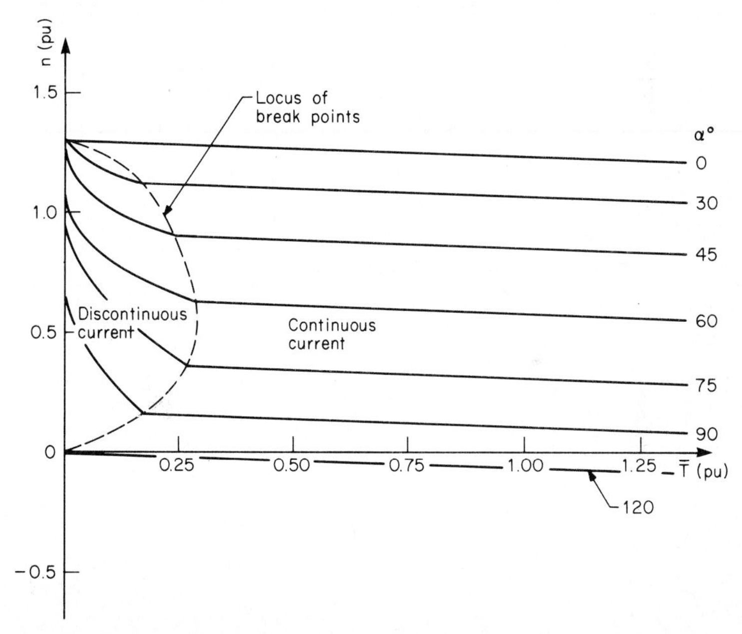


**FIG. 8.26** Speed-torque characteristics for a three-phase rectifier drive with freewheeling.

Operation with freewheeling in the fourth quadrant is illustrated in Fig. 8.28. The overall length of the gating signals is maintained constant at $(4\pi/3 + \omega t_{\text{off}} + \mu)$, but a break is introduced commencing at $\pi/3$ from the end of each gating signal. Delay angle $\alpha$ now signifies the interval from $\omega t = \pi/3$ to the start of the second part of the $i_{G1}$ signal. The length of the first part remains constant at $(\pi + \omega t_{\text{off}} + \mu)$. If the brief interval when $v_t > 0$ is ignored, then

$$\bar{v}_t = \frac{3\sqrt{2}V}{\pi}\left[\cos\left(\alpha + \frac{\pi}{3}\right) - 1\right] \quad \text{V} \qquad \frac{5\pi}{3} < \alpha < 2\pi \tag{8.49}$$

Operation in the fourth quadrant without freewheeling has been illustrated in Fig. 8.23, and Eq. (8.43) applies over the range $2\pi/3 < \alpha < (\pi - \omega t_{\text{off}} - \mu)$.

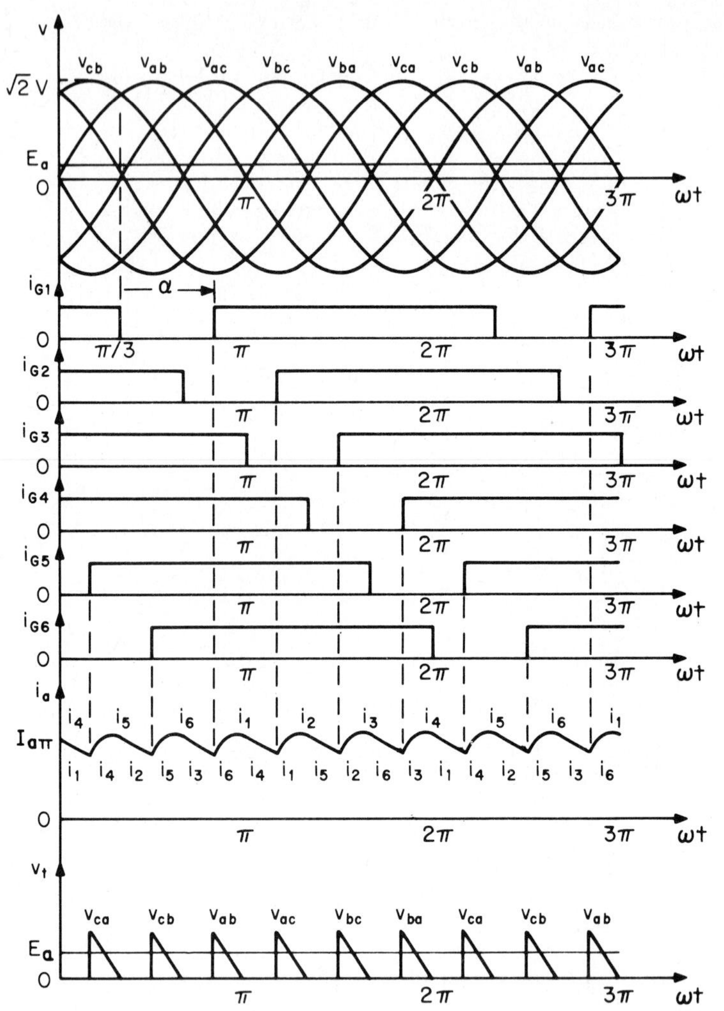


**FIG. 8.27** Three-phase drive with regeneration and freewheeling—first-quadrant operation.

## Power in Load and Source Circuits

For continuous-current operation, the assumptions made are the same as the four listed under "Power in Load and Source Circuits" in Sec. 8.6, except that now the current harmonic of lowest frequency is the sixth, and all others may be ignored.

The amplitude $c_6$ of the sixth harmonic of $v_t$ varies with $\alpha$, as shown in Fig. 8.29. The rms sixth-harmonic current is

$$I_{R6} = \frac{c_6}{6\sqrt{2}\omega L_a} \qquad \text{A} \tag{8.50}$$

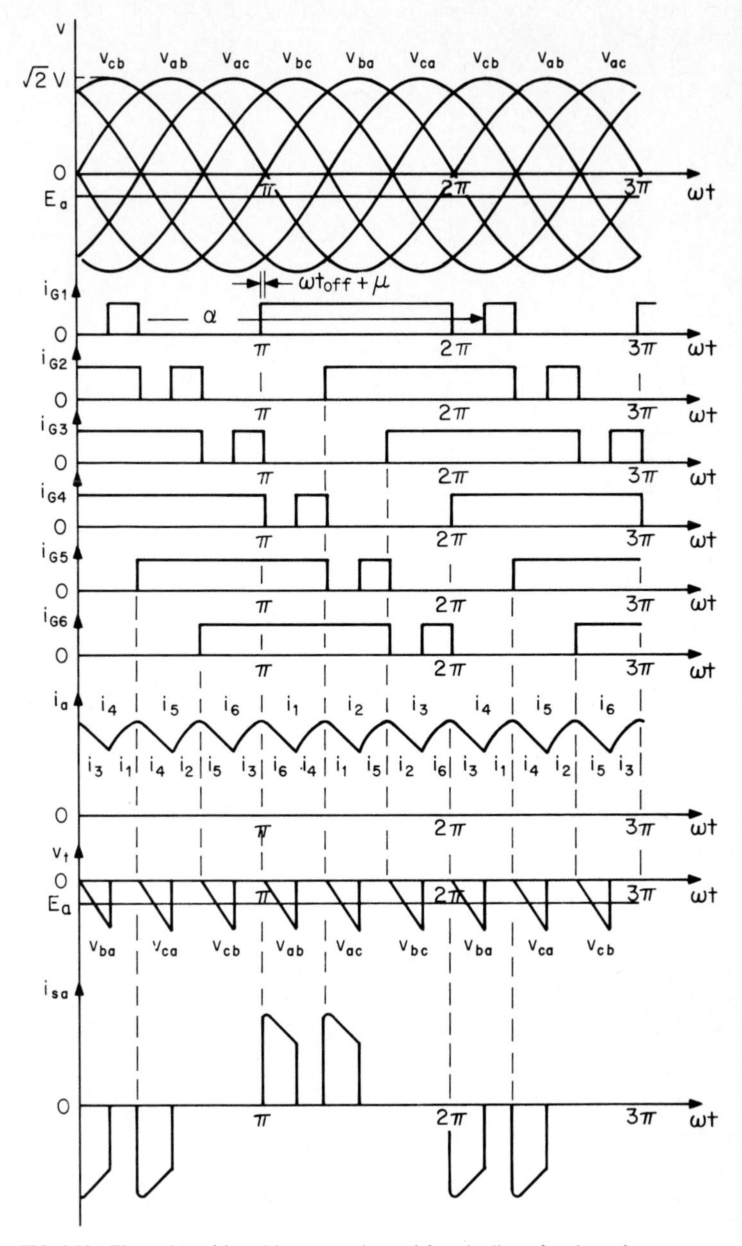


**FIG. 8.28** Three-phase drive with regeneration and freewheeling—fourth-quadrant operation.

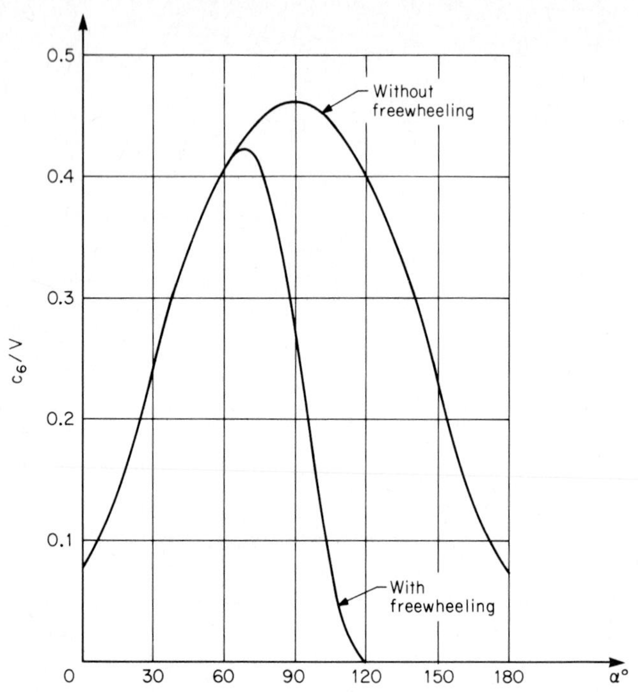


**FIG. 8.29** Variation of $c_6/V$ as a function of $\alpha$ for three-phase bridge rectifier drives.

and the rms armature current is

$$I_R = [(\bar{i}_a)^2 + I_{R6}^2]^{1/2} \qquad \text{A} \tag{8.51}$$

For rectifiers without freewheeling (Fig. 8.22), the rms magnitude of the approximate waveform of $i_{sa}$ is

$$I_{sa} = \sqrt{\frac{2}{3}}\,\bar{i}_a \qquad \text{A} \tag{8.52}$$

The power input to the rectifier is

$$P_{\text{in}} = \frac{3\sqrt{2}}{\pi} V \bar{i}_a \cos\alpha = R_a I_R^2 + k\phi\Omega_m \bar{i}_a \qquad \text{W} \tag{8.53}$$

and the apparent power factor at the rectifier ac terminals is

$$\text{pf} = 0.955 \cos\alpha \tag{8.54}$$

For rectifiers with freewheeling (Fig. 8.27),

$$I_{sa} = \left[\frac{2}{\pi}\left(\frac{2\pi}{3} - \alpha\right)\right]^{1/2} \bar{i}_a \qquad \text{A} \tag{8.55}$$

$$P_{\text{in}} = \frac{3\sqrt{2}V\bar{i}_a}{\pi}\left[1+\cos\left(\alpha+\frac{\pi}{3}\right)\right] = R_a I_R^2 + k\phi\Omega_m\bar{i}_a \qquad \text{W} \tag{8.56}$$

$$\text{pf} = \frac{\bar{v}_t\bar{i}_a}{\sqrt{3}V I_{sa}} \tag{8.57}$$

## *8.8 CHOPPER (DC-TO-DC CONVERTER) DRIVES*

A chopper is supplied from a dc source, which may possibly be a rectifier. Choppers may be designed to operate with a dc motor in one, two, or four quadrants of the diagram of Fig. 8.6. They may be classified according to the number of quadrants of the $\bar{v}_t - \bar{i}_a$ diagram in which they are capable of operating, and a convenient classification is shown in Fig. 8.30.

### Class A Chopper

The basic circuit is shown in Fig. 8.31*a*, where the thyristor requires forced commutation. This is a *step-down* chopper, since the average terminal voltage of the armature circuit is less than (or equal to) that of the source. The operation of the circuit is shown for discontinuous- and continuous-current operation by the waveforms in Figs. 8.31*b* and *c*, respectively. Equations (8.19) and (8.21) apply to this system, and

$$\bar{v}_t = \frac{t_{\text{on}}V}{T_p} \qquad \text{V} \tag{8.58}$$

The limit to $\bar{v}_t$ is reached when $t_{\text{on}} = T_p$. The operation of a chopper-driven motor thus corresponds to the speed-torque curves shown in Fig. 8.4, when $V_t$ is replaced by $\bar{v}_t$. When $t_{\text{on}} = T_p$, operation by field weakening is possible, as illustrated in Fig. 8.5.

Since discontinuous-current operation occurs only at low torque, only continuous-current operation will be discussed in this and following sections.

The degree of cyclic variation of $i_a$ is governed by the time constant of the armature circuit

$$\tau_a = \frac{L_a}{R_a} \qquad \text{s} \tag{8.59}$$

and the upper and lower limits of $i_a$ are

$$I_{a1} = \frac{V}{R_a}\,\frac{1-e^{-t_{\text{on}}/\tau_a}}{1-e^{-T_p/\tau_a}} - \frac{E_a}{R_a} \qquad \text{A} \tag{8.60}$$

$$I_{a2} = \frac{V}{R_a}\,\frac{e^{t_{\text{on}}/\tau_a}-1}{e^{T_p/\tau_a}-1} - \frac{E_a}{R_a} \qquad \text{A} \tag{8.61}$$

The frequency of the fundamental harmonic of $i_a$ is equal to the chopping frequency $\omega_o$, and this is made as high as practicable (ca. 500 Hz) to reduce interference with communication circuits. In addition, a filter is fitted at the chopper input; the filter usually consists of a series inductance in one line and a parallel capacitance across the chopper input terminals. Harmonics can be further reduced by operating two or more choppers with staggered pulses from a common filter.

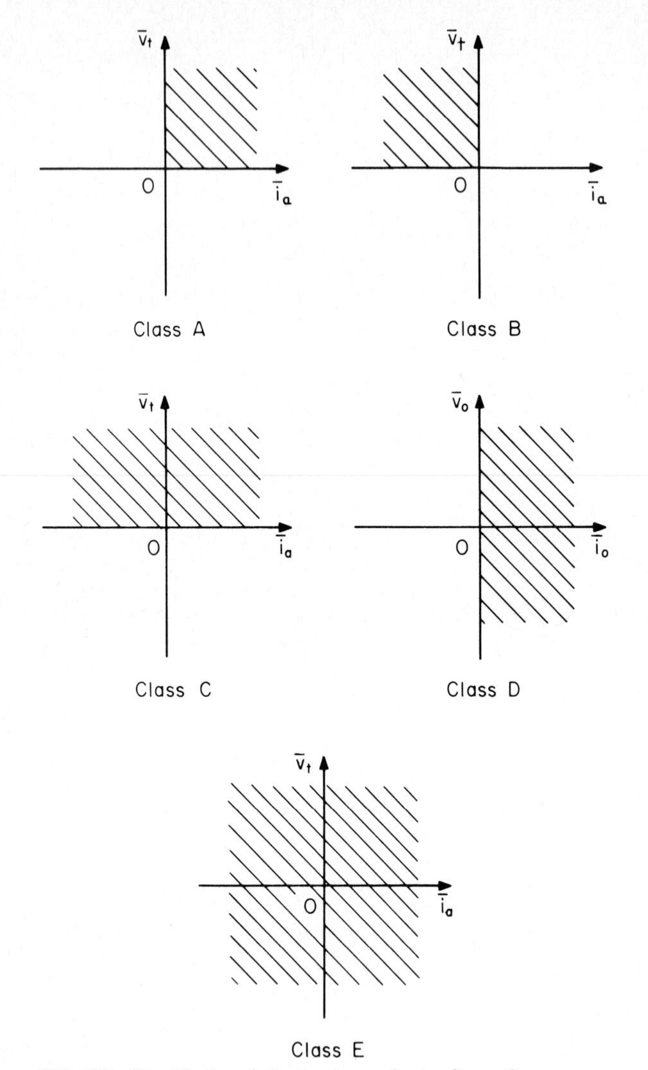


**FIG. 8.30** Classification of choppers by quadrants of operation.

In the following, only the fundamental component of $i_a$ will be taken into consideration. This is

$$i_{a1} = \frac{2V}{\pi \omega_o L_a} \sin \frac{\omega_o t_{on}}{2} \sin \left( \omega_o t - \frac{\omega_o t_{on}}{2} \right) \quad \text{A} \tag{8.62}$$

of which the rms magnitude is

$$I_{R1} = \frac{\sqrt{2}V}{\pi \omega_o L_a} \sin \frac{\omega_o t_{on}}{2} \quad \text{A} \tag{8.63}$$

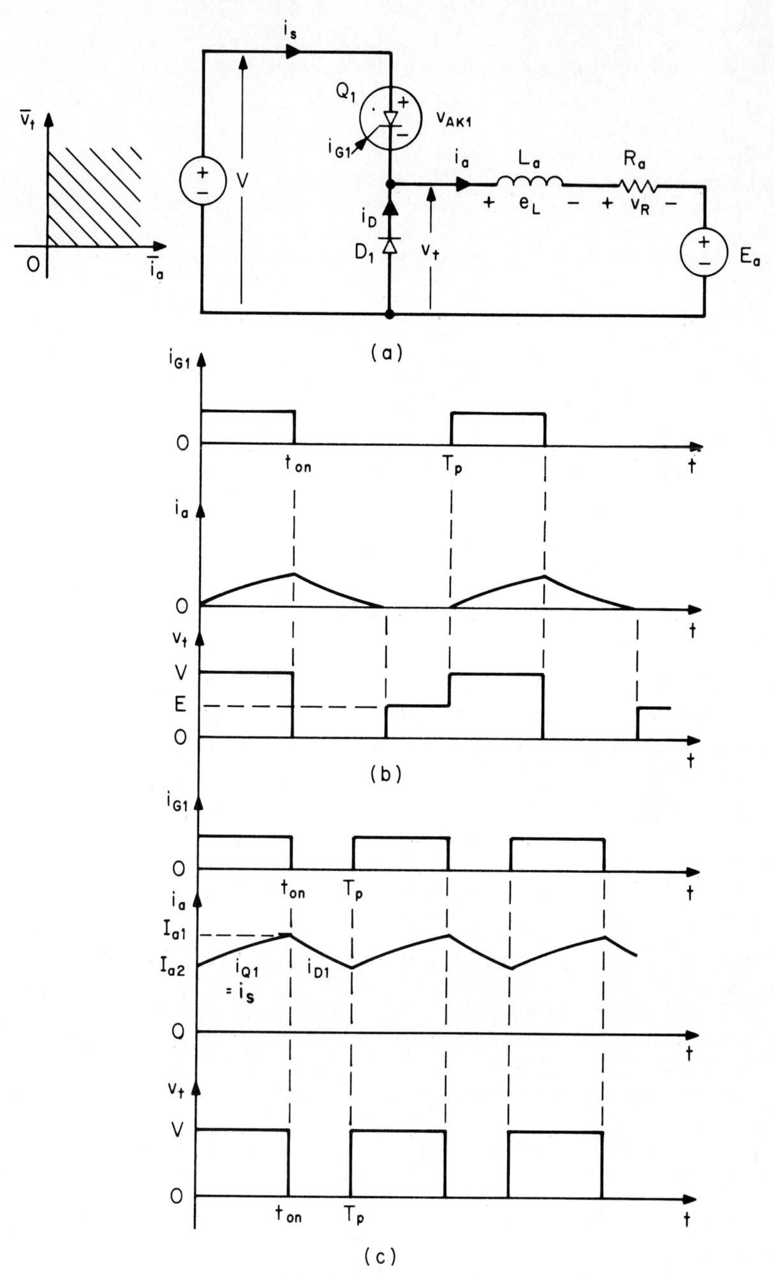


**FIG. 8.31** Basic principles of a class A step-down single-quadrant chopper: ***(a)*** chopper circuit; ***(b)*** and ***(c)*** current and voltage waveforms.

and, to a good approximation, the rms armature current is

$$I_R = [(\bar{i}_a)^2 + I_{R1}^2]^{1/2} \qquad \text{A} \tag{8.64}$$

The harmonics of the waveform of the line current $i_s$ may be obtained to a good approximation by assuming that, since $\omega_o L_a >> R_a$, the armature current is constant at magnitude $\bar{i}_a$. The source current is thus assumed to be a series of pulses of magnitude $\bar{i}_a$, duration $t_{\text{on}}$, and periodic time $T_p$. If once more only the fundamental harmonic component is taken into account, then

$$i_s = \frac{t_{\text{on}}}{T_p}\bar{i}_a + \frac{2\bar{i}_a}{\pi}\sin\frac{\omega_o t_{\text{on}}}{2}\sin\left(\omega_o t + \frac{\omega_o t_{\text{on}}}{2}\right) \qquad \text{A} \tag{8.65}$$

of which the rms magnitude is

$$I_s = \bar{i}_a\left[\left(\frac{t_{\text{on}}}{T_p}\right)^2 + \left(\frac{\sqrt{2}}{\pi}\sin\frac{\omega_o t_{\text{on}}}{2}\right)^2\right]^{1/2} \qquad \text{A} \tag{8.66}$$

The power input to the chopper or motor is then

$$P_{\text{in}} = \frac{t_{\text{on}}}{T_p}V\bar{i}_a = \bar{v}_t\bar{i}_a = R_a I_R^2 + k\phi\Omega_m\bar{i}_a \qquad \text{W} \tag{8.67}$$

## Class B Chopper

A class B chopper steps up the terminal voltage of a regenerating motor to feed energy back to the dc source. In a system in which a switching interval between driving the motor by class A action and regenerating is acceptable, the components of the class A circuit may be rearranged to form the class B chopper. The firing circuit is the same for both modes of operation. The basic circuit is shown in Fig. 8.32*a*. The symbols and reference directions of the variables are the same as in Fig. 8.31, and thus the system operates in the second quadrant of the $\bar{v}_t - \bar{i}_a$ diagram. Only continuous-current operation will be considered.

If thyristor $Q_2$ is never turned on and $V > E_a$, then the circuit is completely inactive. If $Q_2$ is turned on and commutated during regular intervals of period $T_p$, then emf $E_a$ stores energy in inductance $L_a$ while $Q_2$ is conducting, and part of that stored energy is delivered to source $V$ by current through diode $D_2$ when $Q_2$ is commutated. The operation of the system is illustrated by the waveforms in Fig. 8.32*b*. The interval during which $D_2$ conducts is designated $t_{\text{on}}$ for this chopper, and a cycle of operation commences at $t = 0$ when $Q_2$ is commutated.

Equations (8.58) to (8.67) may be applied to this chopper if it is borne in mind that currents $i_a$ and $i_s$ are always negative.

## Class C Two-Quadrant Chopper

If switching from class A to class B configuration is not an acceptable method of obtaining regenerative braking, then a smooth transition from class A to class B operation, or vice versa, may be obtained by combining the circuits of Figs. 8.31 and 8.32 to give that of Fig. 8.33*a*. Here the thyristors are turned on and commutated alternately. A short interval (about 100 $\mu$s) is allowed to elapse between the

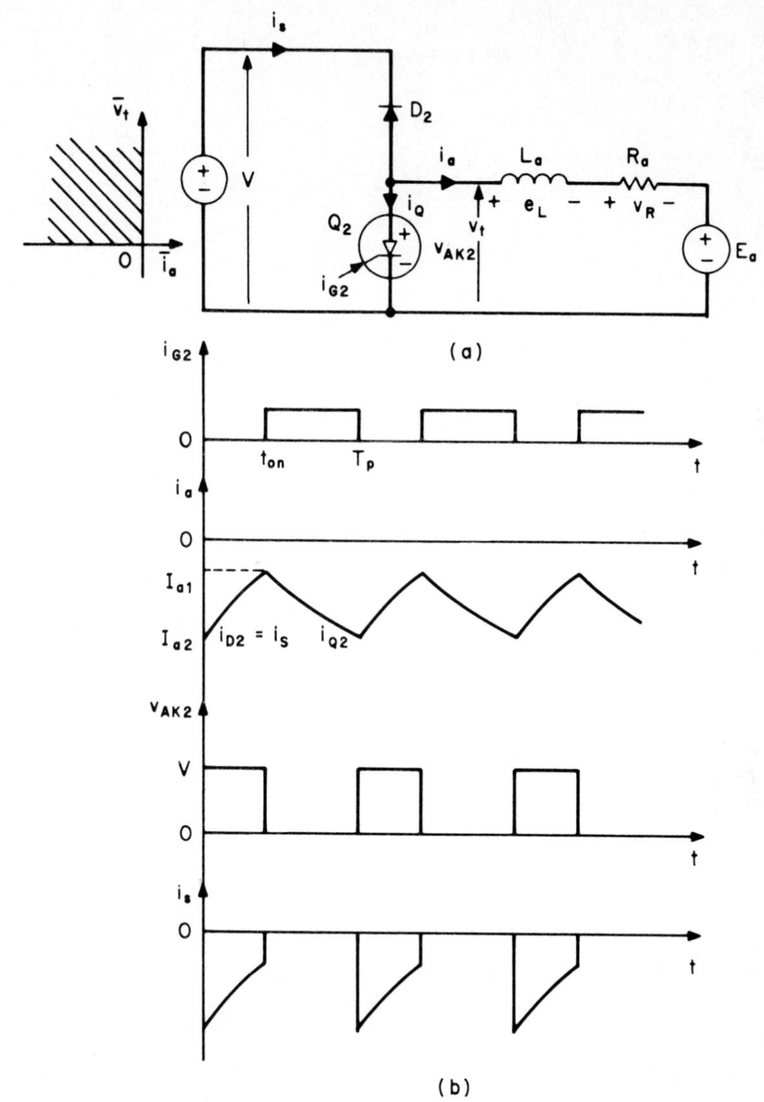


**FIG. 8.32** Basic principles of a class B step-up single-quadrant chopper: ***(a)*** circuit and ***(b)*** waveforms of various currents and the terminal voltage.

commutation of one thyristor and the gating of the other so that the dc source may not be short-circuited.

If the conditions of loading are such that either the class A or the class B chopper would operate with discontinuous current, then the cycle illustrated in Fig. 8.33*b* arises. Equations (8.58) to (8.67) may be applied to this chopper. The quadrant of operation is determined by means of Eq. (8.58).

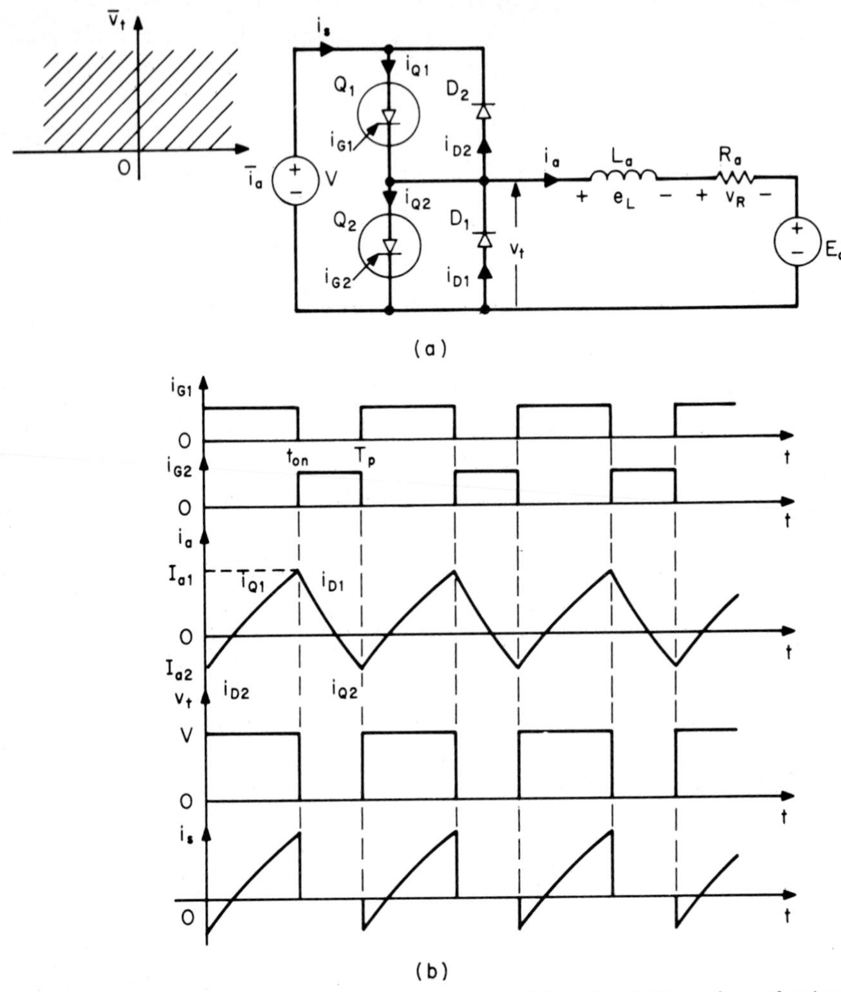


**FIG. 8.33** Basic principles of a class C two-quadrant chopper: *(a)* circuit and *(b)* waveforms of various currents and the terminal voltage.

## Class D Two-Quadrant Chopper

The power circuit of this chopper is shown in Fig. 8.34. There is no advantage in employing it as a source for a dc motor armature, since first- and second-quadrant or four-quadrant operation is required for that purpose. It is useful for controlling the field current of a dc or synchronous machine, however, for when a rapid change of that current is required, it can short-circuit the field winding to reduce the current or force the field to increase the current. It is also of use as a source for other converters, notably the current-source inverter. Thus, although a source of emf is included in the load circuit of Fig. 8.34, there may not be such an emf present.

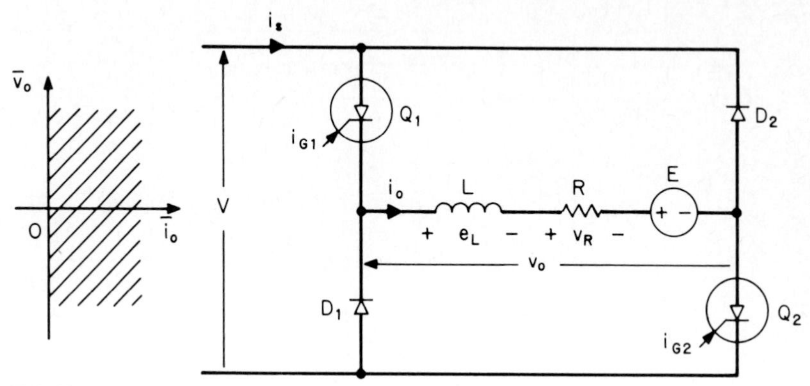


**FIG. 8.34** Power circuit of a class D two-quadrant chopper.

The operation of the circuit is illustrated in Fig. 8.35. If both thyristors are turned on, the load current approaches the magnitude

$$i_o = \frac{V - E}{R} \qquad \text{A} \tag{8.68}$$

If only one thyristor is turned on, the load circuit is shorted through that thyristor and a diode. The load current then approaches zero. Only continuous current in the two modes of operation will be considered.

For steady-state operation in mode 1, where $t_a < T_p/2$, it is necessary that $V > E$, where possibly $E = 0$. The current varies between the limits

$$I_{o1} = \frac{V}{R} \frac{1 - e^{-(T_p/2 - t_a)/\tau}}{1 - e^{-T_p/2\tau}} - \frac{E}{R} \qquad \text{A} \tag{8.69}$$

$$I_{o2} = \frac{V}{R} \frac{e^{(T_p/2 - t_a)/\tau} - 1}{e^{T_p/2\tau} - 1} - \frac{E}{R} \qquad \text{A} \tag{8.70}$$

where

$$\tau = \frac{L}{R} \qquad s \tag{8.71}$$

For steady-state operation in mode 2, where $T_p/2 < t_\alpha < T_p$, it is necessary that $E < 0$ and $-E > V$; thus, a source of emf in the load circuit is required. The current varies between the limits

$$I_{o1} = -\frac{V}{R} \frac{e^{(t_\alpha - T_p 2)/\tau} - 1}{e^{T_p/2\tau} - 1} - \frac{E}{R} \qquad \text{A} \tag{8.72}$$

$$I_{o2} = -\frac{V}{R} \frac{1 - e^{-(t_\alpha - T_p/2)/\tau}}{1 - e^{-T_p/2\tau}} - \frac{E}{R} \qquad \text{A} \tag{8.73}$$

Since one cycle of $v_o$ takes place in interval $T_p/2$ s, the angular frequency is

$$\omega_o = \frac{4\pi}{T_p} \qquad \text{rad/s} \tag{8.74}$$

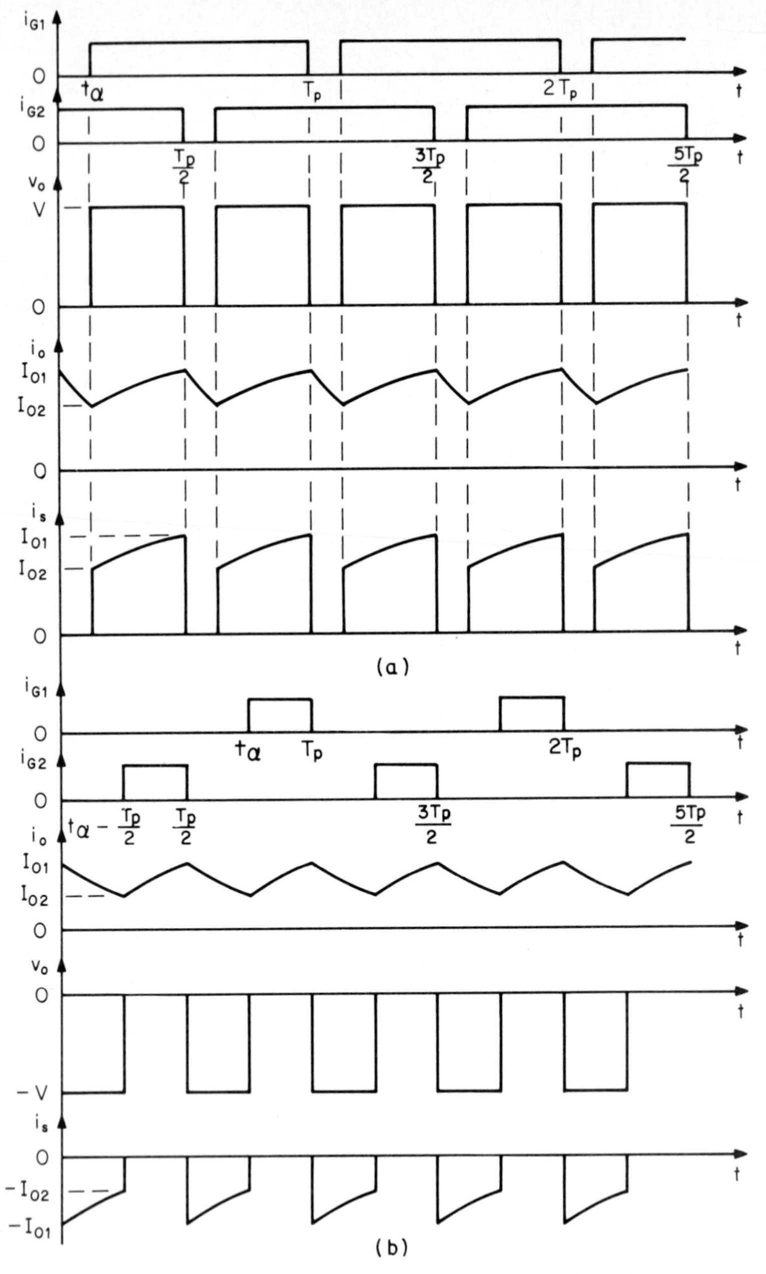


**FIG. 8.35** Waveforms of the variables in Fig. 8.34: ***(a)*** $0 < t_\alpha < T_P/2$ ; ***(b)*** $T_P/2 < t_\alpha < T_P$.

The average value of $v_o$ is

$$\bar{v}_o = \left[1 - \frac{2t_\alpha}{T_p}\right]V \qquad \text{V} \tag{8.75}$$

and the average load current is

$$\bar{i}_o = \frac{\bar{v}_o - E}{R} \qquad \text{A} \tag{8.76}$$

The rms magnitude of the fundamental harmonic component of the load current is

$$I_{R1} = \frac{2\sqrt{2}V}{\pi\omega_o L} \sin \frac{2\pi t_\alpha}{T_p} \qquad \text{A} \tag{8.77}$$

so that, to a good approximation, the rms load current is

$$I_R = \left[(\bar{i}_o)^2 + I_{R1}^2\right]^{1/2} \qquad \text{A} \tag{8.78}$$

On the assumption that $\omega_o L >> R$ and therefore that the waveform of source current $i_s$ consists of pulses of magnitude $\bar{i}_o$, the average source current is

$$\bar{i}_s = \left[1 - \frac{2t_\alpha}{T_p}\right]\bar{i}_o \qquad \text{A} \tag{8.79}$$

and the load power is

$$P_{\text{in}} = \bar{v}_o \bar{i}_o = V\bar{i}_s \qquad \text{W} \tag{8.80}$$

### Class E Four-Quadrant Chopper

The power circuit of this chopper is shown in Fig. 8.36. If thyristor $Q_4$ is turned on continuously, then the antiparallel-connected pair of devices $Q_4$ and $D_3$ constitutes a short circuit. Thyristor $Q_3$ may not be turned on at the same time as thyristor $Q_4$, because that would short-circuit source $V$. Moreover, since under these conditions the terminal voltage of diode $D_4$ is always negative, the pair of devices $Q_3$ and $D_4$ is equivalent to an open circuit. Continuous turn-on of $Q_4$ and inhibition of $Q_3$ thus produces a circuit equivalent to that of Fig. 8.33, and operation in the first and second quadrants is possible. Continuous turn-on of $Q_3$ and inhibition of $Q_4$ permits operation in the other two quadrants.

## *8.9 THREE-PHASE INDUCTION MOTORS*

Until recently, induction motors had been little used for speed control, since with the classical methods employed they became increasingly inefficient as speed was reduced below the synchronous value. They are now employed efficiently with a number of different converters according to the requirements of the drive system and the available power source.

It is convenient to employ a variety of models of the induction motor for prediction of system performance. All of these models are derived from the basic per-phase model [Fig. 4.7*a* (see Chap. 4)]. For this model,

$$\omega_r = \omega_s - \frac{p}{2}\omega_m \qquad \text{rad/s} \tag{8.81}$$

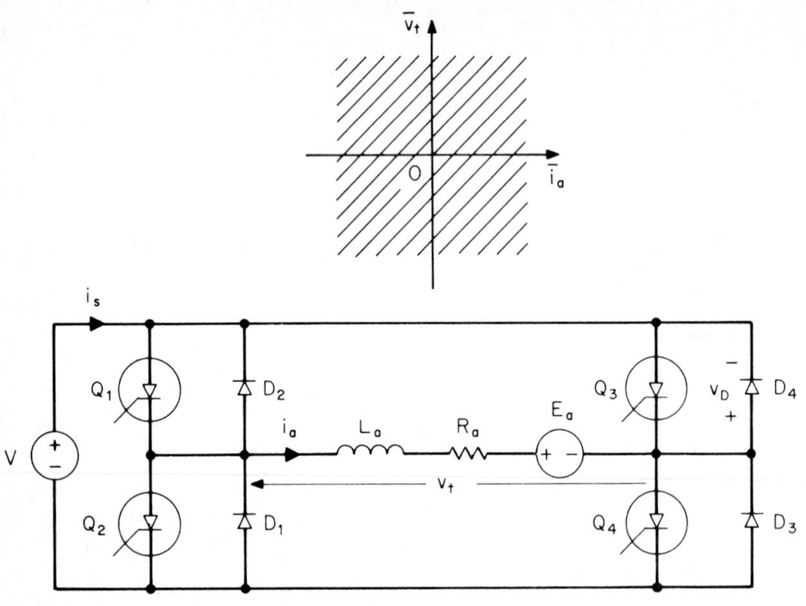


**FIG. 8.36** Power circuit of a class E four-quadrant chopper.

where $\omega_r$ = angular frequency of rotor currents
$\omega_s$ = angular frequency of stator currents
$\omega_m$ = angular speed of rotation
$p$ = number of poles

Slip $s$ is defined by

$$s = \frac{\omega_r}{\omega_s} = \frac{\omega_s - (p/2)\omega_m}{\omega_s} \quad \text{per unit} \tag{8.82}$$

and the ratio of per-phase power in the rotor and stator circuits is

$$\frac{P_{mA}}{P_{ma}} = s = \frac{\text{per-phase power in rotor}}{\text{per-phase power in stator}} \tag{8.83}$$

The three-phase power converted to mechanical form is

$$P_{\text{mech}} = 3(P_{ma} - P_{mA}) = 3(1-s)P_{ma} = 3\frac{p}{2}\frac{\omega_m}{\omega_s}P_{ma} \quad \text{W} \tag{8.84}$$

The airgap torque exerted on the rotor is

$$T = \frac{P_{\text{mech}}}{\omega_m} = 3\frac{p}{2}\frac{P_{ma}}{\omega_s} \quad \text{N}\cdot\text{m} \tag{8.85}$$

In some systems employing wound-rotor motors, it is convenient to use an equivalent circuit incorporating the ideal machine. The turns ratio of the ideal machine, as opposed to the ratio of the actual numbers of turns on the windings, is given by the effective ratio $N_{se}/N_{re}$. In the circuit of Fig. 4.7*a* (with $R_c$ omitted),

$$P_{ma} = \frac{R'_r}{s}(I'_a)^2 = R'_r(I'_a)^2 + \frac{(1-s)}{s}R'_r(I'_a) \quad \text{W} \tag{8.86}$$

and $$P_{\text{mech}} = \frac{3(1-s)}{s} R'_r (I'_a)^2 \qquad \text{W} \tag{8.87}$$

For wound-rotor motors, as well as for class A and class D squirrel-cage motors, the equivalent circuit may be considered to have constant parameters at any speed or slip. Class B and class C squirrel-cage motors, on the other hand, are designed to have parameters that are functions of rotor frequency. In speed-control systems, such motors are frequently operated at or near rated rotor frequency at all speeds, so that equivalent-circuit parameters under rated operating conditions are required. (See also Chap. 4, "Induction Machines.")

## 8.10 AC POWER-CONTROLLER DRIVES

A range of speed control of a loaded induction motor can be obtained by varying its terminal voltage. For this purpose a class D squirrel-cage motor with a high-resistance rotor, or a wound-rotor motor with some fixed external resistance in the rotor circuit, must be employed. A family of speed-torque curves, such as shown in Fig. 8.37, can then be obtained. The intersections of the motor curves with the curve of $T_{\text{loss}} + T_L$ indicate stable points of operation for a series of values of per-phase terminal voltage $V_a$.

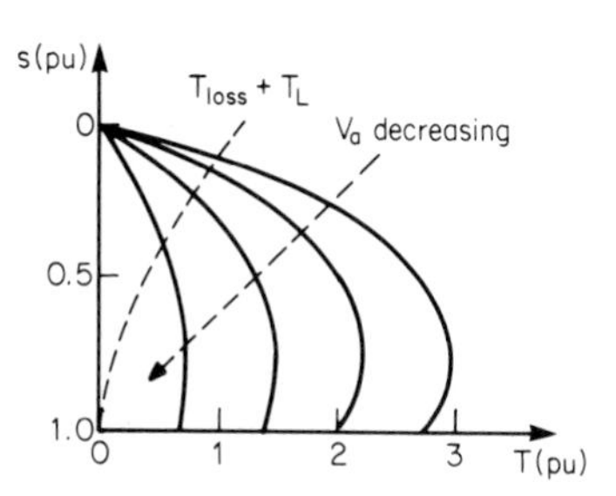


**FIG. 8.37** Speed control by variation of source voltage difference.

A circuit for a balanced three-phase power controller and motor are shown in Fig. 8.38. Waveforms of line-to-line supply voltages and thyristor gating signals for a particular setting of delay angle $\alpha$ are shown in Fig. 8.39.

Figure 8.40 shows the controller with the addition of two pairs of thyristors, which can be brought into operation to reverse the phase sequence of the motor excitation and provide braking by plugging or reverse driving. If the driven mechanism overhauls the motor, as in a vehicle running downhill, regenerative braking occurs; thus, for a particular value of $\alpha$, the four-quadrant relationship of Fig. 8.41 would be obtained. The speed-torque curves illustrated there would not correspond to those obtained with a three-phase source of constant sinusoidal voltage unless $\alpha$ were so small that the controller would be inoperative.

It can be shown that if $\alpha$ is equal to or less than the angle

$$\varphi = \tan^{-1} \frac{\omega_s L}{R} \qquad \text{rad} \tag{8.88}$$

where $\omega_s$ is the angular frequency of the source, while $L$ and $R$ are the per-phase inductance and resistance seen from the stator terminals of the motor, the system will operate as a normal three-phase motor drive with no thyristors in the lines. As each thyristor ceases to conduct current in one direction, its inverse-parallel connected partner begins to conduct current in the reverse direction. An increase in $\alpha$, or an increase in speed causing an increase in $R$ and a reduction in $\varphi$, disturbs this pattern of behavior and results in an interval between the turn-off of one thyristor and the turn-on of its partner.[1] Consequently, the line current flows in alternating pulses. Current harmonics are then present in the motor.

## Pump or Fan Drives

The ideal efficiency of an ideal induction motor is expressed by

$$\eta = 1 - s \qquad \text{per unit} \tag{8.89}$$

Speed control of a motor by a power controller is therefore suitable only when a narrow range of speeds at low slip is required. Drives for fans and centrifugal pumps are of this type, and for these, to a close approximation,

$$T_L = k\omega_m^2 \qquad \text{N} \cdot \text{m} \tag{8.90}$$

where $k$ is a constant. For such a drive, a speed range of 2:1 provides a range of throughput of 8:1. Thus, operation at low efficiency occurs at low power.

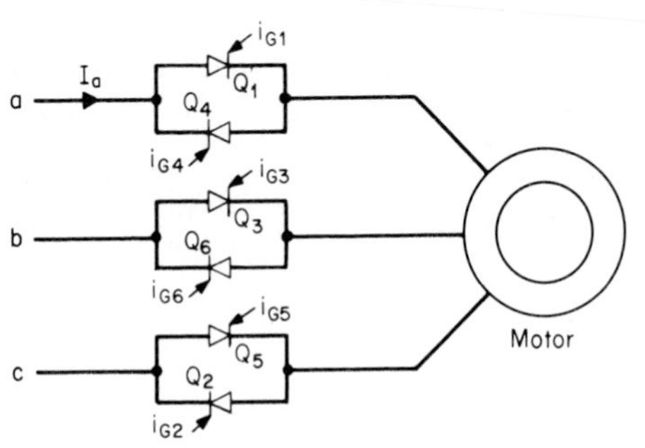


**FIG. 8.38** Symmetrical ac power controller.

The effect of harmonics in the system may be ignored to give approximate performance prediction, and this justifies the further approximation of omitting the magnetizing branch in the equivalent circuit. As a consequence,

$$I_a = I_a' \qquad \text{A} \tag{8.91}$$

and

$$P_{\text{mech}} = 3P_{ma} \qquad \text{W} \tag{8.92}$$

or

$$\frac{3(1-s)}{s} R_r' I_a^2 = \frac{2}{p}\omega_s(1-s)T \qquad \text{W} \tag{8.93}$$

from which

$$T = \frac{3p}{2} \frac{R_r' I_a^2}{s\omega_s} \qquad \text{N} \cdot \text{m} \tag{8.94}$$

It can be shown that maximum current occurs when $s = 1/3$ and that, for a motor with a full-load slip of 0.15, the maximum current is 1.17 times the rated current. Thus, allowing for the approximations made in reaching this conclusion, a derating factor of 0.75 to 0.8 should be applied to the motor.

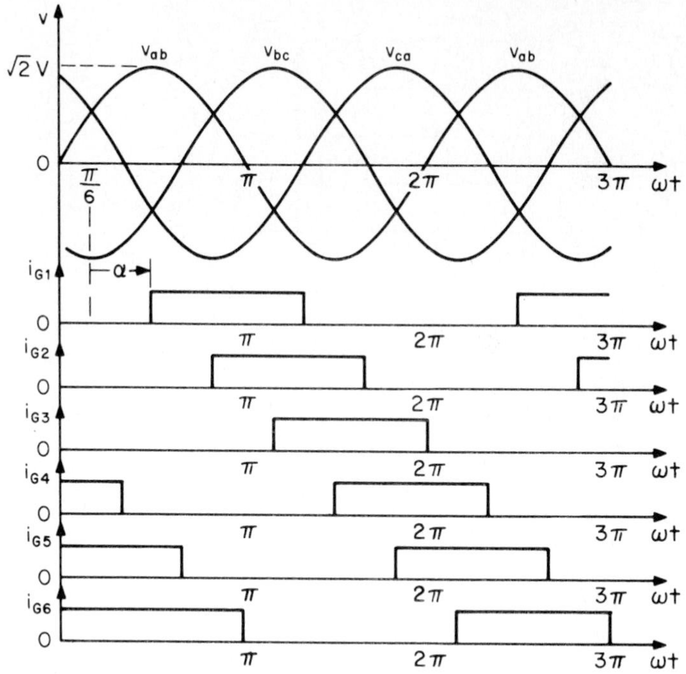


**FIG. 8.39** Waveforms for circuit of Fig. 8.38.

An approximate range of the controller delay angle $\alpha$, as also of the line current, may be obtained from the experimentally determined curves in Fig. 8.42. The base current employed in determining the normalized line current $I_{aN}$ is

$$I_{\text{base}} = \frac{V}{\sqrt{3}Z} \qquad \text{A} \tag{8.95}$$

where $Z\underline{/-\varphi^\circ}$ is the motor impedance at the speed considered.

## 8.11 SPEED CONTROL BY SLIP-ENERGY RECOVERY

This type of drive is suitable for driving centrifugal pumps and fans, but since it can be made highly efficient, it is frequently employed for very large power installations.

Figure 8.43 shows an arrangement of a diode bridge and single resistor that may be used in conventional speed control of a wound-rotor motor in place of the more familiar balanced three-phase variable resistor. This principle is employed in the slip-energy recovery circuit of Fig. 8.44; here, however, the single resistor is replaced by a three-phase controlled bridge operating as an inverter, as illustrated in Fig. 8.23. The energy wasted in the rotor resistor of the conventional control is now fed back by the inverter to the supply system. The inductor in the dc link smoothes the rectified output voltage at the rotor terminals. The effective power ratings of

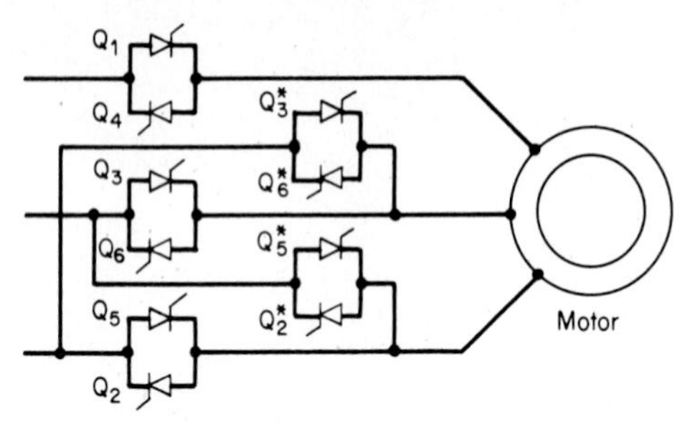


**FIG. 8.40** Reversing power controller.

the converters must be well in excess of the power fed back to the supply system to allow for the wide ranges of rotor terminal voltage and current that occur as the speed and torque are varied.

Figure 8.45 shows the relationship between source voltage $v_{ab}$ and $\bar{v}_o$ appearing at the dc terminals of the controlled bridge when the motor is stationary. If the rotor turns ratio is not such that $|\bar{v}_o < \sqrt{2}V|$, then a transformer must be introduced into

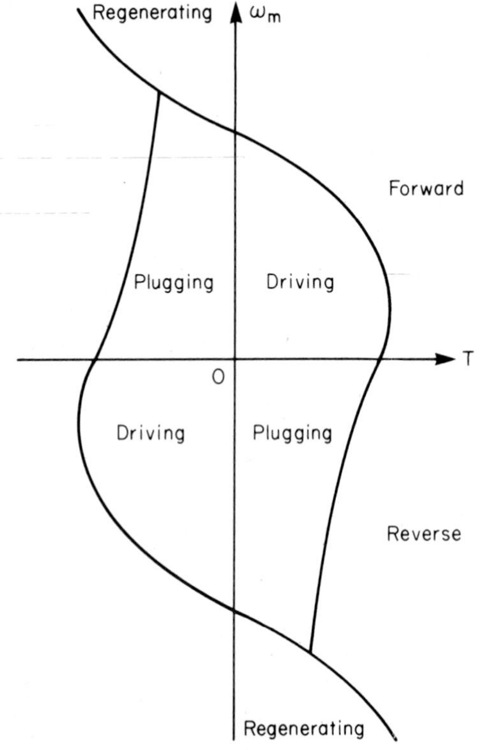


**FIG. 8.41** Reversing system speed-torque diagram.

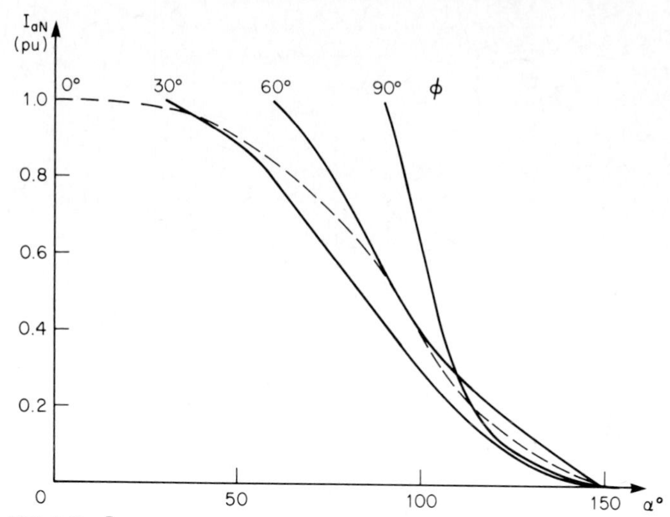


**FIG. 8.42** $I_{aN}$ versus $\alpha$ for the motor and converter.

the ac lines to the controlled bridge. Any reduction in $\alpha$ from the value shown in Fig. 8.45 will result in rotor current and a starting torque. If $\alpha$ is set to some value $90° < \alpha < (\eta - 60°)$, where $\eta$ is defined in Fig. 8.45, the rotor will accelerate to a constant speed. For $\alpha < 90°$, the diode rectifier would short-circuit the dc terminals of the controller bridge.

If converter losses are neglected and it is assumed that $i_o$ is effectively constant at magnitude $I_o$, then Fig. 8.46 shows the phase relationships between the ac terminal variables of the controlled bridge for $\alpha = 90°$, the maximum speed setting, where $i_r$ is in quadrature with $v_a$. An increase in $\alpha$ then results in inversion, with energy delivered to the ac power source. Since the fundamental component of $i_r$ lags $v_a$ by more than 90°, the fundamental power factor at the source terminals will be lower than that at the stator terminals.

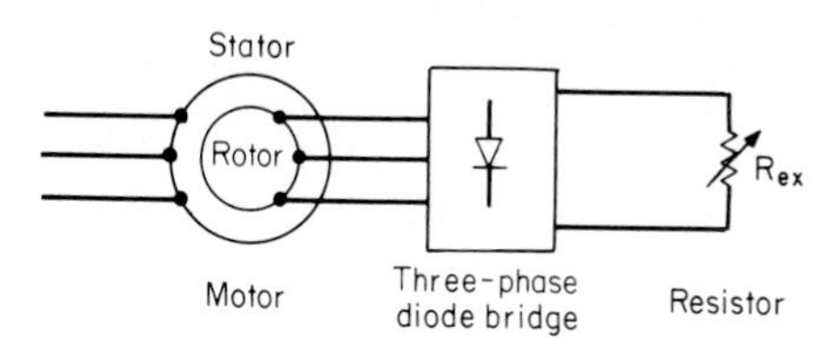


**FIG. 8.43** Alternative method of varying rotor-circuit resistance.

A reasonable approximation to the system performance may be obtained by employing a motor equivalent circuit consisting simply of the magnetizing reactance. It can then be shown that, if the inductor is large and effectively resistance-free,

$$\bar{v}_{LK} = \frac{3\sqrt{2}}{\pi} s \frac{N_{re}}{N_{se}} V = -\bar{v}_o = \frac{-3\sqrt{2}V}{\pi} \cos\alpha \quad \text{V} \tag{8.96}$$

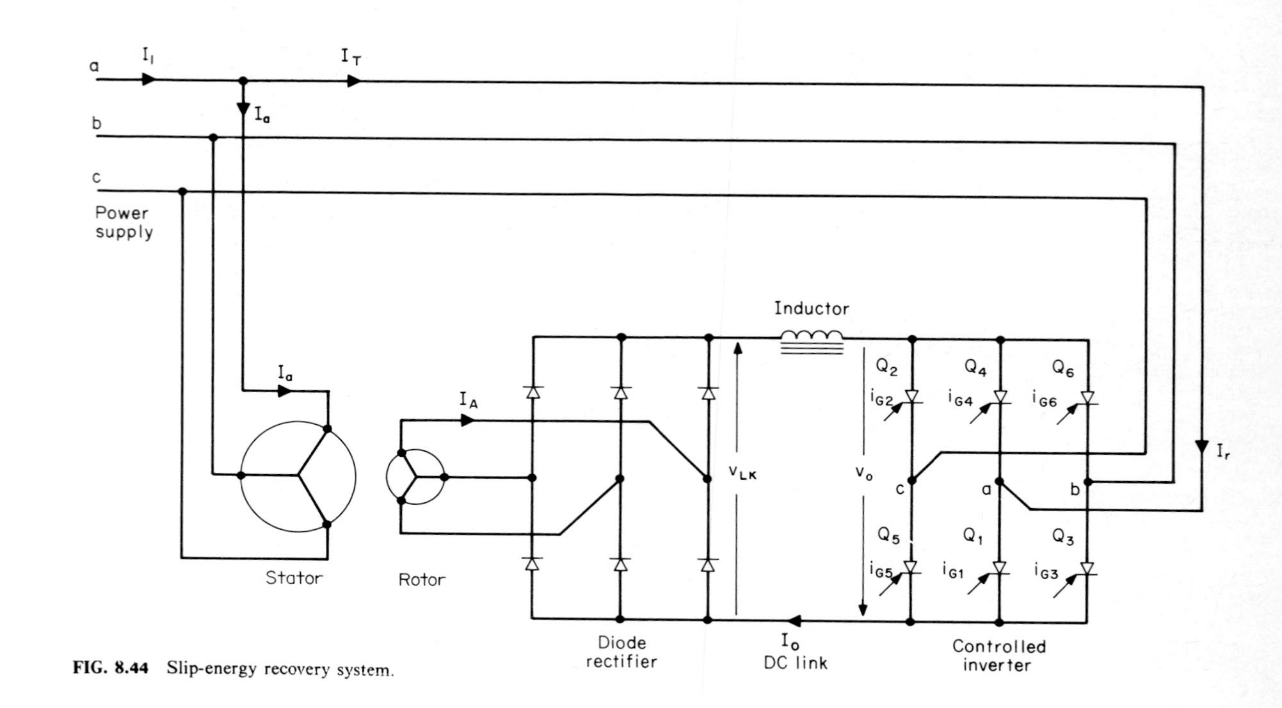


**FIG. 8.44** Slip-energy recovery system.

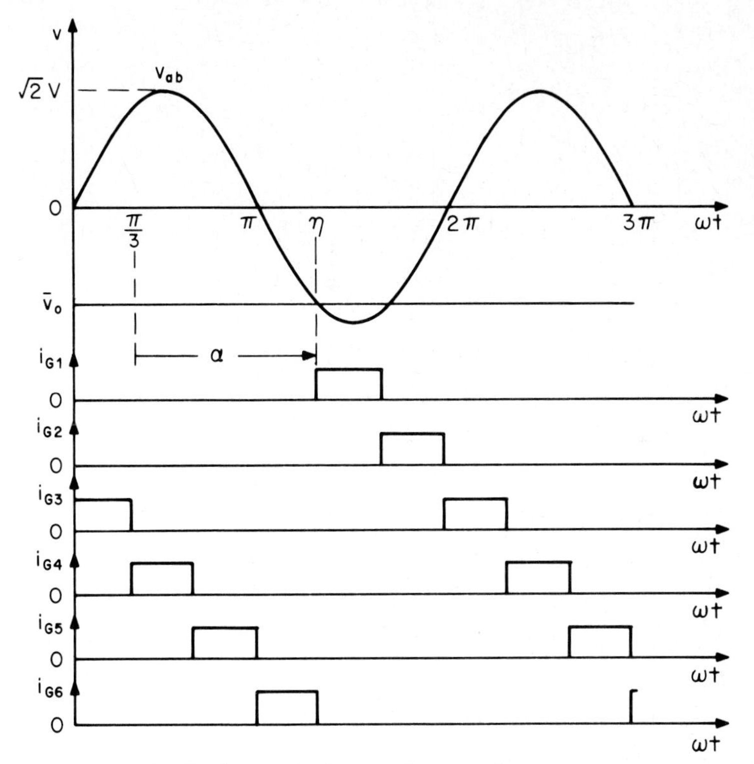


**FIG. 8.45** Relationship of gating signals to $v_{ab}$ for $\omega_m = 0$.

and
$$\omega_m = \frac{2}{p}\omega_s\left(1 + \frac{N_{se}}{N_{re}}\cos\alpha\right) \qquad \text{rad/s} \tag{8.97}$$

where $N_{se}$ = effective number of stator turns
$N_{re}$ = effective number of rotor turns

Thus, speed appears to be a function of $\alpha$ only and is independent of torque. In practice, with full-load torque the motor will have a speed regulation somewhat greater than that occurring under rated operating conditions. It can also be shown that

$$I_o = 0.740 \times \frac{2}{p}\omega_s \frac{N_{se}}{N_{re}}\frac{T}{V} \qquad \text{A} \tag{8.98}$$

and is dependent only on the airgap torque.

To allow for overlap and turn-off time of the thyristors, it is advisable to keep $\alpha$ below 175°. The maximum permissible value for the effective turns ratio of the motor is then

$$\frac{N_{re}}{N_{se}} = \frac{0.985}{s_{\max}} \tag{8.99}$$

If this value is exceeded, a transformer must be introduced between the ac source and the controlled bridge. Equation (8.95) then becomes

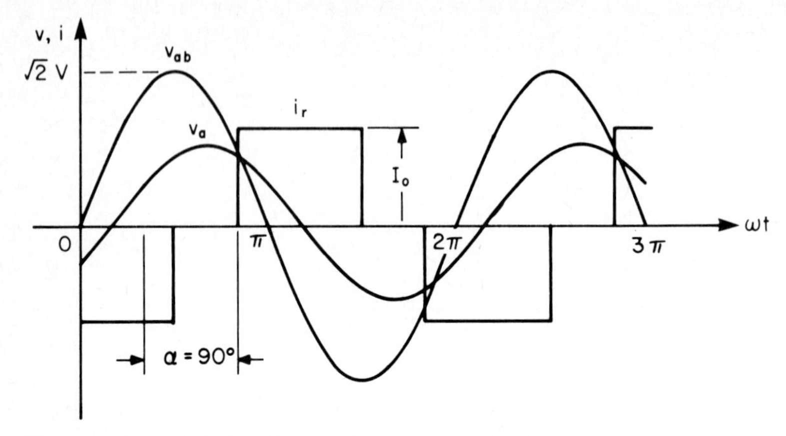


**FIG. 8.46** Phase relationship of controlled inverter terminal variables.

$$\bar{v}_{LK} = \frac{3\sqrt{2}}{\pi} s \frac{N_{re}}{N_{se}} V = -\bar{v}_o = -\frac{3\sqrt{2}}{\pi} \frac{N_p}{N_s} V \cos\alpha \qquad \text{V} \tag{8.100}$$

where $N_s$ = transformer turns on the source side
$N_p$ = transformer turns on the bridge side

The required control range is then

$$90° < \alpha < \cos^{-1}\left(-s_{\max}\frac{N_{re}}{N_{se}}\frac{N_s}{N_p}\right) \qquad \text{deg} \tag{8.101}$$

## Power in Load and Source Circuits

For a given set of values of $T$ and $\alpha$, $\omega_m$ and $s$ may be determined from Eq. (8.96). The magnetizing current $I_{ma}$ is constant at magnitude

$$I_{ma} = \frac{V}{\sqrt{3}\omega_s L_{ms}} \qquad \text{A} \tag{8.102}$$

where $L_{ms}$ is the magnetizing inductance. The airgap flux wave has a sinusoidal space distribution and interacts with the fundamental component of rotor current $I_A$ to give a torque. The rms value of this current component is

$$I_{A1} = 0.780 I_0 \qquad \text{A} \tag{8.103}$$

and this is also the rms value of the fundamental component of $i_r$. The fundamental motor input current is then

$$\bar{I}_{a1} = I_{a1}\,\underline{/-\varphi_{a1}} \qquad \text{A} \tag{8.104}$$

where

$$I_{a1} = \left[\left(\frac{N_{re}}{N_{se}} I_{A1}\right)^2 + I_{ma}^2\right]^{1/2} \qquad \text{A} \tag{8.105}$$

$$\varphi_{a1} = \tan^{-1}\left(\frac{N_{se}}{N_{re}}\frac{I_{ma}}{I_{A1}}\right) \qquad \text{rad} \tag{8.106}$$

$$\bar{I}_{r1} = 0.780 I_o \underline{/-\alpha} \qquad \text{A} \tag{8.107}$$

and the fundamental component of line current is

$$I_{l1}\underline{/-\varphi_{l1}} = \bar{I}_{l1} = \bar{I}_{a1} + \bar{I}_{r1} \qquad \text{A} \tag{8.108}$$

The fundamental source power factor is thus

$$\text{pf}_{l1} = \cos\varphi_{l1} \tag{8.109}$$

The power developed by the source is then

$$P_l = \sqrt{3} V I_{l1} \cos\varphi_{l1} \qquad \text{W} \tag{8.110}$$

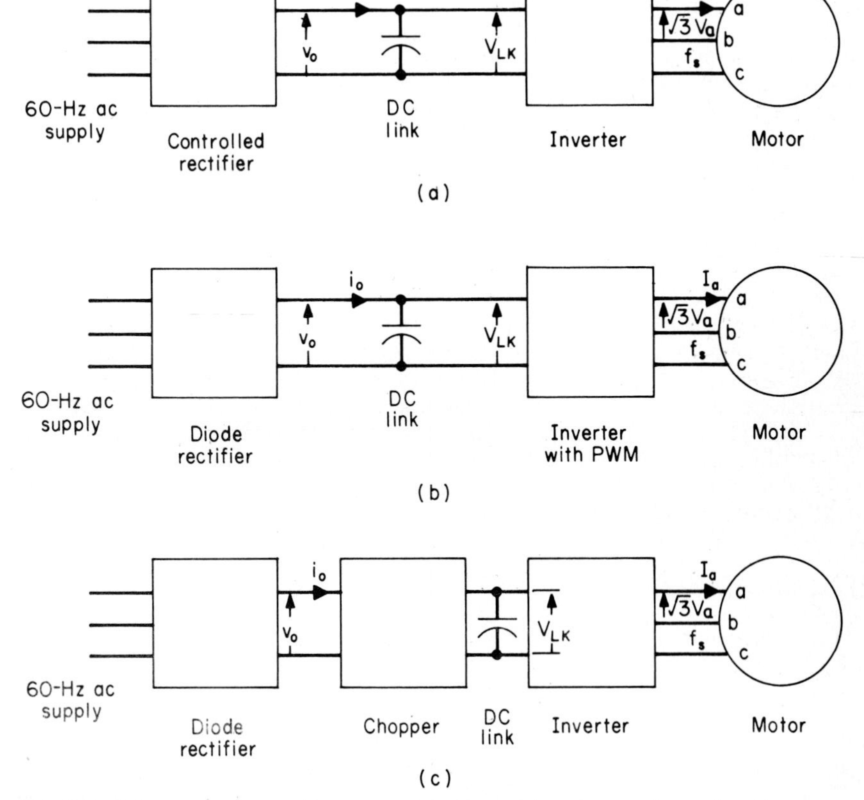


**FIG. 8.47** Voltage-source inverter drive systems: *(a)*, *(b)*, and *(c)* are three possible configurations.

## 8.12 INDUCTION MOTORS WITH VOLTAGE-SOURCE INVERTERS

A voltage-source inverter is supplied from a direct voltage source and may be used to excite a three-phase squirrel-cage induction motor at controllable frequency and therefore at controllable speed. If the motor is not to become magnetically saturated and if the frequency of excitation is to be varied, the terminal voltage must also be varied. Three converter systems that satisfy this requirement are shown in Fig. 8.47. These systems are unable to regenerate, but replacement of the rectifier by a dual converter and provision of a class C chopper in the system of Fig. 8.47*c* would make regeneration possible.

Figure 8.48*a* shows the power circuit of a three-phase inverter supplied via a dc link from a voltage source. The thyristors are numbered in the sequence in which the gating signals are applied, and each thyristor is forced-commutated at the end of its gating signal. A possible set of gating signals is shown in Fig. 8.48*b*, and the resulting waveforms of line-to-line and line-to-neutral voltages are shown in Figs. 8.49*b* and *c*. The amplitudes of these waveforms are directly related to the link voltage $V_{LK}$, so that only by varying $V_{LK}$ can the terminal voltage applied to the motor be controlled. The systems of Figs. 8.47*a* and *c* are able to vary $V_{LK}$, giving *amplitude modulation* of the pulses of line-to-line terminal voltage.

The fixed relationship between $V_{LK}$ and the motor terminal voltage may be avoided by replacing each of the gating signals by a train of shorter signals. The pulse of line-to-line voltage is split up correspondingly, and the actual motor terminal voltage may be taken as the average value over 120° of the pulses in one half-cycle. The amplitude of the waveform of terminal voltage in the motor is then varied with constant $V_{LK}$ by varying the width of the gating pulses in the train, giving pulse-width modulation (PWM). The inverter in Fig. 8.47*b* operates in this way. In the following, only pulse-amplitude modulation (PAM) will be discussed.

While the inverter output contains odd harmonics of voltage and current, the torque produced by the fundamental components essentially determines the behavior of the system. The discussion of system operation will therefore be continued in terms of the fundamental components of the motor variables.

There will be an upper limit to the fundamental line-to-neutral voltage $V_{a1}$ that can be applied to the motor. This may be determined by the strength of the motor insulation or, more probably, by the maximum available value of $V_{LK}$. To obtain the maximum torque per ampere of input current $I_{a1}$, the magnetizing current $I_{ma1}$ should be held at a value equal to that of $I_{ma}$ under rated conditions. Above the limit of $V_{a1}$ it is still possible to increase the speed by increasing $\omega_s$, but this will cause $I_{ma1}$ to decrease, and the motor will then operate with field weakening. The top speed will be determined by mechanical factors.

### Constant-Flux Operation

From the approximate equivalent circuit,

$$\bar{I}_{ma} = \frac{\bar{E}_{ma}}{j\omega_s L_{ms}} = -j\frac{k}{L_{ms}} \quad \text{A} \qquad (8.111)$$

Thus, if up to the limiting value of $V_{a1}$

$$\bar{I}_{ma1} = \frac{\bar{E}_{ma1}}{j\omega_s L_{ms}} = \bar{I}_{ma} \quad \text{A} \qquad (8.112)$$

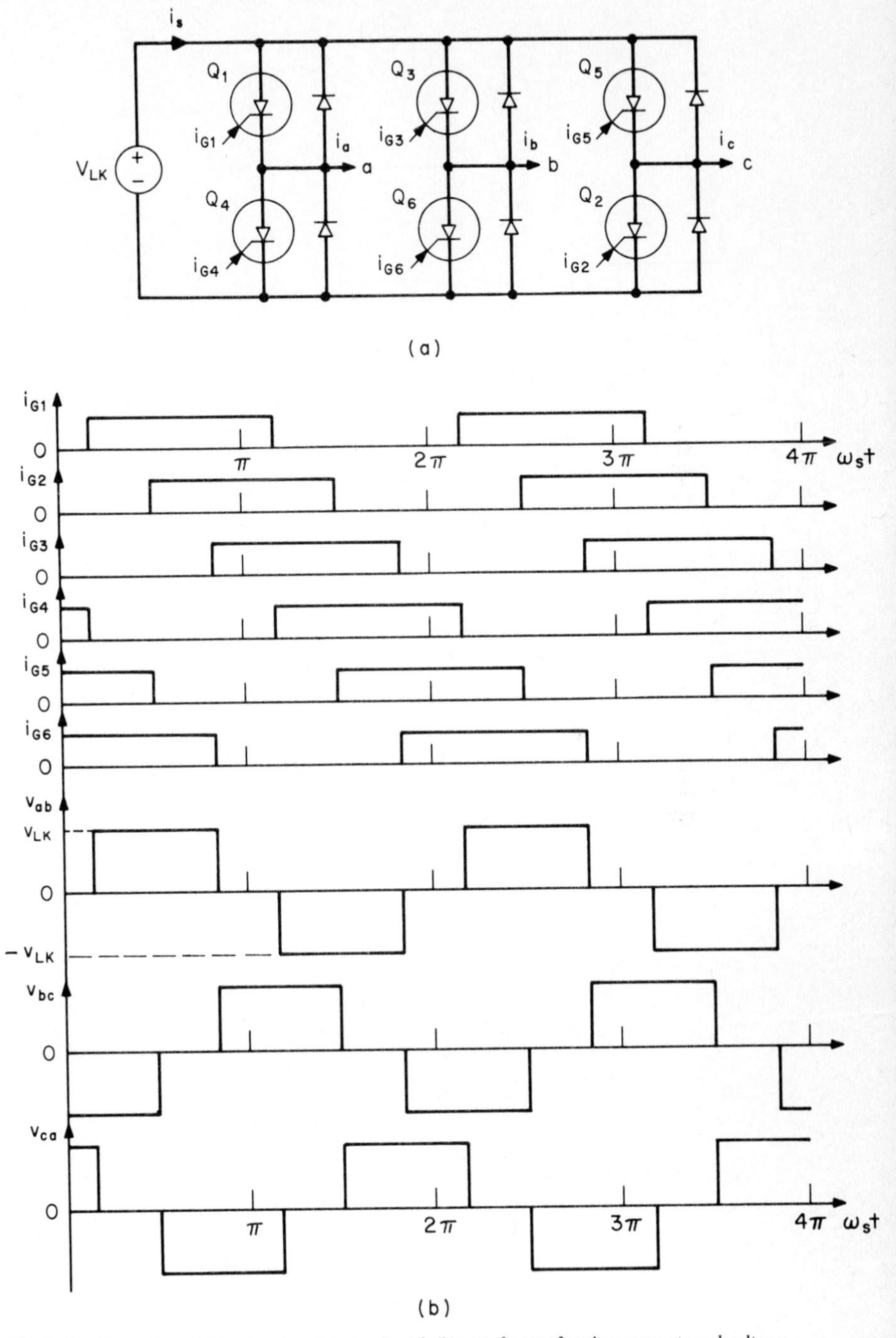


**FIG. 8.48** Three-phase bridge inverter: ***(a)*** circuit and ***(b)*** waveforms of various currents and voltages.

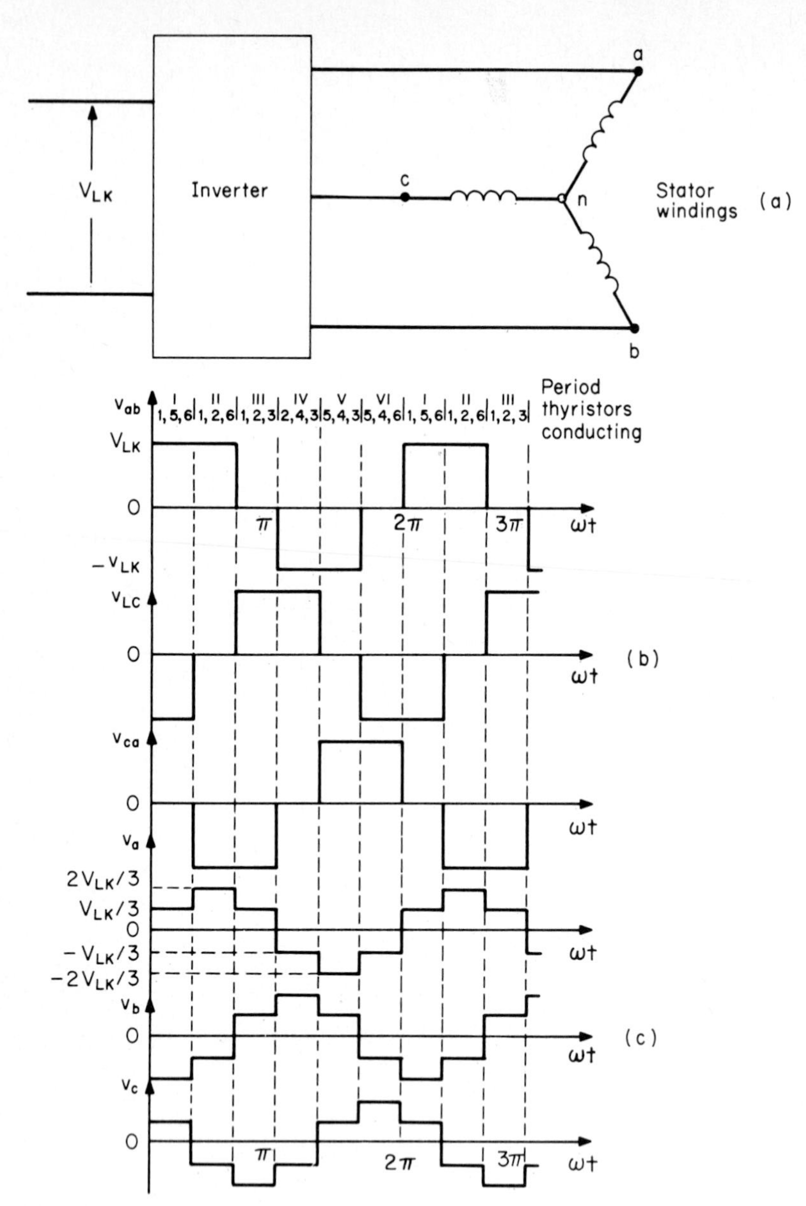


FIG. 8.49 Waveforms of motor line-to-neutral voltages: *(a)* a three-phase inverter-fed motor; *(b)* line-to-line voltage waveform; *(c)* line-to-neutral voltage waveform.

then it is necessary that

$$\frac{E_{ma1}}{\omega_s} = k \qquad \text{V/(rad} \cdot \text{s)} \tag{8.113}$$

and an appropriate control loop is required to satisfy this equation. The mechanical power developed is, from Eq. (8.86),

$$P_{mech} = \frac{3(1-s)}{s} R_r'(I_a')^2 = \omega_m T \qquad \text{W} \tag{8.114}$$

From Eq. (8.82) and the equivalent circuit, it can then be shown that

$$T = \frac{3R_r' k^2 [\omega_s - (p/2)\omega_m]}{(R_r')^2 + \{[\omega_s - (p/2)\omega_m] L_{lr}'\}^2} \qquad \text{N} \cdot \text{m} \tag{8.115}$$

From this equation a speed-torque curve may be obtained for any value of $\omega_s$.

From the equivalent circuit, $\bar{I}_{a1}$ and $\bar{V}_{a1}$ may also be calculated and the fundamental power factor obtained.

By analysis of the waveform of $V_a$ in Fig. 8.49,

$$V_{a1} = 0.450 V_{LK} \qquad \text{V} \tag{8.116}$$

Then, assuming that the inverter is lossless, from Eq. (8.86)

$$3P_{ma} = V_{LK} I_o \qquad \text{W} \tag{8.117}$$

and $I_o$ may be determined.

## Operation with Field Weakening

With field weakening, Eq. (8.113) no longer applies, so that it is not possible to derive a relationship similar to that in Eq. (8.115). Also, if the magnitude of $V_{a1}$ must be limited to some value less than that determined by the available magnitude of $V_{LK}$, then a control loop is required to limit the motor terminal voltage to what is acceptable.

To obtain a point on a speed-torque curve for a given inverter frequency $\omega_s$, first assume a value of slip $s$. Then from Eq. (8.82)

$$\omega_m = \frac{2}{p}\omega_s(1-s) \qquad \text{rad/s} \tag{8.118}$$

For this chosen value of slip and the known limiting value of $V_{a1}$, all variables in the equivalent circuit can be calculated on the assumption that the circuit parameters remain constant. Equation (8.114) may then be applied, after which

$$T = \frac{P_{\text{mech}}}{\omega_m} \qquad \text{N} \cdot \text{m} \tag{8.119}$$

The magnitude of $I_o$ is obtained from Eq. (8.117).

The speed-torque curves in Fig. 8.50 were obtained by the procedures described in the foregoing. The curves for the field-weakening range are approximate, however, since the slips are considerably greater than the rated value.

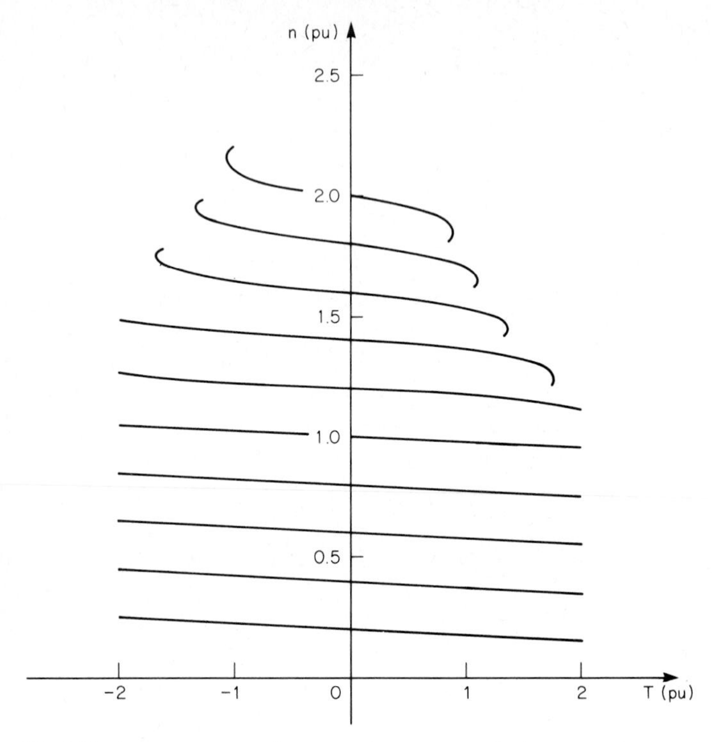


**FIG. 8.50** Speed torque curves for a squirrel-cage motor and voltage-source inverter.

## 8.13 INDUCTION MOTORS WITH CURRENT-SOURCE INVERTERS

A current-source inverter is supplied from a dc source. It is normally used to drive only a single motor or a number of similar motors mechanically coupled together, as in a rapid-transit system drive. Two alternative converter combinations are shown in Fig. 8.51, in which input current to the inverter is kept constant by a feedback loop. The phase-controlled rectifier in Fig. 8.51*a* has a low power factor at the ac terminals when load is low and has a slow dynamic response. The diode rectifier in Fig. 8.51*b* has a high power factor at all loads, but forced commutation and fast-response thyristors are required in the chopper. When a dc distribution network is available, the system in Fig. 8.51*b* can be employed without the rectifier.

### Three-Phase Current-Source Inverter

Figure 8.52 shows the power circuit of a three-phase inverter. The thyristors are numbered in the sequence in which they are turned on to give phase sequence *ABC* at the load terminals. Six identical commutating capacitors are connected in delta across the three phases of the circuit. The gating signals and ideal waveforms of the resulting motor line currents and line-to-neutral terminal voltages are shown in

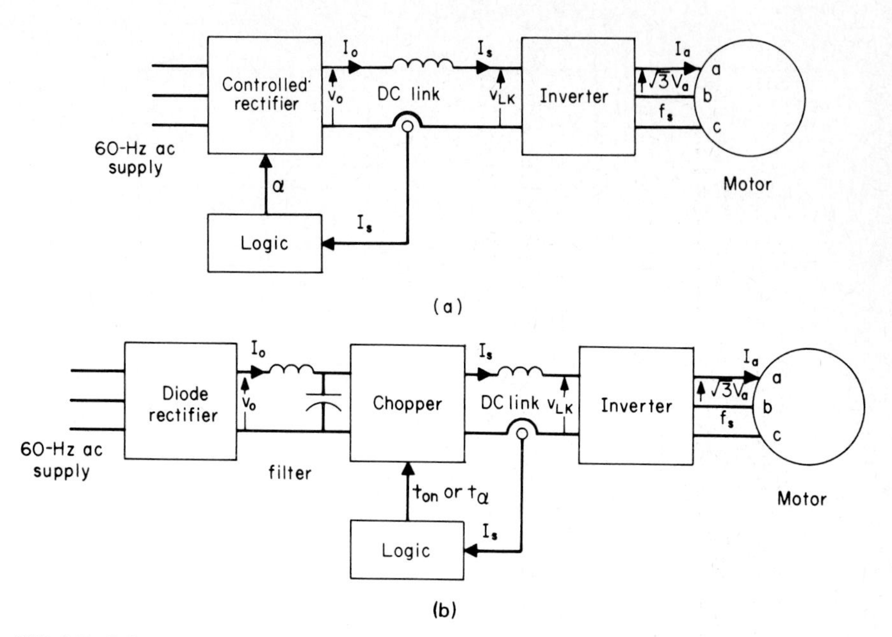


**FIG. 8.51** DC current-source inverter drive systems: *(a)* and *(b)* are two possible configurations.

Fig. 8.53. The voltage curves are approximately sinusoidal, but superimposed upon them are spikes due to the high emf's induced in the motor by the rapid changes of current at the beginning and ending of each rectangular (in fact, trapezoidal) pulse of line current. Reversal of the phase sequence of the inverter output currents is obtained by reversal of the sequence of the gating signals.

## Operation of an Induction Motor on a Current Source

The operation of an induction motor with constant impressed line current differs markedly from that of the same motor with constant impressed terminal voltage. The equivalent circuit of Fig. 4.7*a* may be employed to illustrate this point. From the circuit it can be shown that

$$T = \frac{3p}{2}\frac{R_r'}{s}\frac{\omega_s(L_{ms}I_a)^2}{(R_r'/s)^2 + \omega_s^2(L_{ms} + L_{lr}')^2} \qquad \text{N} \cdot \text{m} \tag{8.120}$$

where $I_a$ is the line current. This equation gives a relationship between torque $T$ and slip $s$ for fixed values of $\omega_s$ and $I_a$.

Curves of per-unit torque and terminal voltage versus per-unit speed of the motor operating with unit line current at fixed rated frequency are shown in Fig. 8.54, assuming constant parameters. Actually the motor saturates with $V_a$ about 1.2 per unit, the magnetizing current increases, and the torque is less than that shown. The rated values of torque, terminal voltage, current, and speed have been used as the base values. The torque-speed curve differs very greatly from the familiar form of that for an induction motor operating from a voltage source. In particular, the rated operating point occurs on the unstable part of the torque-speed curve, so that a

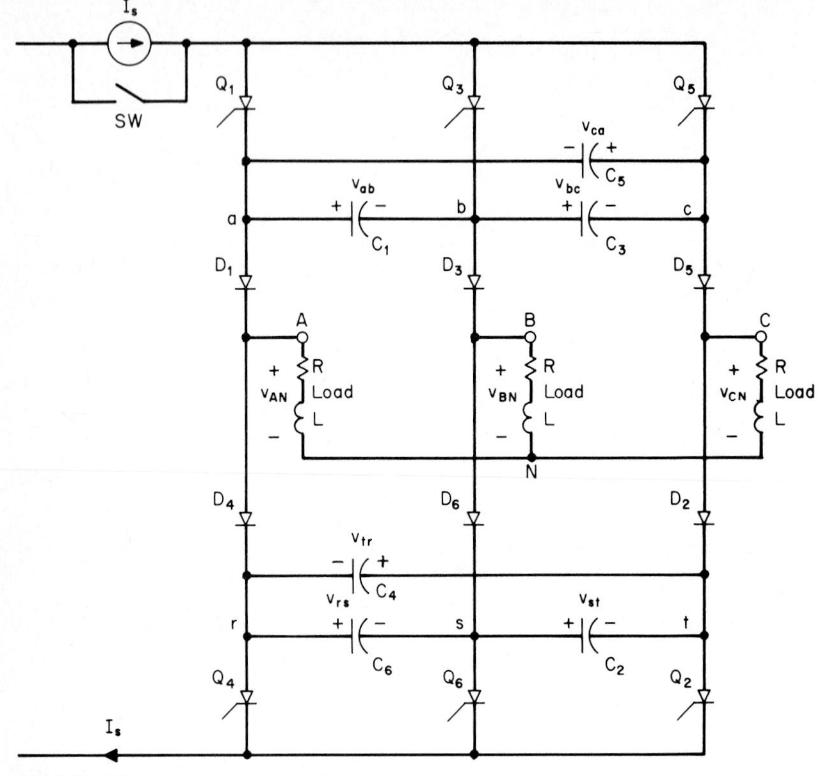


**FIG. 8.52** Three-phase current-source inverter.

speed feedback loop is required in the control system to maintain operation at such a point.

## Operation on a Current Source of Variable Frequency

Once again the ratio $E_{ma}/\omega_s$ must be kept constant if the motor is not to saturate. From Fig. 4.7*a*.

$$I_{ma} = \left[ \frac{(R_r')^2 + (\omega_r L_{lr}')^2}{(R_r')^2 + \omega_r^2 (L_{ms} + L_{lr}')^2} \right]^{1/2} \times I_a \quad \text{A} \tag{8.121}$$

This equation shows that if $\omega_r$ is held constant, so also is $I_{ma}$ for any given value of $I_a$, but if $I_a$ varies, then $I_{ma}$ also varies. If $\omega_r$ is held constant at the rated value, the motor operates with field weakening at currents below the full-load value and will saturate if, for short periods, the motor draws current above the full-load value. Such an arrangement is preferable to the complication of making $\omega_r$ a function of $I_a$. Provision must therefore be made in the control system for holding $\omega_r$ constant.

The angle by which the motor line-to-neutral voltages lead the fundamental components of the line currents depends upon the angle of the motor impedance.

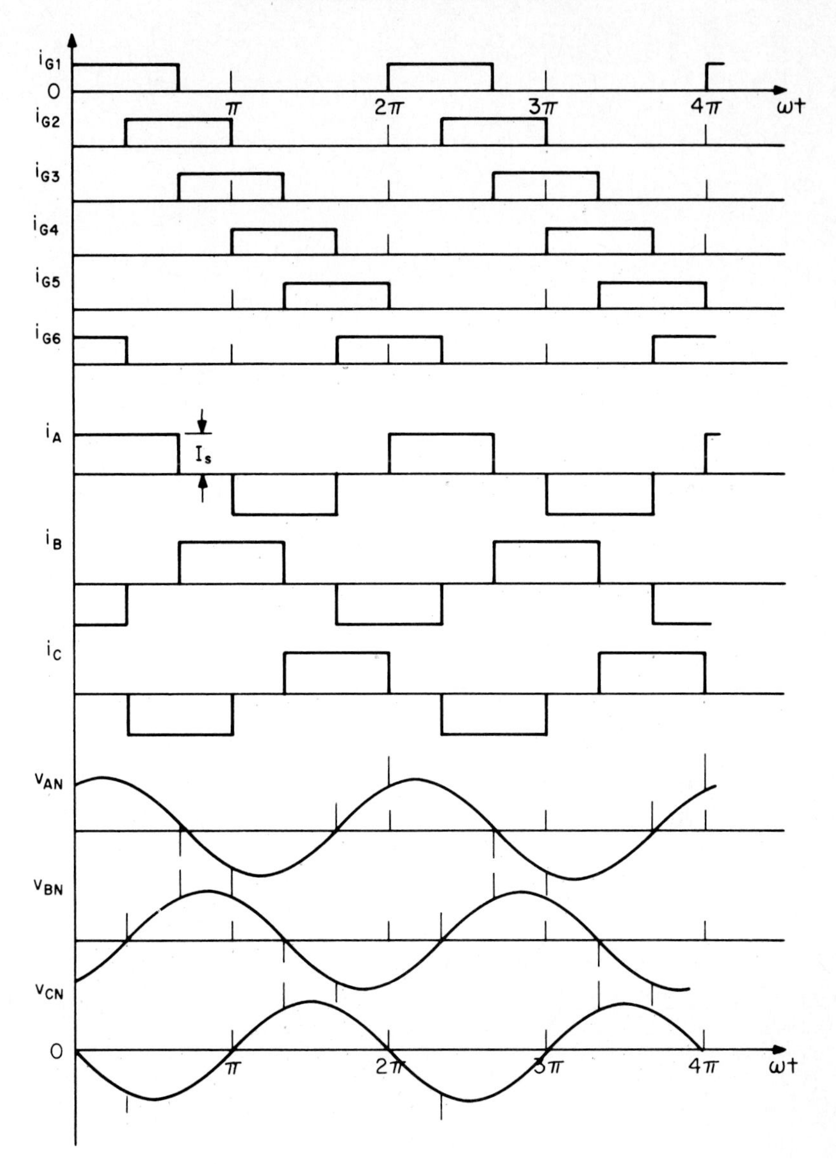


**FIG. 8.53** Waveforms of the circuit variables in Fig. 8.52.

This angle depends upon the chosen rotor frequency and the motor speed. With constant $\omega_r$, the motor impedance is substantially constant except at low speed when the effect of $R_s$ is significant. When regeneration takes place, the waveforms of voltage in Fig. 8.53 advance in phase to lead the fundamental components of the current waveforms by more than 90°. Since the direction of source current $I_s$ is unchanged when this takes place, the input voltage of the inverter becomes negative.

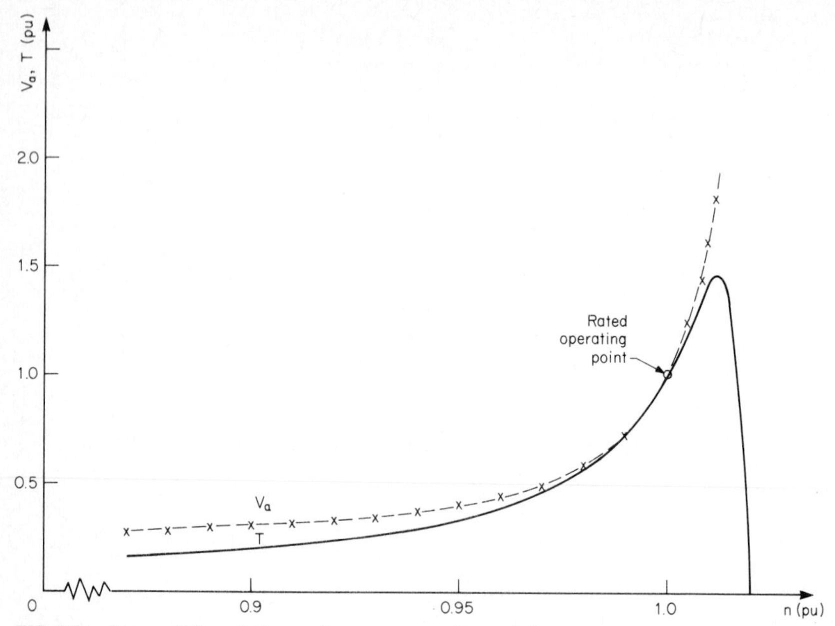


**FIG. 8.54** Curves of $V_a$ and $T$ per unit versus $n$ per unit at unit line current.

A unidirectional current at positive or negative voltage can be supplied by the single controlled rectifier in the system of Fig. 8.51*a*. This system can therefore provide regeneration, which, in conjunction with reversal of the sequence of the inverter gating signals, gives a four-quadrant drive. In the system of Fig. 8.51*b*, regeneration requires a class D chopper and replacement of the diode rectifier by a controlled rectifier in which $\alpha$ changes from zero to the value $(\pi - \omega t_{\text{off}} - \mu)$ when the input voltage of the chopper is required to go negative.

When the motor is driven from the current-source inverter, $I_a$ in Eqs. (8.120) and (8.121) must be replaced by the fundamental component of the line current $I_{a1}$. The rms value of the line current is

$$I_a = 0.817 I_s \qquad \text{A} \tag{8.122}$$

and

$$I_{a1} = 0.780 I_s \qquad \text{A} \tag{8.123}$$

Current harmonics other than the fundamental are considered to flow in the rotor branch of the equivalent circuit, and their rms value is

$$I_{ah} = 0.242 I_s \qquad \text{A} \tag{8.124}$$

From Eq. (8.120)

$$T = \frac{3p}{2} \frac{R_r' \omega_r (L_{ms} I_{a1})^2}{(R_r')^2 + \omega_r^2 (L_{ms} + L_{lr}')^2} \qquad \text{N} \cdot \text{m} \tag{8.125}$$

and

$$I_{a1}' = \frac{\omega_r L_{ms} I_{a1}}{[(R_r')^2 + \omega_r^2 (L_{ms} + L_{lr}')^2]^{1/2}} \qquad \text{A} \tag{8.126}$$

The mechanical power developed is

$$P_{\text{mech}} = T\omega_m = 3(I'_{a1})^2 R'_r \frac{(\omega_s - \omega_r)}{\omega_s} \qquad \text{A} \tag{8.127}$$

The resistive losses in the motor are

$$P_R = 3[R_s I_a^2 + R'_r(I'_{a1})^2 + R'_r I_{ah}^2] \qquad \text{W} \tag{8.128}$$

Thus, if rotational losses are neglected,

$$P_{\text{in}} = P_{\text{mech}} + P_R \qquad \text{W} \tag{8.129}$$

If the inverter is assumed to be lossless, then

$$\bar{v}_{LK} = \frac{P_{\text{in}}}{I_s} \qquad \text{V} \tag{8.130}$$

The magnitude of $V_{a1}$ at a known speed $\omega_m$ may be expressed by employing $\bar{E}_{ma1}$ as the reference phasor. The phase angle of $\bar{I}_{a1}$ can be obtained from

$$I_{m1a}\angle{-90^\circ} = \frac{(R'_r + j\omega_r L'_{lr})\bar{I}_{a1}}{R'_r + j\omega_r(L_{ms} + L'_{lr})} \qquad \text{A} \tag{8.131}$$

and

$$\bar{V}_{a1} = \omega_s L_{ms} I_{ma1}\angle 0 + (R_s + j\omega_s L_{ls})\bar{I}_{a1} \qquad \text{V} \tag{8.132}$$

To compensate for extra heating due to harmonics, the motor should be derated by about 10 percent. If rated torque is required at below about 30 percent of rated speed, further derating or a separately driven fan becomes necessary.

## Operation with Field Weakening at High Speed

At some point, as the motor speed is increased, the average inverter input voltage $\bar{v}_{LK}$ reaches a limit. If, for a controlled value of motor current $I_a$, the speed is to be increased above that at which this voltage limit is reached, then inverter frequency $\omega_s$ must continue to increase while $\bar{v}_{LK}$ and $I_s$ remain fixed. Since resistive losses in the motor are virtually constant, this means that a range of operation at constant mechanical power is entered; that is,

$$P_{\text{mech}} = T\omega_m \simeq \text{const.} \qquad \text{W} \tag{8.133}$$

If it is assumed that

$$\omega_r(L_{ms} + L'_{lr}) >> R'_r \qquad \Omega \tag{8.134}$$

it can be shown that

$$T\omega_r \simeq \text{const.} \qquad \text{W} \tag{8.135}$$

and

$$\omega_r \propto \omega_m \qquad \text{rad/s} \tag{8.136}$$

The actual relationship can be shown to be

$$\omega_r^2 - \frac{3pR'_r\omega_m(L_{ms}I_{a1})^2\omega_r}{2P_{\text{mech}}(L_{ms} + L'_{lr})^2} + \frac{(R'_r)^2}{(L_{ms} + L'_{lr})^2} = 0 \tag{8.137}$$

This, to a close approximation, yields a straight line. If the maximum required value of $\omega_m$ is substituted in Eq. (8.137), the maximum required value of $\omega_r$ may be obtained, and this gives the upper end of the straight line describing the approximate relationship between $\omega_r$ and $\omega_m$. The lower end is obtained by substituting the value of $\omega_m$ at which $\bar{v}_{LK}$ reaches its limit. From these two values the equation

$$\omega_r = a + b\omega_m \qquad \text{rad/s} \tag{8.138}$$

may be obtained, in which $a$ and $b$ are constants. This function must be provided by the control system.

As speed is increased, the breakdown torque of the motor falls. There is therefore a limit to the upper end of the constant power region at which the torque demanded is equal to the breakdown torque.

## 8.14 SYNCHRONOUS MOTOR DRIVES

Accurate speed control may be obtained by using synchronous motors powered by converters. Several synchronous motors may be driven from a single voltage-source inverter, but, as in the case of induction motors, current-source inverters can usually supply only a single motor or a number of similar, mechanically coupled motors. If wound-field or permanent magnet machines are operated at leading power factor, they are able to provide load commutation for a current-source inverter.

### The Wound-Field Synchronous Motor

Per-phase equivalent circuits for a wound-field synchronous motor are shown in Fig. 8.55. In these circuits emf $E_o$ is called the excitation voltage, $i_f$ is the actual direct current in the field winding, and the factors $n'$ and $n''$ are defined as

$$n' = \frac{\sqrt{2}}{3}\frac{N_{re}}{N_{se}} \tag{8.139}$$

$$n'' = \frac{L_{ms}}{L_{ms} + L_{ls}}\, n' \tag{8.140}$$

where $N_{re}$ = effective turns of the field winding
$N_{se}$ = effective turns per phase of the stator winding

If $v_{an}$ is defined as

$$v_{an} = V_a \sin \omega t \qquad \text{V} \tag{8.141}$$

then $\beta_{oe}$ is the angle of the rotor in electrical radians with respect to phase $a$ of the stator winding at instant $t = 0$. The position of the rotor at any instant in electrical radians is then given by

$$\beta_e = \omega_s t + \beta_{oe} \qquad \text{rad} \tag{8.142}$$

The sum of the stator leakage inductance and the magnetizing inductance gives the synchronous inductance $L_s$, so that

$$L_s = L_{ms} + L_{ls} \qquad \text{H} \tag{8.143}$$

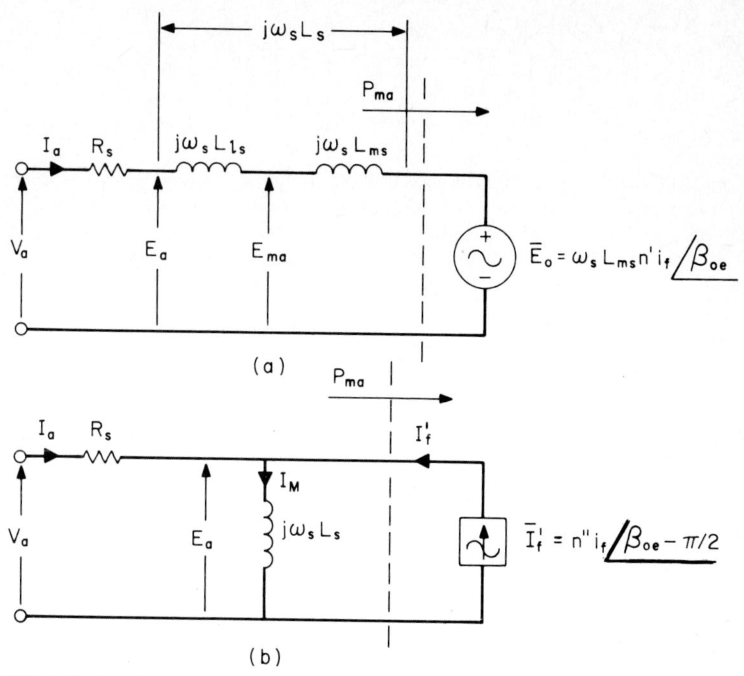


**FIG. 8.55** Steady-state equivalent circuits of a cylindrical-rotor synchronous machine: ***(a)*** and ***(b)*** are two forms of equivalent circuits.

The angular speed of rotation is

$$\omega_m = \frac{2}{p}\omega_s \qquad \text{rad/s} \tag{8.144}$$

## Voltage-Source Inverter Drives

The converter combinations illustrated in Fig. 8.47 may be employed. As for the induction motor, the line-to-neutral waveforms of the motor terminal voltage may be considered to be those in Fig. 8.49. Apart from causing some extra heating that necessitates derating of the motor, the effect of harmonics on the steady-state behavior is negligible except at very low speed, where sinusoidal PWM may be desirable.[3] The following discussion will therefore be conducted in terms of the fundamental components of the variables, and Eq. (8.116) may be modified to the form

$$V_a = 0.450 V_{LK} \qquad \text{V} \tag{8.145}$$

where $V_a$ signifies the fundamental component of the waveform $v_a$ in Fig. 8.49. To avoid saturation, the magnetizing current $I_{ma}$ must be kept at or below the value for rated operation by control of the ratio $E_{ma}/\omega_s$.

If rotational losses are neglected, then the airgap torque of the motor is equal to the coupling torque, and from Fig. 8.55*b*

$$T = \frac{3p}{2\omega_s} E_a n'' i_f \sin \beta_{oe} \qquad \text{N} \cdot \text{m} \tag{8.146}$$

$$\bar{I}_a = \bar{I}_M - \bar{I}'_f \qquad \text{A} \tag{8.147}$$

$$\bar{V}_a = \bar{E}_a + R_s \bar{I}_a \qquad \text{V} \tag{8.148}$$

It may be desirable to drive a synchronous motor at speeds in excess of that at which the stator emf can be made proportional to speed because of a limit on the supply voltage. It may then be assumed that $V_a$ will be fixed while $\omega_s$ increases and ratio $E_a/\omega_s$ falls, giving field weakening. Current $I_a$ must be maintained at or less than the value permissible during operation at rated speed so that excessive heating due to resistive losses does not take place.

If the maximum torque per ampere of line current is to be obtained, then the power factor must be held at or near unity. Thus, the maximum permissible input power to the machine $P_{in}$ will be constant over the field-weakening range.

## Permanent Magnet Synchronous Motor

A permanent magnet (PM) synchronous motor may be envisaged simply as a normal synchronous motor operating with fixed field current. Its equivalent circuit may be taken to be that of Fig. 8.55*b*.

The synchronous reactance of a PM motor is usually low (below 0.5 per unit) compared with that of a wound-field machine. This is due to the low permeability of the permanent magnet material. A motor may be designed to be overexcited or underexcited, that is, to operate with a leading or lagging power factor under rated conditions. In the equivalent circuit of Fig. 8.55*b*, $I_M$ is considerably larger than the rated value of $I_a$, and the phasor diagram of Fig. 8.56 shows that the stator current will decrease in magnitude as torque is reduced.

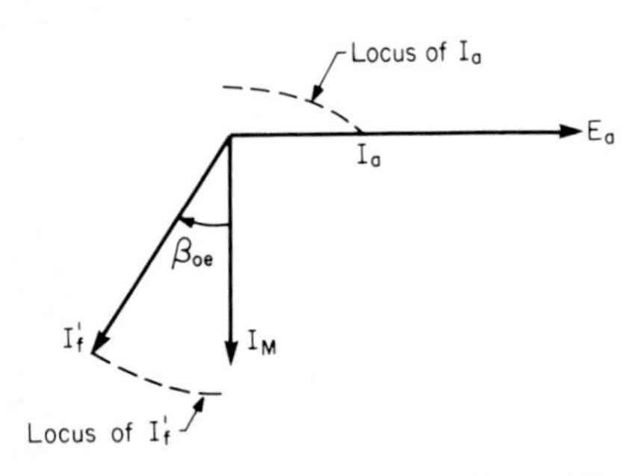


**FIG. 8.56** Locus of stator current for variable torque and constant field current.

The PM motor may also be operated with field weakening over a limited speed range. For constant power, the component of stator current $I_a$ in phase with stator voltage $V_a$ is constant. Figure 8.57 shows a locus of stator current as the speed is increased from a base value of $\omega_{mb}$ to $2\omega_{mb}$ at constant power while the angle $\beta_{oe}$ is maintained constant by the control system. By appropriate choice of a lagging power factor at the base speed, an equal value of a leading power factor is achieved at twice the base speed.

## Three-Phase Synchronous Reluctance Motor

The reluctance motor has no rotor mmf from a field winding or permanent magnets. It can operate only at a lagging power factor. However, careful design of the rotor can raise the full-load power factor to above 0.75. Like any other electromagnetic machine, the motor can also regenerate. The equivalent circuit of the motor is shown

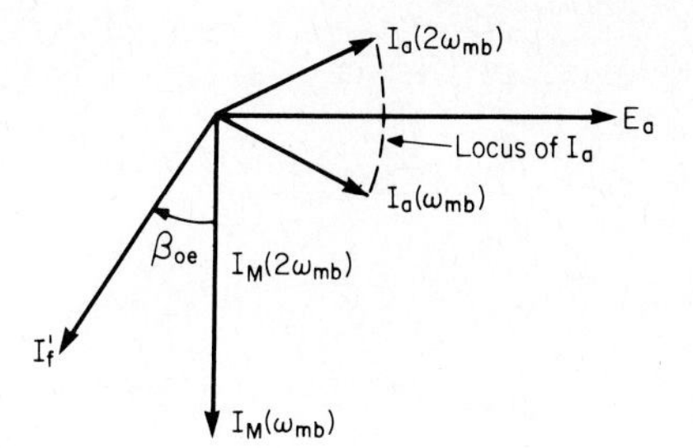


**FIG. 8.57** Operation of a PM motor in the field-weakening region.

in Fig. 8.58, where the resistive element accounts for the per-phase energy converted to or from a mechanical form.

As with any other synchronous machine, $\beta_{oe}$ is the angle between the axis of phase $a$ of the stator and the direct axis of the rotor at instant $t = 0$. When $\beta_{oe}$ is zero, the resistive element in the equivalent circuit becomes infinite; thus, no energy conversion takes place, and only a magnetizing current flows into the machine. When $\beta_{oe}$ becomes negative, the machine motors; when it becomes positive, the machine regenerates.

For a $p$-pole motor,

$$T = -\frac{3p}{4\omega_s}\frac{L_d - L_q}{L_d L_q} V_a^2 \sin^2 \beta_{oe} \qquad \text{N} \cdot \text{m} \tag{8.149}$$

It can be shown that the relationship between $\bar{V}_a$ and $\bar{I}_a$ can be represented by the circle diagram in Fig. 8.59, in which

$$Oa = \frac{V_a}{2\omega_s}\frac{L_d + L_q}{L_d L_q} \qquad \text{A} \tag{8.150}$$

$$ab = \frac{V_a}{2\omega_s}\frac{L_d - L_q}{L_d L_q} \qquad \text{A} \tag{8.151}$$

The power output is given by

$$P_o = 3V_a I_a \cos\varphi \qquad \text{W} \tag{8.152}$$

and

$$T = \frac{P_o}{\omega_m} = \frac{p}{2}\frac{P_o}{\omega_s} \qquad \text{N} \cdot \text{m} \tag{8.153}$$

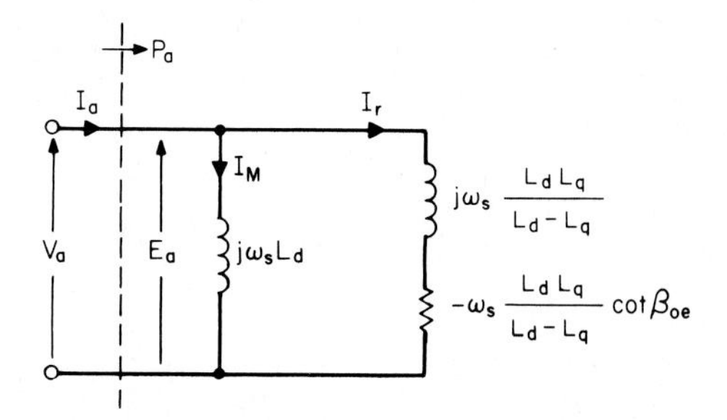


**FIG. 8.58** Equivalent circuit of a reluctance motor.

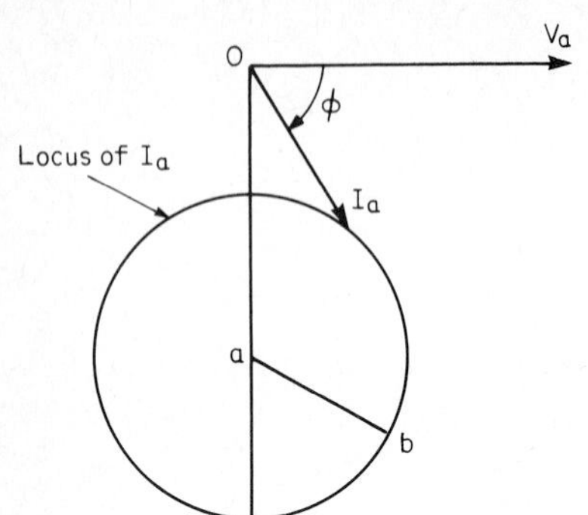


**FIG. 8.59** Circle diagram for a reluctance machine.

## Current-Source Inverter Drive

For a synchronous motor, there is no essential difference between driving it from a voltage-source inverter at a controlled line current and driving it from a current-source inverter that delivers the same current. The current-source inverter drive may produce higher losses in the motor than the voltage-source inverter drive as a result of current harmonics, but it has the advantage of a less expensive combination of converters.

With a current-source drive, it is desirable to express the torque and other motor quantities in terms of the stator current. The equivalent circuit of Fig. 8.55*b* and the phasor diagram of Fig. 8.60 give

$$\bar{E}_a = j\omega_s L_s(\bar{I}_a + I'_f) \qquad \text{V} \tag{8.154}$$

$$T = -\frac{3p}{2} L_s I_a I'_f \sin\delta \qquad \text{N}\cdot\text{m} \tag{8.155}$$

Thus, if the field current and angle $\delta$ are held constant by the control system, the torque is directly proportional to the line current $I_a$ at all speeds. Figure 8.61 shows a favorable locus for $\bar{E}_a$ as the torque varies from zero to maximum. This mode of control is particularly suitable for PM motors where the field cannot be controlled.

Operation with field weakening is required for speeds above the value at which the limit of $v_{LK}$ is reached. For a PM motor, this may be accomplished over a limited range by increasing the angle $\delta$. For a wound-field machine, the field current can be varied by the control system to control the stator emf and power factor.

The synchronous motor can be overexcited to draw a leading current and permit reduction of the size of the commutating capacitors required in the inverter of Fig. 8.52, where the *RL* circuit branches represent the effect of an induction motor load. The effect of a load consisting of an overexcited synchronous motor may

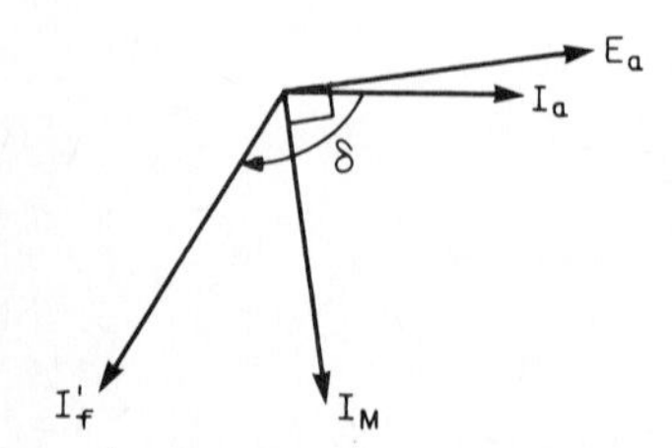


**FIG. 8.60** Phasor diagram for current-source drive.

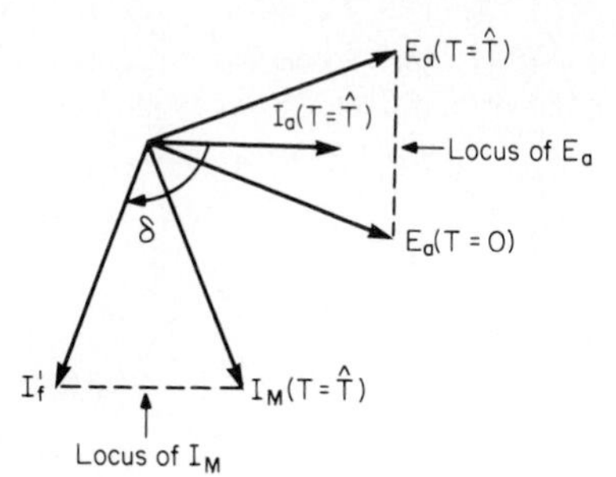


**FIG. 8.61** Phasor diagram for control with constant field current and constant angle $\delta$.

be represented as in Fig. 8.62 by three wye-connected resistance elements $R_{eq}$ in parallel with three delta-connected capacitances $C_{eq}$. The reactive components of the motor currents have the effect of additional commutating capacitors, and this may permit a reduction in the size of the actual commutating capacitors included in the inverter circuit.

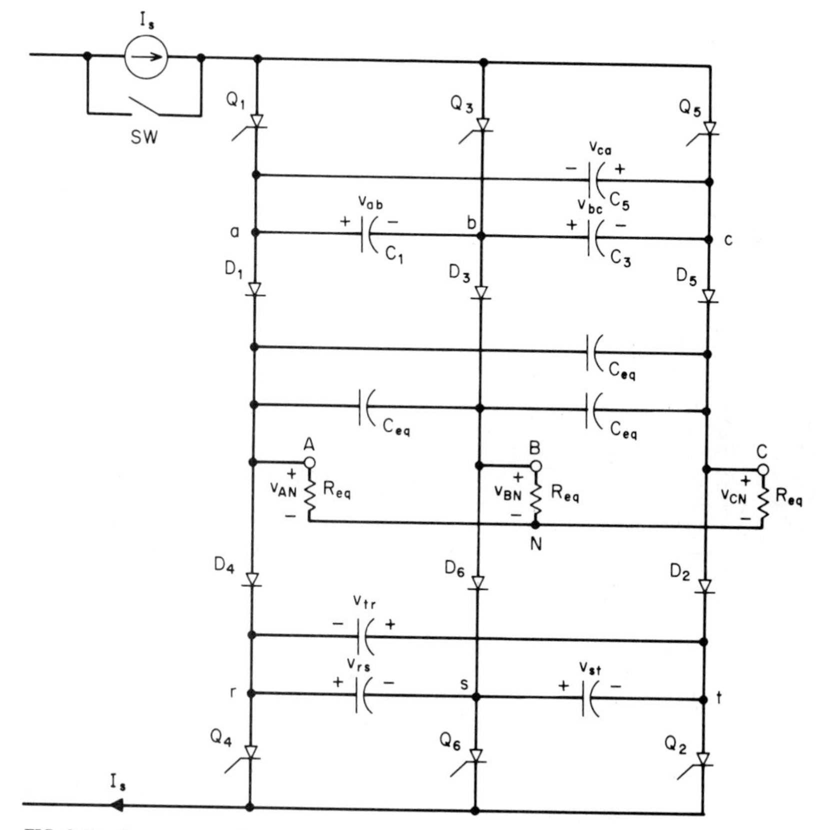


**FIG. 8.62** Current-source inverter with overexcited synchronous motor.

If the per-phase equivalent circuit of the motor is represented by a resistance and a capacitance connected in parallel, the fundamental current in the capacitance will be

$$I_c = I_a \sin \varphi \qquad \text{A} \tag{8.156}$$

so that in Fig. 8.62

$$C_{\text{eq}} = \frac{I_a \sin \varphi}{3\omega_s V_a} \qquad \text{F} \tag{8.157}$$

At standstill, $V_a \rightarrow R_s I_a$ and $\varphi \rightarrow 0$. Capacitance $C_{\text{eq}}$ therefore approaches zero. Thus, in determining the reduction (if any) that may be made in the size of

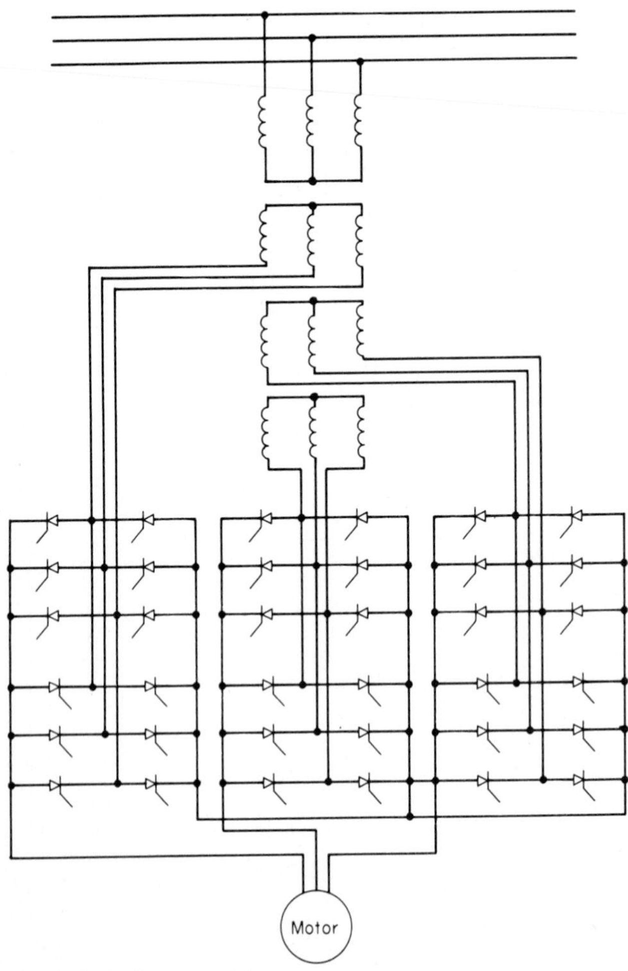


**FIG. 8.63** Cycloconverter drive.

the commutating capacitors, the variation of $C_{eq}$ over the entire operating area in the speed-torque diagram must be considered.

### Cycloconverter Drive

The cycloconverter is a direct ac-to-ac converter in which no dc link exists. One circuit arrangement is shown in Fig. 8.63, but others are possible.[4] That shown in Fig. 8.63 can be seen to be made up of three dual converters, each supplying one phase of the load. By switching each of the dual converters cyclically through all four quadrants of operation, a waveform of per-phase output terminal voltage of the form shown in Fig. 8.64 may be synthesized. The output frequency is lower than the input frequency and, because of the high harmonic content of both the output voltages and the input currents, must be limited to about 45 percent of the input frequency.

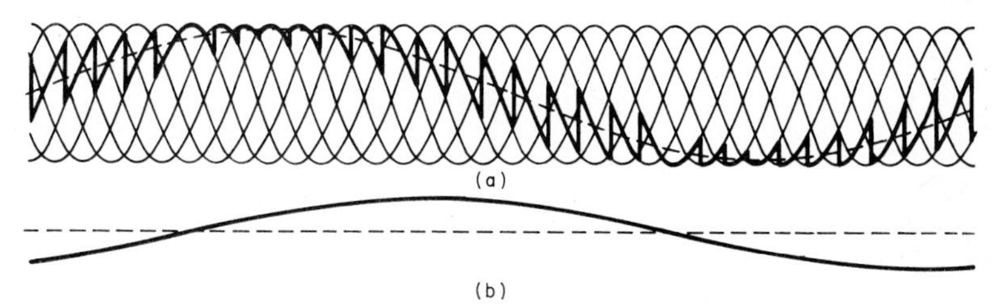


**FIG. 8.64** Cycloconverter output waveforms: ***(a)*** unfiltered output voltage (--- wanted component); ***(b)*** output current.

The converter can operate in all four quadrants and is very suitable for supplying multipole motors, providing a low-speed direct drive without the use of gears. The drive may be considered to be a voltage-source inverter drive in which the rectifier and inverter are combined in a single converter.

In summary, Table 8.1 gives an indication of possible applications of the electronic control of motors.

## REFERENCES

1. S. B. Dewan and A. Straughen, *Power Semiconductor Circuits*, Wiley-Interscience, New York, 1975.
2. S. B. Dewan and W. G. Dunford, *Improved Power Factor Operation of the Three-phase Rectifier Bridge through Modified Gating*, IEEE IAS Conference, 1980, Conference Record pp. 830–837.
3. S. B. Dewan, G. R. Slemon, and A. Straughen, *Power Semiconductor Drives*, Wiley-Interscience, New York, 1984.
4. B. R. Pelley, *Thyristor Phase-Controlled Converters and Cycloconverters*, Wiley-Interscience, New York, 1971.

**TABLE 8.1** Applications of the Electronic Control of Motors

| Converter | Motor | Typical output range, kW | Practical speed range for a one- or two-quadrant drive | Quadrants of operation | Supply | Variables governing speed | Applications |
|---|---|---|---|---|---|---|---|
| DC drives | | | | | | | |
| Single-phase rectifier with center-tapped transformer | DC separately excited or permanent magnet | 0–5 | 1:50 | Quad 1 driving<br>Quad 2 dynamic or mechanical braking<br>Reversal or armature connections or dual converter for four-quadrant operation | 120-V single-phase ac | Armature terminal pd<br>Field current | Small variable-speed drives in general |
| Single-phase fully controlled bridge rectifier | DC separately excited or permanent magnet | 0–10 | 1:50 | Quad 1 driving<br>Quad 2 dynamic braking<br>Reversal of armature connections or dual converter for four-quadrant operation | 240-V single-phase ac | Armature terminal pd<br>Field current | Processing machinery, machine tools |
| Single-phase half-controlled bridge rectifier | DC separately excited or permanent magnet | 0–10 | 1:50 | Quad 1 driving<br>Quad 2 dynamic braking<br>Reversal of armature connections for four-quadrant operation | 240-V single-phase ac | Armature terminal pd<br>Field current | Processing machinery, machine tools |
| Three-phase bridge rectifier | DC separately excited or permanent magnet | 10–200 (10–50 for PM motor) | 1:50 | Quad 1 driving<br>Quad 2 dynamic braking<br>Dual converter for regenerative braking with four-quadrant operation | 240- to 480-V three-phase ac | Armature terminal pd<br>Field current | Hoists, machine tools, centrifuges, calendar rollers |
| Three-phase bridge rectifier | DC separately excited | 200–1000 | 1:50 | Quad 1 driving<br>Quad 2 dynamic braking<br>Dual converter for regenerative braking with four-quadrant operation | 460- to 600-V three-phase ac | Armature terminal pd<br>Field current | Winders, rolling mills |
| Chopper (dc-to-dc converter) | DC separately excited | | 1:50 | Alternatives are: Quad 1 only; quads 1 and 2; quads 1 and 4; all four | 600- to 1000-V dc | Armature terminal pd<br>Field current | Electric trains, rapid transit, streetcars, trolley, buses, cranes |

| AC drives | | | | | | | |
|---|---|---|---|---|---|---|---|
| AC power controller | Class D squirrel-cage induction motor | 0–25 | 1:2 | Quad 1 driving<br>Quad 4 plugging with source phase-sequence reversal<br>Quad 3 driving<br>Quad 2 plugging | 240-V three-phase ac | Stator terminal pd | Centrifugal pumps and fans |
| Slip-energy recovery system | Wound-rotor induction motor | Up to 20,000 | 1:2 | Quad 1 driving, with source phase-sequence reversal<br>Quad 3 driving | Up to 5000-V three-phase ac | Rotor terminal pd | Centrifugal pumps and fans |
| Voltage-source inverter and additional converters | Class B or C squirrel-cage induction motors or synchronous motors | 15–250 | 1:10 | Quad 1 driving<br>Quad 2 regenerative braking, with appropriate additional converters<br>Quad 3 driving<br>Quad 4 regenerative braking | 240- to 600-V three-phase ac | Stator terminal pd and frequency | Group drives in textile machinery and run-out tables |
| Current-source inverter and additional converters | Class B or C squirrel-cage induction motors | 15–500 | 1:10 | Quad 1 driving<br>Quad 2 regenerative braking, with inverter control signal sequence reversal<br>Quad 3 driving<br>Quad 4 regenerative braking | 240- to 600-V three-phase ac | Stator current and frequency | Single-motor drives, centrifuges, mixers, conveyors, etc. |
| Current-source inverter and additional converters | Three-phase synchronous motors | Up to 15,000 | 1:50 | Quad 1 driving<br>Quad 2 regenerative braking, with inverter control signal sequence reversal<br>Quad 3 driving<br>Quad 4 regenerative braking | Up to 5000-V three-phase ac | Stator current and frequency | Single-motor drives, processing machinery of all kinds |

# CHAPTER 9
# PERMANENT MAGNET MACHINES

**K. J. Binns**

## 9.1 INTRODUCTION

The use of permanent magnet field systems in electric machines has found increasing acceptance in recent years. This trend is partly due to the need for inexpensive and reliable field systems, for which ferrite magnets are particularly well suited. A more important development, however, is that the application of new permanent magnet materials (such as the rare earths) in new configurations has resulted in a high specific output or in other characteristics that are hard to match in nonpermanent magnet machines.[1,2] The elimination of the *risk of demagnetization* under fault conditions and the need for a *high torque/volume ratio* have both helped focus attention on the merits of permanent magnet machines. Permanent magnet materials are now available with a wide range of characteristics, allowing considerable scope in the choice of magnet composition.

One of the limiting factors in the development of permanent magnet generators was the possibility of demagnetization with the use of alnico magnets, which operate normally on a minor loop. Some recent designs make use of ceramic and rare earth magnets; with careful design, these neither require stabilization nor are prone to minor-loop operation. There is now a wide choice of configurations which have been explored, and one needs to choose the geometry that will best meet a particular specification.

Permanent magnet machines tend to have higher efficiencies than wound-field machines because they have no excitation losses; in addition, of course, they have no slip-dependent copper losses. Perhaps one of the most important aspects of synchronous machines is that they can operate at or close to unity power factor, which is impossible with an induction motor.

The use of permanent magnets in conventional dc machines with commutators has resulted in an important new class of machines. The magnets take up less space

Professor K. J. Binns wishes to acknowledge the work of his colleagues, Drs. Jabbar, Low, and Wong, whose contributions are of the highest value.

The figures in this chapter are reproduced by permission of the IEE.

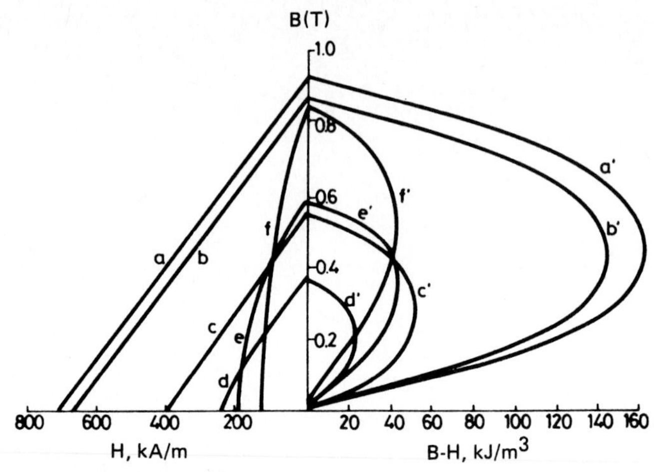


**a** = H22-A (rare earth)
**b** = H18-B (rare earth)
**c** = HERA (polymer-bonded rare earth
**d** = Ceramic 8 (anisotropic ferrite)
**e** = MnAlC
**f** = Alnico VIII

**FIG. 9.1** *B-H* characteristics and *B-H* curves of some permanent magnets.[17]

than a wound field does, and for small frame sizes this advantage is significant. Furthermore, the lack of field-winding loss increases the motor's efficiency, and the reduced cost when ceramic magnets are used tends to make such motors very cost-competitive. Clearly, field control is not practical, but dc chopper control is quite suitable for many applications.

Some designs have used relatively crude geometries that waste the magnet material, which in the case of rare earth magnets is expensive. The use of airgap magnets in synchronous machines is an example of this; the mean gap density is inevitably less than that in the magnet, roughly in the ratio of the pole arc to the pole pitch. This limitation is quite unnecessary for many applications.

## 9.2 THE PROPERTIES OF PERMANENT MAGNETS *

The available range of permanent magnet materials has increased in recent years. It now includes not only the alnico and ferrite materials but also rare earth compounds involving samarium or neodymium as well as such composites as manganese-aluminum-carbon. Figure 9.1 shows the characteristics of some commonly available materials, and Table 9.1 lists certain properties of these and similar materials.

*See also Chap. 2, "Magnetic Circuits."

The most important magnetic properties are the coercivity $H_c$ and the remanence $B_r$. The magnet will operate along a characteristic between these points depending on the reluctance of the magnetic circuit and the level of any applied field, such as the armature reaction field. If this graph is linear over the working region, there is no loss of stored energy as the reluctance of the circuit changes. When the graph is nonlinear, as with alnico magnets, stored energy is lost as the operating point moves up and down the characteristic; the magnet no longer works on the major loop characteristic and is said to operate on a minor loop. Attention will be concentrated in this chapter mainly on magnet materials that can be operated on a linear characteristic since these are the most important materials.

**TABLE 9.1** Magnetic and Physical Properties of Some Commercially Available Permanent Magnets

| Magnet material | $B_r$, T | $-H_c$, kA/m | $B$-$H$ max, kJ/m$^3$ | Density, kg/m$^3$ × 10$^3$ | Max. operating temp, °C | Reversible temp. coeff., % per °C | Curie temp., °C |
|---|---|---|---|---|---|---|---|
| Alnico V | 1.28 | 50 | 41 | 7.3 | 550 | −0.016 | 870 |
| Alnico VIII | 0.90 | 127 | 48 | 7.3 | 550 | −0.005 | 870 |
| Ceramic 7 | 0.36 | 255 | 23 | 4.9 | 350 | −0.200 | 450 |
| Ceramic 8 | 0.40 | 247 | 29 | 4.9 | 350 | −0.200 | 450 |
| Polymer-bonded rare earth cobalt, B2 | 0.55 | 360 | 48 | 5.1 | 120 | −0.040 | * |
| Rare earth cobalt, H-90A | 0.82 | 597 | 122 | 8.2 | 250 | −0.042 | 710 |
| Rare earth cobalt, H-90B | 0.87 | 653 | 142 | 8.3 | 250 | −0.042 | 710 |
| Rare earth cobalt, H-99A | 0.97 | 477 | 176 | 8.5 | 250 | −0.035 | 870 |
| MnAlC | 0.58 | 190 | 45 | 5.0 | 300 | −0.140 | 350 |

* This material disintegrates structurally at temperatures exceeding 120°C as a result of the breakdown of the bonding media.

*Source:* T. M. Wong, "The Finite-Element Analysis of a High-Field Permanent Magnet Machine," Ph.D. thesis, University of Southampton, Hampshire, England, 1983.

It is important to note the maximum permissible operating temperature and also the temperature above which reversible changes take place (Table 9.1). The reversible temperature coefficient gives a measure of the change in strength of a magnet with temperature over the range for which the magnet properties revert to the initial values when the magnet temperature falls to a normal working level. Irreversible temperature coefficients apply outside a normally acceptable working range. Stabilization at a temperature slightly above the specified maximum working temperature is effective in preventing additional irreversible loss of stored energy.

The new neodymium-iron-boron alloys can have energy products exceeding 300 kJ/m$^3$. There is, however, a problem still unresolved concerning the maximum operating temperature, which is at present below the normal rated temperature of most electric machines. Researchers are investing considerable effort in attempts to provide a material that combines high stored energy and a reasonable range of working temperature.

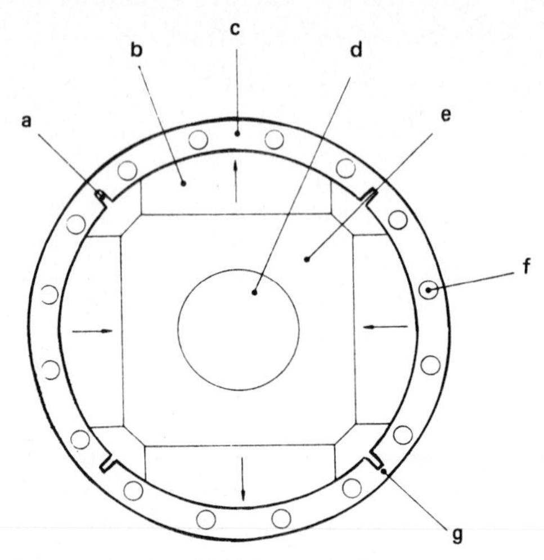


⟶ Direction of initial magnetization

**a** Slot introduced to reduce leakage
**b** Magnet
**c** Laminated ring
**d** Magnetic shaft
**e** Magnetic hub
**f** Cage bar
**g** Iron bridge

**FIG. 9.2** A four-pole rotor configuration of the Permasyn design.[19]

## 9.3 HISTORICAL DEVELOPMENT

The earliest modern work on permanent magnet ac machines was done by Merrill, who developed a motor called the Permasyn.[3] Two very real problems were encountered with this design (Fig. 9.2): the leakage of flux from one pole to another, and the presence of very little reluctance action. A metallic alnico magnet was used, and the motor was very sensitive to increases in the voltage supply during starting, with a risk of partial demagnetization. In addition, the mechanical construction was complicated since the outer ring of lamination had to have a very good interference fit on the permanent magnets.

The development of the Statexyn motor involved the incorporation of permanent magnets into the rotor to increase the useful flux, though at the expense of extra complexity.[4] Honsinger proposed the inclusion of permanent magnets into the regions normally used as flux barriers in reluctance motors so as to reduce the quadrature-axis flux.[5]

A design by Siemens made use of ferrite magnets positioned between separate core segments (Fig 9.3).[6] The rotor was virtually held together by the die-cast cage, and this rotor configuration was the first of the manufactured "buried magnet" designs. There was little or no magnet material between the magnet and the rotor shaft, the aim being to minimize magnet leakage flux and have a separate magnetic circuit for the flux from each magnet. At first glance this seems a reasonable idea, but it limits the performance capability when a choice of possible magnet materials is considered.

Another version used a similar concept but had the rotor external to the stator (Fig. 9.4), which had the advantage of providing more space for magnet material.[7] A novel configuration for a machine started and run from an inverter was one in which ferrite magnets were used as salient magnets and were surmounted by steel pole pieces.[8] Because of the limitation in the flux-carrying capacity of ferrite magnets, the direct-axis flux was severely limited and the machine behaved like a segmental rotor reluctance motor.

Configurations of hybrid permanent magnet reluctance motors have been developed that can make use of ferrite or rare earth magnets.[9,10] One-piece laminations

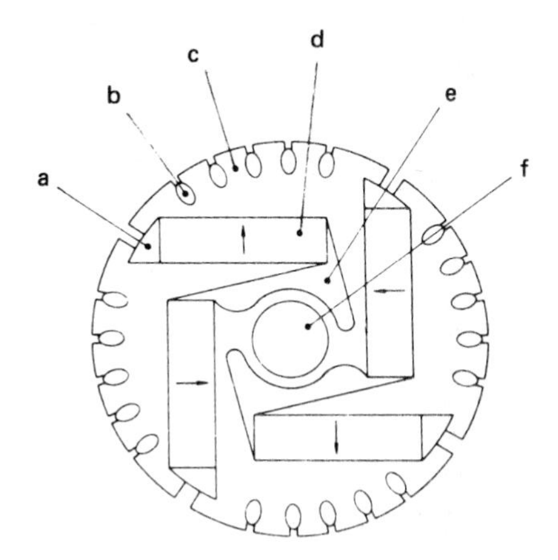


⟶ Direction of initial magnetization
**a** Nonmagnetic spacer
**b** Cage bar
**c** Laminated segments held together by die-cast cage
**d** Magnet
**e** Nonmagnetic spacer
**f** Nonmagnetic shaft

**FIG. 9.3** Rotor configuration of the Siemens permanent magnet motor.[19]

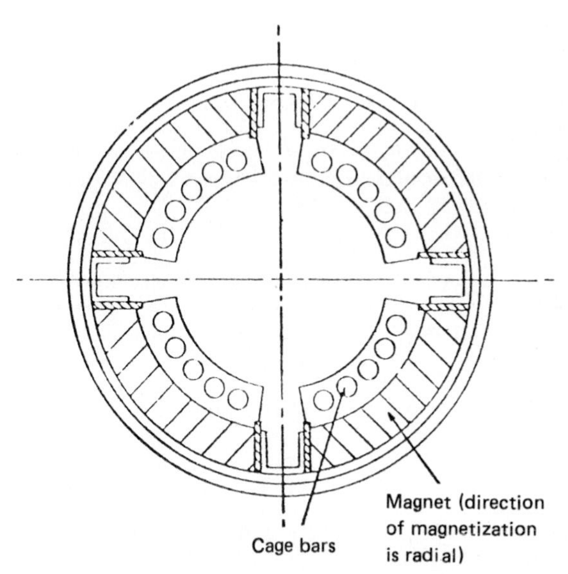


**FIG. 9.4** Siemens' alternative design of a permanent magnet motor.[19]

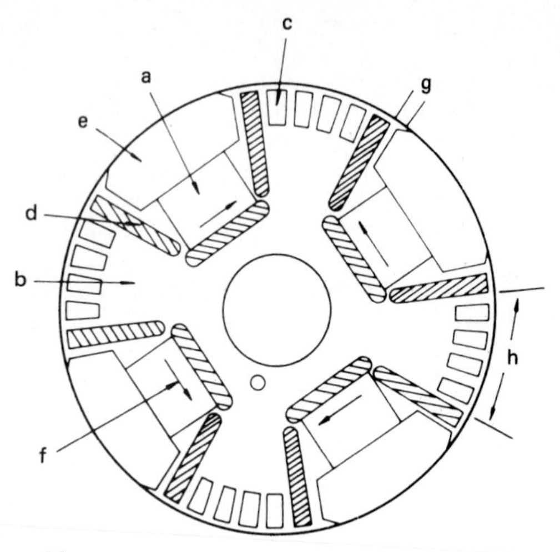


**a** Magnets
**b** Rotor iron
**c** Cage bars
**d** Flux barriers
**e** Channel (filled with copper-aluminum)
**f** Directions of initial magnetization
**g** Pole-shoe width (0.25 cm)
**h** Pole arc (46.74°)

**FIG. 9.5** Schematic diagram of the geometry of a typical hybrid motor.[10]

are used, and they are designed to give a combined action, the paths for the permanent magnet flux being essentially in parallel with the paths of the reluctance action flux. Additionally, the flux path at starting is of low reluctance and is shared with that of the reluctance action when synchronized. The purpose of the permanent magnets is to improve the polarities on the corners of the rotor poles, which greatly improves synchronization and also serves to supply additional useful flux.

**TABLE 9.2** Effect on Performance of Width of Pole Shoes

| Pole-piece width, cm | No-load current, A | Pull-out power, W | Maximum efficiency, % | Maximum power factor |
|---|---|---|---|---|
| 0.24 | 4.50 | 2980 | 78 | 0.63 |
| 0.48 | 4.08 | 2630 | 76 | 0.61 |
| 0.95 | 3.47 | 1970 | 72 | 0.59 |
| 1.59 | 3.53 | 1410 | 68 | 0.54 |

*Source:* K. J. Binns, M. A. Jabbar, and G. E. Parry, "Choice of Parameters in the Hybrid Permanent Magnet Synchronous Motor," *Proc. IEE*, vol. 126, no. 8, 1979, p. 742.

Figure 9.5 shows the preferred form of construction for this type of machine. The magnets are of simple shape and fit into slots in the one-piece laminations. The rotor can be die-cast not only to provide a squirrel cage for self-starting but also to fill the flux barriers separating the reluctance circuit from the permanent magnet circuit.

The reluctance action clearly depends on the degree of saliency of the poles and the width of the flux barriers. The pole arc is critical in its influence on pull-out torque, but it also determines the space available for the permanent magnet and its iron circuit. For a given space, one should attempt to maximize the magnet flux reaching the airgap. A less obvious point is the need to concentrate the flux over a small peripheral section of airgap in order to make the rotor field's form steep-sided at the pole edges. This requires narrow pole pieces to the flux from the magnets, as shown in Table 9.2. For still narrower pole pieces, mechanical problems occur; also, the rise in no-load current can become disadvantageous.

A very different design of machine is shown in Fig. 9.6.[11] This machine does not incorporate a cage and thus is not self-starting on a constant-frequency supply. The machine is appropriate for a fairly large pole number: for example, 12 (as shown in Fig. 9.6). Designs for pole numbers less than six are impractical, but there is virtually no clear upper limit. It can readily be deduced that if the slots can be sufficiently deep, the airgap flux density can be considerably higher than that in the magnet. This configuration of rotor is particularly suited to applications in which a

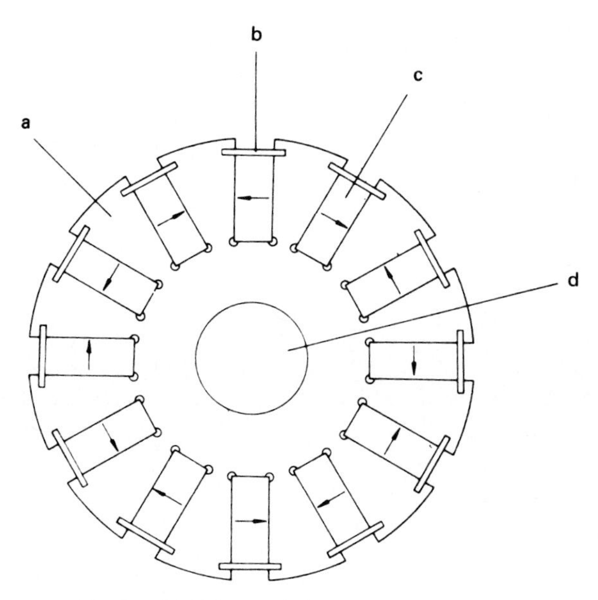


⟶ Direction of initial magnetization

**a** Rotor-iron lamination

**b** Magnet retaining plate

**c** Magnet

**d** Nonmagnetic shaft

**FIG. 9.6** A 12-pole rotor with tangentially magnetized magnets between its pole segments.[16]

stable, slow running speed is desirable. The machine can also serve as an efficient generator, particularly when frequencies well above 50 Hz are required. The major drawback of this design is the mechanical strength of the rotor. It is clear from Fig. 9.6 that the section of the base of a rotor tooth is critical and the basic design has the challenging problem that the optimization of the mechanical strength is in conflict with the maximization of rotor flux.

Recently, a new configuration of self-starting permanent magnet machine has been proposed and intensively developed for a range of applications.[12,13] It can make use either of the inexpensive barium or strontium ferrites or of any of the rare earth composites. A typical four-pole configuration is shown in Fig. 9.7. The magnets are arranged in a T formation for any pair of adjacent poles and are of

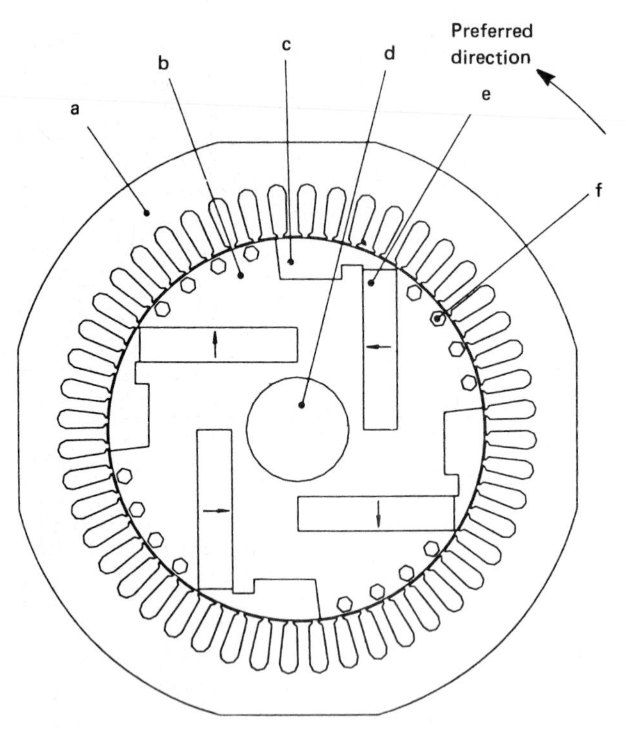


⟶ Direction of magnetization

**a** Stator-iron lamination

**b** Rotor-iron lamination

**c** Nonmagnetic material

**d** Nonmagnetic steel shaft

**e** Magnet

**f** Rotor-cage bar (circular)

**FIG. 9.7** The configuration of the initial design of a 45-kW permanent magnet synchronous motor.[13]

a simple rectangular section. By careful design, it is possible to eliminate almost completely the magnet leakage flux, and the gap density would normally exceed the density in the magnet. This configuration superficially resembles that of Fig. 9.3, but there are important differences in the way the flux produced by the magnets is disposed through the rotor. In the Siemens machine the rotor has a nonmagnetic spacer between the magnet and the shaft to form effectively a flux barrier, and this also serves to prevent flux from an adjacent magnet from passing through another magnet. The rotor of Fig. 9.7, on the other hand, has an iron bridge between the magnet and the nonmagnetic shaft. Some of the flux leaving a given pole passes through one magnet only, while the rest passes through two magnets. The total flux per pole therefore exceeds the total flux per magnet. This is shown in Fig. 9.8, from which it is evident that a high gap density can be achieved.

Another important difference is the presence of a salient pole in this design, with a pole-arc/pole-pitch ratio not much greater than 60 percent; in contrast, the Siemens rotor has a very high value (close to 90 percent). The flux focusing achieved in the new design gives a gap density considerably higher than that in the magnet itself and also a "sharp" edge to the field form.

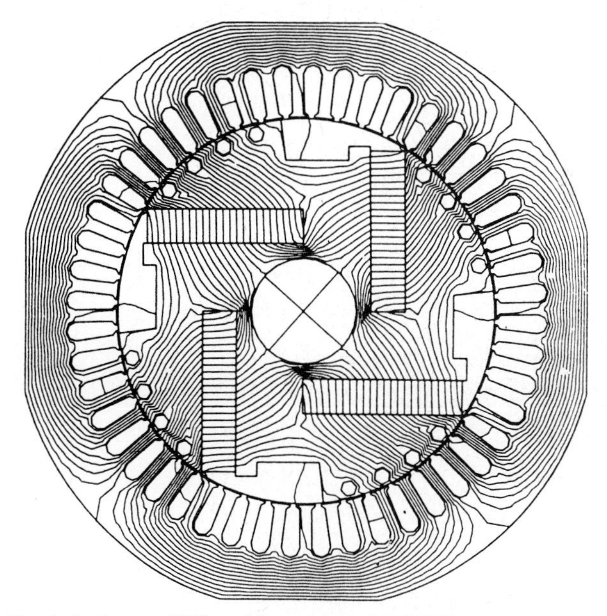


| | |
|---|---|
| Terminal voltage = 476 V | Magnet material = H-90B |
| Stator current = 50 A | Radial airgap = 0.838 mm |
| Output power = 39.7 kW | Magnet thickness = 19.8 mm |
| Power factor = 0.965 lag | Magnet width = 92 mm |
| Frequency = 60 Hz | |

**FIG. 9.8** Flux distribution of the initial design of the motor with a thicker magnet on load in the preferred direction of rotation.[13]

Early designs of permanent magnet generators made use of integrally cast magnets involving isotropic alnico magnets.[14] Poles were produced on the rotor surface with this one-piece construction, as shown in Fig. 9.9. For large machines, block magnets were mounted on the rotor hub and steel pole pieces and magnets were supported by nonmagnetic bolts, as in Fig. 9.10. The magnets were radially polarized, and the whole rotor assembly could be die-cast in aluminum.

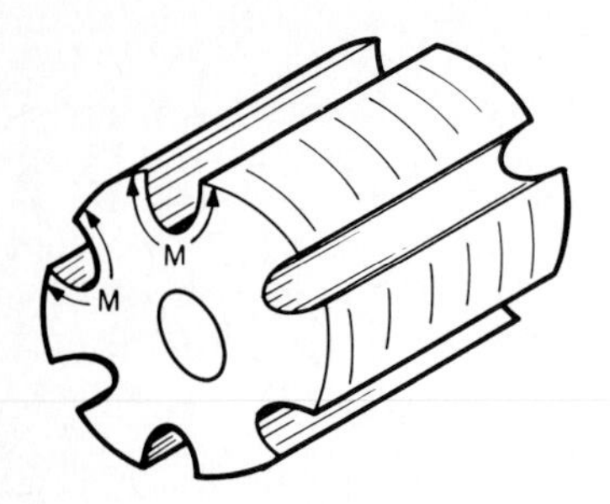


**FIG. 9.9** Integrally cast magnet for small machines.[16]

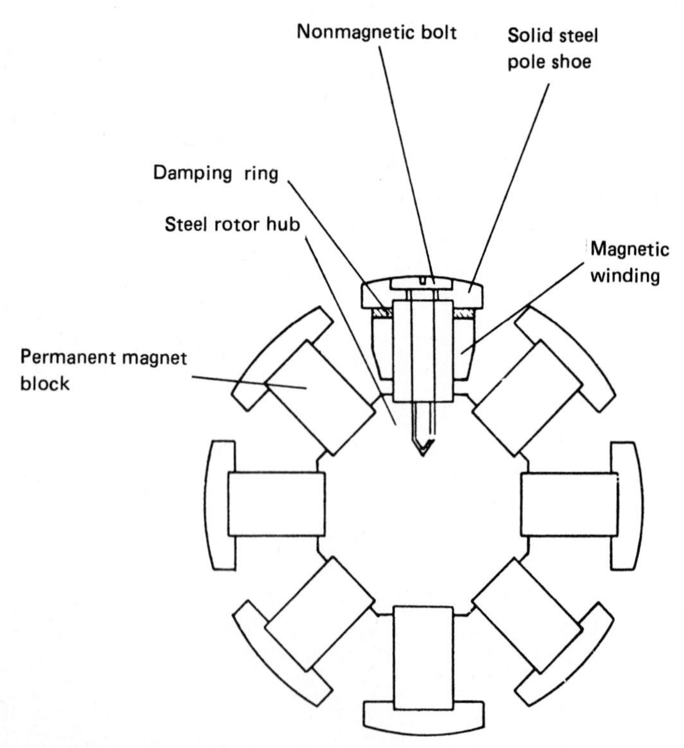


**FIG. 9.10** Design using salient magnets and steel pole shoes.[16]

Another class of permanent magnet generator uses axially magnetized magnets; an early form of this is commonly referred to as the Lundell, or claw-type, form of construction. Axially magnetized magnets have also been used in another form of construction.[15] The rotor has a disk magnet with a central hole and a number of radially projecting parts on its periphery, as seen in Fig. 9.11. Pole shoes are arranged on the projecting pole part and overlap the latter in the peripheral direction so that the polarity of the radial projections on the first differs from the second, thereby achieving alternating polarity.

The rotor already discussed in its role as a motor field system (Fig. 9.6) serves also as a good multipole generator. Designs involving the use of ferrites and rare

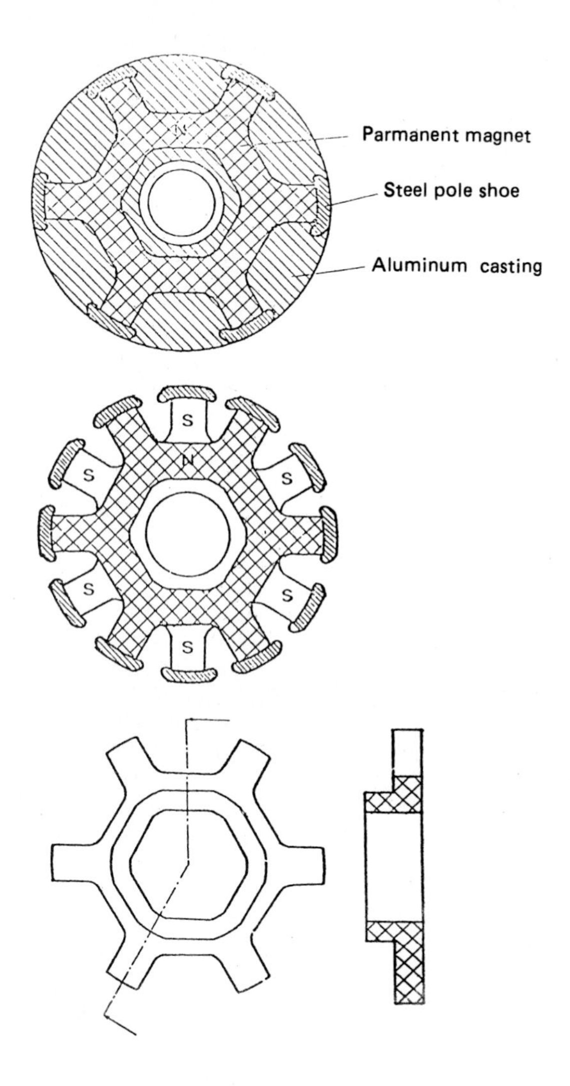


**FIG. 9.11** Rotor using axially magnetized magnets.

earth magnets have been widely investigated. This configuration seems simple to implement but has the disadvantage that the airgap flux density cannot exceed the density in the magnet.

The axial rotor configuration has been neglected because of the inherent difficulty in guiding the axial flux radially across the airgap. More recently, a form of generator has been designed with a disk-magnet rotor and a multistacked arrangement, as shown in Fig. 9.12.[16] The multistacked generator is ideally suited to the use of ceramic magnets, but other magnets can be used, including the rare earth composites.[17] It is made up of a stack of rotor units, each unit comprising two steel

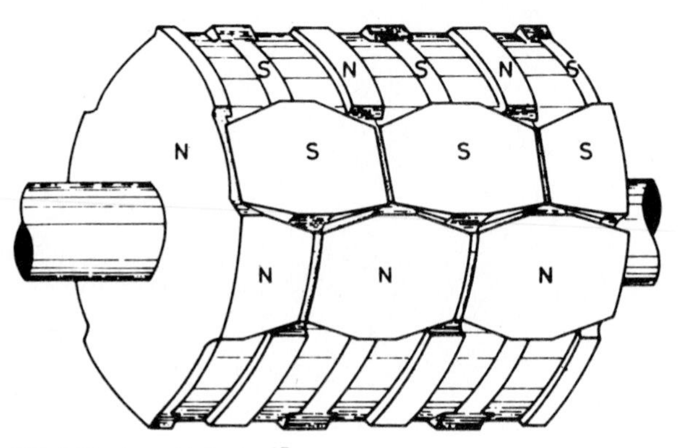


**FIG. 9.12** Assembled rotor.[17]

flux guides with an axially magnetized disk magnet sandwiched between them. The steel disk guides the flux radially to the poles, which are shaped like a trapezoid and extend from the plate, which is attached to one pole face of the magnet. Figure 9.12 shows an assembled rotor, and Fig. 9.13 shows an individual steel flux guide. Figure 9.14 shows a cross section through the rotor axis of a four-unit machine.

DC machines with conventional commutators have recently made extensive use of permanent magnets for field excitation. Motors for the automobile industry are a prominent example of their commercial use. Essentially, the wound-field system is replaced by that of a set of permanent magnets (Fig. 9.15). Chopper control systems can be used for speed control, and ceramic magnets give an inexpensive form of machine that has applications in many areas. Good field forms can be obtained, and radially polarized anisotropic magnets tend to predominate.

## 9.4 COMPUTER-AIDED DESIGN OF PERMANENT MAGNET MACHINES

Increasingly, electric machines are being designed with the assistance of computers, which not only solve the field equations but also process the display of geometry and data.

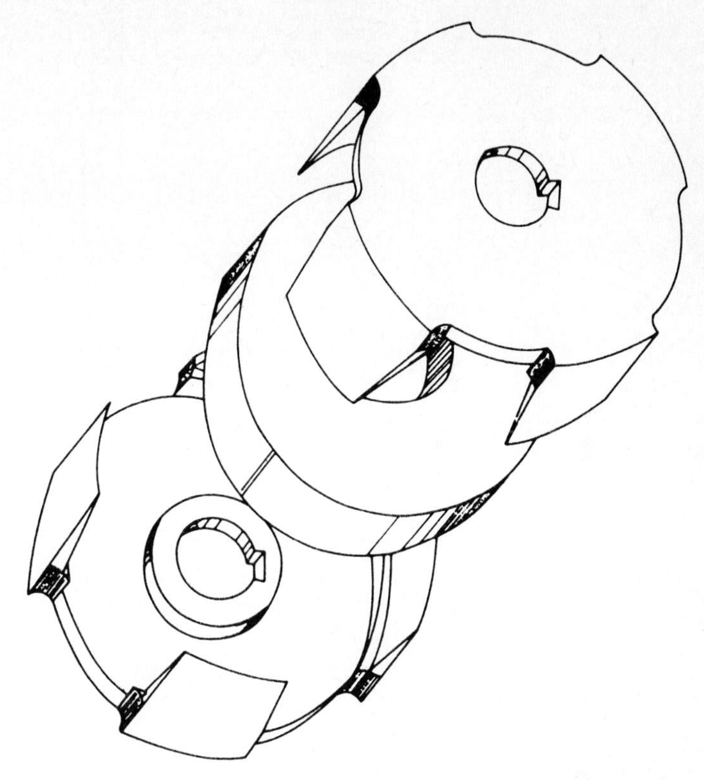

**FIG. 9.13** Rotor unit.[17]

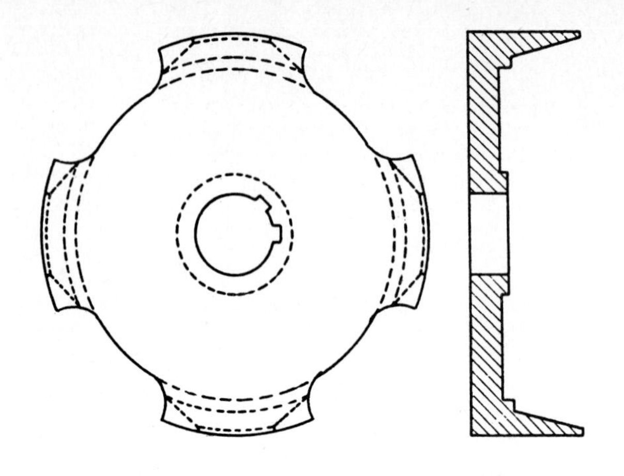

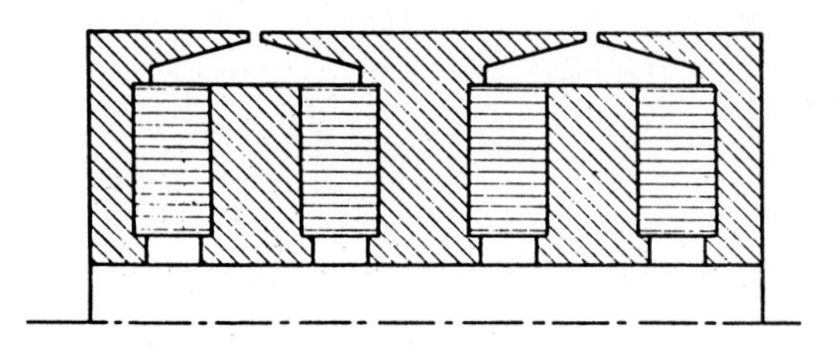

Permanent-magnet material

Steel

**FIG. 9.14** Cross section and assembly of new rotor.[16]

One of the most valuable methods for the fundamental design analysis of electric machines is the finite-element method.[13,16,18] The magnetic field distribution is evaluated by a discretization process, which involves replacing the field equation by a set of simultaneous equations that apply to a network of elements distributed throughout the machine or a cross section of it. Such an arrangement of elements is shown in Fig. 9.16 for a repeatable section of the type of configuration shown in Fig. 9.7.

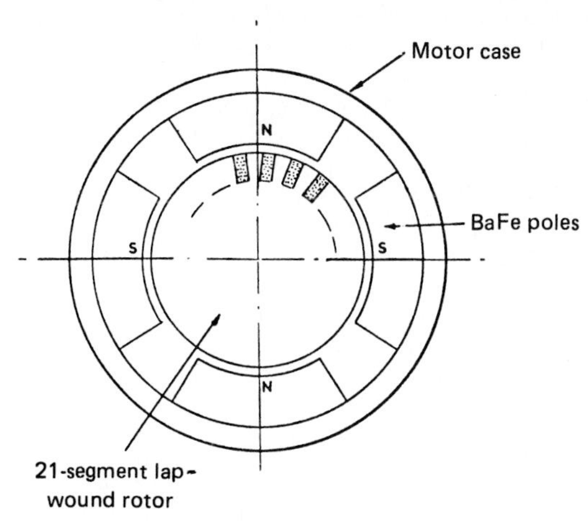


**FIG. 9.15** Cross section of permanent magnet dc motor with segmental poles.

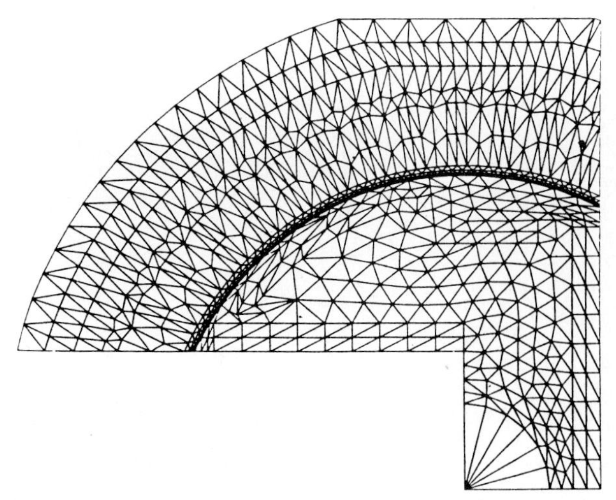

**FIG. 9.16** Triangular finite-element mesh of the initial design of a 45-kW permanent magnet synchronous motor over a pole pitch (1500 elements).[13]

As a result of the application of finite-element methods, conclusions can be drawn and design criteria established for permanent magnet machines. A number of important examples will now be discussed over a wide range of useful geometries.

## 9.5 SYNCHRONOUS PERFORMANCE OF A HYBRID PERMANENT MAGNET RELUCTANCE MOTOR

A configuration of a hybrid motor is shown in schematic form in Fig. 9.17, and Fig. 9.18 gives a flux distribution at a point close to full load for a machine making use of ferrite magnets.[10]

The pole arc (Fig. 9.17) is clearly an important design parameter, particularly in regard to power factor, efficiency, no-load current, and pull-out torque. The pole-arc/pole-pitch ratio affects the torque because of reluctance action, but it also influences the space available for the magnet and its associated flux guides and barriers. The pull-out torques for three values of pole arc are shown in Fig. 9.19 for different load-current levels. The trend is clear: The pull-out torque rises as the arc narrows. Tests on machines show that this continues into a configuration for which other parameters suffer unacceptably for practical designs.

The effect of pole-shoe width is also a critical parameter. Its effect on pull-out

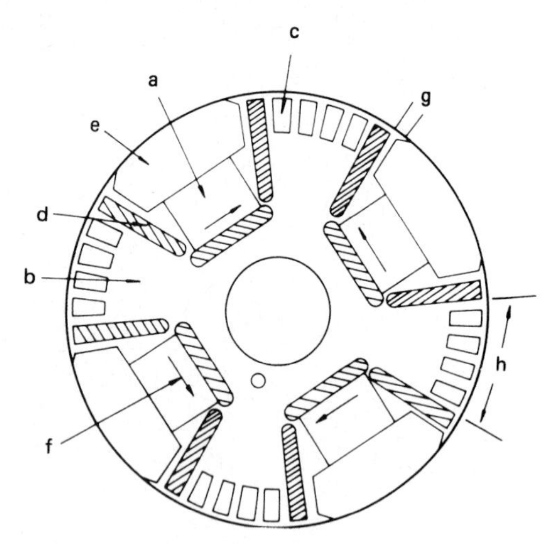


**a** Magnets
**b** Rotor iron
**c** Cage bars
**d** Flux barriers
**e** Channel (filled with copper-aluminum)
**f** Directions of initial magnetization
**g** Pole-shoe width (0.25 cm)
**h** Pole arc (46.74°)

**FIG. 9.17** Schematic diagram of the geometry of a typical hybrid motor.[10]

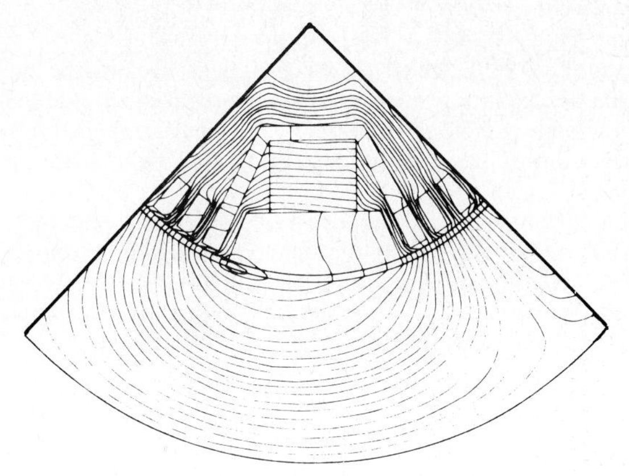

FIG. 9.18 Flux plot at near full load.[10]

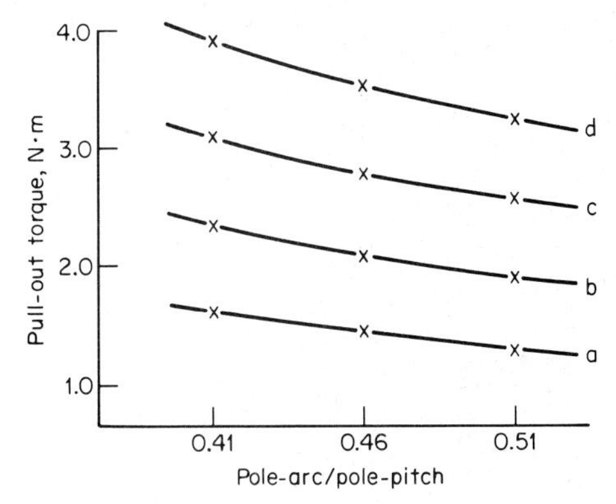


**FIG. 9.19** Variation of pull-out power with pole-arc/pole-pitch ratio.[10]

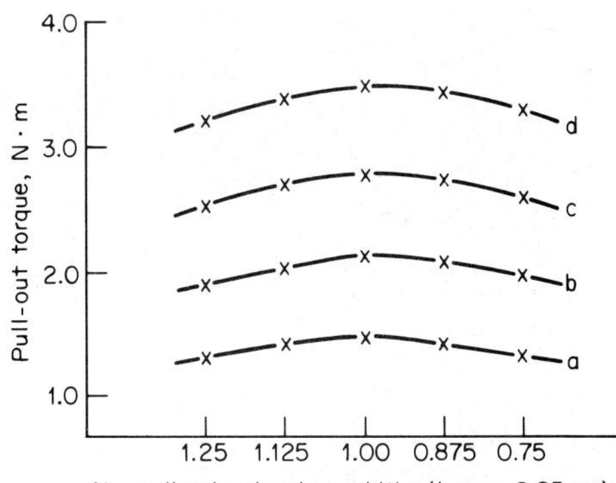


**a** 4 A **b** 5 A **c** 6 A **d** 7 A

**FIG. 9.20** Variation of pull-out torque with pole-shoe width as a function of stator currents.[10]

torque is seen in Fig. 9.20, which shows variations around the maximum torque position. The maximum torque condition can be realized in practice, but narrower peaks are impractical as well as undesirable. A significant degree of saturation is needed, and it is desirable to maintain a high flux density at the pole edge to counteract armature reaction.

The flux barrier plays a vital role in guiding the permanent flux and improving the reluctance action. It helps limit the demagnetizing forces on the magnets and concentrate the flux into the required pattern. However, the actual width of the barrier above a certain level is unimportant; frequently, the choice of width is affected by mechanical considerations. The barrier's effect on pull-out torque is shown in Fig. 9.21.

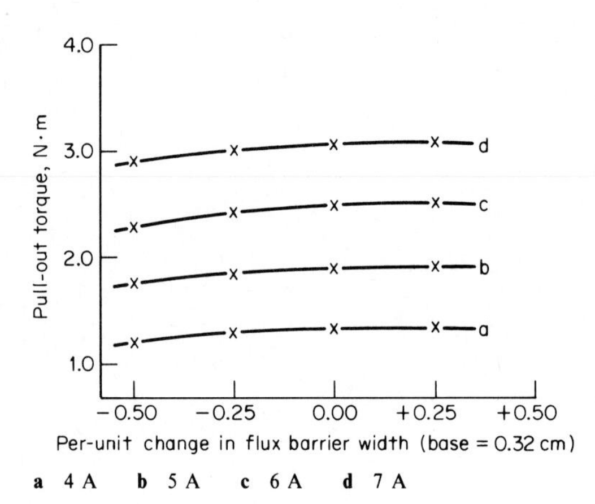


**a** 4 A **b** 5 A **c** 6 A **d** 7 A

**FIG. 9.21** Variation in pull-out torque with changes in flux-barrier width as a function of stator currents.[10]

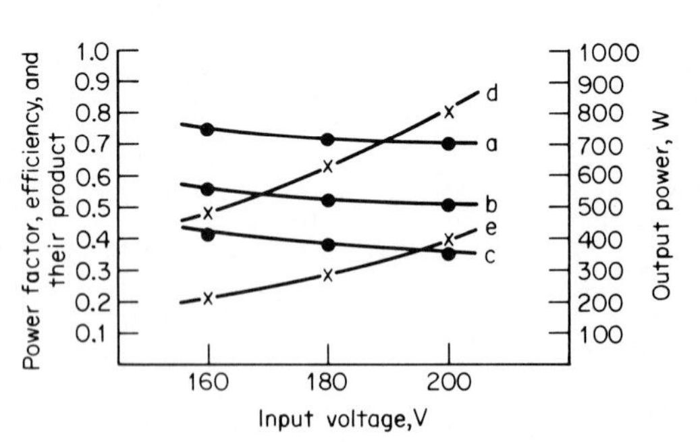


**a** Efficiency

**b** Power factor

**c** Product of power factor and efficiency

**d** Pull-out power

**e** lem Pull-in power (load inertia = 12 × rotor inertia)

**FIG. 9.22** Synchronous performance characteristics of a typical hybrid motor at various flux levels.[10]

Both power factor and efficiency are load-dependent, and a performance factor of importance is the product of efficiency and power factor. Some typical performance figures are indicated in Fig. 9.22 for a four-pole motor with anisotropic ferrite magnets. Despite the high connected inertia, the synchronizing capability of the machine is good.

## 9.6 IMBRICATED PERMANENT MAGNET GENERATORS

The significance of design parameters is here examined for the configuration shown in Fig. 9.12. This type of rotor is ideally suited to the use of cheap ceramic magnets,

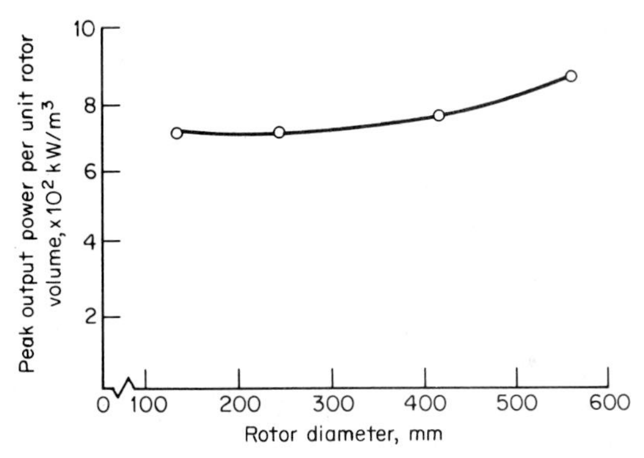


**FIG. 9.23** Influence of rotor diameter on the output power density for generators using ferrite magnets (ceramic 8).[17]

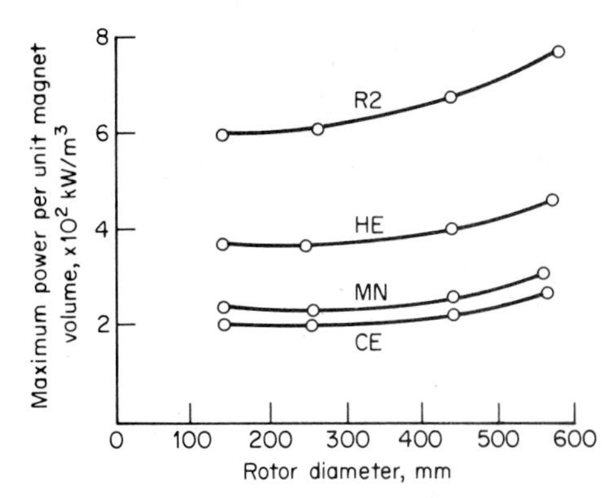


**FIG. 9.24** Relationship between output, magnet material, and rotor size.[17]

but high-energy magnets can also be used to advantage. The most appropriate magnet materials are characterized by a high coercivity, which is utilized to prevent demagnetization even under the application of a short circuit. Another feature of this rotor is its ability to achieve an airgap flux density in excess of the magnet's remanence by using a flux concentration obtained by careful design of the steel configuration.[17,20]

The influence of the rotor diameter on the output power is shown in Fig. 9.23. It relates the maximum power output per unit rotor volume with the rotor diameter. The power density can be seen to rise with an increase in diameter.

A comparison of the outputs of rotors of different sizes employing different magnet materials is shown in Fig. 9.24 in terms of output per unit volume of magnet. The curves give a measure of the power density that can be achieved for each material used for excitation. The increase in output with size increases with the maximum energy product of the magnet. The increase in output with material falls behind the improvement in $B\text{-}H_{max}$.

## 9.7 HIGH-FIELD PERMANENT MAGNET MACHINES

A configuration of high-field machines has been shown in Fig. 9.7 for a machine in which the gap density would normally exceed that of the existing magnets. A flux

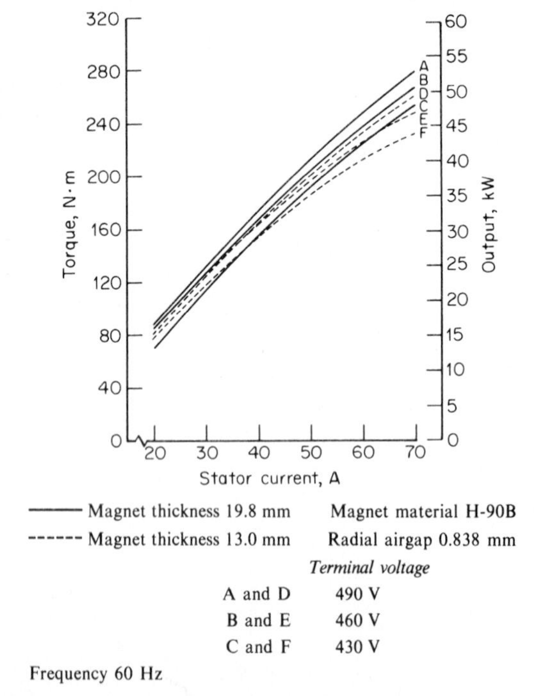


**FIG. 9.25** Computed torque-current characteristics for the two different sizes of magnet.[19]

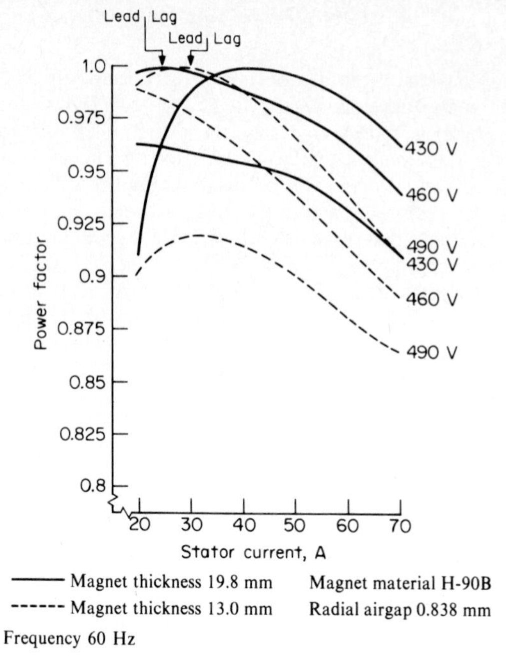


**FIG. 9.26** Computed power-factor–current characteristics for the two different sizes of magnet.[15]

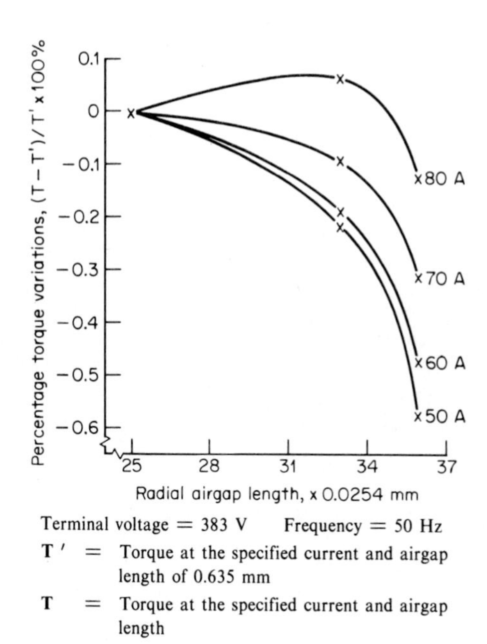


Terminal voltage = 383 V    Frequency = 50 Hz

**T'** = Torque at the specified current and airgap length of 0.635 mm

**T** = Torque at the specified current and airgap length

**FIG. 9.27** The effect of the radial airgap length of the torque in the preferred direction of rotation.[19]

density distribution has also been demonstrated for a four-pole configuration. Some of the more fundamental design aspects will now be considered.

The choice of magnet thickness across the direction of polarization is clearly important. Figure 9.25 demonstrates the variation in torque as a function of current variation for two different thicknesses of magnet. Figure 9.26 shows the variation in power factor under the same conditions. It can be seen that a better performance can be obtained at leading power factors with the thinner magnet, while at a lagging power factor the thicker magnet is superior.

The effect of the radial airgap length is clearly important. Three airgap lengths (0.63 mm, 0.84 mm, and 0.91 mm) are here examined. Figures 9.27 and 9.28 illustrate the variation of torque and power factor, respectively, as functions of the airgap for a range of currents. As would be expected, both the torque and the power factor decrease as the airgap is increased. The power-factor varies almost linearly with the airgap over the range considered, but the torque decreases more rapidly as the airgap increases from 0.84 to 0.91 mm than it does in the range from 0.63 to 0.84 mm.

The iron bridge (between the magnets and the nonmagnetic shaft) through which any flux passing through more than one magnet must flow is an interesting parameter. This is illustrated in Fig. 9.29. A decrease in the depth of the bridge produces an increase in rotor field for each pole. The power factor for a predefined voltage and current input has a trend to move from lagging to leading Fig. 9.30 as a function of the reduction of the bridge width.

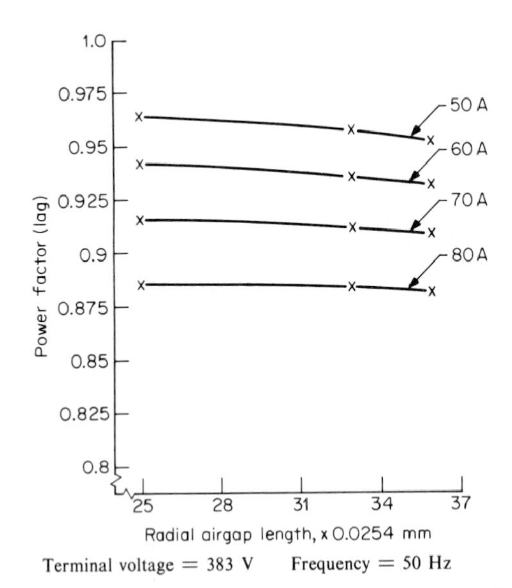


**FIG. 9.28** The effect of the radial airgap length on the power factor in the preferred direction of rotation.[19]

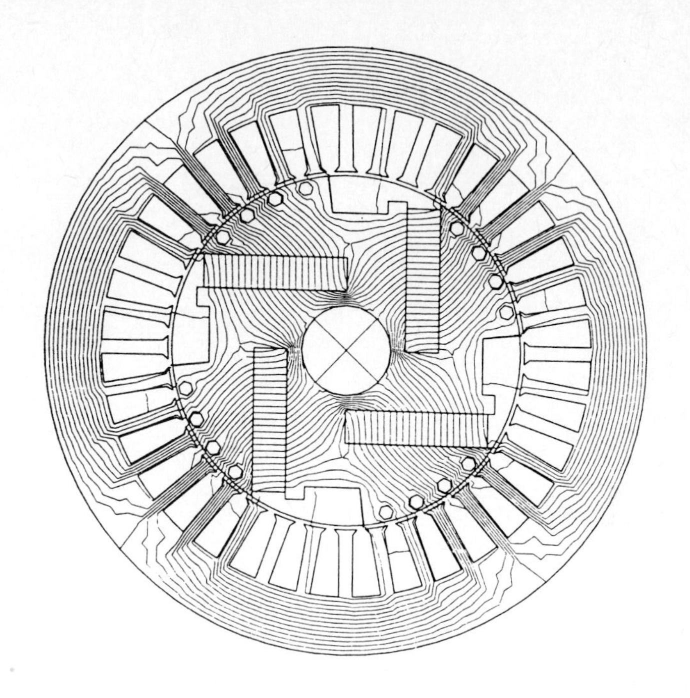

Scale 1:1

Minimum potential contour = −0.01443 Wb/m
Maximum potential contour = 0.01443 Wb/m
Potential contour increments = 0.00111 Wb/m

**FIG. 9.29** Flux distribtuion of the base machine at full load in the preferred direction of rotation.[19]

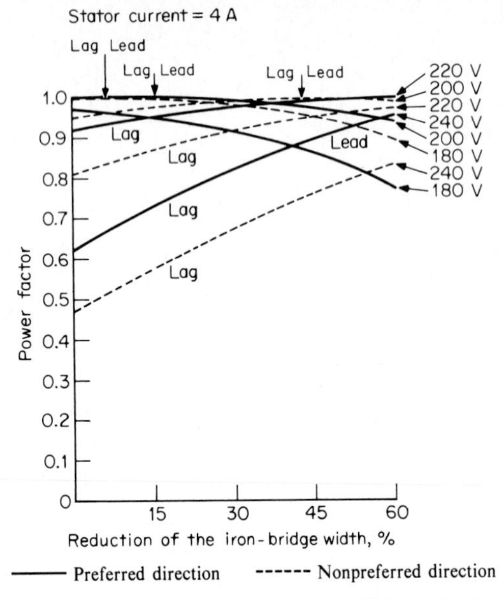


**FIG. 9.30** The variation of the power factor with the iron-bridge width as a function of terminal voltage.[19]

## REFERENCES


1. K. J. Binns and M. A. Jabbar, "A High-Field Self-Starting Permanent Magnet Synchronous Motor," *Proc. IEE*, vol. 128, pt. B, no. 3, May 1981, pp. 157–160.
2. K. J. Binns, "Permanent Magnet Motors for Inverter-Fed Drives," *Conference on Drives, Motors, Controls*, Brighton, England, October 1981.
3. F. W. Merrill, "Permanent Magnet Excited Synchronous Motors," *AIEEE Transactions*, vol. 73, pt. 3, 1954, pp. 1754–1760.
4. Heemaf, "Synchronous Electric Machine," U.K. Patent 1 366 949.
5. V. R. Honsinger, "Permanent Magnet Motors," U.K. Patent 1 126 493.
6. Siemens, "An Electric Machine Having Permanent Magnets Mounted in the Rotor between Its Pole Segments," U.K. Patent 1 177 247.
7. Siemens, "Synchronous Machines," U.K. Patent 1 056 605.
8. M. Lajoie-Mazenc, R. Carlson, J. Hector, and J. J. Pesque, "Characterization of a New Machine with Ferrite Magnets by Resolving the Partial Difference Equation for the Magnetic Field," *Proc. IEE*, vol. 124, no. 8, 1977, pp. 697–701.
9. K. J. Binns, W. R. Barnard, and M. A. Jabbar, "Hybrid Permanent Magnet Synchronous Motors," *Proc. IEE*, vol. 125, no. 3, 1978, pp. 203–208.
10. K. J. Binns, M. A. Jabbar, and G. E. Parry, "Choice of Parameters in the Hybrid Permanent Magnet Synchronous Motor," *Proc. IEE*, vol. 126, no. 8, 1979, pp. 741–744.
11. A. Hameed, K. J. Binns, A. Vandenput, and W. Geysen, "Finite-Element Computation of the Field in a Permanent Magnet Machine," *Int. Sym. on Mach. and Converters*, Liège, Belgium, May 1984.

12. K. J. Binns and M. A. Jabbar, "Comparison of Performance Characteristics of a Class of High-Field Permanent Magnet Machines for Different Magnet Materials," *IEE Conference (SSEM)*, London, 1981.

13. K. J. Binns and T. M. Wong, "Analysis and Performance of a High-Field Permanent Magnet Synchronous Machine," *Proc. IEE*, vol. 131, pt. B, no. 6, November 1984, pp. 252–258.

14. C. J. Mole, "Permanent Magnet Generators," *Electrical Times*, December 1956, pp. 893–898.

15. Y. Zumazawa, "Improvements in or Relating to Permanent Magnet Generators," U.K. Patent 1 204 844, May 1970.

16. K. J. Binns and A. Kurdali, "Permanent Magnet AC Generators," *Proc. IEE*, vol. 126, no. 7, July 1979.

17. K. J. Binns and T. S. Low, "Performance and Application of Multistacked Imbricated Permanent Magnet Generators," *Proc. IEE*, vol. 130, pt. B, no. 6, November 1983.

18. K. J. Binns, M. A. Jabbar, and W. R. Barnard, "Computation of the Magnetic Field of Permanent Magnets in Iron Cores," *Proc. IEE*, vol. 122, no. 12, December 1975.

19. T. M. Wong, "The Finite-Element Analysis of a High-Field Permanent Magnet Machine," Ph.D. thesis, University of Southampton, England, 1983.

20. T. S. Low, "The Optimal Configuration of a Multistacked Permanent Magnet Generator," Ph.D. thesis, University of Southampton, England, 1982.

# CHAPTER 10
# SUPERCONDUCTING MACHINES

**Carl Flick**

## *10.1 PRINCIPLES AND APPLICATIONS OF SUPERCONDUCTIVITY*

### General

The application of superconductors to electric machine windings overcomes the major factor limiting both output capability and efficiency, namely, the "ohmic" (or resistive) losses. Losses in superconducting windings approach zero with dc and can be kept small even with ac by appropriate conductor design. The refrigeration power and other provisions required by the use of superconductors slightly offset these advantages, but the net result remains beneficial for certain types and sizes of machines.

*Superconductivity* is the absence of electrical resistance. H. Kamerlingh Onnes, the Dutch physicist (1853–1926), found that at a certain critical temperature the resistance of mercury dropped to a value too small to be measured. However, practical superconductors did not emerge until the early 1960s. The evolution in this field has been very rapid since that time and has taken a prominent place in the literature on electric machines, especially during the 1970s and 1980s. In the initial phases of this evolution, superconductivity was hailed as promising solutions to numerous, previously insuperable design problems. As of 1986 a great deal of development toward commercialization had taken place, but prospects for early commercialization had receded; active development in specialized areas, such as defense applications in earth and space environments, continued at an increased pace.

---

*Acknowledgment*: The permission of the Westinghouse Electric Corporation to publish this material is gratefully acknowledged.

Thanks are due to the membership of the "superconducting community," many of whom have been most generous with information and advice. My special gratitude goes to those who shared their valuable time to review this chapter and provide comments—in particular, Mr. J. S. Edmonds of EPRI; Dr. J. L. Kirtley, Jr., of MIT; and Drs. J. K. Hulm and J. H. Parker, Jr., of Westinghouse.

Not least, my thanks to Ms. Ilene Kelley of Westinghouse who provided invaluable and greatly appreciated help in processing the manuscript.

For most applications, superconductivity offers a useful alternative design principle rather than a unique approach that must be utilized in order to achieve the design objective. As an example, superconducting ac generators for electric utility use have been shown to be more efficient than their normal conducting counterparts as well as offering better performance. The potential efficiency improvement for a large generator is illustrated in Fig. 10.1. Figure 10.2 shows that the improvement in critical fault clearing time, an important indicator of stable performance, is significant.[1,2] To convert the efficiency and stability improvements into cash revenue requires, however, a very high degree of reliability and availability, which must be considered a design requirement ab initio. At the same time, superconducting machine construction is necessarily quite complex. When these conflicting considerations are sorted out, superconductivity is most likely to be preferred to normal conductors in cases in which it offers either a unique solution or a major advantage. At this time, an example would be the construction of two-pole ac generators with ratings of the order of 1500 MVA.

In the main, superconductor applications are most successful for use with dc, as in the field windings of synchronous machines. Superconductors are not as yet economically feasible for ac windings of electric machines, because ac losses cannot be avoided. Removal of these losses from the cold zone requires an input power of several hundred to a thousand watts for each watt of loss. Thus, to be competitive, ac superconductors must be designed and built for lower levels of losses than have yet been commercially accomplished.

In short, even while all the fundamental problems in applying superconductors to electric machines have been solved, innovative engineering may produce better

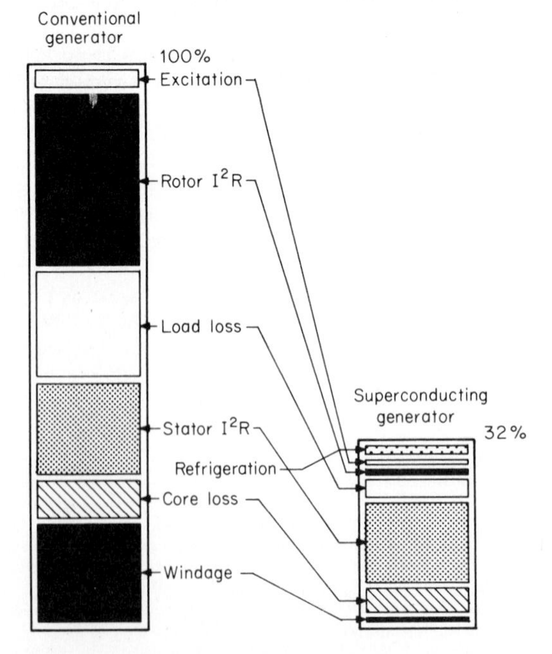


**FIG. 10.1** Loss comparison for conventional versus superconducting turbine generators, 1200-MVA rating.

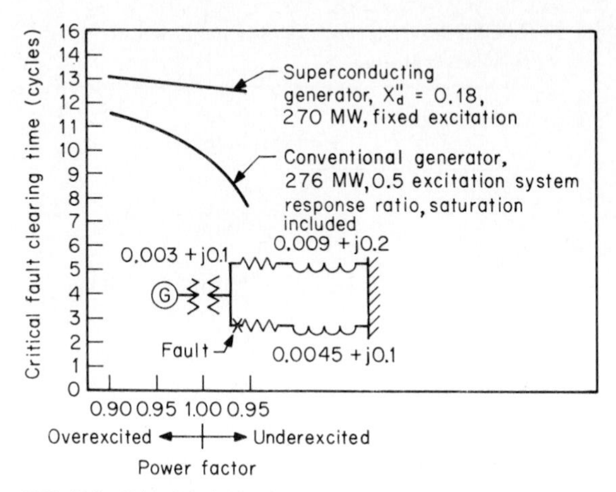


**FIG. 10.2** Critical fault clearing time versus prefault power factor for conventional and superconducting generators.

solutions. The widespread application of superconducting devices appears to be only a matter of time.

## Superconducting Materials

Superconductivity is exhibited by many metals, alloys, and intermetallic compounds and has also been reported for some nonmetallic materials. In pure metals, the transition to the superconducting state occurs spontaneously and suddenly as soon as the critical temperature is reached and results in perfect conductivity. Experiments with closed superconducting circuits have shown that the resistance is sufficiently low to produce time constants of many thousands of years.

In the ideal superconducting state typical of pure metals, perfect conductivity is accompanied by perfect diamagnetism, so that no magnetic flux can exist in the interior of the superconductor below a very thin depth of penetration. This thin layer, of the order of $10^{-7}$ m thick, carries permanent shielding currents that exclude flux from the interior. At a limiting value of magnetic field intensity $H_{c1}$, however, this current suddenly collapses and the conductivity reverts to normal. For typical pure metals, this critical field is of the order of 0.1 T; consequently, such metals are not useful for practical electric machines. These low-field (soft) superconductors are known as type I.

With both type I and the "magnetically hard" type II superconductors, reversion to the normal conducting state also occurs when the temperature exceeds the critical value $T_c$ or when the current or field intensity exceed their own critical values $I_c$ and $H_c$, respectively. Return to the normal conducting state is known as *normalization.* The normalization of an entire winding, or of a significant part of one, is known as a *quench.*

## Type II Superconductors

Magnetically "hard" superconductors were first discovered in 1961. In these, superconductivity is not lost abruptly at the critical field $H_{c1}$; rather, although perfect

**TABLE 10.1** Properties of Some Practical Type II Superconductors

| Compound or alloy | Critical temperature, K | Critical field, T |
|---|---|---|
| NbTi | 10 | 11.4 |
| NbZr | 10.8 | 11 |
| $Nb_3Al$ | 18 | 32 |
| $Nb_3(Al_{0.7}Ge_{0.3})$ | 20.6 | 41 |
| $Nb_3Ga$ | 14.5 | 34 |
| $Nb_3Ge$ | 22.5 | 36.3 |
| $Nb_3Sn$ | 18 | 26.3 |
| $V_3Ga$ | 16.8 | 22 |
| $PbMo_{5.1}S_6$ | 14 | 50 |
| $PbMo_6S_8$ | 14 | 45 |

*Note:* Properties of superconductors prepared in commercial quantities may be expected to show variability and variations from published data. For design purposes, it is best to obtain performance data for specific conductors from the manufacturer.

***Sources:*** J. K. Hulm and B. T. Matthias, "High-Field, High-Current Superconductors," *Science*, vol. 208, no. 4446, May 23, 1980, p. 881; T. H. Geballe and J. K. Hulm, "Superconductors in Electric Power Technology," *Scientific American*, vol. 243, no. 5, November 1980, p. 138; J. W. Ekin, "Superconductors," in R. P. Reed and A. F. Clark (eds.), *Materials at Low Temperatures*, American Society for Metals, Metals Park, Ohio, 1983.

surface shielding is lost, the material continues to be superconducting (even with some flux penetration below the surface) up to a much higher field $H_{c2}$. Of interest is that these materials may exhibit superconductivity at flux densities in excess of 10 T, or far above those encountered in conventional electric machines with magnetic flux paths. Type II superconductors include niobium-titanium (NbTi) and niobium-tin ($Nb_3Sn$), among others. Some practical type II superconductors and their critical temperatures and fields are listed in Table 10.1. As illustrated chronologically in Fig. 10.3, progress in raising critical temperatures was achieved by continued materials development.

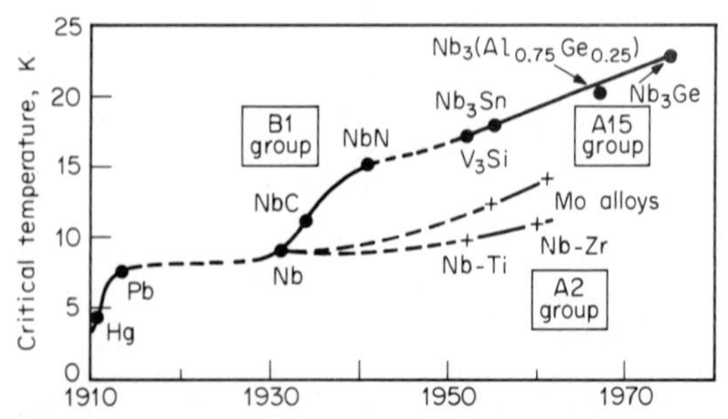


**FIG. 10.3** Critical temperature for various superconductors.

Note in Fig. 10.3 that the critical temperatures of practical superconductors are limited to under 25 K. This is below the liquefaction temperature of such common gases as nitrogen. Hydrogen liquefies at 20 K, but this is not enough temperature "margin." The cooling of superconducting windings must therefore be accomplished with the rarer and more expensive liquefied helium, which boils at 4.2 K at atmospheric pressure.

Practical type II superconductors are either alloys or intermetallic compounds. Some of the important superconductors, such as $Nb_3Sn$, are brittle as normally prepared and thus require specialized winding techniques to attain their full properties in the completed windings.

At low fields, type II superconductors act about the same as type I in that flux is excluded from the interior by surface shielding currents. Type II materials respond to increasing field not by the collapse of the surface current but by the formation of a microscopic array of flux tubes of normal material, each encompassed by superconducting material that carries a shielding current around each tube. This "mixed state" is exhibited only by type II materials and results in their ability to handle much larger critical fields. Complete flux penetration occurs at still higher fields and results in normalization.

## Conductor Stability

Superconductors normalize when any one of the three critical parameters $T_c$, $I_c$ (or $J_c$), or $H_c$ is exceeded. The region of stability may be described by a surface in three-dimensional space, as illustrated for three practical superconductors in Fig. 10.4. For each superconductor (e.g., NbTi), there is a surface in this three-dimensional space above which the superconductor is in its normal resistive state. Below this surface (i.e., closer to the origin), it is in the superconducting (zero-resistance) state.

Figure 10.4 shows the three superconductors—NbTi, $Nb_3Sn$, and $Nb_3Ge$—to be in order of increasing performance. The first two, NbTi and $Nb_3Sn$, are commercially available practical conductors today. $Nb_3Sn$ gives higher performance than NbTi but is quite brittle; to accommodate its special mechanical properties, special techniques

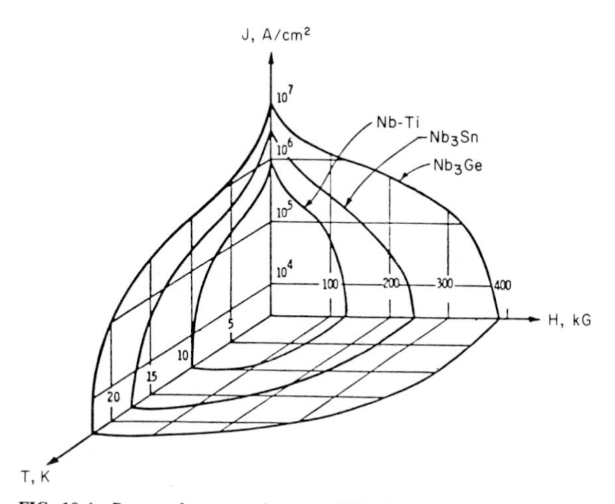


**FIG. 10.4** Range of superconductor stability for three materials.

must be used in its manufacture and in winding coils. The third superconductor, $Nb_3Ge$, outperforms the other two, but it has not as yet been fabricated into a conductor that is practical for forming into a generator field winding.

Superconductor stability is essential for successful application. Devices that incorporate superconducting windings must generally remain fully functional through and after disturbances. Special methods have been developed to ensure stable performance of even large and complex superconducting windings.

Quench of a superconducting winding can be the result of either local or global temperature, current, or flux density transients. In power-system devices, these tend to occur either simultaneously or in close sequence. Current and flux transients, moreover, tend to produce wire motions, and these can cause energy dissipation. Additionally, loss mechanisms within the superconductor, such as flux jumps, can assist in driving it normal. (*Flux jumps* are discontinuous movements of a group or bundle of flux lines. Especially when operating near the $T_c$-$J_c$-$H_c$ stability limit, small magnetic or mechanical disturbances may be enough to initiate a thermal runaway and a quench.)

In order to achieve a practical high-field conductor, the total superconductor cross section must be subdivided into a large number of fine filaments, usually less than 100 $\mu$m in diameter, embedded in a matrix of high-conductivity normal metal, such as copper. This construction enables current transfer to occur between the matrix and any superconducting filaments that may partly or wholly revert to normal conductivity, thereby ensuring stability against magnetic-thermal transitions that could lead to excess heating and the eventual normalization of the entire superconducting wire.

The degree of stability required for a superconducting winding is determined by its functional application. The criteria that distinguish among different degrees of stabilization are:

1. *Adiabatic or enthalpy criterion*: The local energy release is absorbed and does not propagate.
2. *Dynamic criterion*: There is sufficient matrix material to confine the local energy release.
3. *Cryogenic stabilization, or cryostability*: Local normalization of the superconductor may occur, but it will not occasion a thermal runaway or quench. In the case of full cryostability, an entire section of a winding may normalize, but it will spontaneously recover to the superconducting state.

Cryostability is an important characteristic of a winding that must operate stably through a variety of disturbances. In stationary magnets, it can be achieved by providing a fairly large amount of matrix material. Such an approach is not feasible for the windings of electric machines, where space is at a premium. Investigations have, however, shown that the exceptionally high heat-transfer rates that characterize even natural convection in highly accelerated frames, such as high-speed rotors, afford the ability to design cryostable windings by providing sufficient coolant and heat-transfer area.[3]

## Application Requirements

The application of superconductors to rotating electric machines and other apparatus constitutes a means to an end, which is the more efficient, more economical, and at least equally reliable performance of the equipment. While superconductors perform optimally under conditions of pure dc excitation, these conditions almost never prevail in power apparatus, which must start, stop, and operate under a variety of transients. The application of superconductors under such conditions requires close attention to ac loss mechanisms; it also requires a careful choice between sufficient protection of the superconducting windings and excessive shielding, which can lead to other difficulties.

A good deal of application experience is available in the field, especially with respect to superconducting magnet windings, which include magnets with dimensions ranging from a few centimeters to several meters. Magnet technology is directly applicable to rotating-machine technology, but the latter requires additional considerations which are not usually encountered in magnet design and which may be ascribed to armature reaction, including transients and steady-state asynchronous effects that can penetrate to the superconducting winding.

The overall application process can be diagramed as shown in Fig. 10.5. Having selected a physical configuration of parts, the designer selects a trial superconductor for which the ac and transient losses are known or measured. By calculation, the heating effects of disturbances that penetrate past the shielding provisions are estimated and compared to the superconductor's stability profile for the specific ambient conditions, such as winding current, local steady-state temperature, and local field strength. If the conductor retains its stability during and after the disturbance, the trial is judged successful; if it does not, the process must be repeated with a different choice of conductor or cooling arrangement.

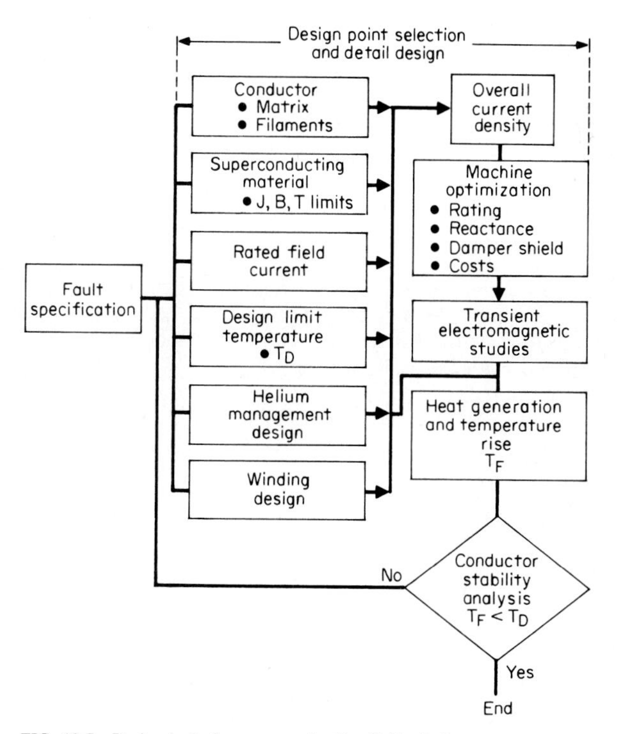


**FIG. 10.5** Design logic for superconducting field winding.

In fact, superconductor application to rotating-machine windings is a more complex process than is made explicit by Fig. 10.5. The designer must constantly be aware of the possible degradation of winding performance by physical strain and wire movement. Even small strains measured in fractions of a percent can seriously degrade the current capability of some superconductors, especially the brittle materials, such as $Nb_3Sn$. Wire movement produces localized heating through friction and electrical losses. Constraint against wire movement may be improved by providing more structural support, which tends to reduce cooling volume or area, or both, thus reducing stability margins. The designer must balance the conductor constraint and cooling requirements such that stable performance will result.

It has been a common experience with magnet windings that they *train*, i.e., quench at successively higher values of excitation current as they are brought through a series of charging and quench sequences. The quenches are usually attributed to wire motion, and the increase in current capability without quench is attributed to the settling of the wire into locations that are not as amenable to further motion. In the case of rotating machines intended for practical applications, any training will have to be completed before the equipment is launched into commercial service.

The simultaneous and mutually conflicting requirements for conductor constraint and cooling have been approached with various winding designs and techniques, ranging from complete impregnation with a suitable resin to pure mechanical clamping by spacers that also provide flow passages. The first has been applied to numerous magnets and can work well, given the proper selection of external support provisions and impregnating resins.* In this arrangement, heat generated in the superconductors must be dissipated to the coolant at least in part by conduction through the impregnant. With mechanical clamping, there is usually more opportunity for direct contact between the coolant and the conductor surface. In rotating windings, advantage may be taken of the substantial enhancement of heat transfer in the acceleration field produced by the rotation. It is possible to achieve cryostable performance in such cases without the use of a great deal of normal conductivity shunting material.

The advantages of superconductivity are purchased at the price of having to refrigerate a major machine component to approximately 4 K, the temperature of liquid helium. This imposes design requirements upon the structure of the machine member that supports the superconducting winding, and it of course requires circulation of helium between that member and a suitable helium refrigerator-liquefier. The structural requirements are occasioned by the need for thermal isolation between the cryogenic zone and the portions in contact with the normal-temperature surroundings. Thermal distance pieces are usually employed for this purpose. If metallic, these still constitute considerable heat leaks and must be very carefully proportioned to avoid excessive refrigeration requirements.

### Loss Generation, Heating, and Quench

As Fig. 10.5 shows, the determination of heat generation and temperature rise is central to ascertaining that stable performance can be ensured. Studies of loss generation in superconductors have been a major topic of research in this field. Testing is costly and cumbersome because it requires the establishment of superconductivity under suitable cryogenic conditions. The development of loss calculation methods

*Concern about cracking of the impregnating resins or fillers persists, however. Some recent papers attribute the persistence of training behavior and premature quenches to filler-material failures.

has been a difficult problem, now largely solved mainly through the dedicated efforts of Carr and his coworkers.[4]

Losses in superconductors are found to be of two kinds: eddy-current losses in the normal-conductivity matrix, and hysteresis losses in the superconducting filaments. In the analysis developed by Carr et al., the composite superconducting wire is treated by applying Maxwell's equations as if it were a continuum with anisotropic properties. Both hysteresis and eddy-current losses are thus determined.

For the practical stabilized multifilamentary composite superconductors, the following variables enter into the loss calculations:

1. Rate of change of field or current: i.e., frequency of alternating current or ramping rate of unidirectional current.
2. Presence of ambient field.
3. Orientation of the field vector(s) with respect to the conductor axis.
4. Conductor design: e.g., matrix/superconductor ratio, resistivity of matrix parallel to superconductor filaments, resistivity of matrix perpendicular to filaments, presence or absence of high-resistivity barriers, filament size, twist pitch, and presence or absence of a normal-conductivity sheath.

***Hysteresis Losses.*** For magnetic field changes ($\dot{B} \triangleq dB/dt$) large enough to penetrate the filaments fully, the hysteresis loss per unit volume in a cylindrical multifilamentary superconductor is given by:[5]

$$\frac{P_h}{V} = \frac{2|\dot{B}|\lambda j_c d}{3\pi} \tag{10.1}$$

where $\lambda$ = volume fraction of the filaments
$\dot{B}$ = rate of change of the field
$d$ = filament diameter

For sinusoidally varying fields of amplitude $B_m$ and frequency $f$, the quantity $|\dot{B}|$ is replaced by $4fB_m$ in Eq. (10.1). Meter-kilogram-second (mks) units are used throughout.

The field change necessary for full penetration is given by[6]

$$\Delta B_p = 4 \times 10^{-7} d j_c \tag{10.2}$$

The case of solid rectangular filaments has also been studied. For full field penetration, the result is

$$\frac{P}{V} = \frac{\lambda j_c |\dot{B}| v}{2} F\left[\theta, \frac{s}{v}, \frac{j}{j_c}\right] \tag{10.3}$$

This equation was derived by treating the filaments as elliptical, with $2s$ and $2v$ the lengths of the major and minor axes, respectively, and with $\theta$ the angle between the magnetic field and the major axis. The function $F$ has been derived for $\theta = 0$ and $\theta = \pi/2$:

$$F = \frac{s}{v}\left[1 + \left(\frac{j}{j_c}\right)^2\right] \qquad \theta = \frac{\pi}{2} \tag{10.4a}$$

and

$$F = 1 + \left(\frac{j}{j_c}\right)^2 \qquad \theta = 0 \tag{10.4b}$$

In the case in which the field change is not large enough to penetrate the filaments fully, the hysteresis loss for round filaments is given by

$$\frac{P_h}{V} = \frac{256}{9\pi} \frac{\lambda B_m^3 f}{\mu_o^2 d j_c} \tag{10.5}$$

This formula will apply to the calculation of losses for the ac fields with amplitudes less than $\Delta B_p$ given by Eq. (10.2).

***Eddy-Current Losses.*** The loss formulas usually overestimate the actual losses but are generally within a factor of two of the measured value. Thus, the calculated results tend to be conservative.

The model assumes anisotropic conductivity and permeability, with components parallel to and transverse to the superconducting filaments. These properties are designated as:

Parallel to filaments : $\sigma_{\|},\ \mu_{\|}$

Transverse to filaments : $\sigma_{\perp},\ \mu_{\perp}$

The transverse conductivity is given by the following:

$$\sigma_{\perp} = \frac{1+\lambda}{1-\lambda}\sigma_m \tag{10.6}$$

or

$$\sigma_{\perp} = \frac{1-\lambda}{1+\lambda}\sigma_m \tag{10.7}$$

where $\sigma_m$ = matrix conductivity
$\lambda$ = fraction of high-conductivity material in wire

The following equations apply to the case in which the matrix material is homogeneous.

The eddy-current losses are influenced by the path the current takes through the matrix and the filaments. High contact resistance at the filament surfaces causes the current to flow primarily through the matrix and bypass the filaments; otherwise, the current enters each filament. Equation (10.6) applies to the case in which no significant interfacial resistance is present, whereas Eq. (10.7) applies to the opposite case.

For magnetic field changes large enough to penetrate the filaments fully ($\Delta B > 4 \times 10^{-7} d j_c$), both permeability values may be taken as 1; otherwise,

$$\mu_{r\|} = 1 - \lambda \quad \text{and} \quad \mu_{r\perp} = \frac{1-\lambda}{1+\lambda} \tag{10.8}$$

The model recognizes three frequency regions of loss behavior for applied alternating magnetic fields. The low-frequency region is limited by

$$f < f_{t1}, f_{t2}$$

where

$$f_{t1} = \frac{4\pi}{\mu_o \mu_{r\perp} \sigma_{\perp} L^2} \tag{10.9a}$$

and

$$f_{t2} = (\pi R_o^2 \mu_o \mu_{r\|} \sigma_{\perp})^{-1} \tag{10.9b}$$

where $L$ is the twist pitch length and $R_o$ is the radius of the active area of the conductor. In the low-frequency regime, only the outermost filaments, to a depth of

$w_o$, are saturated by the eddy current, and for this regime to apply, it is necessary that $w_o << R_o$, where $w_o$ is given by

$$w_o = \frac{2\mu_{r\perp}}{\mu_{r\perp} + 1} \left(\frac{L}{2\pi}\right)^2 \frac{\sigma_\perp |\dot{B}|}{\lambda j_c} \tag{10.10}$$

Many of the field changes that produce superconductor losses are transient rather than steady-state. For transient fields to be in the low-frequency limit, the field changes should occur in times long compared to the relaxation time of the eddy currents: i.e., in times long compared to $\tau_1 = (2\pi f_{t1})^{-1}$ and $\tau_2 = (2\pi f_{t2})^{-1}$. The eddy-current loss for round conductors (total wire radius $R_t$) in the low-frequency limit has been shown to be approximated by

$$\frac{P_e}{V} = \left(\frac{R_o}{R_t}\right)^2 \dot{B}^2 \sigma_\perp \left[\left(\frac{L}{2\pi}\right)^2 + \frac{R_o}{4}\right]^2 \tag{10.11}$$

Note that the difference $(R_t - R_o)$ is the thickness of a normal-conductivity outer sheath, if one is present. For the case in which $\Delta B < B_p$, $\mu_r \neq 1$ and Eq. (10.11) becomes

$$\frac{P_e}{V} = \left(\frac{R_o}{R_t}\right)^2 \dot{B}^2 \sigma_\perp \left(\frac{2\mu_{r\perp}}{1 + \mu_{r\perp}}\right)^2 \frac{L^2}{4\pi^2} \tag{10.12}$$

At intermediate frequencies for which $f_{t1} < f < f_{t2}$, the inner filaments are shielded by the currents induced in outer layers of filaments and the eddy-current loss is given by

$$\frac{P_e}{V} = \left(\frac{R_o}{R_t}\right)^2 \frac{B_m^2 \times 10^{14}}{2\sigma_\perp L^2} \tag{10.13}$$

Note that the loss is inversely proportional to $\sigma_\perp$ and $L^2$. Thus, if a conductor is subjected to more than one time-varying field, designing for a low loss due to one will yield a higher loss from the other if they occur on different sides of $f_{t1}$.

At very high frequencies, $f > f_{t2}$, the skin-effect region occurs and the loss is given by

$$\frac{P_e}{V} = \left(\frac{R_o}{R_t}\right)^2 \frac{R_o}{L^2} \left(\frac{64\pi^5 \mu_{r\parallel} f}{\sigma_\perp \mu_0^3}\right)^{1/2} B_m^2 \tag{10.14}$$

Eddy-current losses in rectangular conductors have been treated by using the anisotropic continuum model and approximating the conductor cross section as an ellipse with major and minor axes of length $2a$ and $2b$, respectively. In the low-frequency limit for full field penetration ($\mu_{r\perp} \simeq 1$), the loss is given by

$$\frac{P_e}{V} = \frac{\sigma_\perp \dot{B}^2}{4} \left\{ \left(\frac{L}{\pi}\right)^2 \left[\left(\frac{b}{a}\right)^2 \cos^2\theta - \frac{a}{b}\sin^2\theta\right] + b^2 \cos^2\theta + a^2 \sin^2\theta \right\} \tag{10.15}$$

where $\theta$ is the angle between the transient field and the major axis of the conductor cross section.

For linearly polarized sinusoidally varying fields, the quantity $\dot{B}^2$ in the eddy-current loss equations is averaged over a cycle and $\dot{B}^2$ is replaced by $2\pi^2 B_m^2 f^2$.

For rotating fields, the result must be multiplied by 2. Since the hysteresis loss is linear in $B_m$, a rotating field produces the same loss as an oscillating one.

***Quench.*** Analysis of the conditions that can produce quench (a normalization of the entire winding) is difficult, for it requires a knowledge of loss distributions, heat-transfer rates, and heat storage that must consider not only the two-dimensional problem of the winding cross section but also axial heat flow along the conductor.

While the losses due to electromagnetic disturbances can be estimated with fair accuracy, those relating to winding motion are often more difficult to assess. The most effective countermeasures are:

1. Minimizing loss generation by using low-loss conductors
2. Providing adequate locally distributed supplies of helium
3. Ensuring adequate support provisions consistent with (**2**)
4. Keeping the ratio of conductor cooling surface to conductor volume large
5. Avoiding surface coatings that increase thermal resistance to the helium coolant

Consistent use of these methods, especially with the high heat-transfer rates afforded by high-speed rotation, can result in very stable and quench-resistant winding performance.

### Commercial Development of Superconducting Materials and Devices

Development efforts have been directed to applications in which superconductors are substituted for normal conductors and to innovative applications in which superconductivity plays a crucial role. On the other hand, the rather complex and sophisticated technology (including cryogenics and vacuum containments) required for practical applications has tended to offset the economic benefits of efficiency increases and material volume reductions; consequently, the present range of commercial applications still remains substantially below the potential of this advanced technology.

Superconducting magnet technology has found a place in power devices in the form of field magnets for rotating and linear electric machines, of magnets for magnetohydrodynamic and fusion experiments, and of energy-storage coils. Related industrial applications are magnetically levitated trains, magnetic separation (e.g., for mining), and magnets for testing and diagnostic purposes. Magnets for imaging by nuclear magnetic resonance (NMR) have reached major commercial significance.

With respect to superconductors proper (i.e., materials such as wires for magnets), the last two decades have witnessed the development of manufacturing methods to satisfy essentially any of the viable applications. Research efforts in recent years have been aimed at reducing costs rather than at new material technology.

## *10.2 SUPERCONDUCTING DC MACHINES*

### General Description and Types

The earliest applications of superconductivity to electric machines featured dc machines (Chap. 5), including both homopolar and heteropolar types; practical development efforts, however, have concentrated on homopolar machines virtually to the exclusion of heteropolars. In conventional dc machines, magnetic iron provides a path of high magnetic permeability to help the magnetomotive force (mmf) of the field winding produce a high flux density. This flux density is, however, limited by saturation of the magnetic materials. The major advantage of superconductivity is that it lifts this limitation, so that significantly higher flux densities and power densities can be obtained at no sacrifice—in fact, with an improvement—in efficiency.

Superconductors are presently applied in dc and ac machines only to windings that carry direct current, i.e., the field winding. This implies that the armature of a superconducting, commutator-type heteropolar machine would have to be a room-temperature conventional-conductor design. The advantages of this approach would be limited to the increased power density offered by the higher flux densities and to the reduced excitation loss that a superconducting field winding provides. Some benefit would also be derived from the reduction in inductance of the airgap armature winding (as compared to one contained in the slots of a magnetic structure), which would improve the commutation. The disadvantages of this approach would include the need for cryogenic containment of the field winding, for which compensating windings might also have to be provided,[11] and the concerns about the dynamics of a nonmetallic armature support structure. The power output of this design would still be limited by the commutator, particularly in terms of the voltage that might be permitted across the insulation between bars. Careful shaping of the distribution of the high-intensity magnetic field would be required in order to prevent unacceptable losses in the portions of the armature that must remain conducting, such as the commutator and support rings, and in the shafting. The combination of these features has not proven economically attractive. A possible alternative, in which a rotating superconducting dipole field excites an ac armature connected to a solid-state rectifier, is really an ac machine. For ac generators, the economic break-even point that justifies the use of superconducting field windings has been shown to be of the order of several hundred megawatts, well above the usual range for dc machines.

Homopolar construction eliminates these concerns, but at the cost of introducing another rather difficult problem: the collection of high currents. Much of the work in the area of homopolar superconducting machines has been devoted to the development of satisfactory collector arrangments, and the problem cannot yet be regarded as completely resolved. Disk-type and drum-type machines are the two basic arrangements for homopolar machines. These are amenable to a range of variations in the location of field magnets relative to the armature, in the shaft arrangements, and in the means for current collection. The use of superconductors for the field windings introduces the need for cryogenic isolation of a part of the machine. Although for purposes of maintainability it might appear desirable to locate the cryogenic vessel near the outer envelope of the machine, some designs have been made with the field magnets within an inner cavity of the rotor.[7]

## Field and Armature Configurations

The geometry of the field winding of a homopolar dc machine is among the most regular of the types of superconducting windings encountered in practice, being essentially solenoidal and of compact dimensions. Examples may be seen in Refs. 7, 8, 9, and 10. The field distributions for this form of exciting winding are given in Ref. 11.

The engineering challenges in designing such field windings reside in the need to ensure stable performance of the winding thermally and mechanically, as well as electrically, and in providing the necessary support for this winding with minimum heat leak. Design of the cryogenic containment may be complicated by application specifications, such as the attitude in service being other than horizontal. Detailed design parameters for a 3000-hp motor developed as a prototype for a motor technology valid up to 40,000 hp may be found in Ref. 10.

Armature configurations for homopolar superconducting machines are based on the same principles as those for conventional design, but the recent developments in this field have emphasized innovative concepts that exploit the advantages of the higher power densities offered by superconducting field coils. The basic types of armatures remain the disk and the drum. Machines have been designed, however, in which drums are connected in series through appropriate slip-ring arrangements so as to increase the terminal voltage available for a given current.[12] The same idea can be applied to disk armatures.[13] Alternately, disk armatures can be segmented, with the brushes arranged so as to create series or series-parallel paths that increase the terminal voltage while keeping it within the limit set by the capabilities of the insulation between adjacent segments.[11]

## Cryogenic Design and Isolation

Since the basic configuration of the homopolar machine relieves the superconducting winding of torque reactions and since this winding is also stationary, the design of a suitable cryogenic enclosure is, in comparison with other superconducting devices, not an overly demanding task. It generally consists of an enclosing vessel of a cryogenically strong, easily fabricated, helium-tight material, such as stainless steel. Although the outer vessel is not exposed to cryogenic fluids, except for local areas, it must be able to survive the possible effects of an internal rupture that could result from the overpressures that might accompany a quench. High strength is also required in order to resist transient stresses, such as during cool-down.[11]

Winding supports between the outer vessel and the field coils can be made of a good thermal insulator, such as fiberglass-reinforced synthetic resin, and must be thin or slender in comparison to the length of path for heat conduction. Radiant-heat shields (one or more) are used to intercept thermal radiation from the room-temperature enclosure; they may be conduction-cooled or gas-cooled with boil-off helium, depending on the design of the field-winding cooling system. These shields will be thermally "anchored" so that they operate efficiently at an average temperature of about 100 K.

The options for superconductor cooling include forced-flow cooling and natural convection.

## Current Collection

Given the low armature voltages and high load currents typical of homopolar machines, current collection constitutes a major design challenge. The two options for addressing this issue are:

- Solid brushes
- Liquid-metal collection

The option chosen and the subsequent design must consider (1) slip-ring speed, brush drop, and contact resistance, which affect efficiency; and (2) brush temperature, brush wear, and replacement, which affect life-cycle costs, disposition of brush dust, etc. The atmosphere (type of gas, humidity, etc.) can have a major influence on brush wear. It has been found, for example, that both low humidity and the proximity of silicone resins are inimical to good brush performance.

Solid brushes can be applied at current densities ranging from about 100,000 $A/m^2$ for ordinary electrographitic brushes, such as are used with turbine-generator (turbogenerator) collector rings, to nearly 10 times that value.[11] To achieve the latter high density, a special metal-coated carbon fiber brush has been developed.

Liquid metal offers the possibility of reducing the objectionable features of solid brushes. Liquid-metal arrangements have in fact been applied both to superconducting and to more conventional machines. They can operate at extremely low brush drop as well as at low friction. They do present other practical problems of safe containment and operation, however; these result from the basic properties of the usable metals, which include the alkali metals: e.g., sodium and potassium (and their low-melting eutectic NaK), gallium, gallium-indium, and mercury. None of these is user-friendly. The alkali metals, for example, are reactive and require containment in an inert atmosphere. Mercury is heavy and tends to cause erosion, and its vapor is poisonous. Only mercury remains liquid to subzero temperatures (−38.9°C); thus, other choices require heaters or low-temperature protection during shutdowns.

Methods of calculating the losses with liquid-metal current collection are summarized in Ref. 14.

### Design Considerations and Applications

The need for dc machine applications involving high currents, pulsing, wide speed range, and/or other specialized requirements—combined with the availability of superconductors—has caused a resurgence of interest in such machines in the last two decades. Recent concepts feature the use of liquid cooling for the heat-producing portions of the machine and liquid-metal current collection.

An innovation in homopolar machines is described in Ref. 15. The armature elements are cone-shaped and are connected at appropriate points through the liquid metal, which floods the interior. The field magnet is located in a toroidal cryostat centered about the axis of the shaft on which the conical armature elements are centered.

An interesting range of application for homopolar machines is the generation of high-current pulses. Fusion reactor research, for example, requires pulsed energy systems. Homopolar machines are well suited to these applications by virtue of their very short armature time constants and their ability to store energy inertially in the rotating mass of the armature.[16]

## 10.3 SUPERCONDUCTING AC MACHINES

### General Description and Types

The emphasis in superconducting ac machine development has been on high-speed generators, which include both large turbine generators for electric utility central station use and other types aimed at special applications, which include defense and marine uses.

The development of superconducting ac machines began in the mid-1960s, with early work at AVCO as well as important contributions at MIT. The first laboratory demonstration device was a synchronous machine with a stationary superconducting field winding and a rotating armature winding that used normal conductors. Later machines have uniformly been constructed with rotating superconducting field windings. A good summary of the early conceptual, theoretical, and prototypical developments is found in Smith and Keim's article in the compendium, *Superconducting Machines and Devices.*[17]

The decision to place the superconducting field winding on the rotor rather than the stator is not a priori obvious. As recalled by Smith, a continuous and prolific investigator of the subject, considerable effort was devoted early in the analysis of superconducting ac machines to defining the optimum configuration.[18] Smith concludes that the optimum configuration of a superconducting ac machine can be described as having the following characteristics:

**1.** Central rotor containing a superconducting field winding and having:
   **a.** Connections to external excitation-current supply
   **b.** Shielding system using normally conducting eddy-current shields
   **c.** Liquid-helium coolant with provisions for circulation
   **d.** High-vacuum cryogenic insulation
**2.** Surrounding stator containing:
   **a.** Airgap-type normal-conductivity armature winding operating with ambient-temperature coolant
   **b.** Laminated flux shield surrounding armature winding

In addition to a helium refrigerator-liquefier system, superconducting ac generators require the usual auxiliaries, including an excitation system. Since steady-state excitation losses are limited to lead and brush-drop losses, the excitation system power requirements are small and easily satisfied by a static excitation system, which must, however, be designed to operate stably at quite a low voltage (on the order of 5 V). A more stringent requirement for the excitation system is that the harmonic content of its output voltage must also be extremely low so as to keep the ac losses in the superconductor and the rest of the cryogenic zone within acceptable bounds. This requirement sets an upper limit to the exciter ceiling voltage that can be used for forced excitation for stability considerations. Field forcing is not of significant value for superconducting generators, since the field and damper time constants are large; therefore, a low value of maximum ceiling is acceptable. In a typical case, it will still be from 20 to 30 times the steady-state full-load excitation voltage.

## Shield, Field, and Armature Configurations

A simple schematic of a superconducting ac generator is shown in Fig. 10.6. As shown, the magnetic materials are limited to the environmental shield and the outer structure, and the entire space occupied by the rotor and the stator windings is devoid of magnetic substances. Both the field winding and the armature winding are airgap windings and are separated in space by "electrothermal" shielding provisions, which normally incorporate both electromagnetic and radiant-heat shielding. These shields are mounted on the same rotating structure as the field winding and normally travel synchronously with the fundamental flux wave.

***Shields.*** As it turns out, the electrothermal shield design exerts a major influence on all aspects of the overall design, for it determines the degree to which the

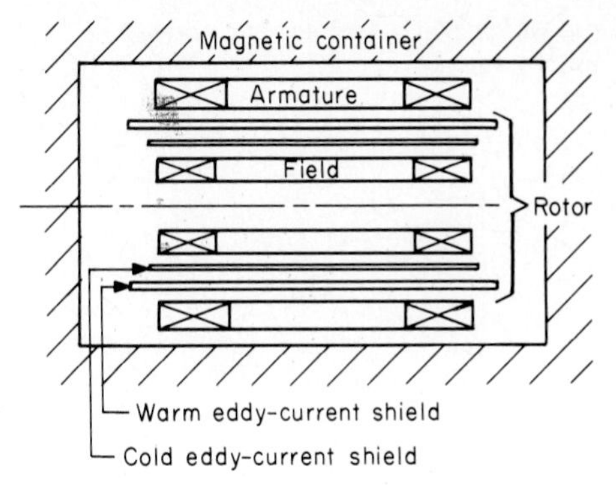


**FIG. 10.6** Cross-sectional schematic of superconducting generator.

field winding is isolated from external disturbances that can potentially cause it to quench and thus disable the machine. At the same time, the dimensions and physical properties of the shield have a major influence on the machine reactances and on the damping of rotor oscillations, which are of major significance to the power system to which the machine is connected. Alternate approaches to the design of these shields include:

1. Electromagnetic and thermal shielding served by separate structures:
   a. Electromagnetic shield at room temperature.
   b. Electromagnetic shield at cryogenic temperature.
2. Electromagnetic and thermal shielding combined in the same structure. Shield configurations described in the literature include the following:
   a. Room-temperature composite* electromagnetic shield integral with outer-rotor support structure, with either conduction-cooled or gas-cooled radiant-heat shield in the vacuum space between the inner and outer rotors. Figure 10.7 illustrates such a configuration as well as the general arrangement of other rotor parts.
   b. Room-temperature monolithic electromagnetic shield; otherwise, the same as (**2*a***). Note that the use of magnetic material for such a shield has also been proposed. Due to the high magnetic fields in which it operates, however, it is highly saturated and its permeability is greatly reduced.
   c. Compound shield, including the room-temperature shield of (**2*a***) as well as an intermediate cold electromagnetic shield on the inner rotor just outside the field winding.
   d. Shields of the above types, but mounted with bearing surfaces at the rotor ends so as to be free to rotate. Such inertial shields have dynamic properties different from solidly mounted shields and can function to reduce the short-circuit torques on rotor elements. To date, this approach has not been applied in a machine.

*Concentric shell (or "sandwich") construction of high-conductivity shielding and high-strength structural material.

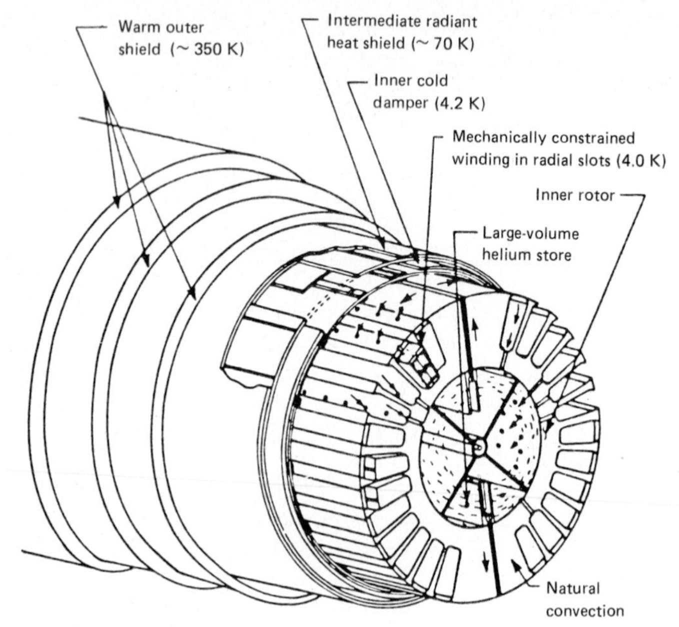


**FIG. 10.7** Cutaway view of rotor.

**e.** Electromagnetic shielding at cryogenic temperature, consisting of one or more concentric shielding shells mounted on the rotor. The shells may be either of solid material or, to reduce the heating on the cold rotor, composed of discrete conductors. In the latter case, end connections may be made to an external resistor that absorbs most of the energy required to damp out rotor swings and thus improves the dynamic performance.

Cryogenic temperature shields can be made relatively thin if pure copper is used. The resistivity of copper at temperatures near absolute zero can be up to three orders of magnitude lower than at room temperature but depends greatly on the chemical purity of the material and, to a lesser extent, on its state of stress. Good progress has been made in applying thick copper electroplates that offer the requisite purity without objectionable residual stress.

***Field Windings.*** Options for the configuration of field windings include the basic shape and conductor arrangement and the support provisions. Alternatives for coil shape correspond to conventional parallel- and radial-slot windings.

For the airgap windings used for superconducting fields, the parallel-slot arrangement is implemented as a stack of "racetrack" (or "pancake") coils, which are individually formed and stacked on each other to make the complete winding. Windings of this type can be made using the technology developed earlier for stationary superconducting magnets, and they have functioned well in small machines. The largest machine known to have this type of field winding is a 20-MVA unit.[19]

Designers for most of the more recent, usually larger, superconducting turbine-generator prototypes have selected windings in which the conductors are distributed along arcs of the rotor periphery. These radial-slot arrangements used in the various designs differ mainly in the means adopted to support the conductors. The design of

**FIG. 10.8** Exterior view of a helical armature winding. *(Courtesy of Massachusetts Institute of Technology.)*

any superconducting winding represents a delicate compromise between cooling and support. In rotating-machine design, a major motivation for using superconductors is the increase in power density. The high magnetic flux density at the field winding required to achieve this results in high levels of magnetic self-forces of a magnitude similar to that of the centrifugal-force loading.* Moreover, the conductors must be supported against these so as to prevent conductor motion, which could produce localized losses and raise the possibility of a quench.

***Armature Windings.*** Armature windings for superconducting generators are airgap windings. An impressive variety of types of airgap windings have been proposed. While a few designers have been satisfied to make use of the same types and forms of coils as are used in conventional generators (i.e., drum windings, with conical involute ends), a larger number have utilized the design freedom offered by the removal of stator teeth to devise novel arrangements. These include, prominently:

- Helical winding
- Spiral pancake winding

Major motivations for innovation in armature windings include the objective of high power density and the potential of being able to design the armature to generate directly at transmission line voltage level. The elimination of the stator teeth not only liberates space for conductors and insulation but also moves the ground potential plane outside of the winding proper. Conductors and insulation can be redistributed to the maximum advantage. The spiral pancake winding[20] was invented to favor the integrity of high-voltage insulation and proposed as a possible means to the eventual development of transmission line voltage armatures.[21] Development of the helical winding has also had this objective as an ultimate goal.

The helical winding is illustrated in Fig. 10.8. It can be considered to be the result of a conventional winding using cylindrical involute end turns from which

*See Table 10.6 for an example.

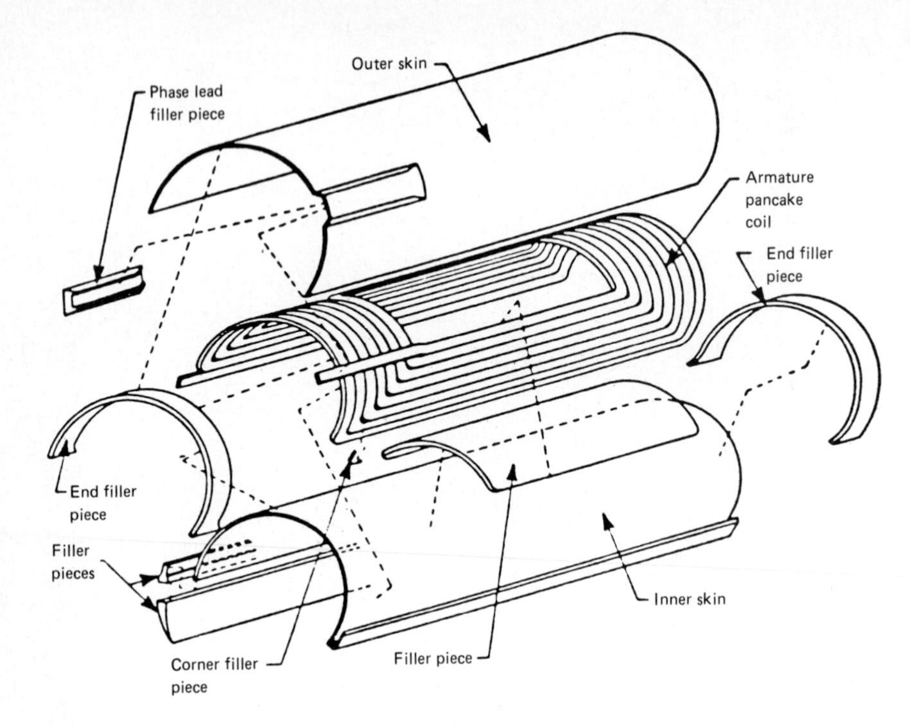


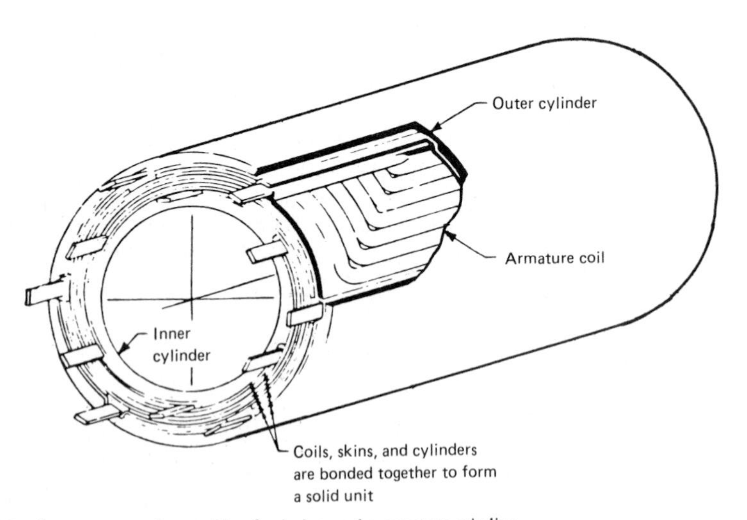


**FIG. 10.9** Components and assembly of spiral pancake armature winding.

the straight central section has been removed. In such a winding, the changes in coil direction and support conditions, which are imposed by the straight part-to-end transitions in conventional windings, are eliminated.

The spiral pancake winding is illustrated in Fig. 10.9 for a three-phase armature. It consists of six identical pancake coils with concentrically wound turns. Each coil is formed to the shape of a spiral so that all parts of a given conductor are at the same radius, but adjacent conductors are situated at different radii, which increase with the angle of the conductor from the start turn. The complete winding is formed by interleaving these coils equidistantly around a base circle and then bringing them radially inward to be in close contact. Figure 10.9 illustrates a winding design in which the available space has been used for both conductors and insulation, but not in the conventional manner of insulating each conductor from every other one for full overvoltage test withstand. Instead, ground insulation is applied to each of the six coils as a whole. Within each coil, except for the phase lead, only turn-to-turn insulation is applied. The insulating "skins" serve simultaneously to provide structural support for the conductors.

## Cryogenic Design and Isolation

Utilization of superconductors for the field windings of ac synchronous machines (e.g., turbine generators) requires the creation of a cryogenic environment aboard the rotor so that its inner parts operate at or near the atmospheric pressure boiling point of helium (4.2 K); the external parts of the rotor, or at least its ends, operate at or above room temperature. Thus, suitable provisions must be made to accommodate the large temperature differences between the inner and outer structures. This must be accomplished so as to maintain the mechanical integrity and helium tightness of all components as well as to ensure smooth operation at the requisite high speed. The design must also accommodate thermal stresses incident to operation, cool-down, and warm-up as well as to abnormal operation, such as balanced or unbalanced terminal faults.

Thermal isolation of the inner rotor is generally accomplished by a combination of (1) vacuum isolation in the annular space between the inner and outer cylindrical structures and (2) thermal distance pieces at one or both ends. The thermal distance pieces must carry torque between the inner and outer rotors and are therefore referred to as torque tubes. These provisions may be augmented by radiant-heat shields in the annular gap, as well as at appropriate locations at the ends, and by gas cooling with boil-off helium for the torque tubes; both methods are used to achieve more effective isolation and reduce the heat leak or flow of thermal energy from the outer to the inner rotor zones.

An example of the thermal isolation system designed for a certain machine is shown in Fig. 10.10. Helium flow connects between the shaft-mounted transfer coupling and a delivery tube that acts as a self-regulating flow control by maintaining a balance between the internal pressure developed by rotation and the external pressure applied by the helium supply. The rotor bore accepts the incoming helium and stores it in an annular layer. Centrifugal force transports supercritical helium toward the outer periphery and maintains its separation from helium vapor. The entire winding is immersed in supercritical fluid, in which no vaporization occurs. Fluid spaces in the winding provide enough local fluid to absorb transient heating effects. The warmed fluid is less dense, so convection loops are formed, and these transport the heat inward. The warmer fluid vaporizes at the surface of the supercritical pool and is replenished by cold fluid.

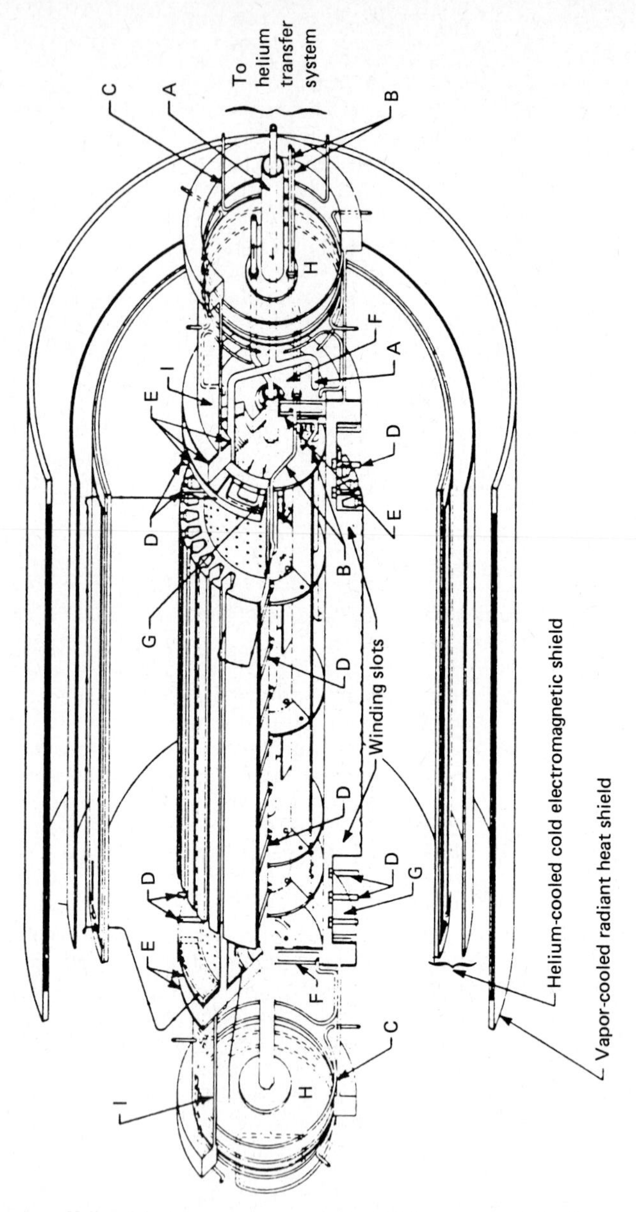


| | | | |
|---|---|---|---|
| **A** | Helium inlet tube | **F** | Helium pump with liquid trap |
| **B** | Vapor-cooled field leads | **G** | Hollowed-out ground-wall insulation |
| **C** | Torque-tube cooling tubes | **H** | Axial radiant-heat shields |
| **D** | Radial copper heat exchangers | **I** | Vapor-cooled torque tubes |
| **E** | Copper electromagnetic shields | | |

**FIG. 10.10** Diagram of rotor cryogenic cooling (helium management) system.

The evaporated helium enters pumplike chambers, from which it is circulated through cooling coils in the torque tubes and flow passages in the radiant-heat shield. A normal rate of flow corresponding to steady-state operation achieves torque-tube temperature distributions that are optimized by design with respect to thermal stress. It also establishes the temperature of the radiant-heat shield so that it will intercept the radiant heat incident from the outer rotor and reject it back to the refrigerator with reasonable efficiency. Electrical transients, such as system faults, produce transient increases in the flow of vaporized helium, as well as reductions in its temperature, and tend to change the temperature distribution in the torque tubes. The radiant-heat shield will usually have sufficient mass that its temperature is not greatly affected.

Vaporized helium leaves the rotor at the transfer coupling and is returned to the refrigerator for cooling back to liquefaction temperature.

When the outer rotor structure is at or above room temperature, the large temperature difference between the outer and inner structures results in a differential contraction, which must be accommodated by providing either a spring member or a sliding joint between these members. Both approaches have been applied, but the use of a spring appears to be more common in current designs. The use of a sliding joint incorporated in a double bearing may, however, prove indispensable for long rotors.

The design of the cryogenic system must provide for the losses and heat liberated in all parts in contact with the coolant, not only during normal operation but especially during fault conditions. Very precise determination of the losses is required in order to ensure a design that is able to operate stably through and following a severe system fault. As noted above, stability enhancement is a major motivation for the development and eventual application of superconducting turbine generators. Industry needs have been defined in terms of longer critical fault clearing times and hence of greater tolerance for fault conditions. During such faults, heat deposition occurs throughout the rotor. The specific heat of solid materials (e.g., metals, including superconductors) is extremely small at cryogenic temperature; thus, even small amounts of heat deposition become potential risks of quench or normalization. The cooling system must be designed to prevent this; moreover, it must prevent normalization without major changes in the temperatures of associated structural parts in order to prevent excessive thermal stress. Figure 10.10 is an example of a system designed to provide a high degree of superconductor stability and to satisfy the other constraints imposed by control of thermal stresses and refrigeration capabilities.[22]

The calculations made for the design of this system were subject to verification by a series of experiments for the modeling of important performance aspects and characteristics. The cryostability of the superconductor application, in particular, was confirmed by measurement of the heat-transfer coefficients in a rotating frame at the same peripheral speed as under actual operation; the experimental results, shown in Figs. 10.11 and 10.12, contributed significant new knowledge about natural convection in high-acceleration fields.[23,24] The application of the data to the field winding of the machine demonstrated its cryostability in both the straight central part and the curved end turns.

For practical applications of continuously operating superconducting generators, the helium coolant must be circulated from the refrigerator-liquefier to the cryogenic confinement aboard the rotor and back again. Liquid helium is supplied at about 4 K and may be returned, after it is used to cool the torque tubes, at approximately room temperature. The transfer coupling that performs this function requires both vacuum insulation and close clearances in order to operate at reasonable efficiency. It is possible for the helium stream to experience thermal oscillations, making for supply instabilities aboard the rotor.

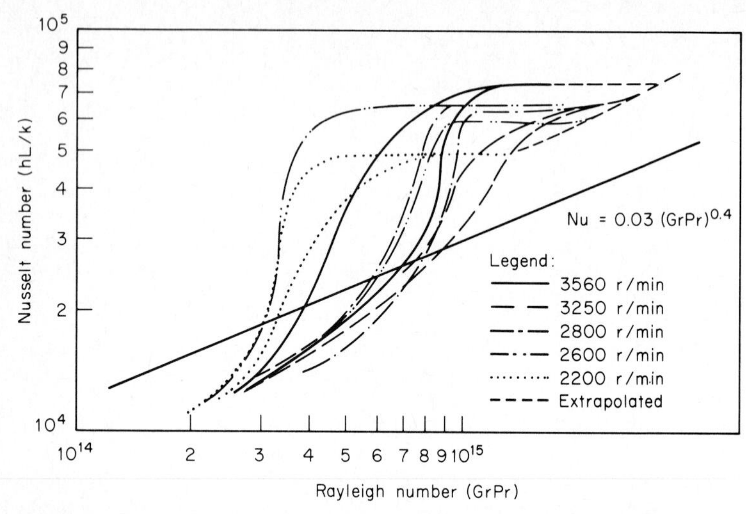


**FIG. 10.11** Correlation of heat transfer in the winding straight section (Coriolis vector normal to the heated surface).

Transient operating conditions of superconducting generators must be taken into account in specifying refrigerator requirements.

## Electromechanical Design and Performance

The electromechanical design of a superconducting ac machine is a process that occurs in a sequential and iterative order.

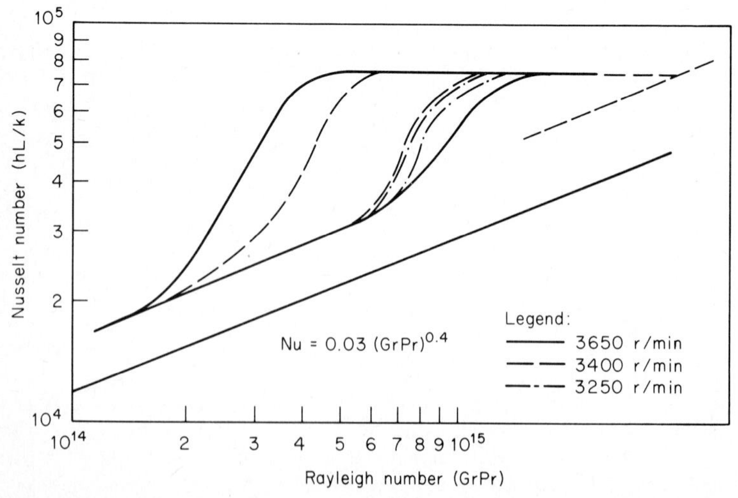


**FIG. 10.12** Correlation of heat transfer in the winding end turns (Coriolis vector parallel to the heated surface).

**TABLE 10.2** Major Specifications of 300-MVA Superconducting Generator

| | |
|---|---|
| Line voltage | 24 kV |
| Frequency | 60 Hz |
| Speed | 3600 r/min |
| Power factor | 0.90 |
| Negative sequence capability | $I_2 \geq 0.1$† <br> $I_2^2 t \geq 15$† |
| Reactances | $X_d \leq 50\%$‡ <br> $X'_d \leq 35\%$‡ <br> $X''_d \leq 20\%$‡ |
| Efficiency | 99.4% |

† $I_2$ in per-unit of rated current.
‡ In per-unit on rated MVA base.

As suggested in Fig. 10.5, the process begins with a specification. In the usual case of the specification for, say, a turbine generator, the rating and normal operating capability will be defined; certain of the transient performance parameters may also be specified. The major aspects of a particular specification are summarized in Table 10.2. In addition, this specification requires a minimum value of critical fault clearing time (CFCT) for a fault of specified severity; the system for determination of this CFCT was indicated in Fig. 10.2.

***Configuration Design.*** The major configurations selected for meeting such requirements have been discussed in the literature. (See the References and Bibliography at the end of this chapter.) Once these configurations have been selected, an initial determination of the major dimensions can be undertaken. Although not sufficient for final design purposes, the initial sizing can be done on the basis of a two-dimensional field analysis for the central part of the machine.[25] The end-region effects are, for this purpose, taken into account by an empirical correction factor. The equations applicable to a two-pole generator are given in Tables 10.3 and 10.4; these tables afford initial estimates for the fields in the regions of major interest as well as calculations for the inductances. The symbols used in these tables are defined in the Nomenclature at the end of this chapter.

It is also convenient to evaluate the expressions for rated voltamperes

$$P_r = 3V_t I_a \tag{10.16}$$

per unit synchronous reactance

$$x_s = \frac{X_a I_a}{V_t} \tag{10.17}$$

as well as the transient and subtransient reactances, using the formulas in Table 10.4. In order to do this, evaluate $V_t$ in terms of the internal machine voltage

**TABLE 10.3** Magnetic Fields for a Two-Pole Generator

$np \neq 2$

| Field-winding fields |
|---|
| $r < R_{fi}$ |
| $H_{rf} = -\sum_{n_{odd}} \frac{2J_f \sin(n\theta_{wfe}/2) \sin np(\theta - \phi)}{n\pi(2-np)} r \left(\frac{r}{R_{fo}}\right)^{np-2} \left[1 - y^{2-np} \pm \left(\frac{2-np}{2+np}\right)\left(\frac{R_{fo}}{R_s}\right)^{2np} (1 - y^{2+np})\right]$ |
| $H_{\theta f} = -\sum_{n_{odd}} \frac{2J_f \sin(n\theta_{wfe}/2) \cos np(\theta - \phi)}{n\pi(2-np)} r \left(\frac{r}{R_{fo}}\right)^{np-2} \left[1 - y^{2-np} \pm \left(\frac{2-np}{2+np}\right)\left(\frac{R_{fo}}{R_s}\right)^{2np} (1 - y^{2+np})\right]$ |
| $R_{fi} < r < R_{fo}$ |
| $H_{rf} = -\sum_{n_{odd}} \frac{2J_f \sin(n\theta_{wfe}/2) \sin np(\theta - \phi)}{n\pi(4-n^2p^2)} r \left[-2np - (2-np)\left(\frac{R_{fi}}{r}\right)^{np+2} + (2+np)\left(\frac{r}{R_{fo}}\right)^{np-2} \pm (2-np)\left(\frac{r}{R_s}\right)^{np-2}\left(\frac{R_{fo}}{R_s}\right)^{np+2}(1 - y^{np+2})\right]$ |
| $H_{\theta f} = -\sum_{n_{odd}} \frac{2J_f \sin(n\theta_{wfe}/2) \cos np(\theta - \phi)}{n\pi(4-n^2p^2)} r \left[-4 + (2-np)\left(\frac{R_{fi}}{r}\right)^{np+2} + (2+np)\left(\frac{r}{R_{fo}}\right)^{np-2} \pm (2-np)\left(\frac{r}{R_s}\right)^{np-2}\left(\frac{R_{fo}}{R_s}\right)^{np+2}(1 - y^{np+2})\right]$ |
| $R_{fo} < r < R_s$ |
| $H_{rf} = -\sum_{n_{odd}} \frac{2J_f \sin(n\theta_{wfe}/2) \sin np(\theta - \phi)}{n\pi(2+np)} r \left(\frac{R_{fo}}{r}\right)^{np+2} (1 - y^{np+2}) \left[1 \pm \left(\frac{r}{R_s}\right)^{2np}\right]$ |
| $H_{\theta f} = -\sum_{n_{odd}} \frac{2J_f \sin(n\theta_{wfe}/2) \cos np(\theta - \phi)}{n\pi(2+np)} r \left(\frac{R_{fo}}{r}\right)^{np+2} (1 - y^{np+2}) \left[1 \mp \left(\frac{r}{R_s}\right)^{2np}\right]$ |

**TABLE 10.3** Magnetic Fields for a Two-Pole Generator *(Cont.)*

Armature-winding phase A fields

$r < R_{ai}$

$$H_{ra} = -\sum_{n_{\text{odd}}} \frac{2J_a \sin(n\theta_{wae}/2)\sin np\theta}{n\pi(2-np)} r \left(\frac{r}{R_{ao}}\right)^{np-2} \left[1 - x^{2-np} \pm \left(\frac{2-np}{2+np}\right)\left(\frac{R_{ao}}{R_s}\right)^{2np}(1 - x^{2+np})\right]$$

$$H_{\theta a} = -\sum_{n_{\text{odd}}} \frac{2J_a \sin(n\theta_{wae}/2)\cos np\theta}{n\pi(2-np)} r \left(\frac{r}{R_{ao}}\right)^{np-2} \left[1 - x^{2-np} \pm \left(\frac{2-np}{2+np}\right)\left(\frac{R_{ao}}{R_s}\right)^{2np}(1 - x^{2+np})\right]$$

$R_{ai} < r < R_{ao}$

$$H_{ra} = -\sum_{n_{\text{odd}}} \frac{2J_a \sin(n\theta_{wae}/2)\sin np\theta}{n\pi(4-n^2p^2)} r \left[-2np - (2-np)\left(\frac{R_{ai}}{r}\right)^{np+2} + (2+np)\left(\frac{r}{R_{ao}}\right)^{np-2} \pm (2-np)\left(\frac{r}{R_s}\right)^{np-2}\left(\frac{R_{ao}}{R_s}\right)^{np+2}(1 - x^{np+2})\right]$$

$$H_{\theta a} = -\sum_{n_{\text{odd}}} \frac{2J_a \sin(n\theta_{wae}/2)\cos np\theta}{n\pi(4-n^2p^2)} r \left[-4 + (2-np)\left(\frac{R_{ai}}{r}\right)^{np+2} + (2+np)\left(\frac{r}{R_{ao}}\right)^{np-2} \pm (2-np)\left(\frac{r}{R_s}\right)^{np-2}\left(\frac{R_{ao}}{R_s}\right)^{np+2}(1 - x^{np+2})\right]$$

$R_{ao} < r < R_s$

$$H_{ra} = -\sum_{n_{\text{odd}}} \frac{2J_a \sin(n\theta_{wae}/2)\sin np\theta}{n\pi(2+np)} r \left(\frac{R_{ao}}{r}\right)^{np+2}(1 - x^{np+2})\left[1 \pm \left(\frac{r}{R_s}\right)^{2np}\right]^{\dagger}$$

$$H_{\theta a} = -\sum_{n_{\text{odd}}} \frac{2J_a \sin(n\theta_{wae}/2)\cos np\theta}{n\pi(2+np)} r \left(\frac{R_{ao}}{r}\right)^{np+2}(1 - x^{np+2})\left[1 \mp \left(\frac{r}{R_s}\right)^{2np}\right]$$

† Wherever double signs (± or ∓) are used, the upper sign refers to the ferromagnetic environmental shield case (as illustrated in Fig. 10.6). The lower sign is appropriate if a conductive image or reflecting shield is used.

**TABLE 10.4** Inductances for a Two-Pole Generator

$np \neq 2$

$$L_f = \sum_{n_{odd}} \frac{16 l_f \mu_o N_{ft}^2 \sin^2(n\theta_{wfe}/2)}{n^3 p \pi \theta_{wfe}^2 (n^2p^2 - 4)(1-y^2)^2} \left[ (np-2) + 4y^{np+2} - (np+2)y^4 \pm 2\left(\frac{np-2}{np+2}\right)(1-y^{np+2})^2 \left(\frac{R_{fo}}{R_s}\right)^{2np} \right]$$

$$L_a = \sum_{n_{odd}} \frac{16 l_a \mu_o N_{at}^2 \sin^2(n\theta_{wae}/2)}{n^3 p \pi \theta_{wae}^2 (n^2p^2 - 4)(1-x^2)^2} \left[ (np-2) + 4x^{np+2} - (np+2)x^4 \pm 2\left(\frac{np-2}{np+2}\right)(1-x^{np+2})^2 \left(\frac{R_{ao}}{R_s}\right)^{2np} \right]$$

$$L_{ab} = \sum_{n_{odd}} \frac{16 l_a \mu_o N_{at}^2 \sin^2(n\theta_{wae}/2)\cos(2n\pi/3)}{n^3 p \pi \theta_{wae}^2 (n^2p^2 - 4)(1-x^2)^2} \left[ (np-2) + 4x^{np+2} - (np+2)x^4 \pm 2\left(\frac{np-2}{np+2}\right)(1-x^{np+2})^2 \left(\frac{R_{ao}}{R_s}\right)^{2np} \right]$$

$$L_a - L_{ab} = \frac{16 l_a \mu_o N_{at}^2}{p\pi\theta_{wae}^2(1-x^2)^2} C_{sx}$$

where $C_{sx} = \sum_{n_{odd}} \frac{\sin^2(n\theta_{wae}/2)[1-\cos(2n\pi/3)]}{n^3(n^2p^2-4)} \left[ (np-2) + 4x^{np+2} - (np+2)x^4 \pm 2\left(\frac{np-2}{np+2}\right)(1-x^{np+2})^2 \left(\frac{R_{ao}}{R_s}\right)^{2np} \right]$

$$M_a = \sum_{n_{odd}} \frac{32 l_m \mu_o N_{ft} N_{at} C_{mn} \cos np\theta}{p\pi\theta_{wfe}\theta_{wae}(1-y^2)(1-x^2)} \left(\frac{R_{fo}}{R_{ao}}\right)^{np}$$

where $C_{mn} = \frac{\sin(n\theta_{wae}/2)\sin(n\theta_{wfe}/2)(1-y^{np+2})}{n^3(4-n^2p^2)} \left[ 1 - x^{2-np} \pm \frac{2-np}{2+np}(1-x^{2+np}) \left(\frac{R_{ao}}{R_s}\right)^{2np} \right]$

$$L_{dn} = \frac{\mu_o l_d \pi N_{np}^2}{8np} \left[ 1 \pm \left(\frac{R_t}{R_s}\right)^{2np} \right]$$

$$L_{adn} = \frac{2\mu_o l_{ad} N_{at} N_{np} \sin(n\theta_{wae}/2)\cos np\theta}{n^2 p \theta_{wae}(1-x^2)(2-np)} \left(\frac{R_t}{R_{ao}}\right)^{np} \left[ 1 - x^{2-np} \pm \frac{2-np}{2+np}(1-x^{2+np}) \left(\frac{R_{ao}}{R_s}\right)^{2np} \right]$$

$$L_{fdn} = \frac{2\mu_o l_{fd} N_{ft} N_{np} \sin(n\theta_{wfe}/2)}{\theta_{wfe}(1-y^2)n^2p} \left(\frac{R_{fo}}{R_t}\right)^{np} \left(\frac{1-y^{np+2}}{np+2}\right) \left[ 1 \pm \left(\frac{R_t}{R_s}\right)^{2np} \right]$$

See footnote table at the top of page 10-29.

| To get | Use | Replace | By |
|---|---|---|---|
| $M_b$ | $M_a$ | $\phi$ | $\phi - 2\pi/3$ |
| $M_c$ | $M_a$ | $\phi$ | $\phi + 2\pi/3$ |
| $L_{bdn}$ | $L_{adn}$ | $\phi$ | $\phi - 2\pi/3$ |
| $L_{cdn}$ | $L_{adn}$ | $\phi$ | $\phi + 2\pi/3$ |
| $L_{aqn}$ | $L_{adn}$ | $\cos np\phi$ | $\sin np\phi$ |
| $L_{bqn}$ | $L_{adn}$ | $\cos np\phi$ | $\sin np(\phi - 2\pi/3)$ |
| $L_{cqn}$ | $L_{adn}$ | $\cos np\phi$ | $\sin np(\phi + 2\pi/3)$ |

$$E_f = \frac{\omega M I_f}{\sqrt{2}} \tag{10.18}$$

From the relationship of machine voltages and currents and the law of cosines, one may write

$$\frac{V_t}{E_f} = \sqrt{1 - x_a^2 \cos\Psi} - x_a \sin\Psi \tag{10.19}$$

where cos $\Psi$ is the power factor and where

$$x_a = \frac{X_a I_a}{E_f} \tag{10.20}$$

is the per-unit synchronous reactance calculated with internal voltage $E_f$ as a base. The per-unit synchronous reactance based on terminal voltage is then

$$x_s = \frac{I_a X_a}{V_t} = x_a \frac{E_f}{V_t} \tag{10.21a}$$

with

$$x_a = \frac{\omega(L_a - L_{ab})I_a}{E_f} \tag{10.21b}$$

The subtransient reactance is calculated by assuming that the flux linked by the electrothermal shield is constant. The equivalent damper windings of the shell play the same role as the field winding plays in transient reactance. Further, since this shell is symmetrical, the direct- and quadrature-axis subtransient reactances are the same:

$$x_d'' = x_q'' = x_s \left[1 - \frac{3}{2} \frac{L_{ad1}^2}{(L_a - L_{ab})L_{d1}}\right] \tag{10.22}$$

Because of the very long transient time constants (see Table 10.6), the subtransient reactance also plays the role of the transient reactance of conventional machines in stability studies.

Initial assessments can be made on this basis also for the forces of electromagnetic origin incident upon the field winding, the armature winding, and the shield or shields. In particular, the ovalizing forces on the rotor shield (assumed for this purpose to be at the outer surface of the rotor and of negligible thickness) can be calculated by extending the work of Furuyama.[26]

Preliminary calculation of the short-circuit forces on the armature winding and of the electromagnetic torques can be accomplished on similar grounds. Closed form expressions are available for the torques experienced by the rotor during the different types of short circuit;[27] these are sufficiently accurate for preliminary sizing of the shaft sections and internal connections (e.g., the spring support used to accommodate thermal contraction). The electromagnetic forces on at least the straight part of the armature can also be found from closed form expressions by forming the $\mathbf{J} \times \mathbf{B}$ vector product.[28]

The use of the design equations in Tables 10.3 and 10.4 requires the field- and armature-winding current densities $J_f$ and $J_a$ as input items. In conventional design practice, these densities can usually be estimated from previous designs based on the same or similar cooling methods. For superconducting generators, cooling remains an important criterion, but other criteria, such as the critical current of the superconductor, must also be taken into account. Also note that the densities $J_f$ and $J_a$ are averaged, or "smeared out," over the annular segments occupied by the windings. These spaces include not only active conductors but also materials required for structural support; thus, the average achievable densities are well below the densities in the conducting materials.

In order to utilize the generator volume economically and to obtain an efficient partition between active conductor area and supporting material, the designer may resort to design optimization. A number of obstacles to the use of classical methods of optimization, such as steepest descent, may be encountered, including the large number of variables, some of which are discrete rather than continuous. Subopti-mization of separable design features, such as the field winding, may then be a more effective approach.[29] Optimization by means of multiparametric analysis has also proven useful. One such analysis identified armature current density as an important "figure of merit" that has ramifications for the technical and economic viability of the design concept.[30]

Preliminary cooling studies for all parts of the machine must also be started at about this time. If values for $J_f$ and $J_a$ have been assumed, some confirmation of these assumptions is needed.

***Detail Design.*** Refer again to Fig. 10.5. The six functional blocks in the left part of the design process are, for simplicity, shown in parallel, but this is not strictly correct. For example, the helium management design and the winding design are not independent efforts; rather, they constitute an iterative subloop, which may be exited only when a proper match has been achieved between the cooling capability of the helium supply system and the cooling requirements of the winding. This must be accomplished not only for the steady state but also to provide enough tolerance for fault conditions to prevent the precipitation of a quench.[31]

An overview of the detail design process is shown in Table 10.5, in which, for simplicity and clarity, only the highest-level tasks have been displayed. It will be noted that a key task basic to almost all other analyses and design is the precise determination of the three-dimensional distribution of the magnetic fields in the interior of the machine, not only for the steady-state conditions but especially for all transient conditions of interest. The need for this precision arises largely from the sensitivity of the superconducting field winding to electromagnetic disturbances that retain significant magnitude even after attenuation by the compound electrothermal shielding system.

The absence of magnetic material from the interior of a superconducting ac generator invalidates the design approach used for iron-core machines and requires the mathematical modeling of the machine as an entity. Moreover, the model must be constructed so as to account with sufficient accuracy for the diffusion problem posed by the conducting shields and to afford a reliable assessment of the magnetic field disturbances at the field winding. Since even small field changes can cause significant energy liberation, highly accurate numerical results are required. The ovalizing forces cause the conducting shields to deform under the impact of fault loading,

**TABLE 10.5** The Design Process

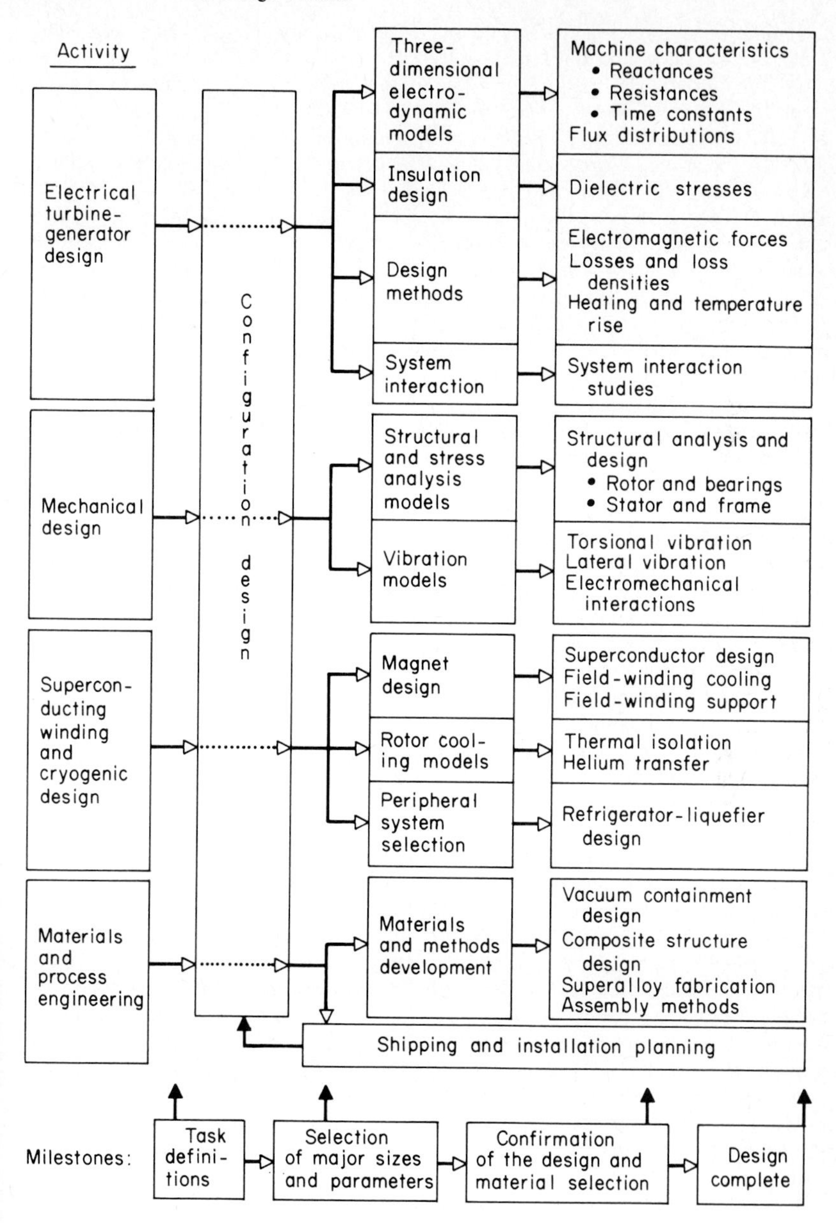

which results in another mechanism of loss generation in the superconductors. The more elaborate models account for this effect as well.

For detail design, the method of three-dimensional analysis is used to predict transient system interactions, superconductor losses, and temperature rise as well as the overall effect on the power system–cryogenic system stability issue.[32–34] Some results for a specific case are shown below. The design for which these performance calculations were made is summarized in Table 10.6 (p. 10-33).

***Design Example.*** Figure 10.6 shows the configuration to which the analysis was applied.[35,36] An ideal magnetic container and windings developed on cylindrical surfaces are assumed. Magnetic diffusion through a solid shield is modeled by replacing the thick shield with a set of thin concentric subshells, which are inductively coupled but resistively isolated. Linearity and superposition are assumed to hold. A coupled-circuit approach, using an extension of the Park $d$-$q$ analysis of the synchronous machine, is employed to resolve the discrete windings and the conducting shields into series of coupled circuits that carry currents modeling the actual currents in the physical windings and to calculate their transient values.

The results of this modeling process can readily be interpreted to yield both an equivalent-circuit model of the complete machine and the corresponding set of extended Park's equations, as suggested by Fig. 10.13 (p. 10-34). This model is thus amenable to the types of transient interaction analyses that have become standard in studying the system performance aspects of synchronous machine application, e.g., stability and fault analyses. Inasmuch as the basic motivation for so complex a model is the accurate representation of the fluxes and currents in the interior of the machine, rather careful attention must be paid to computational accuracy, and the model is not readily amenable to inclusion in system representations that require many machines to be modeled simultaneously. For these purposes, the model has been adapted to yield the usual stability study constants shown in Table 10.6. To obtain these values, the model is excited with currents of variable frequency, and the standard constants are obtained by curve-fitting the frequency-domain operational impedances.

The complete representation of the machine with acceptable accuracy calls for the manipulation of matrices of significant size ($\simeq 200 \times 200$), which must be inverted with high numerical precision. Efficient methods to achieve solutions of acceptable accuracy and computer cost have been developed.[33] The following figures show various important performance characteristics of this machine.[33–35]

***Performance Characteristics.*** Although these figures illustrate concepts basic to all rotating-machine theory, they exhibit features in which the characteristics of superconducting generators differ markedly from machines of more conventional construction.

The fundamental factor leading to these differences is the effective isolation or shielding of the field winding from the armature and its electromagnetic reaction. It was found that the shielding calculations made with increasing numbers of axial harmonics converge rather rapidly with harmonic order and that six harmonics are enough to get good results. Figure 10.14 (p. 10-35) shows that the cold shield must be extended to very near the ends of the armature winding to prevent all leakage of flux around the shield ends. The distribution of the shielding currents among the successive rotor shells as a function of frequency is illustrated in Fig. 10.15 (p. 10-36). It will be noted that for disturbances with frequencies higher than 1 to 2 Hz, the field winding is substantially decoupled from the armature. As a result of this decoupling, field forcing is not effective for stability enhancement of superconducting generators.

Losses produced by the shielding currents are plotted in Fig. 10.16 (p. 10-36) for a severe disturbance (three-phase fault). As may be expected, nearly all the energy in the warm structure appears in the brief initial period of the fault. The energy loss

**TABLE 10.6** Design Parameters for 300-MVA Superconducting Generator

| Rotor dimensions* | | Stator dimensions* | |
|---|---|---|---|
| Field-winding inner radius | 242 | Radius to innermost conductor | 589 |
| Field-winding outer radius | 312 | Radius to outermost conductor | 848 |
| Cold-shield outer radius | 367 | Flux-shield inner radius | 925 |
| Cold-shield thickness | 5.1 | Flux-shield outer radius | 1372 |
| Warm-damper outer radius | 448 | Single end-turn width | 897 |
| Warm-damper thickness | 25.4 | Single pancake thickness | 63 |
| Outer radius | 476 | Mean length of turn | 9449 |
| Active length | 1981 | Active length | 1981 |
| Total number of slots | 24 | Connection | three-phase |
| Total number of turns | 2556 | Series turns per phase | 20 |
| Superconductor size | 2.03 × 3.84 | Uninsulated conductor | 37.8 × 73.2 |
| Copper/superconductor ratio | 2.5:1 | Copper cross section | 1051 |
| NbTi filament diameter, microns | 50 | Inner phase-belt width | 69° |
| Filament twist pitch | 40 | Outer phase-belt width | 56° |
| Critical current at 5 T, 4.2 K | 3300 A | Winding span | 210° |

| Rotor slot and turns distribution | | | | Field-winding electromagnetic forces | | | |
|---|---|---|---|---|---|---|---|
| | | | | Normal force, MN/m | | Radial force, MN/m | |
| Slot no. | Turns | Angle, pole CL† to slot CL, deg | Maximum flux density | In slot | On end | In slot | On end |
| 1 | 198 | 28.0 | 6.4 | 1.54 | 1.77 | 0.05 | 0.11 |
| 2 | 216 | 39.7 | 5.9 | 1.10 | 1.62 | 0.10 | 0.18 |
| 3 | 216 | 51.4 | 5.3 | 0.89 | 1.41 | 0.13 | 0.23 |
| 4 | 216 | 62.9 | 5.1 | 0.60 | 1.16 | 0.15 | 0.25 |
| 5 | 216 | 74.1 | 4.7 | 0.44 | 0.77 | 0.17 | 0.24 |
| 6 | 216 | 84.7 | 4.8 | 0.16 | 0.37 | 0.19 | 0.22 |

Standard generator stability constants

| Reactances, per-unit on 300-MVA base | | | Time constants, s | | |
|---|---|---|---|---|---|
| Direct axis | | | | | |
| Synchronous | $X_d$ | 0.335 | Transient open circuit | $T'_{do}$ | 4040 |
| Transient | $X'_d$ | 0.161 | Subtransient open circuit | $T''_{do}$ | 0.30 |
| Subtransient | $X''_d$ | 0.175 | Transient short circuit | $T'_d$ | 3270 |
| | | | Subtransient short circuit | $T''_d$ | 0.20 |
| Quadrature axis | | | | | |
| Synchronous | $X_q$ | 0.335 | Transient open circuit | $T'_{qo}$ | 3.70 |
| Transient | $X'_q$ | 0.224 | Subtransient open circuit | $T''_{qo}$ | 0.07 |
| Subtransient | $X''_q$ | 0.175 | Transient short circuit | $T'_q$ | 2.47 |
| | | | Subtransient short circuit | $T''_q$ | 0.06 |
| Negative sequence | $X_2$ | 0.174 | Armature | $T_a$ | 0.23 |
| Zero sequence | $X_o$ | 0.044 | Generator inertia constant | $H$ | 0.62 |

* Linear dimension in mm; areas in $mm^2$; flux densities in T (tesla).
† CL: centerline.

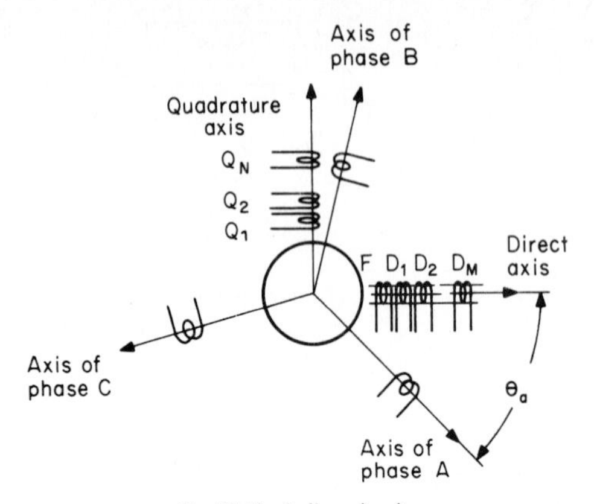


F Field-winding circuit

$D_1, D_2, \ldots, D_M$ Lumped parameter circuits on direct axis for eddy-current shields

$Q_1, Q_2, \ldots, Q_N$ Quadrature-axis circuits for shields

**FIG. 10.13** Schematic representation of a three-phase superconducting generator.

in the cold shield is delayed by diffusion effects. As shown by Fig. 10.17 (p. 10-37), the multiple shielding of the field winding (warm and cold shields acting in concert) achieves virtually complete attenuation of all higher-frequency disturbances. The time response shown is the gradual rise of field current required to oppose the synchronous component of the increased armature reaction. This smooth and gradual response may be contrasted to the field response in a conventional generator, in which, even with an effective amortisseur cage, a considerable fundamental-frequency component may be observed.

Despite this excellent isolation, energy dissipation in the cryogenic zone of the rotor will be significant during faults, and especially during the slow rotor swings that accompany recovery to a new steady state. The case of faults that persist right up to the critical fault clearing time (CFCT) has been studied in great detail for the case of a 300-MVA machine, inasmuch as increased CFCT has been an important program objective. The results are illustrated in Figs. 10.18, 10.19, and 10.20 (pp. 10-37 and 10-38); in this order, they show the losses induced in the superconductor, the density of energy dissipation in the same, and the cold-shield loss. These losses are then dissipated to the helium coolant. As a result of extremely effective heat transfer mediated by the high gravity (g) acceleration of these rotating parts, the release of the energy to the coolant is completed in a period of seconds. However, the high rate of energy release produces a considerable increase in pressure of the helium system, and this must be provided for in the mechanical design of the rotor parts. The pressure increase is illustrated in Fig. 10.21 (p. 10-39).

Finally, Fig. 10.22 (p. 10-39) shows the diminution in the volume of helium stored in the central cavity of the rotor to provide the requisite degree of fault tolerance. Since much of this stored helium is depleted by a long-duration fault, the time to recover full fault tolerance is determined by the rate at which helium is replenished through the transfer mechanism.

***System Stability.*** Electrical design includes the study of normal as well as abnormal interactions between the superconducting generator and the power system

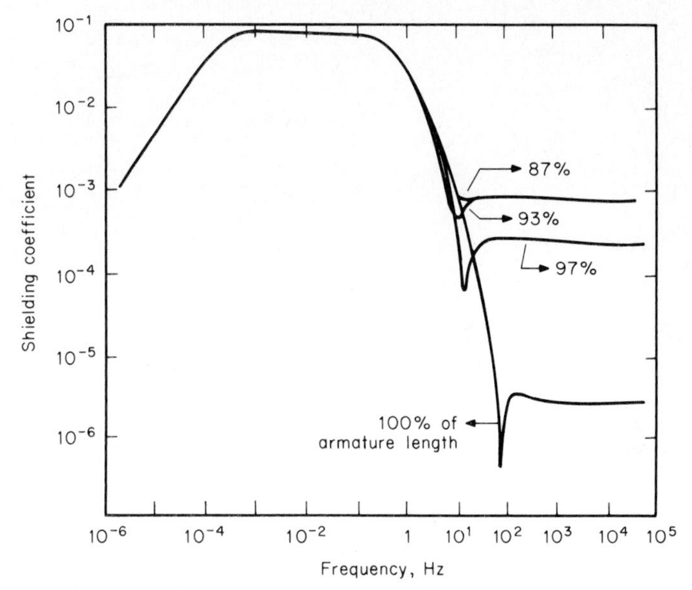


**FIG. 10.14** Shielding coefficient for different lengths of cold shield.

to which it is connected. Determination of the stability of the unit under various sorts of disturbances is of particular interest.

As illustrated in Fig. 10.2, superconducting generators can be designed for longer CFCTs than can conventional machines. The improvement is attributable to a major reduction in reactances, which more than compensates for the reduction in the inertia constant $H$. Further, the long subtransient time constants resulting from the damping-circuit design for sufficient shielding cause subtransient reactance, rather than transient reactance, to dominate these system interactions. In superconducting generator designs in which end-zone flux is controlled by means of a flux shield extending the full length of the stator winding, the capability for capacitive loading can be maintained at full MVA output, as shown in Fig. 10.23 (p. 10-40). The low synchronous and transient reactances permit this capability to be utilized.

***Mechanical Design Considerations.*** Superconducting field windings, as noted in Table 10.6, can achieve significantly greater power densities than conventional designs can, but at the same time they impose greater electrodynamic force levels. As a rule of thumb, the increase in forces can be taken to be proportional to the square of the current density. The mechanical design must accommodate these higher loads as well as the limitations imposed by the necessity to maintain a cryogenic temperature zone within the rotor. The high-strength steels typically used in conventional rotors cannot be exposed to these temperatures. The choice is between some of the stainless steels and the nickel-based "superalloys." Desired levels of strength at both room temperature (at which some running at speed must normally occur) and cryogenic temperature will determine the choice. Since design margins must be kept thin if the advantages offered by superconductivity are not to be sacrificed, detailed stress calculations are needed for all highly stressed parts.[37]

Airgap stator windings also constitute a design element that demands thorough analysis. In such a winding, the armature conductors are exposed to the full main flux field, and the resulting electromagnetic forces must be transferred to the

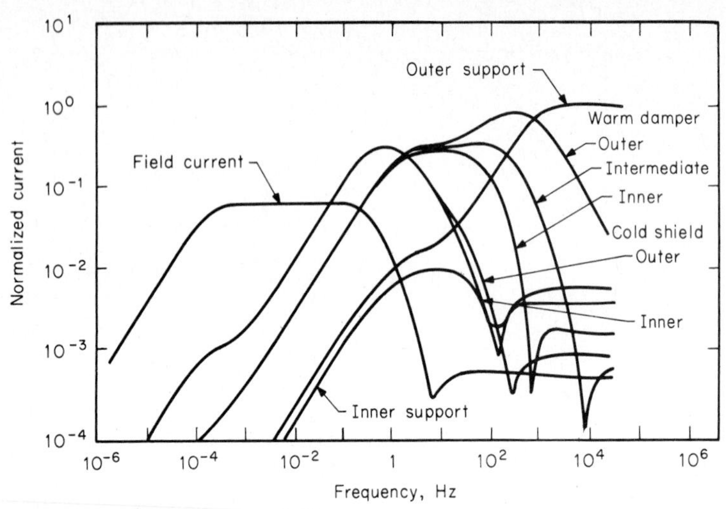


**FIG. 10.15** Shield-shield support layer current distribution and field current versus armature excitation frequency, normalized with respect to armature current increment.

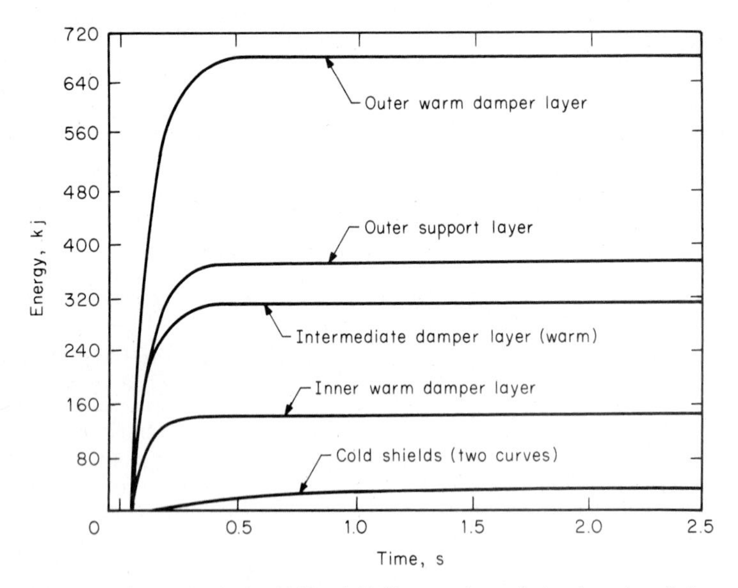


**FIG. 10.16** Energy loss in the shield and shield support layers during three-phase fault.

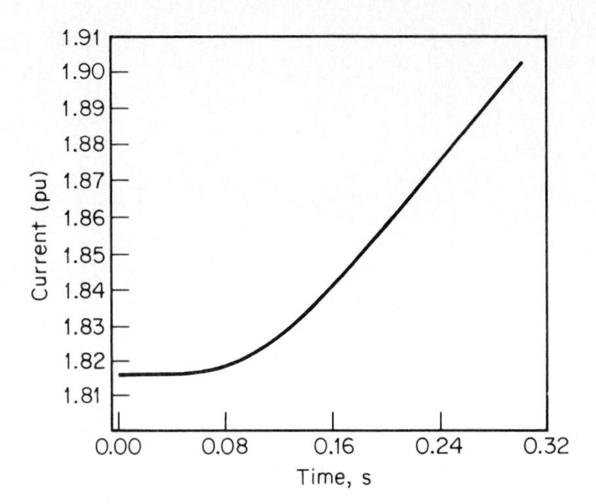


**FIG. 10.17** Field current during three-phase fault, nearly perfect shielding.

supporting structure through nonmetallic materials so as to control motions and vibration within acceptable limits as well as within the strengths of the insulating and support materials. Studies of armature and support structure design have been reported by several authors.[38,39]

An important aspect of high-speed machinery design is vibration analysis of the shaft system with respect to both lateral and torsional oscillations. For purposes of analysis, the rotor structure is represented as a collection of masses (or inertias) and springs. Dashpots may be used to represent damping effects. Vibration analyses for superconducting generators must take account of the fact that the rotor is made up of several concentric shells rather than being solid.

## Development Status and Examples

Development of superconducting generators has been essayed in nearly all the world's industrialized countries. An excellent overview of the development status was given

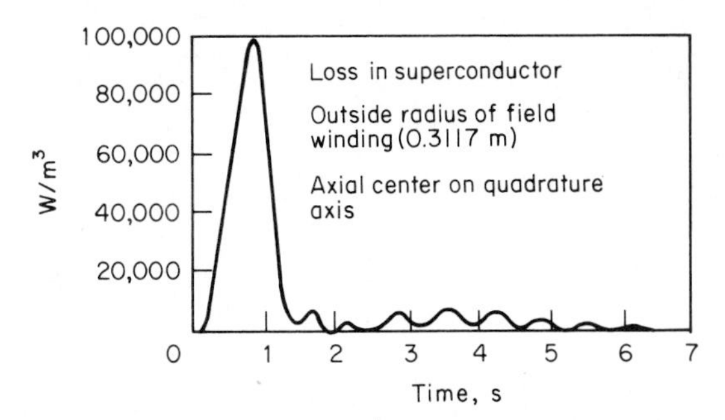


**FIG. 10.18** Superconductor loss for three-phase 15-cycle fault.

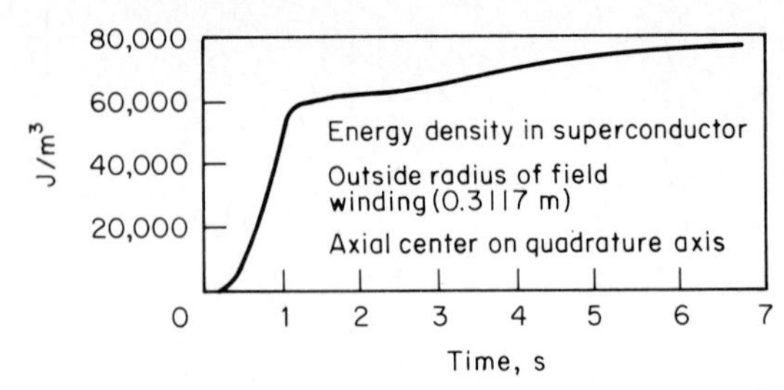


**FIG. 10.19** Superconductor energy density for three-phase 15-cycle fault.

by one of the recognized leaders in the field in 1983.[18] Since that time, the status has changed remarkably with the completion or termination of two major U.S. programs (a 20-MVA generator shop-tested and a 300-MVA generator program terminated). Work continues in Germany and Japan and possibly in the Soviet Union.

The programs oriented toward large machines for application to electric power systems all appear to be based on the same group of motives: to achieve a better fit to user needs by reason of reduced size and weight, improved efficiency, better performance with respect to system needs and exigencies, and the possibility of achieving greater outputs and at higher voltage ratings than afforded by conventional designs. Likewise, a considerable convergence of design and construction approaches and features has taken place as the programs have advanced from conceptualization and feasibility demonstrations in laboratory settings to large-scale construction and test in factory environments. Although the disciplines imposed by limitations of obtainable materials and manufacturing methods have taken some of the wind out of the sails of enthusiasts, the engineering successes of these programs are not to be underestimated. There is every reason to believe that enough work has now been done that construction of superconducting generators on a commercial basis is completely feasible and only a matter of time.

The convergence of features can perhaps best be demonstrated by citing specific examples that go beyond the description of the optimum configuration[18] into more detailed construction features.

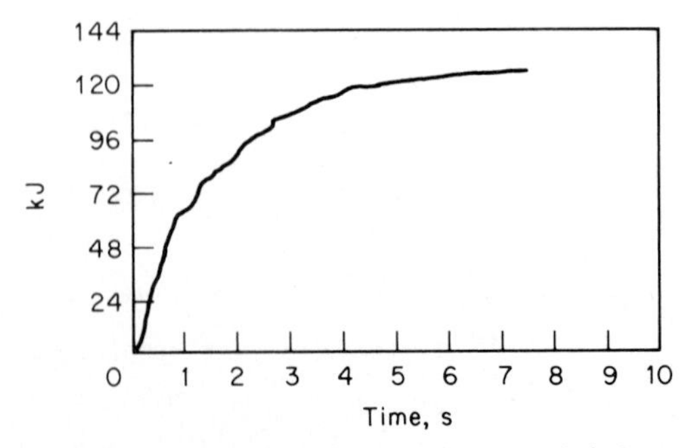


**FIG. 10.20** Cold-shield loss for three-phase 15-cycle fault.

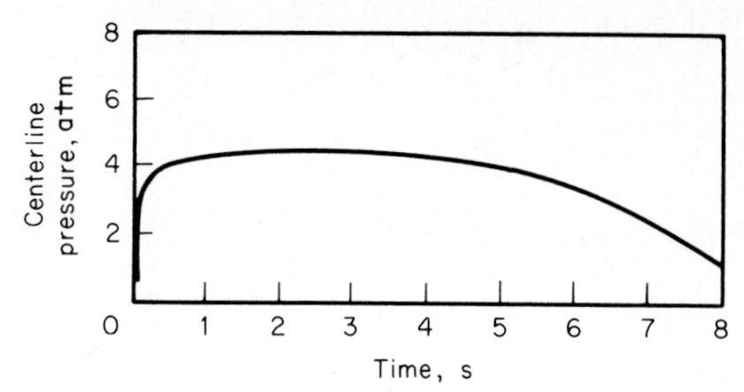


**FIG. 10.21** Helium system pressure rise for three-phase 15-cycle fault.

One example is the field-winding configuration and supporting structure. Although pancake coils were featured in some of the earlier designs and were considered for large machines until as recently as 1977, designers now appear to opt uniformly for saddle-shaped coils, which also seem in all recent designs to be housed in slotted metallic structures. Note that while the toothed structure is not strictly necessary, the support provided by it is—and would otherwise have to be provided by arrays of smaller features requiring precision machining and possibly depending in part on frictional fits.

Another interesting convergence is in the design of the shielding provisions. Compound shielding, using an outer warm shielding structure in conjunction with a low-temperature thermal shield and a cold electromagnetic shield, appears to have found its way into most current designs.

Thermal isolation of rotor cryogenic zones is being accomplished by means of vacuum spaces, usually (but not in all cases) between the cold inner and the warm outer structures, as shown in Fig. 10.7. The isolation remains effective only as long as a hard vacuum persists. Cryopumping (condensation of gas molecules on the cold surfaces) can keep up with slow leaks but is not useful for helium leaks; for this reason, all walls separating the vacuum from the helium must be constructed to be leakproof. An important design decision concerns a permanently sealed versus a continuously pumped vacuum. The latter requires a more complex helium-transfer coupling as well as an additional auxiliary vacuum pump, but the former requires extreme care in the manufacture and joining of the helium circuit. A recent trend appears to favor continuous pumping.

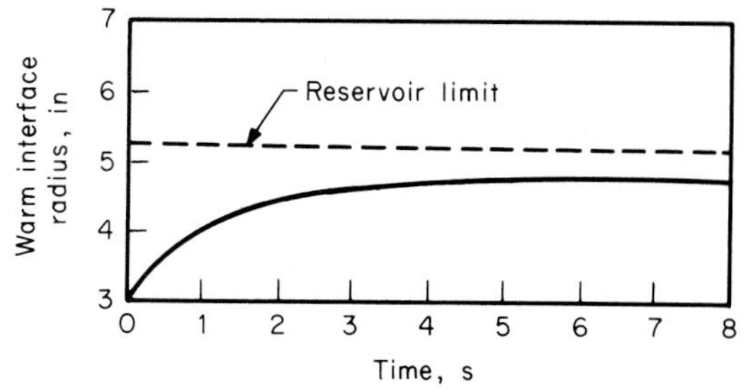


**FIG. 10.22** Helium reservoir depth reduction for three-phase 15-cycle fault.

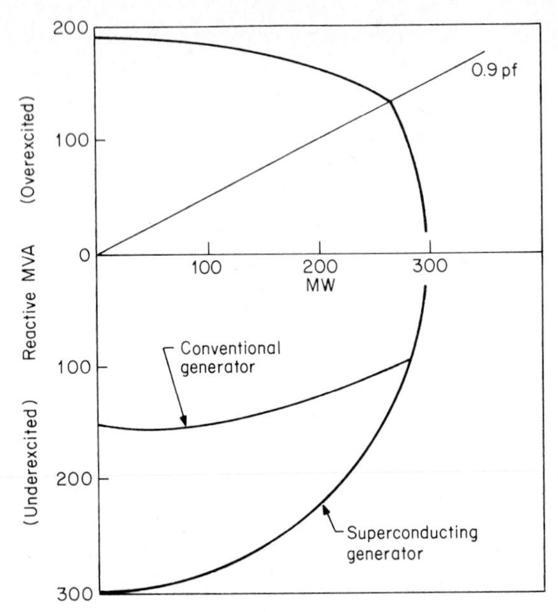


**FIG. 10.23** Improvement in steady-state capability.

In respect to the type of armature winding, the literature indicates that all programs are predicated on airgap windings, but the constructional details differ. Helical windings have been adopted for several programs. Their potential as prototypes for high-voltage armatures is believed to be comparable to that of the spiral pancake type.[40] Other programs rely upon coils with diamond-shaped ends, similar to those used in conventional armatures. Water cooling of the stator coils appears to be uniformly accepted for the larger machines.

The performance expected of such machines in power systems also appears to be converging toward a fairly definite set of objectives. Some of the early anticipations of rocklike stability secured through design for extremely low reactances, which had adverse consequences in terms of high fault currents and torques, have fortunately receded. The current trends appear to peg the synchronous reactance $X_d$ in the 0.4 to 0.7 per-unit range, with the other reactances in proportion. The precise level selected within this range for a particular application will depend upon the importance given to an increase in critical fault clearing time. System studies based on mixes of superconducting and conventional generators indicate that the full potential of system stability enhancement can be realized only when superconducting machines begin to form a significant part of the total population.

The operational characteristics of superconducting generators as power-plant apparatus are also receiving more general recognition and understanding. One significant aspect of operating these machines is the cool-down of the field winding and the other parts of the cryogenic zone. A large amount of heat must be extracted from these parts to bring them from room temperature to 4.2 K, and the coolant to accomplish this must be fed through a transfer coupling of a size limited by various rotor design considerations, not the least of which is that cool-down must proceed slowly enough that thermal stresses due to contraction remain within acceptable bounds. A consensus is being reached that time periods of about 3 days are required for safe and economical cool-down.

## REFERENCES


1. E. W. Kimbark, *Power System Stability*, Wiley, New York, 1948, vol. 1: "Elements of Stability Calculations," p. 159.
2. J. S. Edmonds, "Superconducting Generator Technology—An Overview," *IEEE Transactions on Magnetics*, vol. MAG-15, no. 1, January 1979, p. 673.
3. P. W. Eckels et al., "Heat Transfer Correlation for Cryostable Alternator Field Winding," *Advances in Cryogenic Engineering*, vol. 27, Plenum Press, New York, 1982, p. 357.
4. W. J. Carr, Jr., *AC Loss and Macroscopic Theory of Superconductors*, Gordon and Breach Science Publishers, New York, 1983.
5. J. H. Murphy et al., "Alternating Field Losses in $Nb_3Sn$ Multifilamentary Superconductor," *IEEE Transactions on Magnetics*, vol. MAG-11, no. 2, March 1975, p. 317.
6. W. J. Carr et al., "Alternating Field Loss in a Multifilamentary Superconducting Wire for Weak AC Fields Superposed on a Constant Bias," *J. Appl. Physics*, vol. 46, no. 9, September 1975, p. 4048.
7. A. Arkkio et al., "A 50-kW Homopolar Motor with Superconducting Field Windings," *IEEE Transactions on Magnetics*, vol. MAG-17, no. 1, January 1981, p. 900.
8. A. D. Appleton, "Status of Superconducting Machines at IRD, Spring 1972," *Proceedings of the 1972 Applied Superconductivity Conference*, IEEE Publication 72CH0682-5-TABSC, 1972.
9. A. D. Appleton, "Design and Manufacture of a Large Superconducting Homopolar Motor (and Status of Superconducting AC Generator)," *IEEE Transactions on Magnetics*, vol. MAG-19, no. 3, May 1983, p. 1047.
10. R. A. Marshall, "3000-Horsepower Superconductive Field Acyclic Motor," *IEEE Transactions on Magnetics*, vol. MAG-19, no. 3, May 1983, p. 876.
11. A. D. Appleton, "Superconducting DC Machines," in S. Foner and B. B. Schwartz (eds.), *Superconducting Machines and Devices*, Plenum Press, New York, 1973, p. 236.
12. J. T. Doyle, "Shaped Field Superconductive DC Ship Drive Systems," *Advances in Cryogenic Engineering*, vol. 19, Plenum Press, New York, 1974, p. 162.
13. A. D. Appleton and R. M. McNab, "A Model Superconducting Motor," *International Conference on Low Temperatures and Electric Power*, London, March 1969.
14. J. R. Bumby, *Superconducting Rotating Electrical Machines*, Clarendon Press, Oxford, England, 1983, p. 51.
15. J. P. Chabrerie et al., "Flooded Rotor, Direct-Current Acyclic Motor, with Superconducting Field Winding," *Proceedings of the 1972 Applied Superconductivity Conference*, IEEE Publication 72CH0682-5-TABSC, 1972.
16. K. I. Thomassen, "Conceptual Engineering Design of a One-GJ Fast-Discharging Homopolar Machine for the Reference Theta-Pinch Fusion Reactor," *Electric Power Research Institute Research Project 469*, July 1975.
17. J. L. Smith, Jr., and T. A. Keim, "Applications of Superconductivity to AC Rotating Machines," in S. Foner and B. B. Schwartz (eds.), *Superconducting Machines and Devices*, Plenum Press, New York, 1973, p. 279.
18. J. L. Smith, Jr., "Overview of the Development of Superconducting Synchronous Generators," *IEEE Transactions on Magnetics*, vol. MAG-19, no. 3, May 1983, p. 522.
19. T. E. Laskaris and K. F. Schoch, "Superconducting Rotor Development for a 20-MVA Generator," *IEEE Transactions on Power Apparatus and Systems*, vol. PAS-99, no. 6, November–December 1980, p. 2031.

20. G. Aichholzer, "New Solutions for the Design of Large Turbogenerators up to 2 GVA, 60 kV," *Elektrotechnik und Maschinenbau*, vol. 92, June 1975, p. 249.
21. C. Flick, "New Armature Winding Concepts for EHV and High CFCT Applications of Superconducting Generators," *IEEE Transactions on Power Apparatus and Systems*, vol. PAS-98, November–December 1979, p. 2190.
22. M. Ashkin et al., "Superconducting Generator Field Winding Design for High Fault Tolerance," *IEEE Transactions on Magnetics*, vol. MAG-19, no. 3, May 1983, p. 1035.
23. D. C. Litz et al., "High Tip Speed Test Rig to Study Natural Convection in Liquid Helium," in *Advances in Cryogenic Engineering*, vol. 27, Plenum Press, New York, 1982, p. 799.
24. P. W. Eckels et al., "Heat Transfer Correlations for a Cryostable Alternator Field Winding," in *Advances in Cryogenic Engineering*, vol. 27, Plenum Press, New York, 1982, p. 357.
25. J. L. Kirtley, Jr., "Basic Formulas for Air-Core Synchronous Machines," *IEEE Paper 71CP155-PWR*, presented at IEEE Winter Power Meeting, New York, Jan. 31–Feb. 5, 1971.
26. M. Furuyama, "A Design Concept of a Damper Shield of a Superconducting Alternator," Master's thesis, MIT, 1974.
27. H. S. Kirschbaum, "Transient Electrical Torques of Turbine Generators During Short Circuits and Synchronizing," *AIEE Transactions*, vol. 64, February 1945, p. 65.
28. M. V. K. Chari and T. R. Haller, "Steady-State and Short-Circuit Force Analysis on the Stator Windings of a Superconducting Generator," *IEEE Transactions on Power Apparatus and Systems*, vol. PAS-99, no. 3, May–June 1980, p. 928.
29. J C. White and W. R. McCown, "Progress on the Development of a Prototype 300-MVA Superconducting Generator," *AIM Conference*, Liège, Belgium, October 1981.
30. C. Flick, "Multiparameter Study for Optimization of Superconducting Generator Design under Consideration of Rotor Design Selections," *International Conference on Electric Machines*, Budapest, Hungary, September 1982.
31. M. Ashkin et al., "Stability Criteria for Superconducting Generators—Electrical System Modeling and Cryostability Considerations," *IEEE Transactions on Power Apparatus and Systems*, vol. PAS-101, no. 12, December 1982, p. 4578.
32. A. G. Koronides, "Transient Simulation of Superconducting Synchronous Machines," Ph.D. thesis, University of Pittsburgh, 1980.
33. A. G. Koronides et al., "Superconducting Generator Transient Interaction Analysis Using Three-Dimensional Models," *IEEE Transactions on Power Apparatus and Systems*, vol. PAS-100, no. 6, June 1981, p. 2880.
34. M. Ashkin et al., "Projecting AC Losses in a Superconducting Generator," *Proceedings of the Ninth International Cryogenic Engineering Conference*, Kobe, Japan, May 1982.
35. M. Ashkin, "A Theoretical Analysis of Finite-Length Electromagnetic Shields of Superconducting Turbine Generators," *IEEE Transactions on Power Apparatus and Systems*, vol. PAS-100, no. 3, March 1981, p. 1049.
36. S. D. Umans et al., "Three-Dimensional Transient Analysis of Superconducting Generators," *IEEE Transactions on Power Apparatus and Systems*, vol. PAS-98, no. 6, November–December 1979, p. 2055.
37. W. G. Moore et al., "Development of a 300-MVA Electric Utility Superconducting Generator," presented at South Eastern Electric Exchange Conference, New Orleans, 1983.
38. M. Watanabe et al., "Experimental Study of a Practical Airgap Winding Stator Arrangement for Large Turbine Generators," *IEEE Transactions on Power Apparatus and Systems*, vol. PAS-100, no. 4, April 1981, p. 1901.
39. R. D. Nathenson and M. R. Patel, "Designing an Airgap Armature of a Large Superconducting Generator for Electromagnetic and Thermal Loads," *IEEE Transactions on Power Apparatus and Systems*, vol. PAS-102, August 1983, p. 2629.

40. "High-Voltage Stator-Winding Development," *EPRI Report EL-3391*, Research Project 1716-1 D. R. Albright, program manager, April 1984.

## BIBLIOGRAPHY

The literature on superconductivity and its applications to electric machines is voluminous. The following is a small selection, which is largely based on the author's personal experience and on availability in English. It is hoped that the items included in this bibliography are representative of the state of the art, and omissions should not be construed as judgmental. Current work is regularly published, notably in *Cryogenics*, *Advances in Cryogenic Engineering*, and *IEEE Transactions on Magnetics*.

Items listed above under "References" have not been repeated.

### 1. Superconductivity, Superconductors, and Losses

Bardeen, J., et al.: "Theory of Superconductivity," *Phys. Rev.*, vol. 108, 1957, p. 1175.

Carr, W. J., Jr.: "Conductivity, Permeability, and Dielectric Constant in a Multifilamentary Superconductor," *Journal of Applied Physics*, vol. 46, 1975, p. 4043.

Ekin, J. W.: "Superconductors," in R. P. Reed and A. F. Clark (eds.), *Materials at Low Temperatures*, American Society for Metals, Metals Park, Ohio, 1983.

Gregory, E.: "Multifilamentary Superconducting Materials for Large-Scale Applications," *Cryogenics*, May 1982, p. 203.

Gregory, W. D., et al. (eds.): *The Science and Technology of Superconductivity*, Plenum Press, New York, 1973.

Hulm, J. K., and D. W. Deis: "Applications of Superconductivity," *Electro-Technology*, July 1969, p. 57.

Hulm, J. K., and B. T. Matthias: "High-Field, High-Current Superconductors," *Science*, vol. 208, no. 4446, May 23, 1980, p. 881.

Kunzler, J. E., et al.: "Superconductivity in $Nb_3Sn$ at High Current Density in a Magnetic Field of 88 k Gauss," *Phys. Rev. Letters*, vol. 6, 1961, p. 89.

Minervini, J. V.: "Analysis of Loss Mechanisms in Superconducting Windings for Rotating Electric Generators," Ph.D. thesis, MIT, 1981.

Powell, R. L., et al.: "Definition of Terms for Practical Superconductors," *Cryogenics*, pt. 1, December 1977, p. 697; pt. 2, March 1978, p. 137; pt. 3, June 1979, p. 327.

### 2. General Applications

Cohen, M. H. (ed.): *Superconductivity in Science and Technology*, University of Chicago Press, 1968.

Foner, S., and B. B. Schwartz (eds.): *Superconducting Machines and Devices—Large Systems Applications*, Plenum Press, New York, 1973.

Geballe, T. H., and J. K. Hulm: "Superconductors in Electric Power Technology," *Scientific American*, vol. 243, no. 5, November 1980, p. 138.

Heinz, W.: "Status and Trends of S.C. Magnet Development in Europe," *IEEE Transactions on Magnetics*, vol. MAG-19, no. 3, May 1983, p. 167.

Yasukochi, K.: "Superconducting Magnet Development in Japan," *IEEE Transactions on Magnetics*, vol. MAG-19, no. 3, May 1983, p. 179.

## 3. Superconducting DC Machines

Appleton, A. D.: "Superconducting DC Machines—Concerning Mainly Civil Marine Propulsion but with Mention of Industrial Applications," *IEEE Transactions on Magnetics*, vol. MAG-11, no. 2, 1975, p. 633.

Fox, G. R., and B. D. Hatch: "Superconductive Ship Propulsion Systems," *Proceedings of the 1972 Applied Superconductivity Conference*, IEEE Publication 72CH0682-5-TABSC, 1972, p. 33.

Levedahl, W. J.: "Superconductive Naval Propulsion Systems," *Proceedings of the 1972 Applied Superconductivity Conference*, IEEE Publication 72CH0682-5-TABSC, 1972, p. 26.

## 4. Superconducting AC Generators

### A. Scoping studies

Concordia, C.: "Future Developments of Large Electric Generators," *Philosophical Transactions of the Royal Society of London*, ser. A, vol. 275, 1973, p. 39.

Joyce, J. S., et al.: "Will Large Turbine-Generators of the Future Require Superconducting Field Turbogenerators?" *Proceedings of the American Power Conference*, Chicago, April 1977.

### B. State of the art

Appleton, A. D.: "Status of Superconducting AC Generators in the U.K." *AIM Conference*, Liège, Belgium, October 1981.

Bykov, V. M., et al.: "20-MVA Superconducting Generator and Its Test Results," *AIM Conference*, Liège, Belgium, October 1981.

Flick, C., et al.: "General Design Aspects of a 300-MVA Superconducting Generator for Utility Application," *IEEE Transactions on Magnetics*, vol. MAG-17, no. 1, January 1981, p. 873.

Fujino, H.: "Technical Overview of Japanese Superconducting Generator Development Program," *IEEE Transactions on Magnetics*, vol. MAG-19, no. 3, May 1983, p. 533.

Glebov, I. A., and V. N. Shaktarin: "High Efficiency and Low Consumption Material Electrical Generators," *IEEE Transactions on Magnetics*, vol. MAG-19, no. 3, May 1983, p. 541.

Intichar, L., and D. Lambrecht: "Technical Overview of the German Program to Develop Superconducting AC Generators," *IEEE Transactions on Magnetics*, vol. MAG-19, no. 3, May 1983, p. 536.

Iwamoto, M., et al.: "A 6250-kVA Superconducting Generator," *IEEE Power Engineering Society Paper A 79 013-4*, presented at 1979 Winter Power Meeting, New York, Feb. 5–9, 1979.

Kirtley, J. L., Jr.: "Supercool Generation," *IEEE Spectrum*, vol. 20, no. 4, April 1983, p. 28.

Kösler, H., H. Fillunger, and S. Gründorfer: "SUSI und SMG—ein Entwicklungsprogramm für Synchrongeneratoren mit supraleitender Erregerwicklung" (SUSI and SMG—A Development Program for Synchronous Generators with Superconducting Field Windings), *ELIN* Zeitschrift, vol. 36, no. 1/2, 1985, pp. 34–43.

Kumagai, M., et al.: "Development of Superconducting AC Generator," *IEEE Paper 85 SM 333-0*, presented at 1985 Power Engineering Society Summer Meeting, Vancouver, B.C., Canada, July 14–19, 1985.

Maki, N., et al.: "Design and Component Development of a 50-MVA Superconducting Generator," *IEEE Transactions on Power Apparatus and Systems*, vol. PAS-99, no. 1, January–February 1980, p. 185.

McCown, W. R., and J. S. Edmonds: "300-MVA Superconducting Generator—Plan for Design, Testing, and Long-Term Operation," *CIGRE 1980*, Subject 11-08.

McCown, W. R., et al.: "Superconducting Turbogenerators—Current Situation and Prospects," *CIGRE 1982*, Subject 11-4.

Nakamura, S., et al.: "30-MVA Superconducting Synchronous Condenser: Design and Its Performance Test Results," *IEEE Transactions on Magnetics*, vol. MAG-21, no. 2, March 1985, p. 783.

Parker, J. H., Jr., and R. A. Towne: "Design of Large Superconducting Generators for Electric Utility Applications," *IEEE Transactions on Power Apparatus and Systems*, vol. PAS-98, no. 6, November–December 1979, p. 2241.

Sabrie, J. L., and J. Goyer: "Technical Overview of the French Program," *IEEE Transactions on Magnetics*, vol. MAG-19, no. 3, May 1983, p. 529.

Sato, K., et al.: "An Approach to Optimal Thermal Design of Superconducting Generator Rotor," *IEEE Paper SM 85 334-8*, presented at Power Engineering Society 1985 Summer Meeting, Vancouver, B.C., Canada, July 14–19, 1985.

Smith, J. L., Jr.: "Superconducting Generators: From Conception to Power Station," *Mechanical Engineering*, April 1981, p. 44.

———and A. G. Liepert: "Construction of MIT-DOE 10-MVA Superconducting Generator," *IEEE Transactions on Magnetics*, vol. MAG-21, no. 2, March 1985, p. 791.

Yamada, T., et al.: "$Nb_3Sn$/NbTi Superconducting Windings for 30-MVA Synchronous Rotary Condenser," *IEEE Transactions on Magnetics*, vol. MAG-17, no. 5, September 1981, p. 2194.

Ying, A. S., et al.: "Mechanical and Thermal Design of the EPRI-Westinghouse 300-MVA Superconducting Generator," *IEEE Transactions on Magnetics*, vol. MAG-17, no. 1, January 1981, p. 894.

### C. Design

Bejan, A.: "Improved Thermal Design of the Cryogenic Cooling System for a Superconducting Generator," Ph.D. thesis, MIT, 1974.

"Superconducting Generator Design," *EPRI Report EL-577*, Research Project 429-1, prepared by Westinghouse Electric Corporation (J. H. Parker, Jr., and R. A. Towne, principal investigators), 1977.

"Superconducting Generator Design," *EPRI Report EL-663*, Research Project 429-2, prepared by General Electric Company (M. J. Jefferies and P. A. Rios, principal investigators), 1978.

"Superconductors in Large Synchronous Machines," *Electric Power Research Institute Final Report TD-255*, Project 672-1, prepared by Massachusetts Institute of Technology School of Engineering (J. L. Smith, Jr., principal investigator), August 1976.

Tepper, K. A.: "Mechanical Design of the Rotor of a Fault-Worthy, 10-MVA Superconducting Generator," Ph.D. thesis, MIT, September 1980.

### D. Cryogenic cooling

Bejan, A.: *Entropy Generation Through Heat and Fluid Flow*, Wiley, New York, 1982, chap. 10: "Low-Temperature Applications."

## NOMENCLATURE

### Machine Rating and Densities

$p$ = Number of pole pairs

$P$ = Machine rating (in voltamperes)

$E_f$ = Internal voltage

$V_t$ = Terminal voltage
$I_a$ = Rated terminal current
$J_a$ = Armature-winding current density
$J_f$ = Field-winding current density
$N_{at}$ = Number of turns per phase in the armature
$N_{ft}$ = Number of turns in the field winding (approximately equal to $l + \Delta l$)

## Major Dimensions

$l$ = Length of straight section of the armature
$\Delta l$ = Length of one armature end turn
$l_a$ = Effective length of armature for calculating self-inductance
$l_f$ = Effective length of field winding for calculating self-inductance
$l_m$ = Effective length for calculating field-to-armature mutual inductance
$l_d$ = Effective length of damper for calculating self-inductance
$l_{ad}$ = Effective length for calculating armature-to-damper mutual inductance
$l_{fd}$ = Effective length for calculating field-to-damper mutual inductance
$r$ = Radius
$R_{fi}$ = Inner radius of field winding
$R_{fo}$ = Outer radius of field winding
$R_t$ = Radius of electrothermal shield
$R_{ai}$ = Inner radius of armature winding
$R_{ao}$ = Outer radius of armature winding
$R_s$ = Radius of magnetic shield
$x = R_{ai}/R_{ao}$ = Ratio of inner to outer radii of armature winding
$y = R_{fi}/R_{fo}$ = Ratio of inner to outer radii of field winding
$\theta_{wf}$ = Winding angle of field winding
$\theta_{wfe} = p\theta_{wf}$ = Electrical winding angle of field winding
$\theta_{wa}$ = Winding angle of each phase belt of the armature
$\theta_{wae} = p\theta_{wa}$ = Electrical winding angle of each armature phase

## Fields

$H_{rf}$ = Radial field due to field winding
$H_{\theta f}$ = Azimuthal field due to field winding
$H_{ra}$ = Radial field due to armature winding
$H_{\theta a}$ = Azimuthal field due to armature winding

### Inductances

$L_a$ = Self-inductance of one armature phase winding
$L_f$ = Self-inductance of the field winding
$L_{ab}$ = Mutual inductance between two armature phase windings
$M_a$ = Mutual inductance between phase A of the armature and field windings
$M_b$ = Mutual inductance between phase B of the armature and field windings
$M_c$ = Mutual inductance between phase C of the armature and field windings
$L_d$ = Self-inductance of direct-axis damper
$L_{ad}$ = Mutual inductance between phase A and direct-axis damper
$L_{aq}$ = Mutual inductance between phase A and quadrature-axis damper

### Reactances

$X_a$ = Synchronous reactance (in ohms)
$x_a$ = Synchronous reactance (in per-unit, with $E_f$ as base voltage)
$x_s$ = Synchronous reactance (in per-unit, with terminal voltage as base)
$x'_d$ = Transient direct-axis reactance (in per-unit)
$x'_q$ = Transient quadrature-axis reactance (in per-unit)
$x''_d$ = Subtransient direct-axis reactance
$x''_q$ = Subtransient quadrature-axis reactance

### Miscellaneous

$\theta$ = Angle (a variable)
$\phi$ = Angular displacement of rotor
$\Psi$ = Power-factor angle

CHAPTER 11

# MOTORS FOR CONTROL APPLICATIONS

B. C. Kuo

## 11.1 INTRODUCTION

Motors play an important role in control systems. Depending on the type of control functions, various types of motors can be used for control purposes. However, not all motors used in the industry are suitable for control system applications.

The control systems considered in this chapter are mainly closed-loop systems; that is, feedback is incorporated in one form or another, so that only certain ac and dc motors are appropriate. For ac motors, the two-phase induction motors represent a class of rugged motors especially popular in airborne or aerospace systems. DC motors represent another class of motors that has become popular in recent years as a result of the advancements made in brush technology and magnet material. There is another subclass of motors called step motors, which are typically driven by dc pulses. Because of their incremental motion and direct digital control, step motors have become quite popular for digital computer peripherals, including disk drives, printers, and plotters, among many other applications.

Figure 11.1 illustrates the block diagram of a typical control system with a motor as the prime mover. The motor is powered by the driver, which acts as a power amplifier. The controller usually contains the logic (or compensation) circuit that governs the behavior of the system. The objective of the system is to achieve certain control functions in accordance with the reference input signal $r(t)$.

Figure 11.2 gives a specific example of a printwheel control system of an electronic printer using a dc motor. In this case, the motor is directly coupled to the printwheel. The dc power amplifier receives the signal from the microprocessor through a digital-to-analog (D/A) converter. The direction and distance-to-travel commands are generated by the microprocessor depending on the specific character to be printed. The minor feedback loop with the tachometer is used for the purpose of stabilizing the closed-loop system.

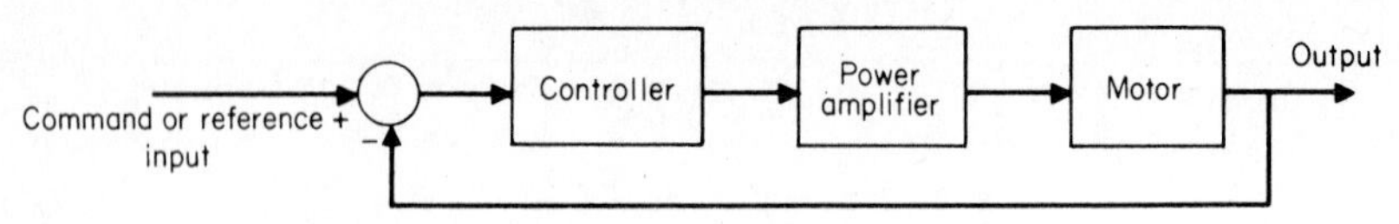


**FIG. 11.1** Block diagram of a typical motor control system.

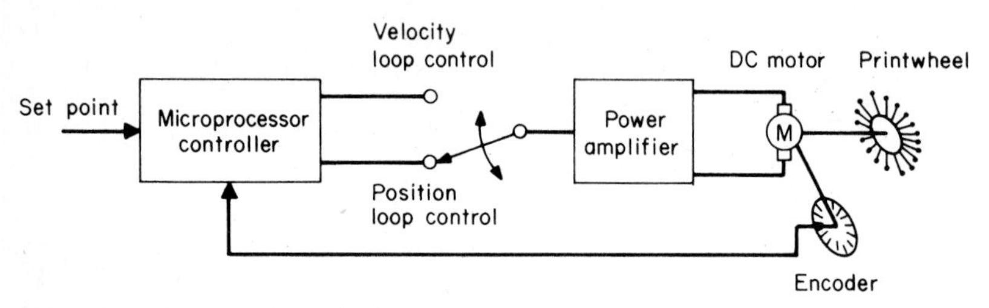


**FIG. 11.2** A dc-motor printwheel control system.

## 11.2 STEP MOTORS

*Step motors* are electromagnetic actuators that convert digital pulse inputs to analog output motion. These motors differ from the conventional motors in that, when energized by a voltage or current pulse train, the step motor moves in an incremental fashion. Each pulse advances the motor shaft one step increment.

Step motors are used in many types of control systems in the industry. By far the highest-volume users of step motors are found in the computer industry. Such peripheral equipment as printers, tape drives, capstan drives, and disk drives for computer memories all have step motors in them. In addition, step motors are used in numerical control systems, machine tool controls, process control systems, robots, clocks and watches, and numerous other applications.

**1.** Advantages of step motors

- **a.** Can be driven open-loop without feedback; no stability problems
- **b.** No accumulative positional error
- **c.** More natural for digital control
- **d.** Mechanically simple; no commutator or brushes to maintain
- **e.** Can be repeatedly stalled without damage
- **f.** For electronic drives, are simpler and more economical than other motors

**2.** Disadvantages of step motors

- **a.** Fixed step angle or increment of motion; lack flexibility in step resolution
- **b.** Low efficiency with ordinary driver
- **c.** Relatively high overshoot and oscillation in step response
- **d.** Limited ability to handle large inertia loads
- **e.** Frictional loads increase position error with open-loop control
- **f.** Limited power output and size available

**3.** Types of step motors

**a.** Solenoid-ratchet type
**b.** Variable-reluctance (VR) type
**c.** Permanent magnet (PM) type
**d.** Hybrid permanent magnet type
**e.** Electromechanical type
**f.** Electrohydraulic type
**g.** Piezoelectric type

## Principles of Operation of Step Motors

The following discusses the principles of operation of the three most popular types of step motors: the VR, PM, and hybrid PM types.

***VR Step Motor.*** The VR-type step motor operates on the principle of variable reluctance. Figure 11.3 illustrates the end view of the stator lamination and the rotor configuration of a single-stack VR step motor. In this case, the stator has 12 poles and the rotor has eight teeth. The stator is wound as a three-phase motor with four poles in each phase, although in Fig. 11.3 only the windings in one phase (phase A) are shown. When a dc current is sent through the windings of phase A, four of the rotor teeth will line up with the four stator poles of the phase. The rotor in Fig. 11.3 is now at the detent position of phase A.

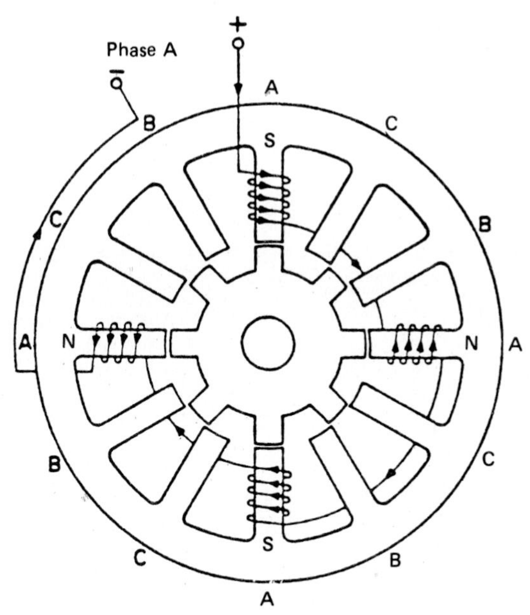


**FIG. 11.3** Schematic diagram of a three-phase single-stack VR step motor.

When phase B is energized next and phase A is de-energized, the rotor will rotate clockwise by 15°, and thus the other four rotor teeth will line up with the four stator poles of phase B. If, from the energized phase A condition, phase C is energized, the rotor will rotate counterclockwise by 15°. This illustrates the basic principle of operation of a VR step motor in that the rotor teeth seek the minimum-reluctance path with the stator teeth of the energized phase.

In general, the step resolution of a single-phase VR step motor with the configuration shown in Fig. 11.3 is given by

$$R = 360\frac{|N_s - N_r|}{N_r N_s} \qquad \text{degrees} \tag{11.1}$$

where $N_s$ = number of stator poles
$N_r$ = number of rotor teeth

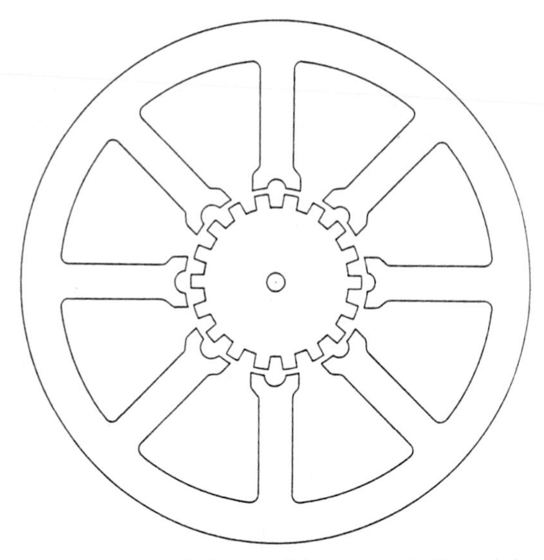

**FIG. 11.4** Schematic diagram of the stator lamination and the rotor of a high-resolution four-phase VR step motor.

For VR step motors with small step angles, such as 1.8° per step, the stator poles have multiple numbers of teeth. In Fig. 11.4, for instance, the motor has four phases with two poles per phase; each stator pole has two teeth, resulting in a total of 16 teeth. In this case, the step resolution of the motor is

$$R = \frac{360}{nN_r} \qquad \text{degrees} \tag{11.2}$$

where $n$ = number of phases

The number of stator teeth is related to the number of rotor teeth by

$$kN_s = N_r \pm q \tag{11.3}$$

where $k$ = positive integer
$q$ = number of poles per phase

Thus, for the configuration shown in Fig. 11.4, $N_s = 16$, $N_r = 18$, and $n = 4$. Thus, the step resolution is

$$R = \frac{360}{4 \times 18}$$
$$= 5^\circ \text{ per step} \tag{11.4}$$

Similarly, for a four-phase VR step motor with a step resolution of 1.8°, the rotor must have

$$N_s = \frac{360}{nR}$$
$$= \frac{360}{4 \times 1.8}$$
$$= 50 \text{ teeth} \tag{11.5}$$

For two poles per phase, $q = 2$; Eq. (11.3) gives

$$N_s = N_r - 2 = 48 \qquad k = 1 \tag{11.6}$$

or 48 teeth on the stator. However, we must leave space between the stator poles for the windings to be slipped in place, so in reality we may find only four or five teeth on each of the eight stator poles.

***PM Step Motors.*** There are many different types of PM step motors, but the most common type has a cylindrical permanent magnet rotor, with poles magnetized alternately in the radial direction. There are no distinctive teeth structures on the rotor. Figure 11.5 illustrates a PM rotor with six alternate poles. The stator assembly usually consists of two or more sections with offsetting teeth, and each section is equipped with a bobbin-wound coil. Figure 11.6 shows the cutaway view of a two-section, or two-phase, PM step motor.

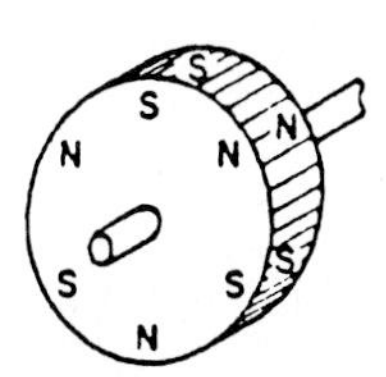


**FIG. 11.5** The rotor of a permanent magnet step motor.

Figure 11.7 shows the schematic diagram of a simple motor with two phases and a two-pole rotor. Each of the stator phases has a double-start wound coil so that the current in the phase can be reversed by simple switchings of the switches $S_1$, $S_2$, $S_3$, and $S_4$ from a single-ended, or unipolar, dc power supply. This type of winding is referred to as a *bifilar-type* winding. As an alternative, each of the motor phases can be equipped with a single coil, a *unifilar-type* winding. Then, to reverse the direction of current flow in the winding, one has to use a double-ended, or bipolar, dc power supply.

As illustrated in Fig. 11.7, when switches $S_1$ and $S_3$ are closed, the stator teeth are magnetized with the polarities as shown, and the PM rotor will be aligned with the pointer at position 1. When the stator windings are energized according to the sequence indicated in Fig. 11.7, the rotor will rotate clockwise or counterclockwise with a step angle of 90°. Another "unwrapped" diagram of a two-phase PM step motor is shown in Fig. 11.8, which can be used to illustrate further the motor's

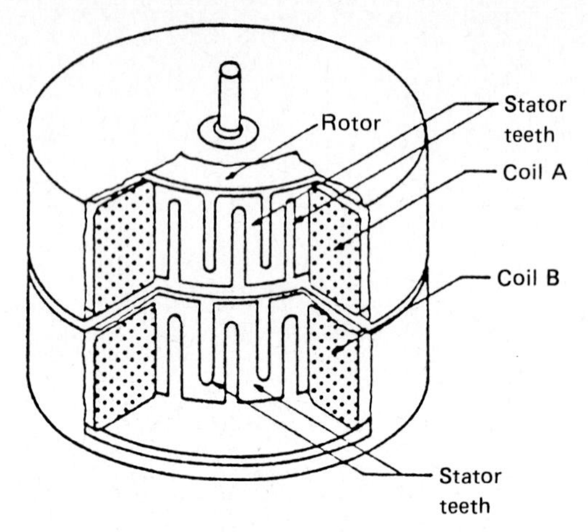


**FIG. 11.6** Cutaway view of a two-stator permanent magnet–rotor motor.

principle of operation. The bifilar windings on each phase are shown as two separate windings, whereas in reality they are wound on one bobbin.

In general, the step angle of the PM step motor discussed above is given by

$$R = \frac{360}{2PN} \tag{11.7}$$

where $P$ = number of pole pairs on PM rotor
$N$ = number of stator sections, or phases

***Hybrid PM Step Motors.*** The name "hybrid" is due to the fact that the principle of operation of the motor is a combination of the VR and PM step motors. The original version of the hybrid PM step motor is a low-speed synchronous motor, or synchronous inductor motor. This type of synchronous motor runs on multiphase ac power, and with many teeth on the permanent magnet rotor, the motor can run at a very low speed.

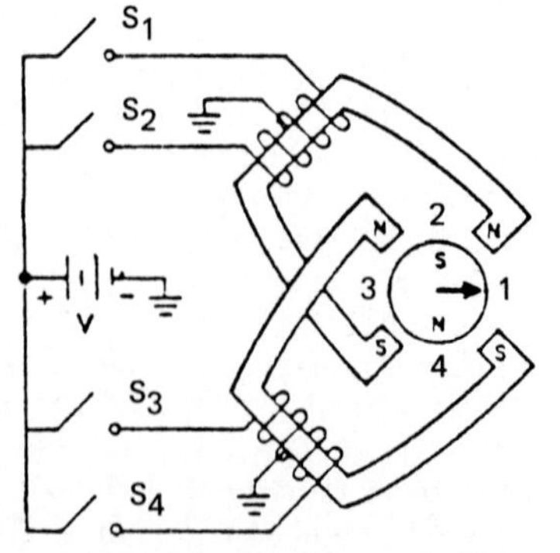


| Switches closed | Step position |
|---|---|
| $S_1, S_3$ | 1 |
| $S_2, S_3$ | 2 |
| $S_2, S_4$ | 3 |
| $S_1, S_4$ | 4 |

**FIG. 11.7** Schematic diagram of a PM-rotor step motor.

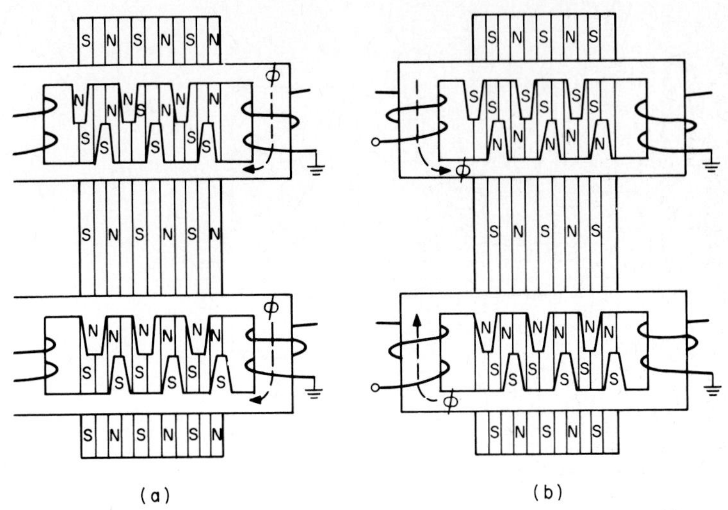


**FIG. 11.8** "Unwrapped" diagram of a PM step motor: (*a*) reference position; (*b*) one-step move to the right.

For step-motor operation, the motor phases are energized sequentially with dc currents. Most hybrid PM motors have two phases, although with the bifilar windings in each phase they are sometimes called four-phase motors. The most popular step angle of a two-phase hybrid PM step motor is 1.8°, or 200 steps per revolution. This is due to some earlier applications of the motor in which it was coupled to a lead screw of 0.2-in pitch, with one motor step corresponding to a 0.001-in linear movement.

Figure 11.9 illustrates the axial view of the hybrid PM step motor. The rotor has a permanent magnet positioned in the axial direction to produce a unidirectional magnetic field. The permanent magnet flux paths are shown by the dotted lines in the diagram. The rotor contains two end sections made of magnetic iron or iron laminations; one section is magnetized as south and the other as north by the permanent magnet. The teeth on each rotor section have a uniform tooth pitch, but the teeth on the two rotor sections are misaligned with respect to each other by one-half of a rotor tooth pitch. Figure 11.10 shows the two end views of the motor. The stator in this case is shown to have eight poles, and the rotor has 10 teeth on each of its two end sections. In this case, four alternate poles of the stator form one phase of the motor.

The stator can be wound with a unifilar two-phase four-pole configuration or with a bifilar "four-phase" four-pole configuration. Figure 11.11 shows the bifilar windings of one phase of a two-phase hybrid PM step motor. Similar windings are found on the remaining four poles.

***The Static-Holding-Torque Curve.*** When one phase of a step motor is energized, then if an external torque is applied to the rotor shaft, causing the rotor teeth and the stator teeth of that phase to be misaligned, a counteracting torque is developed that tends to restore the rotor to the *detent*, or stable equilibrium, position of that phase. The torque developed as described is called the *static holding torque* of the motor, and its value varies with the amount of misalignment between the stator and rotor teeth of the energized phase.

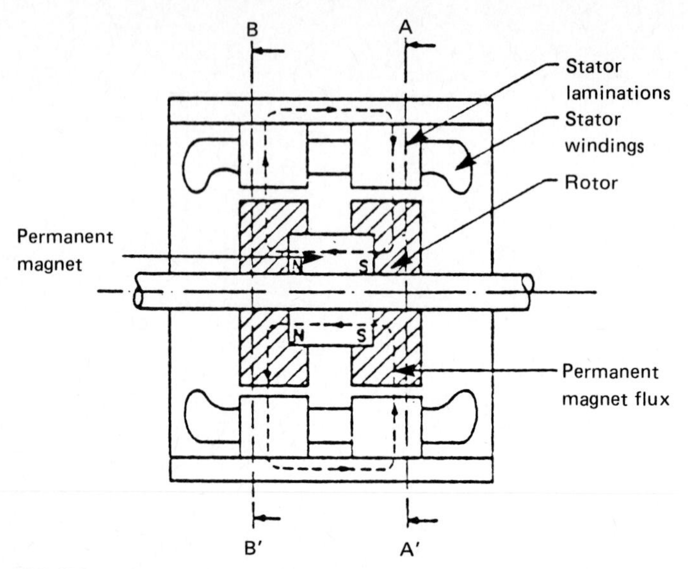


**FIG. 11.9** Axial view of a hybrid PM step motor.

Figure 11.12 shows an ideal static-holding-torque curve of a step motor. It should be pointed out that static-holding-torque curves are generally not sinusoidal (as is the case in Fig. 11.12). The peak value of the static holding torque is normally given as an important motor rating; however, one cannot rely on the peak holding torque as an absolute guide concerning a motor's quality, for the shape of the torque curve is also important in evaluating the motor's overall performance.

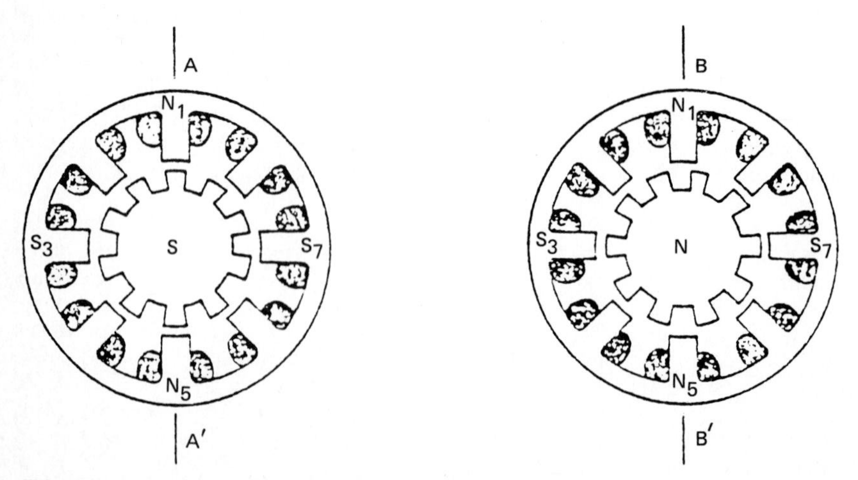


**FIG. 11.10** Two end views of a hybrid PM step motor.

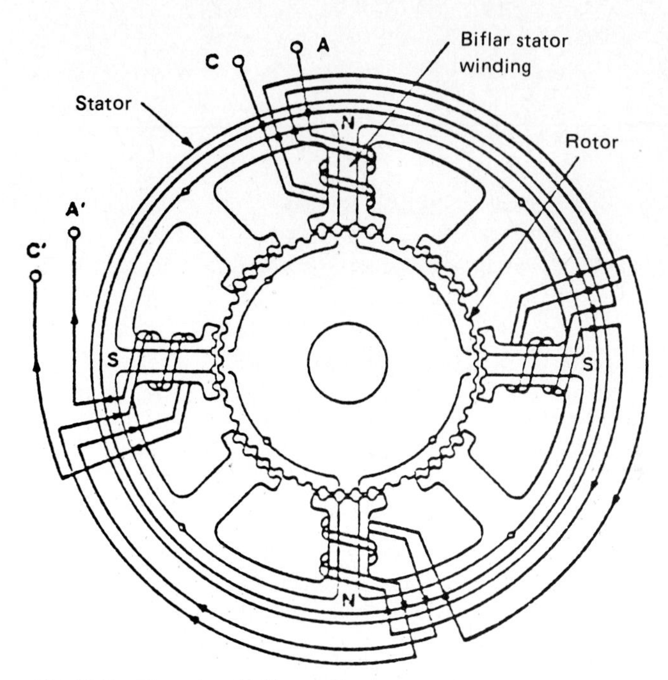


**FIG. 11.11** Illustration of bifilar windings.

When all the phases of the motor are excited separately, the static-holding-torque curves are displayed relative to each other with respect to the rotor position. Figure 11.13 illustrates the individual holding-torque curves developed by each phase of a four-phase step motor. The detent (stable equilibrium) positions of the various phases are displaced by the step angle of the motor.

## Modeling of Step Motors

***VR Step Motors.*** The voltage equation of the $k$th phase of an $N$-phase VR step motor can be written as

$$v_k = R_k i_k + \frac{d\lambda_k}{dt} \tag{11.8}$$

where $v_k$ = voltage applied to phase $k$
$i_k$ = current in phase $k$
$R_k$ = resistance of phase $k$
$\lambda_k$ = flux linkages of phase $k$

The flux linkages can be expressed in terms of the inductances and current as

$$\lambda_k = \sum_{j=a}^{N} L_{kj} i_j \tag{11.9}$$

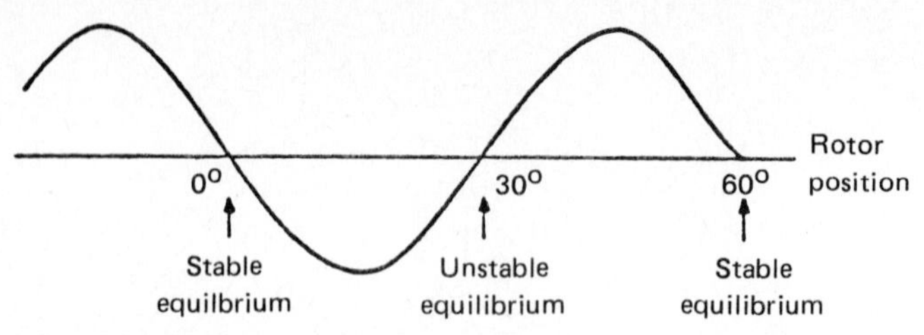


FIG. 11.12 An ideal static-holding-torque curve of one phase of a step motor.

where $L_{kj}$ = average self-inductance of phase $k$, $k = j$
$L_{kj}$ = average mutual inductance between phases $k$ and $j$, $k \neq j$

Then Eq. (11.8) is written as

$$v_k = R_k i_k + \sum_{j=a}^{N} \left( \frac{\partial L_{kj}}{\partial i_j} i_j + L_{kj} \right) \frac{di_j}{dt} + \frac{\partial L_{kj}}{\partial \theta} \frac{d\theta}{dt} ij \qquad (11.10)$$

For a VR step motor, the inductances are functions of the rotor position, and with magnetic saturation they are also functions of the phase currents.

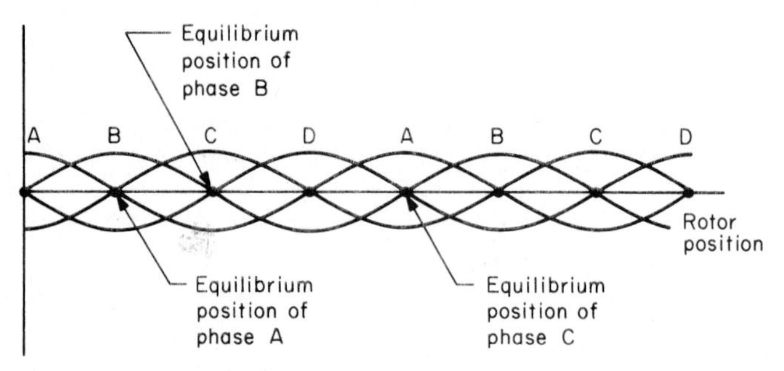


FIG. 11.13 Idealized static-holding-torque curves (four-phase motor).

The torque developed in phase $k$ can be written as

$$T_k = -K_T i_k^2 \sin \left[ N_r \theta + 2(k-1) \frac{\pi}{N} \right] \qquad (11.11)$$

where $K_T$ = torque constant
$N_r$ = number of rotor teeth
$\theta$ = rotor displacement in mechanical radians

The total torque developed by the motor is

$$T = \sum_{j=a}^{N} T_j \qquad (11.12)$$

The output equation is

$$T = J\frac{d^2\theta}{dt^2} + T_L \tag{11.13}$$

where $J$ = total inertia at motor shaft
$T_L$ = load torque

***PM and Hybrid PM Step Motors.*** The voltage equations of a PM or hybrid PM step motor can also be modeled by Eq. (11.8), except that the flux linkage should also include the flux of the permanent magnet. For a four-phase motor that is a two-phase motor with bifilar windings, the flux linkages can be written as

$$\lambda_k = \lambda_{ka} + \lambda_{kb} + \lambda_{kc} + \lambda_{kd} + \lambda_{kf} \tag{11.14}$$

where $\lambda_{kj}$ = flux linking phase $k$ due to current flowing in phase $j$, $j = a,\ b,\ c,\ d$
$\lambda_{kf}$ = flux linking phase $k$ due to the permanent magnet

The final state equations are written as

$$\frac{d\lambda_a}{dt} = \frac{1}{2}(V_a - V_c) - \frac{R}{2L}\lambda_a + \frac{R}{2L}k_b \cos N_r\theta \tag{11.15}$$

$$\frac{d\lambda_b}{dt} = \frac{1}{2}(V_b - V_d) - \frac{R}{2L}\lambda_b + \frac{R}{2L}k_b \cos\left(N_r\theta - \frac{\pi}{2}\right) \tag{11.16}$$

where $L$ = inductance of motor phases
$k_b$ = back-emf constant due to the permanent magnet

The total torque is written as

$$T = -K_a(i_a)\sin N_r\theta - K_b(i_b)\sin\left(N_r\theta - \frac{\pi}{2}\right) - K_c(i_c)\sin(N_r\theta - \pi) - K_d(i_d)\sin\left(N_r\theta - \frac{3\pi}{2}\right) \tag{11.17}$$

The inductances of the motor phases are relatively constant with respect to the rotor position in a PM step motor. The reason for this is that the teeth on the two end stacks of the rotor are offset by one-half of a tooth pitch, so that the variations of the inductances due to each stack are essentially canceled.

From the above discussions we see that the mathematical models of all types of step motors are highly nonlinear. The nonlinear characteristics are due to the variation of the torque as a function of the rotor displacement and to the variation of the inductances as functions of the rotor position and the current. In addition, step motors are usually designed to operate at or near magnetic saturation. The conclusion is that it is impractical to use a transfer function to describe the properties of a step motor.

### Control of Step Motors

Figure 11.14 shows the block diagram of a step-motor control system that contains most of a typical system's vital components. The power driver sequentially energizes the phase windings of the motor. The sequence logic accepts input step and direction commands and converts them into base-driven signals for the power transistors in the power driver; the power transistors amplify the drive signals to energize the

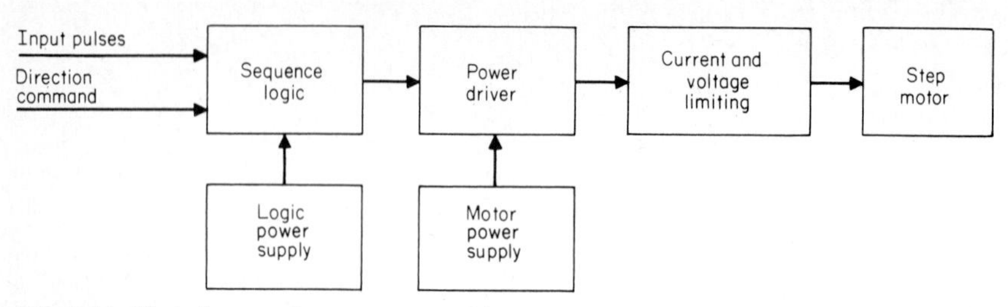


**FIG. 11.14** Block diagram of a step-motor control system.

motor windings. When overexcitation is used to increase the motor performance, the current-limiting circuit is used.

Figure 11.15 illustrates a typical driver stage of a three-phase VR step motor. Figure 11.16 shows the basic components of a bipolar driver, which is used for a unifilar-wound PM step motor.

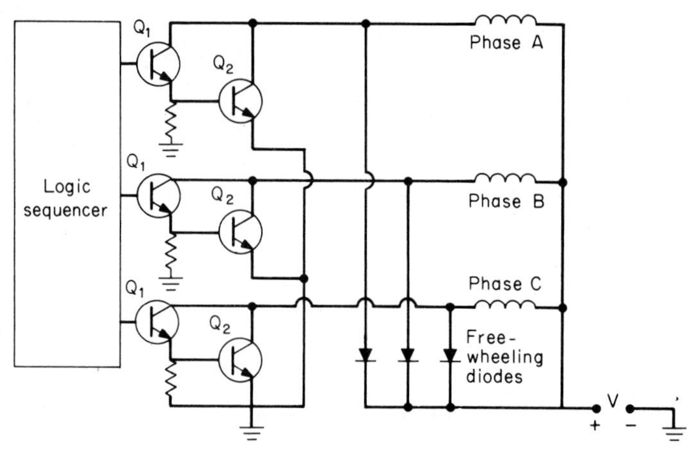


**FIG. 11.15** A typical driver stage of a three-phase VR step motor.

The simplest method of driving a step motor is to energize the phases one at at time—the one-phase-on scheme. For three- and four-phase motors, there may be advantages in driving the motor two phases on at a time. Figure 11.17 illustrates the timing diagram of the one-phase-on drive scheme, and Fig. 11.18 illustrates that of the two-phase-on drive scheme. Figure 11.19 illustrates the timing diagrams of the one-phase-on and two-phase-on drive schemes of a bifilar driver.

## 11.3 TWO-PHASE SERVOMOTORS

Two-phase servomotors are used in low-power control applications. The output power of the two-phase servomotor for control systems usually varies from a fraction of

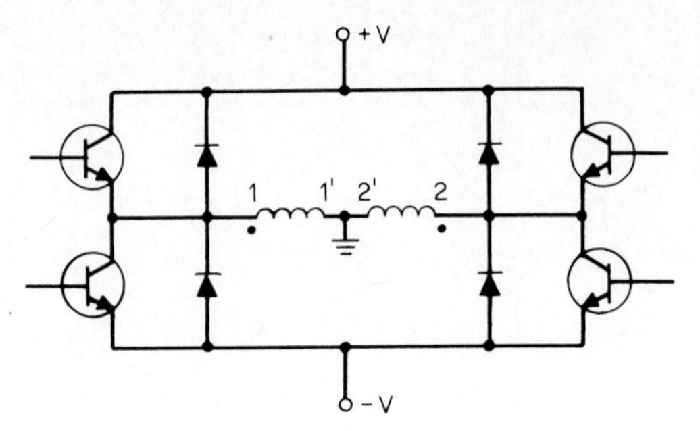


**FIG. 11.16** Basic components of a bipolar driver.

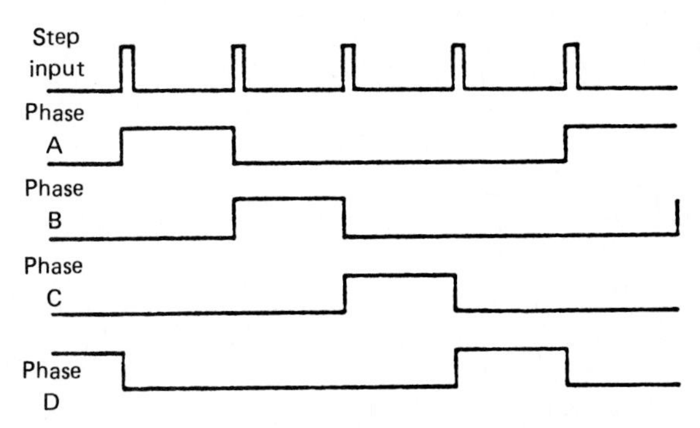


**FIG. 11.17** Timing diagram for the four-phase one-phase-on driver.

a watt up to a few hundred watts. The frequency of operation is normally either 60 Hz or 400 Hz.

A schematic diagram of a two-phase induction motor is shown in Fig. 11.20. The motor consists of a stator with two distributed windings displaced 90 electrical degrees apart. Normally, a fixed voltage from a constant ac source is applied to one

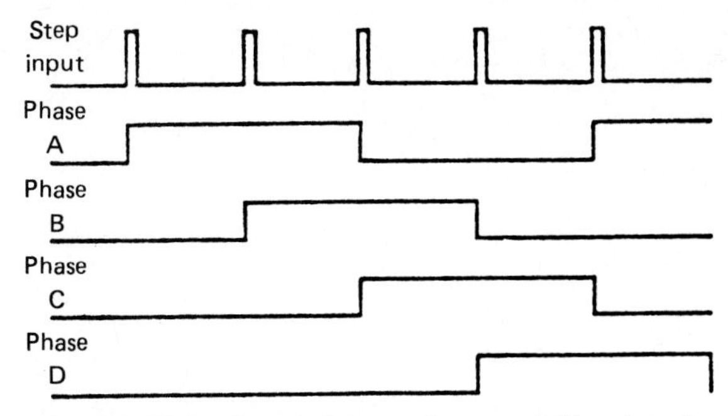


**FIG. 11.18** Timing diagram of the two-phase-on switching scheme for a four-phase step motor.

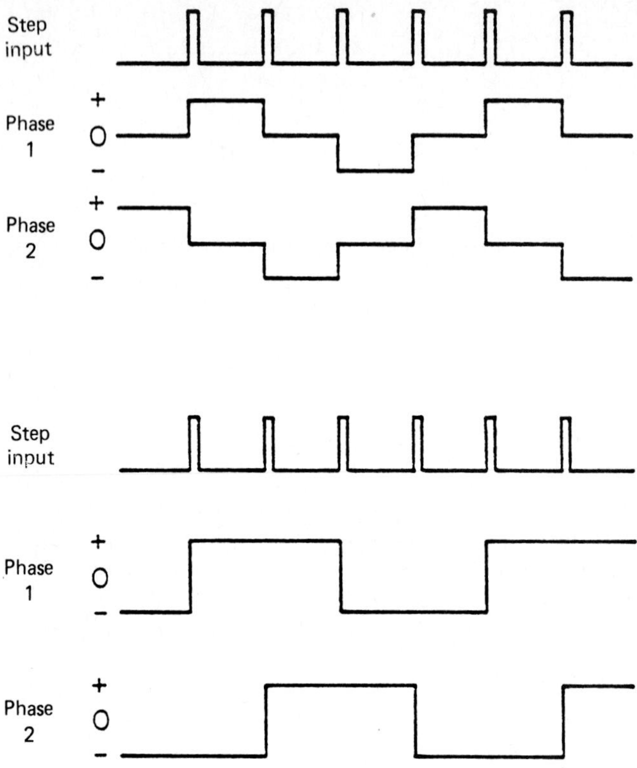


**FIG. 11.19** One-phase-on and two-phase-on bipolar drive schemes.

phase, the reference phase. The other phase, called the control phase, is energized by a variable voltage that is 90 electrical degrees out of phase with respect to the voltage of the fixed phase. The control phase voltage is usually supplied by an ac servo amplifier. The direction of rotation of the motor depends on the phase of the signal of the control phase.

The rotor of a two-phase servomotor is usually of the squirrel-cage or drag-cup type with no windings. The radius of the rotor is usually made small in order to reduce the inertia for fast acceleration control.

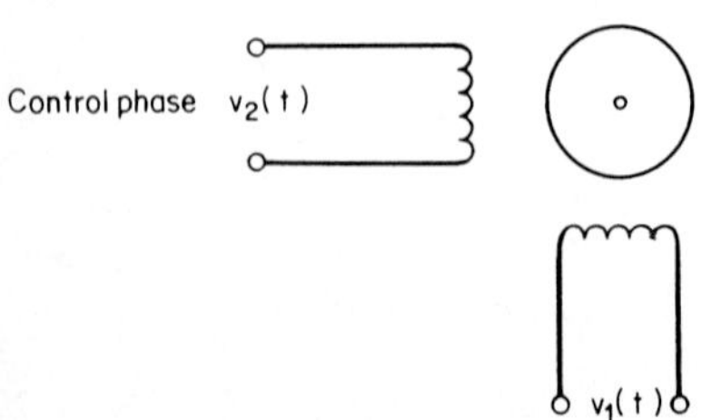


**FIG. 11.20** Schematic diagram of a two-phase induction motor.

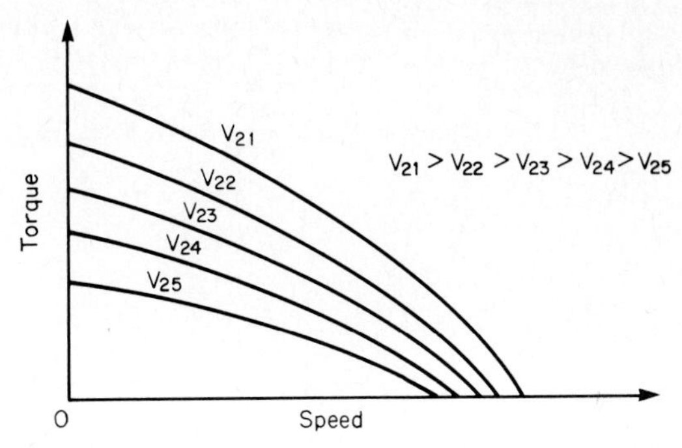


**FIG. 11.21** Typical torque-speed curves of a two-phase induction motor.

Typical torque-speed curves of a two-phase servomotor are shown in Fig. 11.21. These have slopes that monotonically decrease as the speed increases. Such a characteristic is desirable because the slope of the torque-speed curves affects the stability of the closed-loop system, and a positive slope is undesirable. Since the shape of the torque-speed curve depends on the ratio of $X/R$ (reactance/resistance) of the rotor, the rotor of the two-phase servomotor is usually built with high resistance.

The torque-speed curves in Fig. 11.21 show that the two-phase servomotor is a nonlinear device. For linear analysis, we approximate the torque-speed curves by parallel lines, as shown in Fig. 11.22. The torque as represented by the family of straight lines shown in Fig. 11.22 can be expressed as

$$T = kV_2 + m\frac{d\theta}{dt} \tag{11.18}$$

where

$$k = \frac{T_o}{V_1} \tag{11.19}$$

and

$$m = -\frac{T_o}{n_o} \tag{11.20}$$

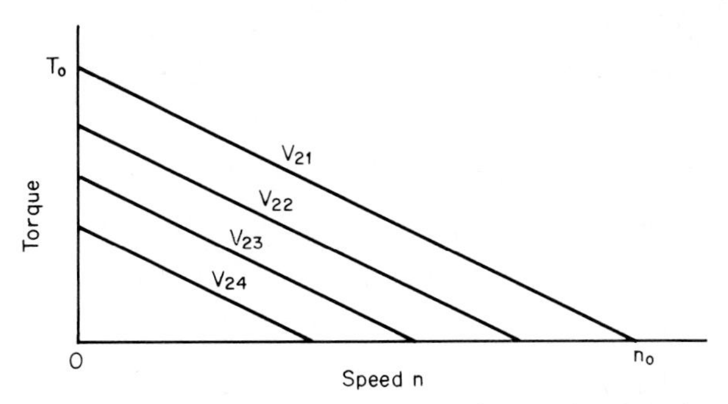


**FIG. 11.22** Linearized torque-speed curves of a two-phase induction motor.

Consider that the total inertia and viscous friction coefficient at the motor shaft are $J$ and $B$, respectively; then

$$T = kV_2 + m\frac{d\theta}{dt} = J\frac{d^2\theta}{dt^2} + B\frac{d\theta}{dt} \tag{11.21}$$

Hence, the motor transfer function is

$$\frac{\theta_m(s)}{V_2(s)} = \frac{K_m}{s(1+\tau_m s)} \tag{11.22}$$

where

$$K_m = \frac{k}{B-m} \tag{11.23}$$

and

$$\tau_m = \frac{J}{B-m} \tag{11.24}$$

Since $m$ is a negative number, these equations show that the effect of the slope of the torque-speed curve is to add more friction to the motor, which improves the damping of the motor. However, if $m$ is positive and $m > B$, negative damping occurs and the motor becomes unstable. Judging from the true torque-speed curves in Fig. 11.21, the value of $m$ is not constant; thus, $K_m$ and $\tau_m$ are also variable with respect to the control voltage. As the value of $V$ increases, the values of $K_m$ and $\tau_m$ both decrease.

## 11.4 TACHOMETERS

When driven mechanically, a tachometer (or tachogenerator) generates an output voltage proportional to the speed of rotation. A tachometer is generally used in control systems to indicate the speed of a rotating shaft, or the output voltage can be used for stability improvements in closed-loop control systems. Figure 11.23 illustrates a typical application of a tachometer in an inner-loop feedback for stabilization.

### AC Tachometer

The ac tachometer is very similar to a two-phase induction motor (Fig. 11.20). A schematic diagram of an ac tachometer is shown in Fig. 11.24. A sinusoidal voltage is applied to the reference winding of the tachometer, setting up an

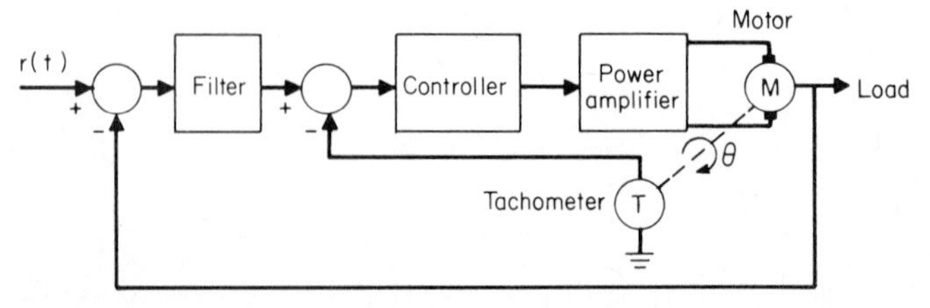


**FIG. 11.23** Application of a tachometer in an inner-loop feedback for control.

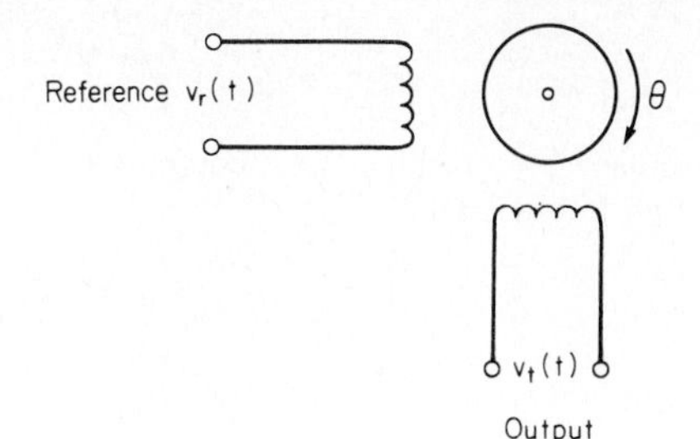


**FIG. 11.24** Schematic diagram of an AC tachometer.

alternating flux in the field. The secondary winding is placed at a 90° angle mechanically with respect to the reference winding; thus, when the rotor shaft is stationary, the output voltage at the secondary winding is zero. When the rotor shaft is driven, the output voltage across the terminals of the secondary winding is closely proportional to the rotor speed. The polarity of the voltage depends on the direction of rotation. Thus, the output voltage of the tachometer can be written as

$$v_t = K_t \frac{d\theta}{dt} \tag{11.25}$$

where $v_t$ = output voltage
$\theta$ = rotor position of the tachometer
$K_t$ = sensitivity or gain of the tachometer, V/(rad·s) or V/r/min

When the tachometer is operated in the linear region, $K_t$ can be regarded as a constant; then the transfer function of the device is

$$\frac{V_t(s)}{\theta(s)} = K_t s \tag{11.26}$$

## DC Tachometer

The basic difference between an ac tachometer and a dc tachometer is that the output voltage of the former is an ac-modulated signal, whereas that of the latter is dc-modulated. Practically all dc tachometers available commercially are of the permanent magnet type. Because of the commutator, there are ripples in the output signal of a dc tachometer.

The output signal of a dc tachometer consists of two components: the desired signal $V_g$ and the ripple $n(t)$. The output signal can be expressed as

$$V = V_g + n(t) \tag{11.27}$$

The voltage $V_g$ is directly proportional to the angular velocity of the shaft; thus

$$V_g = K_t \frac{d\theta}{dt} \tag{11.28}$$

where $K_t$ = tachometer gain constant, V/(rad·s) or V/r/min

Just as in the case of the ac tachometer, for linear operations the dc tachometer can be described by the transfer function

$$\frac{V_g(s)}{\theta(s)} = K_t s \tag{11.29}$$

The ripple, or tachometer noise, is normally caused by several factors: commutator and brush noise, armature eccentricity, and high-frequency noise.

## 11.5 DC MOTORS FOR CONTROL APPLICATIONS *

A dc control motor is basically a torque transducer with its output torque directly controlled by the armature current. The torque produced by the motor is given by

$$T = K\phi i \tag{11.30}$$

where $T$ = torque, N·m
$\phi$ = magnetic flux, Wb
$i$ = current, A
$K$ = a proportionality constant

Thus, if the flux is produced by a permanent magnet, Eq. (11.30) is simplified to

$$T = K_i i \tag{11.31}$$

The simple mathematical relationship that exists between the electrical and mechanical variables in the permanent magnet (PM) dc motor makes it very popular for use in control system applications.

Several basic configurations of PM dc motors exist. The various configurations are aimed at satisfying a wide range of performance requirements, most of which deal with the torque/inertia ratio. This ratio is especially important when motor inertia is a critical system design parameter.

PM dc motors can be classified according to their armature design and commutation structure:

1. According to armature design
   a. Iron-core motors
   b. Surface-wound motors
   c. Moving-coil motors
2. According to commutation structure
   a. Mechanical commutation with brushes
   b. Brushless commutation

### Iron-Core DC Motors

The rotor and stator construction of a typical iron-core dc motor is shown in Fig. 11.25. The magnet (stator) structure could be made of ferrite, alnico, or a rare earth compound. The characteristics of this type of motor are that the rotor inertia and inductance are relatively high; thus, the performance of the motor is moderate.

*See also Chap. 5, "Direct-Current Machines."

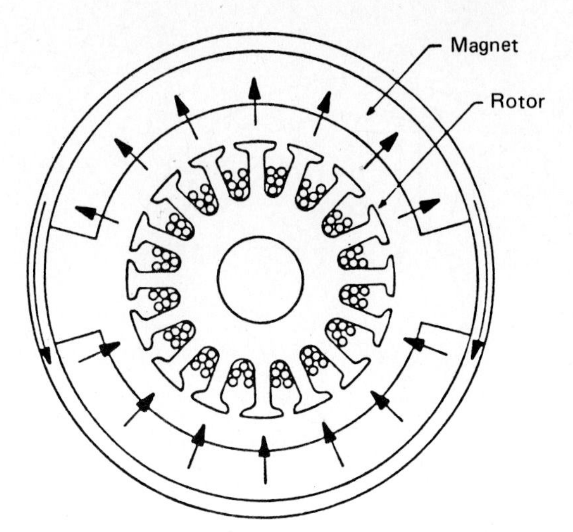


**FIG. 11.25** Cross-sectional view of a PM iron-core dc motor.

## Surface-Wound DC Motors

Figure 11.26 shows the rotor and stator construction of a surface-wound dc motor. In this case, the armature conductors are bonded to the surface of a cylindrical rotor structure, which may be made of laminated disks fastened to the motor shaft. Since no slots are used on the rotor, the rotor does not have any preferential detent positions. The rotor inertia and inductance are also lower for this configuration.

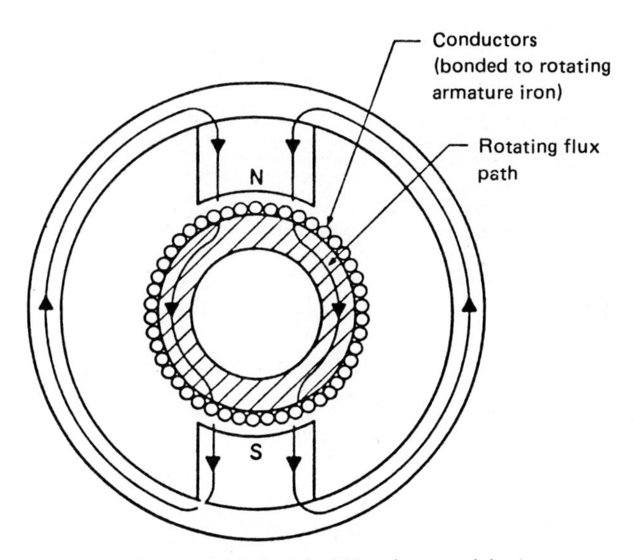


**FIG. 11.26** Cross-sectional view of a PM surface-wound dc motor.

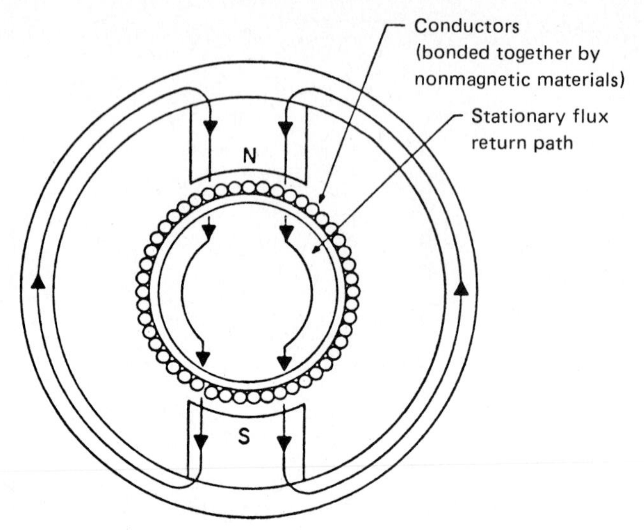


**FIG. 11.27** Cross-sectional view of a moving-coil dc motor.

## Moving-Coil DC Motors

For very low rotor inertia and inductance for high performance, the conductors in the armature are placed in the airgap and supported by a cylindrical structure that is made with nonmagnetic material. The cross-sectional view of the motor is shown in Fig. 11.27. A stationary cylindrical iron piece is placed inside the rotor to carry the flux from the airgap. Since the rotating armature is a thin cylinder without iron, the inertia and inductance of the motor are very low. However, a stronger magnet is needed to push the magnet flux through the airgap.

Another version of the moving-coil motor is shown in Fig. 11.28. This type of motor has a disk-shaped armature, with conductors laid out on the surface. Since

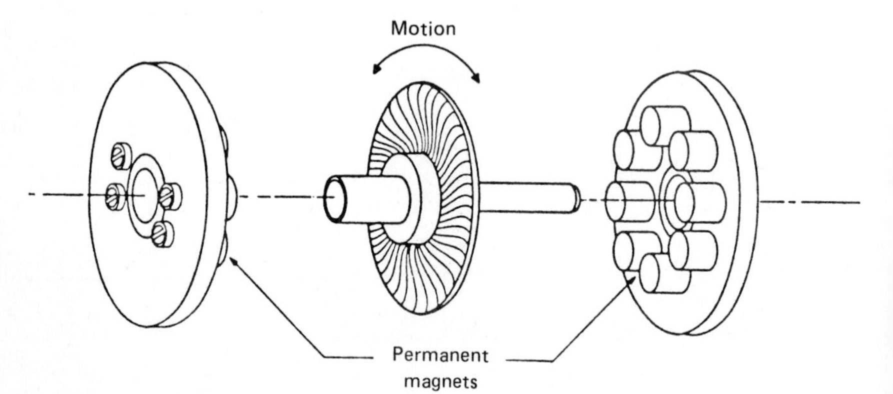


**FIG. 11.28** Cross-sectional view of a disk-rotor moving-coil dc motor.

the conductors are usually placed on the surface of the rotor disk by printed-circuit technology, this motor is also known as the "printed-circuit" motor.

## Brushless DC Motors

Brushless dc motors are constructed without a mechanical commutator; commutation of the armature current is done electronically. Figure 11.29 illustrates the cross-sectional view of a PM brushless dc motor. The rotor contains the permanent magnet and the back-iron support, and the windings are located external to the rotating part.

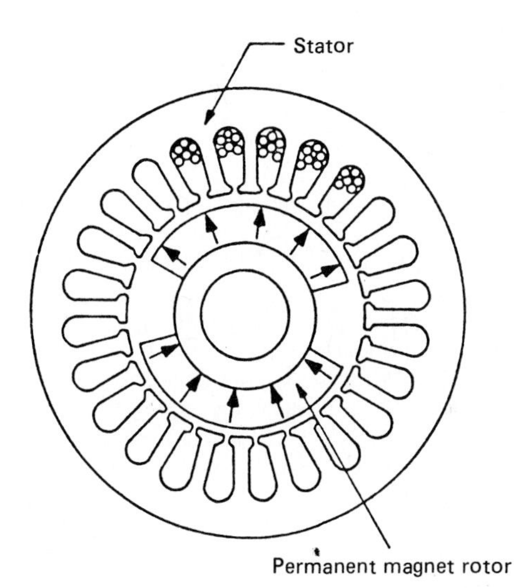


**FIG. 11.29** Cross-sectional view of a brushless dc motor.

Since a mechanical commutator has many sections, it would be impractical to construct an equivalent electronic counterpart because this would require many transistors, each used only for a small time duration for commutation. Figure 11.30 shows a three-phase half-wave commutation controller; the conduction angle is 120 electrical degrees, and only unidirectional currents are utilized. This type of circuit is simple but does not have high winding efficiency, since only one-third of the available copper in the motor is utilized at any time. Figure 11.31 shows a two-phase full-wave circuit, which has a 90° conduction angle and a 50 percent utilization of copper; the circuit requires a double-ended power supply. For

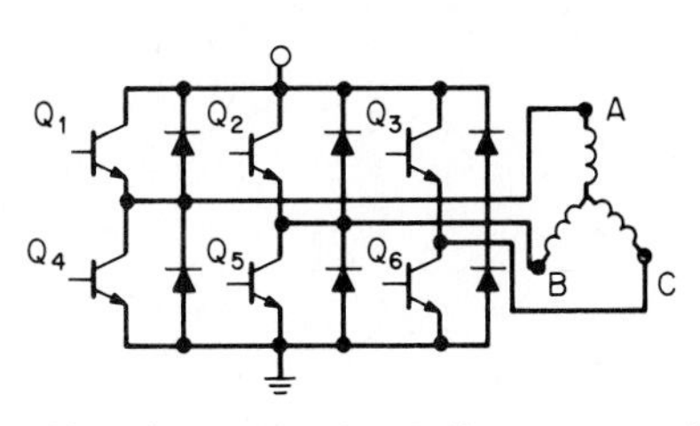


**FIG. 11.30** A three-phase half-wave commutation controller.

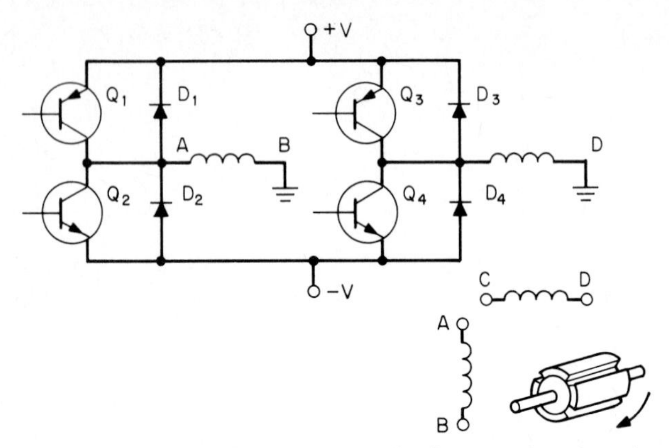


**FIG. 11.31** A two-phase full-wave controller for a two-phase four-pole brushless dc motor.

still higher efficiency, we can use a three-phase full-wave controller, as shown in Fig. 11.32. In this case, six power transistors are connected in a bridge configuration, and only a single-ended power supply is needed. Each transistor conducts for 120 electrical degrees, and the system is switched in intervals of 60°. This system uses two-thirds of the available copper in the motor.

In addition to the power-transistor circuit for switching, we also need to synchronize the commutation switchings with the angular position of the rotor. Several methods are available for the angular-position-sensing system. The most commonly used devices are the Hall-effect sensors; these are usually mounted on the stator structure so as to sense the polarity and magnitude of the permanent magnet field in the airgap. Hall-effect sensors can control the logic functions of the controller to provide current to the proper control member in the stator, and the control can provide some compensation for armature reaction effects, which are prominent in some motors.

## Mathematical Modeling of DC Motors

It is desirable to establish mathematical models for dc motors for control system applications.

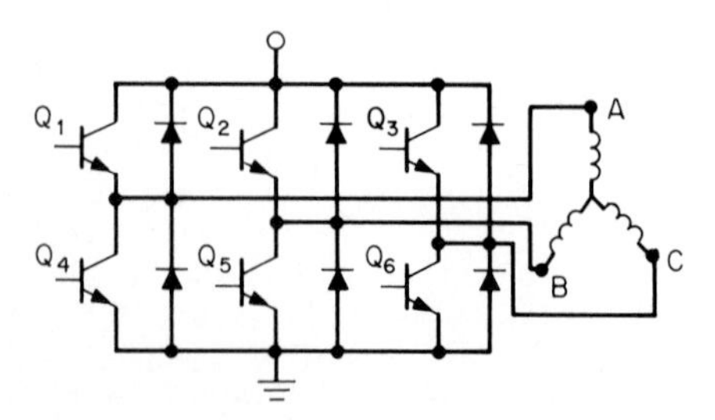


**FIG. 11.32** A three-phase full-wave controller.

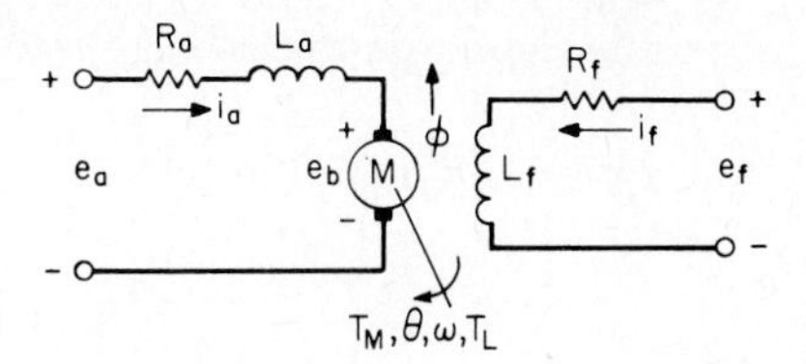


**FIG. 11.33** Equivalent circuit of a separately excited dc motor.

***Separately Excited DC Motor.*** The circuit diagram of such a motor is shown in Fig. 11.33. The armature is modeled as a circuit with a resistance $R_a$ in series with an inductance $L_a$. The voltage source $e_a$ represents the back emf generated by the armature when the rotor rotates. The separate field winding is represented by the resistance $R_f$ connected in series with the inductance $L_f$. The airgap flux is designated by $\phi$. In addition, the following variables and parameters are defined:

$e_a$ = applied voltage
$e_f$ = field voltage
$i_a$ = armature current
$i_f$ = field current
$K_i$ = torque constant
$K_b$ = back-emf constant
$T_m$ = torque developed by the motor
$J_m$ = rotor inertia
$B_m$ = viscous friction coefficient
$\theta$ = rotor displacement
$\omega$ = rotor velocity
$T_L$ = load torque

Assuming that the airgap flux is directly proportional to the field current, then

$$\phi(t) = K_f i_f(t) \tag{11.32}$$

The torque developed by the motor is proportional to the airgap flux and the armature current, and with $\phi$ being constant, we have

$$T_m(t) = K_i i_a(t) \tag{11.33}$$

The voltage equation of the armature is

$$e_a(t) = R_a i_a(t) + L_a \frac{di_a(t)}{dt} + e_b(t) \tag{11.34}$$

The back emf is related to the motor speed as

$$e_b(t) = K_b \omega(t) \tag{11.35}$$

Finally, the torque-output relationship is

$$T_m(t) = J_m \frac{d^2\theta(t)}{dt^2} + B_m \frac{d\theta(t)}{dt} + T_L \tag{11.36}$$

From Eqs. (11.32) to (11.36) we can derive the transfer function of the dc motor as

$$\frac{\theta(s)}{E_a(s)} = \frac{K_i}{s[L_a J_m s^2 + (R_a J_m + B_m L_a)s + (K_b K_i + R_a B_m)]} \tag{11.37}$$

where $T_L$ has been set to zero.

***PM DC Motor.*** For a PM dc motor, the field is set up by the magnet; thus, the airgap flux is constant and we can start directly with Eq. (11.33). The rest of the equations are identical to those of the separately excited motor, so the transfer function of a PM dc motor is still given by Eq. (11.37).

### Power Dissipation of DC Motors

A dc motor can be treated as an energy conversion device in which electric energy is converted to mechanical output energy. A portion of the input electric energy is converted to heat in the windings of the armature. The rest of the electric power is converted to mechanical power. Some of the mechanical power is lost to overcome friction and windage losses, and what is left is the output to drive the mechanical load.

The total electric input power is the product of the instantaneous applied voltage and current in the armature; i.e.,

$$p_i(t) = e_a(t) i_a(t) \tag{11.38}$$

Combining Eqs. (11.33) to (11.36) and (11.38), we have the following power expression:

$$P_i(t) = R_a i_a^2(t) + \frac{K_b B_m}{K_i}\omega^2(t) + \frac{K_e}{K_i}\omega(t) T_L(t) + \frac{K_b}{K_i} J_m \omega(t)\frac{d\omega(t)}{dt} + L_a i_a(t)\frac{di_a(t)}{dt} \tag{11.39}$$

The first term on the right-hand side of Eq. (11.39) represents the $I^2R$ lost in the armature winding as a result of the current flow. The second term represents the power loss due to the viscous friction, and the third term represents the power loss due to the load torque. The fourth term represents the rate of change of the kinetic energy stored in the magnetic field, and the last term represents the actual mechanical power output of the motor.

## 11.6 CRITERIA FOR MOTOR SELECTION

This section discusses some of the important criteria for selecting step motors and dc motors.

### Selection Criteria for Step Motors

***Step-Angle Resolution.*** Step motors are designed to have fixed step positions as the rotor rotates. The number of degrees between adjacent step positions is called the *step angle*. Step motors are available commercially in a wide range of step angles.

In general, if a particular application requires the control of motion in a certain fixed increment, then there may be a certain flexibility in selecting the step-angle resolution of the motor. For instance, if a step-to-step motion control of 15° is required, then any motor having a resolution that is a common divisor of 15° could be used (e.g., we can use three steps of a 5° step-angle motor to realize a 15° motion).

However, there are advantages and disadvantages in using several steps of a step motor to meet the incremental-motion requirement. Since the step-position accuracy of a step motor is noncumulative, a 5° motor with a 5 percent step accuracy will have a smaller absolute position error than a 15° motor with the same step-accuracy percentage. On the other hand, for the same speed in revolutions per minute, the 5° motor would have to run 3 times as fast in steps per second as the 15° motor.

***Torque Requirements.*** One of the most important tasks in selecting a step motor is the specification of the torque requirement. As was mentioned in Sec. 11.2, it is not adequate simply to specify the peak value of the static holding torque of the motor, for the performance of the motor depends not only on the peak value of the static holding torque but also on the shape of the torque curve. In addition, the dynamic performance of the motor also depends on the dynamic torque of the motor. For step motors, the *dynamic torque curve* is defined as the pull-out torque characteristics. In other words, the *pull-out torque* of the motor is defined as the maximum frictional torque that the motor can drive without falling out of synchronism at a specific speed. Therefore, it is important to distinguish the dynamic torque curve of a step motor from that of a dc motor. The dynamic torque (pull-out torque) curve of a step motor is illustrated in Fig. 11.34. A given torque curve is significant only for the motor driver with which the curve is determined.

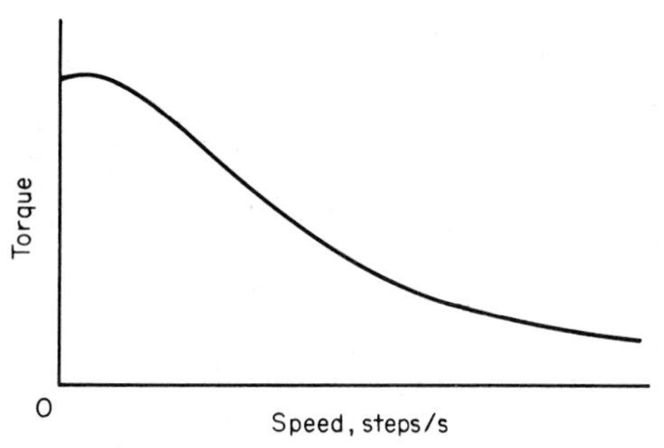


**FIG. 11.34** A typical dynamic torque (pull-out torque) curve of a step motor.

The quickest way to estimate the torque requirement of a step motor is to use the desired acceleration rate and the load parameters. For example, if a 1.8° step motor is to accelerate from standstill to 1000 steps/s in 10 ms, the average acceleration rate necessary to do this is

$$\alpha = \frac{100}{0.01}$$
$$= 100,000 \text{ steps/s}^2$$
$$= \frac{100,000(2\pi)}{200}$$
$$= 3142 \text{ rad/s}^2 \qquad (11.40)$$

For a pure inertia load with inertia $J$ in oz·in·s$^2$, the estimated average torque required to accomplish such an acceleration is

$$T = J\alpha = 3142J \qquad \text{oz} \cdot \text{in} \qquad (11.41)$$

Therefore, if the load inertia is 0.0005 oz·in·s$^2$, the required average torque is

$$T = 1.57 \text{ oz} \cdot \text{in} \qquad (11.42)$$

over the range of speed from 0 to 1000 steps/s. When load friction exists, the frictional load torque must be added to the above torque value since the motor must first overcome the load friction before motion can begin.

***Step-Angle Accuracy.*** Each step motor has a step-position variation due to manufacturing tolerances. This variation is called the *no-load step accuracy*, which is defined as the maximum deviation from the absolute step position under the no-load condition. Since the final position of a step motor determines the positional accuracy of the control system, selecting the accuracy of the step motor is vitally important. Most commercially available step motors have a step accuracy of +5 percent, and some have +3 percent. In actual applications, the frictional load of the motor contributes an additional error to the positional error of the motor system.

## Selection Criteria for DC Motors

Some of the important performance characteristics required of the dc motor in order to meet the demands of control system applications are discussed in the following.

***Armature Inertia.*** For high-performance control systems, fast acceleration and deceleration are important; therefore, the motor's contribution to the system's total inertia should be as close to being optimal as possible. When choosing among the various types of dc motors, note that the moving-coil type has the lowest rotor inertia. However, iron-core motors are lower in cost and can sustain more prolonged overload without damage.

***Armature Inductance.*** A motor's inductance affects its current rise; therefore, in selecting a motor for fast response, it is important to place a maximum limit on the inductance of the armature. Actually, the time constant of the motor's electric circuit controls the current rise time; thus, the $L/R$ ratio is the important parameter to determine. However, when using pulse-width-modulated drivers, low inductance may cause high ripple in the current; thus, a compromise in the inductance value becomes necessary.

***Linear Current-Torque Relationship.*** Ideally, the torque constant of the motor should not be a function of the armature current. The motor should be so selected that it operates in the linear magnetic region.

***Motor Peak-Current Capability.*** Depending on the type of magnet used in the motor, the magnet may be demagnetized if the applied current exceeds a certain level. The fact is that most permanent magnets have a lower resistance to demagnetization at high operating temperatures. Thus, when selecting a motor, it is important to examine the motor specifications in order to determine whether the danger of demagnetization exists.

## 11.7 EXAMPLES OF APPLICATIONS

### Step-Motor Applications

Step motors are generally applied in systems in which the acceleration time and response time are not very demanding. Given the simplicity of their drives and the advantage of direct digital control, step-motor systems are generally more economical than dc motor control systems.

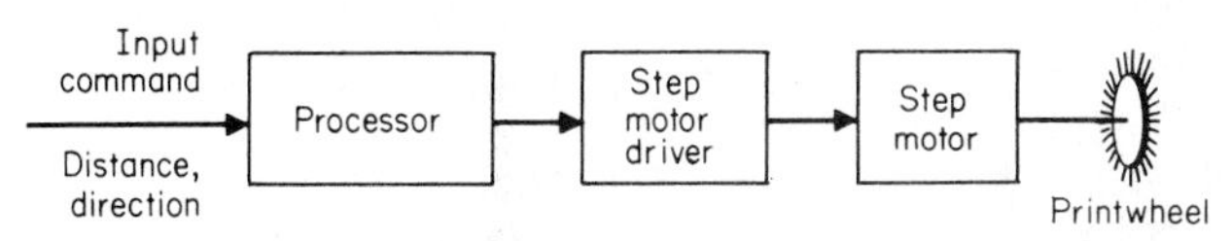


**FIG. 11.35** A printwheel control system using a step motor.

Figure 11.35 shows the basic components of a printwheel control system of an electronic typewriter or printer using a step motor. The system can be driven open-loop as shown, although in practice a feedback encoder is often used for the purpose of step confirmation. Assuming that the printwheel has 96 characters, this means that the step motor may have 96 steps per revolution, or a step resolution of 3.75°. The control objective is to drive the printwheel from standstill to any one of the 48 positions in either direction and stop in a fully damped manner within a certain time limit. If the printer is to have an average access time of 30 characters per second, then the average access time is 33 ms. Assuming that the carriage which moves the printwheel is driven simultaneously with the printwheel drive and that it takes 10 ms for the printing to be done after the printwheel has settled, then the average motion time for the printwheel is only 23 ms.

There is also a step-position accuracy requirement on the motor to ensure that the printing is of high quality. This tolerance may be $\pm 0.1°$. For this application, either a VR or a hybrid PM step motor may be suitable. Some type of electronic damping is necessary at the end of the motion to damp out the natural oscillations of the motor shaft. Figure 11.36 illustrates a typical trace of the output position of the motor of a six-step motion as a function of time.

Another popular application of step motors is in the head drive of computer memory disk packs, both floppy disks and Winchester (hard) disks. Figure 11.37 shows the basic components of a disk drive using a step motor for the read-write head drive. In this case, the spindle drive utilizes a brushless dc motor driven under the constant-speed mode. The step motor is directly coupled to a lead screw. Some drives also use a steel band to couple the motor to the read-write head. As the density of tracks on the memory becomes higher, this poses a severe challenge to the step motor, since it is difficult to get high step resolutions with a small step-motor

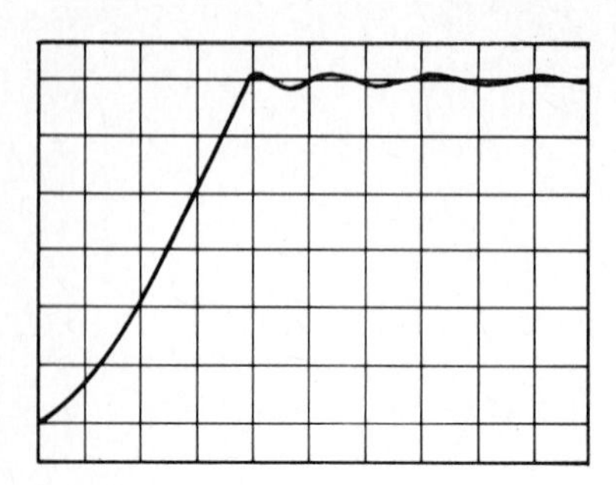

**FIG. 11.36** Six-step motion of the carriage drive system with back-phase damping (10 ms/cm).

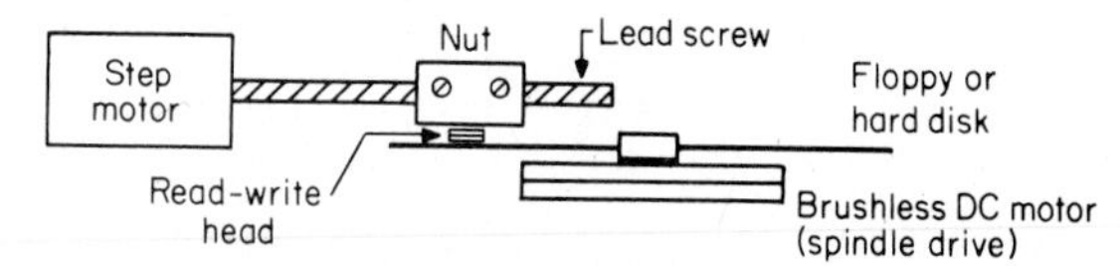


**FIG. 11.37** A disk-head control system using a step motor.

package. Most of the floppy-disk and $5\frac{1}{4}$-in Winchester disk drives now use step motors of size 16 (1.6-in diameter), so it would be difficult to make motors with a step resolution of less than 0.9°.

Figure 11.38 shows a small robot with its axes driven by step motors. Again, since step motors can be driven directly by digital computers, there are definite advantages to using step motors on robots.

## DC Motor Applications

Practically all the step-motor applications described in the previous section can be accomplished with dc motors. The only difference is that for accurate position and speed control with dc motors, feedback must be incorporated. In general, dc motors are preferred when extremely stable and smooth speed control is required.

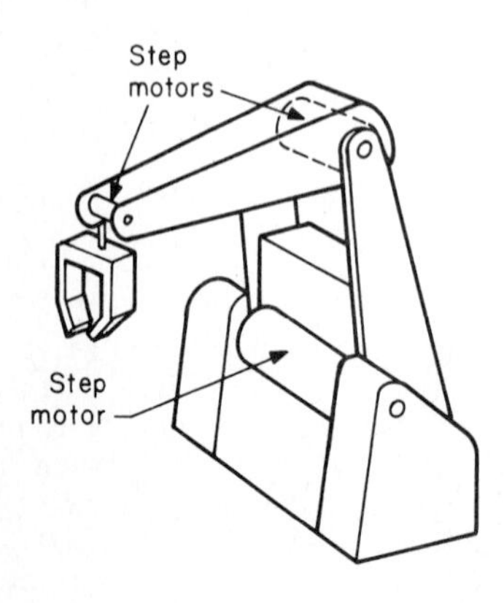


**FIG. 11.38** A robot arm driven by step motors.

For instance, the digital magnetic tape transport has been an established method for bulk data storage for digital computers. The data on magnetic tape are written at densities of 200, 556, 800, 1600, and 6250 bytes/in. These data are formatted into blocks, which theoretically can be as short as 0.01 in and as long as 2400 ft. The data are written or read a block at a time, with the block lengths being of random length. Thus, the capstan system is incremental but must respond in a random fashion, as dictated by the information being written onto or read from the tape, while maintaining certain positional accuracy requirements.

## BIBLIOGRAPHY

Cavanaugh, R. J.: "Timing and Stepper Motors," *Machine Design*, Apr. 9, 1970, pp. 62–67.

Engineering Handbook: *DC Motors Speed Controls, Servo Systems*, Electro-Craft Corporation, Hopkins, Minn, 1985.

Fitzgerald, A. E., C. Kingsley, and A. Kusko: *Electric Machinery*, 3d ed., McGraw-Hill, New York, 1971.

Kordik, K. S.: "The Step Motor—What It Is and Does," in B. C. Kuo (ed.), *Proceedings of the Third Annual Symposium on Incremental Motion Control Systems and Devices*, Champaign, Ill., 1974, pp. A1–A9.

Kuo, B. C. (ed.): *Incremental Motion Control*, vol. 2: *Step Motors and Control Systems*, SRL, Champaign, Ill., 1979.

———and J. Tal (eds.): *Incremental Motion Control*, vol. 1: *DC Motors and Control Systems*, SRL, Champaign, Ill., 1978.

Mea, Anthony N.: "Servomotors," *Machine Design*, Apr. 9, 1970, pp. 55–58.

Persson, E. K.: "Brushless DC Motors in High-Performance Servo Systems," in B. C. Kuo (ed.), *Proceedings of the Fourth Annual Symposium on Incremental Motion Control Systems and Devices*, Champaign, Ill., 1975, pp. T1–T16.

# CHAPTER 12
# HEATING, COOLING, AND VENTILATING

**A. J. Spisak**

## *12.1 INTRODUCTION*

Electric machines are composite structures. Their materials range from conductors to insulators; metal components are selected for their electrical and thermal conductivity, magnetic or physical properties, or some combination of these characteristics. This diverse combination of material requirements gives rise to myriad thermal problems unique to electric machinery. The solutions to these problems have a great impact on the cost and size and to some extent on the feasibility of a particular design.

All electric machines generate losses manifested in the form of heat. Turbine generators used for the production of electric power in blocks of 200 to 1200 MW boast efficiencies of 98.5 percent or better, but even at this high efficiency a 1000-MW generator must dissipate 15 MW of heat. This is equivalent to the capacity of about 500 home heating furnaces. Obviously, this quantity of heat must be effectively removed to prevent damage to the machine. Smaller motors and generators are no less susceptible to thermal problems. Although the losses are lower in absolute quantity, they are generally higher on a per-unit output basis.

In most electric machines the governing thermal limitation is the insulation. Great efforts have been made to qualify electrical insulating materials thermally and to improve their thermal capabilities; modern machines using inorganic materials such as glass and mica in a matrix of polyester, epoxy, or silicone resins exhibit long life at elevated temperatures. The materials used as insulators are classified thermally as A, B, F, and H. The allowable temperatures, which are expressed as a rise above a specified ambient, are defined in NEMA Standards MG 1-12.41, MG 1-12.42, MG 1-20.40, and MG 1-21.40. These standards are specific relative to machine type, size, and voltage rating. The table below provides a guide based on 40°C ambient and observable continuous temperatures, as allowed in the standards.

The method of measuring temperature is also specified in the standards: resistance, embedded detector, or thermometer. In general, embedded detectors are preferred for larger machines; in this case the detector is located to measure the hot-spot temperature and can also serve as a monitor when the machine is in service.

| Insulation class | Typical total winding temperature, °C | |
|---|---|---|
| | 1.15 service factor | 1.0 service factor |
| Class A | 115 | 105 |
| Class B | 140 | 130 |
| Class F | 165 | 155 |
| Class H | — | 180 |

Resistance methods measure the average winding temperature, which is detectable as a change of resistance from that at a known temperature from the formula

$$T_h = \frac{R_h}{R_c}(K + T_c) - K$$

where $T_h$ = hot temperature of winding, °C
$R_h$ = winding resistance at $T_h$
$T_c$ = cold temperature of winding, °C
$R_c$ = winding resistance at $T_c$
$K$ = 234.5 for copper
$K$ = 225 for aluminum

The methods for classifying insulation materials are also defined in standards such as IEEE 275. Insulation life is based on an adaptation of the Arrhenius equation, which states:

$$\log(\text{life}) = \log y = a + b\frac{1}{T}$$

This equation has the algebraic form

$$Y = a + bX$$

where $Y = \log y$
$X = 1/T$
$a, b$ = constants

The constants $a$ and $b$ can be estimated by fitting the experimental data in the form of $\log y$ versus $1/T$ to the above simple linear equation. The straight line thus formed is known as the regression line. A great many electrical insulating systems have been tested in this manner and displayed on curves, such as those in Fig. 12.1. The results shown may be viewed as typical for insulating systems rated per the classifications defined above.

The magnetic iron is also subject to thermal limitations. This component is a stacked structure comprised of laminations that are insulated from each other. The insulating materials vary from organic varnishes to inorganic coatings, all of which have some thermal limit. Furthermore, since the iron is in direct contact with the electrical insulation on the windings, its temperature must be maintained below that to which the insulation is limited; this temperature difference is especially important in machines in which the heat generated in the windings must be transmitted through the magnetic iron.

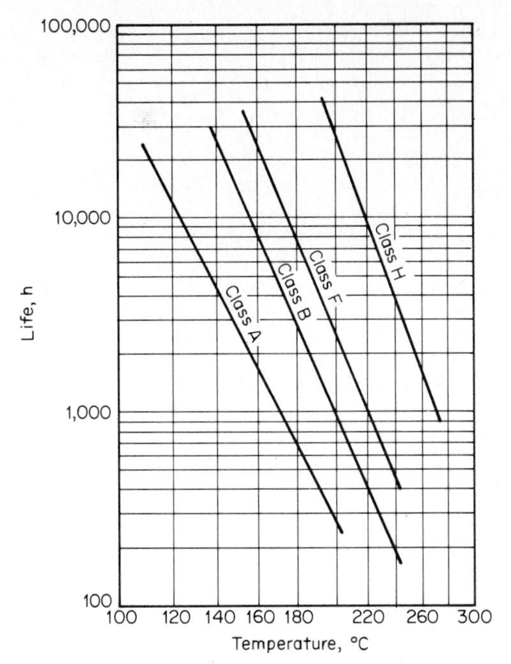


**FIG. 12.1** Insulation life as a function of operating temperature.

Other thermal factors can affect structural components: temperature variations can give rise to high mechanical stresses. For example, the differences in the temperature between the iron and copper, combined with their differences in thermal expansion rates, can cause stresses in the insulation, and these can lead to mechanical failure, such as tape separation. Table 12.1 provides a tabulation of coefficients of linear expansion for various materials commonly found in electric machinery.

The temperature limitation for the conductors, which are normally copper or aluminum, is usually predicated on the thermal capabilities of the adjacent materials. This would normally be the insulation, but consideration must also be given to the connections where solder or brazing alloys are present. Mechanical failures can occur in these joints considerably below the melting point.

## 12.2 HEAT-TRANSFER FUNDAMENTALS

### Heat Generation

In electric machinery there are basically three areas where heat is generated. These are generally referred to as copper loss, core loss, and windage and friction. For the heat-transfer analysis, load losses should be added as appropriate—such that the copper and core losses reflect the total heat generated in those components.

**TABLE 12.1** Coefficients of Linear Expansion, in/(in·°C) × $10^6$

| Material | | Coefficient |
|---|---|---|
| Copper | | 18 |
| Bronze | | 18 |
| Cast iron | | 12.6 |
| Carbon steel | | 11.4 |
| 18-8 stainless steel | | 17 |
| Silicon steel (stator iron) | | 11.7 |
| | Composite | |
| Resin | Laminate | Coefficient |
| Polyester | Unfilled | 100 |
| | Glass roving | 7 |
| | Glass cloth | 10 |
| | Chopped strand | 25 |
| Phenolic | Paper | |
| | Lengthwise | 1.05–2.1 |
| | Crosswise | 1.2–3.3 |
| | Nylon | |
| | Lengthwise | 0.56 |
| | Crosswise | 0.75 |
| | Cotton fabric | |
| | Lengthwise | 1.4–2.4 |
| | Crosswise | 1.8–3.2 |
| | Asbestos | |
| | Lengthwise | 0.9–1.0 |
| | Crosswise | 1.3–1.35 |
| | Glass | |
| | Lengthwise | 1.0–1.5 |
| | Crosswise | 1.5–1.8 |
| Silicone | Glass | |
| | Lengthwise | 1.0 |
| | Crosswise | 1.0 |
| Melamine | Glass | |
| | Lengthwise | 1.0–1.3 |
| | Crosswise | 1.1–2.1 |
| Epoxy | Glass | |
| | Lengthwise | 0.4–0.5 |
| | Crosswise | 0.8–1.0 |
| | Paper | |
| | Lengthwise | 0.40 |
| | Crosswise | 0.80 |

The heat generated in the core and copper is best viewed in terms of watts per unit volume or unit length of conductor. These losses represent the major sources of heating and must be removed effectively to prevent overheating.

The other source of heat generation is windage and friction. Its major impact on ventilating the machine is to raise the temperature of the cooling medium, consequently providing less effective cooling. Bearing losses can usually be isolated from the cooling medium by providing them with their own cooling system (forced lubrication or self-cooling).

The ventilating medium is heated not only by the losses transferred to it but also by the power required to move it through the machine and by the frictional heating due to contact with the rotating parts. As a result, the temperature of the ventilating medium varies throughout the machine. For gross or bulk temperature estimates, average temperatures may be sufficiently accurate; however, to estimate hot-spot temperatures accurately, as required by NEMA Standards, the medium's temperatures throughout the cooling process are required (which will be discussed further in this section under "Heat Transport").

## Heat Conduction

Heat must be moved from the area where generated to the cooling medium. In a typical motor or generator this means conducting the heat through the conductor's ground wall into the iron. There, the core losses are added to the heat flow. The conduction process then transfers the heat to surfaces where the cooling medium can carry them away. The heat transfer through the iron is complicated by the anisotropy resulting from the insulated laminated structure; thus, the heat transfer varies appreciably in the interlaminar and cross-laminar directions.

Table 12.2 lists thermal conductivity heat coefficients for materials most typically found in electric machinery. The units are in W/(in·°C); although these units may seem a bit inelegant, they are quite practical for thermal analysis of electric machines. Table 12.3 provides a conversion table for commonly used thermal conductivity units dealing with electric machinery.

## Convective Heat Transfer

The ultimate heat-removal mechanism is convection from the surfaces to which the heat is conducted. The heat-transfer coefficient $h$ is defined by the expression

$$q = hA(T_s - T_g)$$

where $q$ = heat transferred, W
$A$ = area, in$^2$
$h$ = heat-transfer coefficient, W/(in$^2$·°C)
$T_s$ = temperature of heated surface, °C
$T_g$ = temperature of cooling medium, °C

The cooling medium most commonly used in electric machinery is air. Other important media are hydrogen, helium, water, and oil. The choice of air is obvious since it requires no special containment in most applications. To quantify the merits of these fluids, consider the generalized heat-transfer expression

**TABLE 12.2** Thermal Conductivities

| Material | $k$, W/(in·°C) | Temperature range, °C |
|---|---|---|
| Copper | 9.85–9.32 | 0–315 |
| Aluminum | 5.14–5.23 | 0–315 |
| Brass | 2.46–2.90 | 0–315 |
| Cast iron | 1.21–1.18 | 54–102 |
| Mild steel | 1.1 | |
| Si steel laminations (radial) | 0.5–1.0 | |
| Si steel laminations (axial) | 0.05 | |
| Mica sheet | 0.014 | |
| Micanite | 0.0104 | |
| Asbestos cloth phenolic | 0.0168 | |
| Glass roving polyester | 0.0110 | |
| Glass melamine | 0.0062–0.0088 | |
| Asbestos paper phenolic | 0.0084 | |
| Nylon phenolic | 0.0081 | |
| Cotton fabric phenolic | 0.0081 | |
| Glass cloth polyester | 0.0073 | |
| G-3 glass phenolic | 0.0066 | |
| Glass epoxy | 0.0066 | |
| Shellac | 0.0063 | |
| Paper phenolic | 0.0062 | |
| Chopped strand polyester | 0.0062 | |
| Varnish-impregnated cotton | 0.0061–0.0068 | |
| Paper epoxy | 0.0059 | |
| G-2 glass phenolic | 0.0051 | |
| Unfilled polyester | 0.0044 | |
| Mica folium | 0.0044 | |
| Varnished cambric | 0.004 | |
| Asbestos | 0.0038–0.0049 | 0–100 |
| Glass silicone | 0.0037 | |
| Organic varnishes | 0.0035–0.0045 | |
| Wood | 0.0035–0.0053 | |
| Untreated cotton | 0.0014–0.0017 | |

$$\mathrm{Nu} = C(\mathrm{Re})^m(\mathrm{Pr})^n$$

where Nu = Nusselt number (dimensionless) = $hd/k$
Re = Reynolds number (dimensionless) = $du\rho/v$
Pr = Prandtl number (dimensionless) = $vc/k$
$C, m, n$ = constants determined from test
$d$ = geometric definer (diameter or length)
$k$ = thermal conductivity
$u$ = fluid velocity
$\rho$ = fluid density
$v$ = fluid viscosity
$c$ = thermal capacity of the fluid

**TABLE 12.3** Thermal Conductivity Conversion Factors

| | $\frac{W}{in \cdot °C}$ | $\frac{cal}{s \cdot cm^2 \cdot °C}$ | $\frac{W}{cm \cdot °C}$ | $\frac{W}{m \cdot °C}$ | $\frac{Btu}{h \cdot ft \cdot °F}$ | $\frac{Btu \cdot in}{h \cdot ft^2 \cdot °F}$ |
|---|---|---|---|---|---|---|
| $\frac{W}{in \cdot °C}$ | 1.0 | 0.09406 | 0.3937 | 39.37 | 22.75 | 273 |
| $\frac{cal}{s \cdot cm \cdot °C}$ | 10.63 | 1.0 | 4.186 | 418.6 | 241.9 | 2903 |
| $\frac{W}{cm \cdot °C}$ | 2.54 | 0.2389 | 1.0 | 100 | 57.78 | 693.3 |
| $\frac{W}{m \cdot °C}$ | 0.0254 | $2.389 \times 10^{-3}$ | 0.01 | 1.0 | 0.5778 | 6.933 |
| $\frac{Btu}{h \cdot ft \cdot °F}$ | 0.04396 | $4.134 \times 10^{-3}$ | 0.01731 | 1.731 | 1.0 | 12 |
| $\frac{Btu \cdot in}{h \cdot ft^2 \cdot °F}$ | $3.663 \times 10^{-3}$ | $3.445 \times 10^{-4}$ | $1.4423 \times 10^{-3}$ | 0.1442 | 0.8333 | 1.0 |

***Example:*** $100 \frac{Btu \cdot in}{h \cdot ft^2 \cdot °F} \times \frac{3.663 \times 10^{-3} \ W/(in \cdot °C)}{1.0 \ (Btu \cdot in)/(h \cdot ft^2 \cdot °F)} = 0.366 \ W/(in \cdot °C)$

Thus, the heat-transfer coefficient $h$ depends on the velocity of the fluid, the geometry of the part being heated or cooled, and the fluid properties. For turbulent-flow convection heat transfer, the constants $m$ and $n$ have been determined to be about 0.8 and 0.4, respectively. Thus, for a given geometry and fluid velocity, it can be shown that the heat-transfer coefficient for various fluids 1 and 2 can be compared as follows:

$$\frac{h_1}{h_2} = \left(\frac{\rho_1}{\rho_2}\right)^{0.8} \left(\frac{c_1}{c_2}\right)^{0.4} \left(\frac{k_1}{k_2}\right)^{0.6} \left(\frac{v_2}{v_1}\right)^{0.4}$$

Table 12.4 gives a comparison of these properties for various fluids. It also provides the surface heat-transfer coefficient of the fluid compared to air at atmospheric pressure and room temperature, $h_f/h_a$. For gases, it is obvious that hydrogen is an excellent heat-transfer medium. Even though it is only $^1/_{14}$th as dense as air, it is about 1.5 times as good from a heat-transfer standpoint at standard conditions. At higher pressures this quality is even further enhanced. The combined qualities—convective heat transfer, low density, and high thermal capacity—make hydrogen the gas most generally used in large high-speed generators in which high efficiency and reliability are paramount. Unfortunately, even small amounts of impurities (such as air or carbon dioxide) have significant effects on these characteristics. (Note the properties of a 93 percent mixture of hydrogen in air.)

Helium is also a very good heat-transfer medium compared to air for somewhat the same reasons as hydrogen is. In addition, it is safer to use since it does not form an explosive mixture with air. However, its high cost and lesser properties have limited its use in electric machinery. Liquids are better than gases, as can be seen in Table 12.4, and water ranks among the best. Their high densities, however, are not conducive to using them as the surrounding fluid for the rotating parts since this would give rise to high frictional losses.

Although it is generally recommended that the heat-transfer coefficient be determined on the bases of geometry, fluid velocity, and fluid properties, Table 12.5 gives typical heat-transfer coefficients for air as a cooling medium on various components.

**TABLE 12.4** Properties of Commonly Used Gases and Liquids

| Fluid | Density, $lb/ft^3$ | Specific heat, Btu/(lb·°F) | Conductivity, Btu/(ft·h·°F) | Viscosity, lb/(h·ft) | $h_f/h_a$ |
|---|---|---|---|---|---|
| Air (STP) | 0.075 | 0.240 | 0.0148 | 0.0440 | 1.0 |
| $CO_2$ | 0.1135 | 0.198 | 0.009 | 0.357 | 1.04 |
| He | 0.0103 | 1.26 | 0.0867 | 0.473 | 1.11 |
| $H_2$ (1 atm) | 0.00521 | 3.41 | 0.104 | 0.0214 | 1.47 |
| $H_2$ (2 atm) | 0.0104 | ↓ | ↓ | ↓ | 2.56 |
| $H_2$ (3 atm) | 0.0156 | ↓ | ↓ | ↓ | 3.54 |
| $H_2$ (4 atm) | 0.0208 | ↓ | ↓ | ↓ | 4.45 |
| $H_2$ (5 atm) | 0.0261 | ↓ | ↓ | ↓ | 5.34 |
| 93% $H_2$ | 0.0101 (1 atm) | 1.77 | 0.090 | 0.0268 | 1.61 |
| $H_2O$ | 62.4 | 1 | 0.344 | 2.36 | 515 |
| Light oil | 57.4 | 0.44 | 0.0766 | 140 | 27.5 |

Table 12.6 provides conversion factors for commonly used surface heat-transfer coefficients and the parameters that affect them.

## Heat Transport

In addition to transferring the heat from the surface to be cooled, the cooling medium must also transport the heat away by either exhausting it to the outside or conveying it to a heat exchanger. As the heat is transferred to the medium, its temperature increases per the expression

$$\Delta T_f = \frac{Q}{mc}$$

where $\Delta T_f$ = temperature rise of fluid
$Q$ = heat transferred
$m$ = mass flow rate of fluid
$c$ = thermal capacity of fluid

**TABLE 12.5** Surface Heat-Transfer Coefficients for Air as a Cooling Medium

| Component | $h_a$, W/(in·°C) | Air velocity, ft/min |
|---|---|---|
| End windings and parts of the winding in radial ducts | 0.025–0.10 | 600–6000 |
| Laminations in radial vents | 0.025–0.027 | 800–4000 |
| Laminations in axial channels, 2-in length | 0.03–0.11 | 1200–5000 |
| Laminations in axial channels, 20-in length | 0.02–0.08 | 1200–5000 |
| Field windings | 0.008–0.025 | 800–8000 |

***Source:*** M. Liwschitz-Garik and C. Whipple, *D.C. Machines*, 2d ed., D. Van Nostrand, New York, 1947, p. 275.

**TABLE 12.6** Conversion Factors: Surface Heat-Transfer Coefficients and Their Parameters

| | $\frac{W}{in^2 \cdot °C}$ | $\frac{W}{cm^2 \cdot °C}$ | $\frac{W}{m^2 \cdot °C}$ | $\frac{cal}{s \cdot cm^2 \cdot °C}$ | $\frac{kg\ cal}{h \cdot m^2 \cdot °C}$ | $\frac{Btu}{h \cdot ft^2 \cdot °F}$ |
|---|---|---|---|---|---|---|
| $\frac{W}{in^2 \cdot °C}$ | 1.0 | 0.1550 | $1.55 \times 10^3$ | 0.03703 | $1.333 \times 10^3$ | 273.0 |
| $\frac{W}{cm^2 \cdot °C}$ | 6.452 | 1.0 | 10,000.0 | 0.2389 | $8.60 \times 10^3$ | 1761.0 |
| $\frac{W}{m^2 \cdot °C}$ | $6.452 \times 10^{-4}$ | 0.0001 | 1.0 | $0.2389 \times 10^{-4}$ | 0.86 | 0.1761 |
| $\frac{cal}{s \cdot cm^2 \cdot °C}$ | 27.01 | 4.186 | $4.186 \times 10^4$ | 1.0 | $36.0 \times 10^3$ | 7373.0 |
| $\frac{kg\ cal}{h \cdot m^2 \cdot °C}$ | $0.7502 \times 10^{-3}$ | $0.1163 \times 10^{-3}$ | 1.163 | $0.2778 \times 10^{-4}$ | 1.0 | 0.2048 |
| $\frac{Btu}{h \cdot ft^2 \cdot °F}$ | 3.663 $10^{-3}$ | $0.568 \times 10^{-3}$ | 5.68 | $0.1356 \times 10^{-3}$ | 4.882 | 1.0 |

***Example:*** $100 \frac{Btu}{h \cdot ft^2 \cdot °F} \times \frac{3.663 \times 10^{-3} W/(in^2 \cdot °C)}{1.0 Btu/(h \cdot ft^2 \cdot °F)} = 0.3663\ W/(in^2 \cdot °C)$

Specific heat*

| | $\frac{W \cdot s}{kg \cdot °C}$ | $\frac{W \cdot min}{lb \cdot °C}$ | $\frac{cal}{gm \cdot °C}$ | $\frac{Btu}{lb(mass) \cdot °F}$ |
|---|---|---|---|---|
| (W·s)/(kg·°C) | 1.0 | $7.567 \times 10^{-3}$ | $2.390 \times 10^{-4}$ | $2.388 \times 10^{-4}$ |
| (W·min)/(lb·°C) | 132.16 | 1.0 | 0.03159 | 0.03156 |
| cal/(gm·°C) | 4184 | 31.66 | 1.0 | 0.9993 |
| Btu/[lb(mass)·°F] | 4186 | 31.69 | 1.0007 | 1.0 |

Dynamic viscosity†

| | $\frac{lbf \cdot s}{in^2}$ | $\frac{lbf \cdot s}{ft^2}$ | $\frac{lb(mass)}{s \cdot ft}$ | Poise (P) | cP | $\frac{lb(mass)}{h \cdot ft}$ | Pa·s |
|---|---|---|---|---|---|---|---|
| $\frac{lbf \cdot s}{in^2}$ | 1.0 | 144 | $4.633 \times 10^3$ | $6.895 \times 10^4$ | $6.895 \times 10^6$ | $1.668 \times 10^7$ | $6.895 \times 10^3$ |
| $\frac{lbf \cdot s}{ft^2}$ | $6.944 \times 10^{-3}$ | 1.0 | 32.17 | $4.788 \times 10^2$ | $4.788 \times 10^4$ | $1.158 \times 10^5$ | 47.88 |
| $\frac{lb(mass)}{s \cdot ft}$ | $2.158 \times 10^{-4}$ | 0.03108 | 1.0 | 14.88 | $1.488 \times 10^3$ | $3.600 \times 10^3$ | 1.488 |
| P | $1.450 \times 10^{-5}$ | $2.089 \times 10^{-3}$ | 0.06719 | 1.0 | 100.0 | 241.9 | 0.1 |
| cP | $1.450 \times 10^{-7}$ | $2.089 \times 10^{-5}$ | $6.719 \times 10^{-4}$ | 0.01 | 1.0 | 2.419 | 0.001 |
| $\frac{lb(mass)}{h \cdot ft}$ | $5.996 \times 10^{-8}$ | $8.634 \times 10^{-6}$ | $2.778 \times 10^{-4}$ | $4.134 \times 10^{-3}$ | 0.4134 | 1.0 | $4.134 \times 10^{-4}$ |
| Pa·s | $1.450 \times 10^{-4}$ | $2.089 \times 10^{-2}$ | 0.6719 | 10.0 | 1000.0 | 2419.0 | 1.0 |

* 1 cal (international table)/(gm·°C) = 1.0007 cal/(gm·°C).
† Acceleration of gravity $g = 32.17\ ft/s^2$.

This temperature rise is very important in electric machinery since components are often cooled in series. Thus, the temperature rise of some critical component located downstream in the fluid path through the machine must include the temperature rise of the fluid prior to contacting that component; therefore, the fluid flow must be adjusted to limit this temperature rise. Since the power required to pump the fluid through the machine varies directly with about the cube of the flow, excessive flow requirements can greatly increase these power requirements and detract from the overall machine efficiency. The determination of the exact amount of fluid flow is often very difficult, and precise fluid-flow management is generally reserved for only the largest machines. The following table provides a rough guide for determining the requirements for air-cooled machines.

| Machine speed, r/min | Air required, $ft^3/min$ | | | |
|---|---|---|---|---|
| | Class A | Class B | Class F | Class H |
| Above 1800 | 90 | 70 | 60 | 50 |
| 900–1799 | 100 | 80 | 70 | 60 |
| Below 899 | 110 | 90 | 80 | 70 |

From the standpoint of heat transport, hydrogen again has excellent properties. Its high thermal capacity (over 14 times that of air) greatly reduces the mass flow required to limit the temperature rise to what it would be in air. At atmospheric pressure, the same volumetric flow would be required for hydrogen as for air since hydrogen's higher thermal capacity is almost exactly offset by its lower density; however, the pumping power required for hydrogen would only be about $^1/_{14th}$ of that for air.

## Summary of Fundamentals

The thermal limit of electric machinery is predicated on the temperature rating of the materials. This is normally defined in terms of the temperature rise of the copper windings above ambient. This temperature rise is the summation of the following:

$$\Delta T = \Delta T_{\text{cond}} + \Delta T_{\text{conv}} + \Delta T_{\text{fluid}}$$

That is, $\Delta T$ is (1) the rise required to conduct the heat generated in the windings to the convective surface, plus (2) the temperature difference between the convective surface and the fluid, plus (3) the temperature rise of the fluid prior to reaching the heat transfer surface. Various degrees of design sophistication are employed in an effort to minimize the factors affecting these components of the total temperature rise.

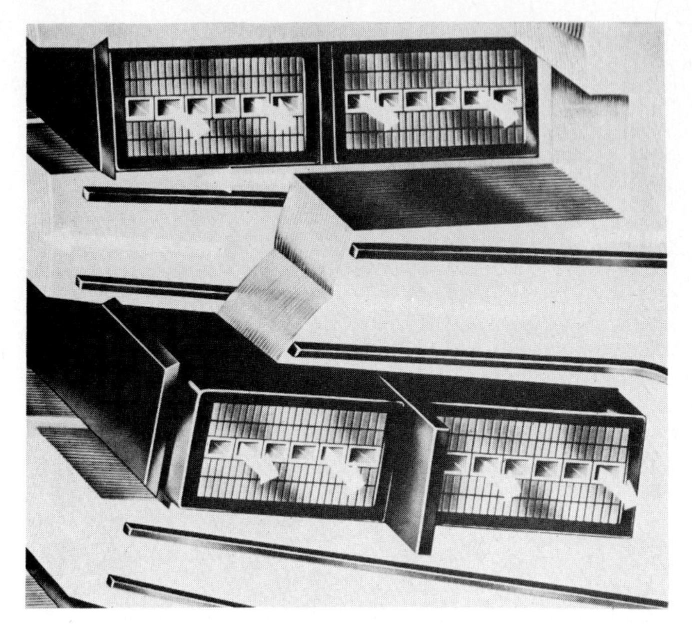

**FIG. 12.2** Direct gas-cooled stator coil section. (*Courtesy of Westinghouse Electric Corporation.*)

To minimize the temperature rise due to conduction air, space in the winding slots must be eliminated to the greatest extent possible. Thus, today's vacuum-post-impregnated windings tend to operate cooler than did their predecessors wound with finished coils. The ultimate in reducing this portion of the temperature rise is achieved by direct cooling of the windings. This is done by providing passages through which the fluid can pass within the conductors, as demonstrated in Figs. 12.2 and 12.3, which depict turbine-generator windings directly cooled with hydrogen gas and water, respectively.

Convective heat transfer is enhanced through fluid management, extended surfaces, and choice of fluid. The effect of the fluid temperature rise can be reduced not only through increased flow and choice of fluid, but also through fluid-path design. Figures 12.4 and 12.5 depict an induction motor of conventional flow path and a reverse-ventilated design. In the conventional arrangement the air is pumped forward into the machine; thus, the pumping losses and their accompanying increase in air temperature are introduced prior to cooling the windings. In the reverse-ventilated design the blowers are arranged to draw the air through the machine, and the added temperature rise at the blower has no effect on the thermal rating. Another approach is to provide a short parallel path for the fluid; this is done in large turbine generators, such as the one depicted in Fig. 12.6.

The growth of central station generators has to a large extent been attributable to improved methods of ventilation and cooling. The high rotational speeds of these machines limit their physical size; therefore, in order to achieve the unit ratings needed, the power density has had to be increased severalfold. Figure 12.7 shows this growth as a function of ventilation improvements.

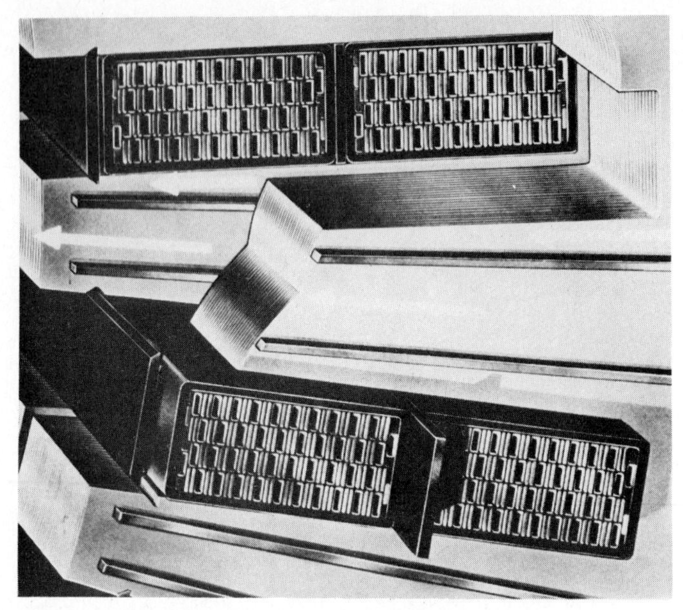

**FIG. 12.3** Water-cooled stator coil section. (*Courtesy of Westinghouse Electric Corporation.*)

## 12.3 FUNDAMENTALS OF VENTILATION

The ventilation of electric machines takes many forms and has varying degrees of complexity. For the most part, the ventilation may be categorized as follows when viewed from outside the machine:

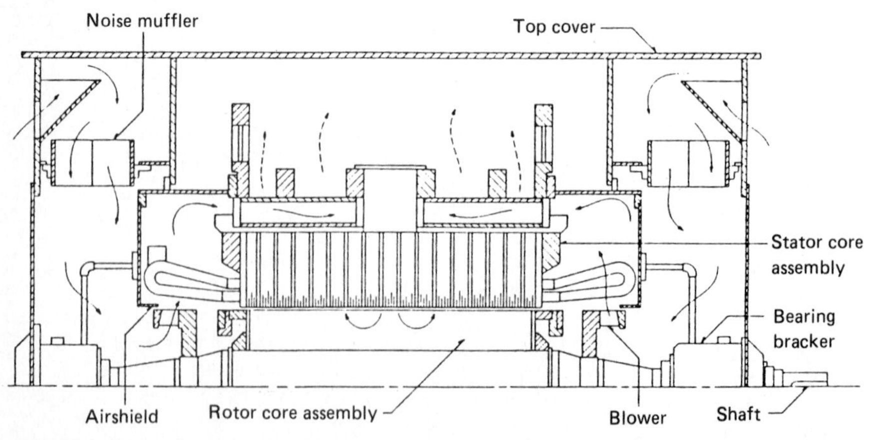


**FIG. 12.4** Conventionally ventilated induction motor. (*Courtesy of Westinghouse Electric Corporation.*)

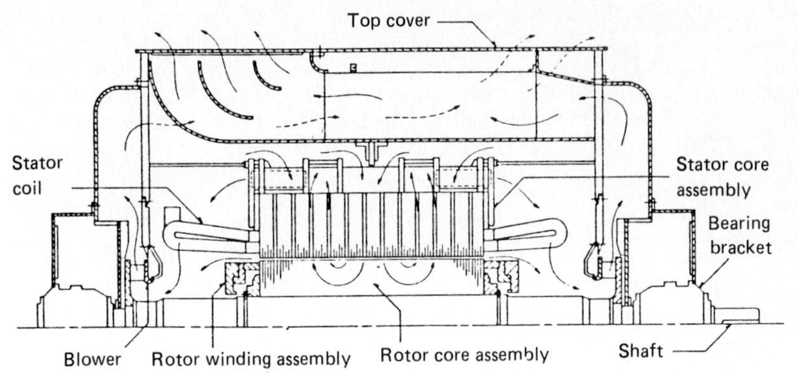


**FIG. 12.5** Reverse-ventilated induction motor. (*Courtesy of Westinghouse Electric Corporation.*)

Externally forced ventilation

Self-ventilated (open, drip-proof, weather-protected types I and II)

Totally enclosed, fan-cooled (TEFC)

Totally enclosed, water-cooled (TEWC)

When viewed from inside the machine, even more variations occur. In most machines the heat generated must be conducted to a convection surface through the electrical insulation; this is referred to as conventional, or indirect, cooling. Direct

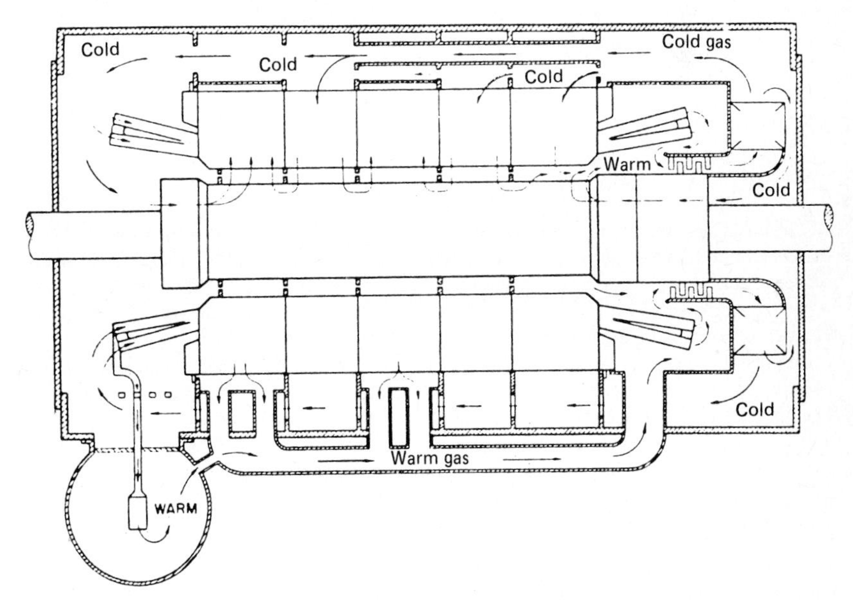


**FIG. 12.6** Ventilation diagram of a zone-cooled generator. (*Courtesy of Westinghouse Electric Corporation.*)

cooling implies that the cooling medium is in direct contact with the conductors.

Rotors and stators may be treated differently. For example, a direct gas-cooled rotor may be used with a direct water-cooled stator or even with a conventionally cooled stator. Ventilation ducts may be arranged axially or radially or both. The cooling medium may be low-density gas, air, pressurized air, or higher-density gases or liquids. The variations possible are limited only by the designer's imagination.

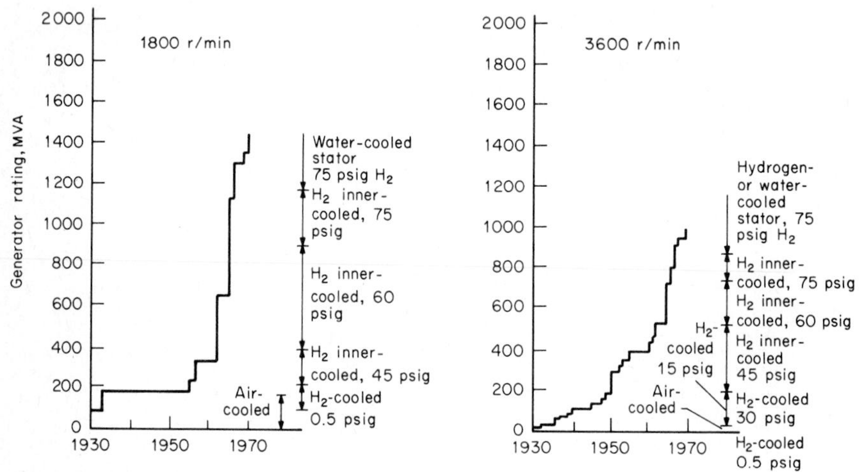


**FIG. 12.7** History of turbine-generator ratings.

The choice of ventilation depends upon many considerations, including the environment in which the machine is to operate and certain machine characteristics—primarily power density and rotational speed.

## Environmental Considerations

Some environments may be hazardous to reliable operation due to dust, dirt, moisture, or chemical contamination. To preclude such hazards, air that has been brought in via ducts from an uncontaminated source is often forced through the machine. Another approach is totally to enclose the machine, thus preventing any interchange between the ventilating medium and the external air.

In other cases the machine may present a problem to the environment. Electric arcs resulting from machine failure or commutation could be serious problems in certain areas; for this case, total enclosure of the machine is usually the only answer. In some cases the enclosure must be designed to be "explosion-proof," thus preventing any explosion occurring inside the machine from igniting surrounding gases or vapors.

More often the problem is the heating that results from the dissipation of the losses. Usually this is solved through forced ventilation, whereby the exhaust is transported to an area where heating is not a concern, or through a TEWC design. Another approach is simply to cool the exhaust from the machine with an air-to-water heat exchanger.

## Machine Characteristics

Rotational speed is an important factor. Very low-speed or variable-speed machines may develop insufficient blower action to provide adequate ventilation, thus requiring externally mounted blowers to force sufficient air through the machine; a solution less often used but also effective would be to pressurize the enclosure, thus increasing the air density and its cooling effectiveness. On the other hand, high-speed machines may generate so much windage loss in air as to make air impracticable as the ventilation medium. This leads to the use of lower-density gases, such as hydrogen or helium, and the accompanying sealed enclosure.

The power density, as defined by the horsepower or kilowatt rating divided by the active machine volume ($D^2L$) at a given speed, dictates the internal ventilation design. The higher the power density, the closer the ventilating medium must come to the point of heat generation and/or the better the heat transfer must be to the cooling medium. This leads to the attractiveness of direct water cooling.

## Elements of Ventilation

With all the variation, there remain a relatively few essential elements. These include:

- Pressure required to pump the ventilating medium through the machine
- Method for developing the necessary pressure (pumping requirements)
- Management of the medium to direct it to the heat-transfer surfaces
- Losses involved in pumping the ventilating fluid
- Frictional losses due to contact with rotating part
- Removal of heat from medium (if recirculated)

***Pressure Requirements.*** The back pressure is generally defined in terms of force per unit area (psi, $N/m^2$, etc.) or head of fluid or, in the case of most air-ventilated machines, inches of water. In electric machines the ventilation paths tend to be tortuous and at first glance not amenable to calculation. However, by using relatively simple techniques it is possible to estimate the pressure drops through a machine with reasonable accuracy. This is done by analyzing each entrance, exit, abrupt change in velocity, abrupt change in direction, and unobstructed path (unobstructed paths can be analyzed as flow through a duct). In most machines it is the abrupt changes in velocity or direction that predominate.

For air-cooled machines, a convenient way of representing the pressure drops is in terms of inches of water, $H_w$, which in turn can generally be expressed as

$$H_w = C\left(\frac{V}{4030}\right)^2 = C\left(\frac{Q}{4030A}\right)^2 \qquad \text{for air at STP}$$

where $C$ = a constant for a particular case
$V$ = velocity at section, ft/min
$Q$ = flow through section, $ft^3/min$
$A$ = area of section, $ft^2$

Cases of particular interest are shown in Figs. 12.8 to 12.13. (Note in Fig. 12.13 that 0.03 is a conservative estimate of the friction factor. For a more refined analysis, refer to a fluids mechanics test or a hydraulics handbook.) For noncircular ducts, the hydraulic diameter $D_h$, which is defined as

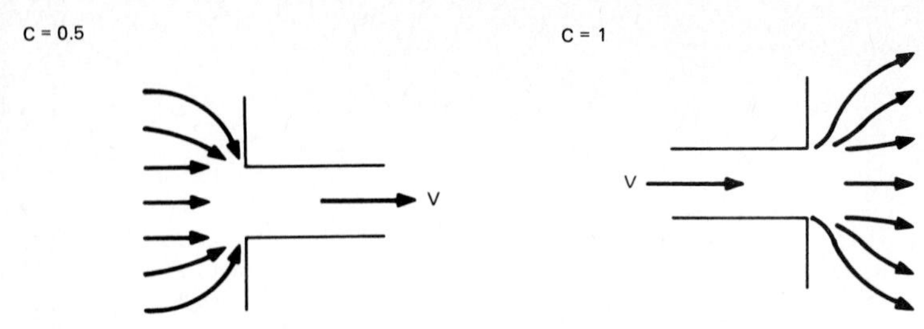


**FIG. 12.8** Entrance into a machine.

**FIG. 12.9** Exit from a machine.

$$D_h = 4 \times \frac{\text{cross-sectional area}}{\text{wetted perimeter}}$$

may be used in place of $D$ in Fig. 12.13. For circular ducts, the hydraulic diameter reduces to $D$. Most complex ventilation paths can be analyzed as components by using the above equations and Figs. 12.8 to 12.13. For components in series, the pressure drops are simply added. For parallel paths, use the electrical analogy shown in Fig. 12.14:

$$R_i = \frac{C_i}{(4030A_i)^2}$$

$$H_w = R_e Q$$

For this case, it can be shown that the equivalent resistance $R_e$ is

$$R_e = \frac{1}{[\sqrt{1/R_1} + \sqrt{1/(R_2 + R_3)}]^2}$$

And in general

$$R_e = \frac{1}{(\Sigma\sqrt{1/R_i})^2}$$

Thus, by adding the component resistances in series or parallel as shown above, the back-pressure characteristic can be represented as shown in Fig. 12.15.

For liquids, it is usually more convenient to express the circuit resistance in terms of pipe or duct length such that

$$p = f\frac{L}{D}\frac{\rho V^2}{2g} \qquad \text{or} \qquad H_f = f\frac{L}{D}\frac{V^2}{2g}$$

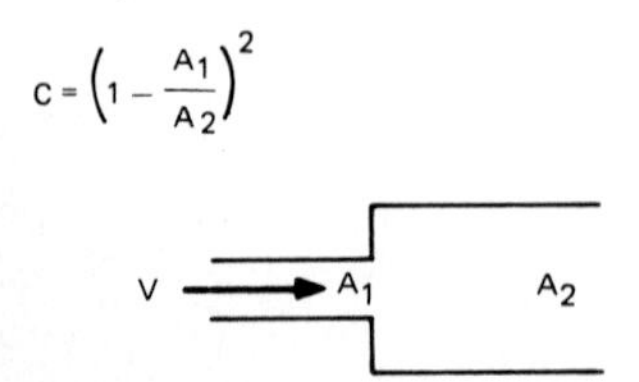


**FIG. 12.10** Sudden expansion.

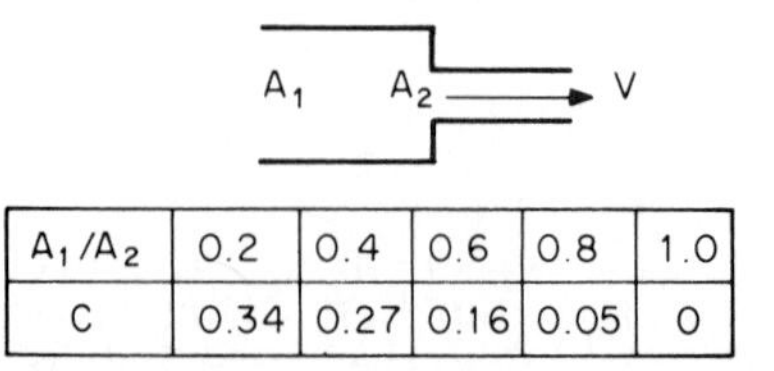


| $A_1/A_2$ | 0.2 | 0.4 | 0.6 | 0.8 | 1.0 |
|---|---|---|---|---|---|
| C | 0.34 | 0.27 | 0.16 | 0.05 | 0 |

**FIG. 12.11** Sudden contraction.

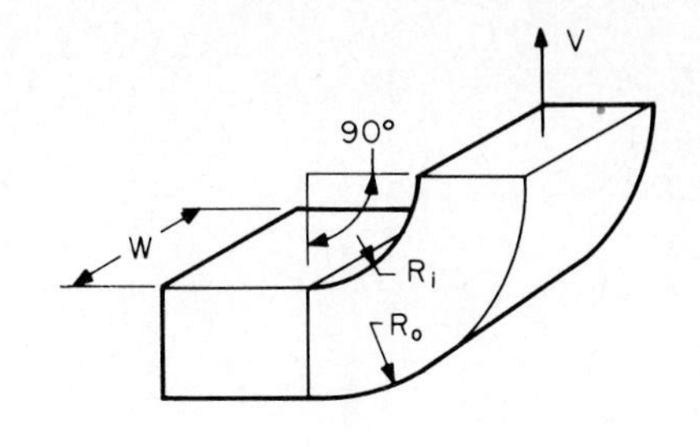


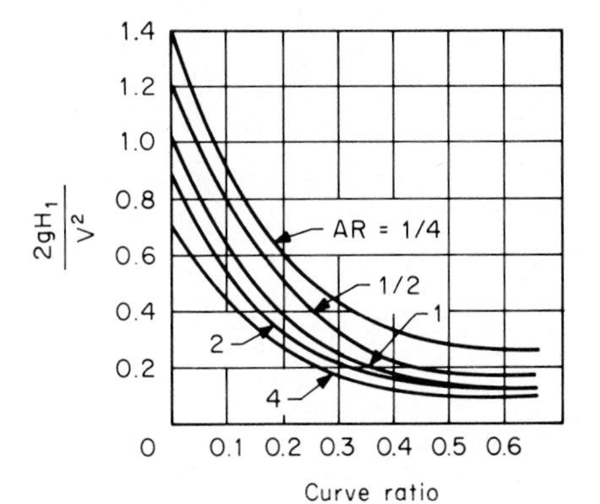


**FIG. 12.12** Flow through a 90° bend, showing loss coefficient for a smooth 90° bend of rectangular cross section. (For a 45° bend, $C_{45} = 0.5C_{90}$.) Curve ratio = inner radius/outer radius. $AR$ = width normal to plane of bend/(outer radius − inner radius).

where $p$ = pressure drop
$H_f$ = head loss
$f$ = friction factor (see Fig. 12.13)
$L$ = length of pipe or duct
$D$ = diameter of pipe or duct (hydraulic diameter for noncircular cross sections)
$V$ = fluid velocity
$g$ = acceleration of gravity
$\rho$ = fluid density

Figure 12.16 provides equivalent lengths for various valves and fittings in terms of the pipe diameter. Various pipe diameters can then be normalized for some

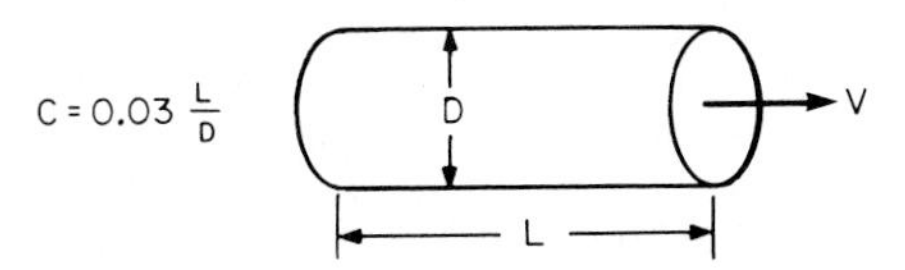


**FIG. 12.13** Flow through a duct.

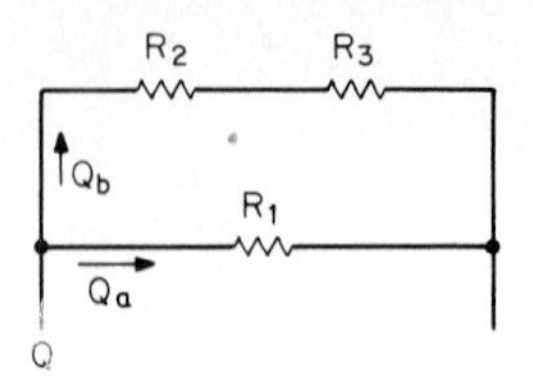


**FIG. 12.14** An electrical analogy.

predominant or convenient pipe diameter $D_e$ by calculating an equivalent length $L_e$ for the pipe in terms of $D_e$:

$$L_e = L\left(\frac{D_e}{D}\right)^5 \frac{f}{f_e}$$

For a given path, the components in series are simply added to determine the overall equivalent length of that path. For parallel paths, the equivalent length of $i$ number of parallels is

$$L = \frac{1}{(\Sigma\sqrt{1/L_i})^2}$$

***Pumping Requirements.*** Knowing the flow rate for the ventilating fluid and the pressure required, the pumping requirements can be defined. This is generally expressed graphically, as shown in Fig. 12.17. Where the pumping is provided from an external blower, it is usually a simple matter of referring to a catalog and determining the appropriate blower and motor for the application. Where the blower is shaft-mounted, the electric machine designer is generally responsible for the design of the blower(s). Dimensionless curves of the type shown in Fig. 12.18 are quite useful in designing these blowers for particular applications; the curve shown is based on characteristics of a blower design familiar to the author. Any blower whose characteristics are known from tests can be represented in a similar fashion. The use of this curve is demonstrated in the following example.

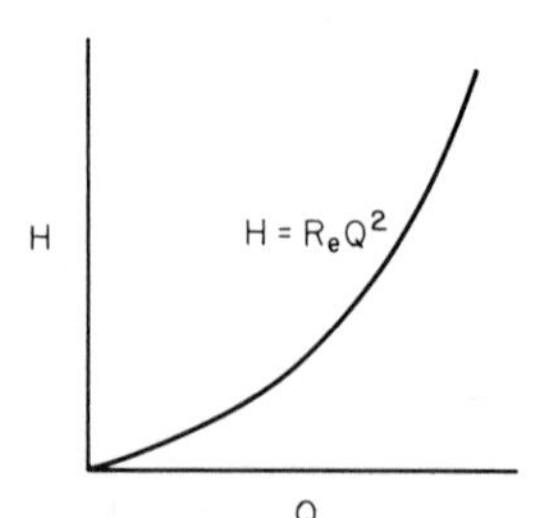


**FIG. 12.15** Back-pressure characteristic.

EXAMPLE: A large 1800-r/min motor requires 600 ft$^3$/min of ventilating air. The head required to pump this volume is 4 in of water. At maximum blower efficiency, $\eta = 0.65$, and the pressure coefficient $\psi$ and the flow coefficient $\phi$ are 0.40 and 0.19, respectively. In order to develop sufficient head, the diameter must be

$$D = \frac{15,400(H_w/\psi)^{0.5}}{\text{r/min}}$$

$$= \frac{15,400(4/0.4)^{0.5}}{1800}$$

$$= 27 \text{ in}$$

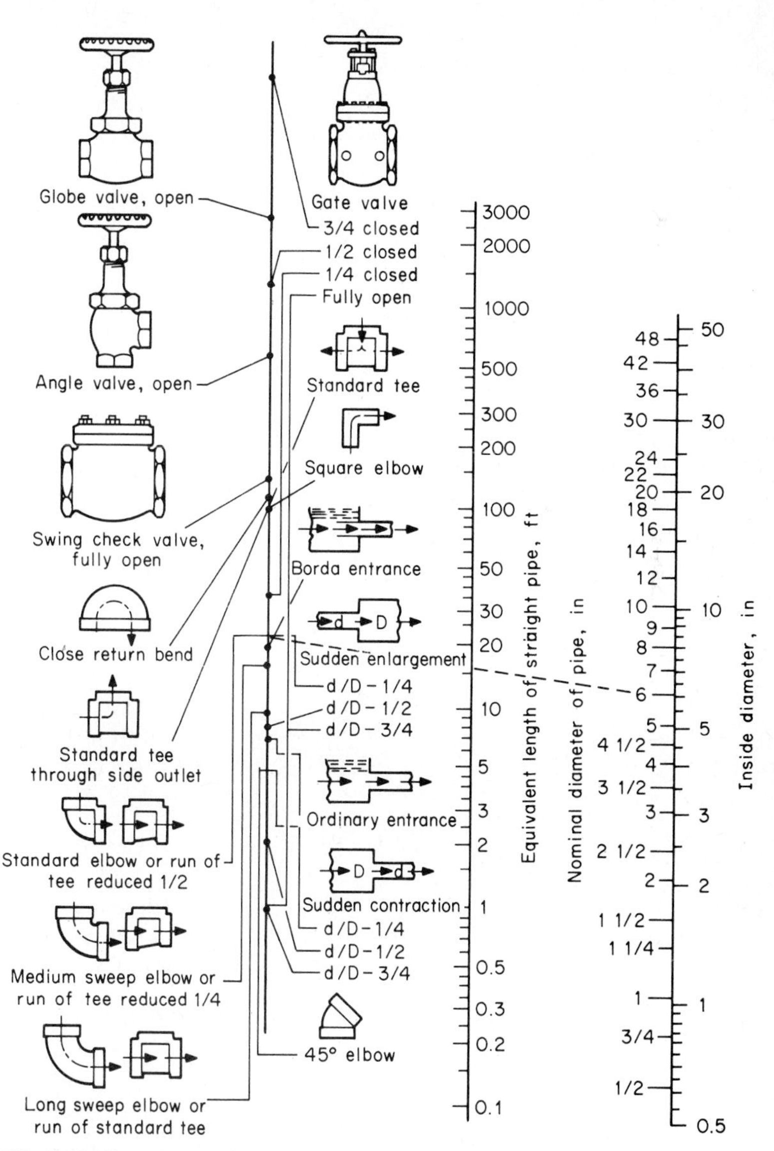


**FIG. 12.16** The resistance of valves and fittings. (*Courtesy of Crane Company.*)

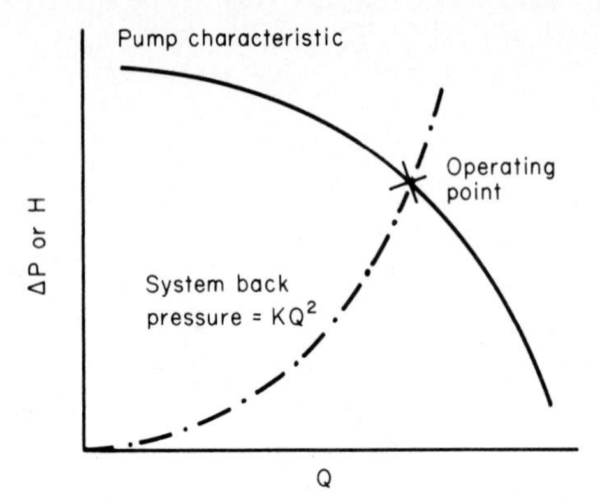


**FIG. 12.17** Pressure/volume curve.

and the width is determined from

$$W = \frac{175Q}{\phi \times D \times \mathrm{r/min}}$$

$$= \frac{175(6000)}{0.19 \times 27 \times 1800}$$

$$= 4.2 \text{ in}$$

The power required to drive the blower, $P$, is determined from

$$P = \frac{Q \times H_f \times \rho}{\eta} = \frac{Q \times p}{\eta}$$

and for any fluid whose pressure rise is measured in inches of water, the horsepower is

$$\mathrm{hp} = \frac{1.576 \times 10^{-4} \times Q \times H_w}{\eta}$$

Thus, the power needed to drive the blower is

$$\mathrm{hp} = \frac{1.576 \times 10^{-4}(6000)(4)}{0.65}$$

$$= 5.8 \text{ hp}$$

All other dimensions are then simply determined as a ratio of $D$.

Some care and judgment must be used in determining the face width $W$ when the ratio $W/D$ varies substantially from that of the blower tested. Generally speaking, large $W/D$ ratios tend to be less effective than predicted, while smaller $W/D$ ratios tend to be more effective unless the flow through the blower approaches laminar.

Another situation that often arises in electric machines is that the rotating parts themselves are used to drive the fluid through the machines. Radial ducts or vents in the rotor, as shown in Fig. 12.19, are utilized for that purpose. For this case, the head $H_f$ available may be estimated as

$$H_f = \frac{[\pi \times (\mathrm{r/min})/60]^2 (D_o^2 - D_i^2)}{2g \times 144} \qquad g = 32.2 \text{ ft/s}^2$$

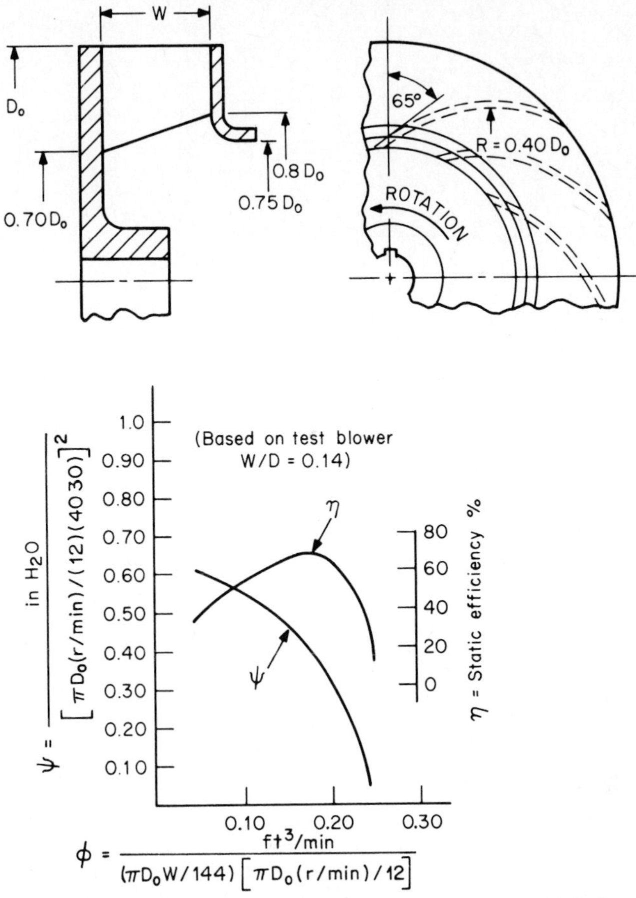


**FIG. 12.18** Pressure/volume characteristics of a typical blower with backward-curved blades.

where $D_o$ = outside diameter of rotor, in
$D_i$ = inside diameter of rotor, in

or in terms of inches of water and with the diameters expressed in inches,

$$H_w = [(r/min)/15,400]^2(D_o^2 - D_i^2)$$

This head may be viewed as being in series with any external blower for ventilating the rotor and stator. Blower vanes attached to the end rings, as shown in Fig. 12.19, are often required to ventilate the ends of the windings.

***Ventilation Management.*** The amount of ventilating fluid circulated through the machine can have an important effect on efficiency. As previously stated, the head (or pressure) required for a given machine varies directly with the square of the volumetric flow. Thus, since the power requirement is proportional to the product of the heat and volumetric flow, the power varies with the cube of the flow.

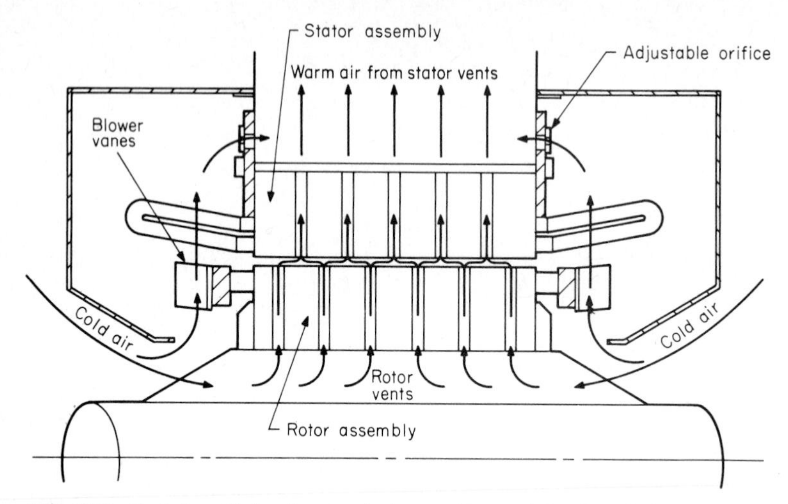


**FIG. 12.19** Typical ventilation diagram of a motor with rotor-induced ventilation.

At the same time, sufficient ventilating fluid must be directed to areas to cool these parts properly. With the complicated series and parallel paths often involved in electric machines, it is tempting to overventilate some areas in order to ensure adequate ventilation for critical components. This leads to excessive overall flow and consequently to an excessive pumping requirement, which necessarily detracts from the machine's efficiency. This can be avoided through management of the ventilating fluid.

The management is accomplished by determining the resistance to flow and the available head through the various paths either by calculation or tests. Then the flow distribution can be determined by solving the network of series and parallel circuits. The resistances can then be calculated to get the best flow distribution. Adjustable baffles such as those shown in Fig. 12.19 are often used to make fine adjustments based on tests.

***Pumping Losses.*** All power required for pumping the ventilating fluid must be considered to be part of the thermal burden to be dissipated. This includes the total power input to all external or shaft-mounted blowers and the rotating parts utilized for forcing the fluid through the machine. For the blowers, this loss is readily determined from curves such as those in Fig. 12.18. For the rotating parts, it is generally assumed that the flow-through leaves the rotor at the peripheral velocity. Thus, this pumping loss may be expressed in watts as

$$P_w = Q \times \rho \times \frac{(V/1000)^2}{10.26}$$

where $Q$ = flow-through, $\text{ft}^3/\text{min}$
$\rho$ = density, $\text{lb}/\text{ft}^3$
$V$ = peripheral velocity = $\pi D_r \times \text{r/min}$, ft/min

As the pressure, or head, decreases along the ventilation path as a result of frictional resistance, the fluid increases in temperature. This temperature rise $\Delta T$ due to incremental pumping losses is

$$\Delta T = \frac{\Delta H}{778.3C} = \frac{\Delta P}{5.405 \times \rho \times C}$$

where $\Delta T$ = temperature rise, °F or °C
$\Delta H$ = head loss, ft of fluid
$\Delta P$ = pressure loss, lb/in$^2$
$C$ = specific heat, Btu/(lb·°F) or Btu/(lb·°C)
$\rho$ = fluid density, lb/ft$^3$

For air, where the head is expressed in inches of water, the temperature rise in °C is

$$\Delta T_c = 0.205 \Delta H_w$$

Note that any fluid leaving the blower has already increased in temperature as a result of absorbing the losses

$$\Delta T = \frac{\Delta H(1-\eta)}{778.3 \times \mathrm{C} \times \eta} = \frac{\Delta P(1-\eta)}{5.405 \times \rho \times \mathrm{C} \times \eta}$$

And for air with the head expressed in inches of water,

$$\Delta T_c = \frac{0.205 \Delta H_w (1-\eta)}{\eta}$$

The temperature rise for fluids pumped through radial ducts in the rotor punchings can be estimated from

$$\Delta T = \frac{(V/1000)^2}{108.4C}$$

Thus, for an air-cooled machine running at 3600 r/min with a 24-in-diameter rotor,

$$V = \frac{24(3600)\pi}{12} = 22,620 \text{ ft/min}$$

$$C = \frac{0.241 \text{ Btu}}{\text{lb} \cdot {}^\circ \text{F} \times 1.8} = \frac{0.434 \text{ Btu}}{\text{lb} \cdot {}^\circ \text{C}}$$

$$\Delta T_c = \frac{(22,620/100)^2}{180.4 \times 0.434} = 6.5^\circ\text{C}$$

Note that this temperature rise is independent of the volumetric flow.

***Frictional Losses.*** In addition to the losses incurred in pumping the ventilating fluid through the machine, there are frictional losses caused by the rotating parts shearing the fluid that is in contact. The cases shown in Figs. 12.20 to 12.22 will generally suffice for estimating these losses in kilowatts on most electric machines.

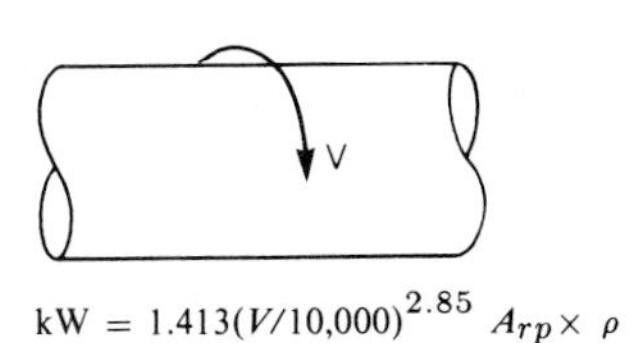


$$\text{kW} = 1.413(V/10{,}000)^{2.85} A_{rp} \times \rho$$

**FIG. 12.20** The open cylinder.

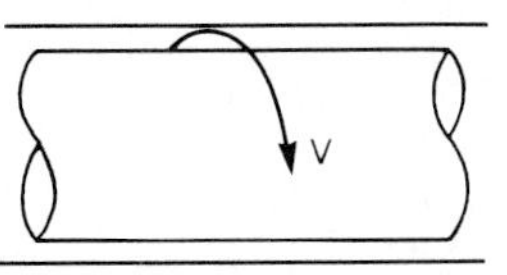


kW = $0.628(V/10{,}000)^{2.85} A_{rp} \times \rho$
$V$ = peripheral velocity, ft/min
$A_{rp}$ = peripheral area, ft$^2$
$\rho$ = fluid density, lb/ft$^3$

**FIG. 12.21** The covered cylinder.

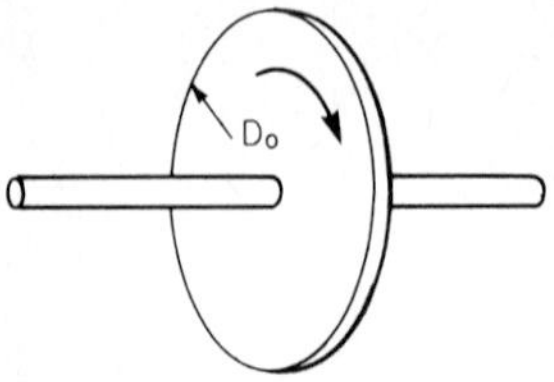

$$kW = 3.702 \times 10^{-5}(D_O)^{4.85}\,[(r/min)/10{,}000]^{2.85}\rho$$

$D_O$ = outside diameter, in

**FIG. 12.22** A thin disk (for both sides).

Although the formulas shown are for relatively smooth surfaces, they have proven to work reasonably well for most rotating surfaces on electric machines.

***Heat Removal.*** In addition to the pumping and frictional losses, the electrical losses expressed in kilowatts, $kW_L$, will all be transported by the ventilating fluid. This will cause a temperature rise of

$$T_c = \frac{32\ kW_L}{Q \times C \times \rho} \quad \text{or} \quad 1780\frac{kW_L}{Q} \quad \text{for air at STP}$$

In a closed system this heat must be removed. Typically, this is done with heat exchangers, which transfer the heat to a secondary fluid. These may be gas-to-liquid or gas-to-gas heat exchangers. In any case, it is this secondary loop that must dissipate the heat to some sink and thereby establish the machine's operating temperature.

Generally, even when part of the motor, the heat exchangers are purchased from a supplier specializing in this equipment. For effective design, the heat-exchanger designer needs the following information:

- Losses to be transferred
- Primary- and secondary-fluid chemistry
- Primary fluid—volumetric flow
- Maximum primary pressure drop across the cooler
- Maximum temperature of the secondary fluid
- Maximum temperature of the primary fluid out of the cooler
- Dimensional constraints
- Maximum pressure drop of the secondary fluid (only if a constraint)
- Maximum volumetric flow of the secondary fluid (only if a constraint)

## 12.4 APPLICATION OF FUNDAMENTALS

From a heat-transfer standpoint, electric machines tend to be somewhat complex. This is due to their geometric configuration, diverse material properties, various heat sources, and tortuous ventilation paths. This is especially true for the most common electric machine design, in which the windings are imbedded in slots in the punchings.

For many years the most reliable design tools for the thermal design was test information obtained from previous machines. This led to a macroscopic approach toward the thermal design—i.e., viewing the machine in terms of overall losses, heat-transfer surface area, average or apparent heat-transfer coefficients, etc. Although practical for machines having a common design philosophy, this approach does not lend itself well to machines outside the bounds of the test information available. Deviations from previous practice involving geometric relationships, materials, ventilation, or loss distribution greatly reduce the reliability in using test data from previous machines.

Today designers rely on a more detailed approach. To understand this method, a simplified model amenable to hand calculations will be utilized. Figure 12.23 portrays a typical stator with the windings imbedded in slots provided in the laminated core. The heat transfer may be represented as a network of resistors, as shown in Fig. 12.24. Current then represents heat flow, and the resistors represent the resistance to heat flow (the reciprocal of conductance).

In Fig. 12.24, note the paths through which heat can flow from the source to the fluid used to remove the heat. The heat generated in the imbedded portion of the windings, $I_1$, has two paths: (1) The most direct is into the air flowing through the vents, but this applies only to the short sections of the winding that cross over the vents; at this area the heat must pass through the insulation $R_1$ and then through the convective "film," which can also be represented as a resistance $R_2$. (2) The other path for the winding losses is into the iron through insulation that has a thermal resistance of $R_3$ and then into the iron teeth $R_4$. In the teeth, the iron losses due to magnetic hysteresis, $I_2$, are added. From the teeth the heat takes several paths—radially toward the airgap ($R_5 + R_6$), axially toward the vents ($R_7 + R_8$), and radially toward the iron below the slots, $R_9$. Resistors $R_6$ and $R_8$ represent the thermal resistance due to convection. In the iron below the slots, more iron losses ($I_3$) may be added to that transferred from the teeth by conduction ($I_e$) and then transferred by conduction to the vent surface ($R_{10}$), from where it is removed by the fluid ($R_{11}$).

In a simplified model the fluid temperature is assumed to remain constant at all the surfaces. In actuality the temperature rises as heat is removed by it, but generally this can be neglected and the fluid temperature be taken as the average temperature of the fluid passing through the component being analyzed (in this case the stator). This average temperature can be determined as: (1) the inlet temperature, plus (2) the temperature rise due to losses dissipated prior to the fluid reaching the component, plus (3) one-half the temperature rise due to the component losses.

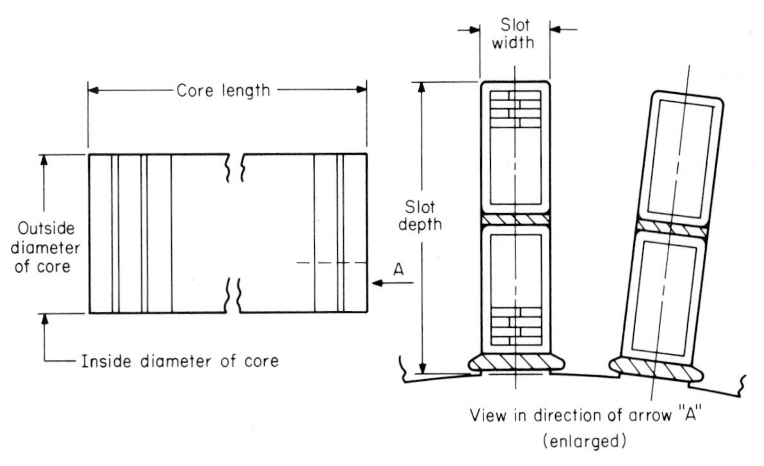


**FIG. 12.23** Typical stator configuration.

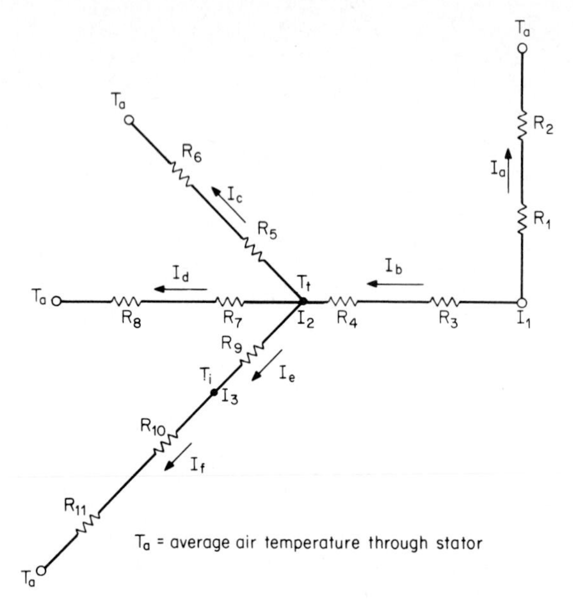


**FIG. 12.24** Resistor network representing typical stator.

Other paths may also be considered, such as the outer radial surfaces of the iron and the top surface of the windings in the airgap. The former are generally neglected since the convective heat transfer is poor and these surfaces tend to be thermally far from the heat source. The latter tends to be a high-resistance path due to the wedges that cover the windings at this point. In any case, by neglecting these paths, the results of the analysis will be conservative.

Even though this model may very well represent a complex heat-transfer problem, it is nonetheless solvable by hand calculations. If we consider one heat source at a time, the resistors can readily be combined to form an equivalent resistance. The current (heat) distribution through the various paths and the potentials (temperatures) at each point can then be determined. If this is done for each heat source, the combined effects can be obtained from superposition.

To demonstrate the applicability of this method to a real machine, consider a 3200-kW wound-rotor induction generator designed for a variable-speed application. In this case, the following apply at rated conditions:

Copper losses in slots = 21 kW
Iron losses between slots = 21 kW
Iron losses below slots = negligible
Inside diameter of stator = 28 in
Outside diameter of stator = 41 in
Length of stator core (including vents) = 33 in
Twenty vents, each 0.38 in wide
Airflow through stator vents = 8000 ft$^3$/min at STP
Air temperature at machine inlet = 40°C

Losses picked up by air before entering stator = 70 kW
Depth of slot = 3 in
Thickness of insulation = 0.085 in
Thermal conductivity of insulation = 0.006 W/(in·°C)
Thermal conductivity of iron = 0.50 W/(in·°C)
Thermal conductivity across laminations = 0.05 W/(in·°C)

The conductive resistances can be calculated as

$$\text{Conductive resistance} = \frac{L}{A \times k}$$

where $L$ = length of conduction path, in
$A$ = mean total conductive area of path, $\text{in}^2$
$k$ = thermal conductivity, W/(in·°C)

This applies directly to $R_1$ $R_3$, $R_4$, and $R_5$. $R_9$ involves a step change from the slotted to unslotted portion of the punchings. Here each step is best calculated separately and then added to determine the effective resistance. Where the thermal input is distributed along the conduction path, it can be shown that a good approximation can be made by setting the path length $l$ equal to one-third the actual length. This applies to $R_6$ and $R_{10}$.

The convective resistances are calculated as

$$\text{Convective resistance} = \frac{l}{A \times h}$$

where $A$ = total convective area off path, $\text{in}^2$
$h$ = convective heat-transfer coefficient, $\text{W/(in}^2\cdot{}^\circ\text{C)}$

Experimental work done by Luke showed that for vents in electric machinery, the heat-transfer coefficient can be determined as*

$$h = 0.0365 \left(\frac{V}{1000}\right)^{0.71} \qquad \text{W/(in}^2\cdot{}^\circ\text{C)}$$

and for surfaces exposed to the airgap,

$$h = 0.0350 \left(\frac{V_r}{2000}\right)^{0.8} \qquad \text{W/(in}^2\cdot{}^\circ\text{C)}$$

where $V$ = local average velocity in vent, ft/min
$V_r$ = peripheral velocity of rotor, ft/min

For the example machine, the network can be represented by the following:

| | |
|---|---|
| $I_1 = 21$ kW | $R_5 = 2.41$°C/kW |
| $I_2 = 21$ kW | $R_6 = 5.61$°C/kW |
| $I_3 = 0$ kW | $R_7 = 0.60$°C/kW |
| $R_1 = 3.84$°C/kW | $R_8 = 2.05$°C/kW |
| $R_2 = 3.76$°C/kW | $R_9 = 3.24$°C/kW |
| $R_3 = 1.54$°C/kW | $R_{10} = 0.23$°C/kW |
| $R_4 = 0.10$°C/kW | $R_{11} = 1.42$°C/kW |

*G. E. Luke, "Surface Heat Transfer in Electric Machines with Forced Air Flow," *AIEE Transactions*, vol. 45, 1927, pp. 1036–1037.

Solving the network, it is found that the temperature rise of the stator copper will be 67°C above the "effective" temperature of the air. Therefore, the total copper temperature will be:

| Temperature factors | Temperature, °C |
|---|---|
| Ambient air temperature | 40.0 |
| Prior air rise (70 × 1780/8000) | 15.5 |
| One-half air rise thru stator [41 × 1780/(2 × 8000)] | 4.5 |
| Copper temperature rise above local air | 67.0 |
| Total copper temperature | 127.0 |

which is very close to the limits for a class B machine.

Besides determining the copper temperature, the model also provides an interesting insight into the thermal behavior of machines of this type. Consider first how the heat is dissipated among the various paths in comparison to the convective heat-transfer area for that path. This comparison is best made on a percentage basis:

| Path | Heat dissipated, % | Total area, % |
|---|---|---|
| Coil section in vent | 21.0 | 12.8 |
| Airgap surface | 13.9 | 3.9 |
| Vent area between slots | 42.2 | 23.4 |
| Vent surface below slots | 22.9 | 59.9 |

Thus, over 77 percent of the heat is dissipated from about 40 percent of the area. This demonstrates the importance of keeping the path short from the heat source to the surface in contact with the fluid.

We can evaluate both the sensitivity of the copper temperature to material conductivities and the convective heat-transfer coefficient by varying the appropriate resistances. A 2:1 variation of the major items results in a percentage effect on the copper temperature rise, as tabulated below. The results of that evaluation for this machine are:

| Item varied over 2:1 range | Effect on copper rise, % |
|---|---|
| Insulation conductivity | 23.0 |
| Through-iron conductivity | 9.2 |
| Across-iron conductivity | 4.5 |
| Convective heat-transfer coefficient | 30.2 |

Thus, at a glance it can be seen what areas are most important. The copper temperature is most dependent on the insulation conductivity and the convective heat-transfer coefficient. On the other hand, 2:1 variations in the iron conductivity will have less than a 10 percent effect on these temperatures. Unfortunately, the insulation and convective heat-transfer information may be the least reliable. The insulation conductivity can be adversely affected by poor contact to the iron in the slot; thus, vacuum-post-impregnated machines tend to run cooler. In large machines wound with fully cured coils, it is recommended that side filler be added as necessary to ensure the best contact possible to the iron. As for the convective heat-transfer coefficient, it is very important not only to get the proper volume of fluid through the stator but also to distribute it as evenly as possible to minimize the hot-spot temperature.

Applying the factors from Table 12.4 for heat-transfer coefficients and using hydrogen as the cooling fluid, the cooling is improved as follows:

| Coolant | Copper rise |
|---|---|
| Air, STP | 67 |
| Hydrogen, 1 atm | 57 |
| Hydrogen, 2 atm | 48 |
| Hydrogen, 3 atm | 44 |
| Hydrogen, 4 atm | 42 |
| Hydrogen, 5 atm | 41 |

This indicates that using hydrogen at over 3 atm (30 psig) has little value for further reducing the copper temperature. The model thus verifies what is common practice—not to exceed 30 psig for a conventionally cooled machine.

Through the application of fundamentals, analytic models can be constructed to provide reasonably accurate estimates of the temperatures throughout electric machines. The approach recommended here is to establish single temperature elements whose heat-transfer and heat-generation characteristics can be well defined. A network of these elements can then be used to represent the component being analyzed. Even coarse grids can be very useful to the experienced designer.

With the computer, finite-element techniques are now available to provide highly refined models of the iron. Unfortunately, the increased accuracy attainable offers little overall improvement in accuracy since the critical temperatures are least sensitive to the iron's conductivity.

In any case, the emphasis of this chapter has been on the application of fundamentals rather than on providing formulas describing the thermal characteristics of specific components. The chief aim has been to point out the most important factors affecting the thermal behavior of electric machinery.

# CHAPTER 13
# ELECTRIC MACHINE INSULATION

**T. W. Dakin**

## *13.1 INTRODUCTION*

The electrical insulation system of rotating machines must be constructed to endure both the electrical stresses and severe mechanical stresses induced by the magnetic field of the moving rotor—a mechanical problem of lesser degree in static apparatus. Also, machines that are designed open to the atmosphere for the purpose of ventilation cooling must resist oxidation, moisture, and contamination, which can seriously reduce their life. Another important feature of the insulation of rotating machines is that the multiple winding turns are constrained to the relatively narrow armature (or stator) slots, which are usually grounded. Although this geometry is not very different from that of a coaxial cable, the cable represents only a single turn; the mechanical forces on the insulation in the machine slot are very much greater.

The difficulty in fabricating a suitable insulation for higher-voltage generators has limited the highest voltages to about 34 kV, and power-generator voltages are more usually 13.8 to 24 kV. Maximum motor operating voltages are usually lower than that, also constrained by the available supply voltage and the power requirements.

Rotating-machine stator insulation construction divides into two principal types, correlated to the machine's physical size, which is related to the power and voltage rating:

1. Random-wound conductor machines, in which the conductors are relatively small and round
2. Form-wound conductor machines, in which the conductors are rectangular and relatively large

### Voltage Requirements

The machine insulation must of course withstand the normally expected operating voltages between conductors and to ground in all parts of the winding(s). Also, the system must withstand (with some prudent margin) the required industry standard overvoltage tests for new machines. The test specified by the American National Standards Institute (ANSI) is twice the rated terminal-terminal voltage plus 1000 V applied for 1 min to the armature windings. For the alternator field windings (in the rotor), the test voltage is 10 times the rated voltage, but not less than 1500 V.

### Temperature Requirements

These requirements are set by the allowable hot-spot temperature due to conductor-resistance heating from the current at rated load. The permissible hot-spot temperature is set by the thermal capability of the organic (resin) insulation used in the machine: i.e., by the chemical stability of the insulation to maintain its mechanical and electrical integrity sufficiently to prevent turn-to-turn or turn-to-ground shorts at operating voltages or expected transient overvoltages.

There is always a slow deterioration of the organic insulation resin(s) by chemical reaction within themselves or with gases, by contamination, or by chemical interactions with other components present. Eventually the deterioration progresses to such a degree that the insulation can no longer resist the mechanical or magnetic forces tending to push the conductors together or to ground. Also, the thermal degradation often develops cracks in the enamel, varnish, or resin, reducing the dielectric strength of the insulation and its resistance to water and contamination.

Since the chemical stability of different types of organic resins varies greatly, depending on their chemical structure, the electric industry has organized them into temperature stability classes based on the temperature at which they are able to endure satisfactorily for their expected service lifetimes.[1] These temperature classes of insulation are as follows:

| | | | |
|---|---|---|---|
| *Class A*: | 105°C | *Class F*: | 155°C |
| *Class B*: | 130°C | *Class H*: | 180°C |

For example, a 5-hp machine with class B insulation would be designed to operate continuously at rated load at a hot-spot temperature of 130°C for a reasonable lifetime under normal ambient conditions.

Industry standards do not specify the life of machines. Partly, this is because their life is affected by various operating conditions. Also, because service does not usually involve operating continuously at maximum load, the machine insulation's actual life may be prolonged by occasional operation at less than the rated hot-spot temperature.

## 13.2 RANDOM-WOUND MACHINE INSULATION

Figure 13.1 shows a typical configuration of the enameled-wire turns and slot insulation (slot liners) for a single-phase motor. The crossover of windings from one slot to another is evident, as is the random compactness of the wires.

The principal insulation components are the enamel insulation on the wire and the insulation between the wire windings and the grounded slot (or between the windings of different phases if it is a multiphase machine). In addition, there may be leads with a different insulation inside the machine housing; these leads might connect from the windings to terminals and to internal switching or protective devices. Another component of the motor's insulation system

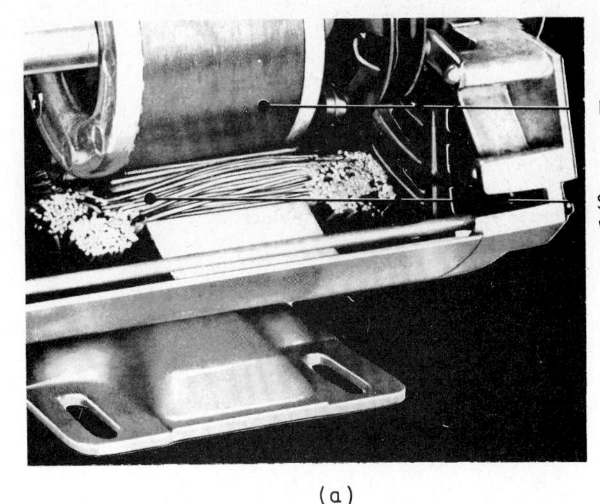


(a)

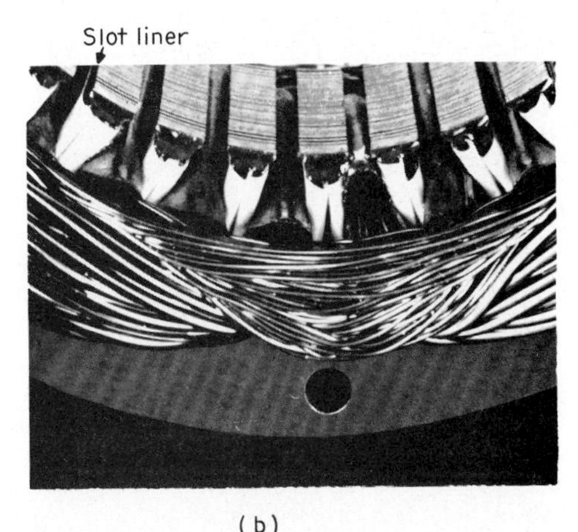


(b)

**FIG. 13.1** Section views of a random-wound motor: (*a*) side view; (*b*) end view. (*Courtesy of Westinghouse Electric Corporation.*)

(with a primarily mechanical function) is the binding cord used to tie down the end windings to restrain their vibration; this cord material has in some cases been incompatible with the other insulating resins and caused reduced insulation life.

## Enameled Wire

Random-wound rotating machines usually have voltage ratings of less than 1000 V root mean square (rms) and carry only moderate currents; thus, the stator-winding current can be handled by relatively small, round, enameled conductors. The enamel on these conductors is a most critical insulation component. The enamel is most commonly applied by passing the wire through a solution of polymerizable resin and directly into a high-temperature curing tower, where it is converted into a thin, solid, flexible coating. Usually several or more passes are required to achieve the desired thickness of 0.025 mm. Double-build coatings are sometimes used.

The American Society for Testing and Materials (ASTM) has standardized a sizable group of tests for testing various qualities of enameled wire for electrical insulation. Many of these qualities are listed in ASTM D-1676; they include: film flexibility and adherence, stiffness, oiliness, continuity, heat shock, unidirectional scrape abrasion, elongation, dielectric breakdown voltage, resistance to softening, solderability, dimensions, cut-through temperature, extractables by refrigerant liquids, and a high-voltage dc (350 to 3000 V in steps) continuity bench test. ASTM Standards Part 39, *Electrical Insulation—Test Methods: Solids and Solidifying Liquids*, should be consulted for others.

Some of the above tests measure qualities that are more important for preserving the integrity of the enamel during insertion of the wire into the machine than during the life of the insulation. For example, the enameled wire is stretched and scraped during winding into the slots and should endure this without serious damage to the enamel.

Some of these standard ASTM tests also deal with the insulating varnish that is applied over the enameled wire after the wire winding of the stator is completed. The function of the varnishing of the stator is twofold: to provide additional protection of the enamel against the environment (moisture, dirt, and chemical contamination) and to provide mechanical support for the wires.

### Slot and Phase Insulation

Another very important component of the random-wound machine insulation is the slot and phase-to-phase insulation (if it is a multiphase machine). This insulation for class A temperatures is most usually a somewhat flexible sheet material (such as cellulosic paper) of the order of 5 to 10 mils (0.125 to 0.25 mm) thick for class A temperature machines, or a combination of a laminate sheet of cellulosic paper and a polyester film, or the film alone. Fused resin coatings are in some cases applied to stators by electrostatic attraction of a fusible polymerizable resin powder. The stator is heated to fuse and cure the resin coating to a smooth, continuous coating providing the slot insulation. This process is used primarily on small motors.

Glass cloth, asbestos, or mica paper treated with varnishes of greater thermal stability are used for higher-temperature machines. High-temperature resin films are also used for slot and phase insulation.

### Thermal Stability

Enamels and varnish overcoatings for wires are produced with temperature ratings in each of the temperature classes listed in Sec. 13.1: for use, of course, in machines of those temperature classes. It is important to be able to test the thermal stability of these enamels and varnishes, and it is particularly important to test any combinations of enamel and varnish overcoat used in the machine. An interaction often occurs between the wire enamel and varnish, and this sometimes reduces the thermal stability below that of either component alone. Wire enamel and varnish tests are usually made on components first, before being tested in more complex models resembling conditions in a part of a complete machine.

For this testing, accelerated-functional-life tests have been developed and standardized by ASTM and the International Electrotechnical Commission (IEC) for individual materials and simple combinations and by the Institute of Electrical and Electronics Engineers (IEEE) for model-motor insulation systems (motorettes) and small motors.[2–4] All of these insulation accelerated-life tests involve the aging of standardized test specimens until they fail at a series of test temperatures well above their planned operating temperatures. The logarithms of these accelerated aging times are then graphed versus their reciprocal kelvin (°C + 273) test temperatures; this Arrhenius graph (as in Fig. 13.2) is extrapolated down to the planned service temperature to predict a lifetime at that temperature.[5,6]

It is prudent to evaluate the data on test failure times (which are usually scattered around a mean extrapolation to the service temperatures) statistically in terms of the minimum probable failure times.[6,7] In some cases, this evaluation is required.

The following test standards for aging apply to wire enamel and varnish insulation:

ASTM D-2307, *Relative Thermal Endurance of Film-Insulated Round Magnet Wire:* Test involves accelerated aging of twisted pairs of insulated wires until the ac breakdown-voltage stress between them declines to about 300 V/mil (1200 V/mm). (Similar to IEC Publication 172.)

ASTM D-1973, *Thermal Endurance of Flexible Electrical-Insulating Varnishes:* Test involves accelerated aging of panels of glass cloth coated with the varnish until the ac breakdown-voltage stress declines to 300 V/mil (1200 V/mm). (Similar to IEC Publication 370.)

ASTM D-3145, *Thermal Degradation of Electrical Insulating Varnishes by the Helical-Coil Method:* Test involves accelerated aging of 0.25-in (6.35-mm)-di-

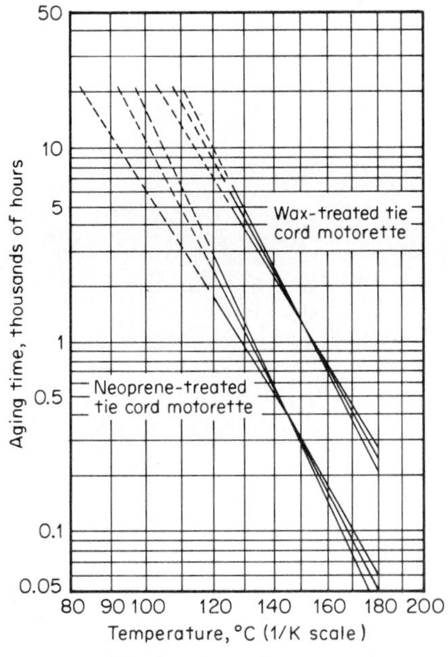


**FIG. 13.2** Arrhenius graph of motorette-life tests.

ameter varnish-coated tight bare-wire coils until the varnish bond breaks under a specified mechanical load. (Similar to IEC Publication 290.)

Accelerated-life tests of models that closely resemble the conditions of the whole insulation system of a random-wound motor have been standardized in IEEE Standard 117. These model motors have become known as motorettes (Fig. 13.3). The Arrhenius graph of motorette-life test results (Fig. 13.2) showed the very deleterious effect of a neoprene-treated tie cord, which reduced the insulation system life by a factor of about 4.[6] Neoprene rubber contains a chlorine atom in each monomer unit of the polymer; the chlorine probably splits out as hydrochloric acid during aging and catalyzes the degradation of the varnish resin or the phase insulation, which in this case was a polyester film–paper laminate. The wire enamel in these tests was Formvar, a class A (105°C) enamel.

The actual test temperatures used in obtaining the life test data of Fig. 13.2 ranged between 120 and 180°C. The failure-time–temperature data (more than 100 test points) were fitted to the Arrhenius equation by linear regression statistical analysis,[7] and then the mean line was graphed with the upper and lower confidence limits for the line; the mean regression line (for the system with wax-treated tie cord) extrapolated to a life of 16,000 h at 105°C, the class A temperature. It should be noted that the environmental conditions in these tests (such as 100 percent humidity with water condensation, and with applied voltage imposed periodically between heating cycles in the procedure) were very severe, perhaps more severe than for most motor applications. Thus, such tests may sometimes predict a conservatively shorter life than can be expected in less severe conditions.

Similarly planned life tests of random-wound motor insulation systems are sometimes run on full-size small or medium motors. Heating to the accelerated higher

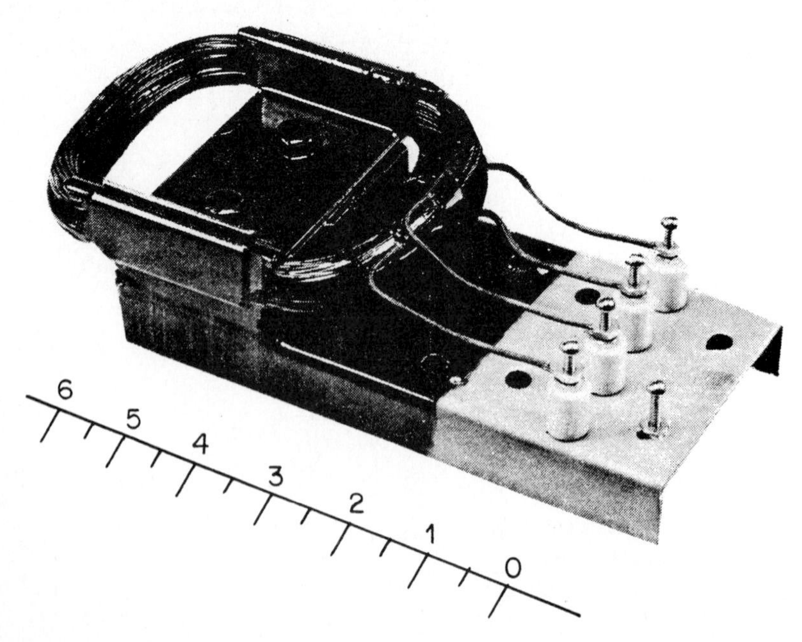


FIG. 13.3 Motorette.

temperatures is accomplished by frequent motor reversals or by thermally enclosing the motors so that they run hotter. The motor reversing also introduces accelerated mechanical aging.

## Voltage Aging

Another degrading effect affecting the life of the electrical insulation is that due to partial discharges—if they occur continuously or regularly for appreciable time. Whenever the applied-voltage stress on any series airgaps or gas gaps in the insulation structure exceeds the breakdown-voltage stress of the gas there, a small discharge occurs between insulation surfaces or between insulation and metal surfaces. Such discharges slowly erode the organic enamel or varnish at a rate dependent on the magnitude of the overvoltage above the threshold stress for occurrence of discharges. Since this threshold is above a few hundred volts, as illustrated in Fig. 13.4, it is usually above the operating voltage of most low-voltage random-wound machines and is a potential problem only for higher-voltage machines.

Note in Fig. 13.4 that the threshold voltage is a function of the ratio of the insulation thickness $t$ divided by the dielectric constant $k$. A graph like Fig. 13.4 would be a guide for estimating the possibility of discharge in any higher-voltage random-wound machine design. One should consider (1) all minimum spacings and (2) the possible voltages between conductors in the same winding, between windings of different phases, and from winding to ground.

Voltage aging by partial discharges is a more serious problem in form-wound machines because their voltages are often higher, so the effects of discharges will be discussed more extensively in Sec. 13.3. Wherever partial discharges occur

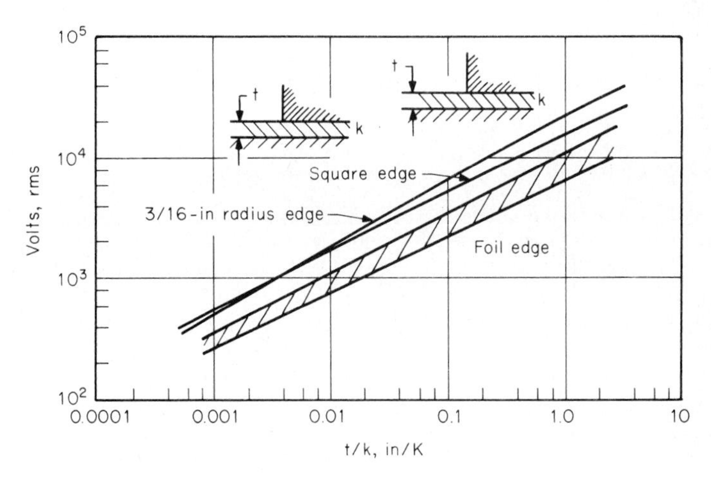


**t** = insulation thickness, in

**k** = relative dielectric constant

**FIG. 13.4** Partial-discharge threshold voltage. [*From D. G. Fink and H. Wayne Beaty (eds.), Standard Handbook for Electrical Engineers, McGraw-Hill, New York, 1978, sec. 4, p. 126.*]

continuously, discharge-resistant insulation like mica should be used, since organic resins can be punctured much more easily by the discharges.

## 13.3 FORM-WOUND MACHINE INSULATION

As indicated above, form-wound windings are usually employed in higher-power machines. The conductors are rectangular in cross section to fit into the rectangular stator slot. The dimensions of the conductor(s) with the applied insulation are coordinated to those of the slot so that they snugly fill the slot. The insulation (not necessarily including impregnation) of the coil turns is applied before winding into the slot. In most high-voltage machines, the coils are completely impregnated with resin and are in a finished form prior to winding.

Higher-voltage rotating machines above a few kilovolts are usually insulated from ground with a resin-bonded mica applied as a wrapper or tape, with a supporting fibrous sheet. The very earliest machines used overlapping mica splittings a few inches in breadth, with a shellac binder and a cellulosic paper support. Later machines employed an asphalt vacuum impregnation of the mica-taped or -wrapped insulation, but this was abandoned when some of these machines failed after some years of service because of a phenomenon variously called tape separation or girth cracking.

Most present-day high-voltage machines employ a vacuum impregnation of the mica tape or wrapper after application to the rectangular conductor. This is done with liquid polymerizable resins, which are subsequently cured to solids by heating. During the curing, the insulated conductor may be constrained to dimensions appropriate to fit the machine slot into which it will be placed. These resins are typically sufficiently elastic that they will contract back into the slot after a heating cycle. The resins used are primarily or partly of an epoxy-type composition proprietary with each manufacturer.

Modern machines employ mica paper in place of large mica splittings. This paper is formed in continuous sheets from tiny flakes of mica. The mica paper is weak in tension, so it must be supported during application to the conductor with a stronger (and very thin) glass cloth or polyester mat.

### Gas-Cooled Conductors

The largest generators produce so much heat from the resistance of their windings at rated currents that the complete generator is enclosed and the housing filled with hydrogen gas at 4- to 5-atm pressure. Hydrogen provides so much better heat transfer than other gases that it is used in spite of the risk in handling it. Secondary benefits accrue to the generator insulation due to the prevention of oxidation of the resin insulation and an increase of the flashover voltage of the gas spaces in the end turns and the connections.

A stepped cross section of a slot of a large generator with inner-cooled hydrogen-gas-cooled conductors is illustrated schematically in Fig. 13.5. With the exception of the central cooling tubes, the cross section of an externally cooled high-voltage coil would be similar. The small arrows from the central cooling tubes in Fig. 13.5 indicate the gas flow axially through the tubes; the vertical arrow represents cooling gas flow through radial ducts through the stator iron. Each conductor is divided into a number of parallel strands, which are often insulated from each other with a thin, varnish-bonded fiberglass insulation to reduce eddy-current heating. The major

**FIG. 13.5** Cross section of a hydrogen inner-cooled conductor slot of a generator. (*Courtesy of Westinghouse Electric Corporation.*)

insulation on the conductor turns is typically mica-tape insulation; it is applied in overlapped fashion and in the required number of layers to get the required thickness for the voltage rating of the machine.

For inner-cooled gas-cooled generators, the cooling tubes at each end of the stator are brought out through the major conductor-turn insulation away from the stator iron. There, the gas vents into the enclosed volume of the generator housing. The housings at each end of the generator are isolated, and a pressure is maintained between them to provide gas flow through the cooling tubes. This explanation is given only to indicate the insulation aspects of this type of machine cooling. The cooling tubes are insulated along the conductors in much the same fashion as the parallel strands are.

## Water-Cooled Conductors

Inner cooling by water flow is sometimes employed in the largest generators. The tubes are located centrally among the strands, somewhat similar to Fig. 13.5, but beyond the end of the stator, at a safe distance, they are connected through the major insulation wall on the coil to insulating tubes that carry the water to grounded heat exchangers. This cooling method presents an interesting insulation problem, since one end of the above-mentioned insulating tubes is at the high potential of the generator coil turn and the other end is at ground, and the tube is filled with water, which is not considered an insulator. The problem is managed by continuously deionizing the cooling water to a resistivity of the order of $10^6$ Ω·cm and making the water-filled tube sufficiently long that the resistance heating of the water column will be negligible.

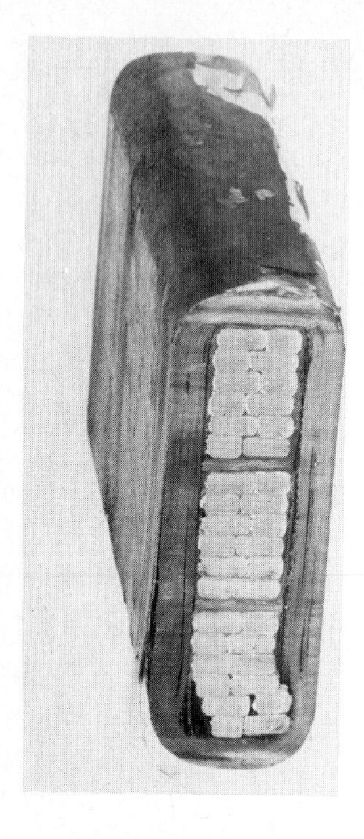

**FIG. 13.6** Cross section of waterwheel generator coil (damaged in service). (*Courtesy of Westinghouse Electric Corporation.*)

The cross section of a coil turn (with insulated strands) of an older waterwheel generator is shown in Fig. 13.6. In this case the strands in the top of the coil turn had loosened and vibrated by magnetic forces part way through the insulation to ground. This condition had produced a breakdown to ground at a nearby location. This example illustrates the importance of mechanically strong and compact insulated coil structures in machines.

## Voltage Aging

When there are partial discharges (sometimes referred to colloquially as corona) impinging continuously against organic insulation, degradation occurs, which can lead to eventual failure. For this reason, mica is incorporated in the major insulation of higher-voltage machine coils. Mica resists the discharges very well, but for practical reasons of construction the mica has to be bonded together with organic resins, so the mica insulations conventionally used suffer a little, and the electric stress must be designed to accommodate this.

Figure 13.7 shows the times to failure in accelerated stresses—much higher than service stresses—for typical high-voltage machine insulations tested as flat sheets and applied to coil bars simulating actual coils. The failure times of a 100 percent organic resin sheet in the same stress range are shown for comparison. The times to failure of organic insulations seem to be principally a function of the electric stress and somewhat independent of insulation thickness. The data of Fig. 13.7 show the dielectric strength of the resin sheet alone declining much more steeply than the

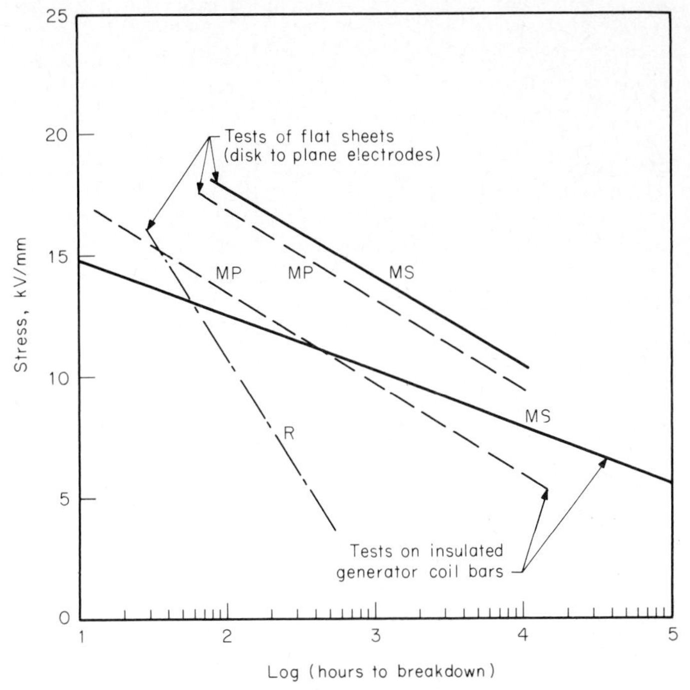


**MP:** Resin-impregnated mica paper

**MS:** Resin-impregnated mica splittings

**R:** Resin alone

**FIG. 13.7** Voltage aging of generator insulation and resin. (*From A. Wichmann and P. Gruenwald, Proceedings IEEE International Symposium on Electrical Insulation, Montreal, IEEE Publication 76CH1088-4-EI, 1976, p. 88.*)

dielectric strength of the mica-paper-filled resin sheet. At the lowest electric stress (about 6 kV/mm) at which the data can be exactly compared, the mica-paper-filled resin failure time is more than 30 times longer than the resin-sheet failure time. The data in this figure also indicate that the insulation with mica splitting insulation applied to coils may be somewhat superior to that with mica paper insulation applied to coils at lower stresses, closer to operating stresses. There are, however, practical advantages that weigh in favor of the mica paper insulation.

Partial discharges are usually avoided on the outside surface of high-voltage major insulation—between the coil surface and the slot iron of high-voltage coils. This is accomplished by applying a conducting paint over the outside surface of individual slot coil turns after their insulation is applied, impregnated, and pressed to size to fit the slot. The conducting paint is applied only over the slot portion of the length. Discharges are avoided at the end edge of the conduction paint by applying a controlled high-resistance semiconducting paint for a selected length extended from the edge of the conducting paint. This gradually reduces the electric stress from the conducting paint toward the conductor to a value below the discharge threshold (gas breakdown) level at that edge. The conducting paint on the coil surface must be securely connected to the grounded slot iron; otherwise, large arclike "slot discharges" will occur between the iron and the conducting paint. Such discharges could discharge the capacitance of the slot portion of the entire length of one or two coils. If the discharges persisted, they would be likely to cause a failure through the major insulation on the coil.

## Thermal Stability

Lower- and medium-voltage coil insulation life is measured in accelerated higher-temperature tests according to IEEE Standard 275 by using model systems called formettes, as illustrated in Fig. 13.8. The test procedures alternate heating cycles with periods of environmental exposure, which include mechanical vibration, 100 percent humidity with normal applied voltage followed by ac proof tests of the ground and phase insulation, and ac and impulse voltage proof tests on the turn insulation. Testing continues until failure occurs on one of the tests. Multiple duplicate tests of each insulation system are run at several aging temperatures and the failure times are plotted on Arrhenius graph paper. IEEE Standard 275 recommends that extrapolation of the accelerated tests be limited to 20 or 30°C.

## Diagnostic Tests

It is of course desirable to test the capability or integrity of larger, more expensive machines by using nondestructive diagnostic tests, and valuable progress has been made toward this goal. These tests are mostly electrical, but a physical-chemical test will be mentioned at the end of this section. The tests are designed to indicate faults or weaknesses in the insulation system.

**1.** *DC tests:* These may sometimes be informative, even on ac machines. ANSI/AIEE Standard 43-1974 recommends for both ac and dc armatures (sta-

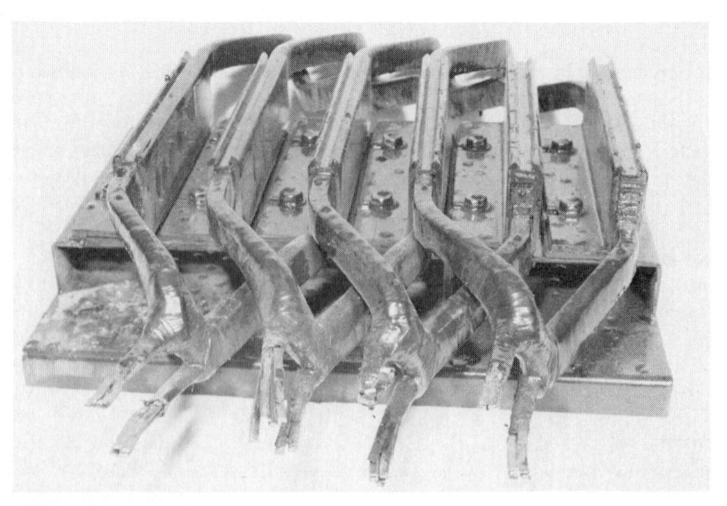

**FIG. 13.8** Formette.

tors) that the minimum dc resistance for the whole winding, $R$, in megohms at 1 min and at 40°C be determined by the expression

$$R = kV + 1$$

where $kV$ is the rated machine voltage, terminal to terminal, in kilovolts.

The polarization index, which is the ratio of dc 10-min (after voltage application) to 1-min resistance, is also measured to indicate the insulation condition. A low value is indicative of surface moisture or possible faults in the insulation. IEEE Standard 43 recommends a minimum value of 1.5 for class A insulation and of 2 for class B and F insulations. This standard suggests preventing moisture in windings during shutdown by keeping the winding warm.

The dc resistance of generator insulation is sometimes measured as a function of voltage up to high voltages in order to detect irregular upward trends, which have sometimes been found to be associated with faults. In such tests, correction must be made for the effect of polarization with time.

**2.** *Capacitance and dissipation-factor tests:* Partial discharges do occur in the cavities of resin-impregnated mica insulation of high-voltage machines in spite of the vacuum-impregnation processing. The resin shrinks during the curing polymerization, leaving thin cavities between the resin and mica. Partial discharges occur in these cavities at electric stresses that often start well below the operating stresses. Also, during thermal and mechanical aging of insulation, original cavities may enlarge and new cavities may develop. (Figure 13.6 illustrates a gross example of a large cavity that developed because of the mechanical vibration of some strands.) The quality of the insulation can be measured by the quantity and magnitude of these partial discharges.

The collective effect of these discharges is to increase the capacitance and dissipation factor of the insulation as the applied voltage is increased above the voltage at which the discharges initiate. The increase is gradual. The increase in the capacitance and dissipation factor occurs because the discharges are internal current pulses partly in phase and partly out of phase with the applied voltage. Measurements of the dissipation factor increase with voltage (popularly referred to as "power-factor tip-up") and were used very early in the industry as a measure of discharges in generator insulation. More recently, a more thorough analysis of the effect and improved instrumentation have made this type of measurement more precise and better understood.[8] It is used most accurately in tests of single coil lengths in the factory, where proper guard rings can be applied to avoid effects at the end of the conducting paint. Even if surface discharges are avoided there by a semiconducting stress grading paint, this resistive paint will cause a superimposed increasing dissipation factor with increasing voltage unless a guard ring is applied here.

It is possible to estimate the volume percentage of the cavities from an accurate measurement of the capacitance increase (due to discharges) with increasing voltage.[8] The measurement of capacitance and dissipation factor with increasing voltage is a sort of average measure of discharges and cavities and does not discriminate between a few large ones versus numerous small ones. A modification of the separate capacitance and dissipation-factor measurement is a rapid recording of the admittance increase with voltage.[9]

Capacitance and dissipation-factor measurements on complete machines incur numerous uncertainties as a result of interfering effects (such as the semiconducting paint). Also, partial-discharge pulse measurements with high-frequency circuits connected to the machine terminals may yield confused results—caused both by the superposition of numerous small pulses and by the attenuation of pulses between their original site and the terminals. High-frequency pulse-detection circuits are very helpful in locating sites of abnormally larger discharges if an inductively coupled pickup probe is placed over a coil slot with the winding energized.[10] If the detector frequency is 5 MHz or higher, the attenuation of the discharge pulse along the winding is high enough to permit location of larger discharges to within one slot of a large generator. The voltage of the winding need not be set above the normal operating voltage to ground in making such tests. Slot discharges are extremely easy to detect and locate with such a probe.

Capacitive probes have also been applied to the end turns, but the spacing of the pickup capacitor electrode to the coil surface must be controlled sufficiently to avoid causing discharges from the coil surface to the probe, which is grounded.

Tests of complete machine windings may be thwarted if a low-current capability high-potential test set is used, since the capacitance current to ground of a complete machine winding may exceed the set's capability. One way of avoiding this limitation is to make tests at low frequency, and special test sets have been made for this. There are some differences from 60 Hz in the insulation behavior that must be considered in the interpretation of the results. Another method of avoiding the high current required at high voltage is to employ a series-adjustable inductor to resonate with the machine capacitance at 60 Hz and to use a moderate voltage supply.

High-frequency inductively coupled monitoring detectors have been applied to the neutral grounding lead of large generators. These are not suitable for detecting small partial discharges but are intended to detect quickly any internal arcing faults in the machine.[11]

**3.** *Gas test:* Thermal deterioration of the organic insulation in large, enclosed, hydrogen-cooled generators can now be detected and measured by periodic analysis of the gas in the generators. The gas is analyzed into its various components using gas chromatography, and the analysis is compared to laboratory tests of the same insulation materials aged in sealed vessels of hydrogen. Another gas test to detect overheating is described in Ref. 13.

A good survey article on high-voltage generator insulation is given in Ref. 14.

## REFERENCES


1. *General Principles for Temperature Limits in the Rating of Electric Equipment and for the Evaluation of Electrical Insulation*, IEEE Standard 1. (See also Refs. 2, 3, and 4.)
2. *IEEE Recommended Practices for the Preparation of Test Procedures for the Thermal Evaluation of Insulation Systems for Electrical Equipment*, IEEE Standard 99.
3. IEC Publication 610, *Principal Aspects of Functional Evaluation of Electrical Insulation Systems—Aging Mechanisms and Functional Procedures*, International Electrotechnical Commission, Geneva, 1978.
4. IEC Publication 611, *Guide for the Preparation of Test Procedures for Evaluation of the Thermal Endurance of Electrical Insulation Systems*, International Electrotechnical Commission, Geneva, 1978.
5. T. W. Dakin, "Electrical Insulation Deterioration Treated as a Chemical Rate Phenomenon," *AIEE Transactions*, vol. 67, pt. 1, 1948, p. 113.

6. IEEE Working Group 117, *Refinements in IEEE 117 Test Procedure for Evaluation of the Life Expectancy of Random-Wound Motor Insulation Systems*, IEEE Transactions Paper 636, 1968.
7. *Guide for Statistical Analysis of Thermal Life Test Data*, IEE Standard 101.
8. T. W. Dakin, "The Relation of Capacitance Increase with High Voltages to Internal Electric Discharges and Discharging Void Volume," *AIEE Transactions*, vol. 78, 1959, p. 790; "A Capacitance Bridge Method for Measuring Integrated Corona Charge Transfer and Power Loss per Cycle," *AIEE Transactions on Power Applications and Systems*, vol. 79, 1960, p. 648.
9. H. Terase, S. Hirabayashi, S. Hasegawa, and K. Kimura, "A New AC Current Testing Method for Non-Destructive Insulation Tests," *IEEE Transactions on Power Applications and Systems*, vol. PAS-99, 1980, p. 1557.
10. T. W. Dakin, J. S. Johnson, and C. N. Works, "An Electromagnetic Probe for Detecting and Locating Discharges in Large Rotating Machine Stators," *IEEE Transactions on Power Applications and Systems*, vol. 88, 1969, p. 251.
11. R. T. Harrold and F. T. Emergy, "Radio Frequency Diagnostic Monitoring of Electrical Machines," *IEEE Electrical Insulation Magazine*, vol. 2, 1986, p. 18.
12. G. Metzger and R. Fournie, "Detection of Local Overheating in the Insulation of Hydrogen-Cooled Turbo Generators" (in French), *Revue générale de l'électricité*, vol. 78, no. 4, April 1976.
13. H. E. Pietsch, E. M. Fort, D. C. Phillips, and J. D. B. Smith, "Sacrificial Coatings for Improved Detection of Overheating in Generators," *IEEE Transactions on Power Applications and Systems*, vol. PAS-96, 1977, p. 1675.
14. J. C. Botts, "High-Voltage Generator Insulation," *Proceedings NEMA-IEEE Electrical Insulation Conference*, Chicago, 1963, pp. 202–207.

# CHAPTER 14
# NOISE AND VIBRATION

S. J. Yang
A. J. Ellison

## 14.1 UNITS

### Sound-Pressure Level

The level of a sound having components within a given frequency band is expressed by the *sound-pressure level*, defined as

$$L_p = 20 \log \frac{p}{p_o} \qquad \text{dB} \tag{14.1}$$

where $p_o$ = root-mean-square (rms) reference sound pressure
$p$ = rms sound pressure for the frequency band

The reference sound pressure is $2 \times 10^{-5}$ N/m$^2$ (i.e., 20 $\mu$Pa), which is approximately the sound pressure of a pure tone of 1000 Hz at the normal threshold of hearing.

A basic sound-level meter often gives a sound-pressure-level reading based on Eq. (14.1) for an octave band, for a one-third-octave band, or in A-weighting. Table 14.1 shows the bandwidths of the internationally standardized octave bands and one-third-octave bands.

An electric machine does not emit noise uniformly into the surroundings, and thus the noise field around it is complicated. Figure 14.1 shows the variation of sound-pressure level along radial lines for a pure-tone component from an electric motor in an anechoic room, i.e., in a room without sound reflection.

### Sound-Power Level

The sound power $W$ in watts emitted from a machine for a given frequency band is expressed by the *sound-power level*, defined as

$$L_W = 10 \log \frac{W}{W_o} \qquad \text{dB} \tag{14.2}$$

**TABLE 14.1** Center and Approximate Cutoff Frequencies for Octave and One-Third-Octave-Band Filters

| Octave bands | | | One-third-octave bands | | |
|---|---|---|---|---|---|
| Center frequency $f_0$, Hz | Approximate lower cutoff frequency $f_1$, Hz | Approximate upper cutoff frequency $f_2$, Hz | Center frequency $f_0$, Hz | Approximate lower cutoff frequency $f_1$, Hz | Approximate upper cutoff frequency $f_2$, Hz |
| 16 | 11 | 22 | 16.0 | 14.1 | 17.8 |
| | | | 20.0 | 17.8 | 22.4 |
| | | | 25.0 | 22.4 | 28.2 |
| 31.5 | 22 | 44 | 31.5 | 28.2 | 35.5 |
| | | | 40.0 | 35.5 | 44.7 |
| | | | 50.0 | 44.7 | 56.2 |
| 63 | 44 | 88 | 63.0 | 56.2 | 70.8 |
| | | | 80.0 | 70.8 | 89.1 |
| | | | 100.0 | 89.1 | 112.0 |
| 125 | 88 | 177 | 125.0 | 112.0 | 141.0 |
| | | | 160.0 | 141.0 | 178.0 |
| | | | 200.0 | 178.0 | 224.0 |
| 250 | 177 | 355 | 250.0 | 224.0 | 282.0 |
| | | | 315.0 | 282.0 | 355.0 |
| | | | 400.0 | 355.0 | 447.0 |
| 500 | 355 | 710 | 500.0 | 447.0 | 562.0 |
| | | | 630.0 | 562.0 | 708.0 |
| | | | 800.0 | 708.0 | 891.0 |
| 1,000 | 710 | 1,420 | 1,000.0 | 891.0 | 1,122.0 |
| | | | 1,250.0 | 1,122.0 | 1,413.0 |
| | | | 1,600.0 | 1,413.0 | 1,778.0 |
| 2,000 | 1,420 | 2,840 | 2,000.0 | 1,778.0 | 2,239.0 |
| | | | 2,500.0 | 2,239.0 | 2,818.0 |
| | | | 3,150.0 | 2,818.0 | 3,548.0 |
| 4,000 | 2,840 | 5,680 | 4,000.0 | 3,548.0 | 4,467.0 |
| | | | 5,000.0 | 4,467.0 | 5,623.0 |
| | | | 6,300.0 | 5,623.0 | 7,079.0 |
| 8,000 | 5,680 | 11,360 | 8,000.0 | 7,079.0 | 8,913.0 |
| | | | 10,000.0 | 8,913.0 | 11,220.0 |
| | | | 12,500.0 | 11,220.0 | 14,130.0 |
| 16,000 | 11,360 | 22,720 | 16,000.0 | 14,130.0 | 17,780.0 |
| | | | 20,000.0 | 17,780.0 | 22,390.0 |

where $W_o$ = reference power (1 pW, or $10^{-12}$ W)

Although the sound-pressure levels around an electric machine vary from point to point, the sound-power level for an electric machine emitting a steady noise under a given operating condition in a free field has a unique value.[1]

## A-Weighted Sound Level

The *A-weighted sound level* is the overall sound-pressure level for the entire audible frequency range (approximately from 20 Hz to 20 kHz), with each component

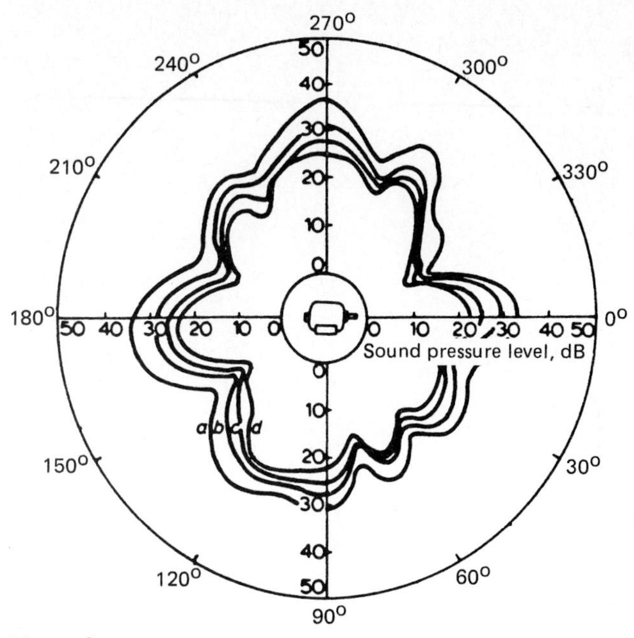


Distance from motor center:

*a* 0.3 m  *c* 0.61 m

*b* 0.58 m  *d* 0.76 m

**FIG. 14.1** Variations of sound-pressure level along four concentric circular paths for a pure-tone component emitted from an electric motor.[1]

weighted by a special A-weighting. The internationally standardized A-weighting is shown in Fig. 14.2. A-weighting takes a frequency of 1000 Hz as the reference and gives positive or negative adjustments to all other frequencies in such a way as to simulate roughly the subjective response of the human ear. The relationship between the A-weighted sound level and the sound-pressure levels of all noise components is given by

$$L_A = 10 \log \sum_{i=1}^{m} 10^{0.1 L_{p,i,A}} \qquad \text{dB(A)} \tag{14.3}$$

where the sum is taken over all the components of the noise and where the subscript $A$ indicates the A-weighted value. The A-weighted sound level is a convenient noise scale for rating noise annoyance and hearing damage. For general noise measurement and control of electric machinery, the mean A-weighted sound level at a reference radius of 3 m and the A-weighted sound-power level are often used in machine specifications.

## A-Weighted Sound-Power Level

The *A-weighted sound-power level* is the overall sound-power level with all individual sound-power levels for the entire audible frequency range (approximately from 20

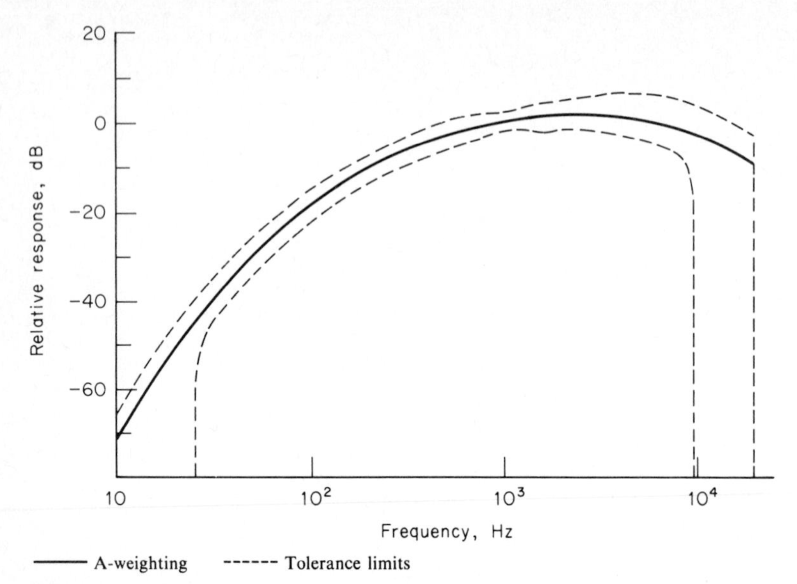


**FIG. 14.2** A-weighting curve.

Hz to 20 kHz) scaled by the A-weighting. The relationship between the A-weighted sound-power level and all individual sound-power levels at all the frequencies concerned is

$$L_{W,A} = 10 \log \sum_{i=1}^{m} 10^{0.1 L_{W,i,A}} \tag{14.4}$$

where the sum is taken over all the frequency components in the audible range and where the subscript $A$ indicates the A-weighted value.

## Vibration Velocity and Amplitude

Vibration of rotating electric machines is expressed by the rms value of the *vibration velocity*, defined as

$$v_{\text{rms}} = \sqrt{\frac{1}{T} \int_{o}^{T} v^2(t)\, dt} \tag{14.5}$$

where $T$ = total measuring period, s
$v(t)$ = instantaneous vibration velocity, mm/s

The vibration of an electric machine usually contains a number of individual frequency components; thus, the rms value of the vibration velocity can also be expressed as

**TABLE 14.2** National Standards on Noise and Vibration of Rotating Electric Machines

| Nation | Identification of standard | Name of standard |
|---|---|---|
| Australia | AS1081 (1975) | Measurement of airborne noise emitted by rotating electric machinery. |
| | AS1359 Part 51 (1978) | General requirements for rotating electric machines: Noise-level limits. |
| Belgium | S 01-200 | Test codes for measurement of airborne noise emitted by rotating electric machinery. |
| Bulgaria | BDS 6011-66 | Measurement of noise emitted by electric rotating machines. |
| Czechoslovakia | CSN 35 0000 | Measurement of noise emitted by electric machines. |
| | CSN 35 0019 Part III | Special testing methods for electric machines: Noise measurement. |
| | CSN 36 1005 | Noise measurement of domestic electric motor–operated appliances. |
| France | S 31-006 (1979) | Test codes for measuring noise emitted by rotating electric machinery. |
| | E 90-310 (1973) | Mechanical vibrations of rotating electric machines with shaft heights between 80 and 400 mm—Evaluation of vibration severity. |
| Germany (DDR) | TGL 50-29034 | Measurement of noise emitted by rotating electric machines, guidelines. |
| | TGL 200-3110 | Electric machines, definitions: Procedures for calculating noise levels. |
| | TGL 22423-2 | Rotating electric machines: Test procedures for measurement of noise. |
| Germany (FRG) | DIN 45635 Part 10 (1974) | Measurement of airborne noise emitted by machines, enveloping measurement method, rotating electric machines. |
| | DIN 45665 (1968) | Vibration of rotating electric machines with shaft heights from 80 to 315 mm. |
| Hungary | MSZ KGST 828 (1977) | Rotating electric machines: Noise-measuring method. |
| | MSZ KGST 1348 (1978) | Rotating electric machines: Permissible noise limits. |
| India | IS: 6098 (1971) | Methods of measurement of the airborne noise emitted by rotating electric machinery. |
| Netherlands | NEN 21 680 (1971) | Test code for measurement of airborne noise emitted by rotating electric machinery. |

**TABLE 14.2** (*Cont.*)

| Nation | Identification of standard | Name of standard |
|---|---|---|
| Japan | JEC-37 (1979) | Induction machines, 12.5 noise-level measurement. |
| | JIS C 0911 | Vibration testing procedures for electric machines and equipment. |
| | JIS C 0912 | Shock testing procedures for electric machines and equipment |
| Poland | PN-81 E-04257 | Rotating electric machines: Determination of noise level. |
| | PN-72 E-06019 | Rotating electric machines: Sound-level limits. |
| | PN-73 E-04255 | Rotating electric machines: Measurement of vibrations. |
| | PN-73 E-06020 | Rotating electric machines: Vibration limits. |
| Romania | STAS 7301-74 | Measurement of noise emitted by electric rotating machines. |
| | STAS 7536-66 | Measurement of vibration from electric rotating machines. |
| Spain | UNE 20121 (1975) | Acceptable noise levels for rotating electric machinery. |
| | UNE 20137 (1978) | Test codes for measuring airborne noise emitted by rotating electric machinery. |
| Great Britain | BEAMA Publ. 225 | BEAMA recommendations for the measurement and classification of acoustic noise from rotating electric machines. |
| | BS 4999 Part 51 (1973) | General requirements for rotating electric machines: Noise levels [Amendments 2328 (1977) and 2659 (1978)]. |
| | BS 4999 Part 50 (1978) | General requirements for rotating electric machines: Mechanical performance—vibration. |
| | BS 5265 Part 1 (1979) | Mechanical balancing of rotating bodies: Recommendations on balance quality of rotating rigid bodies. |
| | BS 5265 Part 2 (1981) | Mechanical balancing of rotating bodies: Recommendations on balance quality of flexible rotors. |
| United States | IEEE Std. 85 (1973) | Test procedure for airborne sound measurements on rotating electric machinery. |
| Soviet Union | GOST 11929 (1981) | Measurement of noise emitted by electric rotating machines. |
| | GOST 16372 (1977) | Rotating electric machines: Noise-level limits. |
| | GOST 160,688-013-71 | Turbogenerators: Vibration limits. |

$$v_{\text{rms}} = \frac{1}{\sqrt{2}} \sqrt{\sum_{i=1}^{n} (2\pi D_i f_i)^2} \tag{14.6}$$

where indices 1 to $n$ refer to all $n$ components, $D_i$ is the vibration amplitude of the $i$th component, and $f_i$ is the frequency of the $i$th component.

For large machines with shaft-center heights exceeding 400 mm, the peak-to-peak vibration displacement in millimeters is used when specifying the vibration limits.[2]

## 14.2 STANDARDS

Table 14.2 lists the names of national standards and test codes directly related to noise and vibration of rotating electric machines. Table 14.3 gives the names of international standards and test codes related to machinery noise and vibration problems. These standards and test codes provide specifications on noise and vibration measurement methods and limits.

A series of standards from the International Organization for Standardization, ISO 3740 to ISO 3746 (see Table 14.3), are applicable to noise test methods for determining the sound-power levels of any type of stationary machinery or equipment. Compared with the noise test methods specified by most national standards in Table 14.2, these ISO standards provide more comprehensive and up-to-date specifications. ISO 3740 presents guidelines for the preparation of noise test codes. Table 14.4 summarizes the applicability of ISO 3741 to 3746, and Table 14.5 gives the uncertainty involved in determining sound-power levels according to the various methods specified. Table 14.6 shows the various factors affecting the selection of an appropriate method.

### Noise-Level Limits[3]

According to BS 4999 Part 51, machines can be divided into three classes in terms of the A-weighted sound-power level. These are (1) normal sound-power class, (2) reduced sound-power class, and (3) especially low sound-power class.

Machines in the normal sound-power class are of the manufacturers' standard design; the output has not been limited, and no special acoustic treatment has been provided to reduce the noise emitted. Table 14.7 gives the limiting A-weighted sound-power-level values for standard machines rated up to 16,000 kW. Limits for machines larger than 16,000 kW should be agreed upon between the manufacturer and the purchaser.

Sound-power-level values for machines in the reduced sound-power class will not exceed the values given in Table 14.7 reduced by 5 dB. Machines in this class are basically of the manufacturers' standard design but may have some modifications (e.g., special fans) to obtain a moderate reduction in noise emission.

Machines in the especially low sound-power class may have special electrical and mechanical design, and the sound-power-level values in this class are a matter for agreement between the manufacturer and the purchaser.

The values in Table 14.7 are the limiting sound-power levels for standard machines (excluding small-power machines and machines for traction vehicles) tested on no-load in a free field over a reflecting plane, which is considered to be its base.

**TABLE 14.3** International Standards Related to Noise and Vibration of Rotating Electric Machines

| Organization | Identification of standard | Name of standard |
|---|---|---|
| International Electrotechnical Commission | IEC 34-9 Part 9 (1972) | Rotating electric machines: Noise limits. |
| | IEC 34-14 Part 14 (1982) | Rotating electric machines: Mechanical vibration of certain machines with shaft heights 56 mm and higher—measurement, evaluation, and limits of the vibration severity. |
| International Organization for Standardization | R 1680 (1970) | Test code for the measurement of the airborne noise emitted by rotating electric machinery. |
| | ISO/DIS 1680/1 Part 1 | Revision of R 1680 (1970), Test Code for the measurement of airborne noise emitted by rotating electric machinery: Engineering method for free-field conditions over a reflecting plane. |
| | ISO/DIS 1680/2 Part 2 | Revision of R 1680 (1970), Test Code for the measurement of airborne noise emitted by rotating electric machinery: Survey method. |
| | ISO 2372 (1974) | Mechanical vibration of machines with operating speeds from 10 to 200 r/s—Basis for specifying evaluation standards. |
| | ISO 2373 (1974) | Mechanical vibration of certain rotating electric machinery with shaft heights between 80 and 400 mm—Measurement and evaluation of the vibration severity. |
| | ISO 3945 (1977) | Mechanical vibration of large rotating machines with speed range from 10 to 200 r/s—Measurement and evaluation of vibration severity *in situ*. |
| | ISO 3740 (1980) | Acoustics: Determination of sound-power level of noise sources—Guidelines for the use of basic standards and for the preparation of noise test codes. |
| | ISO 3741 (1975) | Acoustics: Determination of sound-power sources—Precision methods for broadband sources in reverberation rooms. |
| | ISO 3742 (1975) | Acoustics: Determination of sound-power level of noise sources—Precision methods for discrete-frequency and narrowband sources in reverberation rooms. |
| | ISO 3743 (1976) | Acoustics: Determination of sound-power levels of noise sources—Engineering methods for special reverberation test rooms. |

**TABLE 14.3** (*Cont.*)

| Organization | Identification of standard | Name of standard |
|---|---|---|
| International Organization for Standardization | ISO 3744 (1981) | Acoustics: Determination of sound-power levels of noise sources—Engineering methods for free-field conditions over a reflecting plane. |
| | ISO 3745 (1977) | Acoustics: Determination of sound-power levels of noise sources—Precision methods for anechoic and semianechoic rooms. |
| | ISO 3746 (1979) | Acoustics: Determination of sound-power levels of noise sources—Survey method. |
| | ISO/DIS 3747 | Acoustics: Determination of sound-power levels of noise sources—Survey method using a reference sound source. |
| | ISO/DIS 3748 | Acoustics: Determination of sound-power levels of noise sources—Engineering method for small, nearly omnidirectional sources under free-field conditions over a reflecting plane. |

For all cases in which the A-weighted sound-power level of a machine is greater than 93 dB(A) or one or more tones are prominent, noise measurements should be made in octave bands on frequencies from 63 to 8000 Hz and a correction should be made to the A-weighted sound-power level.[3]

## Vibration Limits

The vibration limits for dc and three-phase ac machines with shaft-center heights between 80 and 400 mm, as specified by BS 4999 Part 50 (1978), are shown in Table 14.8. This table does not apply to electric machines having shaft-center heights below 80 mm or to single-phase ac machines. The recommended limits for the N grade in Table 14.8 apply to normal electric machines. For special applications needing extremely low vibration levels, the vibration limits regarding machine mountings and vibration measuring points should be a matter of agreement between the manufacturer and the user.

For horizontal electric machines with heights exceeding 400 mm, BS 4999 Part 50 gives the vibration limits (see Table 14.9) in terms of the maximum peak-to-peak vibration displacement in millimeters for the radial vibration measured on the bearings of the machine when the completely assembled machine is operating on no-load at rated voltage and rated speed or at the highest speed of its speed range.

Vibration limits for certain machines are also specified in IEC 34-14 (1982), which applies to dc and three-phase ac machines, with shaft-center heights of 56 mm and higher and a rated power up to 50 MW, at nominal speeds from 600 r/min up to and including 3600 r/min.[4]

**TABLE 14.4** International Standards Specifying Various Methods for Determining the Sound-Power Levels of Machines and Equipment

| International Standard no. | Classification of method | Test environment | Volume of source | Character of noise | Sound-power levels obtainable | Optional information available |
|---|---|---|---|---|---|---|
| 3741 | Precision | Reverberation room meeting specified requirements | Preferably less than 1 % of test room volume | Steady, broadband | In one-third-octave or octave bands | A-weighted sound-power level |
| 3742 | Precision | Reverberation room meeting specified requirements | Preferably less than 1 % of test room volume | Steady, discrete-frequency or narrowband | In one-third-octave or octave bands | A-weighted sound-power level |
| 3743 | Engineering | Special reverberation test room | Preferably less than 1 % of test room volume | Steady, broadband, narrowband, discrete-frequency | A-weighted and in octave bands | Other weighted sound-power levels |
| 3744 | Engineering | Outdoors or in large room | Greatest dimension less than 15 m | Any | A-weighted and in one-third-octave or octave bands | Directivity information and sound-pressure levels as a function of time; other weighted sound-power levels |
| 3745 | Precision | Anechoic or semianechoic room | Preferably less than 0.5 % of test room volume | Any | A-weighted and in one-third-octave or octave bands | Directivity information and sound-pressure levels as a function of time; other weighted sound-power levels |
| 3746 | Survey | No special test environment | No restrictions: limited only by available test environment | Any | A-weighted | Sound-pressure levels as a function of time; other weighted sound-power levels |

***Source:*** ISO 3740, *Acoustics: Determination of Sound-Power Level of Noise Sources—Guidelines for the Use of Basic Standards and for the Preparation of Noise Test Codes*, International Organization for Standardization, Geneva, 1980.

**TABLE 14.5** Uncertainty in Determining Sound-Power Levels, Expressed as the Largest Value of the Standard Deviation in Decibels

| International Standard no. | Octave bands, Hz | | | | | |
|---|---|---|---|---|---|---|
| | 125 | 250 | 500 | 1000–4000 | 8000 | A-weighting |
| | One-third-octave bands, Hz | | | | | |
| | 100–160 | 200–315 | 400–630 | 800–5000 | 6300–10,000 | |
| 3741<br>3742 | 3 | 2 | 1.5 | | 3 | — |
| 3743 | 5 | 3 | 2 | | 3 | 2 |
| 3744 | 3 | 2 | | 1.5 | 2.5 | 2 |
| 3745 | 1 (Anechoic room) | 1 | | 0.5 | 1 | — |
| | 1.5 (Semianechoic room) | 1.5 | | 1 | 1.5 | — |
| 3746 | — | — | — | — | — | 5 |

***Source:*** ISO 3740, *Acoustics: Determination of Sound-Power Level of Noise Sources—Guidelines for the Use of Basic Standards and for the Preparation of Noise Test Codes*, International Organization for Standardization, Geneva, 1980.

**TABLE 14.6** Factors Influencing the Choice of the Method

■ Information in accordance with International Standards
□ Optional information

| Factor | | ISO 3741 | ISO 3742 | ISO 3743 | ISO 3744 | ISO 3745 | ISO 3746 |
|---|---|---|---|---|---|---|---|
| Size of source | Large sources—not movable | | | | ■ | | ■ |
| | Small sources—movable | ■ | ■ | ■ | ■ | ■ | ■ |
| Character of noise | Steady—broadband | ■ | | ■ | ■ | ■ | ■ |
| | Steady—narrowband—discrete-frequency | | ■ | ■ | ■ | ■ | ■ |
| | Nonsteady | | | □ | ■ | ■ | ■ |
| Classification of method | Precision | ■ | ■ | | | ■ | |
| | Engineering | | | ■ | ■ | | |
| | Survey | | | | | | ■ |
| Application of data | Noise control work | ■ | ■ | ■ | ■ | ■ | |
| | Type testing | ■ | ■ | ■ | ■ | ■ | |
| | Comparison of machines or equipment: | ■ | ■ | ■ | ■ | ■ | |
| | different types | ■ | ■ | ■ | ■ | ■ | |
| | same type | ■ | ■ | ■ | ■ | ■ | ■ |
| Information obtained | Octave band levels | ■ | ■ | ■ | ■ | ■ | |
| | One-third-octave band levels | ■ | ■ | | ■ | ■ | |
| | A-weighted levels | □ | □ | ■ | ■ | ■ | ■ |
| | Other weightings | | | □ | □ | □ | □ |
| | Directivity information | | | | □ | □ | |
| | Temporal pattern | | | | □ | □ | □ |
| Test environment | Laboratory reverberation rooms | ■ | ■ | | | | |
| | Special reverberation test room | | | ■ | | | |
| | Large rooms, outdoors | | | | ■ | | |
| | Laboratory anechoic rooms | | | | | ■ | |
| | *In situ*, indoors, outdoors | | | | ■ | | ■ |

***Source:*** ISO 3740, *Acoustics: Determination of Sound-Power Level of Noise Sources—Guidelines for the Use of Basic Standards and for the Preparation of Noise Test Codes*, International Organization for Standardization, Geneva, 1980.

## 14.3 NOISE MEASUREMENTS

### Precision Methods in Anechoic and Semianechoic Rooms[5]

For small electric machines, noise measurements can be made in an anechoic room (i.e., in a free field) or in a semianechoic room (i.e., in a free field above a reflecting plane). The quantities to be measured are the sound-pressure levels (A-weighted and in various frequency bands) over a prescribed surface, and the quantities to be calculated are the A-weighted sound-power level, the sound-power level in frequency bands, and the directivity characteristics of the machine.

The sound-power level is calculated from

$$L_W = \bar{L}_p + 10 \log S + C \tag{14.7}$$

**TABLE 14.7** Limiting Mean Sound-Power Level $L_W$ in dB(A) for Airborne Noise Emitted by Rotating Electric Machines*

| | | Protective enclosure | | | | | | | | | | | |
|---|---|---|---|---|---|---|---|---|---|---|---|---|---|
| | | IP 22 | IP 44 | IP 22 | IP 44 | IP 22 | IP 44 | IP 22 | IP 44 | IP 22 | IP 44 | IP 22 | IP 44 |
| | | Rated speed, r/min | | | | | | | | | | | |
| Rating kW (or kVA) | | 960 and below | | 961–1320 | | 1321–1900 | | 1901–2360 | | 2361–3150 | | 3151–3750 | |
| Above | Up to | Sound-power level, dB(A) | | | | | | | | | | | |
| | 1.1 | | 76 | | 79 | | 80 | | 83 | | 84 | | 88 |
| 1.1 | 2.2 | | 79 | | 80 | | 83 | | 87 | | 89 | | 91 |
| 2.2 | 5.5 | | 82 | | 84 | | 87 | | 92 | | 93 | | 95 |
| 5.5 | 11 | 82 | 85 | 85 | 88 | 88 | 91 | 91 | 96 | 94 | 97 | 97 | 100 |
| 11 | 22 | 86 | 89 | 89 | 93 | 92 | 96 | 94 | 98 | 97 | 101 | 100 | 103 |
| 22 | 37 | 89 | 91 | 92 | 95 | 94 | 97 | 96 | 100 | 99 | 103 | 102 | 105 |
| 37 | 55 | 90 | 92 | 94 | 97 | 97 | 99 | 99 | 103 | 101 | 105 | 104 | 107 |
| 55 | 110 | 94 | 96 | 97 | 101 | 100 | 104 | 102 | 105 | 104 | 107 | 106 | 109 |
| 110 | 220 | 98 | 100 | 100 | 104 | 103 | 106 | 105 | 108 | 107 | 110 | 108 | 112 |
| 220 | 630 | 100 | 102 | 104 | 106 | 106 | 109 | 107 | 111 | 108 | 112 | 110 | 114 |
| 630 | 1100 | 102 | 104 | 106 | 107 | 107 | 111 | 108 | 111 | 108 | 112 | 110 | 114 |
| 1100 | 2500 | 105 | 107 | 109 | 110 | 109 | 113 | 109 | 113 | 109 | 113 | 110 | 114 |
| 2500 | 6300 | 106 | 108 | 110 | 112 | 111 | 115 | 111 | 115 | 111 | 115 | 111 | 115 |
| 6300 | 16000 | 108 | 110 | 111 | 113 | 113 | 116 | 113 | 116 | 113 | 116 | 113 | 116 |

*IP22 corresponds generally to drip-proof, ventilated, and similar enclosures. IP44 corresponds generally to totally enclosed fan-cooled, closed air-circuit air-cooled, and similar enclosures (see BS 4999 Part 20). No positive tolerance is allowed on the above sound-power levels.

***Source:*** BS 4999, *General Requirements for Rotating Electric Machines*, pt. 51: *Noise Levels*, BSI, London, 1973.

**TABLE 14.8** Limits of Vibration Severity for Electric Machines Having Shaft-Center Heights between 80 and 400 mm

| Quality grade | Speed, r/min | Maximum allowable rms values of the vibration velocity, in mm/s, for the shaft height $H$, in mm | | |
|---|---|---|---|---|
| | | $80 \leq H \leq 132$ | $132 < H \leq 225$ | $< H \leq 400$ |
| N (normal) | $> 600 \leq 3600$ | 1.8 | 2.8 | 4.5 |
| R (reduced) | $> 600 \leq 1800$<br>$> 1800 \leq 3600$ | 0.71<br>1.12 | 1.12<br>1.8 | 1.8<br>2.8 |
| S (special) | $> 600 \leq 1800$<br>$> 1800 \leq 3600$ | 0.45<br>0.71 | 0.71<br>1.12 | 1.12<br>1.8 |

*Source:* BS 4999, *General Requirements for Rotating Electric Machines*, pt. 50: *Mechanical Performance—Vibration*, BSI, London, 1978.

**TABLE 14.9** Limits of Vibration Amplitude (Peak to Peak, in mm) for Horizontal Machines Having Shaft-Center Heights Exceeding 400 mm

| Speed, r/min | Column* | Column† | Speed, r/min | Column* | Column† |
|---|---|---|---|---|---|
| 3600 | 0.020 | 0.025 | 1100 | 0.051 | 0.064 |
| 3400 | 0.021 | 0.027 | 1000 | 0.055 | 0.069 |
| 3200 | 0.022 | 0.028 | 900 | 0.059 | 0.074 |
| 3000 | 0.024 | 0.030 | 850 | 0.061 | 0.076 |
| 2800 | 0.025 | 0.032 | 800 | 0.063 | 0.078 |
| 2600 | 0.027 | 0.034 | 750 | 0.065 | 0.081 |
| 2400 | 0.029 | 0.036 | 700 | 0.068 | 0.085 |
| 2200 | 0.031 | 0.039 | 650 | 0.071 | 0.089 |
| 2000 | 0.033 | 0.042 | 600 | 0.074 | 0.092 |
| 1800 | 0.036 | 0.045 | 550 | 0.077 | 0.096 |
| 1700 | 0.038 | 0.047 | 500 | 0.080 | 0.100 |
| 1600 | 0.040 | 0.050 | 450 | 0.085 | 0.105 |
| 1500 | 0.041 | 0.052 | 400 | 0.090 | 0.115 |
| 1400 | 0.043 | 0.054 | 350 | 0.094 | 0.120 |
| 1300 | 0.045 | 0.057 | 300 | 0.100 | 0.125 |
| 1200 | 0.048 | 0.060 | 250 | 0.100 | 0.125 |

*Tested at the manufacturer's works.

†Tested at the site where installed.

*Source:* BS 4999, *General Requirements for Rotating Electric Machines*, pt. 50: *Mechanical Performance—Vibration*, BSI, London, 1978.

where $\bar{L}_p$ = mean sound-pressure level over spherical or hemispherical test surface, dB
$S$ = area of test surface ($4\pi r^2$ for spherical, $2\pi r^2$ for hemispherical), $m^2$
$C$ = correction term for influence of atmospheric pressure and temperature, dB

The correction term $C$, which is necessary only if the atmospheric temperature $\theta$ and pressure $p$ differ significantly from 20°C and 980 mbar, respectively, is determined by

$$C = -10\log\left[\left(\frac{423}{400}\right)\left(\frac{273}{273+\theta}\right)^{0.5}\left(\frac{p}{1000}\right)\right] \quad \text{dB} \tag{14.8}$$

where $\theta$ is in degrees Celsius and $p$ is in millibars.

When fixed microphone positions are used and the positions are distributed uniformly over the measuring surface, the mean sound-pressure level is given by

$$\bar{L}_p = 10\log\left(\frac{1}{n}\sum_{i=1}^{n}10^{0.1L_{p,i}}\right) \quad \text{dB} \tag{14.9}$$

where $n$ = number of measuring points over test surface
$L_{p,i}$ = sound-pressure level at $i$th measuring point

Figure 14.3 gives the location of 10 basic measuring points on a hemispherical surface, and Table 14.10 provides the coordinates of 20 basic measuring points on a spherical surface, as recommended by ISO 3745. If the difference in decibels between the highest and lowest sound-pressure levels is numerically greater than half the number of measuring points, additional measuring points are needed.

The radius of the test sphere or hemisphere shall be equal to or greater than twice the major machine dimension, but not less than 1 m. The measuring spherical or hemispherical surface should be centered on the acoustic center of the machine.

The mean sound-pressure level over the test surface can also be measured by moving a microphone at constant speed along a number of prescribed parallel circular paths (Fig. 14.4).

The directivity characteristic of the machine noise is expressed by the directivity index (DI) in decibels, defined as

$$\text{DI} = L_{p,i} - L_{p,\text{sph}} \tag{14.10}$$

where $L_{p,\text{sph}}$ is the mean sound-pressure level over a test spherical surface of radius $r$ m and where $L_{p,i}$ is the sound-pressure level measured at $r$ m from the center of the machine in the particular direction in which the directivity-index value is required.

### Engineering Methods for Free-Field Conditions over a Reflecting Plane[6]

For medium-size and large electric machines, it may be necessary to determine the noise emission in a large room, which provides a free field over a reflecting plane or in a flat outdoor area. In such cases, an environmental correction $K$ (as discussed below) should be considered in determining the sound-power level.

When environmental conditions permit, a hemispherical measurement is preferred. The radius of the hemisphere is preferably 1, 2, 4, 6, 8, 10, 12, 14, or 16 m, and the

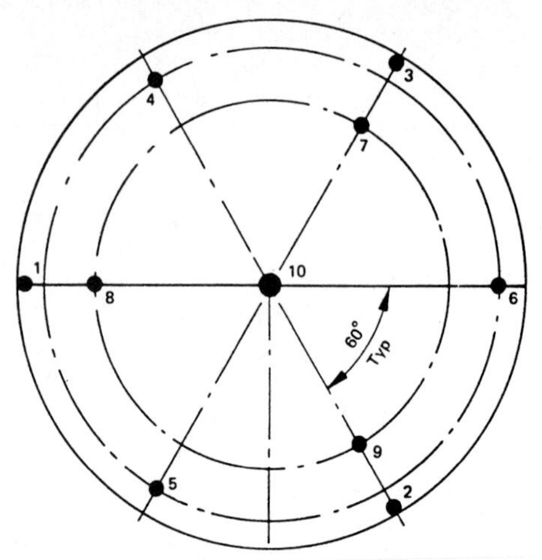


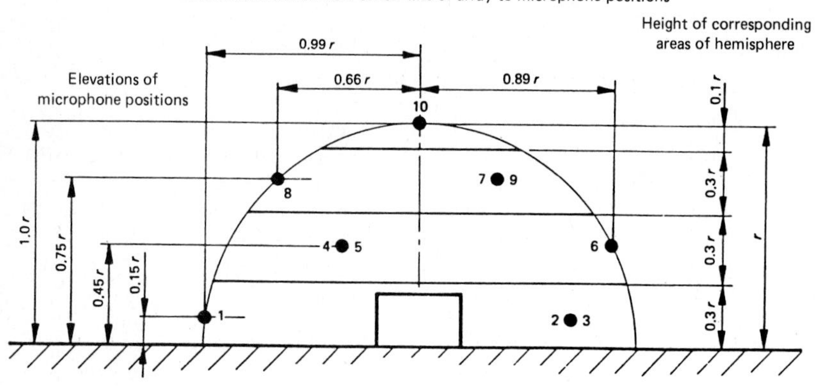


$r$ = radius of hemisphere

**FIG. 14.3** Basic array of microphone positions in a free field over a reflecting plane.[5]

key microphone positions can be the same as those given in Fig. 14.3. If hemispherical measurements cannot be used because of practical constraints, either a rectangular parallelepiped or a conformal surface may be used for the measurement surface (Figs. 14.5 and 14.6).

The sound-power level is calculated from

$$L_W = \bar{L}_p + 10 \log S - K \tag{14.11}$$

where $\bar{L}_p$ = mean sound-pressure level over measurement surface after corrections for background noise, dB

$S$ = surface area, $m^2$

$K$ = mean value of environmental correction over measurement surface, dB

All measured sound-pressure levels should be corrected for background noise according to Table 14.11.

**TABLE 14.10** Recommended Array of Microphone Positions in a Free Field*

| Microphone positions | | | |
|---|---|---|---|
| No. | $x/r$ | $y/r$ | $z/r$ |
| 1 | −0.99 | 0 | 0.15 |
| 2 | 0.50 | −0.86 | 0.15 |
| 3 | 0.50 | 0.86 | 0.15 |
| 4 | −0.45 | 0.77 | 0.45 |
| 5 | −0.45 | 0.77 | 0.45 |
| 6 | 0.89 | 0 | 0.45 |
| 7 | 0.33 | 0.57 | 0.75 |
| 8 | −0.66 | 0 | 0.75 |
| 9 | 0.33 | −0.57 | 0.75 |
| 10 | 0 | 0 | 1.0 |
| 11 | 0.99 | 0 | −0.15 |
| 12 | −0.50 | 0.86 | −0.15 |
| 13 | −0.50 | −0.86 | −0.15 |
| 14 | 0.45 | −0.77 | −0.45 |
| 15 | 0.45 | 0.77 | −0.45 |
| 16 | −0.89 | 0 | −0.45 |
| 17 | −0.33 | −0.57 | −0.75 |
| 18 | 0.66 | 0 | −0.75 |
| 19 | −0.33 | 0.57 | −0.75 |
| 20 | 0 | 0 | −1.0 |

*Gives the cartesian coordinates ($x$, $y$, $z$) with origin at the center of the source. The $z$ axis is chosen perpendicularly upward from a horizontal plane ($z = 0$).

***Source:*** ISO 3745, *Acoustics: Determination of Sound-Power Levels of Noise Sources—Precision Methods for Anechoic and Semianechoic Rooms*, International Organization for Standardization, Geneva, 1977.

For the parallelepiped measuring surface (Fig. 14.5), the surface area is given by

$$S = 4(ab + bc + ca) \tag{14.12}$$

where $a = 0.5l_1 + d$
$b = 0.5l_2 + d$
$c = l_3 + d$
$d =$ measurement distance

and $l_1$, $l_2$, and $l_3$ are the length, width, and height, respectively, of the reference box in Fig. 14.5. The preferred value of $d$ is 1 m or one of the following: 0.25, 0.5, 2, 4, or 8 m.

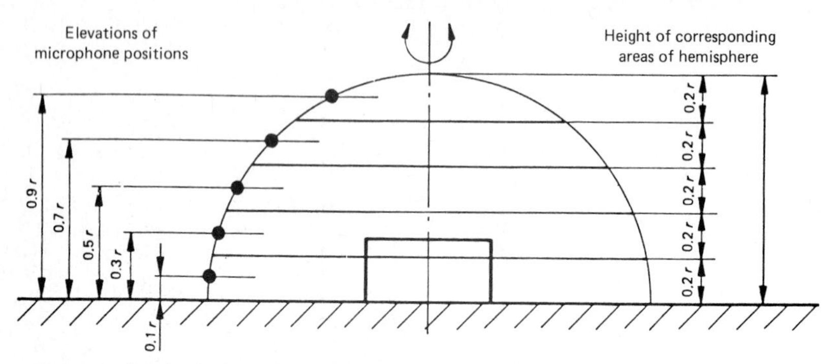


**FIG. 14.4** Coaxial circular paths in parallel planes for microphone traverses in a free field over a reflecting plane.[5]

For the conformal measuring surface (Fig. 14.6), the surface area is given approximately by

$$S = 4(ab + bc + ca) \times \frac{a+b+c}{a+b+c+2d} \tag{14.13}$$

where $a$, $b$, $c$, and $d$ are the same as for the rectangular parallelepiped.

The calculation of $K$ in Eq. (14.11) is performed by one of the following two methods:

**1.** Measure the sound-power level of a reference sound source in the test environment. Place the reference source essentially in the same position as that of the machine under test, and use the same measurement surface as that used for measuring the machine. The environmental correction is given by

$$K = L_W - L_{WR} \tag{14.14}$$

where $L_W$ is the calculated A-weighted or band sound-power level of the reference sound source (using the above procedures) with $K = 0$ in Eq. (14.11) and where $L_{WR}$ is the nameplate A-weighted or band sound-power level of the reference sound source, which should have characteristics that meet the requirements of ISO/DIS 6926.[7] If the value of $K$ is greater than 2 dB, either a smaller measuring surface or a better test environment is required.

**2.** Determine the environmental correction from

$$K = 10 \log \left(1 + \frac{4}{A/S}\right) \tag{14.15}$$

where $A$ = total sound-absorption area of test room, $m^2$
$S$ = area of measurement surface

Equation (14.15) is plotted in Fig. (14.7). The value of $A$ may be estimated by

$$A = \alpha S_V \tag{14.16}$$

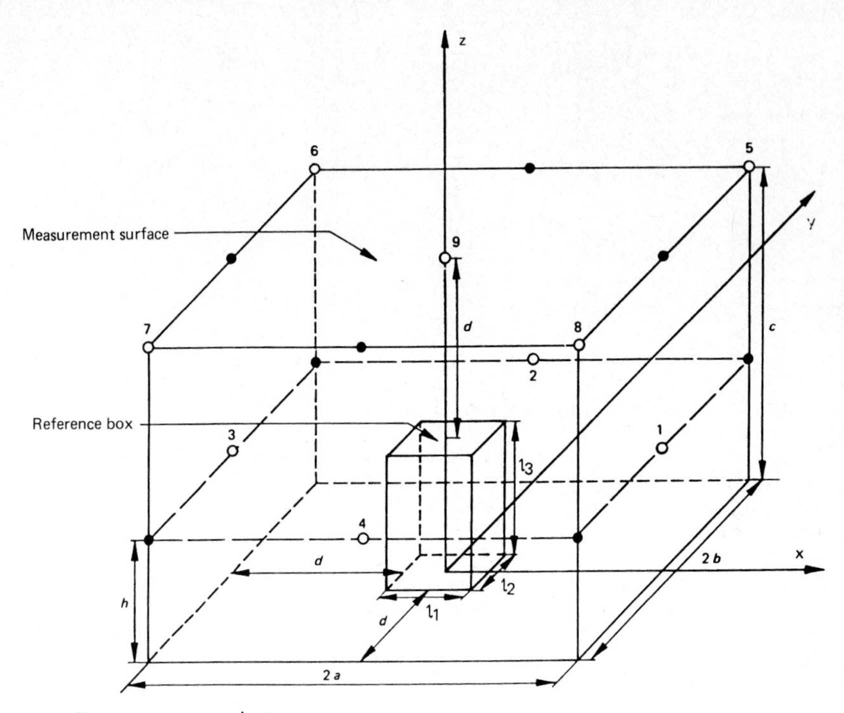


○ Key measurement points

● Additional measurement points

Coordinates of Key Measurement Points

| No. | $x$ | $y$ | $z$ |
|---|---|---|---|
| 1 | $a$ | 0 | $h$ |
| 2 | 0 | $b$ | $h$ |
| 3 | $-a$ | 0 | $h$ |
| 4 | 0 | $-b$ | $h$ |
| 5 | $a$ | $b$ | $c$ |
| 6 | $-a$ | $b$ | $c$ |
| 7 | $-a$ | $-b$ | $c$ |
| 8 | $a$ | $-b$ | $c$ |
| 9 | 0 | 0 | $c$ |

**FIG.14.5** Microphone array on the parallelepiped.[6]

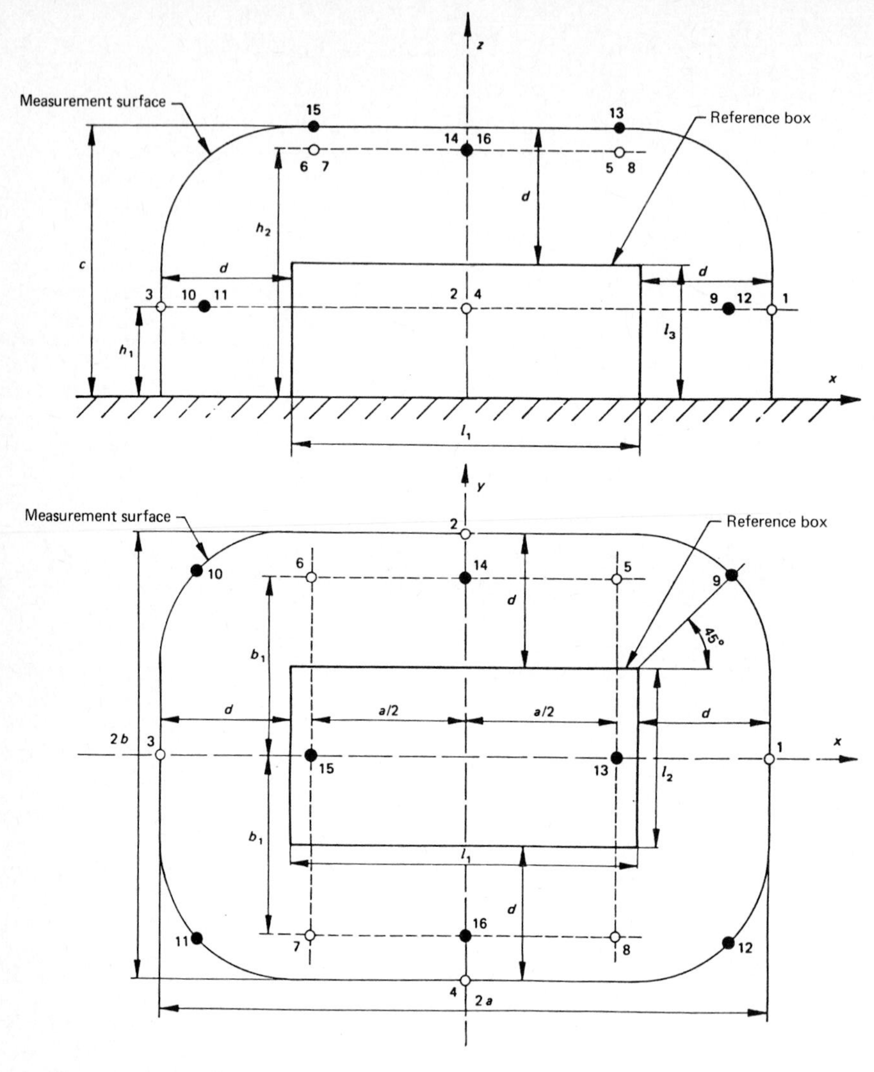


○ Key measurement points

● Additional measurement points

**FIG. 14.6** Microphone array on the conformal surface.[6]

**TABLE 14.11** Correction for Background Noise

| Difference between sound-pressure level measured with sound source operating and background sound-pressure level alone, dB | Correction to be subtracted from sound-pressure level measured with sound source operating to obtain sound-pressure level due to sound source alone, dB |
|---|---|
| 3 | 3 |
| 4 | 2 |
| 5 | 2 |
| 6 | 1.3 |
| 7 | 1.0 |
| 8 | 0.8 |
| 9 | 0.6 |
| 10 | 0.4 |
| 11 | 0.3 |
| 12 | 0.3 |
| 13 | 0.2 |
| 14 | 0.2 |
| 15 | 0.1 |

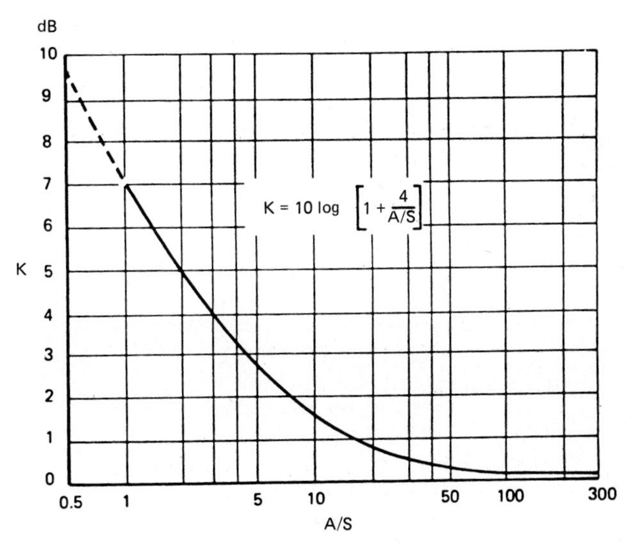


**FIG. 14.7** Environmental correction $K$ in decibels.[6]

where $S_V$ = total area of test room (walls, ceiling, floor), $m^2$
$\alpha$ = approximate mean acoustic absorption coefficient (Table 14.12)

Alternatively, $A$ may be determined by measuring the reverberant time of the test room:

$$A = 0.16\frac{V}{T} \tag{14.17}$$

where $V$ = volume of test room, $m^3$
$T$ = reverberation time, s

To satisfy the free-field requirements specified by ISO 3744, the ratio $A/S$ should exceed 6 (i.e., $A/S > 6$). If this cannot be satisfied, choose a smaller measuring surface and take care about the near-field error. Alternatively, introduce additional sound-absorptive materials into the test room to increase the $A$ value.

## Survey Method[8]

This method may be applied *in situ* to electric machines that cannot be moved to a special test environment.

Figures 14.8 and 14.9 show the measurement surfaces and positions specified by ISO 3746. The radius of the hemispherical surface in Fig. 14.8 is preferably equal

**TABLE 14.12** Approximate Values of the Mean Acoustic Absorption Coefficient $\alpha$

| $\alpha$ | Description of room |
|---|---|
| 0.05 | Nearly empty room with smooth hard walls made of concrete, brick, plaster, or tile |
| 0.1 | Partly empty room; room with smooth walls |
| 0.15 | Room with furniture; rectangular machinery room; rectangular industrial room |
| 0.2 | Irregularly shaped room with furniture; irregularly shaped machinery room or industrial room |
| 0.25 | Room with upholstered furniture; machinery or industrial room with a small amount of acoustic material (e.g., partially absorptive ceiling) on ceiling or walls |
| 0.35 | Room with acoustic materials on both ceiling and walls |
| 0.5 | Room with large amounts of acoustic materials on ceiling and walls |

***Source:*** ISO 3746, *Acoustics: Determination of Sound-Power Levels of Noise Sources—Survey Method*, International Organization for Standardization, Geneva, 1979.

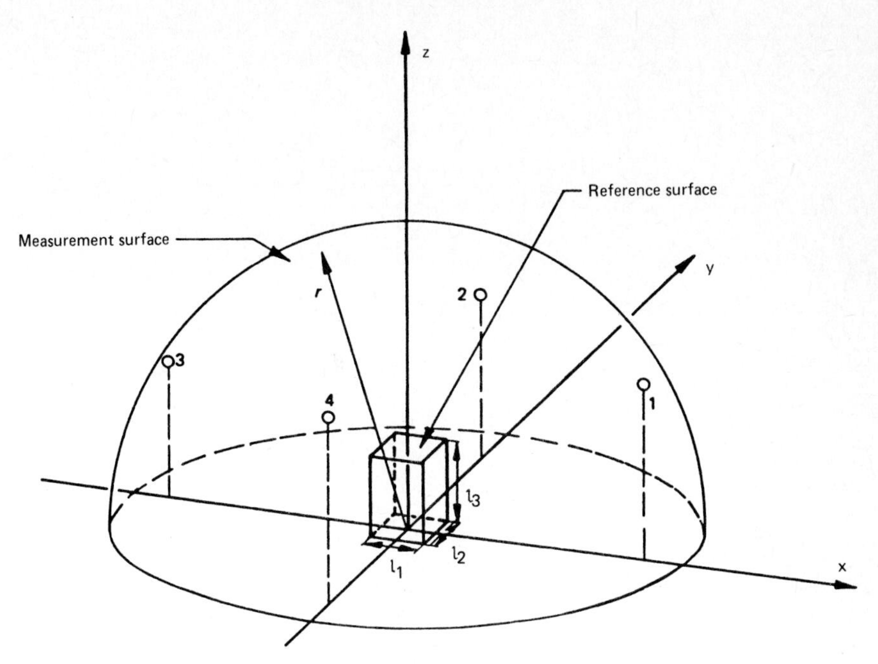


○ Measurement positions

Coordinates of Microphone Positions in Terms of Distances from Center of Hemisphere along Three Mutually Perpendicular Axes ($x$, $y$, $z$)

| Microphone number | Microphone at height $z = 0.6r$ | | | Microphone just above reflecting plane | | |
|---|---|---|---|---|---|---|
| | $x/r$ | $y/r$ | $z/r$ | $x/r$ | $y/r$ | $z$ |
| 1 | 0.8 | 0.0 | 0.6 | 1.0 | 0.0 | < 0.05 m |
| 2 | 0.0 | 0.8 | 0.6 | 0.0 | 1.0 | < 0.05 m |
| 3 | −0.8 | 0.0 | 0.6 | −1.0 | 0.0 | < 0.05 m |
| 4 | 0.0 | −0.8 | 0.6 | 0.0 | −1.0 | < 0.05 m |

**FIG. 14.8** Microphone array on the hemisphere.[8]

to 1, 2, 4, 6, 8, 10, 12, 14, or 16 m. The measurement distance $d$ in Fig. 14.9 is normally 1 m and shall not be less than 0.15 m.

The A-weighted sound-power level is then given by

$$L_{W,A} = \bar{L}_A + 10 \log S - K \tag{14.18}$$

where $\bar{L}_A$ = mean surface A-weighted sound level corrected for background noise, dB

$S$ = measuring surface area, $m^2$

$K$ = environmental correction to account for influence of reflected sound, dB

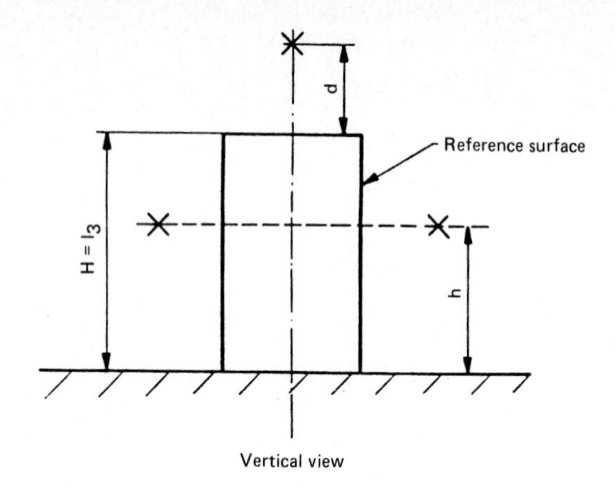


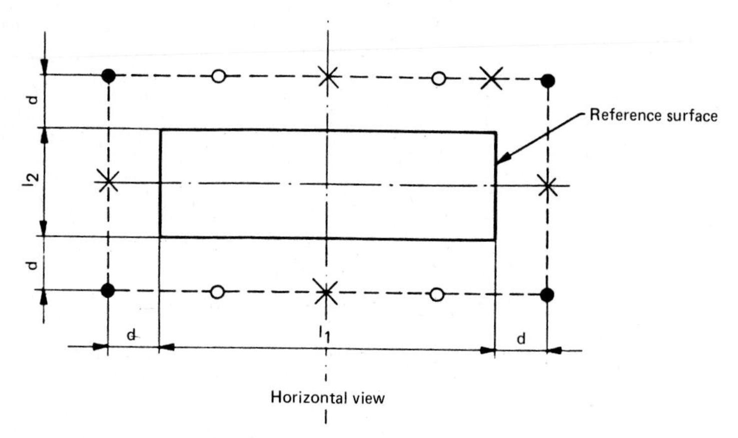


X Key measurement positions

● Corner measurement positions for sources with $l_1$ or $l_2 > 1$ m

○ Other intermediate measurement positions equally spaced between key and corner positions at intervals $\leq 2$ m for $d \leq 1$ m and $\leq 2d$ m for $d > 1$ m

For $H \leq 2.5$ m, $h = (H + d)/2$.
The minimum height of the microphone above the reflecting plane shall be 0.15 m

For $H > 2.5$ m, take instead of $h$ two heights:

$h_1 = (H + d)/2$ (four key microphone positions, as shown)

$h_2 = (H + d)$ (four microphone positions in the corners)

**FIG. 14.9** Location of measurement positions on the parallelepiped.[8]

For the determination of $K$ values, see the previous subsection. To obtain the accuracy specified by ISO 3746, $K$ should be equal to or less than 7. If this requirement cannot be satisfied, use a new, smaller measuring surface or introduce additional sound-absorptive materials into the test room.

## Nominally Identical Small Machines[9]

A considerable variation in noise data between the members of a group of nominally identical small electric machines made from the same production line has been observed. The range of sound-pressure-level data taken at the same points under the same test conditions for mass-produced, nominally identical small machines may exceed 20 dB (Fig. 14.10). It is therefore essential to use statistical techniques for the presentation of the noise data for nominally identical electric machines.

For mass-produced small electric machines, the limited number of machines randomly chosen for noise measurements from a given production line should be considered as a sample from a large population. If $J$ nominally identical machines are each tested at $I$ points on the same measuring surface and $N$ observations are

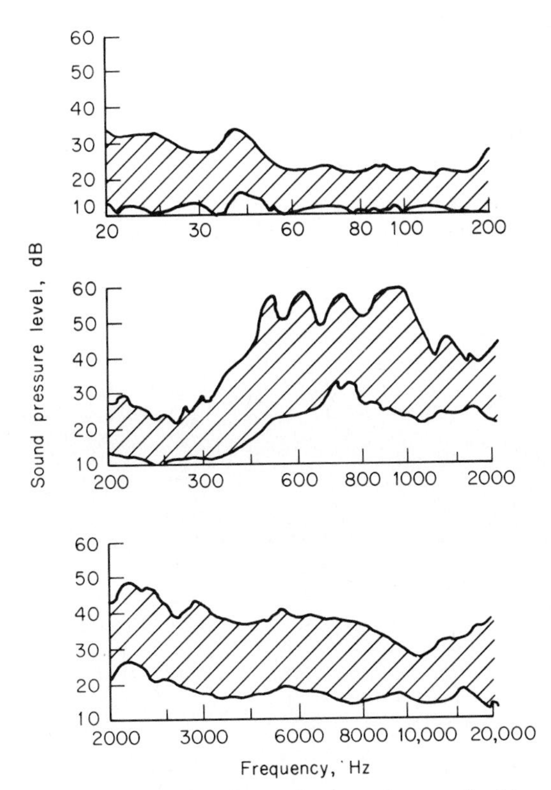


**FIG. 14.10** Variation of narrowband sound-pressure level between 10 nominally identical motors with ball bearings at same point in acoustic field (passband 6 percent of center frequency).[9]

made at each point for each machine, the standard deviation of the total error in $p_{\text{av}}^2/p_{\text{ref}}^2$ referred to a large population is given by

$$\sigma_{\text{tot}} = \left( \frac{\sigma_A^2}{I} + \frac{\sigma_B^2}{J} + \frac{\sigma_{AB}^2}{IJ} + \frac{\sigma_e^2}{IJN} \right)^{1/2} \tag{14.19}$$

where $\sigma_A^2$ = variance for effect of points (factor A)
$\sigma_B^2$ = variance for effect of machines (factor B)
$\sigma_{AB}^2$ = variance for interaction of points and machines
$\sigma_e^2$ = variance for residual errors

For the upper limit of $\sigma_{\text{tot}}$ at a given confidence level, see Ref. 9.

The upper limit of the mean sound-pressure level in a frequency band is given by

$$\bar{L}_{p,u} = 10 \log \left[ \frac{(p^2)_{\text{av,measured}}}{p_{\text{ref}}^2} + \gamma \sigma_{\text{tot},u} \right] \tag{14.20}$$

where $(p^2)_{\text{av,measured}}/p_{\text{ref}}^2$ is the mean $p^2/p_{\text{ref}}^2$ value for all the sound-pressure levels taken, suffix $u$ indicates the upper limit, and $\gamma$ is any positive value. The $\bar{L}_{p,u}$ value is at a confidence level of $P_\gamma$ times the confidence level for $\sigma_{\text{tot},u}$, where $P_\gamma$ is the probability that the standard normal variable $Z$ is less than $\gamma$ [i.e., $P_\gamma = P\ (-\infty \leq Z \leq \gamma)$].

### Measuring On-Load Noise

Care should be taken when measuring the noise from an electric machine operating on load since its load machine usually also emits noise, which can mix with the noise from the test machine.

For machines tested in anechoic and reverberant rooms, the load machine should be placed outside the special test room. For machines tested according to the methods described earlier in this section, the distance of the microphone $d$ from the reference surface of the machine (see Figs. 14.5, 14.6, and 14.9) shall be 0.15 m in order to minimize the noise contribution by the load machine to all measurements. Such a short distance may introduce near-field error in the sound-power-level results. For estimating the near-field error due to magnetic sources, see Ref. 65.

## *14.4 VIBRATION MEASUREMENTS*

### Machine Mounting

For electric machines having shaft-center heights not exceeding 400 mm, the machine shall be installed in a state of "free suspension." This is achieved by suspending the machine on a spring or by mounting it on an elastic support.

The natural frequency of the suspension/machine system, in the six possible degrees of freedom, shall be lower than a quarter of the frequency corresponding to the lowest nominal operational speed. This condition is often fulfilled for the perpendicular direction when the spring travel during suspension or mounting reaches at least the values given in Fig. 14.11 and, in the case of an elastic support, when the elastic baseplate is compressed to a maximum of half its original thickness.

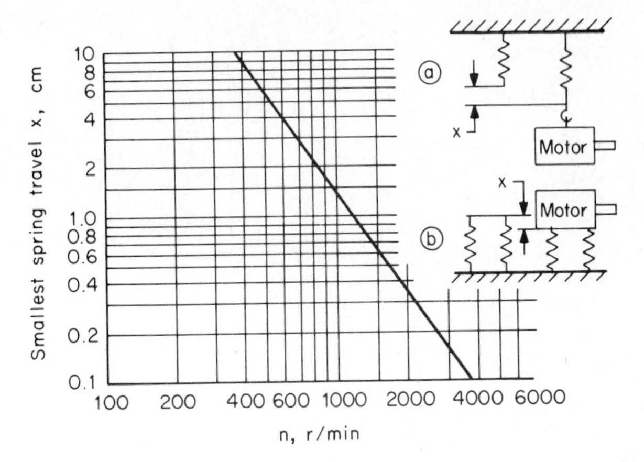


*a* Suspended on a spring

*b* Mounted on an elastic baseplate

**FIG. 14.11** The smallest spring travel for a small spring constant.[10]

Vibrations should be measured when the machine is fed at rated voltage and frequency on no-load. For machines having several operational speeds, the test should be carried out at each operational speed.

For machines having shaft-center heights greater than 400 mm, vibrations shall be measured *in situ*.

### Measuring Points

BS 4999 Part 50 recommends that vibration measurements be made on the bearings, in the neighborhood of the shaft, in three perpendicular directions (Fig. 14.12), the machine operating only in its normal position (shaft horizontal or vertical). For vibration-measuring instrumentation, see Ref. 11.

## 14.5 NOISE AND VIBRATION SOURCES

### Electromagnetic Forces on Stator

The magnetic flux waves passing through the airgap of an electric machine produce radial, axial, and tangential forces acting on the stator and rotor. These forces excite the machine to vibrate and thus to emit noise. The radial forces are usually the most important causes of airborne magnetic noise. The major electromagnetic forces include the following:

- For induction machines, there are radial forces introduced by slotting and by homopolar and other flux waves.[12,13] Homopolar flux passes through the airgap from the rotor to the stator and completes its path through the stator frame, the end shields, the shaft, and back to the rotor.
- Rotor eccentricity gives rise to a series of additional low pole-pair (i.e., mode-

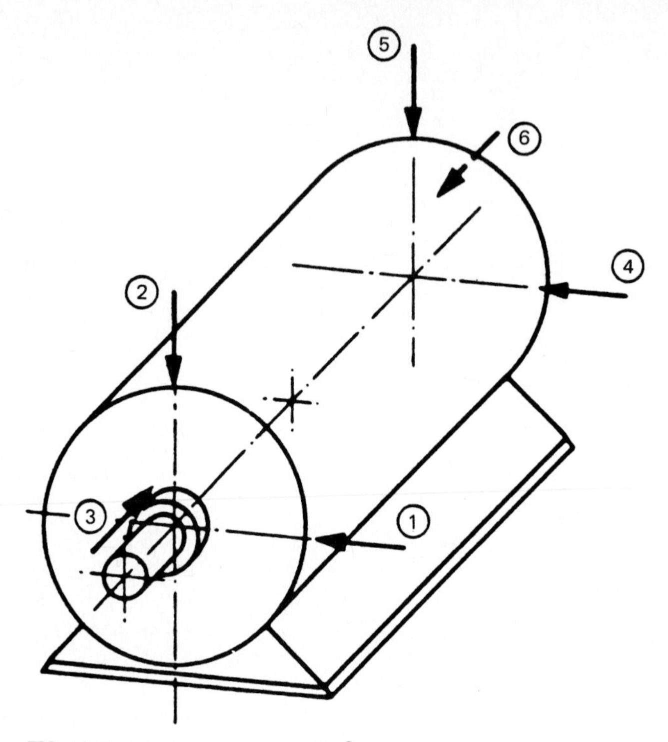


**FIG. 14.12** Vibration measuring points.[2]

number) radial forces (see Refs. 12, 14, and 15) and hence to a higher noise emission (Fig. 14.13). Some of these radial forces are at frequencies of

$$f = (Z_{rt} \pm i_{ec}) \frac{f_1(1-s)}{p} \pm 2f_1 \tag{14.21}$$

where $Z_{rt}$ = rotor slot number
$i_{ec}$ = any integer
$f_1$ = supply frequency
$s$ = slip value
$p$ = pole-pair number

When an ac machine is fed from a nonsinusoidal supply, additional radial and tangential forces are produced by the interaction between various flux waves as a result of current harmonics.[16–18] Some of the radial forces in a three-phase induction machine fed from a nonsinusoidal supply are at frequencies of[17]

$$f = Z_{rt} \frac{f_1(1-s)}{p} \pm q' f_1 \qquad 14.22$$

where $$q' = q_1 \pm q_2$$

where $q_1$ and $q_2$ are integers representing the current harmonic orders (Tables 14.13 and 14.14).

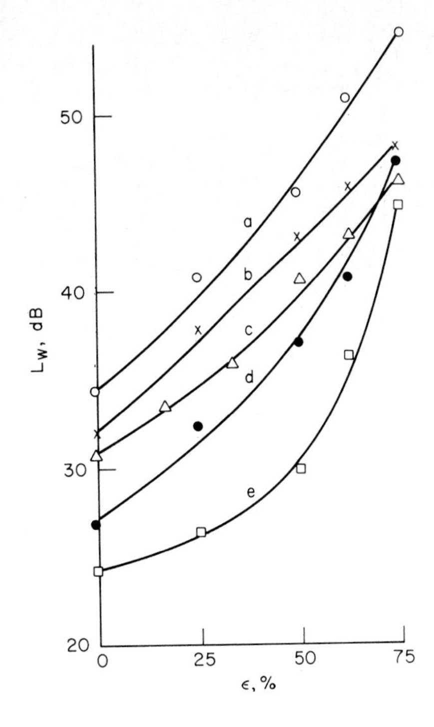


Passband is 6 percent of center frequency. All winding sections are in series, star-connected, 97.5 percent confidence level, with center frequencies of:

***a*** 1200 Hz ***b*** 1000 Hz ***c*** 1300 Hz
***d*** 1100 Hz ***e*** 1400 Hz

**FIG. 14.13** Variation of narrowband sound-power level $L_W$ with relative eccentricity $\epsilon$ for a three-phase 1/2-hp motor.[14]

For determining the accurate radial forces acting on the stator, it is necessary to use finite-element methods, taking into account iron nonlinearity, the actual operating conditions, and the structural details.

## Electromagnetic Forces on Rotor

The electromagnetic forces acting on the rotor consist of the axial forces, the tangential forces (giving rise to the operating torque), and the radial forces. The tangential electromagnetic forces during a start-up or an abnormal operation can cause substantial torsional vibration.[19,20]

The sum of the electromagnetic radial forces acting on the whole rotor periphery of an electric machine is usually zero. However, owing to rotor eccentricity and certain combinations of harmonic waves, the sum of certain radial forces acting on the rotor is nonzero, resulting in an unbalanced magnetic pull.[21]

Chapman showed that two flux waves can produce an unbalanced magnetic pull if the two waves differ by two poles around the whole rotor periphery.[22] In Fig. 14.14 the curve $\phi_4$ represents a flux wave having four poles around the whole periphery of the rotor, and the curve $\phi_6$ represents a flux wave similarly having six poles at a certain instant; $\phi_T$ represents the resultant of these two flux waves at that instant. Bearing in mind that the radial force is proportional to the square of the flux density value at any point, the corresponding radial forces around the periphery of the rotor would be as shown in Fig. 14.15, in which a "strong zone" is immediately opposite a "weak zone." This leads to an unbalanced magnetic pull, which can be represented by a radial force vector.

**TABLE 14.13** Frequencies and Origins of Additional Radial Forces (with Odd Harmonics Only)

| Frequency, rad/s | Origins | | | |
|---|---|---|---|---|
| | Funda-mental | Fifth harmonic | Seventh harmonic | Eleventh harmonic |
| $i'_{rt}Z_{rt}\omega_{rt} \pm 4\omega_1$* | √ | √ | | |
| $i'_{rt}Z_{rt}\omega_{rt} \pm 4\omega_1$ | | | √ | √ |
| $i'_{rt}Z_{rt}\omega_{rt} \pm 6\omega_1$ | √ | √ | | |
| $i'_{rt}Z_{rt}\omega_{rt} \pm 6\omega_1$ | √ | | √ | |
| $i'_{rt}Z_{rt}\omega_{rt} \pm 6\omega_1$ | | √ | | √ |
| $i'_{rt}Z_{rt}\omega_{rt} \pm 8\omega_1$ | √ | | √ | |
| $i'_{rt}Z_{rt}\omega_{rt} \pm 10\omega_1$ | √ | | | √ |
| $i'_{rt}Z_{rt}\omega_{rt} \pm 10\omega_1$ | | √ | | |
| $i'_{rt}Z_{rt}\omega_{rt} \pm 12\omega_1$ | | √ | √ | |
| $i'_{rt}Z_{rt}\omega_{rt} \pm 14\omega_1$ | | | √ | |
| $i'_{rt}Z_{rt}\omega_{rt} \pm 16\omega_1$ | | √ | | √ |
| $i'_{rt}Z_{rt}\omega_{rt} \pm 18\omega_1$ | | | √ | √ |

*$i'_{rt}$ = any integer, including zero
$Z_{rt}$ = rotor slot number
$\omega_{rt}$ = rotor angular frequency
$\omega_1$ = fundamental angular frequency

***Source:*** S. J. Yang and P. L. Timar, "The Effect of Harmonic Currents on the Noise of a Three-Phase Induction Motor," *IEEE Transactions*, vol. PAS-99, no. 1, 1980, pp. 307–310.

The unbalanced magnetic force vector rotates in space at a speed given by[22]

$$n = \tfrac{1}{2}(n_2 y - n_1 x) \tag{14.23}$$

where $n_1$ = rotating speed of $x$-pole flux wave, r/s
$n_2$ = speed of $y$-pole flux wave

If this force-vector speed is close to the rotor's critical speed, severe vibration will occur.

The unbalanced magnetic pull due to a static rotor eccentricity consists of a time-varying component and a static component. The static component of the unbalanced magnetic pull $F$ in the direction of the minimum airgap is approximately given by[23]

$$F = Q\frac{\pi D l}{4\mu_o}B_m^2\varepsilon \quad \text{N} \tag{14.24}$$

where $\varepsilon$ = relative eccentricity (i.e., ratio of eccentricity to mean airgap)
$B_m$ = maximum flux density, T
$D$ = rotor diameter, m
$l$ = rotor length, m
$\mu_o$ = permeability of free space ($4\pi \times 10^{-7}$ H/m)
$Q$ = pole number correction factor, as given in the following table[23]

| Pole number | 2 | 4 | 6 | 8 |
|---|---|---|---|---|
| $Q$ | 0.25 | 0.712 | 0.86 | 0.896 |

**TABLE 14.14** Frequencies and Origins of Additional Radial Forces (with Odd and Even Harmonics)

| Frequency, rad/s | Origins | | | | |
|---|---|---|---|---|---|
| | Fundamental | Second Harmonic | Fourth Harmonic | Fifth Harmonic | Seventh Harmonic |
| $i'_{rt}Z_{rt}\omega_{rt} \pm \omega_1$* | √ | √ | | | |
| $i'_{rt}Z_{rt}\omega_{rt} \pm 3\omega_1$ | √ | √ | | | |
| $i'_{rt}Z_{rt}\omega_{rt} \pm 3\omega_1$ | √ | | √ | | |
| $i'_{rt}Z_{rt}\omega_{rt} \pm 3\omega_1$ | | √ | | √ | |
| $i'_{rt}Z_{rt}\omega_{rt} \pm 3\omega_1$ | | | √ | | √ |
| $i'_{rt}Z_{rt}\omega_{rt} \pm 4\omega_1$ | √ | | | √ | |
| $i'_{rt}Z_{rt}\omega_{rt} \pm 5\omega_1$ | √ | | √ | | |
| $i'_{rt}Z_{rt}\omega_{rt} \pm 5\omega_1$ | | | √ | | √ |
| $i'_{rt}Z_{rt}\omega_{rt} \pm 6\omega_1$ | √ | | | √ | |
| $i'_{rt}Z_{rt}\omega_{rt} \pm 6\omega_1$ | √ | | | | √ |
| $i'_{rt}Z_{rt}\omega_{rt} \pm 6\omega_1$ | | √ | √ | | |
| $i'_{rt}Z_{rt}\omega_{rt} \pm 7\omega_1$ | | √ | | √ | |
| $i'_{rt}Z_{rt}\omega_{rt} \pm 8\omega_1$ | √ | | | | √ |
| $i'_{rt}Z_{rt}\omega_{rt} \pm 9\omega_1$ | √ | | | | √ |
| $i'_{rt}Z_{rt}\omega_{rt} \pm 9\omega_1$ | | | √ | √ | |

*Same symbols as in Table 14.13.

***Source:*** S. J. Yang and P. L. Timar, "The Effect of Harmonic Currents on the Noise of a Three-Phase Induction Motor," *IEEE Transactions*, vol. PAS-99, no. 1, 1980, pp. 307–310.

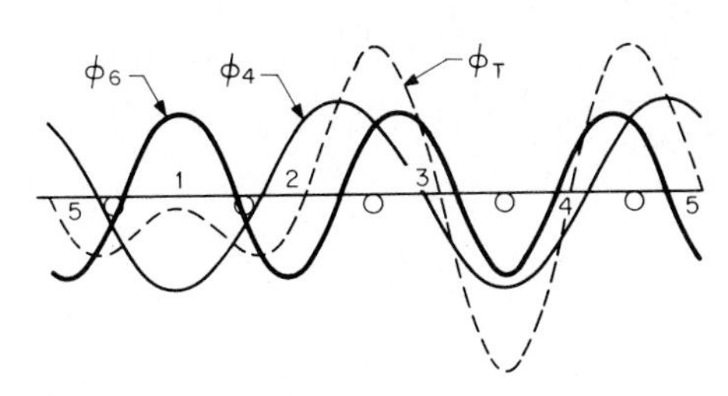


$\phi_4$ Four-pole flux density wave

$\phi_6$ Six-pole flux density wave

$\phi_T$ Resultant of $\phi_4$ and $\phi_6$

**FIG. 14.14** Summation of two flux density waves.[22]

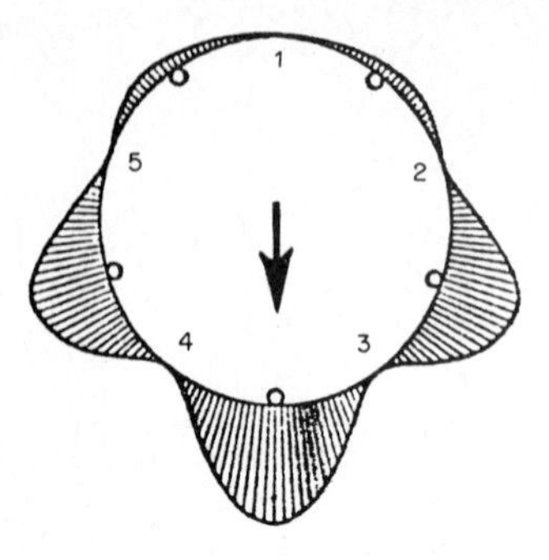


**FIG. 14.15** Radial forces around rotor periphery.[22]

Equation (14.24) shows that the static component is proportional to the relative eccentricity; this has been demonstrated to be true for relative eccentricity values up to about 10 percent.[24] This static component alters the effective shaft stiffness and hence the rotor's critical speed (see Sec. 14.6).

Taking into account the effect of homopolar flux, for a two-pole machine having a static rotor eccentricity $\varepsilon$, the approximate unbalanced magnetic pull $F_x$ in the direction of the minimum airgap is given by[25]

$$F_x = \frac{1}{4}\frac{\pi Dl}{8\mu_o}B_m^2\varepsilon[(1+Q')+Q'\cos 2\omega_1 t] \tag{14.25}$$

where $\omega_1$ is the angular supply frequency, $Q'$ is a factor to allow for the magnetic reluctance of the return path of the homopolar flux (the maximum value of $Q'$ is 1 when the reluctance is assumed to be zero), and the other symbols are the same as those for Eq. (14.24).

The unbalanced magnetic pull $F_y$ at right angles to the direction of the minimum airgap is expressed by[25]

$$F_y = \frac{1}{4}\frac{\pi Dl}{8\mu_o}B_m^2 Q'\sin 2\omega_1 t \tag{14.26}$$

Equations (14.24), (14.25), and (14.26) are based on the assumption that the center of the rotor is stationary.

If the rotor has a dynamic rotor eccentricity of $\varepsilon_d$ and the stator and rotor are concentric at rest (i.e., the center of the rotor is moving in a circular path), the unbalanced magnetic pull will be speed-dependent. For a two-pole induction machine operating with a dynamic rotor eccentricity $\varepsilon_d$, the unbalanced magnetic pull in the vertically downward direction is given by[25]

$$F_{\text{vert}} = \frac{1}{4}\frac{\pi Dl}{8\mu_o}B_m^2\varepsilon_d\sqrt{1+2Q'+2Q'^2+2(1+Q')Q'\cos 2s\omega_1 t}\cos[(1-s)\omega_1 t+\psi] \tag{14.27}$$

where $s$ is the slip value and the other symbols are the same as for Eq. (14.25). This expression shows that the vibration of the rotor bearing's pedestal should contain a component at twice the slip frequency. For the effects on the unbalanced magnetic pull of the magnetic reluctance of the homopolar flux return paths, Ref. 26 is recommended.

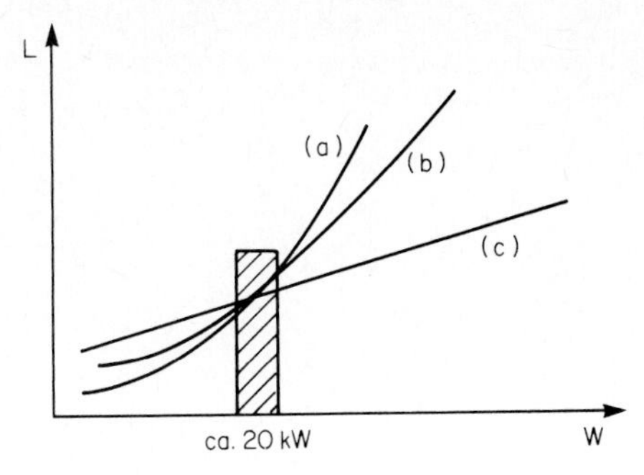


*L* Sound level
*W* Power rating
(*a*) Aerodynamic noise
(*b*) Magnetic noise
(*c*) Mechanical noise (bearings)

**FIG. 14.16** Relative importance of three noise sources.[27]

## Mechanical Sources

The sources of a mechanical nature are: (1) bearings, (2) brush-commutator and brush–slip-ring assemblies, and (3) rotor unbalance.

For small electric machines up to about 20 kW, bearing noise can be predominant, compared with noise of a magnetic or aerodynamic origin (Fig.14.16). Most large and medium-size electric machines have sleeve bearings, which support the rotor on a thin oil film. Since there is no direct metal contact under normal operation, sleeve bearings are usually quiet. However, the ball and roller bearings used in small electric machines generate considerable noise and vibration (Table 14.15). Figure 14.17 gives the ball-bearing excitation frequencies for four types of defect.

The brush-commutator and brush–slip-ring assemblies used in electric machines generate noise and vibration through the sliding contact of current-carrying brushes against commutator segments or slip rings. The noise generated by brush–slip-ring assemblies is usually not significant. However, the brush-commutator assembly can

**TABLE 14.15** Variation of Mean Sound Level A [dB(A)] Referred to 3-m Radius between Nominally Identical Machines* (at 97.5 Percent Confidence Level)

| Type of bearing | No. of motor | | | | | | | | | |
|---|---|---|---|---|---|---|---|---|---|---|
| | 1 | 2 | 3 | 4 | 5 | 6 | 7 | 8 | 9 | 10 |
| Sleeve | 26.1 | 29.9 | 27.9 | 24.2 | 31.3 | 28.7 | 24.6 | 27.8 | 28.7 | 29.9 |
| Ball | 31.8 | 37.5 | 34.2 | 35.1 | 32.4 | 37.1 | 40.4 | 31.5 | 35.9 | 36.8 |

*Three-phase 1/3-hp cage-rotor induction motors.

*Source:* A. J. Ellison and S. J. Yang, "Acoustic-Noise Measurements on Nominally Identical Small Electric Machines," *Proc. IEE*, vol. 117, no. 3, 1970, pp. 555–560.

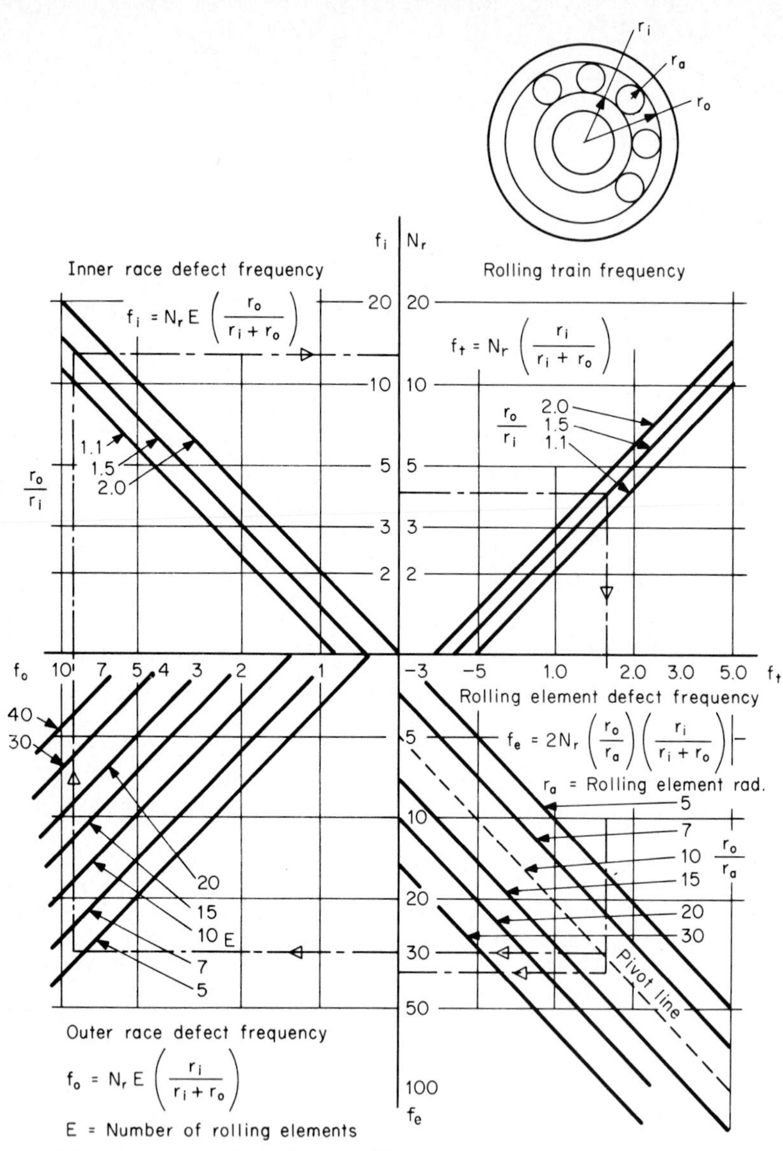


**FIG. 14.17** Ball-bearing excitation frequences.[28] $N_r$ may represent Hz or r/s.

be an important noise and vibration source for universal machines and dc machines. The brush vibrations are at the following frequencies:

$$f_b = iZ_c n_{rt} \tag{14.28}$$

where $i$ = any integer
$Z_c$ = number of commutator segments
$n_{rt}$ = machine speed, r/s

Rotor unbalance affects the dynamic response of the rotor and causes dynamic rotor eccentricity. The rotor-unbalance vibration frequency $f_u$ is at the rotational frequency, i.e.,

$$f_u = n_{rt} \tag{14.29}$$

where $n_{rt}$ = machine speed, r/s

Völler found that for a four-pole 3-kW motor, the vibration velocity at 25 Hz increased from 50 to 900 $\mu$m/s when the rotor unbalance was increased from 0 to 30 gcm.[29] The dynamic rotor eccentricity due to rotor unbalance gives rise to additional electromagnetic radial forces acting on the stator.

### Aerodynamic Sources

Aerodynamic noise in electric machines consists of discrete noise components and broadband noise. The discrete components are produced by periodic disturbances in air pressure due to rotating components (such as fan blades), stationary obstacles in the airstreams, and resonances in cavities. The broadband noise is caused by the eddies and random disturbances in the airstreams.

Figure 14.18 gives the general characteristics of aerodynamic noise sources for electric machines having ratings of from 0.1 to 10 MW. The figure also shows the relative importance of aerodynamic noise sources, compared with electromagnetic and mechanical noise sources.

## *14.6 VIBRATION BEHAVIOR*

### Stator Natural Frequencies and Mode Shapes

In designing electric machines for low noise and vibration levels, the accurate calculation of the natural frequencies of the stator is of considerable importance, since it is essential to avoid coincidence (hence resonance) between these natural frequencies and those of the exciting forces.

There is a vast literature on the calculation of stator natural frequencies. Table 14.16 gives a brief summary of the main analytical methods and the representative references.

Owing to constructional and geometrical dissymmetries in a stator, a phenomenon of dual resonance occurs.[37] This is characterized by two natural frequencies, which differ by a few percent in their numerical values but belong to the same vibration mode.

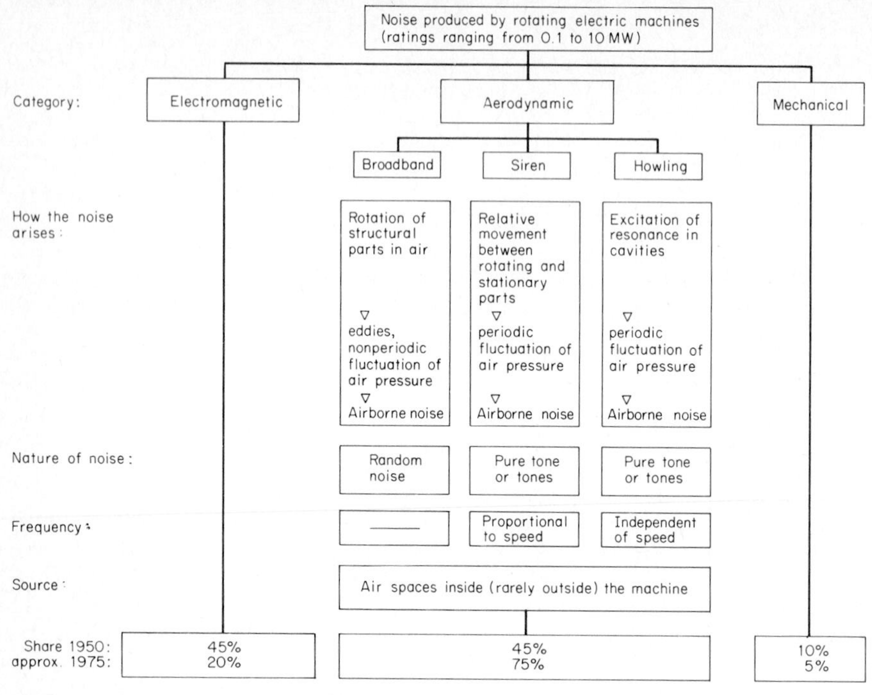


**FIG. 14.18** Acoustic noise components.[30]

**TABLE 14.16** Main Methods for Determining Stator Natural Frequencies

| Stator type | Analytical method | Reference |
|---|---|---|
| Simple single ring | Thin-ring or thick-ring theory | 31, 32 |
| Single ring encased by frame with ribs | Three-dimensional energy method | 33 |
| Single ring with formed coils | Three-dimensional finite-element method | 34 |
| Double ring | Two-dimensional energy method | 35, 36 |

## Force Response of Stators

In order to determine analytically the dynamic response of an electric machine stator subjected to a known forcing function, it is necessary to know the mechanical properties of the stator structure, including its damping characteristics. At present, very little information is available about stator damping characteristics; therefore, it is possible only to estimate the force response of a stator. Reference 12 provides a useful estimation.

A mechanical mobility test is useful in providing experimental data on the force response of a given stator. The mobility **M** at a particular point is defined as

$$\mathbf{M} = \frac{\mathbf{V}}{\mathbf{F}} \tag{14.30}$$

where **V** = vibration velocity
**F** = exciting force at the given point

The frequency of the exciting force should vary from zero to a few kilohertz, covering the important frequency range for noise and vibration considerations. Figure 14.19 gives typical measured mobility values for a stator and shows that the mobility varies with both frequency and temperature.

## End-Winding Vibrations

The experience gained from the use of epoxy-mica insulation for large electric machines over some 20 years indicates that end-winding vibrations constitute one of the main causes for stator-winding insulation failure.[38]

Since large-machine insulation failures lead to expensive repairs and outages, many organizations have devoted considerable effort to the investigation of large-machine end-winding vibrations. The U.K. CEGB[39] and others[40,41] have developed analytical techniques based on finite-element methods for calculating the vibrations and stresses of large electric machine stator end windings under both steady and transient load conditions.

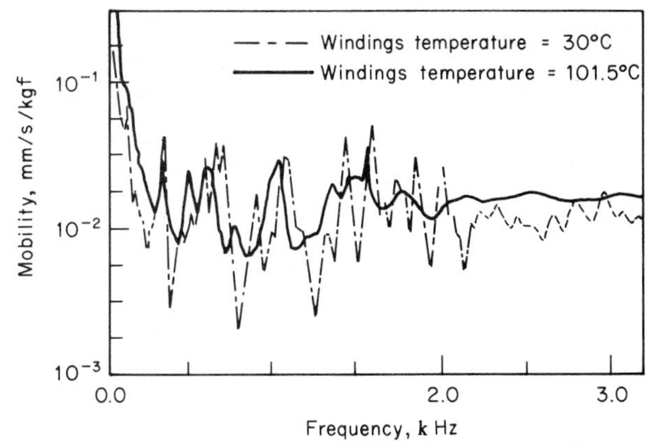


**FIG. 14.19** Measured driving-point mobility values for a 2100-kW motor stator.[34]

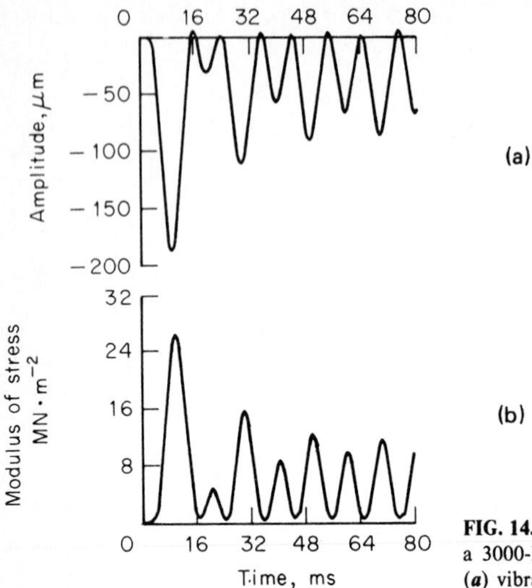


**FIG. 14.20** Calculated transient vibration and stress of a 3000-hp-motor end-winding conductor at start-up:[39] (*a*) vibration displacement; (*b*) stress.

Stator end-winding vibrations are primarily caused by electromagnetic forces acting on the end-winding conductors; these forces are produced by the interaction of the conductor current and the magnetic flux in the end-winding region. Reference 42 gives a comprehensive analysis of the forces on turbogenerator end windings.

Figure 14.20 shows typical variations of both the vibration displacement and stress of a large-motor end-winding conductor during start-up. The vibrations mainly consist of a steady component, a component at the supply frequency, and a component at twice the supply frequency. Figure 14.21 depicts the radial and tangential forces acting on the end-winding conductors for an 800-MVA generator caused by a three-phase short circuit; the variation of the maximum radial forces and deflections of the end winding with the axial position is shown in Fig. 14.22.

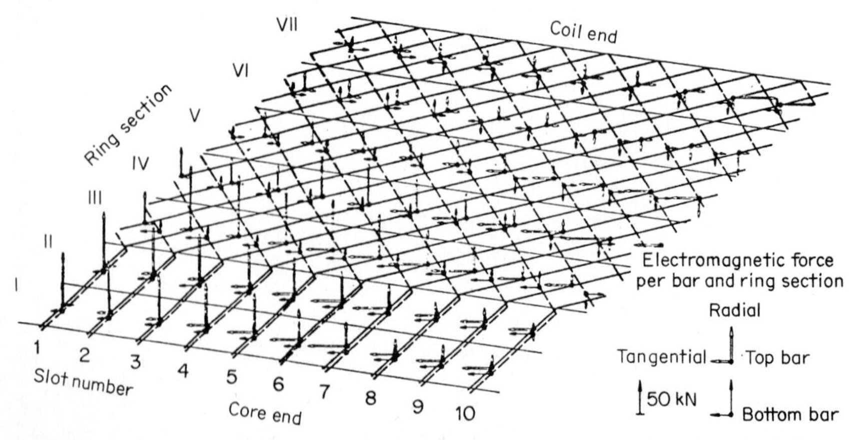


**FIG. 14.21** Electromagnetic forces in the end winding caused by a three-phase short circuit for an 800-MVA 60-Hz generator with 42 stator slots.[41]

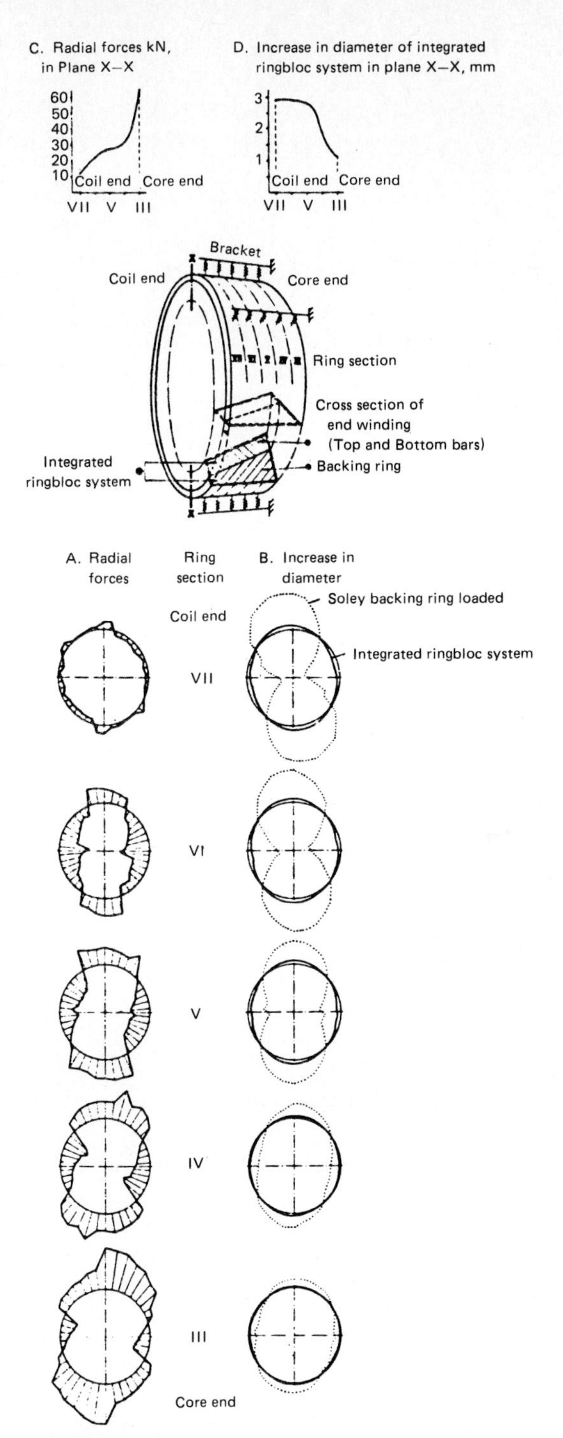


**FIG. 14.22** Calculated maximum radial forces and deflections resulting from a three-phase terminal short circuit at rated voltage of a two-pole 800-MVA 60-Hz generator.[41]

There are several methods for measuring high-voltage end-winding vibrations. These include the use of strain gauges, high-speed film, various optical techniques, and piezoelectric accelerometers.[40,41,43–45]

## Rotor Vibrations

Severe electric machinery rotor vibrations can lead to high noise emission and also cause rotor-bar fracture and damage to the bearings and the shaft-coupling system. In order to avoid excessive rotor vibrations, it is necessary to carry out a rotor dynamic analysis to determine (1) the critical speeds, (2) the force response, and (3) the stability of the rotor. There exists a vast literature on rotor dynamics. References 19, 23, 46, 47, and 48 form part of this, directly related to electric machines. For information concerning general rotor dynamics, Refs. 49, 50, and 51 are recommended.

For rigid-shaft electric machines, the first critical speed should be at a value approximately 30 percent above the nominal operating speed. For flexible-shaft electric machines having an operating speed between the first and second critical speeds, the rotating system should be designed to run some 30 to 40 percent above the first critical speed.

In the rotor dynamic analysis, it is essential to consider the mass and stiffness of the rotor, the bearing stiffness and damping, the bearing support-structure dynamics, and the electromagnetic forces acting on the rotor (see Sec. 14.5), e.g., the unbalanced magnetic pull. Although many electric machinery manufacturers have rotor dynamic analysis computer programs, the magnetic forces are sometimes neglected in this analysis, resulting in erroneous results. In fact, the static component of the unbalanced magnetic pull reduces the critical speed while the time-varying component affects the force response of the rotor system.

For an ideal rotor shaft case, assuming the static unbalanced magnetic pull to be proportional to the relative eccentricity (i.e., $F = K_m \varepsilon$), the critical speed is given by

$$\omega = \sqrt{\frac{K_s - K_m}{M}} \tag{14.31}$$

where $M$ = disk mass
$K_s$ = shaft stiffness constant
$K_m$ = unbalanced magnetic pull constant

The unbalanced magnetic pull acts as a "negative" spring and thus reduces the critical speed. As a first approximation, this constant $K_m$ can be estimated from Eq. (14.24) as

$$K_m = Q \frac{\pi D l}{4\mu_o} B_m^2 \tag{14.32}$$

where the symbols are the same as for Eq. (14.24).

Unbalanced magnetic pull can also cause self-excited radial rotor vibrations. For slip-ring induction motors having roller bearings and winding parallel paths, strong self-excited radial vibrations of the rotor can arise. These oscillations are at a frequency slightly lower than the supply frequency (see Ref. 52) and are caused by the unbalanced radial vibratory forces due to the interaction of the fundamental $p$-pole-pair flux density wave with two rotating flux density waves, one with $(p + 1)$

pole pairs and the other with ($p - 1$) pole pairs. For self-excited torsional vibrations in induction machines, Ref. 53 may be consulted.

In the case of the rotor of a submerged electric motor, the coolant fluid around the rotor forms a substantial "hydrodynamic bearing."[54] Because this "bearing" envelops the entire rotor, very substantial effects of fluid inertia are present. Thus, the effective mass of the rotor is increased substantially, leading to a great reduction of the critical speed. For a submerged rotor of radius 102 mm (4 in), of length 406 mm (16 in), and having a clearance gap of 1 mm (0.04 in) for the coolant water, the critical speed in water is found to be 247.5 rad/s, compared with a critical speed of 500 rad/s in air, and the onset speed of unstable rotation is 1000 rad/s.[54] Thus, the fluid inertia cannot be neglected in rotor dynamic analysis for submerged motors.

## 14.7 SOUND RADIATION

Owing to the complexity of the aerodynamic and mechanical sources in an electric machine, methods are not yet available for determining analytically the sound-pressure value at a given point in the noise field of an electric machine having predominant aerodynamic and/or mechanical noise. However, analytical methods have been presented for determining the sound-pressure value around an electric machine having predominant magnetic noise.[55]

## 14.8 REDUCTION OF NOISE AND VIBRATION

Electric machine design developments have resulted in a smaller weight per unit power and in increased electric and magnetic loadings, leading to higher flux densities, a thinner core and frame, and the need for increased cooling. These have resulted in greater noise and vibration problems.

The noise and vibration sources described in Sec. 14.5 are primarily the by-products of the normal operation of an electric machine, and it is not feasible to eliminate them completely. However, these sources or their effects can be reduced to acceptable levels so that a particular machine can satisfy a given set of noise and vibration specifications. Reference 28 gives useful information regarding noise-source identification with the aid of narrowband-frequency vibration analysis.

Since the noise and vibration of an electric machine are related to the exciting forces, to the dynamic response of the machine structure, and to the sound radiation and vibration transmission from the machine, the control measures can be classified as:

1. Reduction of the exciting forces
2. Reduction of the dynamic response of the machine
3. Reduction of sound radiation and vibration transmission

### Reduction of the Exciting Forces

Reference 12 discusses various measures to reduce the electromagnetic forces. These include:

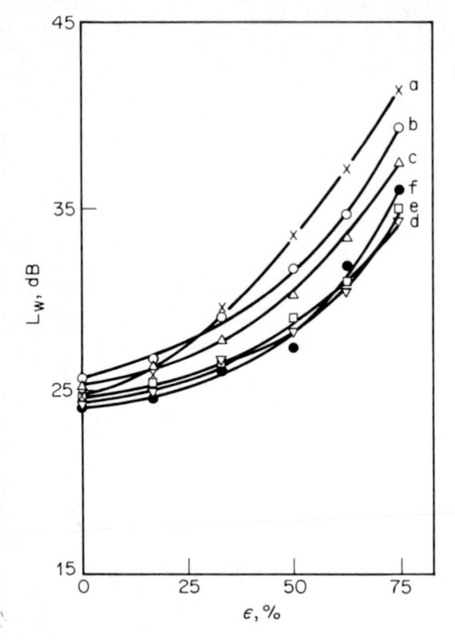


Passband is 6 percent of center frequency, constant applied voltage per winding section, total of 240 V when all in series, star connection; confidence level = 97.5 percent.

***a*** Two parallel paths with opposite pole groups in parallel, but without equalizers

***b*** All pole groups in series

***c*** Two parallel paths with odd-numbered pole groups in series and without equalizers

***d*** Six parallel paths

***e*** Three parallel paths with two equalizers

***f*** Two parallel paths with opposite pole groups in parallel and two equalizers

**FIG. 14.23** Variation of narrowband sound-power level $L_W$ centered at 900 Hz with relative eccentricity $\epsilon$ for various parallel paths and connections for a three-phase 1/2-hp six-pole motor.[14]

1. Enlarging the airgap.
2. Reducing the harmonics (of both space and current).
3. Reducing the magnetic permeance variation in the stator or rotor cores.
4. Skewing the stator or rotor slots (beneficial only for machines of ratings up to about 100 kW).
5. Using parallel paths in the stator winding and reducing the rotor eccentricity (Fig. 14.23).
6. Reducing the homopolar fluxes by using nonmagnetic end shields or nonmagnetic rings between the bearings and the end shields[26] or, for two-pole single-phase motors, by maintaining the phase angle between the main and auxiliary currents to 90° (Fig. 14.24).

The reduction in sound level from enlarging the airgap may be expressed approximately as[12]

$$L_1 - L_2 = 10 \log \left( \frac{g_2}{g_1} \right)^4 \tag{14.33}$$

where $g$ = mean airgap length

The reduction in sound level from skewing the rotor slot may be expressed as[56]

$$L_1 - L_2 = 20 \log \frac{\sin (\lambda \alpha_s / 2)}{(\lambda \alpha_s / 2)} \tag{14.34}$$

where $L_1$ = sound level without rotor skewing
$L_2$ = sound level with rotor skewing

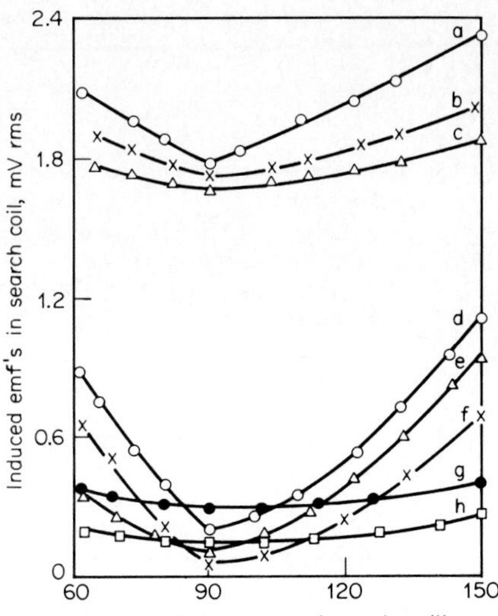


***Specifications:*** 240-V 50-Hz two-pole single-phase machine; main-winding current = 0.56 A; auxiliary-winding current = 0.51 A; effective turns ratio of auxiliary winding to main winding = 1.1; 30-turn search coil.

***Machine 1:*** With dynamic rotor unbalance of 1.5 gcm and unscrambled rotor laminations (rotor weight = 0.68 kg)

***Machine 2:*** Same as machine 1 but with dynamic rotor unbalance of 0.5 gcm

***Machine 3:*** Same as machine 1 but with scrambled rotor laminations

***a*** At 50 Hz for machine 1, 2970 r/min

***b*** At 50 Hz for machine 2, 2970 r/min

***c*** At 50 Hz for machine 3, 2970 r/min

***d*** At $(2 - s)f_1$ = 99.5 Hz for machine 1, 2970 r/min

***e*** At $(2 - s)f_1$ = 99.5 Hz for machine 2, 2970 r/min

***f*** At $(2 - s)f_1$ = 99.5 Hz for machine 3, 2970 r/min

***g*** At $sf_1$ = 5 Hz for machine 1, 2700 r/min

***h*** At $sf_1$ = 5 Hz for machine 2, 2700 r/min

**FIG. 14.24** Variation of induced emf's in search coil for homopolar flux measurement with phase angle between the main- and auxiliary-winding currents.[13]

**TABLE 14.17** Balance-Quality Grades for Various Groups of Representative Rigid Rotors

| Balance-quality grade G | $e\omega$,* mm/s | Rotor types: General examples |
|---|---|---|
| G 16 | 16 | Drive shafts (propeller shafts, cardan shafts) with special requirements |
| G 6.3 | 6.3 | Parts or process plant machines<br>Marine main-turbine gears (merchant service)<br>Centrifuge drums<br>Fans<br>Flywheels<br>Pump impellers<br>Machine-tool and general machinery parts<br>Normal electrical armatures |
| G 2.5 | 2.5 | Gas and steam turbines, including marine main turbines (merchant service)<br>Rigid turbogenerator rotors<br>Rotors<br>Turbocompressors<br>Machine-tool drives<br>Medium and large electrical armatures with special requirements<br>Small electrical armatures |
| G 1 | 1 | Tape recorder and phonograph (gramophone) drives<br>Grinding-machine drives<br>Small electrical armatures with special requirements |
| G 0.4 | 0.4 | Spindles, disks, and armatures of precision grinders |

* $\omega = 2\pi\ n/60 \simeq n/10$ if $n$ is measured in revolutions per minute and $\omega$ in radians per second. In general, for rigid rotors with two correction planes, one-half of the recommended residual unbalance is to be taken for each plane; these values apply usually for any two arbitrarily chosen planes, but the state of unbalance may be improved upon at the bearings. For disk-shaped rotors, the full recommended value holds for one plane.

***Source:*** BS 5265, *Mechanical Balancing of Rotating Bodies*, pt. 1: *Recommendations on Balance Quality of Rotating Rigid Bodies*, BSI, London, 1979.

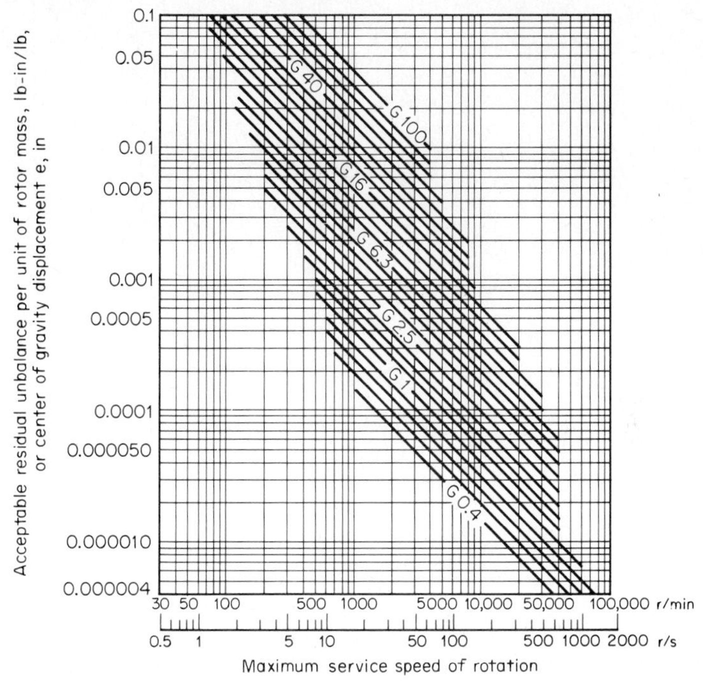


**FIG. 14.25** Maximum residual specific unbalance corresponding to various balance-quality grades (G) (ISO 1940).

$\lambda$ = number of pole pairs of the rotor mmf wave contributing to the force wave concerned
$\alpha_s$ = geometrical skewing angle of rotor slot, rad

Various methods for reducing the noise and vibration excited by rolling bearings and brush-commutator assemblies are described in Refs. 12, 27, and 57. Reference 57 shows that, for small machines, the optimum brush pressure is 400 to 450 g/cm$^2$ for minimum noise emission. The residual unbalance of the rotor should be kept to a minimum, or within the limits specified by BS 5265 (see Table 14.17 and its associated Figs. 14.25 and 14.26).

For the measures suppressing the aerodynamic exciting forces, Refs. 12, 30, 58, 59, and 60 are recommended. These include the use of the smallest acceptable fan, two smaller fan impellers instead of one large one, unidirectional fans, unevenly spaced fan blades, a porous fan-blade surface, a special fan-rotor coupling, a streamlined air circuit, and an enlarged distance between the fan-blade tips and the casing ribs. Reference 60 recommends that the fan-blade peripheral speed not exceed 50 m/s. This peripheral speed is attained by the fan-impeller sizes and speeds shown in Fig. 14.27. The variation of sound level with fan-impeller diameter is given in Fig. 14.28.

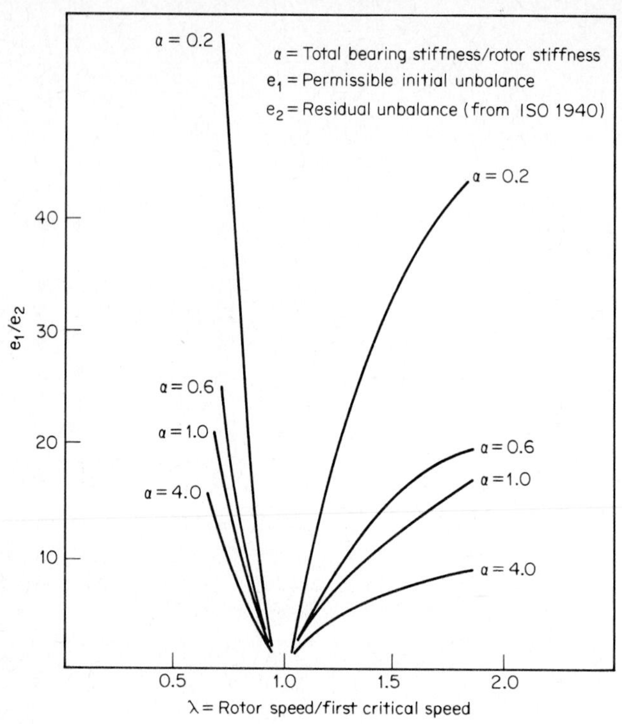


**FIG. 14.26** Permissible initial unbalance for flexible rotors (BS 5265 Part 2).

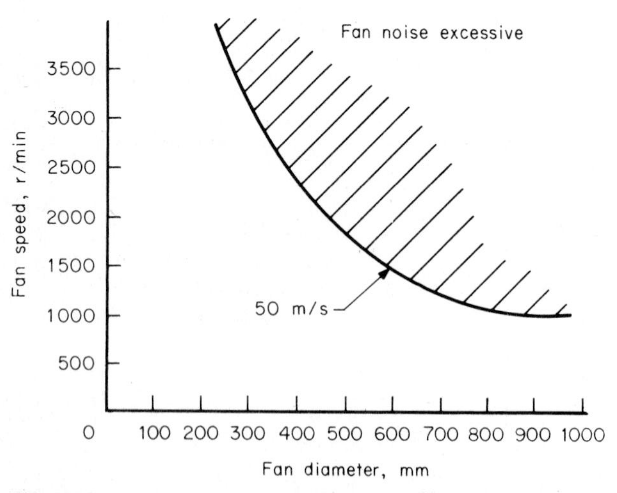


**FIG. 14.27** Fan constant peripheral velocity graph.[60]

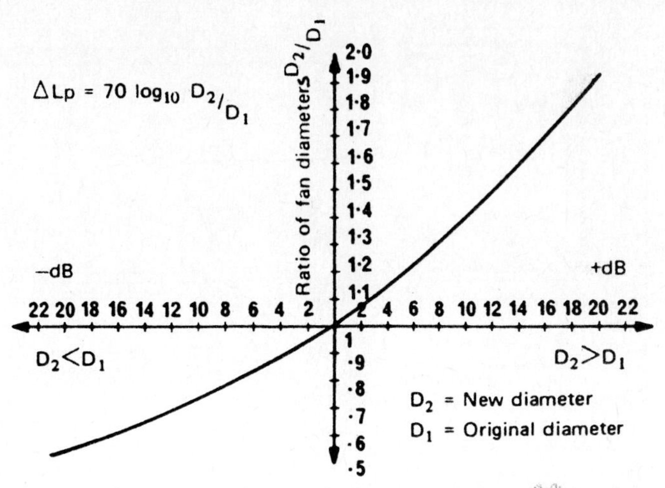


**FIG. 14.28** Change in sound-pressure level $\Delta L_p$ expressed as a function of the ratio of fan diameters for geometrically similar fans at constant speed.[60]

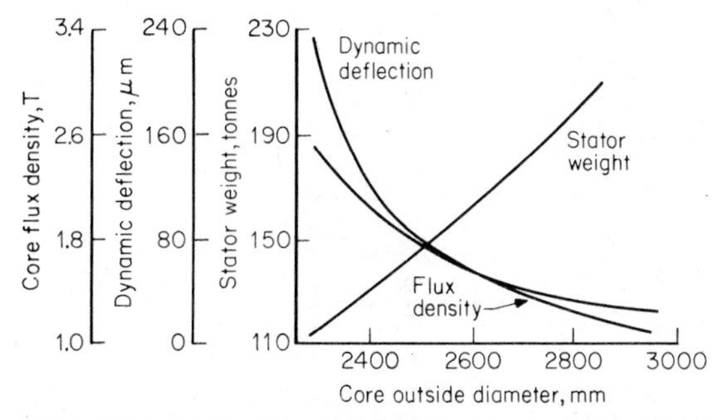


**FIG. 14.29** Variation of dynamic deflection, stator weight, and flux density with outside core diameter for cold-rolled grain-oriented core plate for a 590-MVA generator.[61]

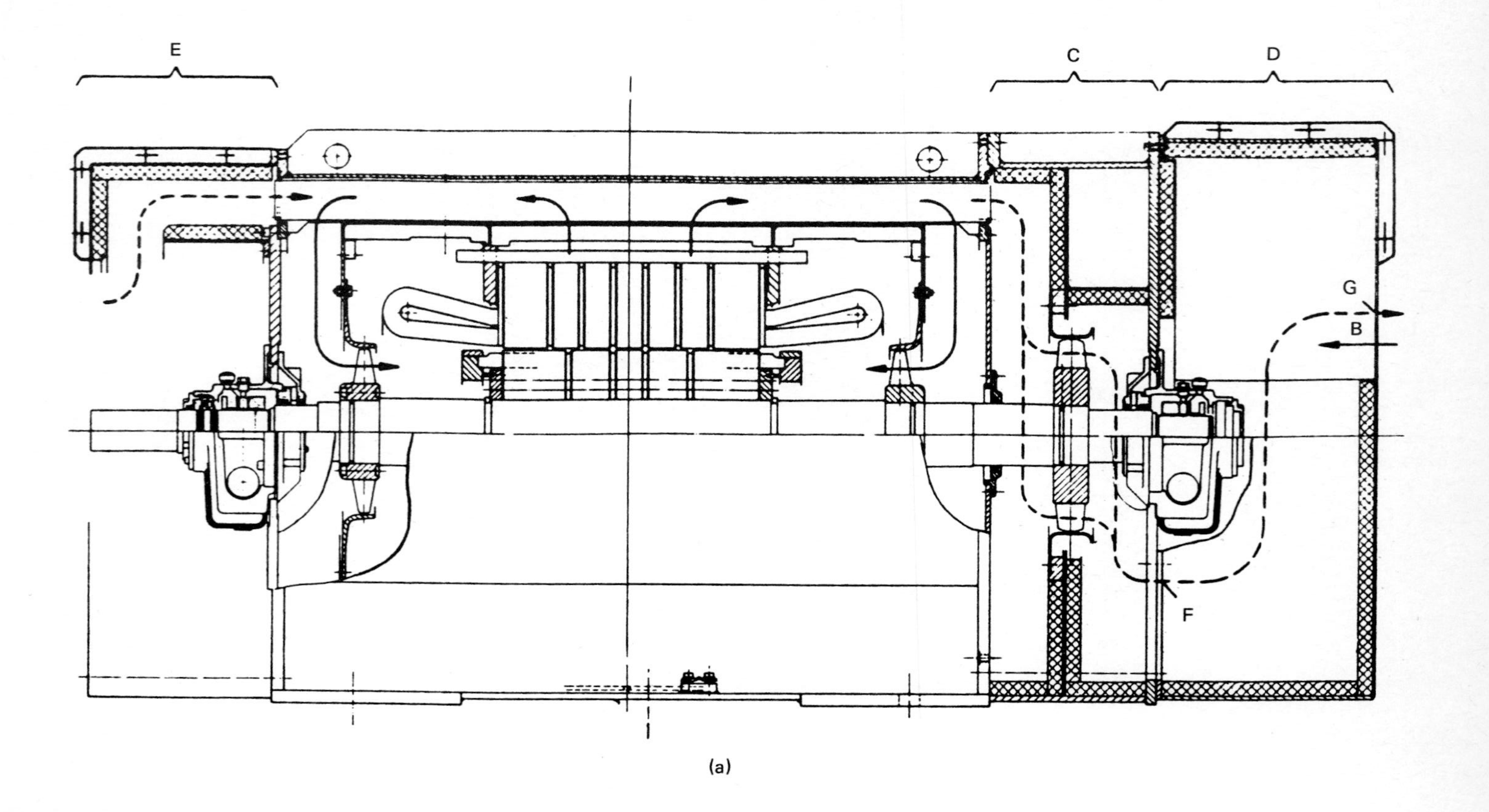


(a)

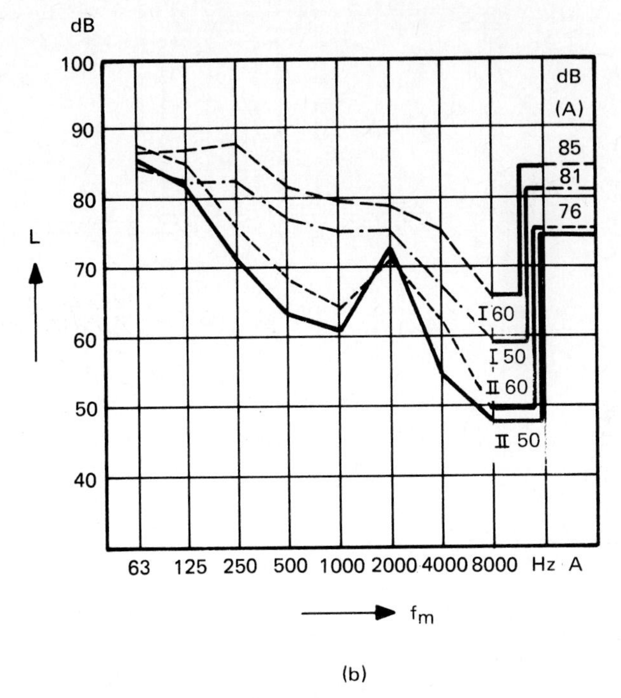


*C* Fan space *D, E* Silencer elements

⟶ Internal air - - → External air

*L* Sound-pressure level in octave bands

$f_m$ Octave-band center frequency

*I* With smaller fan blade, fan space *C* lined, without silencer elements *D* and *E*

*II* As for I, but with silencer elements *D* and *E*

*50* Mains frequency 50 Hz, $\hat{\simeq}$ 3000 r/min

*60* Mains frequency 60 Hz, $\hat{\simeq}$ 3600 r/min

**FIG. 14.30** (***a***) Tube-cooled squirrel-cage motor, type eQRG 560 ia 2, rated 650 kW, 2981 r/min, 6 kV, with noise-muffling elements.[30] (***b***) Mean octave-band spectra and mean sound levels in dB(A) of the motor measured at 1-m distance from the machine's surface.[30]

## Reduction of the Dynamic Machine Response

The measures for reducing the dynamic response of the machine structure are briefly summarized as follows:[12]

1. Mismatching the natural frequencies of the machine structure and the exciting-force frequencies
2. Using flexible links between the stator core and the frame
3. Increasing the exciting-force mode number values by selecting suitable stator/rotor slot numbers

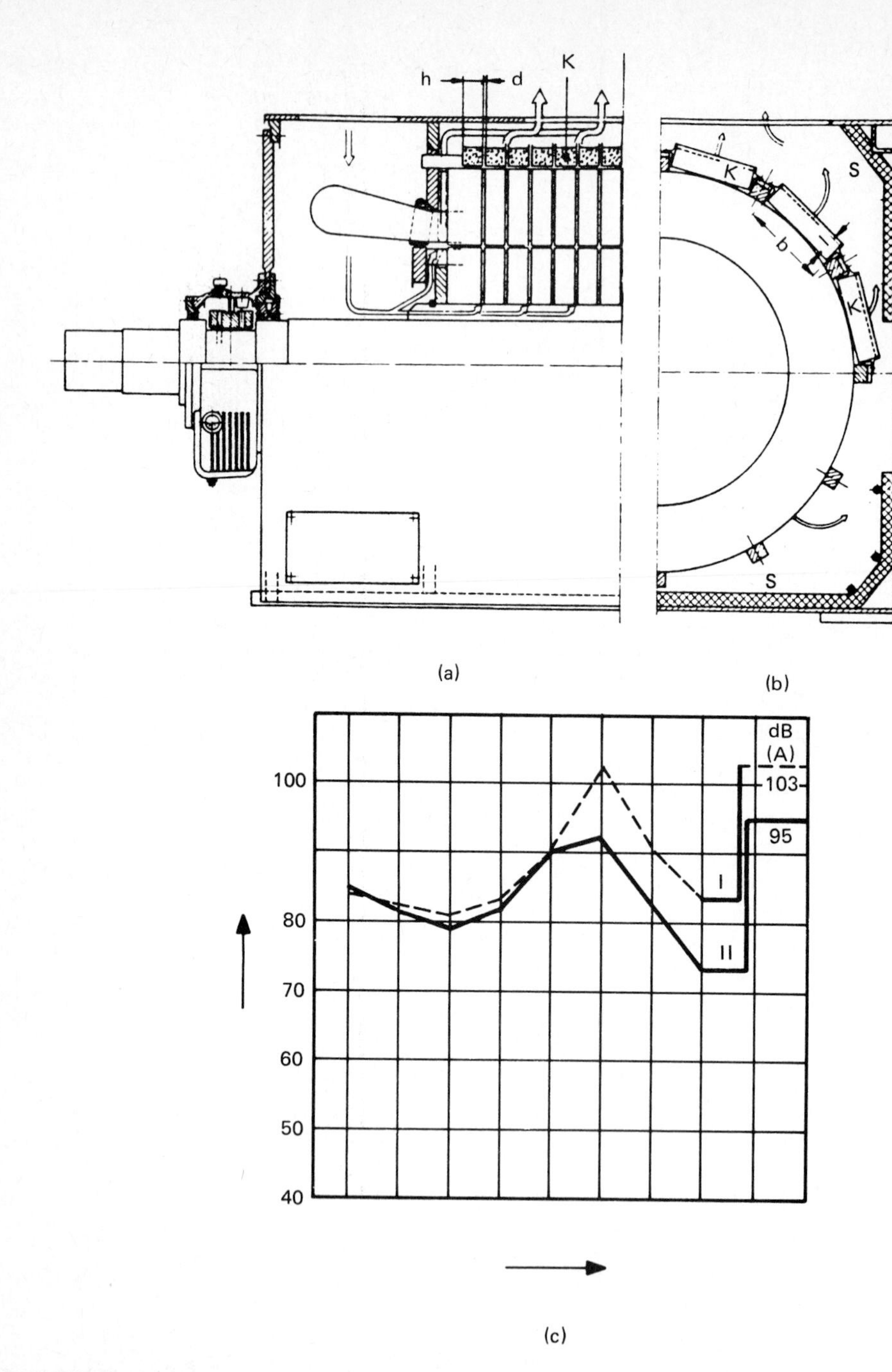


*K* Silencer cassettes
*S* Stator space coating
*L* Sound-pressure level in octave bands
$f_m$ Octave-band center frequency
*I* Without sound-reduction measures
*II* With silencing cassettes and stator space coating

**FIG. 14.31** Basic construction of a pipe-ventilated squirrel-cage motor rated 3000 kW, 1790 r/min, 6.9 kV:[30] (***a***) longitudinal elevation; (***b***) cross section; (***c***) mean octave-band spectra and mean sound levels in dB(A) of the motor, measured at 1-m distance from the machine's surface.

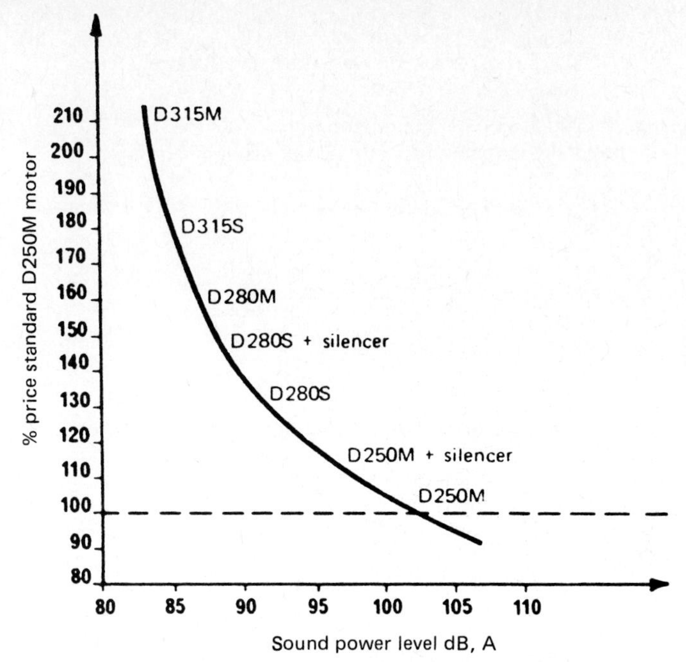


**FIG. 14.32** Cost of low noise level on two-pole 50-Hz TEFC motor.[60]

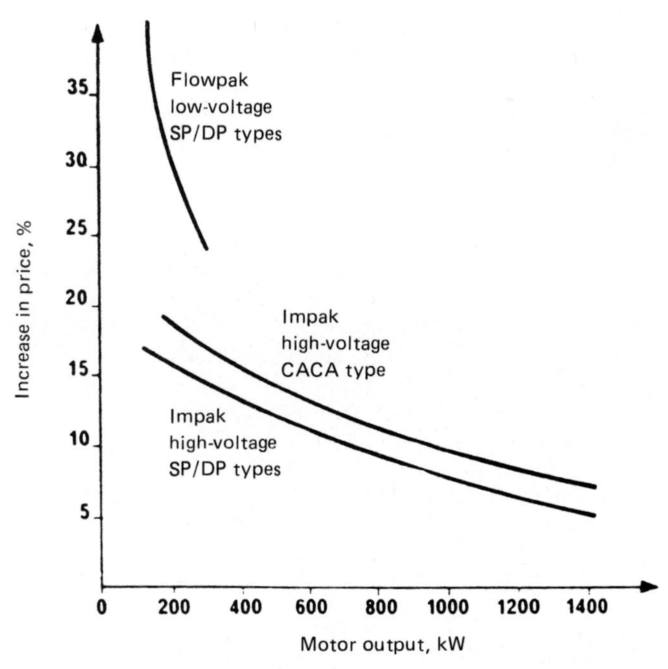


**FIG. 14.33** Percentage increase in price of fitting a silencer to two-pole 50-Hz motors with rolling element bearings to achieve a 15-dB(A) noise-level reduction.[60]

4. Increasing the thickness of the stator core, i.e., the outside core diameter (Fig. 14.29) and the damping capacity of the machine structure[62]

The sound-level reduction due to an increase in stator core thickness is approximately given by[12]

$$L_1 - L_2 = 10 \log \left( \frac{h_2}{h_1} \right)^6 \tag{14.35}$$

where $h$ = core thickness, excluding the teeth.

### Reduction of Sound Radiation and Vibration Transmission

References 12 and 13 show that, for reducing a sound radiation of electromagnetic origin from the machine surface, the combination of a shorter length and a larger diameter is preferred to one with a longer length and a smaller diameter.

Sound-absorbent enclosures and air-inlet and -outlet silencers are often used for medium-size and large machines with predominant aerodynamic noise. Air-inlet and -outlet silencers and closed air-circuit air-cooled enclosures can reduce the predominant aerodynamic noise (A-weighted sound-power level) by 10 dB(A) or more. Figure 14.30 shows the typical effects of air-inlet and -outlet silencers on the octave-band frequency spectra for a large motor. The effect of silencer cassettes placed around the stator-core air ducts on air-duct siren noise is shown in Fig. 14.31.

From the point of view of structure-borne noise and vibration, it is essential to reduce the vibrations transmitted from the machine to its base and to any structures connected to it. For measures to reduce vibration transmission, see Refs. 63 and 64.

### Costs of Noise and Vibration Control

The important factors affecting the costs include:[60]

1. The required output compared with the maximum output of the assigned frame size
2. The type of enclosure of the machine
3. The size of the machine
4. The required noise and vibration levels

Figure 14.32 shows the cost increase for a two-pole totally enclosed, fan-cooled (TEFC) induction motor; the significant cost increase in this figure is due to the need to increase the frame size. Figure 14.33 gives the typical percentage increase in price for fitting a silencer to a two-pole induction machine to achieve a noise reduction of 15 dB(A).

## REFERENCES

1. A. J. Ellison, C. J. Moore, and S. J. Yang, "Methods of Measurement of Acoustic Noise Radiated by an Electric Machine," *Proc. IEE*, vol. 116, 1969, pp. 1419–1431.

2. BS 4999, *General Requirements for Rotating Electric Machines*, pt. 50: *Mechanical Performance—Vibration*, BSI, London, 1978.
3. BS 4999, *General Requirements for Rotating Electric Machines*, pt. 51: *Noise Levels*, BSI, London, 1973.
4. IEC 34-14, *Rotating Electric Machines*, pt. 14: *Mechanical Vibration of Certain Machines with Shaft Heights 56 mm and Higher—Measurement, Evaluation, and Limits of the Vibration Severity*, International Electrotechnical Commission, Geneva, 1982.
5. ISO 3745, *Acoustics: Determination of Sound-Power Levels of Noise Sources—Precision Methods for Anechoic and Semianechoic Rooms*, International Organization for Standardization, Geneva, 1977.
6. ISO 3744, *Acoustics: Determination of Sound-Power Levels of Noise Sources—Engineering Methods for Free-Field Conditions over a Reflecting Plane*, International Organization for Standardization, Geneva, 1981.
7. ISO/DIS 6926, *Acoustics: Determination of Sound-Power Levels of Noise Sources—Characterization and Calibration of Reference Sound Sources*, International Organization for Standardization, Geneva, n.d.
8. ISO 3746, *Acoustics: Determination of Sound-Power Levels of Noise Sources—Survey Method*, International Organization for Standardization, Geneva, 1979.
9. A. J. Ellison and S. J. Yang, "Acoustic-Noise Measurements on Nominally Identical Small Electric Machines," *Proc. IEE*, vol. 117, no. 3, 1970, pp. 555–560.
10. DIN 45665, *Schwingstärke von rotierenden elektrischen Maschinen der Baugrössen 80 bis 315. Messverfahren und Grenzwerte*, Beuth Vertrieb GmbH, Berlin, 1968.
11. J. T. Broch, *Mechanical Vibration and Shock Measurements*, Bruel & Kjaer, Naerum, Denmark, 1980.
12. S. J. Yang, *Low-Noise Electrical Motors*, Oxford University Press, New York, 1981.
13. S. J. Yang, "Acoustic Noise from Small Two-Pole Single-Phase Induction Machines," *Proc. IEE*, vol. 122, no. 12, 1975, pp. 1391–1396.
14. A. J. Ellison and S. J. Yang, "Effects of Rotor Eccentricity on Acoustic Noise from Induction Machines," *Proc. IEE*, vol. 118, no. 1, 1971, pp. 174–184.
15. S. P. Verma and R. Natarajan, "Effect of Eccentricity in Induction Machines," *Proceedings of International Conference on Electric Machines*, Budapest, 1982, pp. 930–933.
16. S. J. Yang, "Noise and Vibration of Inverter-Fed Induction Motors," *Proceedings of International Conference on Electric Machines*, Vienna, 1976, pp. 19-1–19-9.
17. S. J. Yang and P. L. Timar, "The Effect of Harmonic Currents on the Noise of a Three-Phase Induction Motor," *IEEE Transactions*, vol. PAS-99, no. 1, 1980, pp. 307–310.
18. S. Williamson, D. Pearson, and A. M. Rugege, "Acoustic Noise and Pulsating Torques in a Triac-Controlled Permanent-Split-Capacitor Fan Motor," *Proc. IEE*, vol. 128, pt. B, 1981, pp. 201–206.
19. H. Pecken, C. Troeder, and G. Diekhans, "Torsional Vibrations during the Starting Process in Drive Systems with Three-Phase Motors," *Proceedings of the Second International Conference on Vibrations in Rotating Machinery*, Cambridge, England, 1980, pp. 427–435.
20. T. J. Hammons, "Effect of Three-Phase System Faults and Faulty Synchronization on the Mechanical Stressing of Large Turbine Generators," *Revue générale de l'électricité*, vol. 86, no. 7/8, 1977, pp. 558–580.
21. P. von Kaehne, "Unbalanced Magnetic Pull in Rotating Electric Machines," ERA Report Z/T 142, 1963.
22. F. T. Chapman, "The Production of Noise and Vibration by Certain Squirrel-Cage Induction Motors," *Journal of the IEE*, vol. 61, 1922, pp. 39–48.
23. M. T. Wright, D. S. M. Gould, and J. J. Middlemiss, "The Influence of Unbalanced Magnetic Pull on the Critical Speed of Flexible Shaft Induction Machines," *IEE Conference Publication 213*, 1982, pp. 61–64.

24. K. J. Binns and M. Dye, "Identification of Principal Factors Causing Unbalanced Magnetic Pull in Cage Induction Machines," *Proc. IEE*, vol. 120, no. 3, 1973, pp. 349–354.

25. W. G. Crawford, "Unbalanced Magnetic Pull as a Cause of Vibration in Two-Pole Induction Motors," *Proceedings of Electric Machines in the Seventies*, Dundee, 1970, pp. 48-1–48-4.

26. R. Belmans, W. Geysen, H. Jordan, and A. Vandenput, "Unbalanced Magnetic Pull and Homopolar Flux in Three-Phase Induction Motors with Eccentric Rotors," *Proceedings of International Conference on Electric Machines*, Budapest, 1982, pp. 916–921.

27. H. Pittroff, "Reliable and Noiseless Bearings for Small Electric Motors" (in German), *Technica* (Switzerland), vol. 20, 1971, pp. 2499–2504.

28. P. J. Tsivitse and P. R. Weihsmann, "Polyphase Induction Motor Noise," *IEEE Transactions*, vol. IGA-7, no. 3, 1971, pp. 339–358.

29. R. Völler, "Vibration Testing of Electric Motors for Machine Tools," *Eng. Dig.*, vol. 26, 1965, pp. 95–97 and 107.

30. B. Ploner, "Aerodynamic Noise in Medium-Size Asynchronous Motors," *Brown Boveri Review*, vol. 63, no. 8, 1976, pp. 493–499.

31. H. Jordan and H. Frohne, "Ermittlung der Eigenfrequenzen des Ständers von Drehstrommotoren," *Lärmbekämpfung*, vol. 7, 1957, pp. 137–140.

32. H. Pavlovsky, "Calculation of Natural Frequencies of Stator Core of Electric Machines" (in Czech), *Elektrotech. Obz.*, vol. 57, no. 6, 1968, pp. 305–311.

33. R. S. Girgis and S. P. Verma, "Method for Accurate Determination of Resonant Frequencies and Vibration Behavior of Stators of Electric Machines," *Proc. IEE*, vol. 128, pt. B, no. 1, 1981, pp. 1–11.

34. S. Watanabe et al., "Natural Frequencies and Vibration Behavior of Motor Stators," *IEEE Transactions*, vol. PAS-102, no. 4, 1983, pp. 949–956.

35. A. J. Ellison and S. J. Yang, "Natural Frequencies of Stators of Small Electric Machines," *Proc. IEE*, vol. 118, no. 1, 1971, pp. 185–190.

36. E. Erdelyi and G. Horvay, "Vibration Modes of Stators of Induction Motors," *ASME Transactions (E)*, vol. 24, 1957, pp. 39–45.

37. S. P. Verma and R. S. Girgis, "Experimental Verification of Resonant Frequencies and Vibration Behavior of Stators of Electric Machines," *Proc. IEE*, vol. 128, pt. B, no. 1, 1981, pp. 22–32.

38. A. Wichmann, "Two Decades Experience and Progress in Epoxy-Mica Insulation Systems for Large Rotating Machines," Paper 82 WM 235-0, IEEE PES Winter Meeting, New York, 1982.

39. S. Potter and G. D. Thomas, "Calculating Vibration and Stresses of Large Electric Machine Stator End Windings, for Both Steady Load and Transient Conditions," *IEE Conference Publication 213*, 1982, pp. 195–200.

40. M. Ohtaguro, K. Yagiuchi, and H. Yamaguchi, "Mechanical Behavior of Stator End Windings," *IEEE Transactions*, vol. PAS-99, no. 3, 1980, pp. 1181–1185.

41. D. Lambrecht and H. Berger, "Integrated End-Winding Support for Water-Cooled Stator Winding," *IEEE Transactions*, vol. PAS-102, no. 4, 1983, pp. 998–1006.

42. P. J. Lawrence, "Forces on Turbogenerator End Windings," *Proc. IEE*, vol. 112, 1965, pp. 1144–1158.

43. P. E. Clark and I. E. McShane, "Stator End-Winding Movement—Calculation and Measurement," *IEE Conference Publication 213*, 1982, pp. 1–5.

44. A. Futakawa and S. Yamasaki, "Dynamic Deformation and Strength of Stator End Windings during Sudden Short Circuits," *IEEE Transactions*, vol. EI-16, no. 1, 1981, pp. 31–39.

45. S. Nagano et al., "Early Detection of Excessive Coil Vibration in High-Voltage Rotating Electric Machines," *IEEE Transactions*, vol. PAS-101, no. 6, 1982, pp. 1551–1560.

46. R. O. Eis, "Electric Motor Vibration—Cause, Prevention, and Cure," *IEEE Transactions*, vol. 1A-11, no. 3, 1975, pp. 267–275.
47. W. Blase, "Die Auslegung Schwingungsarmer zweipoliger Asynchronmotoren," *AEG-Mitt.*, vol. 52, no. 5/6, 1962, pp. 197–204.
48. T. Hensel, "Critical Speeds and Mechanical Stability Parameters of Asynchronous Motors with Power of 100 to 1800 kW" (in German), *Elektrie*, vol. 20, no. 8, 1966, pp. 315–318.
49. N. F. Rieger, *Vibrations of Rotating Machinery*, 2d ed., pt. 1: *Rotor-Bearing Dynamics*, The Vibration Institute, Clarendon Hills, Ill., 1982.
50. "Rotordynamic Instability Problems in High-Performance Turbomachinery," *NASA Conference Publication 2133*, National Aeronautics and Space Administration, Texas, 1980.
51. "Rotordynamic Instability Problems in High-Performance Turbomachinery," *NASA Conference Publication 2250*, National Aeronautics and Space Administration, Texas, 1982.
52. M. Krondl, "Self-Excited Radial Vibrations of the Rotor of Induction Machines with Parallel Paths in the Winding" (in French), *Bull. Assoc. Suisse Elect.*, vol. 47, pp. 581–588.
53. H. W. Lorenzen, "Self-Excited Torsional Vibrations of Three-Phase Induction Machines," *Brown Boveri Review*, vol. 55, no. 10/11, 1968, pp. 650–663.
54. H. F. Black, "Effects of Fluid-Filled Clearance Spaces on Centrifugal Pump and Submerged Motor Vibrations," *Proceedings of the Eighth Turbomachinery Symposium*, Texas, 1979, pp. 29–34.
55. A. J. Ellison and S. J. Yang, "Calculation of Acoustic Power Radiated by an Electric Machine," *Acustica*, vol. 25, 1971, pp. 28–34.
56. H. Jordan and H. Müller-Tomfelde, "Akustische Wirkung der Schrägung bei DrehstroAsynchronmaschinen mit Käfigläufern," *ETZ(A)*, vol. 82, 1961, pp. 788–792.
57. N. V. Astakhov, "Brush Noise in Miniature Single-Phase Commutator Motors" (in Russian), *Electrichestvo*, vol. 9, 1959, pp. 46–50.
58. P. Francois, "The Generation of Noise and the Response of the Structures in Asynchronous Motors, Particularly as Far as the Flow Is Concerned," *Applied Acoustics*, vol. 3, no. 1, 1970, pp. 23–45.
59. A. Tulleth, "The Design Triangle," *IEE Conference Publication 213*, 1982, pp. 79–83.
60. C. N. Glew, *The Control of Induction Motor Noise*, GEC Machines, Bradford, England, 1973.
61. P. Richardson and R. Hawley, "Generator Stator Vibration," Paper 70 CP186-PWR, IEEE Winter Power Meeting, New York, 1970.
62. S. D. Haddad and M. F. Russell, "Noise Sources of Small Electric Machines and Some Control Methods," *Proceedings of 9th ICA*, Madrid, Paper E-57, 1977, p. 207.
63. C. M. Harris, *Handbook of Noise Control*, 2d ed., McGraw-Hill, New York, 1979.
64. L. L. Beranak, *Noise and Vibration Control*, McGraw-Hill, New York, 1971.
65. S. J. Yang and A. J. Ellison, *Machinery Noise Measurement*, Oxford University Press, New York, 1985.

## ACKNOWLEDGEMENT

Extracts from British Standards are reproduced by permission of the British Standards Institution (BSI). Complete copies can be obtained from BSI at Linford Wood, Milton Keynes, MK14 6LE, U.K.

# APPENDIX 1
# UNIT CONVERSION

## *A1.1 ALTERNATE UNITS*

In addition to the International System of Units (SI), two systems are in relatively common use in certain industrial and scientific applications. These are the English and centimeter-gram-second (cgs) systems.

For the purposes of electromagnetic analysis, the relationships among these systems of units and the SI system are best described by means of the equations in Chap. 2 relating the two fundamental magnetic parameters: Eqs. (2.7) to (2.10). These are repeated below in appropriate representation for the three systems:

$$\mathbf{B} = \mu_o \mu_R \mathbf{H} \qquad \text{SI and English}$$
$$\mathbf{B} = \mu_R \mathbf{H} \qquad \text{cgs}$$

where the units and numerical values are as follows:

| Symbol | SI | English | CGS |
|---|---|---|---|
| $\mu_o$ | $4\mu \times 10^{-7}$ | 3.19 | 1 |
| **B** | tesla | lines per square inch | gauss |
| **H** | ampere per meter | ampere per inch | oersted |

## A1.2 UNIT CONVERSION

| | | One: | is equal to: | |
|---|---|---|---|---|
| Symbol | Description | SI unit | English unit | CGS unit |
| **B** | Magnetic flux density | tesla (= 1 $Wb/m^2$) | $6.452 \times 10^4$ $lines/in^2$ | $10^4$ G |
| **H** | Magnetic field intensity | ampere per meter | 0.0254 A/in | $0.004\pi$ Oe |
| $\phi$ | Magnetic flux | weber | $10^8$ lines | $10^8$ maxwells |
| *D* | Viscous damping coefficient | newton-meter-second | 0.73756 lb·ft·s | $10^7$ dyne·cm·s |
| *F* | Force | newton | 0.2248 lb | $10^5$ dynes |
| *J* | Inertia | kilogram-square meter | 23.73 $lb \cdot ft^2$ | $10^7$ $g \cdot cm^2$ |
| *T* | Torque | newton-meter | 0.73756 ft·lb | $10^7$ dyne·cm |
| *W* | Energy | joule | 1 W·s | $10^7$ ergs |

# APPENDIX 2
# PHYSICAL TABLES

**TABLE A2.1** Density of Industrial Alloys at $T = 20°C$

| Name | Composition, % | Density $\rho$ | |
|---|---|---|---|
| | | $kg \cdot m^{-3}$ | $lb \cdot ft^{-3}$ |
| Aluminum alloy | (99.6 Al) | 2,750±50 | 172±3 |
| Aluminum brass | (76 Cu, 22 Zn, 2 Al) | 8,330 | 520 |
| Aluminum bronze | (86 Cu, 4 Fe, 10 Al) | 7,500 | 468 |
| Beryllium copper | (97 Cu, 2 Be) | 8,230±20 | 514±1 |
| Architectural bronze | (57 Cu, 40 Zn, 3 Pb) | 8,530 | 532 |
| Cable sheet lead | (99.8 Pb, 0.028 Ca) | 11,350 | 709 |
| Carbon steel | (99.2 Fe, 0.2 C, 0.45 Mn, 0.15 Si) | 7,860 | 490 |
| Cast iron | (91 Fe, 3.56 C, 2.5 Si) | 7,080 | 442 |
| Cast steel | (97.40 Fe, 0.3 C, 0.7 Mn, 0.6 Si, 0.5 Ni) | 7,690 | 480 |
| Commercial rolled zinc | (99 Zn, 0.3 Pb, 0.3 Cd) | 7,140 | 446 |
| Commercial bronze | (90 Cu, 10 Zn) | 8,800 | 549 |
| Commercial red brass | (85 Cu, 5 Zn, 5 Pb, 5 Sn) | 8,600 | 537 |
| Commercial yellow brass | (71 Cu, 25 Zn, 3 Pb, 1 Sn) | 8,400 | 524 |
| Electrolytic copper | (99.92 Cu, 0.04 O) | 8,900 | 556 |
| Cobalt alloy | (60 Co, 7 W, 10 Ni, 23 Cr) | 8,610 | 538 |
| Grid metal | (91 Pb, 9 Sb) | 10,660 | 665 |
| Bearing lead | (94 Pb, 6 Sb) | 10,880 | 679 |
| Cast nickel | (97 Ni, 1.5 Si, 0.5 Mn, 0.5 C) | 8,340 | 521 |
| Magnesium alloy | (89.9 Mg, 10 Al, 0.1 Mn) | 1,810 | 113 |
| Stainless steel | (70 Fe, 16 Cr, 12 Ni, 2 Mo) | 7,950±50 | 496±3 |
| Tool steel | (77 Fe, 18 W, 4 Cr, 1 V) | 8,670 | 541 |

**TABLE A2.2** Resistivity $\rho$ of Pure Metals at Temperature $T = 0°C$ or $20°C$

| Name | $\rho$, $10^{-8}$ Ω·m | | Name | $\rho$, $10^{-8}$ Ω·m | |
|---|---|---|---|---|---|
| | $T = 0°C$ | $T = 20°C$ | | $T = 0°C$ | $T = 20°C$ |
| Aluminum | 2.54 | 2.78 | Nickel | 7.00 | 7.82 |
| Antimony | 37.60 | 40.96 | Osmium | 8.47 | 9.24 |
| Arsenic | 26.10 | 28.83 | Palladium | 9.85 | 10.62 |
| Bismuth | 105.10 | 115.00 | Platinum | 9.62 | 10.57 |
| Cadmium | 7.21 | 7.42 | Potassium | 6.25 | 6.92 |
| Calcium | 3.92 | 4.25 | Rhodium | 4.39 | 4.77 |
| Chromium | 12.24 | 13.03 | Silver | 1.47 | 1.60 |
| Cobalt | 8.72 | 9.46 | Sodium | 4.29 | 4.72 |
| Copper | 1.55 | 1.72 | Strontium | 19.90 | 22.80 |
| Gold | 2.23 | 2.44 | Tantalum | 12.15 | 12.95 |
| Iridium | 5.15 | 5.68 | Thorium | 13.80 | 15.20 |
| Iron | 8.68 | 9.75 | Tin | 10.20 | 11.30 |
| Lead | 19.26 | 20.80 | Titanium | 43.42 | 43.85 |
| Lithium | 8.50 | 9.30 | Tungsten | 5.05 | 5.55 |
| Magnesium | 3.96 | 4.45 | Uranium | 24.20 | 25.70 |
| Mercury | 94.10 | 95.80 | Zinc | 5.48 | 5.87 |
| Molybdenum | 5.08 | 5.38 | | | |

**TABLE A2.3** Resistivity $\rho$ of Industrial Alloys at $T = 20°C$

| Name | Composition | $\rho$, $10^{-8}$ Ω·m |
|---|---|---|
| Aluminum | AIEE, hand-drawn | 2.83 |
| Brass | 60% Cu + 40% Zn | 6.80 |
| Brass | 66% Cu + 34% Al | 7.18 |
| Bronze | 96% Cu + 4% Al | 18.25 |
| Bronze | 90% Cu + 10% Al | 16.37 |
| Constantan | 55% Cu + 45% Ni | 49.63 |
| German silver | 65% Cu + 18% Ni + 17% Zn | 28.97 |
| Manganin | 84% Cu + 12% Mn + 4% Ni | 47.88 |
| Nichrome | 60% Ni + 15% Cr + 25% Fe | 112.00 |
| Iron | Cast | 60.00 |
| Iron | Wire | 12.00 |
| Steel | Rail | 18.00 |

# Index

## ABOUT THE EDITOR-IN-CHIEF

Syed A. Nasar, Ph.D., University of California at Berkeley, is Professor of Electrical Engineering at the University of Kentucky. He has been involved in teaching, research, and consulting in electric machine technology for over 30 years and is the author (or coauthor) of 18 books, 12 of which are on electric machines. His work has been translated into seven languages, including Russian and Chinese. Dr. Nasar has also published more than 100 papers relating to electric machines and is the founder and Chief Editor of *The International Journal: Electric Machines and Power Systems.* Both a Fellow IEEE and a Fellow IEE (London), he has served on several IEEE committees relating to electric machines.